Paleontology and Geology of Laetoli: Human Evolution in Context

Vertebrate Paleobiology and Paleoanthropology Series

Edited by

Eric Delson
Vertebrate Paleontology, American Museum of Natural History,
New York, NY 10024, USA
delson@amnh.org

Eric J. Sargis
Anthropology, Yale University
New Haven, CT 06520, USA
eric.sargis@yale.edu

Focal topics for volumes in the series will include systematic paleontology of all vertebrates (from agnathans to humans), phylogeny reconstruction, functional morphology, Paleolithic archaeology, taphonomy, geochronology, historical biogeography, and biostratigraphy. Other fields (e.g., paleoclimatology, paleoecology, ancient DNA, total organismal community structure) may be considered if the volume theme emphasizes paleobiology (or archaeology). Fields such as modeling of physical processes, genetic methodology, nonvertebrates or neontology are out of our scope.

Volumes in the series may either be monographic treatments (including unpublished but fully revised dissertations) or edited collections, especially those focusing on problem-oriented issues, with multidisciplinary coverage where possible.

For other titles published in this series, go to
www.springer.com/series/6978

Paleontology and Geology of Laetoli: Human Evolution in Context

Volume 1: Geology, Geochronology, Paleoecology and Paleoenvironment

Edited by

Terry Harrison

Center for the Study of Human Origins, Department of Anthropology, New York University, 25 Waverly Place, New York, NY 10003, USA

Springer

Editor
Terry Harrison
Center for the Study of Human Origins
Department of Anthropology
New York University
25 Waverly Place
New York, NY 10003
USA
terry.harrison@nyu.edu

ISBN 978-90-481-9955-6 e-ISBN 978-90-481-9956-3
DOI 10.1007/978-90-481-9956-3
Springer Dordrecht Heidelberg London New York

Cover illustration: Photograph of skull of *Serengetilagus praecapensis* (EP 3495/00) from Laetoli Locality 15 (© and courtesy of Alisa Winkler) superimposed on a view of the sunrise over Lemugurut from the base camp at Laetoli (© and courtesy Chris Harrison).

Printed on acid-free paper

Springer is part of Springer Science+Business Media (www.springer.com)

To Terri, Chris and Ben Harrison
So that their contributions and sacrifices to the Laetoli project do not go unheralded

Preface

Laetoli in northern Tanzania is one of the most important paleontological and paleoanthropological sites in Africa. It is renowned for the recovery of early hominin fossils belonging to *A. afarensis* and for the discovery of remarkably well-preserved trails of footprints of hominins, presumably made by *A. afarensis*. In fact, the first Pliocene hominin from East Africa was found at Laetoli, only 10 years after the announcement of *Australopithecus africanus* from Taung in South Africa, although it was not recognized as such at the time.

Given the significance of Laetoli for understanding and interpreting the evolutionary history of early hominins the author recognized that renewed investigations at Laetoli and at other fossil localities on the Eyasi Plateau could produce worthwhile results that might contribute to resolving some long-standing issues in hominin evolution. With this in mind, a long-term program of geological and paleontological investigations, directed by the author, was initiated at Pliocene localities on the Eyasi Plateau, including the hominid-bearing localities in the Laetoli area. The overall objectives were simple; to recover additional fossil hominid specimens and to obtain more detailed contextual information on the paleontology, geology, dating, and paleoecology.

Between 1998 and 2005 the author directed eight expeditions to Laetoli to accomplish these goals. These campaigns produced a wealth of original data on the fossil hominins, their associated fauna, and the paleoecological and paleoenvironmental context. The work presented here is the culmination of that research. It represents the combined effort of a dedicated and experienced field crew who were responsible for collecting the new fossils and samples described and analyzed here, and the subsequent research by a team of international specialists in paleoanthropology, vertebrate paleontology, mammalogy, malacology, entomology, ecology, palynology, paleobotany, taphonomy, geology, geochronology, and stable isotopes.

The present volume provides a compendium of research papers dealing with the geology, geochronology, paleoecology, taphonomy and paleobotany, as well as presenting new information on the modern-day Serengeti ecosystem to help provide baseline data for modeling Laetoli during the Pliocene. The companion volume focuses on the morphology, systematics and paleobiology of the fossil hominins and the associated invertebrate and vertebrate fauna. Together, these two volumes aim is to provide a comprehensive account of the geology, paleontology and paleoecology of Laetoli. It is hoped that the research presented here will provide an important building block in a broader understanding of early hominin evolution, faunal diversity and ecological change in East Africa during the Pliocene, and provide the basis for analyzing early hominin adaptation within the context of broader macroevolutionary models of speciation, diversification and extinction.

A special thanks goes to all of the dedicated, hardworking and resourceful team members who participated in the expeditions to Laetoli that contributed to the recovery of the material discussed and analyzed here. These volumes would not have been possible without them. I would especially like to single out the following individuals who were critical to the success of the field project: Peter Andrews, Remegius Chami, Amandus Kweka, Kaposhi Lasotumboyo, Moses Lilombero, Michael L. Mbago, Charles P. Msuya, Simon Mataro, Simon Odunga, Bill

Sanders and Denise Su. I thank the Tanzania Commission for Science and Technology and the Unit of Antiquities in Dar es Salaam for permission to conduct research in Tanzania. Special thanks go to Norbert Kayombo (Director General), Paul Msemwa (Director), Amandus Kweka and all of the curators and staff at the National Museum of Tanzania in Dar es Salaam for their support and assistance. I thank the regional, district and ward officers in Arusha Region for their support and hospitality. I am grateful to the Ngorongoro Conservation Area Authority for permission to conduct research in the conservation area. Emin Korcelik and Naphisa Jahazi of Hertz International in Dar es Salaam arranged the field transportation, and H. Meghji and A. Esmail helped with logistical support in Dar es Salaam.

Research at Laetoli benefited immeasurably from the advice, discussion, help and support from the following individuals: P. Andrews, R. Blumenschine, A. Deino, P. Ditchfield, C. Feibel, S. Frost, C. Harrison, T.S. Harrison, D. M. K. Kamamba, O. Kileo, J. Kingston, A. Kweka, M. G. Leakey, S. Mataro, G. Ole Moita, E. Mbua, L. McHenry, C. P. Msuya, C. S. Msuya, G. Mollel, M. Muungu, O. Mwebi, J. Pareso, C. Peters, M. Pickford, K. Reed, C. Saanane, W. Sanders, C. Swisher, and S. Waane. Bill Sanders deserves special mention for applying his exceptional talents to preparing and casting some of the Laetoli specimens. The following individuals provided critical comments and advice about the research presented in this volume and its companion: A. Alexandre, P. Andrews, M. Anton, M. Avery, M. Bamford, F. Bibi, L. Bishop, R, Bobe, R. Bonnefille, F, Brown, P. Butler, C. Crumly, A. Deino, P. Ditchfield, P. Duringer, M. Erbajeva, R. Evander, C. Feibel, Y. Fernandez-Jalvo, B. Fine-Jacobs, L. Flynn, S. Frost, T, Furman, J. Genise, A. Gentry, D. Geraads, H. Gilbert, U. Goehlich, J.H. Harris, K. Heissig, A. Hill, P. Holroyd, D. Iwan, N. Jablonski, J. Kappelman, T. Kaiser, R. Kay, J. Kingdon, J. Kingston, W. Kimbel, J. Knott, K. Kovarovic, N. Kristensen, O. Kullmer, F. de Lapparent de Broin, M. Lewis, N. Lopez-Martinez, S. Manchester, I. MacDougall, L. McHenry, S. McNaughton, K. Metzger, P. Meylan, C. Mourer-Chauviré, R. Oberprieler, E. O'Brien, D. Parmley, M. Pavia, C. Peters, M. Pickford, I. Poole, B. Ratcliffe, D. Reed, K. Reed, W.J. Sanders, M. Sponheimer, D. Su, Z. Szyndlar, R. Tabuce, P. Tassy, B. Tiffney, J. van der Made, A. Vincens, C. Ward, H. Wesselman, E. Wheeler, and A. Winkler. Special thanks go to Terri Harrison and Leahanne Sarlo for their assistance with many aspects of the editorial process. I thank Eric Delson, Eric Sargis and the Editorial and Production staff at Springer, especially Tamara Welschot and Judith Terpos. Fieldwork at Laetoli and subsequent research was supported by grants from the National Geographic Society, the Leakey Foundation, and the National Science Foundation (Grants BCS-0216683 and BCS-0309513).

New York Terry Harrison

Contents

Contributors

Katherine A. Adelsberger
Department of Environmental Studies, Knox College,
2 East South Street, Galesburg, IL 61401, USA
kadelsbe@knox.edu

Peter Andrews
Palaeontology Department, Natural History Museum, Cromwell Road,
London, SW7 5B, UK
pjandrews@uwclub.net

Marion K. Bamford
Bernard Price Institute for Palaeontological Research, School of Geosciences,
University of the Witwatersrand, Private Bag 3, WITS 2050,
Johannesburg, South Africa
marion.bamford@wits.ac.za

Laura C. Bishop
School of Natural Sciences and Psychology, Research Centre in
Evolutionary Anthropology and Palaeoecology, Liverpool John Moores University,
Byrom Street, Liverpool, L3 3AF, UK
L.C.Bishop@ljmu.ac.uk

Daniel C. Bowman
Department of Geology, Macalester College, St. Paul, MN 55105, USA
dbowman@macalester.edu

Michael J. Carr
Department of Earth and Planetary Sciences, Rutgers University,
610 Taylor Road, Piscataway, NJ 08854, USA
carr@rci.rutgers.edu

Alan L. Deino
Berkeley Geochronology Center, 2455 Ridge Road, Berkeley, CA 94709, USA
al@bgc.org

Christiane Denys
Department of Systematics and Evolution – CP51, UMR7205 CNRS: Origine structure
& évolution de la Biodiversité, MNHN, 55 rue Buffon, 75005, Paris, France
denys@mnhn.fr

Peter Ditchfield
Research Laboratory for Archaeology and the History of Art, School of Archaeology,
University of Oxford, Oxford, UK
peter.ditchfield@rlaha.ox.ac.uk

Mark D. Feigenson
Department of Earth and Planetary Sciences, Rutgers University,
610 Taylor Road, Piscataway, NJ 08854, USA
feigy@rci.rutgers.edu

Terry Harrison
Center for the Study of Human Origins, Department of Anthropology,
New York University, 25 Waverly Place, New York, NY 10003, USA
terry.harrison@nyu.edu

Fritz Hertel
Department of Biology, California State University, Northridge, CA 91330-8303, USA
fritz.hertel@csun.edu

Thomas M. Kaiser
Biozentrum Grindel und Zoologisches Museum, Universität Hamburg,
Martin-Luther-King-Platz 3, D-20146 Hamburg, Germany
thomas.kaiser@uni-hamburg.de

John D. Kingston
Department of Anthropology, Emory University, 1557 Dickey Drive,
Atlanta, GA 30322, USA
john.kingston@emory.edu

Kris Kovarovic
Department of Anthropology, Durham University, Dawson Building,
South Road, Durham, DH1 3LE, UK
kris.kovarovic@durham.ac.uk

Amandus Kweka
National Museum of Tanzania, P.O. Box 511, Dar es Salaam, Tanzania
kwekason@yahoo.co.uk

Godson Leliyo
Arusha Herbarium, TPRI, PO Box 3024, Arusha, Tanzania

Audax Z.P. Mabulla
Archaeology Unit, University of Dar es Salaam, Dar es Salaam, Tanzania
aumab@udsm.ac.tz

Lindsay J. McHenry
Department of Geosciences, University of Wisconsin-Milwaukee,
3209 N. Maryland Avenue, Milwaukee, WI 53211, USA
lmchenry@uwm.edu

Godwin F. Mollel
Department of Earth and Atmospheric Sciences, University of Alberta,
B-14 Earth Sciences Building, Edmonton, Alberta, T6G 2E3, Canada
gmollel@gmail.com

Efrem-Fred Njau
Arusha Herbarium, TPRI, PO Box 3024, Arusha, Tanzania
efrednjau@yahoo.co.uk

Thomas. W. Plummer
Department of Anthropology, Queens College, CUNY & NYCEP,
65–30 Kissena Blvd, Flushing, NY 11367, USA
Thomas.Plummer@qc.cuny.edu

Denné N. Reed
Department of Anthropology, University of Texas at Austin,
1 University Station C3200, Austin, TX 78712, USA
reedd@mail.utexas.edu

Lloyd Rossouw
Department of Archaeology, National Museum, Bloemfontein,
9300, South Africa
lloyd@nasmus.co.za

Louis Scott
Department of Plant Sciences, University of the Free State, P.O. Box 339,
Bloemfontein, South Africa
scottl.sci@ufs.ac.za

Denise F. Su
Department of Anthropology, Bryn Mawr College, 101 N. Merion Avenue,
Bryn Mawr, PA 19072, USA
dfayshensu@brynmawr.edu

Carl C. Swisher III
Department of Earth and Planetary Sciences, Rutgers University,
610 Taylor Road, Piscataway, NJ 08854, USA
cswish@rci.rutgers.edu

Karl R. Wirth
Department of Geology, Macalester College, St. Paul, MN 55105, USA
wirth@macalester.edu

Chapter 1
Laetoli Revisited: Renewed Paleontological and Geological Investigations at Localities on the Eyasi Plateau in Northern Tanzania

Terry Harrison

Abstract Laetoli, one of the key paleontological and paleoanthropological localities in Africa, is renowned for the recovery of fossil remains of early hominins belonging to *Australopithecus afarensis* and for the remarkable trails of hominin footprints. In addition, the faunas from the Upper Laetolil Beds (3.63–3.85 Ma) and Upper Ndolanya Beds (2.66 Ma) are from time periods that are generally poorly represented at other paleontological sites in East Africa. Fossils from these stratigraphic units provide important insights into the faunal and floral diversity during the Pliocene, and they serve as reliably dated reference faunas for comparison with other Plio-Pleistocene sites in Africa. The paleoecology of Laetoli is unusual for early hominin sites in East Africa in the absence of evidence for extensive or permanent bodies of water, and in having habitats that are reconstructed as being less densely wooded. Therefore, Laetoli provides key evidence for interpreting the possible diversity of hominin habitat preferences and for understanding ecological changes in East Africa during the Pliocene. The main goal of renewed fieldwork at Laetoli, starting in 1998, was to recover additional fossil hominid specimens and to obtain more detailed contextual information on the paleontology, geology, dating, and paleoecology. The substantially expanded fossil collections have added significantly to our understanding of the systematics and paleobiology of Pliocene East African faunas. The recovery of new *Australopithecus afarensis* specimens from the Upper Laetolil Beds has contributed information on the morphology, variation and evolutionary status of this taxon. Fossil hominins have been recovered for the first time from the Upper Ndolanya Beds. These include the first specimen of *Paranthropus aethiopicus* to be recovered from outside the Turkana Basin, and one of the oldest securely dated specimens definitively attributable to this taxon.

Keywords Pliocene • Fossil • Garusi • Laetolil Beds • Ndolanya Beds • *Australopithecus afarensis* • *Paranthropus aethiopicus*

T. Harrison (✉)
Center for the Study of Human Origins, Department of Anthropology, New York University, 25 Waverly Place, New York, NY 10003, USA
e-mail: terry.harrison@nyu.edu

Introduction

Laetoli in northern Tanzania is renowned as one of the most important paleontological and paleoanthropological localities in Africa. The site is significant for a number of reasons: (1) the sample of fossil hominins from the Upper Laetoli Beds (3.63–3.85 Ma) is relatively small (just over 30 specimens), but the site has produced the oldest and the second largest sample of *Australopithecus afarensis*, which includes the lectotype, L.H. 4 (Harrison 2011a); (2) the Upper Ndolanya Beds (2.66 Ma) have produced the only specimen of *Paranthropus aethiopicus* definitively known from outside of the Turkana Basin in northen Kenya and southern Ethiopia, and this find extends the distribution of the species 800 km to the south (Harrison 2002, 2011a); (3) the site is unique for the remarkable preservation of trails of footprints of hominins (see Hay and Leakey 1982; Leakey and Harris 1987); this evidence has been used to confirm earlier inferences based on anatomy that bipedalism was an important component of the terrestrial locomotor behavior of Pliocene hominins (e.g., Johanson et al. 1982; Stern and Susman 1983; Susman et al. 1984, 1985; Latimer et al. 1987; Latimer 1991; Susman and Stern 1991; McHenry 1986, 1991, 1994; Stern 2000; Ward 2002); (4) the Upper Laetolil Beds and Upper Ndolanya Beds sample time periods that are generally poorly-represented at other paleontological sites in East Africa, and the abundance of fossils from these stratigraphic units provide important windows into the faunal and floral diversity during the Pliocene; (5) paleontological collections from the Lower Laetolil Beds (3.85–4.4 Ma), Naibadad Beds (~2.1–2.3 Ma), Olpiro Beds and Ngaloba Beds (late Pleistocene) contribute to a broader understanding of the evolutionary history of African hominins and their associated faunas, as well as to the archaeological record (Leakey and Harris 1987; Ndessokia 1990); (6) the mammalian fauna from Laetoli is remarkably diverse (Leakey and Harris 1987; Harrison 2011b), with 115 species currently identified from the Upper Laetolil Beds and Upper Ndolanya Beds (almost one-third of which are based on type material from Laetoli), along with the remains or traces of fossil amphibians, reptiles, birds,

T. Harrison (ed.), *Paleontology and Geology of Laetoli: Human Evolution in Context. Volume 1: Geology, Geochronology, Paleoecology and Paleoenvironment*, Vertebrate Paleobiology and Paleoanthropology, DOI 10.1007/978-90-481-9956-3_1,

insects, gastropods and plants; it thus serves as a key reference fauna, one that is reliably dated, for comparisons with other Plio-Pleistocene sites in Africa; (7) Laetoli is unusual for sites in East Africa in the absence of sedimentological or paleontological evidence for extensive bodies of water (i.e., rivers or lakes) and in having an inferred paleoecological setting with less extensive woodland than penecontemporaneous sites with fossil hominins; as a consequence, the ecological context at Laetoli has been extensively studied and debated (Leakey and Harris 1987; Andrews 1989, 2006; Cerling 1992; Andrews and Humphrey 1999; Musiba 1999; Kovarovic et al. 2002; Kovarovic 2004; Su 2005; Harrison 2005; Kovarovic and Andrews 2007; Kingston and Harrison 2007; Musiba et al. 2007; Su and Harrison 2007, 2008; Andrews and Bamford 2008; Peters et al. 2008), and (8) the paleoecology at Laetoli provides important information on the possible range of hominin habitat preferences and on ecological changes in East Africa during the Pliocene that may have impacted on the mode and tempo of human evolution.

Given the importance of the site for understanding and interpreting the evolutionary history of early hominins and African faunas in general, the author recognized that renewed investigations at Laetoli and other fossil localities on the Eyasi Plateau could produce worthwhile results that might contribute to resolving some long-standing issues in hominin evolution. Following initial discussions about the feasibility and desirability of renewed fieldwork at Laetoli, Dr. Simon Waane, the Director of the Antiquities Unit in Dar es Salaam at the time, invited the author to direct a new long-term research effort at Pliocene localities on the Eyasi Plateau, including the hominid-bearing localities in the Laetoli area. Fieldwork began during the summer of 1998. Initial results confirmed that the localities were still remarkably productive, that new fossil localities could be discovered, that enlarged fossil collections would add significantly to our understanding of the paleobiology and taxonomic diversity of Pliocene East African faunas, that there was a high likelihood of finding new fossil hominins, and that overall the area had tremendous potential for paleontological and geological research. A long-term project was planned that involved an interdisciplinary team of more than 50 research scientists. The overall objective was to recover additional fossil hominin specimens, and to obtain more detailed contextual information on the paleontology, geology, dating, and paleoecology.

Building on the previous research effort directed by Mary Leakey (1974–1982), which laid an excellent foundation for future research (Leakey and Harris 1987), it was possible to design a more problem-oriented approach to renewed investigations, including a detailed study of the geochronology and stratigraphic context, a better appreciation of the preservation and taphonomy, and a broader multidisciplinary approach to understanding the paleoecology. The project focused on the critical role of understanding the paleoecology at Laetoli, not simply as an adjunct to studies of the fossil hominid finds, but as an integral component of any investigation that aims to answer the much more important how and why questions pertaining to the evolutionary history of humans. The main objectives were: (1) to conduct surface collections of fossils at all of the currently known localities in the Laetoli area; (2) to actively prospect the region for new paleontological occurrences; (3) to conduct geological investigations at both the local and regional levels; (4) to obtain further radiometric age determinations to help constrain the estimated ages of the major stratigraphic units; (5) to attempt to reconstruct more comprehensively the paleoecology at Laetoli and neighboring localities; (6) to better document the taxonomy, phylogenetic relationships, paleobiology and zoogeographic relationships of the fauna; and (7) to recover additional fossil hominins.

Between 1998 and 2005 the author directed eight expeditions to Laetoli to accomplish these goals. The current compendium of chapters, and its companion volume on the Laetoli fauna (Harrison 2011b), represent the culmination of this research effort. The research presented here is the combined effort of a dedicated and experienced field crew who were responsible for collecting the new fossils and samples described and analyzed here, and subsequent research by a team of international scientists (50 researchers from 11 different countries) with expertise in paleoanthropology, vertebrate paleontology, mammalogy, malacology, entomology, ecology, palynology, paleobotany, taphonomy, geology, geochronology, and stable isotopes. A number of key findings from this research have already been published (Harrison 2002, 2005, 2010; Kovarovic et al. 2002; Scott et al. 2003; Darlington 2005; Bamford 2005; Andrews and Harrison 2005; Harrison and Msuya 2005; Andrews 2006; Armour-Chelu et al. 2006; Kovarovic and Andrews 2007; Kingston and Harrison 2007; Su and Harrison 2007, 2008; Andrews and Bamford 2008; Sanders 2008; Mollel et al. 2008; Peters et al. 2008), and four doctoral dissertations have appeared, which focused on clarifying the systematics, paleoecology, and environmental and geological context at Laetoli (Kovarovic 2004; Su 2005; Mollel 2007; Dehghani 2008).

History of Paleontological Research at Laetoli

The paleontological significance of Laetoli was initially recognized during the 1930s. Louis and Mary Leakey first visited the area in 1935, and made a small collection of fossil vertebrates and terrestrial gastropods from localities at Laetoli, Naibadad, Ngai (probably Ngaloba) and Endolele (probably Esere), which were forwarded to the Natural History Museum in London. These included a hominin lower canine, which was the first hominin fossil to be recovered from the Pliocene of East Africa, although the specimen was not identified as such until some decades later (White 1981). The Leakeys were accompanied by Peter Kent, who published a preliminary account of the geology of the area (Kent 1941). Ludwig and Margrethe Kohl-Larsen made collections in the Laetoli region in 1939 (at Garussi = Garusi, Deturi = Olaitole, Lemagrut Korongo, Gadjingero, Oldogom, and Marambu = Olduvai Side Gorge, Locality 1), as part of their multifaceted investigation of the ethnography, archaeology and paleontology of the Lake Eyasi region (Reck and Kohl-Larsen 1936; Kohl-Larsen 1943). This extensive faunal material housed at the Museum für Naturkunde, Berlin, and the Institut für Ur- und Frühgeschichte und Archäologie des Mittelalters and Institut und Museum für Geologie and Paläontologie at Eberhard Karls Universität Tübingen, was described by Dietrich (1941, 1942a, b, 1945, 1950, 1951) and Petter (1963). The 1939 collections included a hominin maxilla with P^3 and P^4 (Garusi I) and an isolated M^3 (Garusi II) (Hennig 1948; Weinert 1950; Remane 1950, 1954; Robinson 1953, 1955). An undescribed occipital fragment, apparently of a fossil hominin (Garusi III), has been lost (Kohl-Larsen 1943; Protsch 1981; Ullrich 2001).

Hennig (1948) described the original Garusi specimens as belonging to a new genus, *Praeanthropus*, but as he did not assign a species name, the *nomen nudum* was unavailable. Weinert (1950) later included the Garusi hominins in a new species, *Meganthropus africanus*. Subsequently, Senyürek (1955) transferred the species to *Praeanthropus*, thereby making Hennig's original generic name available. In the late 1970s, Delson identified a previously unrecognized hominin lower incisor in the collections in Berlin (White 1981; Ullrich 2001; Delson, personal communication).

Louis and Mary Leakey revisited the Laetoli region in 1959 (collecting at Laetoli, Esere, and sites along the Kakesio road), and in 1964 they made a one-day excursion from their base camp at Olduvai Gorge (Leakey 1987a). In addition, as part of his geological study of Olduvai Gorge, Hay (1976) reported the discovery of fossils from the Laetolil Beds at Silal Artum, and the Olduvai Side Gorge near Kelogi. These collections are housed in the National Museums of Kenya in Nairobi.

An intensive program of geological and paleontological investigation was initiated at Laetoli in 1974, under the direction of Mary Leakey, and the campaign continued until 1982. This phase of research laid the foundation for a sound appreciation of the geology, geochronology, and paleontology of the area (Hay 1976, 1978, 1981, 1986, 1987; Hay and Reeder 1978; Leakey 1981; Hay and Leakey 1982; Leakey and Harris 1987). During the course of these investigations Mary Leakey and her team recovered a large collection of fossils (over 10,000 mammal specimens) from the Laetolil Beds, including 25 hominins. In addition, they recovered a hominin cranium (L.H. 18) from the Upper Ngaloba Beds (Day et al. 1980; Magori and Day 1983), referrable to *Homo* cf. *H. sapiens*, and a mandible (L.H. 29) referred to *Homo* cf. *H. erectus* (now included in *A. afarensis*, see Harrison 2011a). A list of hominin specimens recovered prior to 1980 was published by Leakey (1987b: 116–117), and the individual specimens were figured and described in detail (Leakey et al. 1976; White 1977, 1980a, 1981, 1985; Day et al. 1980; Magori and Day 1983; Leakey 1987b). At the time, the collection of hominids from the Laetolil Beds comprised 15 isolated teeth, 9 jaw fragments or associated dentitions, and a partial skeleton of an immature individual. Johanson et al. (1978) combined the hominids from the Laetolil Beds with the more substantial sample from Hadar in Ethiopia, to create the hypodigm of a new species, *Australopithecus afarensis*. A partial mandible from Laetoli (L.H. 4) was designated as the type specimen.

In 1978, the remarkable discovery of fossilized footprints of hominids changed the emphasis of investigations at Laetoli. Renewed efforts focused mainly on the study of the trails and their significance for understanding early hominin locomotor behavior (Leakey and Hay 1979; Clarke 1979; White 1980b; Day and Wickens 1980; Charteris et al. 1981, 1982; Hay and Leakey 1982; Leakey 1978, 1979, 1981, 1987c; Robbins 1987; White and Suwa 1987; Tuttle 1985, 1987, 1990, 1994, 2008; Tuttle et al. 1990, 1991, 1992). In 1979, the hominin footprint trails were reburied to preserve them from weathering and erosion.

Several teams worked at Laetoli in the 1980s and early 1990s, but there was no systematic and sustained program of geological and paleontological investigation. The Institute of Human Origins (IHO), directed by D.C. Johanson, carried out paleoanthropological research at Olduvai Gorge and Laetoli from 1985 to 1988. Research was focused on Olduvai, but small collections of fossils were made at Laetoli, which included an isolated molar of a hominin discovered in 1987 (Kyauka and Ndessokia 1990). As part of this project, Ndessokia (1990) conducted detailed excavations in the Upper Ndolanya Beds and Olpiro Beds, and recovered a small sample of fossil mammals (which has not been relocated). In conjunction with geological research conducted by IHO, Manega (1993) carried out a study of the geochronology and geochemistry of the tuffaceous sediments associated with the Ngorongoro Volcanic Highlands, including those at Laetoli.

A joint Getty Conservation Institute-Government of Tanzania project was undertaken at Laetoli from 1994–1997 in an effort to preserve the hominin footprint trail at Loc. 8 from damage by erosion and root penetration (Anonymous 1995; Agnew et al. 1996; Feibel et al. 1996; Agnew and Demas 1998). The trails were re-excavated and then reburied following conservation and further documentation of the anatomical details of the prints (Feibel et al. 1996; Agnew and Demas 1998). The trails are currently still covered, but there has been recent discussion about whether they should be re-excavated to allow them to be exhibited to the general public and be accessible for study by the scientific community (Mabulla 2000; Dalton 2008).

In 1993–1999 the Hominid Corridor Research Project (HCRP), under the direction of F. Schrenk and T.G. Bromage, surveyed localities at Laetoli and Kakesio (Kaiser et al. 1995; Kaiser 2000). During the mid-1990s, C. Musiba conducted his doctoral dissertation research on the ecomorphology, taphonomy and paleoecology of Laetoli (Musiba 1999). In 1998 he resumed investigations (Musiba et al. 2007, 2008), and since that time he has directed a paleontological and archaeological field school at the site.

It should be noted that there has been much confusion concerning the correct name to apply to Laetoli, as well the meaning of the name. During the mid-1930s, Louis Leakey referred to the area in an unpublished field report as Laetolil (or Litolio) (Leakey 1935). Kent (1941) referred to it as the Vogel River area (using an anglicized version of the early German name, Vogelfluss; Jaeger 1913; Obst 1915). His map identifies the area of exposed Pliocene sediments along the Vogel River as Laetolil, and he refers to the Laetolil Beds as part of the Vogel River Series. Later, Kohl-Larsen (1943) identified the main paleontological localities in the area, including the hominin-bearing localities, as Garussi (or Vogelfluss). This is the local name for the river valley (alternatively Garusi, Ngarusi or Eng'arusi) in which most of the fossil localities are situated. Kohl-Larsen also recovered fossils from several nearby valleys, including Gadjingero, Oldogom, Marambu and Deturi. The last of these corresponds to the river valley known today as Olaitole (and evidently represents an incorrect transliteration of the same name), located just to the south of the Garusi River. The name Laetolil (or Laitole) was subsequently used by English-speaking geologists and paleontologists until Leakey (1987a) corrected the spelling to Laetoli to reflect a more accurate rendering of the name in Maa. However, as noted by Pickering (1969) the type locality of the Laetolil Beds (and most of the paleontological localities) is in the Garusi Valley, rather than the Olaitole Valley. Garusi would be a more appropriate name for the site, but the name Laetoli is retained here because of its historical association with the localities and fossil hominins. Leakey (1987a) attributes the name Laetoli to a Maa word for a local red lily, but I have not been able to confirm this. Instead, local informants tell me that it pertains to the river of that name. According to Ndessokia (1990), however, the word Laitoli means "salty plains" in Maa.

Renewed Fieldwork (1998–2006)

In 1998 the author began work at Laetoli and at other sites on the Eyasi Plateau, and fieldwork continued until 2005. As part of the Laetoli project, the author revisited the Manonga Valley in north-central Tanzania (Harrison 1997), about 170 km southwest of Laetoli, during the summer of 2006 to investigate the relationships between the lacustrine sediments in the Manonga Valley and the tuffaceous deposits at Laetoli.

Fossils were primarily recovered from the surface of the outcrops after they had eroded out of the sediments. All anatomically identifiable fossils were recovered. The same strategy was successfully adopted by the author in the Manonga Valley, starting in 1990, because such collections were considered essential for gathering the kinds of data critical for understanding the taphonomy, ecomorphology, paleobiology and community structure of the fauna. Although criticized at the time, it has since proved to be a very useful strategy, and one that has now become the vogue among teams working in eastern Africa. Excavations were carried out to recover specimens that were discovered *in situ*, but no full-scale program of excavations at targeted sites was implemented. Earlier excavations carried out by Mary Leakey and later by Prosper Ndessokia were not found to be sufficiently productive to merit such labor-intensive efforts. Systematic screening for microvertebrates and invertebrates was not untaken, mainly because of the lack of fossil concentrations, but screening was used to recover associated remains and at localities where fossil hominins were found. Occurrences of fossil footprints and trails were recorded and photographed when exposed, but no effort was made to excavate and expose tuffs known to contain footprints, mainly because of concerns about how to conserve them properly. All material collected was catalogued in the field, and each specimen was marked with a field number, individually bagged with an accompanying field label that included information on locality, stratigraphic provenience and provisional taxonomic identification. The field catalogues are available on-line at http://www.nyu.edu/gsas/dept/anthro/programs/csho/pmwiki.php/Home/Collections. The collections are housed at the National Museum of Tanzania in Dar es Salaam. All holotypes are stored in a separate type cabinet, and the fossil hominins are housed in the Museum strong room.

As noted above, earlier collections by Louis Leakey and Mary Leakey and by Kohl Larsen are housed in museums and repositories outside of Tanzania (i.e., in England, Germany and Kenya). All of the material collected by Mary Leakey from 1974 onwards falls under the authority of the Department of Antiquities in Dar es Salaam, with the National Museum of Tanzania as the repository. Much of the material recovered by Mary Leakey was stored at Olduvai Gorge, but a part of the collection was sent on loan for study to Nairobi and to other international museums and researchers. A small but significant part of the collections, including the hominins and some of the type specimens from Laetoli, are still temporarily on loan at the Kenya National Museum. The Kenyan and Tanzanian authorities intend to repatriate them in the near future. Unfortunately, other parts of the collection sent out on loan cannot be accounted for.

With the assistance of Denise Su, and in cooperation with the National Museums of Tanzania in Arusha and Dar es Salaam and the Department of Antiquities, a major effort was undertaken to reconstitute, rehabilitate and curate the Mary Leakey collections from Laetoli. All of the specimens at Olduvai Gorge, which were temporarily stored in wooden crates and cardboard boxes and wrapped in newspaper, were carefully rewrapped and transported back to the National Museum in Dar es Salaam, where they are now permanently curated. The specimens have now been individually bagged or boxed. Field labels, using information derived from a surviving partial copy of Mary Leakey's field catalogue in the archive of the Kenya National Museum were added to each bag. Data from the catalogue, which are essential to determine the stratigraphic provenance of the specimens, were entered into a database by Denise Su to create a permanent electronic catalogue. From the catalogue it can be determined that more than 80% of the Mary Leakey collections have been relocated and are now housed at the National Museum in Tanzania in Dar es Salaam. Without timely intervention it is very likely that the collections and the pertinent documentation would have been lost to the scientific community. The Leakey and Harrison collections, comprising more than 40,000 fossils, are fully accessible for study, subject to approval by the appropriate Tanzanian authorities. Only a small collection of fossils from the IHO project is retained in the National Museum in Dar es Salaam. The material collected by P. Ndessokia from the Upper Ndolanya Beds has not been relocated and is presumed lost. The location of collections made by other research teams after 1982 is unknown and they are currently not available for study.

Field Seasons

1998 (July 31–August 25)

Field Personnel: E. Baker, C. Dawi, A. Kweka, T. Harrison, T.S. Harrison, M. Lilombero, T. Lubulila, A. Malyango, M. Mbago, C.P. Msuya, D. Munisi, F. Nsinge, C. Robinson, and W.J. Sanders (Fig. 1.1).

The expedition carried out geological and paleontological investigations at Laetoli and Kakesio. The area west of Kakesio was surveyed as far as Itinje in the Meatu District of Shinyanga Region. To the south of Laetoli, the Oldogom and Esere area were explored, largely on foot because of the dense vegetation and rugged terrain. New fossil occurrences were discovered at Oleisusu, Engesha, Olaltanaudo and Laetoli Loc. 22S (see Harrison and Kweka 2011). Locality 22S produced a hominin tibia from the Upper Ndolanya Beds. A total of 1,680 fossil mammal specimens were collected.

1999 (January 3–25)

Field Personnel: P. Abwalo, C. Dawi, T. Harrison, A. Kweka, M. Lilombero, M. Mbago, D. Munisi, and C.P. Msuya.

The field campaign was cut short by heavy rains that made the tracks impassable, and even access to the sites by foot proved extremely difficult. The expedition collected fossils at Laetoli, Kakesio and Esere, as well as several new localities. Mr. Simon Mataro (Department of Antiquities superintendent at Laetoli) directed us to previously unrecorded fossil localities south of Laetoli in the region of Noiti (located not

Fig. 1.1 Laetoli field participants in the 1998 season. Back row (*left to right*): Freddie Nsinge, Amandus Kweka, Cosmas Dawi, Terri Harrison, Dominic Munisi, Tobias Lubulila, Eric Baker, Avelin Malyango, Chris Robinson. Front row (*left to right*): Moses Lilombero, Charles Msuya, Michael Mbago, William J. Sanders, Terry Harrison

far from Engesha, which we had surveyed in 1998). Noiti yielded abundant and well-preserved fossil wood, but no vertebrate remains. Members of the expedition explored the area south of Esere as far as Ngaiborgoso and the peak of Eseketeti, where the Precambrian basement underlying the Laetolil Beds is exposed, and west of Laetoli along the Emboremony River as far as Kakesio, which yielded fossils from the Lower Laetolil Beds. A new fossil occurrence was also discovered in a large valley south of the village of Sinoni, at a locality called Lobileita, which exposes the Lower Laetolil Beds overlying the Precambrian basement. Only 359 fossil mammals were collected.

2000 (January 16–February 7 and July 8–29)

Field Personnel: Winter – P. Andrews, R. Chami, C. Dawi, C. Feibel, T. Harrison, S. Hixson, A. Kweka, M. Lilombero, S. Mataro, C. Msuya, D. Munisi, and F. Nsinge. Summer – R. Chami, C. Dawi, T. Harrison, J. Kingston, A. Kweka, M. Lilombero, M. Mbago, K. McNulty, C. Msuya, D. Munisi, F. Nsinge, C. Robinson, L. Scott, and D. Su (Fig. 1.2).

During the course of the winter and summer field seasons, expedition team members made intensive paleontological collections and geological investigations at all of the Laetoli localities. Important new occurrences of fossils in the Upper Ndolanya Beds were discovered at Loc. 15. Craig Feibel discovered a clay horizon rich in fossil insects, seeds and ruminant dung at Loc. 3. The expedition also revisited Noiti, Emboremony, Lobeleita, Oleisusu, and Kakesio. Exposures of Laetolil Beds located to the southwest of Norsigidok were surveyed on foot, but the area yielded few fossils. Samples for the study of stable isotopes and palynomorphs/phytoliths were collected by John Kingston and Louis Scott. A systematic survey was made of modern bones and carcasses along the Gadjingero River, to provide baseline taxonomic and taphonomic data for interpreting the fossils at Laetoli. The collection of fossil mammals from the combined seasons totals 5183 specimens, including a mandibular fragment and an isolated canine of *A. afarensis* from Loc. 16.

2001 (July 5–28)

Field Personnel: P. Andrews, M. Bamford, Cosmas Dawi, A. Deino, M. Durru, T. Harrison, K. Kovarovic, M. Lilombero, M. Mbago, C.P. Msuya, D. Munisi, F. Nsinge, Shadrack (Fig. 1.3).

Fossils were collected at Laetoli, Emboremony and Kakesio. Samples of fossil wood were collected at Noiti by M. Bamford for thin section preparation at the University of Witwatersrand. Rock samples were obtained by A. Deino from Laetoli, Kakesio and Noiti for geochronology. Additional horizons were sampled for palynology. A new outcrop of Upper Ndolanya Beds was located at Silal Artum, and the fossil material recovered included a maxilla of *Paranthropus aethiopicus*. A total of 1,663 fossil mammals were recovered during the course of the field season. At the end of the season, Laetoli fossils stored at Olduvai Gorge and the Arusha Museum were transported to the National Museum of Tanzania in Dar es Salaam for permanent curation.

Fig. 1.2 Laetoli field participants in the summer 2000 season. Back row (*left to right*): John Kingston, Cosmas Dawi, Louis Scott, Kieran McNulty (*behind*), Amandus Kweka (*front*), Michael Mbago, Freddie Nsinge, Chris Robinson. Front row seated (*left to right*): Denise Su, Dominic Munisi, Moses Lilombero, Remigius Chami. Front on ground: Terry Harrison. Missing from picture: Charles Msuya

Fig. 1.3 Laetoli field participants in the 2001 season. Back row (*left to right*): Shadrack, Amandus Kweka, name unknown (lorry driver), Moses Lilombero, Peter Andrews, Charles Msuya, name unknown (lorry driver), Terry Harrison. Front row (*left to right*): Dominic Munisi, Freddie Nsinge, Michael Mbago, Mohamedi Durru, Kris (Fire) Kovarovic, Denise Su. Missing from picture: Alan Deino and Cosmas Dawi

2003 (July 4–28)

Field Personnel: M. Bamford, C. Dawi, F. Kikwa, A. Kweka, T. Harrison, M. Lilombero, C. Msuya, E. Mwawembe, F. Nsinge, S. Odunga, C. Robinson, and D. Su (Fig. 1.4).

The team made paleontological collections at Laetoli, Noiti, Emboremony, and Kakesio localities south of the Kakesio River. Additional fossil wood samples were collected at Noiti by M. Bamford. A total of 3,001 fossil mammals were recovered.

2004 (June 25–July 21)

Field Personnel: P. Andrews, M. Bamford, S. Cooke, C. Dawi, A. Deino, P. Ditchfield, B. Harrison, C. Harrison, T. Harrison, T.S. Harrison, S. Jones, J. Kingston, F. Kikwa, A. Kweka, G. Leliyo, L. McHenry, G. Mkude, G. Mollel, J. Msowaya, S. Mataro, E. Mwawembe, F. Mwawembe, E.A. Njau, F. Nsinge, S. Odunga, L. Rossouw, W.J. Sanders, C. Swisher, M. Tallman, and S. Worthington (Fig. 1.5).

The team made extensive paleontological collections at all localities at Laetoli and Kakesio. Fossil vertebrates were also recovered from the Lower Laetolil Beds at Noiti 3. A total of 1,909 fossil mammals were collected. Stratigraphic, geomorphological and geochronological investigations were carried out by A. Deino, P. Ditchfield, L. McHenry, G. Mollel, and C. Swisher. An important focus of the field season was mapping and inventorying the modern vegetation in the Laetoli area by P. Andrews, M. Bamford, G. Leliyo and E. Njau. Triplicate samples were deposited in the following herbaria: National Herbarium of Tanzania, Arusha; Royal Botanic Gardens, Kew, London; University of Witwatersrand, Johannesburg. Additional samples were collected for analyses of stable isotopes and phytoliths by J. Kingston and L. Rossouw respectively.

2005 (May 27–June 15)

Field Personnel: J. Darlington, C. Dawi, P. Ditchfield, C. Fellmann, T. Harrison, F. Kikwa, A. Kweka, S. Mbegu, G. Mkude, S. Mataro, E. Mwawembe, F. Mwawembe, S. Nassoro, T. Rein, M. Seselj, and D. Su (Fig. 1.6).

The expedition revisited most of the localities at Laetoli and Kakesio, and collected 1,495 fossil mammals. Geological investigations were continued by P. Ditchfield and T. Harrison. A special focus of the field season was studying the fossil termitaries and termite traces, as well as documenting the distribution and taxonomy of the modern termite fauna at Laetoli.

Results

New Paleontological Collections

The Laetoli localities continue to be productive. During the course of renewed fieldwork, more than 25,000 fossils were collected (Table 1.1). These consisted mainly of fossil mammals (58.1%), but also included the remains of birds (4.9%), reptiles and amphibians (1.9%), invertebrates (33.3%) and plants (1.8%). Most of the fossil mammals (82.4%) were recovered from the Upper Laetolil Beds, but smaller samples came from the Lower Laetolil Beds (1.7%) and Upper Ndolanya Beds (15.8%). Representative fossil vertebrates were also recovered from the Olpiro and Ngaloba Beds, but no systematic collections were made. The fauna from the Lower Laetolil Beds is dominated by bovids and equids, which comprise more than half of the specimens collected, but this frequency is probably skewed by a strong taphonomic bias against smaller vertebrates (Table 1.2). The faunal collections from the Upper Laetolil Beds and Upper Ndolanya Beds are composed primarily of bovids and lagomorphs, which together represent 72.0% and 81.2% of all mammals

Fig. 1.4 Laetoli field participants in the 2003 season. Back row (*left to right*): Amandus Kweka, Ephraim Mwawembe, Cosmas Dawi, Charles Msuya, Terry Harrison, Moses Lilombero, Chris Robinson. Front row (*left to right*): Simon Odunga, Freddie Nsinge, Denise Su, Marion Bamford, Francis Kikwa

Fig. 1.5 Laetoli field participants in the 2004 season. Back row (*right to left*): Amandus Kweka, Simon Odunga, John Kingston, Steven Jones, Godson Leliyo, Carl Swisher, Steven Worthington, Terry Harrison, Moses Lilombero, Terri Harrison, Lloyd Rossouw, Efrem-Fred Njau, Peter Andrews. Middle row (*left to right*): Godwin Mollel, Ben Harrison, Gregory Mkude, Chris Harrison, Freddie Nsinge, Joseph Msowaya, Siobhan Cooke, Melissa Tallman, Francis Kikwa, Felix Mwawembe, Marion Bamford. Front row (*left to right*) Lindsay McHenry, Ephraim Mwawembe, Cosmas Dawi, Peter Ditchfield. Missing from picture: Alan Deino and William J. Sanders

Fig. 1.6 Laetoli field participants in the 2005 season. Back row (*left to right*) Maja Seselj, Gregory Mkude, Denise Su, Cosmas Dawi, Connie Fellmann, Seif Nassoro, Tom Rein, Francis Kikwa, Felix Mwawembe, Amandus Kweka. Front row (*left to right*): Terry Harrison, Peter Ditchfield, Samiyu Mbegu, Ephraim Mwawembe, Johanna Darlington

Table 1.1 Number of fossils collected 1998–2005

Taxon	LLB	ULB	UNB	Total	% of total
Mammals[a]	258	12,383	2,378	15,019	58.1
Birds[b]	3	185	9	197	0.8
Struthio[c]	427	343	289	1,059	4.1
Reptiles and amphibians[d]	103	352	34	489	1.9
Mollusks[e]	290	4,612	282	5,184	20.1
Insects[f]	460	1,857	1,103	3,420	13.2
Plants[g]	7	457	4	468	1.8
Total	1,548	20,189	4,095	25,832	100.0

Specimen counts do not include fossils from the Olpiro or Ngaloba Beds
[a] For more detailed information on fossil mammals see Table 1.2
[b] Includes bones and eggs, except for those assigned to *Struthio*
[c] Egg shell fragments of *Struthio*
[d] Mostly the remains of tortoises, but the count does include snakes, lizards and amphibians
[e] Terrestrial gastropods (for more detailed data on specimen counts see Tattersfield 2011)
[f] Mainly consists of cocoons and brood cells of solitary bees, but also includes casts of insects, termitaries, and brood cells of dung beetles
[g] Includes wood, twigs, leaves, and seeds (see Bamford 2011a, b)
LLB Lower Laetolil Beds, *ULB* Upper Laetolil Beds, *UNB* Upper Ndolanya Beds

Table 1.2 Number of specimens and the frequency of fossil mammals collected at Laetoli and other localities on the Eyasi Plateau from 1998 to 2005

	Lower Laetolil Beds		Upper Laetolil Beds		Upper Ndolanya Beds	
Taxon	N	%	N	%	N	%
Macroscelididae	0	0	4	0.03	0	0
Galagidae	0	0	1	0.01	0	0
Cercopithecidae	1	0.40	111	0.91	1	0.04
Hominidae	0	0	2	0.02	2	0.09
Rodentia	10	3.97	855	7.00	104	4.55
Leporidae	15	5.95	4,640	38.00	398	17.41
Carnivora	13	5.16	424	3.47	54	2.36
Proboscidea	37	14.68	158	1.29	24	1.05
Orycteropodidae	1	0.40	26	0.21	2	0.09
Equidae	55	21.83	330	2.70	110	4.81
Rhinocerotidae	21	8.33	473	3.87	29	1.27
Chalicotheriidae	0	0	3	0.02	0	0
Suidae	12	4.76	244	2.00	27	1.18
Camelidae	0	0	26	0.21	6	0.26
Giraffidae	8	3.17	772	6.32	70	3.06
Bovidae	79	31.35	4,145	33.95	1,459	63.82
Total	252	100.0	12,214	100.01	2,286	99.99

respectively. The mammalian fauna from the Upper Ndolanya Beds differs from that from the Upper Laetolil Beds in having a much higher percentage of bovids and a stronger bias against micromammals (Table 1.2). The 1998–2005 collection is now larger than that made by Mary Leakey in 1974–1979 (Table 1.3). It is interesting to note that the relative proportions of the most common mammalian taxa from the Upper Laetolil Beds (those taxa represented by more than 100 specimens) are remarkably similar between the two collections. This concordance suggests that these proportions are probably quite close to their representation in the fossil record. A key difference is that the 1998–2005 expeditions recovered a smaller percentage of large mammals (i.e., Proboscidea, Rhinocerotidae, Equidae, Suidae and Giraffidae) and a higher percentage of micromammals (i.e. Rodentia and Leporidae) compared with the 1974–1979 collections (Table 1.3). This is almost certainly due to larger specimens having a higher recovery rate than smaller specimens, combined with the slower rate at which larger specimens erode out onto the surface. Although the material recovered by Kohl-Larsen was presumably more selectively sampled than either the Leakey or Harrison collections, his expedition did recovered a high proportion of large mammal finds, which provides additional support for the scenario presented above.

Table 1.3 Comparison of the relative proportions of common mammal taxa from the Upper Laetolil Beds collected by Leakey and Harrison

	Leakey collections (1974–1981)[a]		Harrison collections (1998–2005)[b]	
	N	%	N	%
Cercopithecidae	87	0.9	111	0.9
Rodentia	526	5.7	855	7.0
Leporidae	2,861	30.8	4,640	38.2
Carnivora	261	2.8	424	3.5
Proboscidea	228	2.5	158	1.3
Equidae	265	2.8	330	2.7
Rhinocerotidae	644	6.9	473	3.9
Suidae	324	3.5	244	2.0
Giraffidae	808	8.7	772	6.4
Bovidae	3,281	35.3	4,145	34.1
Total	9,285	99.9	12,152	100.0

Includes only those taxa represented by more than 100 specimens in each collection
[a] Data from Leakey (1987a)
[b] Detailed information presented in Table 1.2

Geological Investigations

Renewed geological research on the Eyasi Plateau focused on a better understanding of the local stratigraphic relationships, sedimentology and geochemistry of the tuffs and fossil-bearing horizons at Laetoli and Kakesio (Mollel et al. 2011; McHenry 2011; Ditchfield and Harrison 2011; Adelsberger et al. 2011; Deino 2011). This has allowed a more refined appreciation of the stratigraphic provenance of fossil specimens, correlation of marker tuffs between localities, and a better understanding of the paleoenvironmental setting. Surveys of much of the Eyasi Plateau, between the villages of Endulen and Kakesio, have allowed a much better understanding of the regional geology, especially the contact between the Precambrian Basement and the Laetolil Beds, and the geographic distribution of the main sedimentary units. In addition, research in the Volcanic Highland Complex to the east of Laetoli has allowed a better understanding of the source of the volcanic materials that contributed to the regional sedimentary sequence. However, the general stratigraphic scheme for the sedimentary sequence follows that developed by Hay (1987) (see Fig. 1.7).

Further afield, the mineralogy of the Manonga Valley in north-central Tanzania, about 170 km to the southwest of Laetoli, has provided evidence of a direct relationship between the two sedimentary basins. Mutakyahwa (1997) showed that the Manonga Valley sediments were very likely derived from the same volcanic source as those at Laetoli. However, the fauna from the Manonga Valley is generally older (~4.0–5.5 Ma) than that at Laetoli, although it probably overlaps in age with the Lower Laetolil Beds (~3.8–4.4 Ma). This implies that Lake Manonga may have acted as a sedimentary basin for the accumulation of tuffs deposited distant from the volcanic centers during the late Miocene and early Pliocene, whereas sediments did not accumulate on the Precambrian basement in the Laetoli area proximal to the volcanic highlands until sometime later.

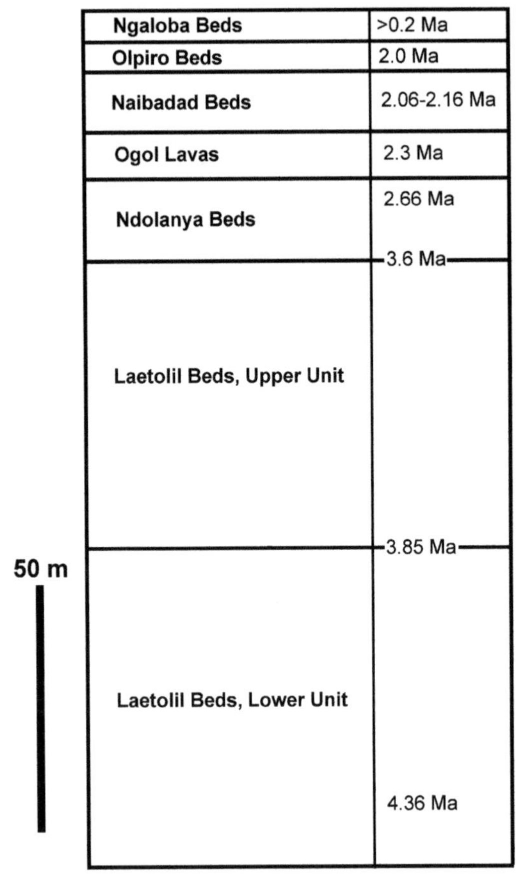

Fig. 1.7 Simplified stratigraphic scheme of Laetoli sediments showing the main stratigraphic units (*left*) and the chronology (*right*, Ma = mega-annum) (after Hay 1987; Ndessokia 1990; Manega 1993; Ditchfield and Harrison 2011; Deino 2011)

Radiometric Age Determinations

Drake and Curtis (1987) published a series of K/Ar dates for Laetoli. These provided age estimates of 3.76 Ma to older than 4.3 Ma for the Lower Laetolil Beds and 3.49–3.76 Ma for the Upper Laetolil Beds. A K-Ar date of 2.41 Ma for the Ogol Lavas provided a maximum age for the underlying Ndolanya Beds. The Naibadad Beds were dated at 2.26 Ma. Ndessokia (1990) provided the first $^{40}Ar/^{39}Ar$ dates, with estimated ages of 2.66 Ma for the Upper Ndolanya Beds, 2.15 Ma for the Naibadad Beds, and 2.0 Ma for the Olpiro Beds. Later, Manega (1993) reported $^{40}Ar/^{39}Ar$ dates of 2.5 Ma for the Ndolanya Beds, 2.10–2.26 Ma for the Naibadad Beds, and 2.04 Ma for the Olpiro Beds. Amino acid racemization of ostrich eggshells provided ages of greater than 200 ka for the Upper Ngaloba Beds and greater than 290 ka for the Lower Ngaloba Beds (Manega 1993). Renewed research at Laetoli since 1998 has allowed a refinement of the geochronological framework at Laetoli (Deino 2011; Mollel et al. 2011; see Fig. 1.7). These new $^{40}Ar/^{39}Ar$ dates confirm the earlier age estimates of the Upper Ndolanya (2.66 Ma) and Naibadad Beds (2.057–2.155 Ma), while the Ogol Lavas (2.3 Ma) are slightly younger. Most importantly, the new dates push back the age of the Upper Laetolil Beds (3.6–3.85 Ma), making *Australopithecus afarensis* and its associate fauna from Laetoli slightly older than previously estimated.

Paleoecology

A major focus of renewed investigations at Laetoli has been on reconstructing the paleoecology. As noted earlier, study of the paleoecology provides critical evidence for understanding the context of early hominin evolution. It allows researchers to pose important questions about hominin habitat preferences, ecology and paleobiology, and to include these data in larger scale macroevolutionary models of speciation, biogeography, diversification and extinction. With these questions in mind, renewed work at Laetoli has attempted to reconstruct the paleoecology using information from a wide diversity of taxa (ranging from plants and invertebrates to mammals) and modern-day ecosystems (ranging from soils and vegetation to micromammals), as well as a range of paleoecological proxies that can be applied to the fossil record (i.e., palynology, phytoliths, stable isotopes, mesowear, ecomorphology, and community structure).

Systematics and Paleobiology of the Fauna

The contributions in Leakey and Harris (1987) provided the last detailed and comprehensive account of the systematics of the Laetoli fauna. Since that time, however, there have been major advances in our understanding of the systematics and paleobiology of late Miocene and Plio-Pleistocene faunas of Africa, with many new localities and faunas reported. Renewed investigations at Laetoli have allowed a thorough revision of the systematics of the mammalian fauna from Laetoli, and a greater emphasis on understanding the paleobiology of the fauna and its paleoecological implications. All of the mammalian taxa have been restudied, with the exception of the Camelidae and Chalicotheriidae (which will be described elsewhere). Equally importantly, study of the non-mammalian fossils was considered essential for a complete understanding and appreciation of the biotic diversity and paleoecology at Laetoli during the Pliocene. These investigations include the first detailed studies to be undertaken of the fossil insects, gastropods, birds, lizards and snakes from Laetoli. Research on the ostriches and birds' eggs has already been published (Harrison 2005; Harrison and Msuya 2005). The new accounts presented in the companion volume (Harrison 2011b) provide the basis for a significant systematic revision of the fauna. The current fauna comprises 115 species of mammals from the Laetolil and Ndolanya Beds (compared with 84 in 1987), and includes five new species of rodents, a new species of proboscidean, one new species of carnivoran, two new species of bovids, and a new genus of galagid. In addition, the non-mammalian fauna has been supplemented by one new species of ostrich (Harrison 2005), two new species of gastropods and four new species of insects. In addition to providing a more complete record of the faunal diversity associated with the early hominins at Laetoli, these studies provide a wealth of critical information that can be used in reconstructing the paleoecology and in better understanding the biogeography and biochronology of African Pliocene faunas.

Additional Fossil Hominins

Renewed investigations at Laetoli have led to the recovery of additional fossil hominins. Two specimens, an isolated lower canine (EP 162/00) and a mandibular fragment with P_3-M_1 (EP 2400/00), recovered from the Upper Laetolil Beds, are referable to *A. afarensis*. These have provided the basis, along with several other previously undescribed specimens, for a reassessment of the morphology and evolutionary status of the *A. afarensis* sample from the type locality of Laetoli. In addition, two hominins, a proximal tibia (EP 1000/98) and an edentulous maxilla (EP 1500/01), were recovered from the Upper Ndolanya Beds, and these represent the first hominins from this stratigraphic unit. EP 1500/01 is important because it represents the only specimen of *P. aethiopicus* recovered from outside the Turkana basin, and is among the oldest securely dated specimens definitively attributable to this taxon. A detailed account of the chronology, anatomy, paleobiology and systematics of the hominins from Laetoli is presented in Harrison (2011a).

Conclusions

Laetoli, one of the key paleontological and paleoanthropological localities in Africa, is renowned for the fossil evidence of early hominins (i.e., *Australopithecus afarensis* and *Paranthropus aethiopicus*) and the remarkable trails of hominin footprints. In addition, the diverse faunas from the Upper Laetolil Beds and Upper Ndolanya Beds, dated to 3.63–3.85 and 2.66 Ma respectively, are from time periods that are generally poorly-represented at other paleontological sites in East Africa. Fossils from these stratigraphic units provide important windows into the faunal and floral diversity during the Pliocene, and they serve as reliably dated reference faunas for comparison with other Plio-Pleistocene sites in Africa. From a paleoecological perspective, Laetoli is also unusual for early hominin sites in East Africa in the absence of evidence for extensive bodies of water, and in having habitats that are reconstructed as being less densely wooded. This evidence provides key information on the possible diversity of hominin habitat preferences and on ecological changes in East Africa during the Pliocene that may have impacted on the mode and tempo of human evolution.

The main aim of renewed fieldwork at Laetoli, starting in 1998, was to recover additional fossil hominid specimens, and to obtain more detailed contextual information on the paleontology, geology, dating, and paleoecology. The results have confirmed that the existing localities are still remarkably productive and that new localities can be found. The substantially expanded fossil collections, numbering more than 25,000 specimens, have added significantly to our understanding of the systematics and paleobiology of Pliocene East African faunas. The recovery of new hominin specimens from the Upper Laetolil Beds has contributed information on the morphology, variation and evolutionary status of *Australopithecus afarensis*. Fossil hominins, recovered for the first time from the Upper Ndolanya Beds, included a partial maxilla of *Paranthropus aethiopicus*. This is the first specimen of *P. aethiopicus* to be recovered from outside the Turkana Basin, and is among the oldest securely dated specimens definitively attributable to this taxon.

The compendium of chapters in this volume, and its companion volume on the Laetoli hominins and their associated fauna (Harrison 2011b), present the results of this renewed research effort.

Acknowledgements A special thanks to all of the dedicated and resourceful team members who participated in the expeditions to Laetoli that contributed to the recovery of the material discussed and analyzed here. This volume and its companion would not have been possible without them. I would especially like to single out the following individuals who were critical to the success of the field project: Amandus Kweka, Michael L. Mbago, Charles P. Msuya, Simon Odunga, Al Deino, Carl Swisher, Peter Ditchfield, Godwin Mollel, Lindsay McHenry, Craig Feibel, Moses Lilombero, Simon Mataro, Denise Su, Peter Andrews, and Bill Sanders. I thank the Tanzania Commission for Science and Technology and the Unit of Antiquities in Dar es Salaam for permission to conduct research in Tanzania. Special thanks go to Norbert Kayombo (Director General), Paul Msemwa (Director), Amandus Kweka and all of the curators and staff at the National Museum of Tanzania in Dar es Salaam for their support and assistance.

Research at Laetoli benefited immeasurably from the advice, discussion, help and support of the following individuals: P. Andrews, R. Blumenschine, C. Harrison, T.S. Harrison, D. M. K. Kamamba, O. Kileo, J. Kingston, A. Kweka, K. Lasotumboyo, M. G. Leakey, S. Mataro, G. Ole Moita, C. P. Msuya, C. S. Msuya, M. Muungu, O. Mwebi, J. Pareso, C. Peters, M. Pickford, K. Reed, C. Saanane and S. Waane. Thanks to M.G. Leakey and C. Robinson for their help in locating and piecing together the Mary Leakey catalogue, and to D. Su for her extraordinary efforts in assembling the digital catalogue. Fieldwork at Laetoli was supported by grants from National Geographic Society, the Leakey Foundation, and the National Science Foundation (Grants BCS-0216683 and BCS-0309513).

References

Adelsberger, K. A., Wirth, K. R., Mabulla, A. Z. P., & Bowman, D. C. (2011). Geochemical and mineralogical characterization of Middle Stone Age tools of Laetoli, Tanzania and comparisons with possible source materials. In T. Harrison (Ed.), *Paleontology and geology of Laetoli, Tanzania: Human evolution in context* (Geology, geochronology, paleoecology and paleoenvironment, vol. 1, pp. 143–165). Dordrecht: Springer.

Agnew, N., & Demas, M. (1998). Preserving the Laetoli footprints. *Scientific American, 279*, 44–55.

Agnew, N., Demas, M., & Leakey, M. D. (1996). The Laetoli footprints. *Science, 271*, 1651–1652.

Andrews, P. J. (1989). Palaeoecology of Laetoli. *Journal of Human Evolution, 18*, 173–181.

Andrews, P. (2006). Taphonomic effects of faunal impoverishment and faunal mixing. *Palaeogeography, Palaeoclimatology, Palaeoecology, 241*, 572–589.

Andrews, P., & Bamford, M. (2008). Past and present ecology of Laetoli, Tanzania. *Journal of Human Evolution, 54*, 78–98.

Andrews, P., & Harrison, T. (2005). The last common ancestor of apes and humans. In D. Lieberman, R. J. Smith, & J. Kelley (Eds.), *Interpreting the past: Essays on human, primate, and mammal evolution in honor of David Pilbeam* (pp. 103–121). Boston: Brill Academic.

Andrews, P. J., & Humphrey, L. (1999). African Miocene environments and the transition to early hominines, climate change and human evolution. In T. Bromage & F. Schrenk (Eds.), *African biogeography* (pp. 282–315). Oxford: Oxford University Press.

Anonymous. (1995). Saving our roots from roots. *Science, 267*, 171.

Armour-Chelu, M., Bernor, R. L., & Mittmann, H.-W. (2006). Hooijer's hypodigm for "Hipparion" cf. ethiopicum (Equidae, Hipparioninae) from Olduvai, Tanzania and comparative material from the East African Plio-Pleistocene. *Beiträge zur Paläontologie, Wien, 30*, 15–24.

Bamford, M. K. (2005). Environmental changes and hominid evolution: What the vegetation tells us. In F. d'Errico & L. Backwell (Eds.), *From tools to symbols. From early hominids to modern humans* (pp. 103–120). Johannesburg: Witwatersrand University Press.

Bamford, M. (2011a). Fossil leaves, fruits and seeds. In T. Harrison (Ed.), *Paleontology and geology of Laetoli: Human evolution in context* (Geology, geochronology, paleoecology, and paleoenvironment, Vol. 1, pp. 235–252). Dordrecht: Springer.

Bamford, M. (2011b). Fossil woods. In T. Harrison (Ed.), *Paleontology and geology of Laetoli: Human evolution in context* (Geology, geochronology, paleoecology, and paleoenvironment, Vol. 1, pp. 217–233). Dordrecht: Springer.

Cerling, T. E. (1992). Development of grasslands and savannas in East Africa during the Neogene. *Palaeogeography, Palaeoclimatology, Palaeoecology, 97*, 241–247.

Charteris, J., Wall, J. C., & Nottrodt, J. W. (1981). Functional reconstruction of gait from the Pliocene hominid footprints at Laetoli, northern Tanzania. *Nature, 290*, 496–498.

Charteris, J., Wall, J. C., & Nottrodt, J. W. (1982). Pliocene hominid gait: New interpretations based on available footprint data from Laetoli. *American Journal of Physical Anthropology, 58*, 133–144.

Clarke, R. J. (1979). Early hominid footprints from Tanzania. *South African Journal of Science, 75*, 148–149.

Dalton, R. (2008). Fears for oldest human footprints. *Nature, 451*, 118.

Darlington, J. P. E. C. (2005). Distinctive fossilized termite nests at Laetoli, Tanzania. *Insectes Sociaux, 52*, 408–409.

Day, M. H., & Wickens, E. H. (1980). Laetoli Pliocene hominid footprints and bipedalism. *Nature, 286*, 385–387.

Day, M. H., Leakey, M. D., & Magori, C. (1980). A new hominid fossil skull (L.H. 18) from the Ngaloba Beds, Laetoli, northern Tanzania. *Nature, 284*, 55–56.

Dehghani, R. (2008). Aspects of carnivoran evolution in Africa. Ph.D. dissertation, University of Stockholm, Sweden.

Deino, A. (2011). $^{40}Ar/^{39}Ar$ dating of Laetoli, Tanzania. In T. Harrison (Ed.), *Paleontology and geology of Laetoli: Human evolution in context.* (Geology, geochronology, paleoecology, and paleoenvironment, Vol. 1, pp. 77–97). Dordrecht: Springer.

Dietrich, W. O. (1941). Die säugetierpaläontologischen Ergebnisse der Kohl-Larsen'schen Expedition 1937–1939 im nördlichen Deutsch-Ostafrika. Zentralblatt für Mineralogie, Geologie und Paläontologie, Abt. B, Stuttgart, pp. 217–223.

Dietrich, W. O. (1942a). Ältestquartäre Säugetiere aus der südlichen Serengeti, Deutsch-Ostafrika. *Palaeontographica, 94A*, 43–133.

Dietrich, W. O. (1942b). Zur Entwicklungsmechanik des Gebisses der afrikanischen Nashörner. Zentralblatt für Mineralogie, Geologie und Paläontologie, Abt. B, Stuttgart, pp. 297–300.

Dietrich, W. O. (1945). Nashornreste aus dem Quartär Deutsch-Ostafrikas. *Palaeontographica, 96A*, 46–90.

Dietrich, W. O. (1950). Fossile Antilopen und Rinder Äquatorialafrikas (Material der Kohl-Larsen'schen Expeditionen). *Palaeontographica, 99A*, 1–62.

Dietrich, W. O. (1951). Daten zu den fossilen Elefanten Afrikas und Ursprung der Gattung Loxodonta. *Neues Jahrbuch für Mineralogie, Geologie und Paläontologie, Stuttgart, 93*, 325–378.

Ditchfield, P., & Harrison, T. (2011). Sedimentology, litho-stratigraphy and depositional history of the Laetoli area. In T. Harrison (Ed.), *Paleontology and geology of Laetoli: Human evolution in context* (Geology, geochronology, paleoecology, and paleoenvironment, Vol. 1, pp. 47–76). Dordrecht: Springer.

Drake, R., & Curtis, G. H. (1987). K-Ar geochronology of the Laetoli fossil localities. In M. D. Leakey & J. M. Harris (Eds.), *Laetoli: A Pliocene site in northern Tanzania* (pp. 48–52). Oxford: Clarendon.

Feibel, C. S., Agnew, N., Latimer, B., Demas, M., Marshall, F., Waane, S. A. C., & Schmid, P. (1996). The Laetoli hominid footprints – a preliminary report on the conservation and scientific restudy. *Evolutionary Anthropology, 4*, 149–154.

Harrison, T. (1997). *Neogene paleontology of the Manonga valley, Tanzania*. New York: Plenum.

Harrison, T. (2002). First recorded hominins from the Ndolanya Beds, Laetoli, Tanzania. *American Journal of Physical Anthropology, Suppl. 32*, 83.

Harrison, T. (2005). Fossil bird eggs from Laetoli, Tanzania: Their taxonomic and paleoecological implications. *Journal of African Earth Sciences, 41*, 289–302.

Harrison, T. (2010). Later Tertiary Lorisiformes (Strepsirhini, Primates). In L. Werdelin & W. J. Sanders (Eds.), *Cenozoic mammals of Africa* (pp. 333–349). Berkeley: University of California Press.

Harrison, T. (2011a). Hominins. In T. Harrison (Ed.), *Paleontology and geology of Laetoli: Human evolution in context* (Fossil hominins and the associated fauna, Vol. 2, pp. 141–188). Dordrecht: Springer.

Harrison, T. (Ed.). (2011b). *Paleontology and geology of Laetoli: Human evolution in context* (Fossil hominins and the associated fauna, Vol. 2). Dordrecht: Springer.

Harrison, T., & Kweka, A. (2011). Paleontological localitier on the Eyasi flateau, including Laetoli. In T. Harrison (Ed.), *Paleontology and geology of Laetoli: Human evolution in context* (Geology, geochronology, paleoecology and paleoenvironment, Vol. 1, pp. 17–45). Dordrecht: Springer.

Harrison, T., & Msuya, C. P. (2005). Fossil struthionid eggshells from Laetoli, Tanzania: Their taxonomic and biostratigraphic significance. *Journal of African Earth Sciences, 41*, 303–315.

Hay, R. L. (1976). *Geology of the Olduvai Gorge*. Berkeley: University of California Press.

Hay, R. L. (1978). Melilitite-carbonatite tuffs in the Laetolil Beds of Tanzania. *Contributions to Mineralogy and Petrology, 17*, 255–274.

Hay, R. L. (1981). Palaeoenvironment of the Laetolil Beds, northern Tanzania. In G. Rapp & C. F. Vondra (Eds.), *Hominid sites: Their geological settings* (pp. 7–24). American association for the advancement of science, Selected symposium 63. Boulder: Westview.

Hay, R. L. (1986). Role of tephra in the preservation of fossils in Cenozoic deposits of East Africa. In L. E. Frostick, R. W. Renaut, I. Reid, & J. J. Tiercelin (Eds.), *Sedimentation in the African rifts* (pp. 339–344). Oxford: Blackwell Scientific. Geological Society special publication No. 25.

Hay, R. L. (1987). Geology of the Laetoli area. In M. D. Leakey & J. M. Harris (Eds.), *Laetoli: A Pliocene site in northern Tanzania* (pp. 23–47). Oxford: Clarendon.

Hay, R. L., & Leakey, M. D. (1982). The fossil footprints of Laetoli. *Scientific American, 246*, 50 57.

Hay, R. L., & Reeder, R. J. (1978). Calcretes of Olduvai Gorge and the Ndolanya Beds of northern Tanzania. *Sedimentology, 25*, 649–673.

Hennig, E. (1948). Quartärfaunen und Urgeschichte Ostafrikas. *Naturwissenschaftliche Rundschau, 1*, 212–217.

Jaeger, F. (1913). Das Hochland der Riesenkrater und die umliegenden Hochländer Deutsch-Ostafrikas. Ergebnisse einer amtlichen Forschungsreise ins abflusslose Gebiet des nördlichen Deutsch-Ostafrika 1906/07. Mitteilungen aus den Deutschen Schutzgebieten. Ergänzungscheft 8.

Johanson, D. C., White, T. D., & Coppens, Y. (1978). A new species of the genus *Australopithecus* (Primates: Hominidae) from the Pliocene of eastern Africa. *Kirtlandia, 28*, 1–14.

Johanson, D. C., Lovejoy, C. O., Kimbel, W. H., White, T. D., Ward, S. C., Bush, M. E., Latimer, B. M., & Coppens, Y. (1982). Morphology of the Pliocene partial hominid skeleton (A.L. 288-1) from the Hadar Formation, Ethiopia. *American Journal of Physical Anthropology, 57*, 403–452.

Kaiser, T. M. (2000). Proposed fossil insect modification to fossil mammalian bone from Plio-Pleistocene hominid-bearing deposits of Laetoli (northern Tanzania). *Annals of the Entomological Society of America, 93*, 693–700.

Kaiser, T., Bromage, T. G., & Schrenk, F. (1995). Hominid Corridor Research Project update: New Pliocene fossil localities at Lake Manyara and putative oldest Early Stone Age occurrences at Laetoli

(Upper Ndolanya Beds), northern Tanzania. *Journal of Human Evolution, 28*, 117–120.

Kent, P. E. (1941). The recent history and Pleistocene deposits of the plateau north of lake Eyasi, Tanganyika. *Geological Magazine, 78*, 173–184.

Kingston, J. D., & Harrison, T. (2007). Isotopic dietary reconstructions of Pliocene herbivores at Laetoli: Implications for hominin paleoecology. *Palaeogeography, Palaeoclimatology, Palaeoecology, 243*, 272–306.

Kohl-Larsen, L. (1943). Auf den Spuren des Vormenschen. Forshungen, Fahrten und Erlebnisse in Deutsch-Ostafrika. 2nd Band. Strecker und Schröder, Stuttgart.

Kovarovic, K. (2004). Bovids as palaeoenvironmental indicators. An ecomorphological analysis of bovid postcranial remains from Laetoli, Tanzania. Ph.D. dissertation, University of London.

Kovarovic, K., & Andrews, P. (2007). Bovid postcranial ecomorphological survey of the Laetoli paleoenvironment. *Journal of Human Evolution, 52*, 663–680.

Kovarovic, K., Andrews, P., & Aiello, L. (2002). An ecological diversity analysis of the Upper Ndolanya Beds, Laetoli, Tanzania. *Journal of Human Evolution, 43*, 395–418.

Kyauka, P. S., & Ndessokia, P. (1990). A new hominid tooth from Laetoli, Tanzania. *Journal of Human Evolution, 19*, 747–750.

Latimer, B. (1991). Locomotor adaptations in *Australopithecus afarensis*: The issue of arboreality. In B. Senut & Y. Coppens (Eds.), *Origine(s) de la Bipédie chez les Hominidés* (pp. 169–176). Paris: CNRS.

Latimer, B. M., Ohman, J. C., & Lovejoy, C. O. (1987). Talocrural joint in African hominoids. Implications for *Australopithecus afarensis*. *American Journal of Physical Anthropology, 74*, 155–175.

Leakey, L. S. B. (1935). East African archaeological expedition. Fourth season 1934–1935. Eighth monthly field report: May 24th-June 23rd. Unpublished report, The Natural History Museum, London.

Leakey, M. D. (1978). Pliocene footprints at Laetoli, northern Tanzania. *Antiquity, 52*, 133.

Leakey, M. D. (1979). 3.6 million years old footprints in the ashes of time. *National Geographic Magazine, 155*, 446–457.

Leakey, M. D. (1981). Tracks and tools. *Philosophical Transactions of the Royal Society of London, B, 292*, 95–102.

Leakey, M. D. (1987a). Introduction. In M. D. Leakey & J. M. Harris (Eds.), *Laetoli: A Pliocene site in northern Tanzania* (pp. 1–22). Oxford: Clarendon.

Leakey, M. D. (1987b). The Laetoli hominid remains. In M. D. Leakey & J. M. Harris (Eds.), *Laetoli: A Pliocene site in northern Tanzania* (pp. 108–117). Oxford: Clarendon.

Leakey, M. D. (1987c). The hominid footprints: Introduction. In M. D. Leakey & J. M. Harris (Eds.), *Laetoli: A Pliocene site in northern Tanzania* (pp. 490–496). Oxford: Clarendon.

Leakey, M. D., & Harris, J. M. (Eds.). (1987). *Laetoli: A Pliocene site in northern Tanzania*. Oxford: Clarendon.

Leakey, M. D., & Hay, R. L. (1979). Pliocene footprints in the Laetolil Beds at Laetoli, northern Tanzania. *Nature, 278*, 317–323.

Leakey, M. D., Hay, R. L., Curtis, G. H., Drake, R. E., Jackes, M. K., & White, T. D. (1976). Fossil hominids from the Laetolil Beds. *Nature, 262*, 460–466.

Mabulla, A. Z. P. (2000). Strategy for Cultural Heritage Management (CHM) in Africa: A case study. *African Archaeological Review, 17*, 211–233.

Magori, C. C., & Day, M. H. (1983). Laetoli Hominid 18: An early *Homo sapiens* skull. *Journal of Human Evolution, 12*, 747–753.

Manega, P. (1993). Geochronology, geochemistry and isotopic study of the Plio-Pleistocene hominid sites and the Ngorongora volcanic highlands in northern Tanzania. Ph.D. dissertation, University of Colorado at Boulder, Boulder.

McHenry, H. M. (1986). The first bipeds: A comparison of the *Australopithecus afarensis* and *Australopithecus africanus* postcranium and implications for the evolution of bipedalism. *Journal of Human Evolution, 15*, 177–191.

McHenry, H. M. (1991). First steps? Analyses of the postcranium of early hominids. In B. Senut & Y. Coppens (Eds.), *Origine(s) de la Bipédie chez les Hominidés* (pp. 133–142). Paris: CNRS.

McHenry, H. M. (1994). Early hominid postcrania: Phylogeny and function. In R. S. Corruccini & R. L. Ciochon (Eds.), *Integrative paths to the past: Paleoanthropological advances in honor of F. Clark Howell* (pp. 251–268). Englewood Cliffs: Prentice Hall.

McHenry, L. J. (2011). Geochemistry and mineralogy of the Laetoli area tuffs: Lower Laetolil through Naibadad Beds. In T. Harrison (Ed.), *Paleontology and geology of Laetoli: Human evolution in context* (Geology, geochronology, paleoecology and paleoenvironment, vol. 1, pp. 121–141). Dordrecht: Springer.

Mollel, G.F. (2007). Petrochemistry and geochemistry of Ngorongoro volcanic highland complex (NVHC) and its relationship to Laetoli and Olduvai Gorge, Tanzania. Ph.D. dissertation, Rutgers University, New Brunswick.

Mollel, G., Swisher, C., Feigenson, M., & Carr, M. J. (2008). Geochemical evolution of Ngorongoro Caldera, northern Tanzania: Implications for crust-magma interaction. *Earth and Planetary Science Letters, 271*, 337–347.

Mollel, G. F., Swisher, C. C., Feigenson, M. D., & Carr, J. D. (2011). Petrology, geochemistry and age of Satiman, Lemagurut and Oldeani: Sources of the volcanic deposits of the Laetoli area. In T. Harrison (Ed.), *Paleontology and geology of Laetoli: Human evolution in Context* (Geology, geochronology, paleoecology and paleoenvironment, vol. 1, pp. 99–119). Dordrecht: Springer.

Musiba, C.M. (1999). Laetoli Pliocene paleoecology: A reanalysis via morphological and behavioral approaches. Ph.D. dissertation, University of Chicago, Chicago.

Musiba, C., Magori, C., Stoller, M., Stein, T., Branting, S., Vogt, M., Tuttle, R., Hallgrimsson, B., Killindo, S., Mizambwa, F., Ndunguru, F., & Mabulla, A. (2007). Taphonomy and paleoecological context of the Upper Laetolil Beds (Localities 8 and 9), Laetoli in northern Tanzania. In R. Bobe, Z. Alemseged, & A. K. Behrensmeyer (Eds.), *Hominin environments in the East African Pliocene: An assessment of the faunal evidence* (pp. 257–278). Dordrecht: Springer.

Musiba, C. M., Mabula, A., Selvaggio, M., & Magori, C. C. (2008). Pliocene animal trackways at Laetoli: Research and conservation potential. *Ichnos, 15*, 166–178.

Mutakyahwa, M. (1997). Mineralogy of the Wembere-Manonga Formation, Manonga Valley, Tanzania, and the possible provenance of the sediments. In T. Harrison (Ed.), *Neogene paleontology of the Manonga valley, Tanzania* (pp. 67–78). New York: Plenum.

Ndessokia, P. N. S. (1990). The mammalian fauna and archaeology of the Ndolanya and Olpiro Beds, Laetoli, Tanzania. Ph.D. dissertation, University of California, Berkeley.

Obst, E. (1915). Das abflusslose Rumpfschollenland in nordöstlichen Deutsch-Ostafrika. Teil 1. *Mitteilungen der Geographischen Gesellschaft in Hamburg, 29*, 4–268.

Peters, C. R., Blumenschine, R. J., Hay, R. L., Livingstone, D. A., Marean, C. W., Harrison, T., Armour-Chelu, M., Andrews, P., Bernor, R. O., Bonnefille, R., & Werdelin, L. (2008). Paleoecology of the Serengeti-Mara ecosystem. In A. R. E. Sinclair, C. Packer, S. A. R. Mduma, & J. M. Fryxell (Eds.), *Serengeti III: Human impacts on ecosystem dynamics* (pp. 47–94). Chicago: University of Chicago Press.

Petter, G. (1963). Étude quelques Viverridés (Mammifères, Carnivores) du Pléistocène inférieur du Tanganyika (Afrique orientale). *Bulletin de la Société Géologique de France*, Serie 7, 5, 265–274.

Pickering, R. (1969). Regional mapping. 1. The geology of the country around Endulen. *Record of the Geological Survey of Tanganyika, 11*, 1–9.

Protsch, R. R. R. (1981). The palaeoanthropological finds of the Pliocene and Pleistocene. In H. Müller-Beck (Ed.), *Die Archäologischen und Anthropologischen Ergebnisse der Kohl-Larsen-Expeditionen in*

Nord-Tanzania 1933–1939, Band 4, 3 (pp. 1–181). Tübingen: Verlag Archaeologica Venatoria, Institute für Urgeschichte der Universität Tübingen.

Reck, H., & Kohl-Larsen, L. (1936). Erster Überblick über die jungdiluvialen Tier- und Menschenfunde Dr. Kohl-Larsen's im nordöstlichen Teil des Njarasa-Grabens (Ostafrika) und die geologischen Verhältnisse des Fundgebietes. *Geologische Rundschau, 27*, 401–441.

Remane, A. (1950). Die Zähne des *Meganthropus africanus*. *Zeitschrift für Morphologie und Anthropologie, 42*, 311–329.

Remane, A. (1954). Structure and relationships of *Meganthropus africanus*. *American Journal of Physical Anthropology, 12*, 123–126.

Robbins, L. M. (1987). Hominid footprints from site G. In M. D. Leakey & J. M. Harris (Eds.), *Laetoli: A Pliocene site in northern Tanzania* (pp. 497–502). Oxford: Clarendon.

Robinson, J. T. (1953). *Meganthropus*, australopithecines and hominids. *American Journal of Physical Anthropology, 11*, 1–38.

Robinson, J. T. (1955). Further remarks on the relationship between "Meganthropus" and australopithecines. *American Journal of Physical Anthropology, 13*, 429–445.

Sanders, W. J. (2008). Review of fossil Proboscidea from the Late Miocene-Early Pliocene site of As Sahabi, Libya. *Garyounis Scientific Bulletin*, Special issue No. 5, 241–256.

Scott, L., Fernandez-Jalvo, Y., Carrion, J., & Brink, J. (2003). Preservation and interpretation of pollen in hyaena coprolites: Taphonomic observations from Spain and southern Africa. *Palaeontologia Africana, 39*, 83–91.

Senyürek, M. (1955). A note on the teeth of *Meganthropus africanus* Weinert from Tanganyika Territory. *Belleten, 19*, 1–55.

Stern, J. T. (2000). Climbing to the top: A personal memoir of *Australopithecus afarensis*. *Evolutionary Anthropology, 9*, 113–133.

Stern, J. T., & Susman, R. L. (1983). The locomotor anatomy of *Australopithecus afarensis*. *American Journal of Physical Anthropology, 60*, 279–317.

Su, D. (2005). The paleoecology of Laetoli, Tanzania: Evidence from the mammalian fauna of the Upper Laetolil beds. Ph.D. dissertation, New York University, New York.

Su, D. F., & Harrison, T. (2007). The Paleoecology of the Upper Laetolil beds at Laetoli: A reconsideration of the large mammal evidence. In R. Bobe, Z. Alemseged, & A. K. Behrensmeyer (Eds.), *Hominin environments in the East African Pliocene: An assessment of the faunal evidence* (pp. 279–313). Dordrecht: Springer.

Su, D., & Harrison, T. (2008). Ecological implications of the relative rarity of fossil hominins at Laetoli. *Journal of Human Evolution, 55*, 672–681.

Susman, R. L., & Stern, J. T. (1991). Locomotor behavior of early hominids: Epistemology and fossil evidence. In Y. Coppens & B. Senut (Eds.), *Origine(s) de la Bipédie chez les Hominidés, Cahiers de Paléoanthropologie* (pp. 121–132). Paris: CNRS.

Susman, R. L., Stern, J. T., & Jungers, W. L. (1984). Arboreality and bipedality in the Hadar hominids. *Folia Primatologica, 43*, 113–156.

Susman, R. L., Stern, J. T., & Jungers, W. L. (1985). Locomotor adaptations in the Hadar hominids. In E. Delson (Ed.), *Ancestors: The hard evidence* (pp. 184–192). New York: Alan R. Liss.

Tattersfield, P. (2011). Gastropoda. In T. Harrison (Ed.), *Paleontology and geology of Laetoli: Human evolution in context* (Fossil hominins and the associated fauna, vol. 2, pp. 567–587). Dordrecht: Springer.

Tuttle, R. H. (1985). Ape footprints and Laetoli impressions: A response to the SUNY claims. In P. V. Tobias (Ed.), *Hominid evolution: Past, Present and Future* (pp. 129–133). New York: Alan R. Liss.

Tuttle, R. H. (1987). Kinesiological inference and evolutionary implications from Laetoli bipedal trails G-1, G-2/3 and A. In M. D. Leakey & J. M. Harris (Eds.), *Laetoli: A Pliocene site in northern Tanzania* (pp. 503–523). Oxford: Clarendon.

Tuttle, R. H. (1990). The pitted pattern of Laetoli feet. *Natural History, 90*, 60–65.

Tuttle, R. H. (1994). Up from electromyography: Primate energetics and the evolution of human bipedalism. In R. S. Corruccini & R. L. Ciochon (Eds.), *Integrative paths to the past* (pp. 251–268). Englewood Cliffs: Prentice Hall.

Tuttle, R. H. (2008). Footprint clues in hominid evolution and forensics: Lessons and limitations. *Ichnos, 15*, 158–165.

Tuttle, R. H., Webb, D. M., Weidl, E., & Baksh, M. (1990). Further progress on the Laetoli trails. *Journal of Archaeological Science, 17*, 347–362.

Tuttle, R. H., Webb, D., & Tuttle, N. (1991). Laetoli footprint trails and the evolution of hominid bipedalism. In Y. Coppens & B. Senut (Eds.), *Origine(s) de la Bipédie chez les Hominidés, Cahiers de Paléoanthropologie* (pp. 203–218). Paris: CNRS.

Tuttle, R. H., Webb, D. M., Tuttle, N. I., & Baksh, M. (1992). Footprints and gaits of bipedal apes, bears, and barefoot people: Perspective on Pliocene tracks. In S. Matano, R. H. Tuttle, H. Ishida, & M. Goodman (Eds.), *Topics in primatology* (Evolutionary biology, reproductive endocrinology, and virology, Vol. 3, pp. 221–242). Tokyo: University of Tokyo Press.

Ullrich, H. (2001). Garusi Hominid 4 – Ein Australopithecinenzahn aus der Sammlung Kohl-Larsen. In: Schultz, M., Atzwanger, K., Bräuer, G., Christiansen, K., Forster, J., Greil, H., Henke, W., Jaeger, U., Niemitz, C., Scheffler, C., Schiefenhövel, W., Schröder, I., Weichmann, I. (Eds.), *Homo* – Unsere Herkunft und Kukunft: Proceedings - 4 (pp. 52–54). Göttingen: Kongress der Gesellschaft für Anthropologie e.V. Cuvillier Verlag.

Ward, C. V. (2002). Interpreting the posture and locomotion of *Australopithecus afarensis*: Where do we stand? *Yearbook of Physical Anthropology, 45*, 185–215.

Weinert, H. (1950). Über die neuen Vor- und Frühmenschenfunde aus Afrika, Java, China und Frankreich. *Zeitschrift für Morphologie und Anthropologie, 42*, 113–145.

White, T. D. (1977). New fossil hominids from Laetolil, Tanzania. *American Journal of Physical Anthropology, 46*, 197–230.

White, T. D. (1980a). Additional fossil hominids from Laetoli, Tanzania: 1976–1979 specimens. *American Journal of Physical Anthropology, 53*, 487–504.

White, T. D. (1980b). Evolutionary implications of Pliocene hominid footprints. *Science, 208*, 175–176.

White, T. D. (1981). Primitive hominid canine from Tanzania. *Science, 213*, 348–349.

White, T. D. (1985). The hominids of Hadar and Laetoli: An element-by-element comparison of the dental samples. In E. Delson (Ed.), *Ancestors: The hard evidence* (pp. 138–152). New York: Alan R. Liss.

White, T. D., & Suwa, G. (1987). Hominid footprints at Laetoli: Facts and interpretations. *American Journal of Physical Anthropology, 72*, 485–514.

Chapter 2
Paleontological Localities on the Eyasi Plateau, Including Laetoli

Terry Harrison and Amandus Kweka

Abstract Sixty paleontological localities are recorded at Laetoli and other areas on the Eyasi Plateau. These include several new localities at Laetoli, and many newly designated localities in the Kakesio and Esere-Noiti areas to the south and southwest of Laetoli. Descriptions of the locations, stratigraphic context and paleontological significance of each locality are presented. The Laetoli localities have produced a rich assemblage of fossils from the Pliocene-aged Upper Laetolil Beds and Upper Ndolanya Beds. The most productive localities are Localities 2, 10E and 18, which have each yielded more than 2,000 fossil mammals since 1974. In addition, smaller samples of fossil vertebrates and stone artifacts have been recovered from the Pleistocene Olpiro and Ngaloba Beds. Fossil hominins are presently known only from localities at Laetoli. The Upper Laetolil Beds have yielded the remains of *Australopithecus afarensis*, and tracks of fossilized footprints of hominins, presumably of *A. afarensis*, are known from Locality 8 (Footprint Site G). Three new specimens of *A. afarensis* have been recovered from Laetoli since 1998, and the provenance of these specimens is described here. In addition, fossil hominin specimens have been recovered from the Upper Ndolanya Beds for the first time. A cranium of an archaic form of *Homo sapiens* is known from the Upper Ngaloba Beds at Locality 2. Localities in the Kakesio area and the Esere-Noiti area have yielded relatively small, but important, collections of fossil vertebrates, invertebrates and plants from the Lower Laetolil Beds. No hominins have yet been recovered from this stratigraphic unit, although it is possible that they may be discovered in the future with more intensive collecting and surveying.

Keywords Garusi • Gadjingero • Emboremony • Kakesio • Esere • Noiti • Laetolil Beds • Ndolanya Beds • Ngaloba Beds • Hominins • Fossil vertebrates

T. Harrison (✉)
Center for the Study of Human Origins, Department of Anthropology, New York University, 25 Waverly Place, New York, NY 10003, USA
e-mail: terry.harrison@nyu.edu

A. Kweka
National Museum of Tanzania, P.O. Box 511, Dar es Salaam, Tanzania
e-mail: kwekason@yahoo.co.uk

Introduction

The Laetolil Beds are widely distributed across the Eyasi Plateau on the northern side of Lake Eyasi, covering an area of more than 1,000 km^2 (Fig. 2.1). They extend as far south as Esere, Olpiro and Sinoni, west beyond the village of Kakesio, north as far as Lake Ndutu, and east just beyond Endulen (Fig. 2.2). The extinct volcanic centers of Lemagurut, Satiman, Oldeani, and Ngorongoro are located to the east, and these represent the potential sources of the tephra and lavas found in the region. The oldest of the volcanic centers, Satiman (previously Sadiman), has been inferred to be the most likely source of the Laetolil Beds (Hay 1987; Mollel et al. 2011).

During the Pliocene, volcanic ashes were blown southwest from the Satiman highlands in the direction of the prevailing winds today, and these settled on the relatively flat terrain of the southern Serengeti. Laetolil tuffs can be traced more than 50 km away in this direction (cognate sediments deposited in lake beds in the Manonga Valley, more than 150 km to the southwest, can be inferred to be derived from the same volcanic source; Mutakyahwa 1997; Mollel et al. 2011). As a consequence, the Laetolil Beds are thickest towards the east, closest to their source in the volcanic highlands, and they become thinner to the southwest. The richly fossiliferous Upper Laetolil Beds and Upper Ndolanya Beds are best exposed in the Laetoli area. West and south of Laetoli, the Laetolil Beds are extensively exposed, but the Upper Laetolil Beds have largely been lost through erosion, and outcrops of the underlying Lower Laetolil Beds predominate. West, beyond the village of Kakesio, and along the margin of the Eyasi Rift escarpment, the Laetolil Beds have been eroded away, presumably washed into the Eyasi basin, to expose the underlying Precambrian basement rocks.

Paleontological localities on the Eyasi Plateau occur in three main geographical areas: Laetoli, Kakesio, and Esere-Noiti (Fig. 2.2). The most productive localities are those associated with the Upper Laetolil Beds and Upper Ndolanya Beds at Laetoli, and this is where previous expeditions have

T. Harrison (ed.), *Paleontology and Geology of Laetoli: Human Evolution in Context. Volume 1: Geology, Geochronology, Paleoecology and Paleoenvironment*, Vertebrate Paleobiology and Paleoanthropology, DOI 10.1007/978-90-481-9956-3_2,

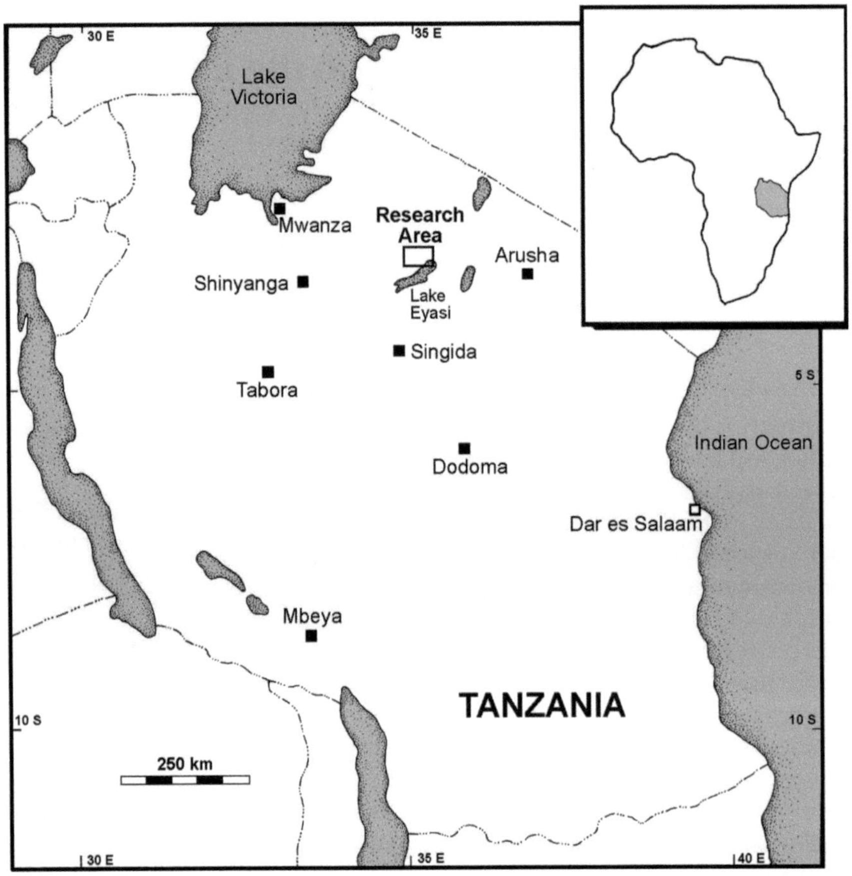

Fig. 2.1 Map of Tanzania showing the Eyasi Plateau-Laetoli research area north of Lake Eyasi. Detail of the research area is shown in Fig. 2.2

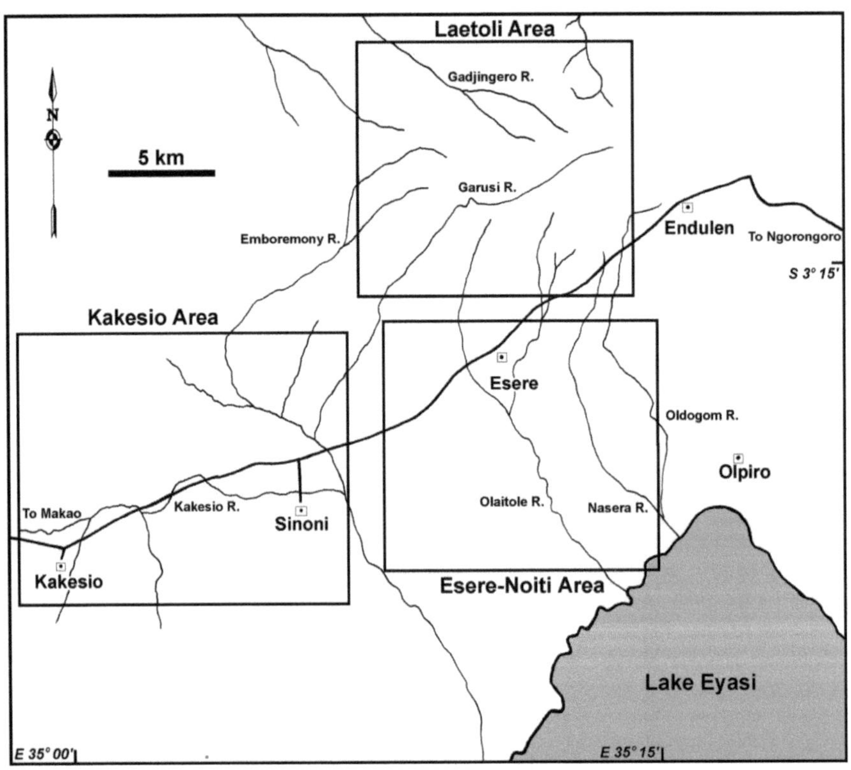

Fig. 2.2 A sketch map of the Eyasi Plateau showing the major rivers and villages, as well as the three main paleontological research areas: Laetoli, Kakesio and Esere-Noiti (see Figs. 2.3–2.5 for detail of insets)

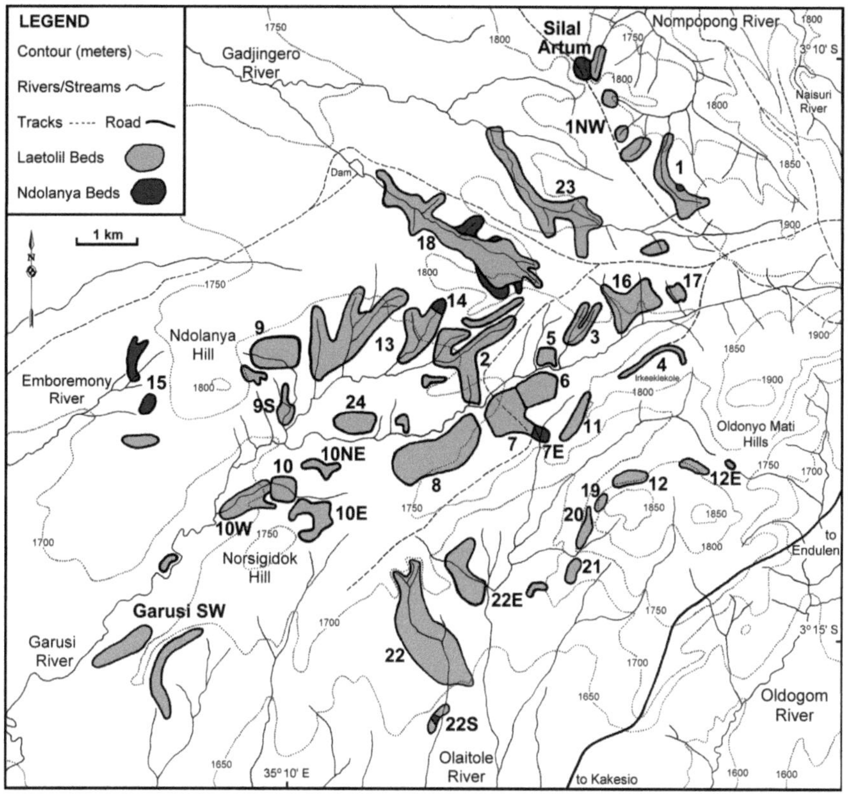

Fig. 2.3 Map of the Laetoli area showing the main outcrops of the Upper Laetolil and Upper Ndolanya Beds and the paleontological collecting localities

concentrated their efforts, where the bulk of the vertebrate fossils have been recovered, including all of the fossil hominins, and where the fossil footprint sites are located (Leakey 1987a, b, c). The localities at Laetoli cover an area of about 100 km^2, with fossiliferous outcrops occurring along the Garusi, Olaitole, Gadjingero and Nompopong River valleys (Fig. 2.3). Extensive outcrops of Lower Laetolil Beds occur in the Kakesio and Esere-Noiti areas, but fossils are much more scanty than at Laetoli, and localities in these areas have so far not yielded any fossil hominins.

Paleontological occurrences at Laetoli were first recorded in 1935. L.S.B. Leakey and other members of the East African Archaeological Expedition (i.e., Peter Kent, Stanhope White, G.T. Bell and Mary Leakey) spent 10 days at Laetoli and localities to the south (Leakey 1935, 1987a; Kent 1941). They made a small collection of fossil vertebrates (including a hominin lower canine) and terrestrial gastropods from exposures along the Garusi valley, as well as at Endulele (probably Esere) and Ngai (probably Ngaloba), which are housed in the Natural History Museum, London. The first major paleontological expedition to Laetoli was directed by Kohl Larsen in 1939. Although Kohl-Larsen did not designate specific collecting localities or apparently keep a field catalogue, he did inscribe on individual specimens the valleys from which they were recovered (i.e., Garussi, Deturi, Vogelfluss, Gadjingero, Marambu and Lemugrut Korongo). By using published sketch maps it is possible to determine the general location of most of the 1939 finds and match them with currently recognized locations (Dietrich 1942; Kohl-Larsen 1943; Protsch 1981; see Table 2.1). Additional

Table 2.1 Place names and collecting locality used by Kohl-Larsen (Dietrich 1942; Kohl-Larsen 1943; Protsch 1981) and Kent (1941), and the currently recognized names and designations (see Figs. 2.2–2.5)

Kohl-Larsen locations	Kent locations	Current place name (Localities)
Deturi	Eilippi River	Olaitole River (Locs. 12, 12E, 19, 20, 21, 22, 22E)
Gadjingero	Lgarja	Gadjingero River (Loc. 18)
Garussi	Vogel River	Garusi River
	Laetoli Springs	Esere Spring
Lanaimuria	Ngarusi Hill	Norsigidok Hill
Lemagrut Korongo		Lemagrut Korongo
Lubiro		Olpiro
Marambu	Oldoway Branch Gorge	Nompopong River (Locs. 1, 1NW)
	Ndulem Hills	Ndonyamati Hills or Oldonyo Mati Hills
Ndulemi		Endulen
Oldogom		Oldogom River (Oleisusu)
Olduwai	Naibadad	Naibadaad
Speer Berg	Speer Mountain	Eseketeti
Vogelfluss	Vogel River	Garusi River

Fig. 2.4 Map of the Kakesio area showing the main outcrops of the Lower Laetolil Beds and the paleontological collecting localities (grey shaded areas)

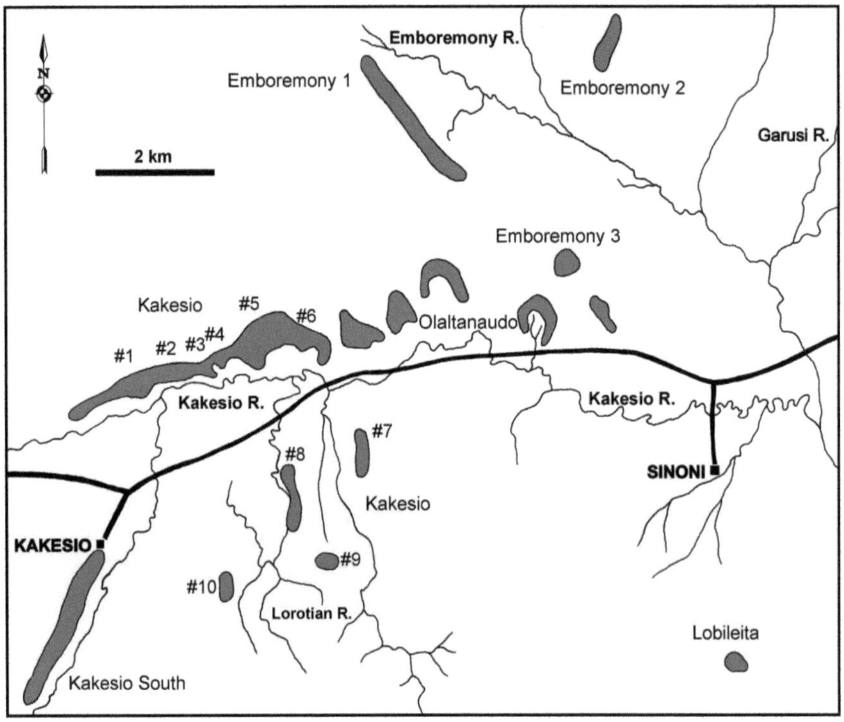

brief excursions to Laetoli were made by the Leakeys and their co-workers in 1959 and 1964 (Leakey 1987a).

Mary Leakey directed full-scale paleontological and geological investigations at Laetoli between 1974 and 1982. During this phase of research the primary focus of attention was on Laetoli, where the bulk of the fossils were collected, but small supplementary collections were made at Kakesio (Leakey and Harris 1987; Harris 1987). Leakey's team designated the various exposures at Laetoli as separate collecting localities, using topographic features and vegetation as natural boundaries. A total of 28 collecting localities and sublocalities were recorded, numbered from 1 to 23 (Leakey 1987a). Most of these localities produced fossils from the Upper Laetolil Beds, while Locs. 7E, 14 and 18 yielded abundant fossils from the Upper Ndolanya Beds (Leakey 1987a).

The Laetoli collecting localities identified by Mary Leakey have been used by subsequent researchers, including teams led by Terry Harrison from 1998 to 2005. However, several new localities have been identified (i.e., Locs. 1NW, 10NE, 22E, 22S, 24, Garusi SW, Silal Artum), bringing the total number of collecting localities at Laetoli to 35. The most important new localities and occurrences are those associated with the Upper Ndolanya Beds (i.e., Locs. 15, 22S and Silal Artum), which have, for the first time, yielded fossil hominins from this stratigraphic unit (Fig. 2.3). In addition, systematic regional surveys on the Eyasi Plateau to the southwest and south of Laetoli have led to the recognition of a number of newly recognized fossil-bearing localities (25 in total), mostly exposing outcrops of the Lower Laetolil Beds. These include localities at Emboremony, Engesha, Esere, Kakesio, Kakesio South, Lobileita, Olaltanaudo, and Noiti (Figs. 2.4 and 2.5).

The aim of this chapter is to briefly describe the location, stratigraphic context, and paleontological significance of the 60 currently recognized localities at Laetoli and elsewhere on the Eyasi Plateau. Particular attention is given to the provenance of the five new hominin specimens recovered since 1998. The main stratigraphic units and horizons producing the fossils at each of these collecting localities are summarized in Table 2.2. Ditchfield and Harrison (2011) provide additional detailed information on the geology of these localities.

Paleontological Localities at Laetoli

Locality 1

This is the easternmost locality at Laetoli (Fig. 2.3). The main fossiliferous exposures of the Upper Laetolil Beds are located at the head of the Nompopong River, a seasonal river course about 3.5 km to the northeast of the Garusi River. The river drains northwards into the Norkuman River, and eventually into the Olduvai Side Gorge. In a series of short steps, the river exposes a ~60 m section of the Upper Laetolil Beds, from the Yellow Marker Tuff to below Tuff 1, as well as more than 60 m of the upper part of the Lower Laetolil Beds. The Upper Laetolil Beds can be traced along the river for a distance of 2.5 km, and

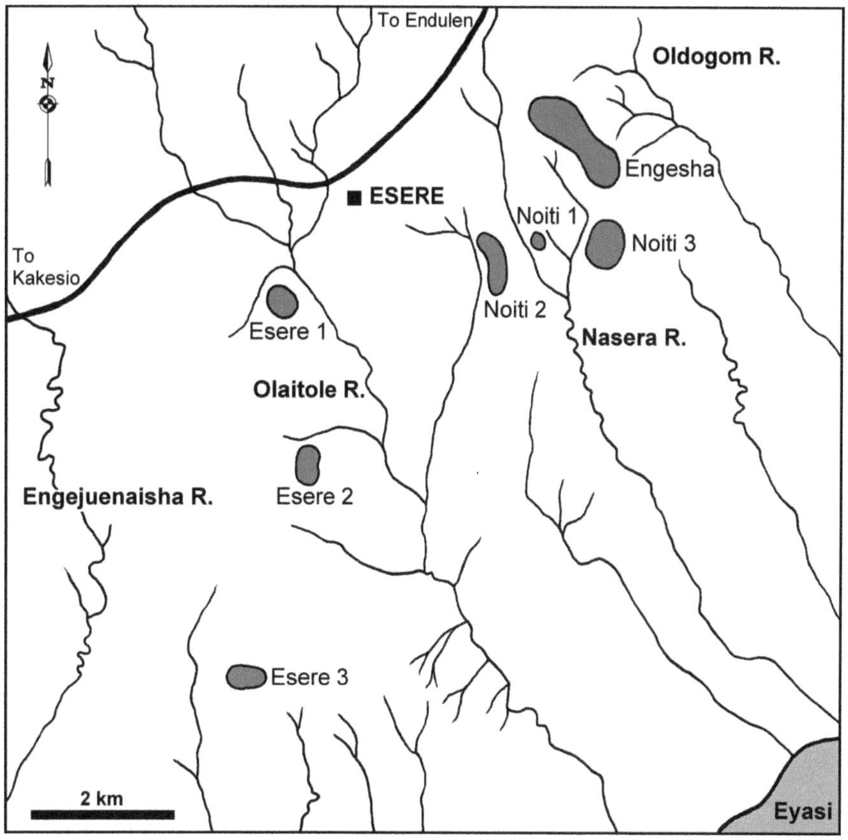

Fig. 2.5 Map of the Esere-Noiti area showing the main outcrops of the Lower Laetolil Beds and the paleontological collecting localities (grey shaded areas)

they represent the most complete and thickest section of these beds at Laetoli. According to Kohl-Larsen's sketch map of the region it is possible to deduce that Loc. 1 is the equivalent of Kohl-Larsen's Marambu (Kohl-Larsen 1943).

Almost all of the fossils from Loc. 1 have been recovered from between Tuffs 6 and just above Tuff 8, mostly from the main gully near the head of the river (Figs. 2.6 and 2.7). It is quite a productive locality, and over 800 fossil mammals have been recovered, including two hominins (L.H. 1 and L.H. 15) collected by Mary Leakey's expeditions (Leakey 1987b). The overlying Yellow Marker Tuff contains abundant fossil wood and root casts. Further downstream, fossils are sparsely represented in the Upper Laetolil Beds, and have not yet been recovered from the Lower Laetolil Beds. A small outcrop of Upper Ndolanya Beds, located in a narrow fault on the eastern side of the gully, has also produced fossils. Overlying the Yellow Marker Tuff is a thick series of Ngaloba Beds, which occasionally yield Pleistocene fossils and stone tools.

Locality 1NW

A series of small, discrete exposures on the western flanks of the Nompopong River valley, located at the origins of several small tributaries and gullies that drain into the main river course. Short sections (less than 10 m) of the Upper Laetolil Beds above and below Tuff 7 are exposed at the locality. The area is heavily vegetated, and fossils are scarce.

Locality 2

A large and complex site on the northern flank of the Garusi River, which extends about 1.5 km E-W and 2 km N-S. The main exposures are associated with seasonal tributaries of the Garusi River. The southern and central areas of the locality are heavily vegetated with bush and tree cover. Two reservoirs were constructed at the locality in 2001, and these now form the western margin of the locality (Fig. 2.8). The locality can be informally subdivided into three main collecting areas: the northwestern arm, the northeastern arm, and the southern arm (Fig. 2.9). Mary Leakey's camp (1976–1981) was located at the junction between the southern and northeastern arms (Leakey 1987a). The most productive areas for fossils are the northwestern and southern arms. Fossils are mainly derived from the Upper Laetolil Beds between Tuffs 5 and 7, including numerous fossil vertebrates, as well as termitaries, insects and abundant seeds. The locality is the second most productive after Loc. 10E, and has yielded over 2,700 fossil mammals. A single hominin specimen, an isolated P_3 (L.H. 25), was found in 1978 from just

Table 2.2 Stratigraphic position of fossil-bearing localities at Laetoli and other sites on the Eyasi Plateau

Locality	Stratigraphic Unit	Main fossiliferous horizons	Comments
1	Upper Laetolil Beds	Between Tuffs 6 and just above Tuff 8	
	Upper Ndolanya Beds		Few fossils
	Olpiro Beds		Few fossils
1NW	Upper Laetolil Beds	Between Tuffs 6 and 8	
2	Upper Laetolil Beds	Between Tuffs 5 and 7	
	Upper Ngaloba Beds		Few fossils and stone tools
3	Upper Laetolil Beds	Between Tuffs 7 and 8	Few fossils between Tuffs 4 and 6
4	Upper Laetolil Beds	Between Tuffs 6 and 8	
5	Upper Laetolil Beds	Between Tuffs 3 and 5	
	Lower Ngaloba Beds		Few fossils
6	Upper Laetolil Beds	Between Tuffs 5 and 7	
7	Upper Laetolil Beds	Between Tuffs 5 and 8	
7E	Upper Ndolanya Beds		
8	Upper Laetolil Beds	Between Tuffs 5 and 7	Few fossils between 7 and just above Tuff 8
9	Upper Laetolil Beds	Between Tuffs 5 and 7	Few between Tuffs 7 and 8
	Upper Ndolanya Beds		Few fossils
9S	Upper Laetolil Beds	Between Tuffs 2 and just below Tuff 1	
10	Upper Laetolil Beds	Between Tuffs 3 and just below Tuff 1	
10E	Upper Laetolil Beds	Between Tuffs 5 and 7	Few fossils between Tuffs 7 and 8
10W	Upper Laetolil Beds	Between Tuffs 3 and just below Tuff 1	
10NE	Upper Laetolil Beds	Uncertain	Few fossils
11	Upper Laetolil Beds	Between Tuffs 7 and 8	
12	Upper Laetolil Beds	Between Tuffs 5 and 8	
12E	Upper Laetolil Beds	Between Tuffs 5 and 7	
13	Upper Laetolil Beds	Between Tuffs 5 and 8	Few fossils between Tuffs 3 and 5
	Upper Ndolanya Beds		Few fossils
	?Ngaloba Beds		Few fossils
14	Upper Ndolanya Beds		
15	Upper Laetolil Beds	Between Tuffs 5 and 7	
	Upper Ndolanya Beds		
16	Upper Laetolil Beds	Between Tuff 7 and Yellow Marker Tuff	
	Olpiro Beds		Fossils and stone tools
17	Upper Laetolil Beds	Between Tuff 7 and Yellow Marker Tuff	
18	Upper Ndolanya Beds		
19	Upper Laetolil Beds	Between Tuffs 5 and 8	
20	Upper Laetolil Beds	Between Tuffs 6 and 8	
21	Upper Laetolil Beds	Between Tuffs 5 and 7	
22	Upper Laetolil Beds	Between Tuffs 5 and 7	Few fossils between Tuffs 1 and 2
22E	Upper Laetolil Beds	Between Tuffs 5 and 7	Few fossils between Tuffs 2 and 4
	Upper Ndolanya Beds		Few fossils
22S	Upper Laetolil Beds	Between Tuffs 1 and 3	Few fossils
	Upper Ndolanya Beds		Restricted to Nenguruk Hill
23	Upper Laetolil Beds	Between Tuffs 7 and 8	Few fossils
	Lower Ngaloba Beds		Fossils in association with stone tools
24	Upper Laetolil Beds	Between Tuffs 6 and 8	Few fossils
Emboremony 1–3	Lower Laetolil Beds		
	?Ngaloba Beds		
Engesha	Lower Laetolil Beds		
Esere 1–3	Lower Laetolil Beds		
Garusi SW	Laetolil Beds	Uncertain	
Kakesio 1–10	Lower Laetolil Beds		
Kakesio South	Lower Laetolil Beds		
Lobileita	Lower Laetolil Beds		
Olaltanaudo	Lower Laetolil Beds		
Oleisusu	Upper Laetolil Beds	Between Tuff 7 and Yellow Marker Tuff	
Ndoroto	Upper Laetolil Beds	Between Tuff 8 and Yellow Marker Tuff	
Noiti 1–3	Lower Laetolil Beds		Fossil vertebrates from Noiti 3 only
Silal Artum	Upper Ndolanya Beds		

Fig. 2.6 Locality 1. View (west) of main exposure of Upper Laetolil Beds at the head of the Nompopong River valley. Marker Tuff 8 and the Yellow Marker Tuff are highlighted. The concrete pedestal (indicated) marks the location of the first fossil hominin find (L.H. 1) made by Mary Leakey's expedition in 1974

Fig. 2.7 Locality 1. View (north) of main exposure of Upper Laetolil Beds at the head of the Nompopong River valley. The concrete pedestal marks the location of the first fossil hominin find (L.H. 1) made by Mary Leakey's expedition in 1974. It is located 1.2 m above Tuff 7. L.H. 20 was initially believed to be a hominin incisor, but was subsequently recognized by Tim White as that of a Cercopithecidae

Fig. 2.8 View (south) of the Garusi Valley showing the location of Locality 2 (northeastern and northwestern arms)

above Tuff 6 (Leakey 1987b). The footprint Tuff, Tuff 7, is exposed along the northwestern margin and southern margin of the locality. Footprint sites O and P are located at the southern end of the locality (Leakey 1987c).

A poorly exposed outcrop of the Laetolil Beds above Tuff 7 is located just north of the main exposure at the foot of Ngirerati Hill, and has yielded few fossils. This horizon is overlain by a thick series of Ndolanya Beds and Naibadad Beds. The Late Pleistocene Upper Ngaloba Beds in the southern part of the locality, close to the Garusi River, have yielded Middle Stone Age tools in association with faunal remains, including a relatively complete cranium of *Homo sapiens* (L.H. 18) (Day et al. 1980; Magori and Day 1983). Mary Leakey (1987b: 116) incorrectly lists the latter specimen as coming from Loc. 25, which is presumably a typographic error for Loc. 2S.

Locality 3

This locality is situated about 1 km east of Loc. 2, on the northern side of the Garusi valley. Outcrops of the Upper Laetolil Beds extend approximately 1 km along a gully running NE-SW that drains into the Garusi River (Fig. 2.10).

Fig. 2.9 Locality 2. View (north) of southern arm

Fig. 2.10 Locality 3. View (north) of main exposures. Norsigidok Hill is visible in the background

A second tributary of the Garusi, just to the east of the main gully, also exposes Upper Laetolil Beds, but these are poorly fossiliferous. Most of the fossils from this locality are derived from between Tuffs 7 and 8, but a few specimens have been recovered from lower in the stratigraphic sequence between Tuffs 4 and 6. Mary Leakey's expedition recovered a single fossil hominin from above Tuff 7, L.H. 2, a partial mandible of a juvenile individual (Leakey 1987b).

Also of interest is a fossiliferous horizon 60–70 cm above Tuff 7, consisting of a calcareous pale brown clay about 17 cm thick, which is rich in fossil dung, twigs, seeds, and insects (see Bamford 2011a; Krell and Schawaller 2011; Kitching and Sadler 2011; Harrison 2011a; Figs. 2.11 and 2.12). The tuffaceous sediments, derived from the surrounding volcanic ashes, were accumulated in a shallow pond, along with a variety of organic debris. The debris was probably washed into the pond from the surrounding land surface by seasonal run-off or accumulated in a shallow depression that was subsequently flooded and inundated with sediment. It may have been a small waterhole that flooded seasonally. Based on the extent of the exposed sediments the pond was at least 40 m wide (Fig. 2.13). After compaction and consolidation of the

Fig. 2.11 Locality 3. The clay horizon yielding fossil plants, coprolites and insects occurs just below the hard tuff running left to right across the center of the picture

deposits, the organic materials decomposed, leaving behind a cavity or void in the sediment. Because the sediments are fine-grained, the void produced a detailed mold of the exter-

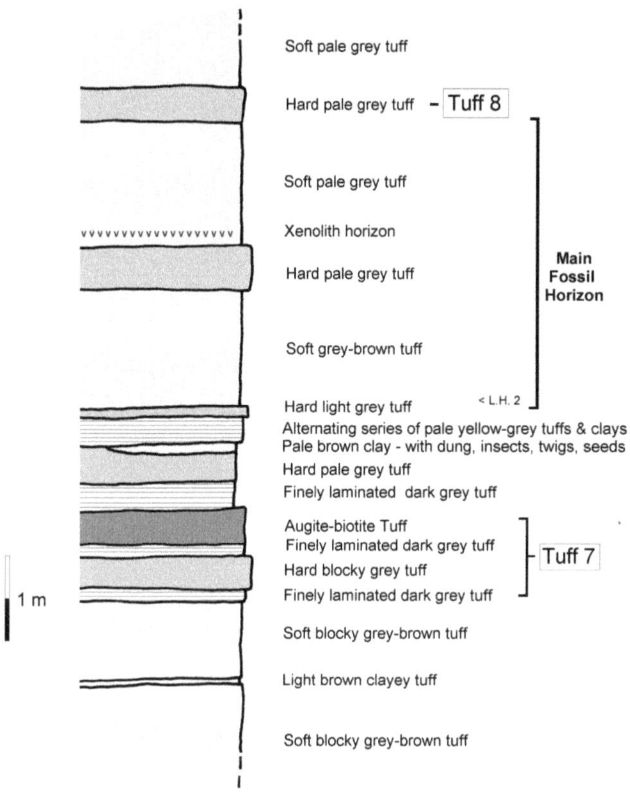

Fig. 2.12 Locality 3. Part of the stratigraphic section that has yielded fossil remains. Most of the vertebrate fossils are derived from the pale grey and grey-brown tuffs below Tuff 8. The stratigraphic position of the L.H. 2 pedestal is indicated. The coprolites, plant material and fossil insects are derived from a pale brown clay about 60–70 cm above Tuff 7

nal structure of the debris. This was subsequently filled with calcite in solution derived from the calcium carbonate rich carbonatite tuffs of the Laetolil Beds. This produced a natural cast of the organic debris, resulting in beautifully preserved fossils (Fig. 2.14).

Locality 4

A series of exposures forming a narrow strip around the base of the northern side of Irkeeklekole Hill, located about 400 m south of the head of the Garusi River (Fig. 2.15). The Kohl-Larsen (1938–1939) and Harrison (1998–2005) camps were both established near this locality. The exposures extend for a length of about 1.3 km, but those along the eastern side of the locality are the most extensive and have produced the majority of fossils. Fossils are not common, and they tend to be fragmentary. All are derived from the Upper Laetolil Beds between Tuffs 6 and 8. The Footprint Tuff (Tuff 7) is exposed along the eastern side of the locality, with a few isolated prints, and the base of the tuff preserves an abundance of calcified plant material, including leaves, twigs (up to 4 cm in diameter) and seeds. Fossil thorns and impressions of compound microphyllous leaves, similar to those of *Acacia*, have been recovered. Higher up the sequence, the poorly exposed Yellow Marker Tuff contains an abundance of calcareous root cast, up to 10 cm in diameter.

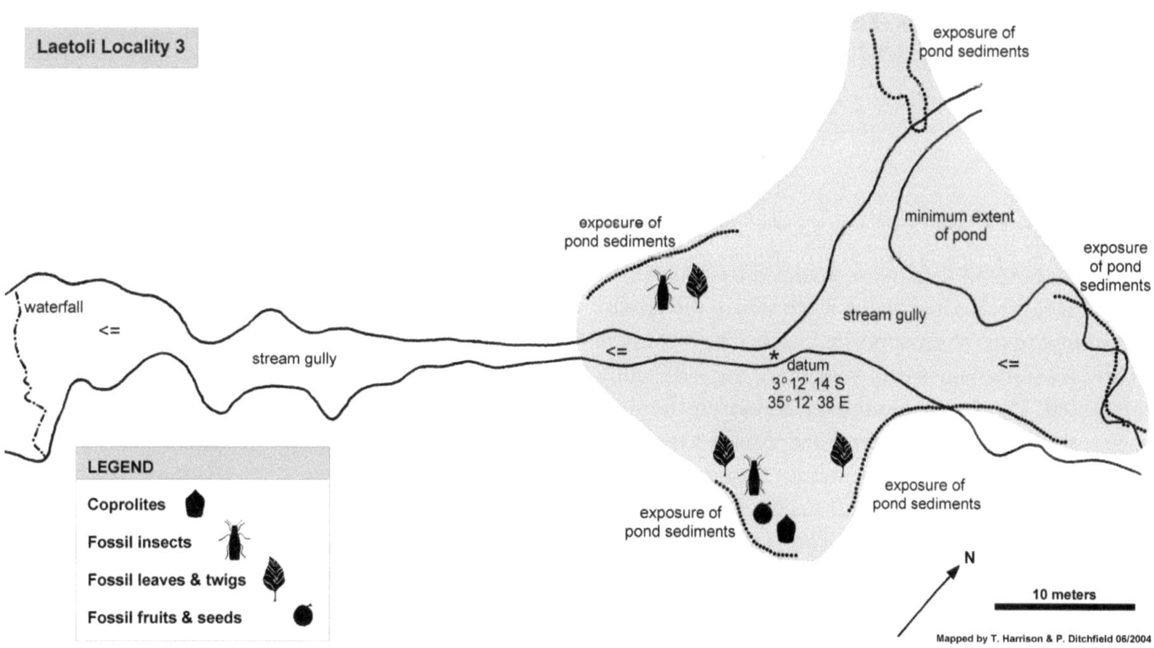

Fig. 2.13 Locality 3. Sketch map of the distribution of the clay horizon that has yielded fossil plants, insects and coprolites, as well as the reconstructed extent (grey shading) of the pond or waterhole

Fig. 2.14 Locality 3. Fragments of fossil wood and seeds that have eroded out of the clay horizon onto the surface of the deposits

Fig. 2.15 Locality 4. View (south) across the Garusi River valley showing the location of Localities 4 and 16, and other major geographical landmarks

Locality 5

This is a relatively small locality, surrounded by thick vegetation (Fig. 2.16). It is located on the north side of the Garusi valley, very close to the river, opposite Loc. 6. The locality is quite richly fossiliferous; small mammals are especially well-represented. The fossils are primarily derived from the Upper Laetolil Beds between Tuffs 3 and 5. Tuff 3 is extensively exposed, and it preserves numerous footprints, especially those of guineafowl and small mammals. Two fossil hominins have been recovered (L.H. 7 and L.H. 12) (Leakey 1987b). A small hillock to the eastern side of the locality is capped by a thin remnant of Ngaloba Beds. The fossils from this horizon are found mixed with those from the Laetoli at the foot of the hillock, but they can be readily separated in most cases based on coloration and degree of mineralization.

Fig. 2.16 Locality 5. View of main exposures between Tuffs 3 and 5

Locality 6

This locality is situated on the southern side of the Garusi Valley, close to the main river course, about 1 km from the western margin of Loc. 4. The fossils are derived from the Upper Laetolil Beds between Tuffs 5 and 7. The site is relatively extensive, with good exposures, and is one of the most productive localities. The combined collections of fossil mammals recovered since 1974 exceed 1,300 specimens. Two isolated molars of fossil hominins were recovered by Mary Leakey's team (L.H. 16 and L.H. 26) in 1976 and 1978 respectively (Leakey 1987b), and a hominin patella (EP 2038/03) was recovered in 2003 (Harrison 2011b). The latter specimen was found in a small side gully (coordinates: 3° 12′ 57″ S, 35° 12′ 12″ E) 30 cm below Tuff 6.

Locality 7

An extensive exposure of Laetolil Beds on the southern side of the Garusi River, located just to the west and south of Loc. 6 and contiguous with it (Fig. 2.17). The motor track that runs through the Laetoli area crosses at a ford onto the southern side of the Garusi Valley and passes through the locality. The main part of the locality exposes the Upper Laetolil Beds between Tuffs 5 and 7. Fossil vertebrates are found throughout the section, but are mainly derived from a 5 m section above and below Tuff 6. On the southeastern side of the locality, near Loc. 7E, there is a series of low cliffs and hillocks, 2–4 m high, which expose a fossiliferous section of the Upper Laetolil Beds above Tuff 7, including Tuff 8 and the Yellow Marker Tuff. Several hominins have been recovered from this locality, including L.H. 3, L.H. 4 (the lectotype of *Australopithecus afarensis*), L.H. 6 and L.H. 30 (Leakey 1987b). There are several good exposures of the Footprint Tuff, including Footprint Sites A, B, and C (Leakey 1987c).

Fig. 2.17 Locality 7. View of main exposures

Locality 7E

This is a small exposure of Upper Ndolanya Beds, extending about 50 m, located just to the southeast of Loc. 7. The Upper Ndolanya Beds are deposited in a minor fault that can be seen as a break in the NE-SW trending ridge that forms the southern boundary of the locality. Due to the faulting, the relatively soft and easily weathered Ndolanya Beds are preserved here, whereas at most other localities these tuffs have been rapidly denuded from the surface of the Laetolil Beds. It is very fossiliferous, having yielded a high percentage of postcranial remains of medium to large-sized bovids, including associated elements and partial skeletons (Leakey 1987a).

Locality 8

This is an extensive locality with good exposures of the Upper Laetolil Beds between Tuffs 4 and the Yellow marker Tuff (Fig. 2.18). It is located on the southern side of the Garusi River, just west of Loc. 7, being separated from it by a low grassy ridge. The western (main) exposures consist of a large open area with sparse tree cover, consisting of a series of low gullies. The eastern exposures are relatively flat, with little topographic relief, and are quite densely vegetated, especially near the Garusi. Locality 8 is one of the most productive localities, with more than 1,800 fossil mammals having been recovered. Most fossils, including the hominins (L.H. 5/27/28, L.H. 13, and L.H. 19), are derived from horizons between Tuffs 5 and 7, but some fossils have also been recovered from between Tuff 7 and just above Tuff 8 (Leakey 1987b). One of the hominin specimens, L.H. 29, a weathered mandibular fragment, was found in 1979 on the surface in the lower part of the sequence. The specimen was assumed to be Pleistocene in age and provisionally attributed to *Homo cf. H. erectus* (Leakey 1987b). However, further comparisons indicate that the specimen is more appropriately attributed to *A. afarensis* (Harrison 2011b). The hominin trail of footprints (Footprint Site G) was excavated at the locality in 1978 and 1979 (Leakey 1987c), and was subsequently covered to conserve the prints in 1996 (Fig. 2.19). Musiba (1999) and Musiba et al. (2007) have conducted a detailed taphonomic and paleoecological investigation at this locality, as well as at Loc. 9.

Fig. 2.18 Locality 8. View of main exposures between Tuffs 6 and 7. The L.H. 5 pedestal is visible center right

Fig. 2.19 Locality 8. View (south) of footprint site G, which was covered in 1996 to conserve the hominin tracks

Locality 9

Locality 9 is located on the northern flank of the Garusi River, East of Ndolanya Hill, and just west of Loc. 13, from which it is separated by a flat grassy area. The locality is quite extensive, with well-exposed Upper Laetolil Beds between Tuffs 5 and 7 (Fig. 2.20). There is little topographic relief, and the beds slope gently towards the Garusi. These exposures have yielded the majority of the vertebrate fossils. Large termitaria

Fig. 2.20 Locality 9. View (northwest) of main exposures of Upper Laetolil Beds. Hard dark grey tuff in the foreground is Tuff 6. Ndolanya Hill is in the background

Fig. 2.21 Locality 9S. View of main exposures at northern end of locality. The cliff is capped by Tuff 1, below which is a series of waterlain tuffaceous clays and silts

are also common above and below Tuff 6 (Darlington 2011). Some fossils occur between Tuffs 7 and 8, but they are not common. Along the heavily vegetated ridge that forms the northern extent of the locality there are exposures of the Upper Ndolanya and Naibadad Beds, but only the former appear to be fossiliferous. In the southern part of the locality, exposures are poor because of heavy vegetation cover, but the Upper Laetolil Beds do extend down to below Tuff 3. The lower part of the section below Tuff 5 has produced few fossils. Calcified plant material is associated with the base of Tuff 3, and several small ovoid termitaries have been observed in situ below Tuff 3. An isolated upper molar of a hominin (L.H. 17) was recovered as a surface find in 1976 (Leakey 1987b). An extensive section of the Footprint Tuff is exposed and is designated as Footprint Site K (Leakey 1987c).

Locality 9S

This locality is situated on the northern side of the Garusi, about 400 m south of Loc. 9, along a river gully that drains into the Garusi River (Fig. 2.21). The boundary between Loc. 9 and Loc. 9S is a 5 m high waterfall. Loc 9S consists of a series of gullies that trend N-S, with steep sided cliffs up to 5 m in height. The surrounding area consists of dense bush and woodland. The section exposes a series of tuffs from the lower part of the Upper Laetolil Beds below Tuff 2. The base of the section consists of finely laminated waterlain tuffaceous beds. Fossil vertebrates are quite common, as are relatively large termitaries (70–100 cm in diameter). Bonnefille and Riollet (1987) analyzed fossil pollen samples obtained from clay horizons 2 m below Tuff 1 and from a termitary 3 m below Tuff 1. Attempts to resample the clay horizons below Tuff 1 have failed to produce definitive fossil pollen (Rossouw and Scott 2011).

Locality 10

Situated on the southern side of the Garusi valley, midway between the highest (western) peak of Norsigidok Hill and the Garusi River. A small seasonal stream flows westwards through the locality to join the Garusi River close to the western border of Loc. 10W. The large gully system is capped by a hard and resistant tuff, Tuff 3, which also forms the upper lip of a waterfall, about 5 m high, at the eastern boundary of the locality. Within the gully system is a series of cliffs and columns, capped primarily by Tuff 2. The stratigraphic sequence consists of about a 10 m section of the Upper Laetolil Beds from Tuff 3 to below Tuff 1. Most of the fossils are derived from between Tuff 2 and the bottom of the section. Fossil vertebrates are common, and include exceptionally well-preserved eggs of ground nesting birds (Harrison 2005). An isolated upper molar of *Australopithecus afarensis* (L.H. 31) was recovered from this locality in 1987 (Kyauka and Ndessokia 1990), but no precise provenance information accompanied the description of the specimen. Large termitaries and gastropods are common (see Darlington 2011; Tattersfield 2011). The Loc. 10 complex of localities (i.e., Locs. 10, 10E and 10W) was the area primarily surveyed by the Leakeys in 1935 and 1959 (Leakey 1987a).

Locality 10W

This locality is contiguous with Loc. 10, and represents the westward extension of the same series of sediments (Fig. 2.22). The exposures extend for about 1 km E-W along a gully system that is cut by the same seasonal river that passes through Loc. 10. As in Loc. 10, fossils are derived from below Tuff 2. These include two hominins (L.H. 10 and

Fig. 2.22 Locality 10W. View of main exposures of Upper Laetolil Beds. The hard dark grey tuff capping the pillars is Tuff 2

L.H., 11), both discovered in 1975 (Leakey 1987b). Fossil termitaries are especially common at this locality (Darlington 2011). In 1975 Mary Leakey set up camp on the western margin of the locality, at the confluence of the river that runs through Locs. 10 and 10W, at the same location as the 1935 and 1959 Leakey camps (Leakey 1987a).

Locality 10E

A large series of exposures in a circular depression about 0.7 km wide, which abuts against the north slope of Norsigidok Hill. Low cliffs that expose sediments between Tuffs 5 and 7 dominate the northeastern part of the locality, and this area has produced most of the fossils (Fig. 2.23). On the flanks of Norsigidok, the sequence of Upper Laetolil Beds continues above Tuff 7 through Tuff 8 to the Yellow Marker Tuff, and these are eventually capped by Ogol Lavas (Fig. 2.24). The section of Upper Laetolil Beds exposed, excluding the Yellow marker Tuff, is about 13–15 m thick. Fossil vertebrates have been recovered from between Tuffs 7 and 8, but these are less common than those between Tuffs 5 and 7. Loc. 10E is the most productive of the localities at Laetoli, with well over 2,500 fossil mammals recovered. Only a single hominin specimen has been recovered; an isolated P_3 (L.H. 24) discovered in 1978 (Leakey 1987b). The Footprint Tuff is well exposed at the locality, and several Footprint sites (F, H, I, J) have been designated (Leakey 1987c).

Locality 10NE

A poorly exposed area of Upper Laetolil Beds north of Loc. 10E and just south of the Garusi River. It is a flat area, covered with eroded sediments, and used heavily as a cattle path. Few fossil mammals have been recovered, and these tend to be fragmentary.

Fig. 2.23 Locality 10E. Section of Upper Laetolil Beds between Tuffs 5 and 7 that has yielded most of the vertebrate fossils. The figure in the white hat (Carl Swisher III) is kneeling on Tuff 7. The backpack of the seated figure (Lindsay McHenry), bottom right, is resting on Tuff 6

Fig. 2.24 Locality 10E. Upper section of the Upper Laetolil Beds, comprising Tuff 8 and the Yellow Marker Tuff, which are capped by Ogol Lavas on top of Norsigidok Hill

Locality 11

A narrow strip of outcrops of Upper Laetolil Beds, about 0.7 km in length, trending NE-SW. It is located on the southern side of the Garusi River valley, less than 0.5 km west of Loc. 4 and just east of Locs. 7 and 7E. It consists of a series of low cliffs (10–15 m high) and rounded hillocks that border the southwestern extension of Irkeeklekole Hill (Fig. 2.25). The surface of the locality is littered with large boulders of Ogol Lava derived from the cap of the hill. A small ephemeral river cuts along its west margin, passes through the break in the ridge south of Loc. 7E, and is eventually captured by the Olaitole River to the south. To the southwest and along the western margin of the locality Tuff 7 is well exposed, and

Fig. 2.25 Locality 11. View (northeast) of main exposures of Upper Laetolil Beds above Tuff 7. The locality is littered with large boulders of Ogol Lava derived from the cap of Irkeeklekole Hill

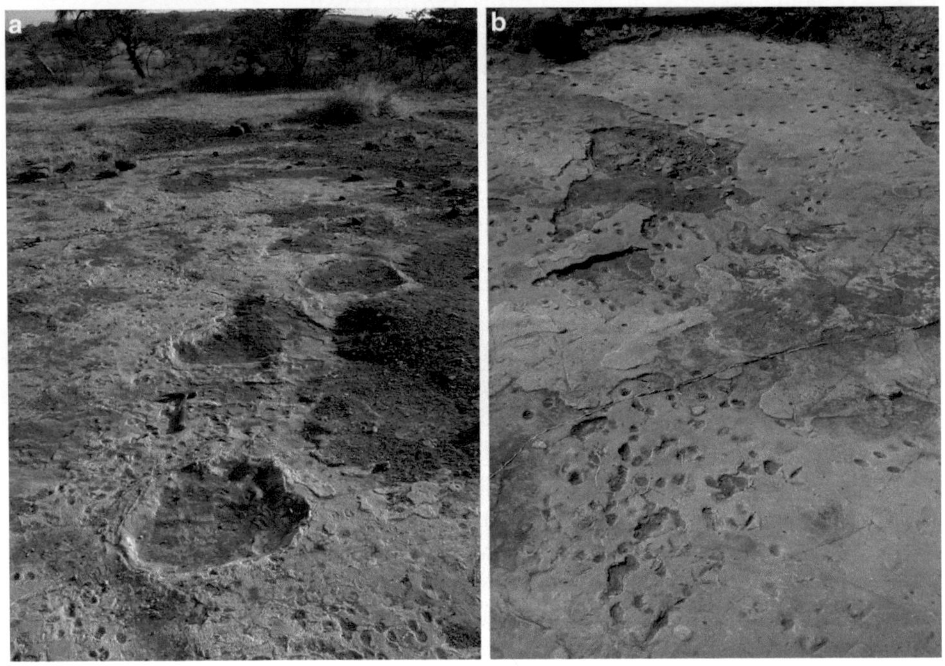

Fig. 2.26 Locality 11. Exposed surface of Tuff 7 with footprints. (**a**) A trail of an elephantid, with numerous small prints of *Serengetilagus* and *Madoqua*. (**b**) Numerous small prints and trails of *Madoqua* and guinea fowl (bottom center and right)

preserves numerous footprints (Footprint Sites D and E; Leakey 1987c) (Fig. 2.26). Fossil vertebrates are derived from the overlying tuffs between Tuffs 7 and 8. It is a relatively productive site, mainly because the tuffaceous sediments are soft and erode quickly. However, trampling by livestock from the nearby bomas (Maasai homesteads) results in a high degree of fragmentation of the fossils. Upper cheek teeth of a single individual of a hominin (L.H. 8 and L.H. 22), found two years apart, were recovered by Mary Leakey's team (Leakey 1987b).

Locality 12

This a narrow strip of exposures around the base of the Ndonyamati (or Oldonyo Mati) Hills overlooking the Olaitole River valley to the north, and about 1.5 km south of the Garusi River. The valley and northern slopes of the Ndonyamati Hills are densely vegetated with tall grass, thick bush and open woodlands. A section of the Upper Laetolil Beds between Tuffs 5 and 8 are exposed, with most of the fossils derived from between Tuffs 5 and 6. Relatively few fossils have been recovered.

Locality 12E

A locality on the southern flanks of the Olaitole River valley, located just to the east of Loc. 12. A short section (about 9 m) of the Upper Laetolil Beds between Tuffs 5 and 7 are exposed, and these have yielded a good number of fossils, mostly from between Tuffs 6 and 7. The base of Tuff 7 is rich in calcified twigs, seeds and roots. L.H. 21, a partial skeleton of an immature individual of *A. afarensis*, was recovered from this locality in 1976 (White 1980; Leakey 1987b; Kyauka 1994).

Locality 13

A series of low gullies and exposures on the northern side of the Garusi River valley, associated with several small channels that drain into the Garusi. It is located immediately west of Loc. 2 and about 0.3 km east of Loc 9, and separated from them by a flat expanse of grass and bush growing on mbuga clay. Although the locality covers a large area, extending about 1.5 km northwards from the Garusi River and E-W for 2 km, most of the exposures are poorly fossiliferous. The majority of the fossils are derived from the Upper Laetolil Beds between Tuffs 5 and 8, but a few have been recovered from exposures close to the Garusi River from between Tuffs 3 and 5. One of the most productive areas is a small E-W oriented drainage channel, about 100 m long, informally named "Snake Gully" (coordinates: 3° 12′ 15.1″ S, 35° 10′ 58.7″ E). Numerous fossil vertebrates have been collected from this gully, including a partial skeleton of a large viper with an immature *Serengetilagus* skeleton preserved in its abdominal cavity (Rage and Bailon 2011), derived from an horizon 2 m below Tuff 7. In addition, a few fossils have been recovered from the Upper Ndolanya Beds, and possibly also from the Ngaloba Beds. The Footprint Tuff (Tuff 7) is extensively exposed along the northern boundary of the locality, and includes Footprint Sites L, M and N (Leakey 1987c).

Locality 14

A small locality, situated between Locs. 2 and 13, about 2 km north of the Garusi River. It exposes tuffaceous channel-fill sediments of the Upper Ndolanya Beds. Excavations by Mary Leakey's team produced numerous fossils (Leakey 1987a). Today, however, the locality is much overgrown and few new fossils have eroded out onto the surface.

Locality 15

This is the westernmost locality at Laetoli, situated at the head of the Emboremony River, about 1 km west of Ndolanya Hill. A short section of Upper Laetolil Beds, about 6 m thick, is exposed on the southern side of the locality (Fig. 2.27). This is overlain by poorly exposed deposits of the Yellow Marker Tuff and Upper Ndolanya Beds, as well as undifferentiated Pleistocene sediments. Upper Laetolil Beds between Tuffs 6 and 8 are exposed, but vertebrate fossils are mostly derived from between Tuffs 6 and 7. A few fossils of Late Pleistocene age, in association with stone tools and ostrich egg shell beads, were found scattered on the surface and derived from the mbuga clays and river alluvium. A newly discovered exposure of Upper Ndolanya Beds occurs on the northern side of the Emboremony along the western flank of Ndolanya Hill (coordinates: 3° 12′ 59.6″ S, 35° 08′ 51.5″ E). A rich fossil assemblage was found eroding out of a series of gullies and low rounded hillocks (Fig. 2.28). The fossils include numerous complete and partial skulls of lagomorphs and rodents, as well as a skull and partial skeleton of a new

Fig. 2.27 Locality 15. Main exposure of the Upper Laetolil Beds on the southern flank of the Emboremony River valley. The thick tuff running left to right across the center of the figure is Tuff 7. The top of the section exposes the Yellow Marker Tuff (the light colored sediment that the middle figure is standing on) and undifferentiated Pleistocene sediments (darker sediments above)

Fig. 2.28 Locality 15. Main exposures of the Upper Ndolanya Beds on the northeast flank of the Emboremony River valley

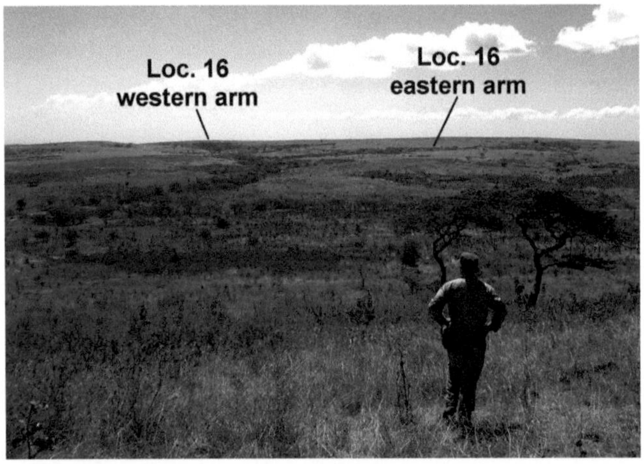

Fig. 2.29 Locality 16. View (north) across the Garusi River valley. The western and eastern arms of Loc. 16 are visible as a line of trees associated with the main gullies

species of medium-size alcelaphine (Gentry 2011). In addition to fossil vertebrates, which are primarily found in the lower part of the section, cocoons and brood cells of solitary hymenopterans are very common, especially in the upper part of the section. A second, smaller outcrop of Upper Ndolanya Beds (coordinates: 3° 12′ 42.0″ S, 35° 08′ 35.3″ E) occurs about 0.6 km to the northeast, but this has yielded few identifiable fossils.

Locality 16

A richly fossiliferous locality on the north side of the Garusi River valley, situated just east of Loc. 3 (Fig. 2.29). It is associated with two ephemeral streams that drain into the Garusi River. The exposed section of Upper Laetolil Beds extends from below Tuff 6 to the Yellow Marker Tuff, but the vertebrate fossils are all derived from between Tuffs 7 and the Yellow Marker Tuff. The western portion of the locality, which extends N-S for about 0.6 km, represents the main gully, and is the most productive. The eastern arm, which extends NE-SW for approximately 0.7 km, is much less fossiliferous. Tuff 7 is well exposed along the floor of the river courses, and it forms a resistant layer capping waterfalls along both the eastern and western tributaries. The base of Tuff 7, which is well exposed in several places, is densely packed with calcified plant material, including wood, twigs (up to 3 cm in diameter), a diversity of seeds, root casts, and fine root mats (Fig. 2.30). A number of small, vertically standing boles penetrate Tuff 7, and these presumably represent the remains of trees that were partially buried by the ash fall. At the eastern end of the locality, a number of leaf impressions are preserved in a coarse-grained dark grey tuff at the top of Tuff 8 (Bamford 2011a). At the northern end of the western arm, there is a section of Yellow Marker Tuff

Fig. 2.30 Locality 16. Fossil seeds, twigs and wood exposed on the basal layer of Tuff 7 in the eastern arm of Loc. 16

about 6 m thick. This is overlain by the Olpiro Beds, which contain large mammal fossils and stone tools (although Leakey [1987a] suggests that most of these may originally have been derived from the Ngaloba Beds) (Fig. 2.31). Excavation by J.W.K. Harris at "Grubb's site" in Loc. 16 produced an Early Stone Age assemblage dominated by core tools and crudely made bifacial tools (Harris and Harris 1981). Ndessokia conducted further excavations at Loc. 16 in 1987, and recovered additional Oldowan artifacts in association with vertebrate fossils (Ndessokia 1990).

In 2000 Michael Mbago found a mandibular fragment with P_3-M_1 (EP 2400/00) of *Australopithecus afarensis* at this locality. This is only the fifth mandibular specimen of a hominin to be recovered from Laetoli, and the first hominin

Fig. 2.31 Locality 16. Main exposure of Olpiro Beds at the northern end of the western arm of Loc. 16

specimen to be found at Loc. 16 (Harrison 2011b). The specimen was found on the surface, near a small mound of soft pale grey calcareous tuff, within a shallow drainage channel on the western flank of the main gully (Fig. 2.32). It was located 12.1 m from the edge of the gully and 49.7 m from the main river course, at a distance of 56 m upstream from the waterfall (coordinates: 3° 12′ 10.9″ S, 35° 12′ 51.7″ E). It is possible that the specimen had been transported a short distance from its original location. There are no fresh breaks, and it is likely that the specimen was initially buried and fossilized in the same fragmentary condition as it was found. A diligent search and screening of the surface sediment failed to yield additional jaw fragments or associated teeth. In terms of stratigraphic position, the hominin was found 51 cm above Tuff 8. Since only 1.32 m of Laetolil Beds are exposed above this horizon, the original provenance of the specimen can be fairly tightly constrained stratigraphically to 0.5–1.3 m above Tuff 8 (Fig. 2.33).

Earlier the same year, Amandus Kweka discovered an isolated left lower canine of *A. afarensis* (EP 162/00) from Loc. 16 (Harrison 2011b). It was found on the surface between Tuffs 7 and 8.

Fig. 2.32 Locality 16. Location of the hominin mandibular fragment (EP 2400/00) discovered in 2000. (**a**) View (southwest) of the western flank of the western arm of Loc. 16. (**b**) View (southeast) of the western flank of the western arm of Loc 16. The river gully is visible as a line of trees in the top right and center. Irkeeklekole Hill is in the background

Locality 17

A small locality situated a short distance east of Loc. 16 on the northern side of the Garusi River valley. A short section of the Upper Laetolil Beds above Tuff 7 is exposed. Although few fossils have been recovered, it has produced examples of some relatively rare taxa, such as *Paracolobus* and *Orycteropus*. One of the tuffaceous horizons contains a dense aggregation of unbranched vertical burrows leading to brood cells of solitary hymenopterans. The tunnels, up to 7 cm deep, are densely packed, with less than 1 cm separating neighboring burrows, and with an estimated density of over 2,000 burrows per square meter (Fig. 2.34).

Locality 18

This is the largest and most productive of the Upper Ndolanya localities. It is located at the head of the Gadjingero River on the southern side of the valley. It is separated from the Garusi valley by a low grassy rise, about 0.5 km north of Loc. 2. Additional small exposures of Upper Ndolanya Beds occur along the northern side of the valley. Upper Laetolil Beds occur for about 2.5 km along the river course, but these are poorly exposed, with a heavy overburden of mbuga clays,

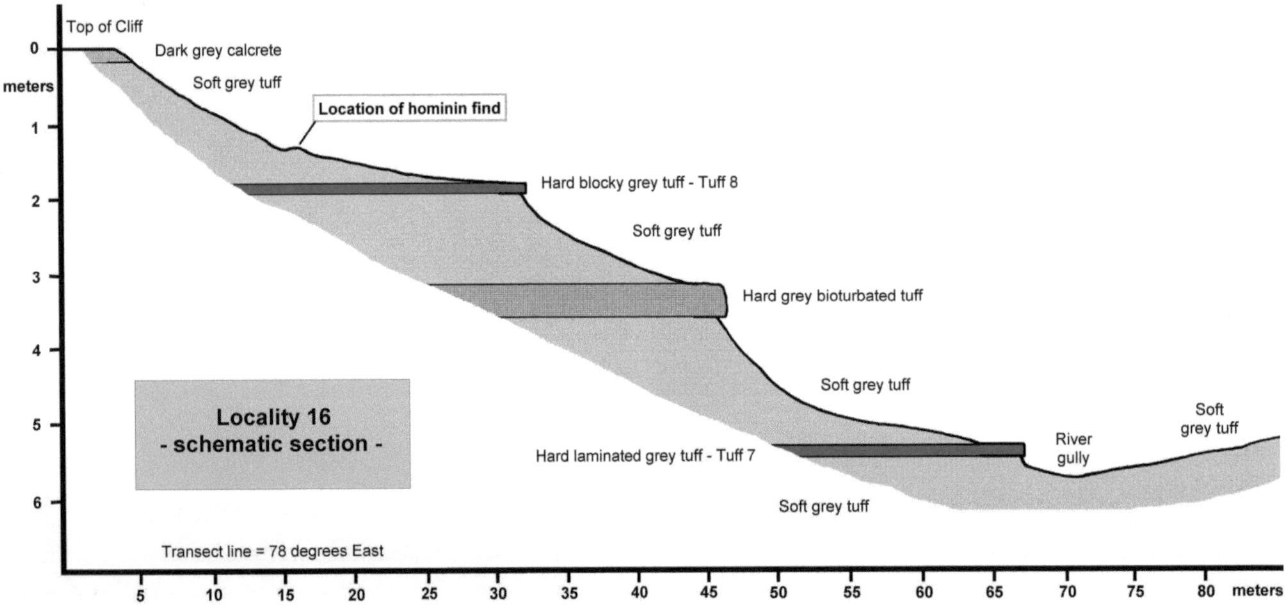

Fig. 2.33 Schematic stratigraphic section at Locality 16 where EP 2400/00 was discovered. The hominin was found 51 cm above Tuff 8. Note that the vertical and horizontal axes are different scales

Fig. 2.34 Locality 17. Tuff of Upper Laetolil Beds preserving an aggregate of burrows and brood cells made by solitary hymenopterans. (**a**) Lateral view; (**b**) Surface view

Fig. 2.35 Locality 18. Main exposures of the Upper Ndolanya Beds

and they are generally poorly fossiliferous. The sides of the valley are mostly gently sloping and covered with grass and open acacia woodland. Good exposures of the Upper Ndolanya Beds are typically found where the cliff faces are steep and actively eroding. The main outcrops extend for more than 200 m, and the exposed section of Upper Ndolanya Beds is at least 8 m thick (Fig. 2.35). A distinctive pedogenic calcrete about midway up the section divides the Beds into two units. Fossil vertebrates are common in both units, consisting predominantly of bovids. Cocoons and brood cells of solitary hymenopterans are common throughout the section, but are most abundant in the upper unit.

This locality was known as Gadjingero by Kohl-Larsen (see Table 2.1), who recovered a large sample of fossils. In addition to extensive surface collections, Mary Leakey conducted systematic excavations in 1976, but few fossils were recovered. Subsequent excavations were carried out by Prosper Ndessokia (in 1987–1988), and more recently by teams led by

Charles Musiba. Kaiser et al. (1995) identified putative artifacts from the surface of the Ndolanya Beds at Loc. 18, but these probably derive from higher in the sequence. Excavations by Leakey and Ndessokia failed to produce any stone tools *in situ*, and subsequent investigations have found no evidence to support the occurrence of stone tools in the Upper Ndolanya Beds, either at Loc. 18 or elsewhere at Laetoli.

Locality 19

This is a small locality on the northwestern flanks of the Ndonyamati Hills overlooking the Olaitole River valley. It is situated about 200 m west of Loc. 12. The Upper Laetolil Beds between Tuffs 5 and 8 are exposed along a series of low cliffs, up to 5 m in height. The surrounding area is thickly vegetated with dense bush and open woodland. The locality has yielded few fossils, but a partial lower dentition of *A. afarensis* (L.H. 14) was recovered in 1975–1976 (Leakey 1987b).

Locality 20

This is another small locality on the southern side of the Olaitole River (Fig. 2.36). It is situated about 250 m west of Loc. 19, on the heavily vegetated northwestern flank of the Ndonyamati Hills. The locality represents a series of discontinuous and poorly exposed outcrops, with a lateral extent of about 100 m. Fossils are derived from the Upper Laetolil Beds between Tuffs 6 and 8, but are not common.

Fig. 2.36 Locality 20. Exposure of the Upper Laetolil Beds. The Ndonyamati Hills are in the background. Blocks of Ogol Lava, derived from the top of the Ndonyamati Hills, are scattered on the Laetolil Beds in the foreground

Locality 21

This is the most productive of the localities on the southern side of the Olaitole River, situated just to the west of Loc. 20. The locality consists of a series of deep gullies, with cliffs up to 6 m in height, which extend E-W for about 120 and 200 m downslope. The surrounding area is thickly vegetated with dense bush and woodland. Fossils are eroding out of the Upper Laetolil Beds between Tuffs 5 and 7, but are most abundant between Tuffs 6 and 7. The Footprint Tuff (Tuff 7) is exposed in the middle of the section, and is associated with calcified plant material.

Locality 22

An extensive locality on the northern side of the Olaitole River, on the opposite side of the valley to Loc. 21. It is separated from the Garusi River valley to the north by a low bush covered ridge capped by Ogol Lavas. It is located about 1 km southeast of Loc. 10E. The locality is situated in a large N-S oriented river gully that drains into the Olaitole River. It is quite heavily vegetated, with small discontinuous exposures of Upper Laetolil Beds. A relatively thick section (about 32 m thick) of Upper Laetolil Beds from Tuff 7 at the northern end of the locality to below Tuff 1 is exposed. Most fossil vertebrates have been recovered from between Tuffs 5 and 7, but a few vertebrates and a good sample of terrestrial gastropods have been recovered from between Tuffs 1 and 2. At the southern end of the gully, the stream cuts through the top part of the Lower Laetolil Beds (about 6 m in thickness), but these appear to be devoid of fossils. At the entrance to the gully at the southern end of the locality, and along a low ridge to the east, the Yellow Marker Tuff is overlain by a relatively thick series of Ndolanya Beds. These beds have yielded a small sample of fossil vertebrates, especially large bovids.

Locality 22E

This locality is situated about 0.5 km to the east of Loc. 22 on the northern side of the Olaitole River valley. It is separated from Loc. 22 by a low dipping ridge with a narrow cattle path running N-S that allows access by car between the Garusi and Olaitole valleys. As at Loc. 22, Loc. 22E is associated with a tributary of the Olaitole River that drains south into the main watercourse. The area is heavily vegetated and outcrops are discontinuous and poorly exposed. The Upper Laetoli Beds are represented by a section between Tuffs 2 and 7. Most of the fossils have been recovered from between Tuffs 5 and 7, but a few fossils have been recovered lower in the section from between Tuffs 2 and 4. Tuff 3 is extensively exposed in the valley floor. The upper part of the tuff preserves a rich concentration of calcified plant remains, especially mats of fine roots, possibly of grasses, with individual roots less than 2 mm in diameter. In one location (coordinates: 3° 14′ 30″ S, 35° 11′ 52″ E) Tuff 3 preserves numerous footprints, especially those of guinea fowl and small mammals. On the eastern flank of the divide between Loc. 22 and Loc. 22E is an extensive series of Upper Ndolanya Beds that have yielded a small sample of fossils.

Locality 22S

This is a newly discovered locality, being first recorded in 1998. It is located south of Loc. 22, further along the same tributary of the Olaitole River, about 4.5 km south of the Garusi River and 4 km north of Esere village. It is the southernmost locality at Laetoli. The topography of the area is relatively flat, and a tributary of the Olaitole meanders across a broad and low valley with dense grass and bush cover. On the western side of the valley is a series of low cliffs, which expose a short section of the Upper Laetolil Beds (about 8 m thick) between Tuffs 1 and 3. Tuff 3 forms the hard resistant cap to the cliffs. The base of Tuff 3 contains an abundance of calcified twigs and root casts. A downfaulted section of the Upper Laetolil Beds with Tuffs 3–5 is exposed in the floor of the valley. Very few fossils have been recovered from these beds. The importance of the locality is that a small outcrop of Upper Ndolanya Beds occurs in a faulted zone of the Upper Laetolil Beds, and this has produced a rich vertebrate fauna, including a proximal tibia of a hominin. This latter specimen (EP 1000/98), found by Chris Robinson in 1998, was the first postcranial specimen of an adult individual to be recovered from Laetoli (Harrison 2011b). The Ndolanya Beds are exposed as a small rounded hillock, about 10 m high and 30 m wide, bordered on either side by steep cliffs formed by the Upper Laetolil Beds (coordinates: 3° 15′ 30″ S, 35° 11′ 9.3″ E) (Fig. 2.37). The area is known locally as Nenguruk, and the hillock represents a distinctive and visible feature within the Olaitole Valley. Eroded sections of Nenguruk Hill shows that the Upper Ndolanya Beds rest unconformably on a downfaulted block of Upper Laetoli Beds capped by Tuff 3, which dips westwards at an angle of 20° towards the center of the valley.

The proximal tibia was found partially buried in loose sediments at the foot of Nenguruk Hill, at a distance of 3.09 m from the base of the slope, within a shallow NE-oriented gully that carries run-off down the northern side of the hill. Although the specimen had clearly been transported from its original location, the context and preservation of the specimen clearly indicates that it was originally derived from the Upper Ndolanya Beds, along with the numerous other vertebrate remains recovered. A thorough search of the locality and sieving of the sediments in the gully produced no additional fragments or further hominin finds. The nature of the breaks indicates that the proximal end of the tibia was all that was preserved when it was originally buried and fossilized (Harrison 2011b).

Locality 23

This locality extends along a shallow gully system cut by a tributary of the Gadjingero River, about 1 km northeast of the main watercourse on which Loc. 18 is situated. It is located 1.5 km north of the Garusi River valley (Loc. 16) and 1.5 km west of the Nompopong River valley (Loc. 1). There is little topographic relief and the main gully cuts through thick deposits of mbuga clay, Ngaloba Beds, and the upper part of the Upper Laetolil Beds, especially the Yellow Marker Tuff. Few fossils were recovered from the river gully itself. Located on the northern flank of the river valley, at the top of a small tree-lined side gully, gently undulating hillocks expose fairly extensive outcrops of Lower Ngaloba Beds (Fig. 2.38). These beds contain numerous fossils vertebrates and Acheulian

Fig. 2.37 Locality 22S. Nenguruk Hill. Exposure of Upper Ndolanya Beds. The location of the hominin proximal tibia (EP 1000/98) found in 1998 is indicated

Fig. 2.38 Locality 23. Exposures of Lower Ngaloba Beds on the northern flank of a tributary of the Gadjingero River

Fig. 2.39 Bifacial tool found on the surface of the Lower Ngaloba Beds at Locality 23

tools (Fig. 2.39). Harris and Harris (1981) presented a preliminary account of archaeological excavations conducted in September 1979. The tools, predominately made of calcrete, were associated with well-preserved remains of large mammals.

Locality 24

Exposures located just to the north of Loc. 10NE, on the northern side of the Garusi River. A short section of the Upper Laetolil Beds above and below Tuff 7 is exposed. These have yielded fossil mammals, but generally the locality is not very productive. The locality was indicated as unnamed outcrops of Upper Laetolil Beds on the locality map of Leakey (1987a, Fig. 1.2), but these were not given a collecting locality designation.

Garusi SW

These are an extensive series of exposures of Upper Laetolil Beds located along the Garusi River Valley west of Loc. 10W. They are situated on the southern side of the Garusi River, on both sides of a promontory, about 2 km southwest of Norsigidok Hill. The outcrops consist of the upper section of the Upper Laetolil Beds above Tuff 8, and primarily consisting of unfossiliferous Yellow Marker Tuff. Fossils do occur in the beds below the Yellow Marker Tuff, but these are scarce. The exposures in this area were initially identified as belonging to the Laetolil Beds by Kent (1941), but fossils were first recovered from this locality in 1998.

Silal Artum

This locality was first discovered in 2001, and is the northernmost collecting locality at Laetoli. It is located 4.5 km north of the Garusi River, on the western flank of the Nompopong River, just to the northwest of Loc. 1NW. The locality is represented by a shallow gully at the base of a grassy hill that forms the divide between the Gadjingero and Nompopong River valleys. Immediately to the east, the Nompopong River exposes a section of Upper Laetolil Beds (16 m thick), which can be traced down to Tuff 5 and is capped by the Yellow Marker Tuff. The latter is overlain by 3 m of Upper Ndolanya Beds, which have produced a rich assemblage of fossils. This was the site of discovery, in 2001, of an edentulous maxilla of *Paranthropus aethiopicus* (EP 1500/01), only the second hominin find from the Upper Ndolanya Beds (the first being the proximal tibia from Loc. 22S found in 1998) (Harrison 2002, 2011b). The specimen was found on the surface, partially buried in loose detrital sediment at the edge of a small acacia thicket, where it was protected from being (further) trampled by livestock (Figs. 2.40 and 2.41). The level at which the specimen was found was 2.55 m below the top of the section and 40 cm above a hard pale brown tuff that forms the base of the exposed section of the Upper Ndolanya Beds. The Upper Ndolanya Beds are capped by a hard dark grey calcrete, which is probably Pleistocene in age (Fig. 2.42). Based on the degree of weathering and abrasion, and on the position of the specimen, it is clear that the fossil had eroded out of and been transported from its original burial location (but there is no doubt that it was originally derived from the Upper Ndolanya Beds). However, assuming that it came from an exposure further upstream, the maximum distance that it could have been transported is 30 m. The entire locality was thoroughly searched and the surface sediment in the immediate vicinity of the discovery was sieved (about 150 m^2), but no

Fig. 2.40 Silal Artum. View (*south*) of main exposure of Upper Ndolanya Beds. The location of the fossil hominin maxilla (EP 1500/01) discovered in 2001 is indicated

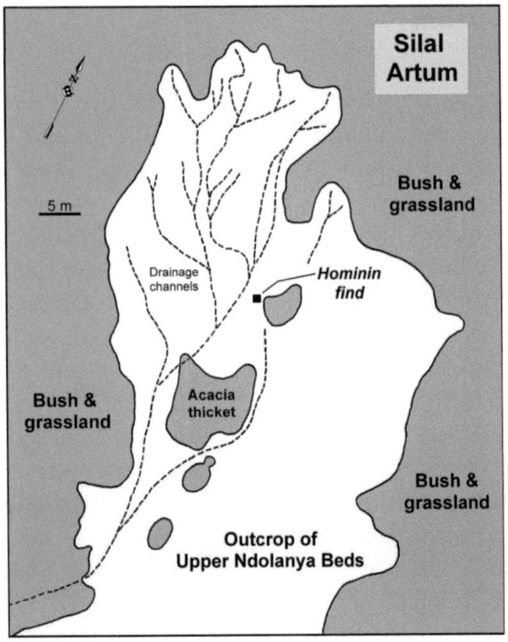

Fig. 2.41 Silal Artum. Plan of the exposure of the Upper Ndolanya Beds at the northern end of the locality (*in white*). Vegetation and recent soil cover is indicated by grey shading. The hominin maxilla (EP 1500/01) was found at the edge of a small clump of acacia and grass (see Fig. 2.40), but the specimen was evidently transported downslope from the northern end of the locality (see text for more detail)

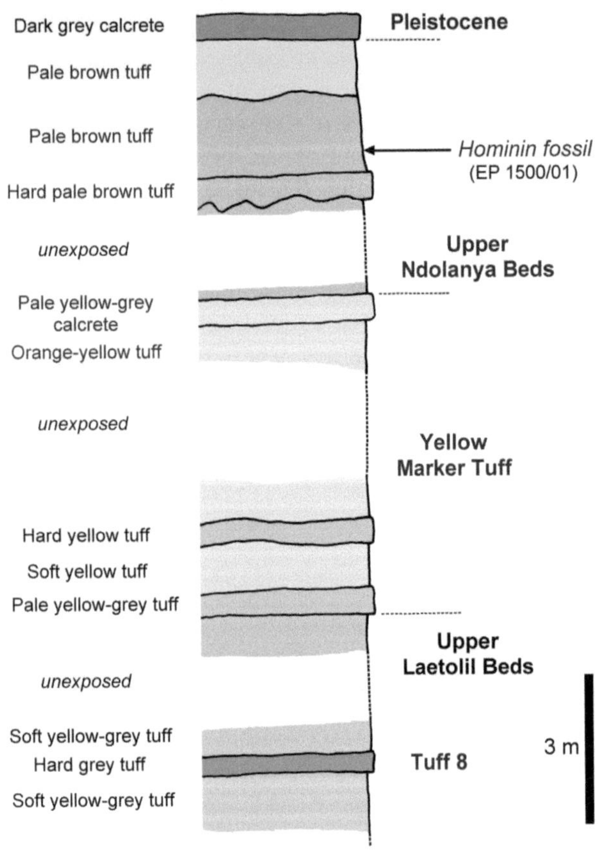

Fig. 2.42 Silal Artum. Upper part of the stratigraphic section at Silal Artum, showing the horizon from which the hominin maxilla (EP 1500/01) was recovered

additional pieces of the specimen were recovered. It is evident from the preservation of the maxilla that the teeth had been lost for a long time prior to its recovery.

Other Paleontological Localities on the Eyasi Plateau

Kakesio

Kakesio is represented by an extensive series of exposures of Lower Laetolil Beds to the east and northeast of Kakesio village, associated with the main course of the Kakesio River and its tributaries. The Kakesio localities are located 40 km southwest of the Laetoli area, and can be reached by road via Endulen. The main outcrops are located about 2 km north of the village of Kakesio, and extend E-W for 5.5 km, about 1 km north of the main Endulen-Kakesio road (Fig. 2.43). This corresponds to Area 5 of Hay (1987; Fig. 2.2). To better constrain the provenance of the fossils collected, this large area has been subdivided, using promontories and gully systems as natural boundaries, into several contiguous collecting localities. These are numbered Kakesio 1 to 6, arranged sequentially from west to east.

The localities consist of a series of E-W trending cliffs facing south towards the Kakesio River. The outcrops are well-exposed along the low slopes and shallow gullies that extend from the cliffs, but closer to the river they are heavily vegetated. The exposed section is 18–20 m thick and consists primarily of massive light yellow-grey to pale brown tuffs, which show variable degrees of bioturbation. The base of the section consists of waterlain tuffs, clays, mudstones, and conglomerates, which rest unconformably on the gneiss basement rocks.

Several smaller outcrops of Lower Laetolil Beds are located east of Kakesio village, south of the Endulen-Kakesio road, and are associated with tributaries of the Kakesio River. The general area is quite densely vegetated bush and open woodland, and the main outcrops consist of eroded cliffs

Fig. 2.43 Kakesio 6. Exposures of Lower Laetolil Beds

along the flanks of the river gullies. These localities are designated as Kakesio 7–10. Kakesio 7 is the easternmost outcrop, located about 1 km south of the Kakesio River, and about 2 km southeast of Kakesio 6. Kakesio 8 is 1.2 km further west, and it represents an extensive series of outcrops along the eastern side of the Lorotian River. Kakesio 9 is located 3 km south of the Kakesio River, at the head of a small ephemeral stream. It consists of a circular depression surrounded by low cliffs. Just south of this locality (about 0.5 km) the Precambrian basement can be observed along the shoulder of the Eyasi Plateau. Kakesio 10 is the southernmost of the Kakesio localities, and it is situated 4.2 km south of the Kakesio River, along the western flank of the west branch of the Lorotian River, 1.7 km further upstream from Kakesio 8. The outcrops are quite extensive, forming a series of high cliffs (Fig. 2.44).

All of these localities have yielded fossil vertebrates, as well as gastropods, termitaria and traces of hymenopterans. The Lower Laetolil Beds are generally much less productive than the Upper Laetolil Beds, and only 144 fossil mammals have been recovered from the Kakesio localities combined since 1998. Nevertheless, more intensive surveys of the Kakesio outcrops may prove rewarding in the future. Kakesio 6 is the most productive locality, having yielded good samples of *Anancus*, *Ceratotherium*, bovids and equids. Fossil vertebrates are found throughout the section, but appear to be most abundant in the lowermost 5 m. The most common fossils in the upper part of the section are small ovoid termitaries (30–50 cm in diameter), bee brood cells and cocoons, terrestrial gastropods, tortoises, and ostrich egg shell fragments. Terrestrial gastropods are dominated by *Limicolaria* and *Achatina*, which are often found in dense concentrations in the subaerially deposited tuffs in the upper part of the section. These aggregations are presumably the result of gastropods that burrowed down into the loosely compacted sediments to aestivate, and were buried by subsequent eruptive events. Calcified wood and root casts (2–5 cm in diameter) are abundant in some levels.

Fig. 2.44 Kakesio 10. Exposures of Lower Laetolil Beds

Vertebrate fossils tend to be scattered and fragmentary, and they commonly show evidence of weathering and rodent gnawing. Kakesio is the only locality on the Eyasi Plateau that has yielded large aquatic vertebrates (i.e., crocodiles), and supports evidence from the geology that fluvial sediments at the base of the Lower Laetolil Beds were associated with permanent or seasonal bodies of water (probably a shallow swampy depression with ephemeral streams, but not a major river or lake system). Dick Hay investigated the geology and paleontology of the Kakesio area in 1982, as part of Mary Leakey's expeditions to the Laetoli area, but the results of his work were never published in detail. The 1982 collections include a small sample of fossil vertebrates, including partial skeletons of *Parahyaena howelli* and aff. *Proteles* sp. (Werdelin and Dehghani 2011).

Kakesio South

A series of low hills and cliffs beginning just south of Kakesio village, and extending southwest for about 3 km. These expose Lower Laetolil Beds comparable to those at Kakesio. Only a few mammalian remains have been recovered, but tortoises, ostrich egg shells, and invertebrate fossils are quite common.

Emboremony

Several fossiliferous outcrops are exposed along the Emboremony River valley, about 10 km to the southwest of Laetoli and 4 km northeast of Kakesio. In this region, the Emboremony River passes through a relatively flat valley, about 4 km wide, with a heavy covering of mbuga clay. The sides of the valley expose Lower Laetolil Beds and Ngaloba Beds. This corresponds to Area 7 of Hay (1987; Fig. 2.2). The best exposures occur along the low sloping cliffs that form the southwestern flank of the valley, about 1.5 km from the main watercourse. Although fossils are found along the entire length of the cliffs, two areas are especially productive and these have been designated at Emboremony 1 and Emboremony 3. On the opposite side of the river valley, about 4 km east of Emboremony 1, there is another productive locality, designated as Emboremony 2.

Emboremony 1 is located 5.5 km north of the Endulen-Kakesio road, 2.7 km west of the Emboremony River. The exposures extend south for about 1 km before low relief and heavy vegetation obscure the outcrops. A section of more than 14 m of Lower Laetolil Beds are exposed, overlain by outcrops of Ngaloba Beds. The former are comparable to the sediments at Kakesio, and consist of massive yellow-grey and pale brown tuffaceous sediments, that grade down to weakly bedded waterlain tuffs and clays. The middle section is the most fossiliferous and contains vertebrates, abundant gastropods *in situ*, and numerous small ovoid termitaries. Fossil bones of large mammals are also common in the Ngaloba Beds, especially bovids and *Equus*, and these are sometimes found mingled together with those from the Laetolil Beds (but in most cases are easily discriminated based on preservation and coloration).

Emboremony 2 is located on the northeast side of the valley. It is visible from Emboremony 1, but separated from it by a flat, sparsely vegetated area, thickly mantled in mbuga clay, on either side of the Emboremony River. The low slopes and gullies that expose the Lower Laetolil Beds extend N-S for a little over 1 km. Once again, surface finds of fossils are derived from two distinct beds, and are distinguished by their contrasting preservation. Heavily mineralized grey-colored fossils, often with bright orange staining, are derived from the Lower Laetolil Beds. The lighter, less heavily mineralized, white to yellow-brown fossils are derived from the overlying dark brown claystones of the Ngaloba Beds. The exposed section is about 10 m thick. The lowermost 5 m represents the Lower Laetolil Beds, consisting of massive pale grey tuffs, heavily bioturbated in the upper section, grading down to pinkish-brown tuffs. In addition to fragmentary bones and isolated teeth of large mammals, small ovoid termitaria are common. Above this is 3.6 m of dark brown to orange claystones and laminated grey tuffaceous clays representing the Ngaloba Beds. In addition to the bones of large mammals, especially *Equus* and large-sized bovids, the surface of these beds has an extensive litter of stone artifacts, including quartzite flakes and hammerstones, as well as cores of lava and chert. The Ngaloba Beds are capped by a thick grey calcrete.

Emboremony 3 is located 5 km south of Emboremony 1 on the southwestern flank of the Emboremony River, about 1 km north of Olaltanaudo. The locality consists of a U-shaped gully with a series of steep cliffs that expose a short section of the Lower Laetolil Beds. Fossil vertebrates are rare, except for ostrich eggshell fragments.

Esere

The localities at Esere represent small fossil-bearing outcrops that occur in the extensive exposures of Lower Laetolil Beds near the village of Esere, south of the Endulen-Kakesio Road. This area was referred to as Laetoli Springs by Kent (1941). Three separate localities have been designated, Esere 1, 2 and 3. Esere village is located 12 km southwest of Endulen, along the main Endulen-Kakesio Road. The localities are all situated on the western flank of the Olaitole River, as it drains southwards from the Laetoli area into Lake Eyasi. Esere is located 3.5 km south of Loc. 22S, the southernmost of the Laetoli localities associated with the Olaitole River. Esere 1 is located 2 km southwest of the village. It has yielded a small collection of fossil vertebrates and terrestrial gastropods. The fossils are derived from a pale grey tuff and underlying red tuff. Both horizons are heavily calcified and resistant to weathering, so fossils erode out of the sediment very slowly. The Lower Laetolil Beds are overlain by mbuga clay, which are associated with a scatter of Pleistocene fossils.

Esere 2 is located 2.5 km south of Esere 1. It consists of a series of low cliffs that expose pale yellow-grey tuffs of the Lower Laetolil Beds, heavily bioturbated by termite burrowing. Fossil vertebrates are not common, but a partial skeleton of a suid was recovered. Brood cells and cocoons of solitary bees are abundant. Esere 3 is 3 km south of Esere 2. It is located at the southernmost limit of the Laetolil Beds, very close to the exposed Precambrian basement rocks on the elevated shoulder of the Eyasi rift scarp, near the peak of Ngaiborgoso. The locality consists of an extensive series of low cliffs, exposing grey-brown tuffs that have been much bioturbated due to termite activity. Few fossil vertebrates have been recovered, but terrestrial gastropods and traces of solitary bees are common.

Kent (1941) made brief reference to "mastodon" (*Anancus kenyensis*) teeth that were collected in this area by Leakey in 1935. The exact location is not recorded, but according to Kent (1941) the specimens were recovered from Laetolil Beds on the southwest slopes of Eseketeti (= Speer Mountain) near the Rift scarp. Additional information is provided on a handwritten note (in L.S.B. Leakey's writing) associated with the specimens in The Natural History Museum in London. The locality is referred to as Endolele or Endulele, and is located near the "Spring" about 8 miles away from the Laetoli area. This would be consistent with the spring at Esere. There is a village called Ndulele, about 7 km southwest of Kakesio village, but there are no Laetolil Beds exposed in the vicinity, and its location does not match that of Kent's description. Leakey (1935), in an unpublished report, indicates that the "the teeth of a mastodon and of a primitive elephant" came from "a place where the base of the Laetolil series was seen resting on granite…". Again, this would be consistent with the outcrops at Esere. Sanders (2011) suggests that the *Anancus* specimens from Endolele are more primitive than those from Kakesio, implying that the Lower Laetolil Beds at Esere might be somewhat older than those at Kakesio.

Engesha

An extensive series of outcrops of the Lower Laetolil Beds about 2.5 km northeast of Esere village, south of the Endulen-Kakesio road, at the head of the Nasero River valley. The gullies expose more than 30 m of Lower Laetolil Beds, consisting of massive grey-brown tuffs with intercalated fluvial facies. A preliminary survey of the area has produced a few vertebrate fossils. Engesha is located just to the north of the Noiti area, and the sediments exposed are a continuity of those found at Noiti 3 (see below).

Lobileita

An extensive outcrop of Lower Laetolil Beds located in a large valley about 3.5 km south of the village of Sinoni, at a place called Lobileita. This corresponds to Area 6 of Hay (1987; Fig. 2.2). The valley is formed by a series of gullies that drain eastwards into the southern end of the Garusi River. The outcrops represent a remnant of the Laetolil Beds. To the southwest, elevated Precambrian inselbergs extend along the Eyasi Rift escarpment. The rocky peak of Enaiborkrum, at an elevation of 1,936 m, is located 3.5 km to the southwest of Lobileita, while 10 km away is Eseketeti (1,965 m) at the edge of the escarpment. To the east of Lobileita are the grassy plains of the Garusi valley, at about 1,600 m elevation, which expose an extensive series of Precambrian rocks. Laetoli Beds can be traced for only about 1 km south of Lobileita, after which they give way to the underlying gneiss basement rocks. A 17 m section of Lower Laetolil Beds is exposed at Lobileita, consisting predominantly of massive yellow-grey and pale brown tuffs, with weakly bedded sandstones and conglomerates at its base. The sediments are capped by a resistant red-orange calcrete, and they rest unconformably on the basement. The bottom half of the section has yielded fossil vertebrates, ostrich egg shell fragments, terrestrial gastropods, termitaries and traces of solitary bees.

Olaltanaudo

An extensive series of outcrops of Lower Laetolil Beds along the Endulen-Kakesio road, about 2.5 km west of the Sinoni junction. The locality is situated just south of Emboremony 3 and 5.5 km east of Kakesio 6. This is equivalent to Area 4 of Hay (1987; Fig. 2.2). The exposures occur along low cliffs to the south and west of the Olaltanaudo Dam. Few fossils have been recovered.

Oleisusu

A locality on the southern flanks of the Oldogom River valley, 2 km south of the village of Endulen. The river gully exposes an extensive series of Upper Laetolil Beds, but the stratigraphy at has not yet been fully investigated. Fossils have been recovered from between Tuffs 7 and the Yellow Marker Tuff.

Ndoroto

A small locality on the east side of the track leading from Endulen to Olduvai Gorge, about 2.3 km northeast of Endulen hospital (coordinates: 3° 11′ 04″ S, 35° 16′ 57″ E). The general area is known as Orumekeke, but according to local informants the specific location is named Ndoroto. The site consists of a small series of steep cliffs, comprising 12 m of Yellow Marker Tuff overlain by Ogol Lavas. At the base of the section are pale grey tuffs of the Upper Laetolil Beds that have yielded a few fossil vertebrates.

Noiti 1-3

The Noiti area is situated 3 km southeast of Esere village, and can be reached by foot from the main Endulen-Kakesio Road. This is the equivalent of Area 2 of Hay (1987; Fig. 2.2). Harrison's team was first directed to Noiti in January 1999 by Mr. Simon Mataro, the Antiquities superintendent at Laetoli. Subsequent surveys of the area have allowed the designation of three separate collecting localities, Noiti 1, 2 and 3.

Noiti 1 was the first locality visited in 1999. It is located on the eastern flank of a small tributary of the Nasera River, with a lateral extent of only about 50 m (Fig. 2.45). There are no fossil vertebrates, but the surface of the locality is littered with well-preserved fossil wood, up to 15 cm in diameter (Bamford 2011b). The fossil wood is associated with a pyroclastic mudflow, which filled a paleochannel that cuts into the underlying Lower Laetolil Beds. The lahar can be seen resting unconformably on a light grey tuff at the base of the sequence that represents the eroded surface of the Lower Laetolil Beds. The massive lahar, 10–13 m thick, is weakly differentiated into a bright red upper unit, an orange-brown tuffaceous agglomerate with poorly-sorted subangular clasts (some up to 1 m in diameter) in the middle of the section, and a pale grey-green agglomerate with smaller inclusions (up to 5 cm) at the base of the section (Fig. 2.46). The lahar is overlain by Ogol Lavas (6–10 m thick), but the contact between the two units is not exposed. Presumably unconsolidated ash and other pyroclastic debris, derived from the apron of the active volcano more than 20 km to the northeast, flowed down river channels, picking up trees and woody detritus along the way, and eventually filling the gullies with carbonate and silica rich volcanic sediments containing large pieces of wood. The wood is potentially derived

Fig. 2.45 Noiti 1. Exposure of lahar that has yielded abundant fossil wood. The lahar rests unconformably on the Lower Laetolil Beds. The dark colored blocks scattered over the surface, especially in the background, are boulders of Ogol Lava derived from the top of the ridge

Fig. 2.46 Noiti 1. (**a**) The figure (Terry Harrison) is standing on the weathered surface of the lahar in the middle part of the section; large boulders of Ogol Lava (dark colored) are present in the foreground derived from the top of the ridge. (**b**) Close up of the lahar showing the subangular inclusions (*above* and *left* of the penknife) and a large piece of wood *in situ* (to the *right* of the penknife). (**c**) Weathered surface detritus with fragments of wood (light colored) and boulders and cobbles of Ogol Lava (dark colored)

from trees living on the slopes of the volcano or from those living more locally along the margins of the paleochannels. The abundance of calcite in the lahar confirms a carbonatite source, similar to the Laetolil Beds. The fossil wood and lahar are inferred to be contemporaneous with the Lower Laetolil Beds, since at other locations in the Esere area the same lahar has been observed interbedded within the Lower Laetolil Beds.

Noiti 2 is located about 0.5 km west of Noiti 1, being separated from it by a low, narrow ridge. The fossiliferous exposures are more extensive than at Noiti 1, with a lateral extent of about 300 m (Fig. 2.47). It is situated on the eastern slope of a small river gully that represents a tributary of the Olaitole River. As at Noiti 1, abundant fossil wood has been recovered in association with the same lahar. Some pieces of wood are very large, implying that some trees had trunk diameters in excess of 50 cm.

Fig. 2.47 Noiti 2. View (*northwest*) of the weathered surface of the lahar covered in large blocks of Ogol Lava (dark colored)

Fig. 2.48 Noiti 3. Exposure of the waterlain Lower Laetolil Beds

Noiti 3 represents an extensive outcrop of Lower Laetolil Beds, located 1 km east of Noiti 1. It was first surveyed in 2003. The same suite of sediments occurs at Engesha, 1.5 km to the north. A series of high cliffs on the east side of the Oloosukutan River, a small tributary of the Nasera River, expose a thick series of Lower Laetolil Beds (Fig. 2.48). The sediments are at least 30 m thick, and consist mainly of a series of waterlain yellow and orange tuffaceous silts, sands and gravels, intercalated with primary airfall tuffs and thin horizons of clays. The sediments were apparently deposited in a shallow lake basin. These sediments are capped in the northeast corner of the locality by a lahar, different in composition from that observed at Noiti 1 and 2. This is overlain in turn by Ogol Lavas. A few vertebrate fossils have been recovered, including teeth and a partial elephant skeleton, from the base of the sequence. No aquatic vertebrates have been recovered, so the lake was probably only seasonally inundated. Fossil wood, similar to that at Noiti 1 and 2, also occurs.

Conclusions

Sixty paleontological localities are currently recorded on the Eyasi Plateau, including those at Laetoli. Of these, seven are newly recognized localities in the Laetoli area, and 25 are formally designated here for the first time in the Kakesio and Esere-Noiti areas to the southwest and south of Laetoli respectively. The localities at Laetoli have produced an abundance of fossils of Pliocene age from the Upper Laetolil Beds and Upper Ndolanya Beds, as well as smaller collections of fossil vertebrates and stone artifacts from the Pleistocene Olpiro and Ngaloba Beds. The most productive localities are Locs. 2, 10E and 18, which have each yielded more than 2,000 fossil mammals since 1974.

Fossil hominins are presently known only from localities in the Laetoli area. The Upper Laetolil Beds have yielded the remains of *Australopithecus afarensis*, which represents a relatively rare component of the vertebrate fauna (Su and Harrison 2008). Tracks of fossilized footprints of hominins, presumably of *A. afarensis*, are preserved at Loc. 8 (Footprint Site G). Three new specimens of *A. afarensis* have been recovered from Laetoli since 1998 (EP 162/00; EP 2400/00; EP 2038/03), and the provenience of these specimens is described here. A weathered mandibular fragment (L.H. 29), previously inferred to be derived from Pleistocene deposits at Loc. 8, and attributed to *Homo* cf. *H. erectus*, is considered here to belong to *A. afarensis*. In addition, fossil hominins assigned to *Paranthropus aethiopicus* (EP 1000/98, EP 1500/01) have been recovered from the Upper Ndolanya Beds for the first time. A relatively complete cranium (L.H. 18) of an archaic form of *Homo sapiens* is known from the Upper Ngaloba Beds at Loc. 2. Localities in the Kakesio area and the Esere-Noiti area have yielded relatively small, but important, collections of fossil vertebrates, invertebrates and plants from the Lower Laetolil Beds. No hominins have yet been recovered from this stratigraphic unit, although it is possible that they may be discovered in the future with more intensive collecting and surveying.

Acknowledgements We are grateful to the Tanzania Commission for Science and Technology and the Unit of Antiquities in Dar es Salaam for permission to conduct research. Thanks to Paul Msemwa (Director) and the staff at the National Museum of Tanzania in Dar es Salaam for their support and assistance. The following individuals provided valuable advice and assistance in the field: P. Andrews, P. Ditchfield, J. Kingston, M. Mbago, C.P. Msuya, C. Swisher and D. Su. W.J. Sanders and D. Su provided very useful comments on an initial draft of the manuscript. We acknowledge with thank the many researchers and support personnel who accompanied us in the field at Laetoli and contributed to the work presented here. Research at Laetoli was supported by grants from the National Geographic Society, the Leakey Foundation, and NSF (Grants BCS-9903434 and BCS-0309513).

References

Bamford, M. (2011a). Fossil leaves, fruits and seeds. In T. Harrison (Ed.), *Paleontology and geology of Laetoli: Human evolution in context* (Geology, geochronology, paleoecology, and paleoenvironment, vol. 1, pp. 235–252). Dordrecht: Springer.

Bamford, M. (2011b). Fossil woods. In T. Harrison (Ed.), *Paleontology and geology of Laestoli: Human evolution in context* (Geology, geochronology, paleoecology, and paleoenvironment, vol. 1, pp. 217–233). Dordrecht: Springer.

Bonnefille, R., & Riollet, G. (1987). Palynological spectra from the Upper Laetolil Beds. In M. D. Leakey & J. M. Harris (Eds.), *Laetoli: A Pliocene site in northern Tanzania* (pp. 23–47). Oxford: Clarendon.

Darlington, J. P. E. C. (2011). Trace fossils interpreted in relation to the extant termite fauna at Laetoli, Tanzania. In T. Harrison (Ed.), *Paleontology and geology of Laetoli: Human evolution in context* (Fossil hominins and the associated fauna, vol. 2, pp. 555–565). Dordrecht: Springer.

Day, M. H., Leakey, M. D., & Magori, C. (1980). A new hominid fossil skull (L.H. 18) from the Ngaloba Beds, Laetoli, northern Tanzania. *Nature, 284*, 55–56.

Dietrich, W. O. (1942). Ältestquartäre Säugetiere aus der südlichen Serengeti, Deutsch-Ostafrika. *Palaeontographica, 94A*, 43–133.

Ditchfield, P., & Harrison, T. (2011). Sedimentology, lithostratigraphy and depositional history of the Laetoli area. In T. Harrison (Ed.), *Paleontology and geology of Laetoli: Human evolution in context* (Geology, geochronology, paleoecology, and paleoenvironment, vol. 1, pp. 47–76). Dordrecht: Springer.

Gentry, A. W. (2011). Bovidae. In T. Harrison (Ed.), *Paleontology and geology of Laetoli: Human evolution in context* (Fossil hominins and the associated fauna, vol. 2, pp. 363–465). Dordrecht: Springer.

Harris, J. M. (1987). Summary. In M. D. Leakey & J. M. Harris (Eds.), *Laetoli: A Pliocene site in northern Tanzania* (pp. 524–531). Oxford: Clarendon.

Harris, J. W. K., & Harris, K. (1981). A note on the archaeology at Laetoli. *Nyame Akuma, 18*, 18–21.

Harrison, T. (2002). First recorded hominins from the Ndolanya Beds, Laetoli, Tanzania. *American Journal of Physical Anthropology, Suppl. 32*, 83.

Harrison, T. (2005). Fossil bird eggs from Laetoli, Tanzania: Their taxonomic and paleoecological implications. *Journal of African Earth Sciences, 41*, 289–302.

Harrison, T. (2011a). Coprolites: Taphonomic and paleoecological implications. In T. Harrison (Ed.), *Paleontology and geology of Laetoli: Human evolution in context* (Geology, geochronology, paleoecology, and paleoenvironment, vol. 1, pp. 279–292). Dordrecht: Springer.

Harrison, T. (2011b). Hominins. In T. Harrison (Ed.), *Paleontology and geology of Laetoli: Human evolution in context* (Fossil hominins and the associated fauna, vol. 2, pp. 141–188). Dordrecht: Springer.

Hay, R. L. (1987). Geology of the Laetoli area. In M. D. Leakey & J. M. Harris (Eds.), *Laetoli: A Pliocene site in northern Tanzania* (pp. 23–47). Oxford: Clarendon.

Kaiser, T., Bromage, T. G., & Schrenk, F. (1995). Hominid corridor research project update: New Pliocene fossil localities at Lake Manyara and putative oldest early Stone Age occurrences at Laetoli (Upper Ndolanya Beds), northern Tanzania. *Journal of Human Evolution, 28*, 117–120.

Kent, P. E. (1941). The recent history and Pleistocene deposits of the plateau north of Lake Eyasi, Tanganyika. *Geological Magazine, 78*, 173–184.

Kitching, I. J., & Sadler, S. (2011). Lepidoptera, Insecta. In T. Harrison (Ed.), *Paleontology and geology of Laetoli: Human evolution in context* (Fossil hominins and the associated fauna, vol. 2, pp. 549–554). Dordrecht: Springer.

Kohl-Larsen, L. (1943). *Auf den Sporen des Vormenschen*. Stuttgart: Strecker und Schröder.

Krell, F.-T., & Schawaller, W. (2011). Beetles (Insecta: Coleoptera). In T. Harrison (Ed.), *Paleontology and geology of Laetoli: Human evolution in context* (Fossil hominins and the associated fauna, vol. 2, pp. 535–548). Dordrecht: Springer.

Kyauka, P. S. (1994). A comparative study of the Laetoli Hominid 21 skeleton and its implications for the developmental biology of *Australopithecus afarensis*. Ph.D. dissertation, University of California, Berkeley.

Kyauka, P. S., & Ndessokia, P. (1990). A new hominid tooth from Laetoli, Tanzania. *Journal of Human Evolution, 19*, 747–750.

Leakey, L. S. B. (1935). East African archaeological expedition. Fourth season 1934–1935. Eighth monthly field report: May 24th-June 23rd. Unpublished report, The Natural History Museum, London.

Leakey, M. D. (1987a). Introduction. In M. D. Leakey & J. M. Harris (Eds.), *Laetoli: A Pliocene site in northern Tanzania* (pp. 1–22). Oxford: Clarendon.

Leakey, M. D. (1987b). The Laetoli hominid remains. In M. D. Leakey & J. M. Harris (Eds.), *Laetoli: A Pliocene site in northern Tanzania* (pp. 108–117). Oxford: Clarendon.

Leakey, M. D. (1987c). Animal prints and trails. In M. D. Leakey & J. M. Harris (Eds.), *Laetoli: A Pliocene site in northern Tanzania* (pp. 451–489). Oxford: Clarendon.

Leakey, M. D., & Harris, J. M. (Eds.). (1987). *Laetoli: A Pliocene site in northern Tanzania*. Oxford: Clarendon.

Magori, C. C., & Day, M. H. (1983). Laetoli Hominid 18: An early *Homo sapiens* skull. *Journal of Human Evolution, 12*, 747–753.

Mollel, G. F., Swisher, C. C., Feigenson, M. D., & Carr, M. J. (2011). Petrology, geochemistry and age of Satiman, Lemagurut and Oldeani: Sources of the volcanic deposits of the Laetoli area. In T. Harrison (Ed.), *Paleontology and geology of Laetoli: Human evolution in context* (Geology, geochronology, paleoecology, and paleoenvironment, vol. 1, pp. 99–119). Dordrecht: Springer.

Musiba, C. M. (1999). Laetoli Pliocene paleoecology: A reanalysis via morphological and behavioral approaches. Ph.D. dissertation, University of Chicago, Chicago.

Musiba, C., Magori, C., Stoller, M., Stein, T., Branting, S., Vogt, M., Tuttle, R., Hallgrimsson, B., Killindo, S., Mizambwa, F., Ndunguru, F., & Mabulla, A. (2007). Taphonomy and paleoecological context of the Upper Laetolil Beds (Localities 8 and 9), Laetoli in northern Tanzania. In R. Bobe, Z. Alemseged, & A. K. Behrensmeyer (Eds.), *Hominin environments in the East African Pliocene: An assessment of the faunal evidence* (pp. 257–278). Dordrecht: Springer.

Mutakyahwa, M. (1997). Mineralogy of the Wembere-Manonga formation, Manonga Valley, Tanzania, and the possible provenance of the sediments. In T. Harrison (Ed.), *Neogene paleontology of the Manonga Valley, Tanzania: A window into the evolutionary history of East Africa* (pp. 67–78). New York: Plenum.

Ndessokia, P. N. S. (1990). The mammalian fauna and archaeology of the Ndolanya and Olpiro Beds, Laetoli, Tanzania. Ph.D. dissertation, University of California, Berkeley.

Protsch, R. R. R. (1981). Die Archäologischen und Anthropologischen Ergebnisse der Kohl-Larsen-Expeditionenin Nord-Tanzania 1933–1939. Band 3: The Palaeoanthropological Finds of the Pliocene and Pleistocene. Verlag Archaeologica Venatoria, Institute für Urgeschichte der Universität Tübingen, Tübingen.

Rage, J. C., & Bailon, S. (2011). Amphibia and Squamata. In T. Harrison (Ed.), *Paleontology and geology of Laetoli: Human evolution in context* (Fossil hominins and the associated fauna, vol. 2, pp. 467–478). Dordrecht: Springer.

Rossouw, L., & Scott, L. (2011). Phytoliths and pollen, the microscopic plant remains in Pliocene volcanic sediments around Laetoli, Tanzania. In T. Harrison (Ed.), *Paleontology and geology of*

Laetoli: Human evolution in context (Geology, geochronology, paleoecology, and paleoenvironment, vol. 1, pp. 201–215). Dordrecht: Springer.

Sanders, W. L. (2011). Proboscidea. In T. Harrison (Ed.), *Paleontology and geology of Laetoli: Human evolution in context* (Fossil hominins and the associated fauna, vol. 2, pp. 233–262). Dordrecht: Springer.

Su, D., & Harrison, T. (2008). Ecological implications of the relative rarity of fossil hominins at Laetoli. *Journal of Human Evolution, 55*, 672–681.

Tattersfield, P. (2011). Gastropoda. In T. Harrison (Ed.), *Paleontology and geology of Laetoli: Human evolution in context* (Fossil hominins and the associated fauna, vol. 2, pp. 567–587). Dordrecht: Springer.

Werdelin, L., & Dehghani, R. (2011). Carnivora. In T. Harrison (Ed.), *Paleontology and geology of Laetoli: Human evolution in context* (Fossil hominins and the associated fauna, vol. 2. 189–232). Dordrecht: Springer.

White, T. D. (1980). Additional fossil hominids from Laetoli, Tanzania: 1976–1979 specimens. *American Journal of Physical Anthropology, 53*, 487–504.

Chapter 3
Sedimentology, Lithostratigraphy and Depositional History of the Laetoli Area

Peter Ditchfield and Terry Harrison

Abstract A review of the stratigraphy and paleoenvironmental context of the main sedimentary units that outcrop at Laetoli is presented here, with a primary focus on the Upper Laetolil Beds. The lithological sequences at many of the numbered paleontological localities, as designated by Leakey (1987a), are described and sedimentary logs for many of these localities are presented. The litho-facies as described by Hay (1987) are reassessed in the light of recent developments in the understanding of volcano-sedimentary processes. Hay (1987) recognized the majority of sediments in the Laetolil Beds as aeolian tuffs, but it is possible that some of these sediments are the product of lahars or hyper-concentrated flows. Gullies and minor channels are relatively common in the Upper Laetolil Beds, indicating that a well-developed drainage system, larger than the present-day Garusi River system, was in place during the Pliocene. Streams and rivers were probably seasonal, but would have been associated with a complex vegetational mosaic, including woodland along the river courses. The general direction of the paleo-flow suggests that run-off originated as precipitation in the elevated areas towards the volcanic highlands in the east and that it flowed southwest across the Laetoli area towards the Eyasi basin. Ponds dotted the landscape during the rainy season, but were dry for much of the year. Primary ash fall deposits periodically blanketed the Laetoli area, forming distinctive marker tuffs. These inundations of volcanic ash would have had an adverse effect on the local ecosystem, leading to a landscape dominated by grasslands and open woodlands. However, these periods of disruption in the climax vegetation were relatively short-lived, with grasslands being quickly replaced by a mosaic of woodland, bushland and grassland.

P. Ditchfield (✉)
Research Laboratory for Archaeology and the History of Art, School of Archaeology, University of Oxford, Oxford, UK
e-mail: peter.ditchfield@rlaha.ox.ac.uk

T. Harrison
Center for the Study of Human Origins, Department of Anthropology, New York University, 25 Waverly Place, New York, NY 10003, USA
e-mail: terry.harrison@nyu.edu

Keywords Eyasi Plateau • Laetolil • Ndolanya • Ogol • Paleoenvironment

Introduction

Laetoli is located on the margin of the Eyasi Plateau at the southern margin of the Serengeti Plains, about 20 km north of Lake Eyasi, Tanzania. The plateau is an uplifted fault block with a general elevation of 1,700–1,800 m, with a maximum elevation of 2,000 m at Eseketeti (formerly Speer Mountain) on the margin of the fault scarp. The Eyasi Plateau is bounded to the south by the Eyasi Fault and to the east by the Ngorongoro volcanic highlands. The Laetoli area straddles the divide between the Lake Eyasi and Olduvai Gorge drainage basins (Kent 1941; Hay 1987).

Outcrops of the Laetolil and Ndolanya Beds are exposed over an area of 1,000 km^2 and fossils from these beds have been recovered from a number of areas, including Oldogom, Endolele, Esere, Kakesio, Laetoli, Lake Ndutu, Silal Artum, Emboremony, and Noiti (Kent 1941; Dietrich 1942; Leakey 1987a; Hay 1976, 1987; Harrison and Kweka 2011). However, the majority of fossils have been recovered from a relatively restricted area (about 100 km^2) along the Garusi, Gadjingero, Nompopong and Olaitole River valleys, which represents the site of Laetoli (see Harrison and Kweka [2011] for a detailed description of the paleontological localities).

Outcrops of fossiliferous sediments within this main area have been divided up into a series of localities and sub-localities, numbered 1 to 24 (Leakey 1987a: Figure 1.2; Harrison and Kweka 2011). An updated map of these localities, based on satellite imagery, is shown in Fig. 3.1.

The geology and geochronology of the sediments in the Laetoli region have been well documented (Kent 1941; Pickering 1964; Hay 1976, 1978, 1987; Hay and Reader 1978; Hay and Leakey 1982; Drake and Curtis 1987; Ndessokia 1990; Manega 1993) and have recently undergone further study and revision (Deino 2011; McHenry 2011; Mollel et al. 2011; Adelsberger et al. 2011; Harrison and Kweka 2011).

T. Harrison (ed.), *Paleontology and Geology of Laetoli: Human Evolution in Context. Volume 1: Geology, Geochronology, Paleoecology and Paleoenvironment*, Vertebrate Paleobiology and Paleoanthropology, DOI 10.1007/978-90-481-9956-3_3,

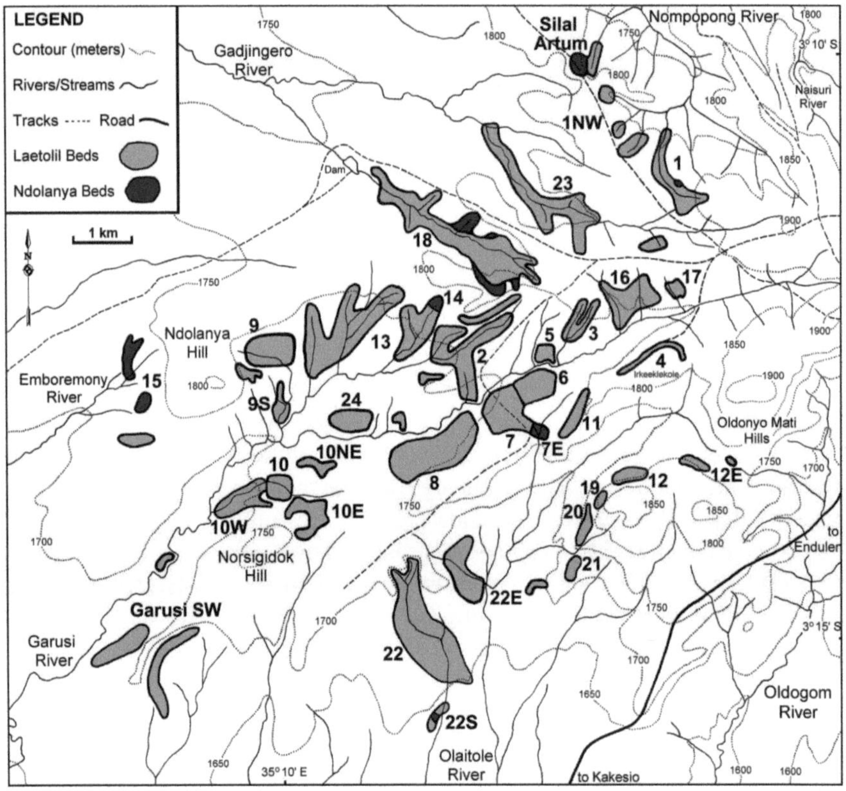

Fig. 3.1 Map showing the numbered paleontological localities in the Laetoli area

The aims of this paper are to review the stratigraphy of the sedimentary units within the Laetoli area and to reassess some of the paleoenvironmental interpretations. It is the upper part of the Upper Laetolil Beds that forms the majority of the outcrops examined in this study and described in detail below with representative logged sections from the various paleontological localities. The UTM coordinates of the sections are summarized in Table 3.1. There is a need to update the stratigraphic nomenclature, to bring it into line with current stratigraphic practice (Hedberg 1976, Whittaker et al. 1991, Murphy and Salvador 1999), but this will be dealt with in a separate publication (Ditchfield et al. in preparation).

Lithostratigraphy

Overview

The Laetolil Beds, which represent the base of the sedimentary sequence, are more than 123 m thick and are divisible into two main lithological units (Hay 1987) (see Fig. 3.2). The lower unit (more than 64 m thick) consists mainly of aeolian tuffs interbedded with primary fall-out tuffs and water-worked tuffaceous sediments. Locally, it is separated from the underlying Precrambrian basement rocks by about 3 m of weathered claystones and sandstones. Dating of the lower unit indicates an age from 3.85 Ma to 4.36 Ma (Drake and Curtis 1987; Deino 2011). Fossil vertebrates have been recovered from this unit, but they are much rarer and more scattered than in the overlying unit.

The upper unit (44–59 m thick) consists of a series of aeolian and fall-out tuffs (Hay 1987). The top of the unit is delimited by the Yellow Marker Tuff (8 m thick), with other tuffs throughout the sequence designated as marker tuffs (Tuffs 1–8). Tuff 8, near the top of the unit, is dated to 3.63 Ma (Deino 2011). The diverse fauna from the Upper Laetolil Beds, including all of the *Australopithecus afarensis* specimens (Harrison 2011a), can be reliably dated, to between ~3.6 Ma and ~3.85 Ma. The Laetolil Beds are probably derived from tephra that was erupted from Satiman, an extinct volcano, located 20 km to the east of Laetoli. The sediments were deposited on a peneplain that was subsequently domally uplifted by more than 250 m (Hay 1987; Mollel et al. 2011).

Disconformably overlying the Laetolil Beds are a series of tuffs and calcretes comprising the Ndolanya Beds (15–20 m thick) (Hay 1987) (Fig. 3.2). These are subdivided into upper and lower units separated by a disconformity. Radiometric dates provide an age of 2.66 Ma for the Upper Ndolanya Beds (Ndessokia 1990; Manega 1993; Deino 2011), and the fauna is consistent with this age. The Upper Ndolanya Beds have yielded a rich vertebrate fauna, including the remains of *Paranthropus aethiopicus* (Harrison 2011a). The Ndolanya Beds are overlain by a series of lavas, the Ogol Lavas, dating to ~2.3 Ma, derived from a range of low volcanic cones to the south and southwest of Lemagurut (Hay 1987, Drake and Curtis 1987; Deino 2011).

Table 3.1 Summary of UTM coordinates for logged sections at Laetoli, Kakesio and Noiti

Locality	UTM zone	UTM E	UTM N
Laetoli localities			
Locality 1 (top)	36	747300	9647130
Locality 1 (graben)	36	747126	9647590
Locality 1 (base)	36	747285	9648580
Locality 2	36	745130	9645204
Locality 3 (top)	36	745739	9645476
Locality 3 (base)	36	745635	9645002
Locality 4	36	741398	9642140
Locality 5	36	745252	9644652
Locality 6 (top)	36	744953	9644234
Locality 6 (middle)	36	744719	9644318
Locality 6 (base)	36	744671	9644396
Locality 7 (top)	36	744929	9643536
Locality 7 (base)	36	744720	9644198
Locality 8	36	743544	9643290
Locality 9 (top)	36	740882	9643980
Locality 9 (base)	36	740951	9644176
Locality 10 East	36	741398	9642140
Locality 10	36	740922	9642586
Locality 10 West	36	740782	9642686
Locality 12	36	747304	9643132
Locality 12 East	36	747402	9643342
Locality 13	36	742642	9645467
Locality 14	36	743063	9645479
Locality 15	36	738244	9643620
Locality 16	36	746102	9645691
Locality 18	36	744261	9645985
Locality 19	36	746007	9642798
Locality 20	36	745790	9642486
Locality 21	36	745552	964286
Locality 22 (top)	36	743022	9641648
Locality 22 (middle)	36	742652	9641380
Locality 22 (base)	36	742667	9641302
Locality 22 East	36	743280	9641620
Other localities			
Kakesio 1	36	722071	9629358
Kakesio 8	36	724251	9626445
Noiti 1	36	746011	9634149
Noiti 2	36	746620	9634827
Noiti 3	36	747400	9634225

A series of tuffs, up to 15 m thick, comprise the Naibadad Beds (Hay 1987). Although contact with the Ogol Lavas has not been observed (Hay 1987), the Naibadad Beds were thought to be somewhat younger than the Ogol Lavas. Hay (1976, 1987) suggested that the Naibadad Beds best correlate lithostratigraphically with lower Bed I and the Naabi ignimbrite (K-Ar dated at ~2.1–2.3 Ma) at the base of the Olduvai sequence. This conclusion is confirmed by more recent radiometric dating of tuffs in the Naibadad Beds, which provide mean ages ranging from 2.057 Ma to 2.155 Ma (Ndessokia 1990; Manega 1993; Deino 2011). Few fossils have been recovered from this unit.

The Olpiro Beds (up to 5 m thick) are disconformable with the underlying Naibadad Beds and are mainly comprised of claystones. A distinctive grey tuff was suggested by Hay (1987) to be correlated with Tuff IID in the upper part of Olduvai Bed II, which provides an estimated age of ~1.5 Ma (Manega 1993). However, $^{40}Ar/^{39}Ar$ dating suggest that the deposits are older, with average ages of ~2.0–2.1 Ma (Ndessokia 1990; Manega 1993; Deino 2011). The overlying Ngaloba Beds consist primarily of claystones, with localized sandstone and conglomerate facies. These are subdivided into an upper unit (up to 5.2 m thick) and a lower unit (up to 6 m thick), of which the former is tentatively correlated with the Ndutu Beds at Olduvai Gorge (Hay 1987). Both units are fossiliferous, and the vertebrate fauna from the Upper Ngaloba Beds includes a hominin (*Homo sapiens*; Magori and Day 1983). Acheulian and Middle Stone Age artefacts have been recovered from the lower and upper units respectively (Hay 1987; Leakey 1987a). A summary of the lithostratigraphic units as described by Hay (1987) is depicted in Fig. 3.2.

Laetolil Beds, Lower Unit

The main exposures of the Lower Laetolil Beds are to the south and east of Laetoli, at Kakesio and Esere-Noiti (see Fig. 3.3). Minor exposures of the upper part of the Lower Laetolil Beds can be found in the Laetoli area at Locs. 1 and 9 (see Fig. 3.3), dominated by a thick and distinctive ignimbrite sequence.

The Lower Laetolil Beds exceed 64 m in thickness, as seen at Kakesio 1 (UTM 36 722071E, 9629358N Alt. 1,692 m). At this location the Lower Laetolil Beds consist of mainly orange to brown aeolian tuffs interbedded with minor fall-out and water-worked tuffs. It is moderately bioturbated with traces of fossil termite activity, which are particularly common in the basal 10 m. It is also moderately fossiliferous at this location containing vertebrates, mollusca and termitaries. It unconformably overlies crystalline basement rocks and is locally separated from the basement by 3 m of highly weathered regolith. Basement topography is generally flat with some gully incision.

The contact with the overlying Upper Laetolil Beds was not observed at Kakesio 1, but can be seen at Laetoli Loc. 1. However, this is some 30 km from the Kakesio outcrops and no marker beds common to both locations were observed. Thus, based on this study it is impossible to give a total thickness for this member or to provide an absolute lithostratigraphic correlation to the Lower Laetolil sediments exposed across the Laetoli area.

The lithologies of the Lower Laetolil Beds, as seen at Kakesio 1, exhibit only minor channel and runnel development. However, 3.6 km to the southeast at Kakesio 8 (UTM 36 724251E, 9626445N), the Lower Laetolil Beds are dominated by fluvially deposited tuffaceous sandstones and conglomerates. These show large-scale fluvial trough cross-bedding with

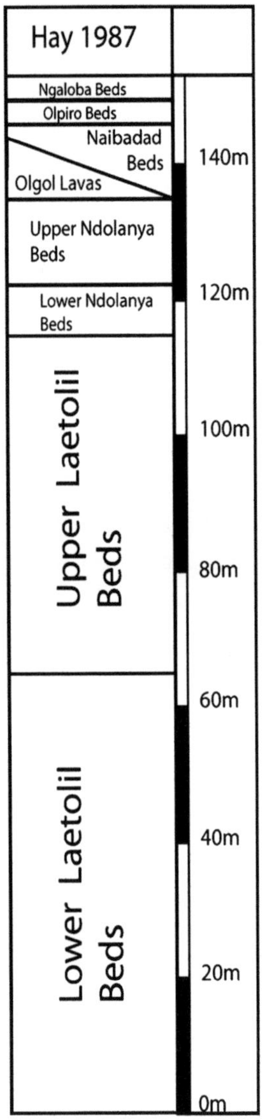

Fig. 3.2 Lithostratigraphic scheme for the Laetoli area (after Hay 1987)

flow directions generally to the west. These are exposed in a channel belt approximately 200 m in width. Individual channel sections where preserved suggest a strongly braided drainage system. Lateral to these channel deposits are overbank sequences of highly pedogenically altered tuffaceous silts, which are incised by the fluvial sediments. Similarly, evidence of a more coherent drainage system can be found in the Lower Laetolil Member at Noiti 3 (UTM 36 747400E, 9634225N), where a similar fluvial sequence can be seen building out into a small lake basin via a series of deltaic deposits.

This contrasts with sequences seen at Noiti 1 and 2 (UTM 36, 746011E, 9634149N, 1,565 m and UTM36, 746620E, 9634827N) where high sediment load lahar deposits can be seen. At least two separate lahar events can be traced, apparently filling and eventually overtopping the local drainage system of relatively small anastomosing channels incised into tuffaceous sandstones. These lahar deposits are rich in fossil wood (Bamford 2011a).

Laetolil Beds, Upper Unit

The Upper Laetolil Beds vary in thickness from 44 m to 59 m thick and consist of a series of aeolian and fall-out tuffs that have undergone variable and often considerable bioturbation, occasional fluvial reworking, and minor pedogenesis. A relatively complete composite section for the Upper Laetolil Beds is exposed at Loc. 1, between approximately UTM 36 747300, 9647130 to UTM 36 747285, 9648580 (Fig. 3.4). Because of its completeness and excellent exposure this section should be taken as a type section for the Upper Laetolil Beds.

There appears to be only a minor disconfomity between the Upper and Lower Laetoli Beds, as seen at Loc. 1. The Upper Laetoli Beds contain distinctive fall-out tuffs, which have been used as lithostratigraphic markers (Hay 1987). McHenry (2011) has shown that these divide into two compositional groups Tuffs 1–4 and Tuffs 5–8. This compositional variation broadly corresponds to sedimentological differences between the upper and lower 'marker tuffs' as described below. The top of the Upper Laetoli Beds has been taken to be a conspicuous yellow tuff (Yellow Marker Tuff). However, McHenry (2011) has shown that this unit has closer compositional affinities to the tuffs in the overlying Ndolanya Beds. This unit makes a convenient lithostratigraphic marker, which is easy to recognize in the field, and thus it seems appropriate to retain it within the Upper Laetolil Beds. Furthermore, because in some localities (i.e., Loc. 10), the Yellow Marker Tuff is overlain by further aeolian tuffs of the Upper Laetolil Beds, it would be inappropriate to define the base of the Ndolanya Beds by this distinctive marker tuff.

Ndolanya Beds

The Ndolanya Beds, which overlie the Laetolil Beds, are subdivided into upper and lower units (Hay 1987). The Ndolanya Beds are generally 15–20 m thick and show a weak angular unconformity with the underlying Upper Laetoli Beds. In addition to this difference, Hay (1987) also describes channels within the basal part of the Ndolanya Beds, which cut into the underlying Laetolil Beds.

Ndolanya Beds, Lower Unit

The Lower Ndolanya Beds are generally 4–12 m thick, although in some places the lower member is locally cut out by an intraformational disconformity. This unit consists of pedogenically altered tuffaceous claystones with occasional crystal and vitric tuffs, which are compositionally distinct

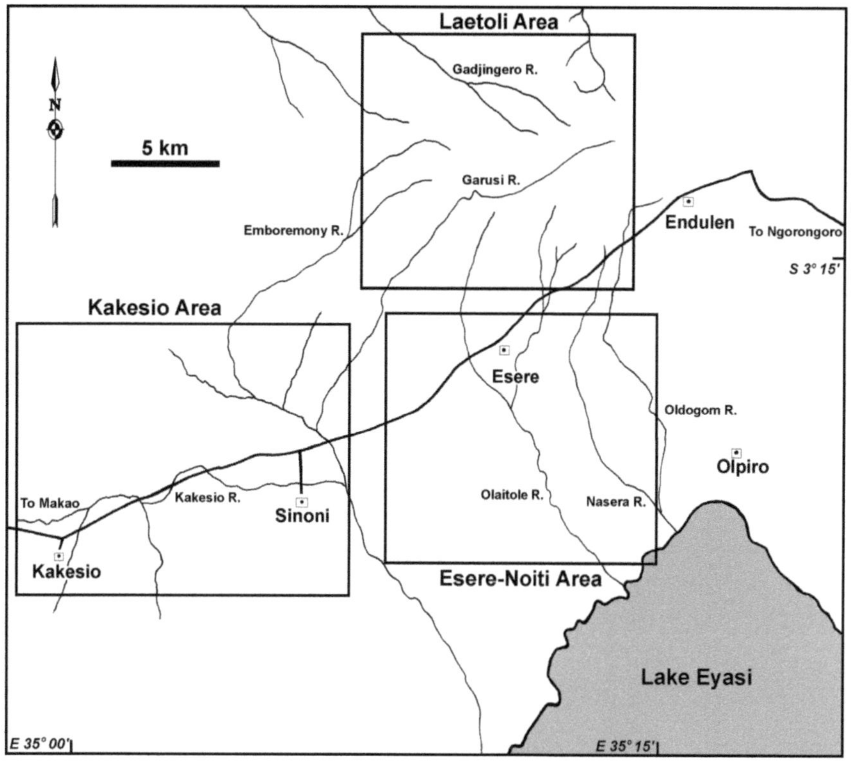

Fig. 3.3 Map of the Eyasi region showing the location of the Laetoli, Kakesio and Esere-Noiti areas. See Fig. 3.1 for detail of the Laetoli area

from the fall-out tuffs of the underlying Laetolil Beds (McHenry 2011).

Ndolanya Beds, Upper Member

The Upper Ndolanya Beds are generally 8–16 m thick. It consists of clay rich aeolian tuff with minor calcretes. It is often richly fossiliferous. The base of the Upper Ndolanya Beds is often marked by a well-developed calcrete (up to 1 m thick) the base of which is erosive into the underlying member and in some cases into the underlying Laetolil Beds. The upper part of the Upper Ndolanya Beds is marked by an increasing frequency and development of tuffaceous calcrete layers, which are probably pedogenic in origin. The Upper Ndolanya Beds are well exposed at Loc. 18 (UTM 36, 744261E, 9645985N).

General Lithostratigraphy

Primary Fall-Out Tuff Correlation

Primary fall-out tuffs ("Air-fall tuffs" of Hay [1987]) were regarded as stratigraphic markers by Hay (1987). These marker tuffs (1 to 8) and the Yellow Marker Tuff were used to correlate between the many separate localities where beds could not be continually traced from one outcrop to the next. This approach is useful in the field, and has been followed in this study, although there are several problems that should be considered. Many of the sections, especially in the lower part of the Upper Laetolil Beds, exhibit many thin (usually less than 5 cm thick), laterally discontinuous air fall tuff beds, which can easily be confused with Tuffs 1 to 4. This is less of a problem in the upper part of the member where the marker tuffs themselves are generally thicker and more lithologically distinctive. Therefore, lithostratigraphic correlation is potentially problematic on its own in the lower part of the Upper Laetoli Beds and should be used in conjunction with tephra chemistry to ensure more reliable correlations between non-contiguous outcrops. In the stratigraphic sections of the various localities presented below these marker tuffs are labelled T1 through T8 (for Tuffs 1–8) and YMT (for the Yellow Marker Tuff).

Color Change

Color of the sediments is generally a poor method of effecting lithostratigraphic correlations due to variations in lighting conditions, moisture content and color perception of the individual. However, as noted by Hay (1987), there are

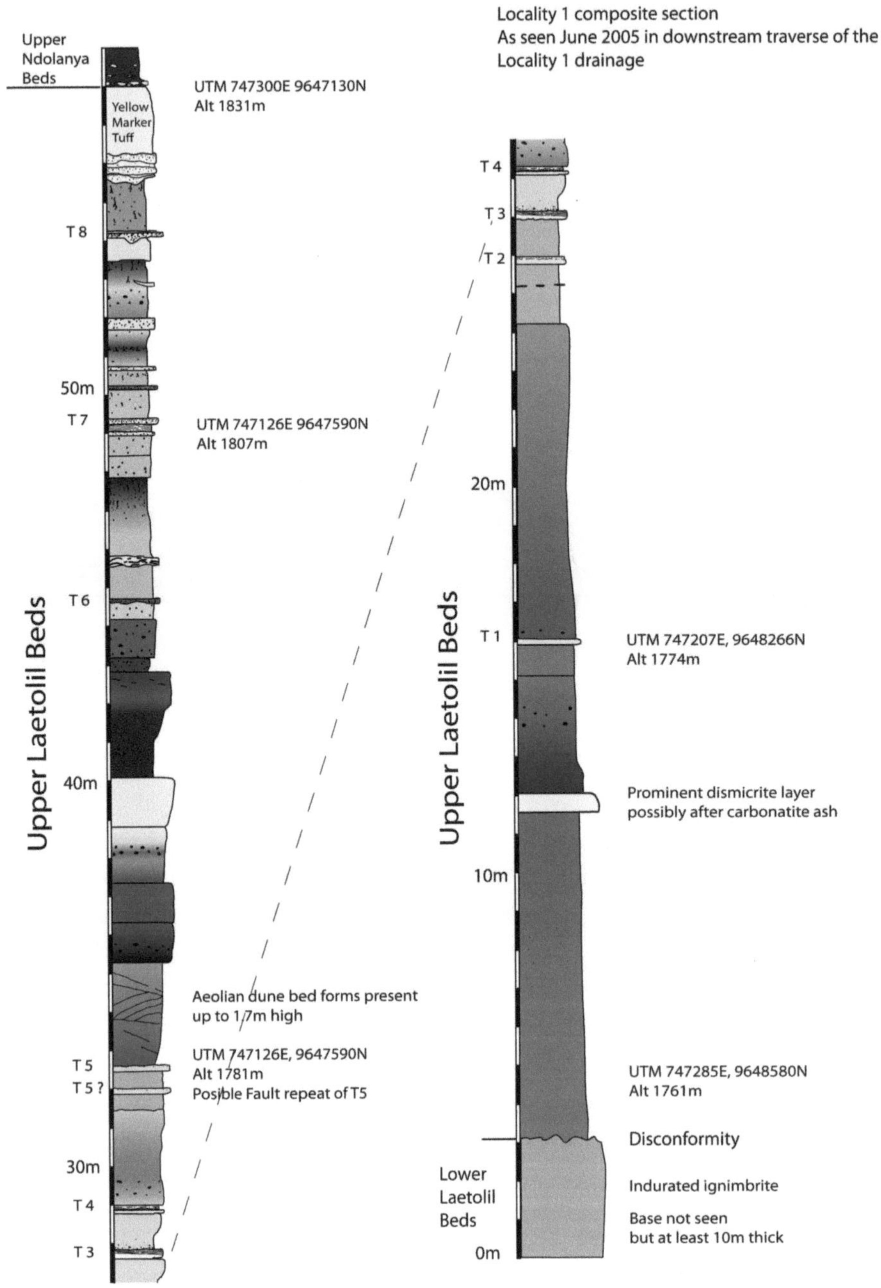

Fig. 3.4 Composite log of the type section of the Upper Laetolil Beds as seen at Laetoli Loc. 1 in a northerly traverse along the Loc. 1 (Nompopong River) drainage from approximately UTM 36, 747300E, 9647130N to UTM 36, 747285E, 9648580N. Marker Tuffs are depicted as T1, T2, T3, T4, T5, T6, T7, T8 and YMT

some consistent color changes within the Upper Laetoli Beds that can be useful in local correlation. The majority of the Upper Laetolil Beds are a light to dark grey in color, with the more carbonate horizons being light and the more clay-rich horizons being darker. However, the interval between Tuffs 5 and 6 is often predominantly brown to grey-brown in color, as is Tuff 6 itself. The sequence above Tuff 8 is often characterized by yellow-grey color, which increases in intensity upwards into the Yellow Marker Tuff.

Description of Logged Sections

Field Methods

Major fossil bearing localities (see Harrison and Kweka 2011) were visited and representative sedimentary sequences for each location were logged using standard field logging techniques: tape measure for vertical sections and Jacob's staff and Abney level for inclined and low-angle sections.

For localities where significant lateral variation was observed, multiple logged sections were recorded. The position of the logged sections was recorded with a Garmin hand held GPS system. All positions are given as UTM coordinates and are recorded relative to WGS84 datum (summarized in Table 3.1).

Sections and Locations

Locality 1

Locality 1 is exposed in a gully system that runs from (UTM 36 747300, 9647130 to UTM 36 747285, 9648580) with the drainage running from south to north. The section is essentially continuous from the Yellow Marker Tuff in the south to the junction with the Lower Laetolil Beds in the north (Fig. 3.4). There is also a section of the Upper Ndolanya Beds exposed in a small graben at (UTM 36 747126E, 9647590N) (Fig. 3.5). Faults relating to this graben cause a repeat of the sequence of the Upper Laetoli Beds at this location.

Locality 2

Locality 2 consists of a series of low-lying exposures near to a reservoir at UTM 36, 74513E, 9645204N. This locality extends to the north as poorly exposed outcrops to the south of and on the flanks of Ngirerati Hill at UTM 36 743847E, 9645366N. The sequence consists of probable Tuff 5 at the southern end of the locality to at least as high as Tuff 8 and the Ndolanya Beds on the flanks of the Ngirerati Hill (see Fig. 3.6). The patchy nature of the outcrop makes sedimentary structures difficult to pick out, except in the middle part of the sequence stratigraphically adjacent Tuff 6 (T6) to where pedogenically altered aeolian tuffs dominates.

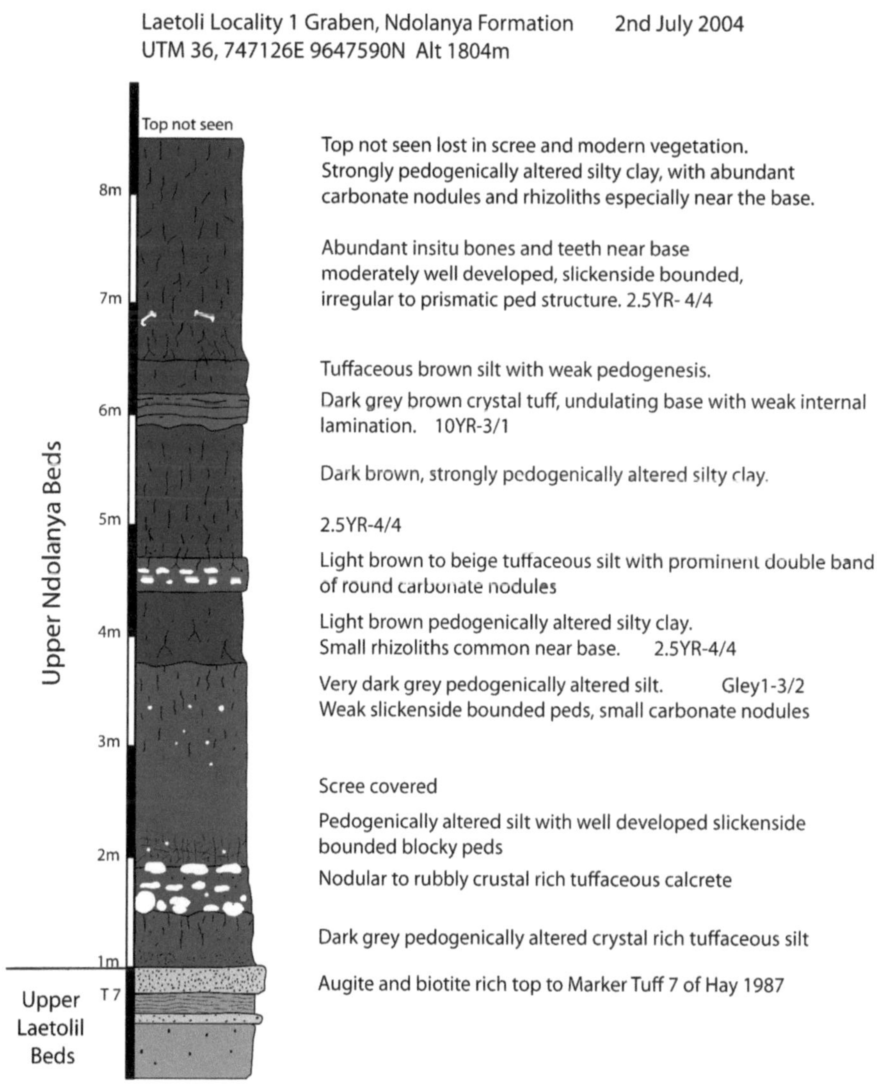

Fig. 3.5 Logged section of the Upper Ndolanya Beds as seen in the Loc. 1 graben feature

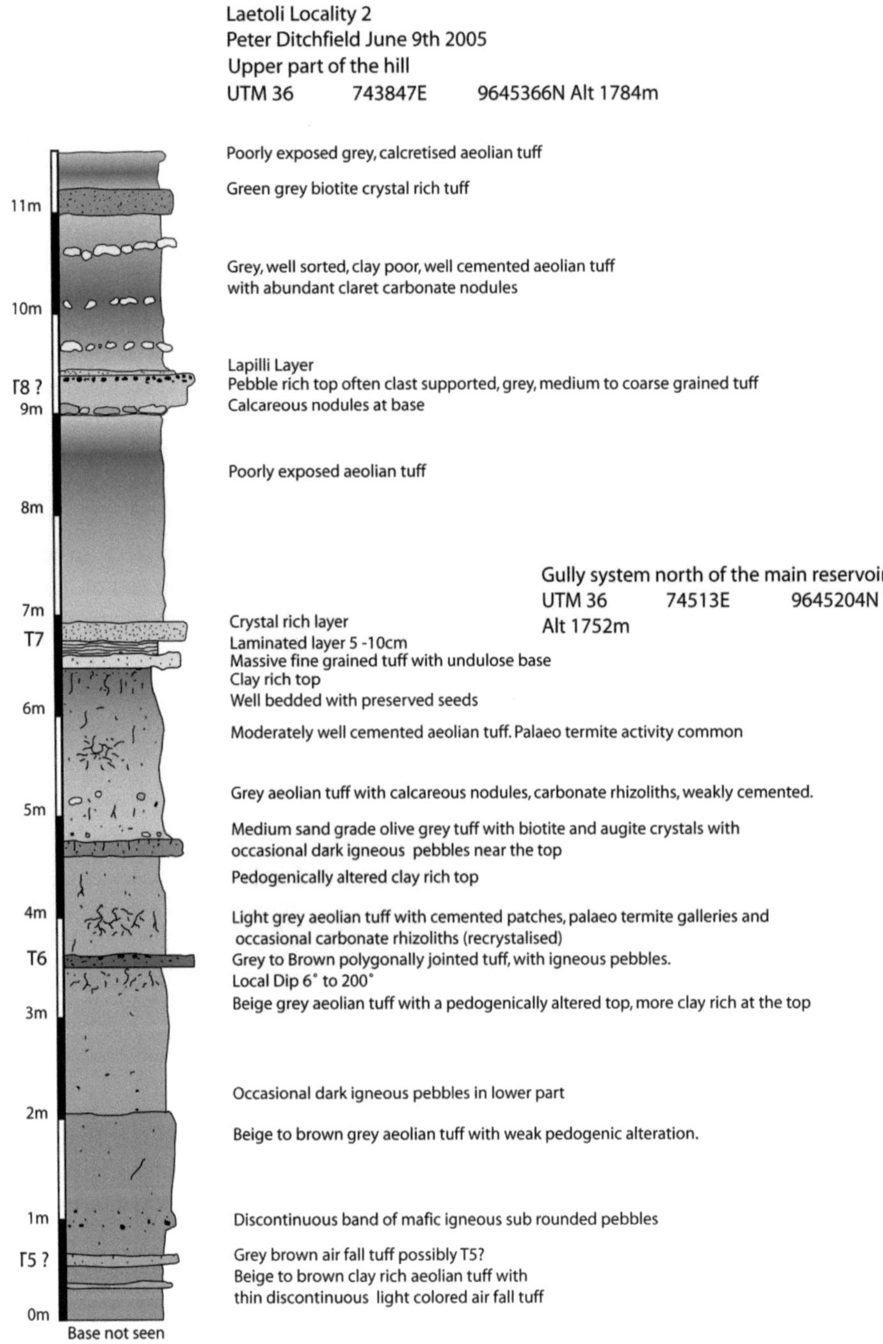

Fig. 3.6 Composite section of the Upper Laetolil Beds sequence exposed in Loc. 2

The middle part of the sequence is cut by a series of small faults striking approximately north to south and steeply dipping to the west. The change from generally grey lithologies to more beige and brown-grey lithologies below Tuff 6 is apparent here as at other localities.

Locality 3

Locality 3 is exposed in a tributary stream of the Garusi River between UTM 36, 745739E, 9645476N for the top of section, and UTM 36, 745635E, 9645002N for the lower part of the section. A composite stratigraphic section for Loc. 3 is shown in Fig. 3.7.

Stratigraphically this section extends down from the Yellow Marker Tuff to below Tuff 5. It is notable for several marked local facies variations, including a localized fine-grained waterlain deposit of reworked tuff, which has yielded very well preserved insect fossils, plant material and ruminant coprolites from between Tuffs 7 and 8 (see Harrison and Kweka 2011; Bamford 2011b; Krell and Schawaller 2011; Harrison 2011b) located at UTM 36, 745664E, 9645210N. This locality also shows an episode of channel cutting and filling approximately 1 m below Tuff 6, with a series of

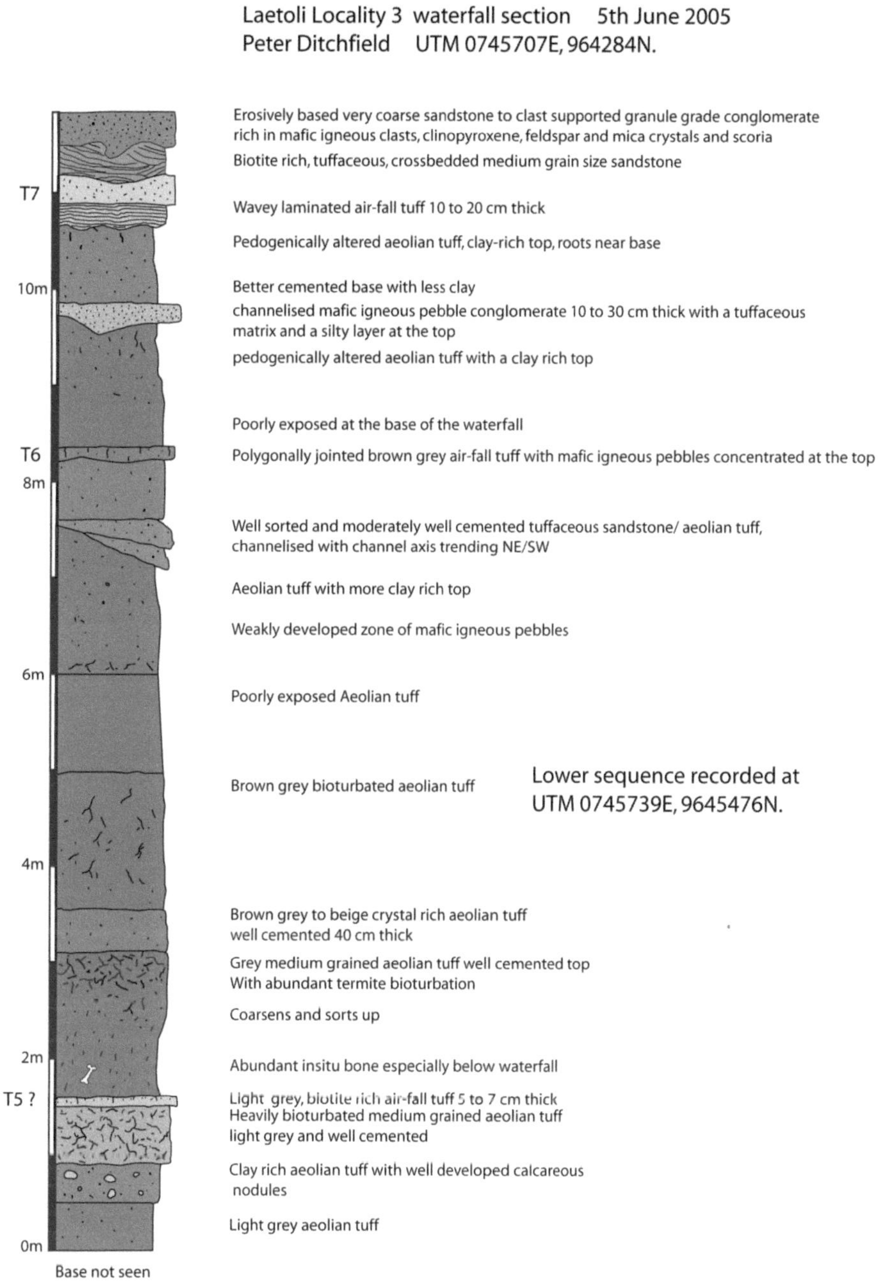

Fig. 3.7 Composite section of the Upper Laetolil Beds sequence exposed in Loc. 3

approximately north-south orientated channels. These are filled by well-sorted tuffaceous sandstones with aeolian tuff intraclasts and occasional bank collapse features. At UTM 36, 745739E, 9645284N, large scale, low-angle, asymmetric cross-bedding can be observed, possibly representing aeolian reworking/deposition of the aeolian tuff.

Locality 4

The sequence at Loc. 4 is well exposed at UTM 36, 741398E, 9642140N. These exposures show an expanded thickness of Tuff 7 and overlying aeolian tuffs, which become progressively more concretized upwards (Fig. 3.8). At this location the Upper Laetoli Member sequence is overlain by scoria and associated Ogol lavas.

Locality 5

Representative sections for Loc. 5 were recorded at UTM 36, 745252E, 9644652N, Alt. 1,749 m and UTM 36, 745207E, 9644536N (Fig. 3.9). Both locations show prominent exposures of Tuff 3, which at the first location has a concrete pedestal marking the location of the L.H. 7 hominin find. Tuff 3 has a clear multi-bed structure at this location and shows occasional footprints on its upper surface. It is overlain by ~5 m of

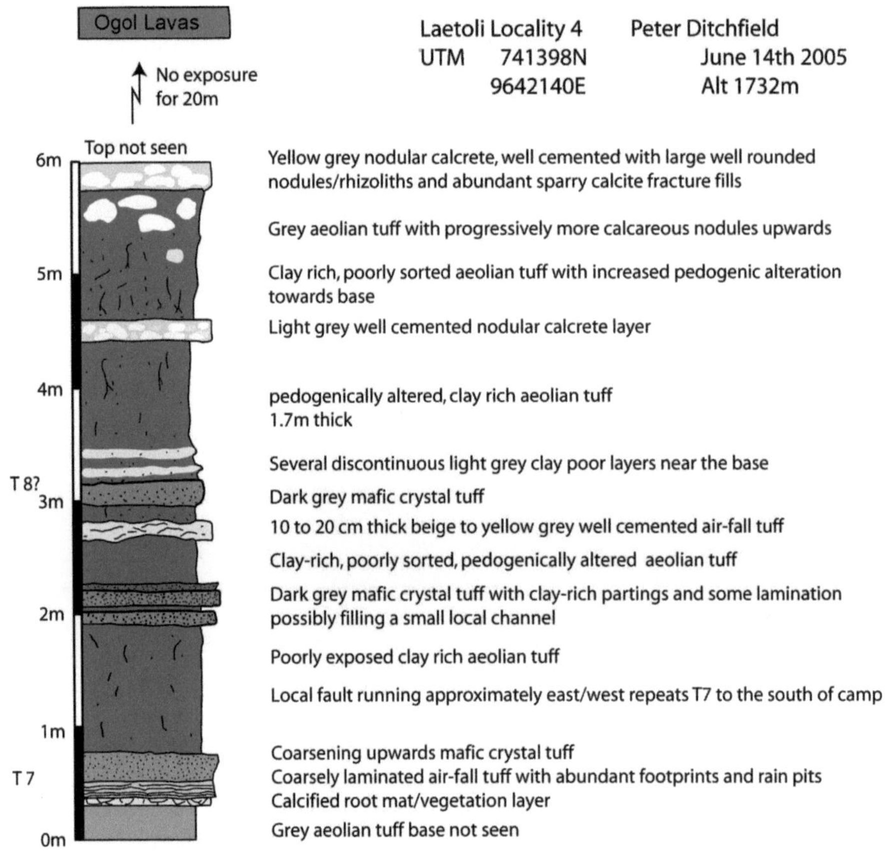

Fig. 3.8 Section of the sequence exposed in Loc. 4

pedogenically altered aeolian tuff, which contains several thin primary fall-out tuff layers, one of which may be an expression of Tuff 4. Below Tuff 3 there are at least 3 m of aeolian tuffs, which are cut by a prominent cross-bedded channel-fill feature, which gives a mean paleoflow direction of 175° ($n = 8$).

Locality 6

The exposures of Loc. 6 are on the south side of the Garusi River extending southwards from the sections exposed along the southern bank. A composite sequence for this locality is shown in Fig. 3.10. The lower part of the sequence was recorded from the Garusi River at UTM 36, 744671E, 9644396N. The middle section, from the hominin pedestal of L.H. 26 to Tuff 6, was recorded at UTM 36, 744719E, 9644318N. Further south, several isolated outcrops of Tuff 7 were seen, as at UTM 36, 744953E, 9644234N.

Localities 7 and 7E

Locality 7 extends southeast from the southern margin of Loc. 6 with good sections being largely restricted to the areas adjacent to Loc. 6. It exposes a similar sequence to that seen in the Loc. 6 with Tuff 5 through Tuff 7 well exposed.

The hominin marker pedestal L.H.3/6 (Leakey 1987b) is located within Loc. 7 at UTM 36, 744720E, 9644198 and another badly weathered hominin marker pedestal for L.H. 4 is located nearby at UTM 36, 744590E, 9644226N. Both of these pedestals are not in the stratigraphic locations identified by Leakey (1987b: Table 3.5.1; see Harrison 2011a). The pedestal L.H. 3/6 is resting directly on Tuff 6 as opposed to its published position 0.5 m above Tuff 7. The pedestal for L.H. 4 is 1.25 m below Tuff 6, not 1.2 m below Tuff 7 as published. We cannot account for these discrepancies, and assume here that the reported stratigraphic provenience of the hominin specimens is sound, and that the pedestals have been inadvertently relocated. Similar discrepancies occur with the placement of hominin pedestals at other localities (see Harrison 2011a for further discussion on this point).

Within Loc. 7 there is evidence of a paleodrainage network within the aeolian tuffs just below Tuff 6. Several small steep-sided paleo-runnels can be traced laterally at this level, and are seen to coalesce to form a larger channel feature approximately 10 m wide and 1 m deep. The channel is filled with fining upwards beds of granule to pebble grade matrix supported conglomerates, with a medium- to coarse-grained, crystal-rich sandy matrix. Occasional bank collapse features can also be observed. This channel system can be traced across several hundred meters, and although moderately sinuous it shows a clear overall direction of the drainage flowing to the east.

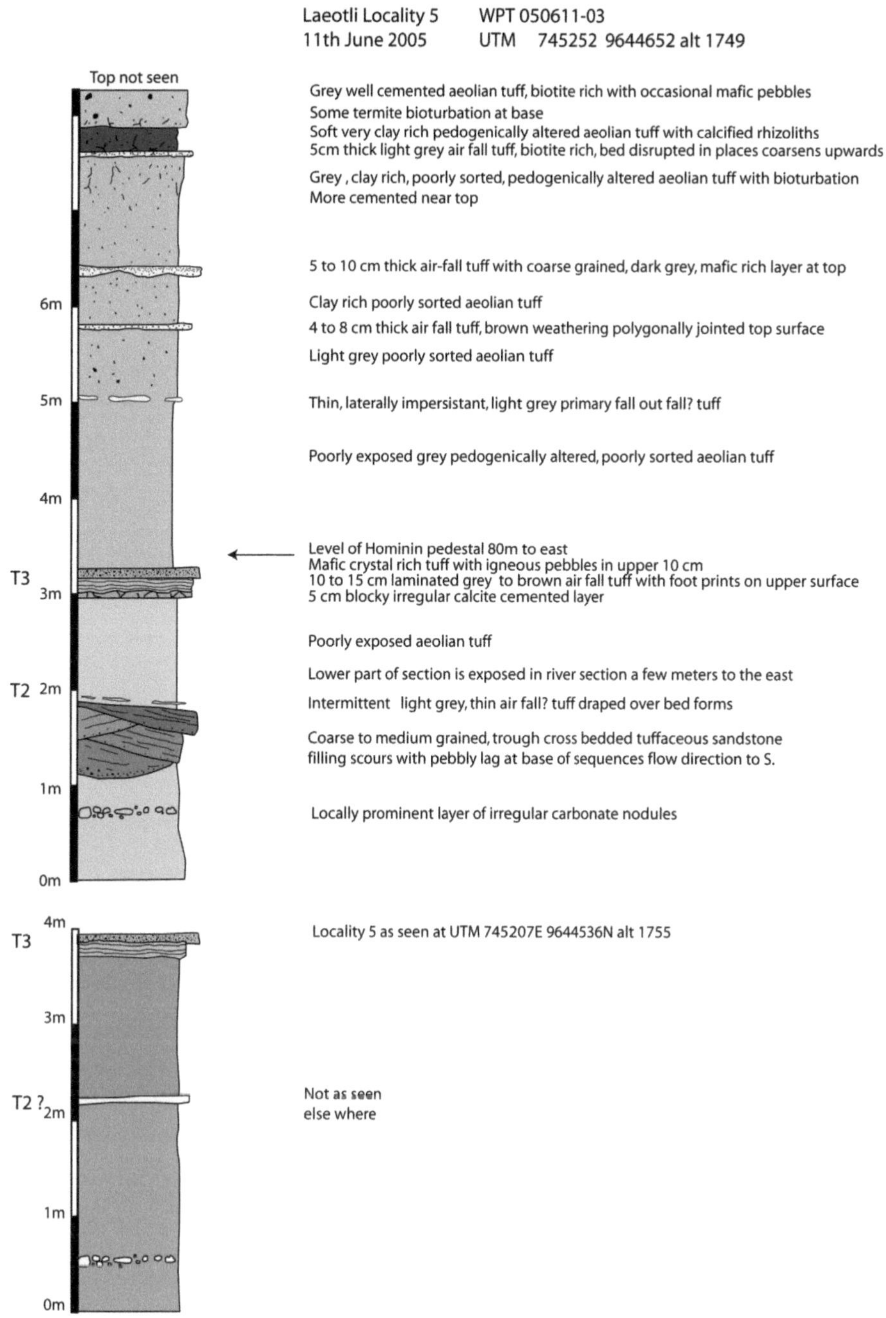

Fig. 3.9 Sections showing the sequence exposed in Loc. 5 and the local facies variation below Tuff 3

To the southeast of Locality 7 the exposures are restricted to a line of low bluffs, with a more coherent series of exposures encountered along the eastern flank of the locality. A thick sequence is exposed, which includes probable Tuff 8 and Yellow Marker Tuff (although exposure of the upper part is poor exposed and partly obscured by overlying Ndolanya Beds as well as boulders of the Ogol Lavas). Locality 7E, just to the south of Loc. 7, exposes the Upper Ndolanya Beds, containing abundant fossil vertebrate remains. This sequence is shown in Fig. 3.11b. At UTM 36 744929E, 9643536N Tuff 8 is displaced approximately 3m by a north/south trending fault, with a down throw to the east. Above Tuff 8 the Upper Laetoli Beds become increasingly calcareous with micritic nodules passing upwards into more tabular calcretes, although it is unclear if this is due to pedogenesis or is diagenetic in origin and related to the prominent calcrete development at the base of the overlying Ndolanya Beds.

Locality 8

Locality 8 comprises a large area of exposures on the southern side of the Garusi River and some 400m to the

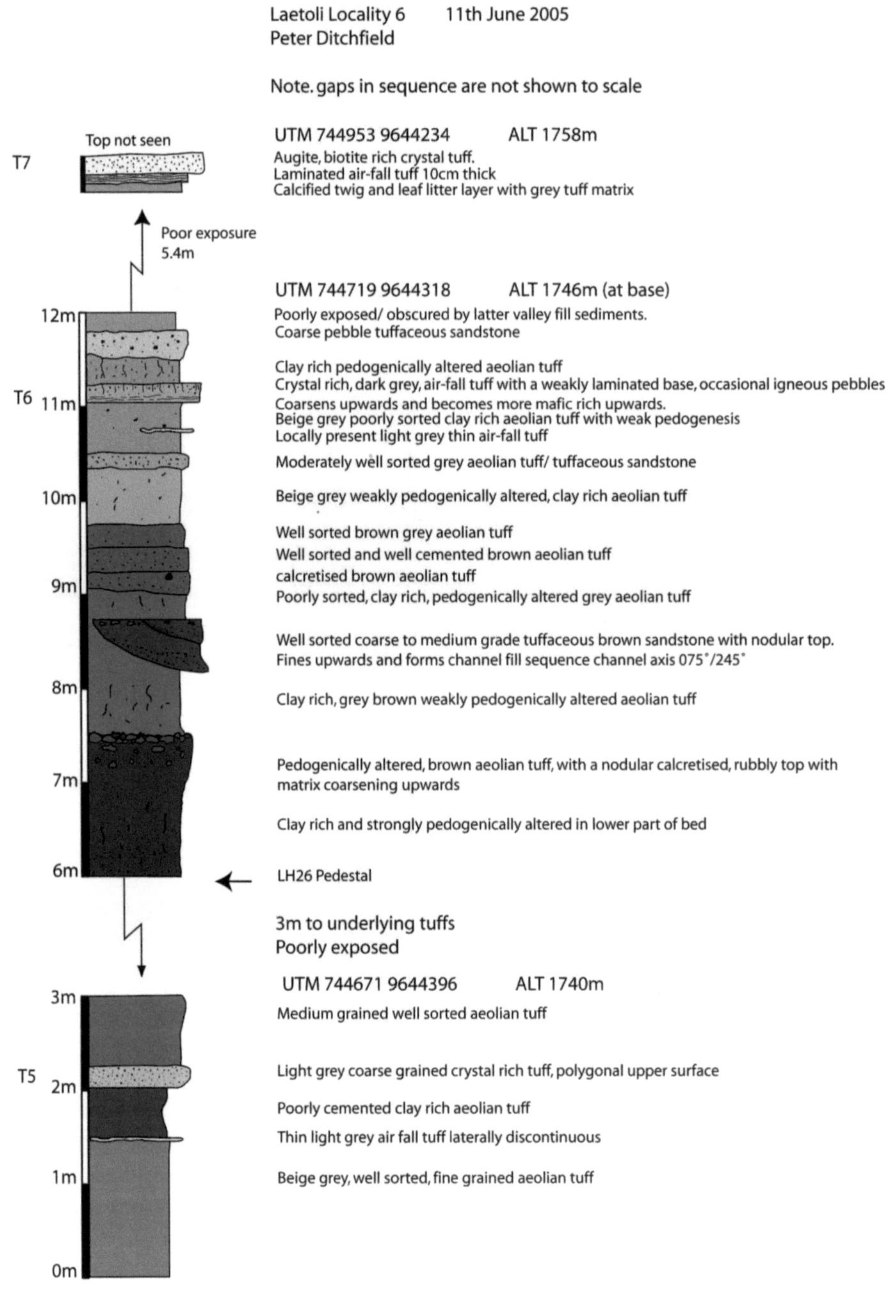

Fig. 3.10 Sections showing the sequence exposed in Loc. 6. Note the stratigraphic gaps in the section are not shown to scale

southwest of Loc. 7 (UTM 36 743544E, 9643290N). The sequence of the Upper Laetoli Beds ranges from Tuff 4 at the base through to the Yellow Marker Tuff at the top and is some 18m thick. Locality 8 contains the hominin footprint trail excavated in 1978 and 1979 (Leakey 1987c). This is now covered as a conservation measure and is no longer exposed, although there are future plans to expose them once again and create a museum exhibit at the locality. The sequence is very similar to that observed in Loc. 7 and, as at Loc. 7, there is evidence of increased channelling and drainage system formation in the aeolian tuffs just below Tuff 6.

Locality 9S

Locality 9S consists of a series of low cliffs, approximately 400 m long, trending SSW to NNE, from UTM 36, 740877E, 9643904N to UTM 36, 740925E, 9644300N. The lithological sequence exposed at Loc 9S is shown in Fig. 3.12a.

The sequence at Loc. 9S shows distinct facies variation form south to north. In the north, the lower part of the sequence consists of in excess of one meter of cross-bedded tuffaceous sands, which show abundant, multidirectional, low angle, low amplitude cross-bedding very similar to hummocky cross-stratification. In the extreme north of the

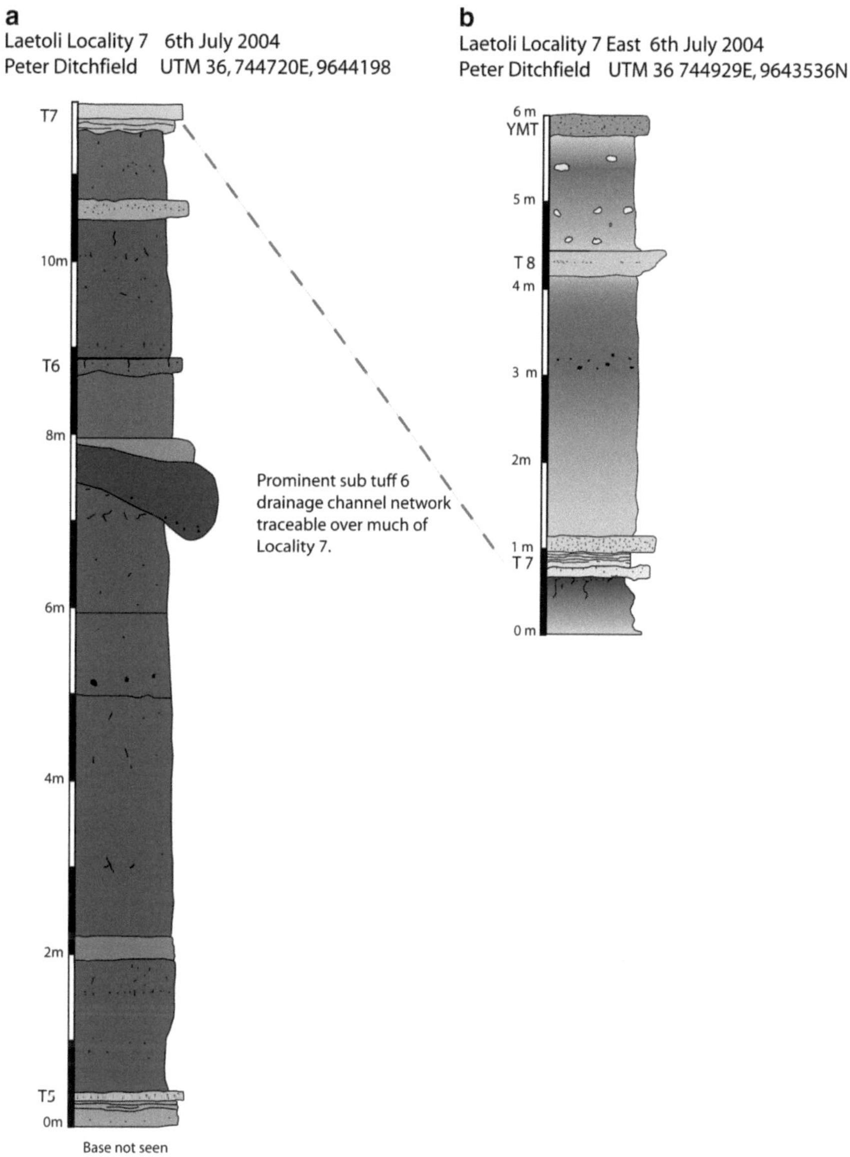

Fig. 3.11 (**a**) Section showing the sequence exposed in Loc. 7. At this locality a sub-Tuff 6, local, paleodrainage system can be traced across much of the outcrop. (**b**) Section showing the exposed sequence within the Upper Laetoli Beds at Loc. 7E adjacent to the fault

outcrop this unit is channelized with trough cross-bedding, indicating flow in a generally southerly direction, whilst in the center of the outcrop it is more tabular in nature. This unit is overlain by a strongly pedogenically altered aeolian tuff unit, which is erosively truncated by a disconformity and progressively cuts down to the south. This disconformity is overlain by a thin, well-cemented fall-out tuff some 20 cm thick, which probably represents Tuff 1. Overlying this tuff is a moderate to poorly pedogenically altered aeolian tuff which itself is overlain by another thin fall-out tuff, Tuff 2. In the middle part of the outcrop this upper aeolian tuff shows prominent small scale point bar deposits covered with clay drapes, which pass southwards into deeply incised minor channels/gullies, which are erosive into more pedogenically altered aeolian tuff (Fig. 3.12b).

The hummocky cross-stratification in the lower part of the sequence may represent the action of waves within a relatively shallow body of standing water, such as a small lake or large pond with a feeder stream coming from the north. This aqueous deposition appears to have been relatively short lived, and is replaced by aeolian tuff deposition and subsequent pedogenesis. This is followed by a period of erosion and tilting of the beds to the north, which is then terminated by the deposition of the first fall-out tuff (probably Tuff 1), which is followed by resumption of aeolian tuff deposition and pedogenesis, which is itself terminated by the deposition of the second fall-out tuff (Tuff 2) at this locality. A spectacular fossil termitary, complete with 'chimney', was exposed in the upper aeolian tuff at the southern end of these exposures in 2005.

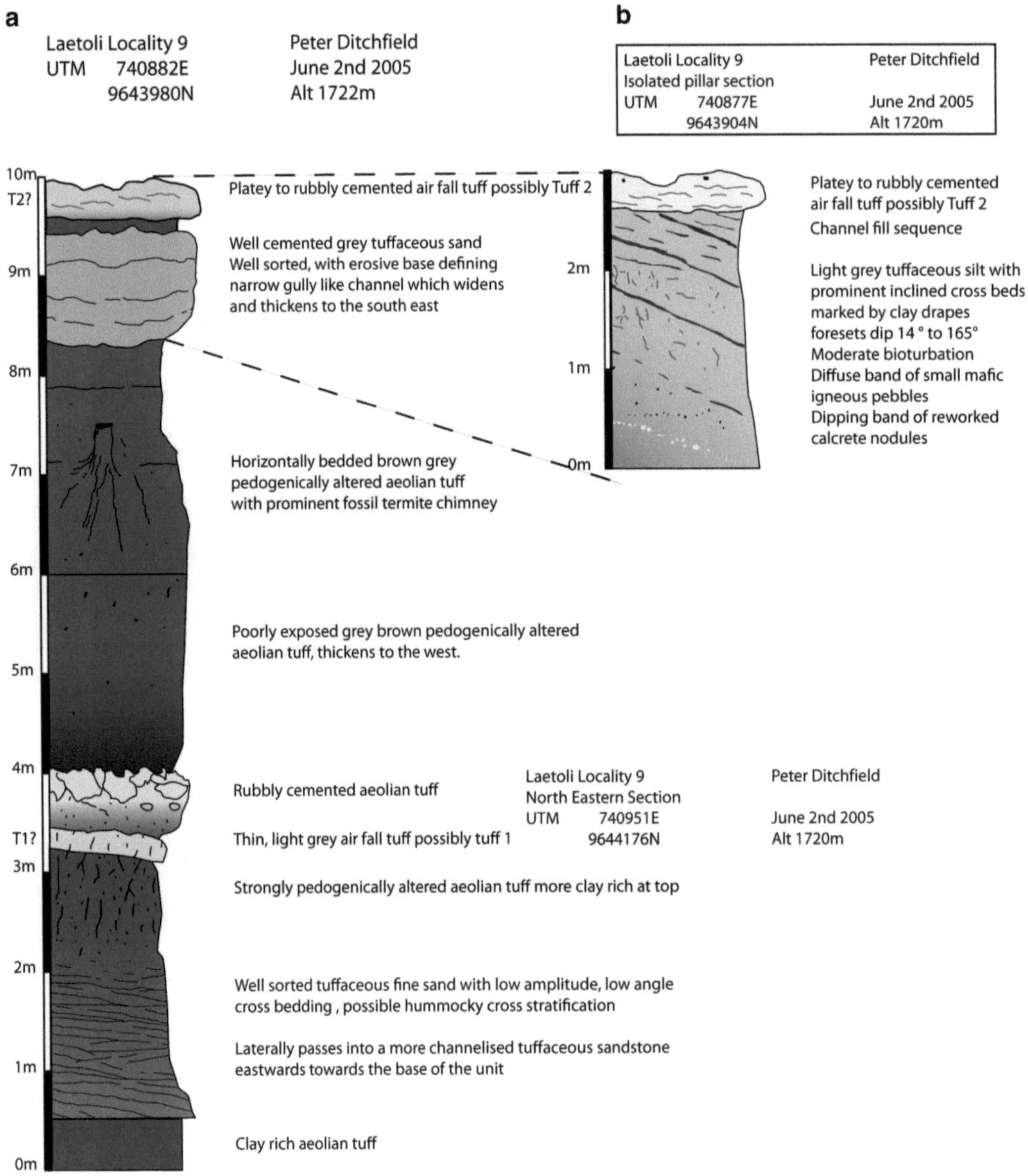

Fig. 3.12 (**a**) Composite logged section showing some of the facies seen at Loc. 9S. (**b**) Shows some of the variation in the interval between Tuffs 1 and 2 towards the southern end of the locality

Localities 10, 10W, 10E and 10NW

This is a large area of exposures between the Garusi River and Norsigidok Hill (Fig. 3.1), and is subdivided into a complex of localities. Outcrop is contiguous between the Locs. 10W, 10 and 10E, and the boundaries between these areas are somewhat arbitrary. The main exposures at Locs. 10E and 10 are separated by a large waterfall on a tributary to the Garusi River, which is capped by Tuff 3.

Representative sections of the exposures at Loc. 10E were recorded at UTM 36, 741398E, 9642140N altitude 1,722 m (Tuff 7 to Yellow Marker Tuff), as well as from UTM 36, 740922E 9642586N, (Tuff 4 up to Yellow Marker Tuff) and from Loc. 10 waterfall section at UTM 36, 740922E, 9642586N, as well as from Loc. 10W at UTM 36 740782E, 9642686N (see Fig. 3.13a–d).

Locality 10E exposes the upper part of the Upper Laetolil Beds, which is overlain to the southeast by poorly exposed Ndolanya Beds. The latter are locally erosive into the underlying Yellow Marker Tuff. The Ndolanya Beds are in turn overlain by olivine bearing vesicular Ogol lavas, which outcrop further to the southeast and form the cap of Norsigidok Hill. The whole sequence is cut by a prominent fault striking approximately north/south across the outcrop and forming a prominent saddle in Norsigidok hill. The fault has a normal sense of movement with a fault plane dipping at 88° to 088°

with a maximum displacement of 5 m (down throw to the East). The two logged sections for Loc. 10E show considerable local differences with Tuff 8 replaced in Section (b) by an interval of cross-bedded and channelized reworked tuffaceous silt. Section (a) shows a much thinner sequence between Tuff 6 and Tuff 7 relative to the sequence observed in Section (b). The local thinning of the sequence in a southwesterly direction may indicate a reduction in accommodation space during the deposition of the Upper Laetolil Beds, which is possibly related to a topographic paleo-high related to earlier movement of the fault noted above.

Locality 10 exposes the middle to lower part of the Upper Laetolil Beds with Tuff 4 through to below Tuff 2 (see Fig. 3.13c). Much of the sequence is well exposed in a waterfall section at UTM36, 740922E, 9642586N. Above this waterfall section in the interval between Tuff 3 and Tuff 5. The 'aeolian tuffs' of Hay (1987) preserve rare evidence of aeolian deposition in the form of rather patchy relict dune bedding which is otherwise absent from most other outcrops of this lithology. This is best observed when the sediments are damp (after rain) and illuminated by low-angle (early evening) light.

Locality 10W exposes the lower part of the Upper Laetolil Beds with Tuff 2 through to below Tuff 1. The northern part of this locality shows striking evidence of fluvial deposition in the form of channel belt sediments with prominent epsilon type cross-bedding after point bar formation in a moderately high sinuosity fluvial system. This channel belt is approximately 50 m wide with paleoflow direction to the southeast (see Fig. 3.13d).

Locality 12 and 12E

These localities lie to the southeast of the Oldonyo Mati Hills on the southeastern side of the Olaitole River drainage. They consist of relatively small and discontinuous

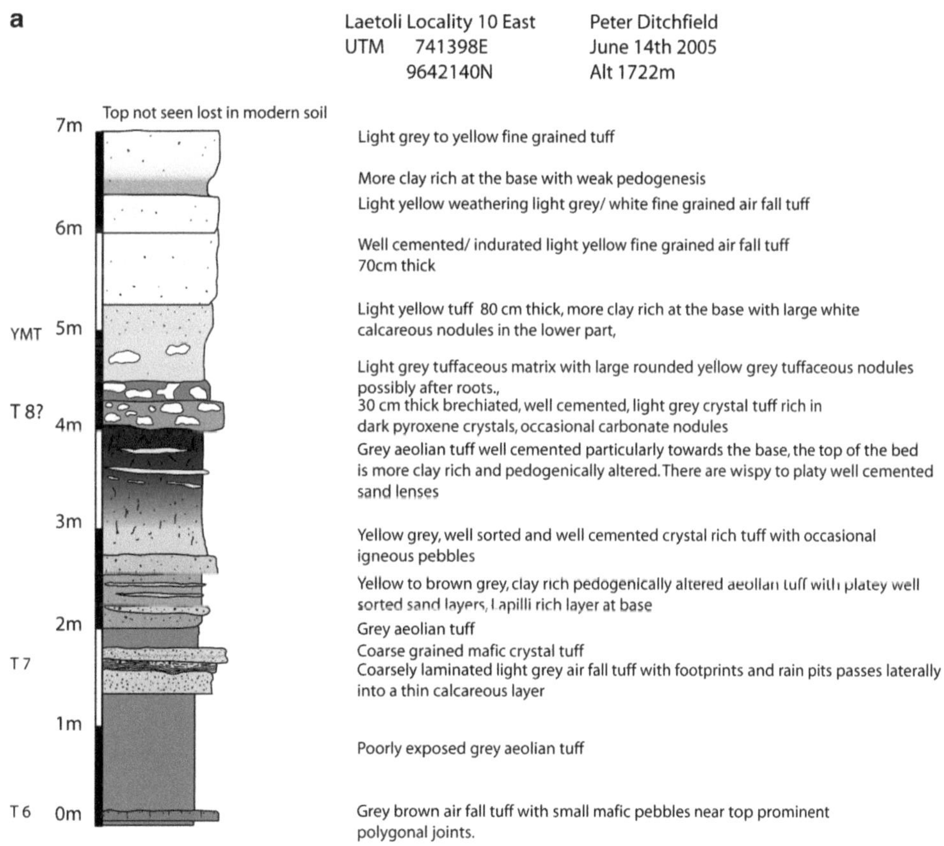

Fig. 3.13 (**a**) Lithological section recorded at Loc. 10E (UTM 36, 741398E, 9642149N Altitude 1,722 m). The Upper Laetolil Beds from the Yellow Marker Tuff (*YTM*) down to Tuff 6 (*T6*) are exposed. At this locality Tuff 8 is present, but has a cemented and nodular texture. (**b**) Lithological section recorded at Loc. 10E (UTM 36, 741477E 9642280N Altitude 1,720 m). This section is to the northeast of the section shown in Fig. 3.13a. It is further from the fault bounded hill which forms the south western margin of this locality and shows a greater thickness of sediments, although at this section location Tuff 8 is cut out by an interval of channelized cross-bedded and tuffaceous silts. (**c**) Lithological section recorded at Loc. 10 (UTM 36, 741625E, 9642538N). Tuff 3 forms the lip of a prominent waterfall at the edge of the Loc. 10 outcrop. This waterfall sections show spectacular fossil termitaries. (**d**) View of a representative outcrop in the Loc. 10W area at UTM 36, 740783E, 9642683N, looking southwest. The photograph shows channel deposits between Tuffs 1 and 2, which display prominent low angle point bar surfaces formed in a sizable fluvial system

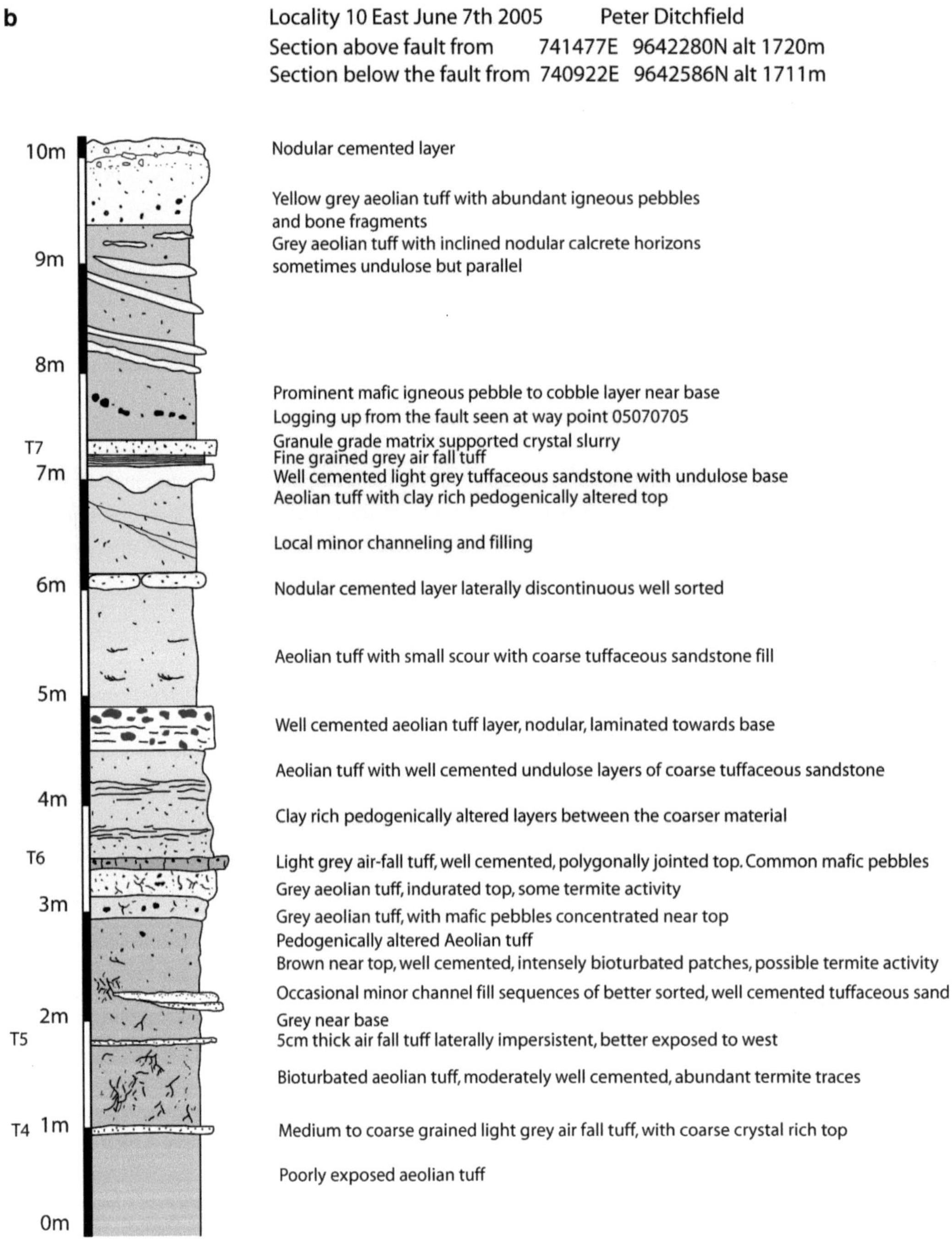

Fig. 3.13 (continued)

exposures. Locality 12E exposures were observed at UTM 747402E, 9643342N and show a section through the Upper Laetolil Beds from Tuff 6 through to above Tuff 7, which are locally erosively cut out by Ndolanya Beds channel facies (Fig. 3.14a). Locality 12, seen at UTM 747304E, 9643132N, contains poorly exposed Tuff 7 at the base overlain by a locally thickened sequence of Upper Laetolil Beds which are also erosively truncated by overlying beds (Fig. 3.14b). The Yellow Marker Tuff was not seen at this location.

Locality 13

Locality 13 is located on the northern side of the Garusi River drainage (UTM 36 742642E, 9645467N). It comprises a series of shallow gullies that feed into the Garusi River. Outcrops are rather poor and overgrown with vegetation, but a sequence of the Upper Laetoli beds are exposed from Tuff 3 through Tuff 8. The Upper Ndolanya Beds also exposed at this locality, but are poorly fossiliferous.

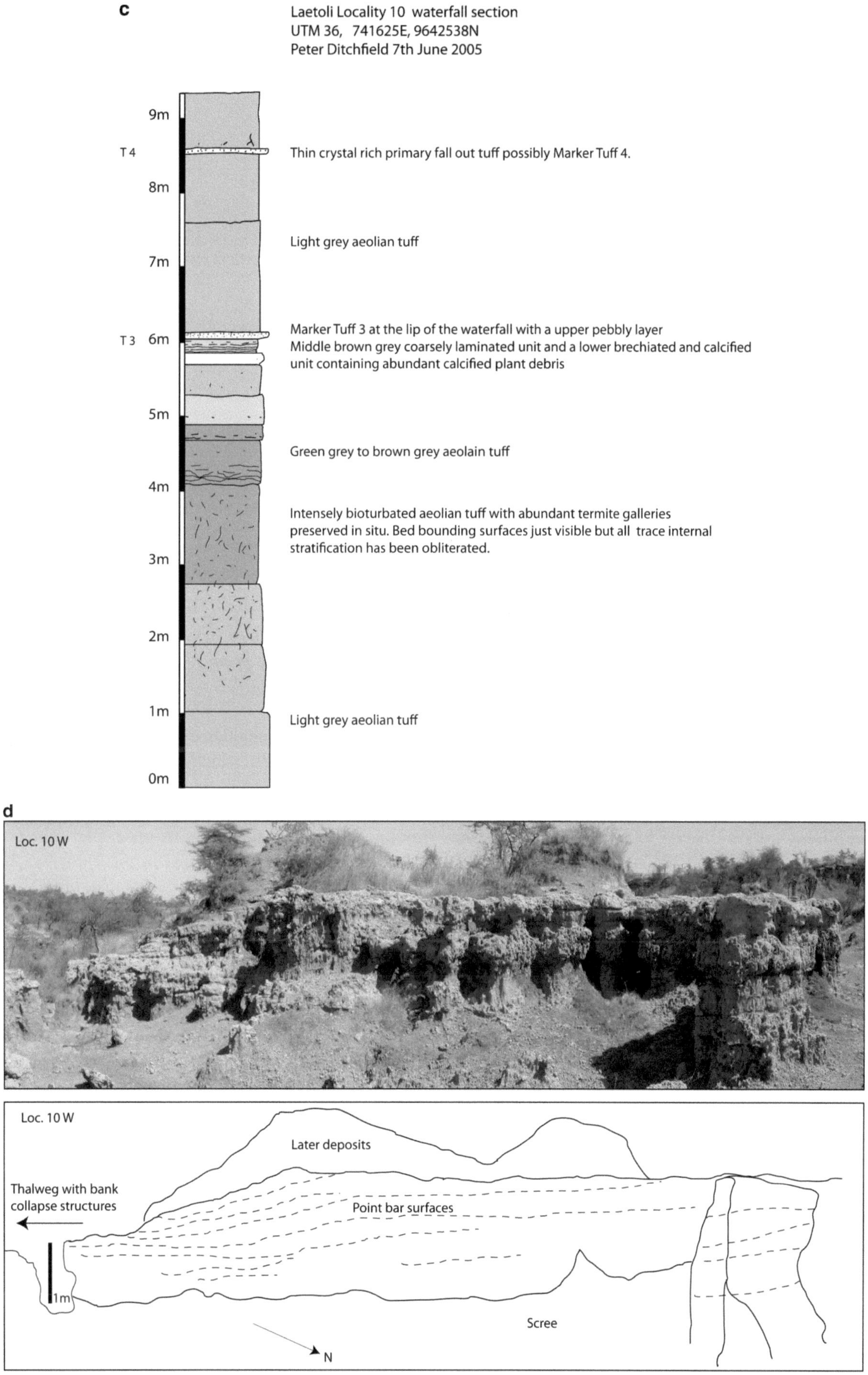

Fig. 3.13 (continued)

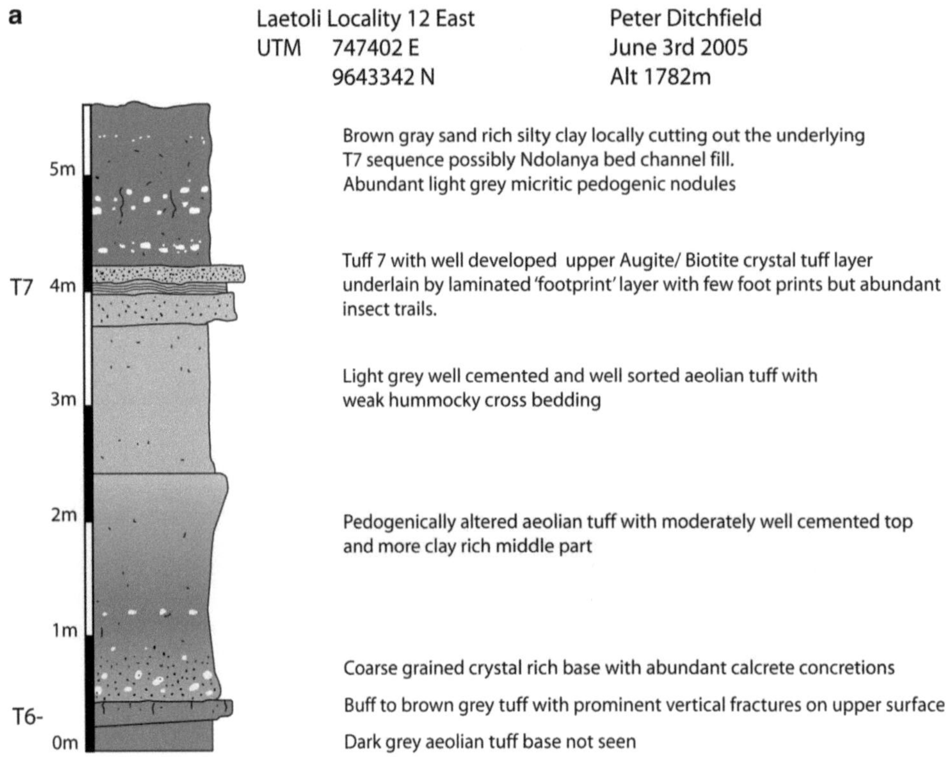

Fig. 3.14 (**a**) Lithological section recorded from Loc. 12E at UTM 36, 747402E, 9643342N, Altitude 1,782 m, showing Upper Laetolil Beds from Tuff 6 to above Tuff 7. (**b**) Lithological section recorded from Loc. 12 at UTM 36, 747304E, 9643132N, Altitude 1,789 m. showing a discontinuous sequence in the Upper Laetolil Beds from Tuff 7 upwards

Locality 14

Locality 14 is located north of the Garusi River between Locs. 2 and 13. The sequences exposed here is an unusual tuffaceous channel-fill facies of the Upper Ndolanya Beds. A representative sequence from UTM 36 743063E, 9645479N is shown in Fig. 3.15. This consists of two, stacked, fining upwards, clast-supported, pebble to granule grade conglomerate units. The sediments occur in a WSW orientated channel with cross-bedding flow directions ranging from 200° to 270°. This channel facies incises into, and is overlain by, more typical Upper Ndolanya lithologies of strongly pedogenically altered, tuffaceous silts, with crystal tuffs and diagenetic calcrete horizons.

Locality 15

Locality 15 is located at the western margin of the main Laetoli area approximately 1.5 km west of Ndolanya Hill. Sections were recorded at UTM 738244E, 9643620N and UTM 738636E, 96444192N. In both sections the Upper Laetolil Beds are erosively truncated by the overlying Ndolanya Beds, which has a relatively steep dip at this location of 10° to 265° (Fig. 3.16). At UTM 38, 738816E, 9644290N a channel fill feature with westerly flow directions can be seen within the Ndolanya Beds. This channel fill contains abundant angular cobbles of partially cemented Yellow Marker Tuff, which have been eroded from the underlying Upper Laetolil Beds. This feature is also unusual as it preserves *in situ* calcified tree remains (horizontal root networks and stumps) in its upper part. The presence of abundant vertebrate remains in the Ndolanya Beds in the second of these sections suggests that this is the Upper Ndolanya Beds. Outcrops of the Upper Laetolil Beds contain exposures of Tuffs 6 and 7. However, the exposures extend downwards following the slope of the drainage, so it is likely that lower marker tuffs are also present to the west.

Locality 16

Locality 16 is situated on the north side of the Garusi River 500 m northeast of Loc. 3 (UTM 36 746102E, 9645691N). The sequence exposed at Loc. 16 ranges from below Tuff 6 through to the Yellow Marker Tuff and, like the nearby Loc. 3 exposures, it shows an expanded sequence of the footprint tuff, Tuff 7. The lower layer of Tuff 7 at this locality is extremely rich in calcified plant material ranging from root mats, to broken stems and twigs, and occasionally fruits and leaves. Some of this debris has undergone minor amounts of transport and in some places shows alignment of long axis of fragments. However, in most places the calcified plant fragments show no preferred orientation and calcified root mats appear to have been preserved

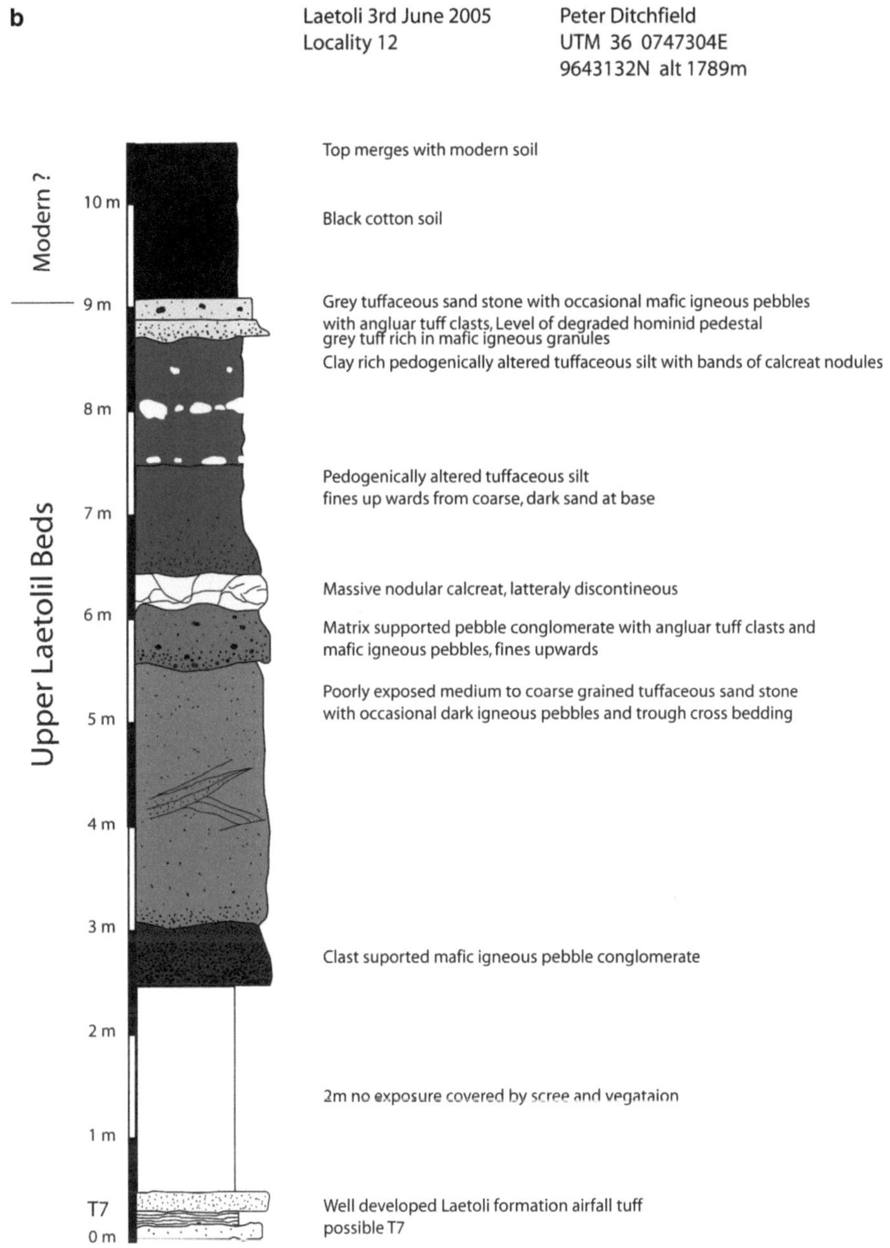

Fig. 3.14 (continued)

in situ. Some of these root mats are well developed and quite laterally extensive. The underlying tuffaceous silts show little, if any, evidence of pedogenesis. This lack of well-developed paleosols is common throughout much of the Upper Laetolil Beds. Clays and silts of the Olpiro Beds, which are locally erosive into the Yellow Marker Tuff and exhibit minor channel development, overlie the Yellow Marker Tuff at this locality.

Locality 18

Locality 18 is a large series of outcrops extending northwest-southeast along the Gadjingero River drainage. A representative section was recorded at UTM 744261E, 9645985N. The section consists entirely of the Ndolanya Beds with the upper few meters of the Lower Ndolanya Beds and some 8 m of the Upper Ndolanya Beds exposed (Fig. 3.17). The Upper Ndolanya Beds are very rich in fossil vertebrate remains and hymenopteran cocoons.

Locality 19

Locality 19 is located on the steep slope of the south side of the Olaitole River drainage downstream from Loc. 12. A representative section was recorded at UTM 746007E, 9642798N (Fig. 3.18).

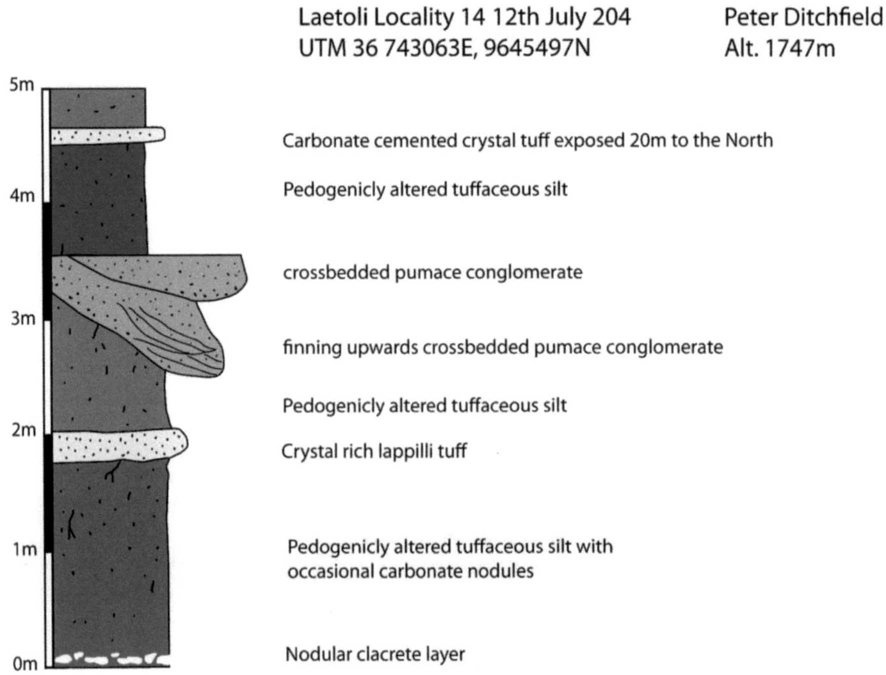

Fig. 3.15 Lithological sequence recorded from Loc. 14 at UTM 36, 743519E, 9645590N, Altitude 1,747 m, showing a channel facies within the Upper Ndolanya Beds

Although this locality exposed fossiliferous aeolian tuffs of the Upper Laetolil Beds no clear marker tuffs were observed so it is not possible to be certain where it occurs in the sequence. Hay (1987) has identified the fossiliferous exposures at Loc. 19 as between Tuff 6 and Tuff 8. The recorded altitude was 40 m lower than that recorded at the adjacent Loc. 12 and Loc. 12E. Assuming no major faulting or folding between the two localities (which are some 700 m apart), it would seem likely that exposures recorded here are stratigraphically lower in the sequence than Tuff 6.

Locality 20

Locality 20 is located on the western side of the Olaitole River between Locs. 19 and 21. The exposures are on the steep slopes below the exposures of Ogol lavas forming the southwestern portion of the Oldonyo Mati Hills. A fossiliferous section in the Upper Laetolil Beds was recorded at UTM 745790E, 9642486N. The exposures range from 2 m below Tuff 7 up to Tuff 8 and are unconformably overlain by poorly exposed tuffaceous sediments of the Ndolanya Beds (Fig. 3.19).

Locality 21

Locality 21 is situated approximately 500 m southwest of Loc. 20 on the west side of the Olaitole River. The locality consists of a series of steep-sided gullies with dense bush and scrub. The sequence exposed ranges from Tuff 3 to above Tuff 7 (Fig. 3.20).

Localities 22 and 22E

Locality 22 and 22E are located on the eastern side of the Olaitole River and south of the northeasterly extension of Norsigidok Hill. These exposures are extensive, but dissected and fragmented by thick bush. At Loc. 22 several sections spanning the stratigraphic interval from Tuff 1 through to Tuff 7 were recorded (Fig. 3.21a–c). Tuffs 3, 6 and 7 are sufficiently well developed and consistent at this location to allow good correlation between the scattered outcrops.

A long section at Loc. 22E was recorded at UTM 743280E, 9641620N (Fig. 3.22). This area contains a sequence, which extends from in excess of 3 m below Tuff 3 to Tuff 7. Tuff 7 is separated from the top of the logged section as seen in Fig. 3.22 by a steep vegetated slope with a vertical separation from the top of the logged section of 4.5 m (Tuff 7 is exposed at UTM 743269E, 9641672N).

Tuff 3 forms the lip to a prominent waterfall in the main part of the section and is easily recognized. Some of the other marker tuffs within the section are problematic. There is a possible Tuff 2 approximately 1 m below marker Tuff 3. However, above Tuff 3 there are at least three thin and laterally discontinuous light grey primary fallout tuffs, which may represent Tuffs 4 and 5. The upper part of the main section is dominated by a series of, stacked channel fill features. These erosively based features may have cut out marker tuffs from the interval below. The sediments that the channel fills cut into have a distinctive brown-grey color, which is suggestive of the interval between Tuffs 5 and 6, as seen at other

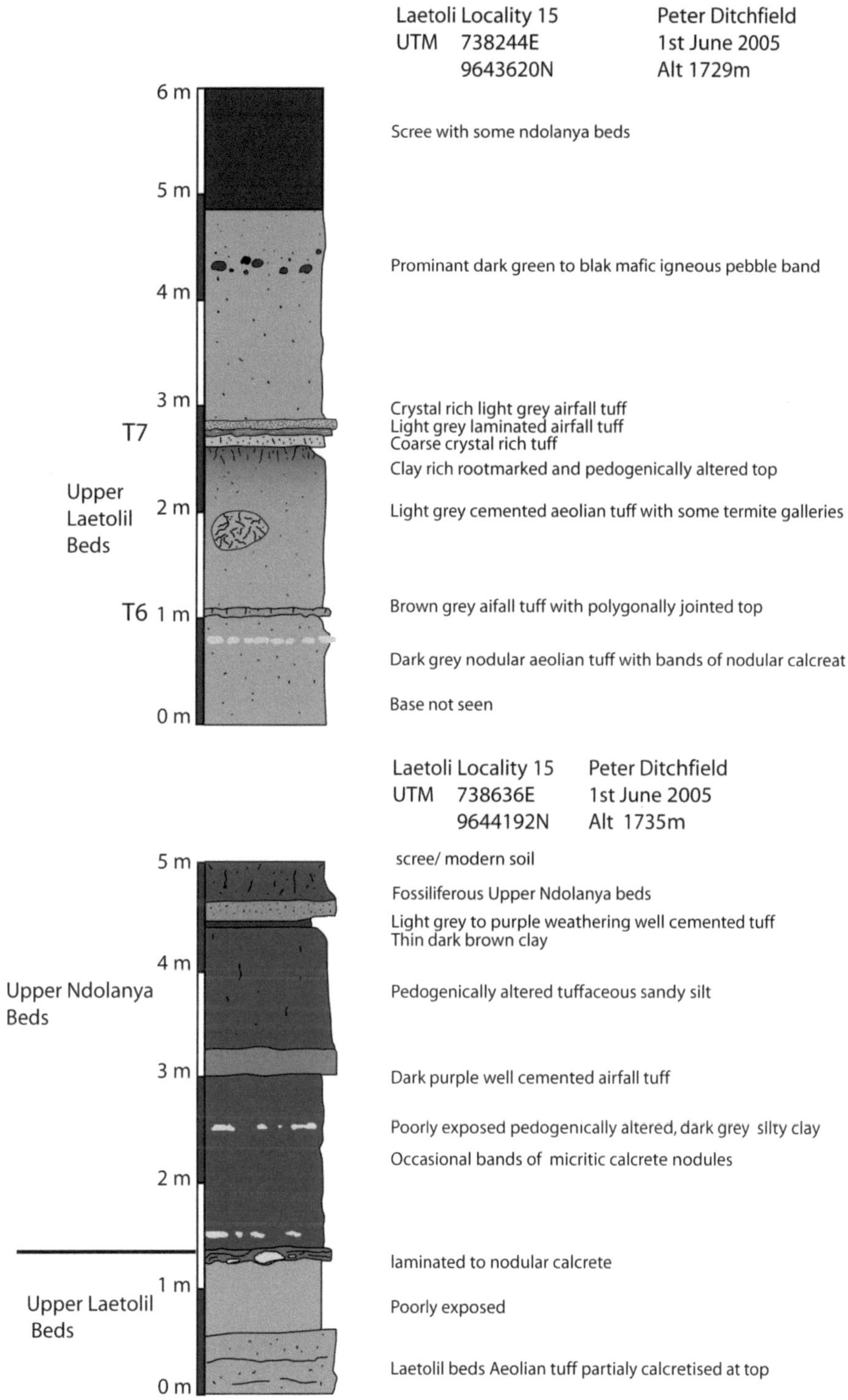

Fig. 3.16 Representative lithological sections recorded from Loc. 15 at UTM 738244E, 9643620N and UTM 738636E, 96444192N. In both cases outcrops of the Upper Laetolil Beds are erosively truncated by the Upper Ndolanya Beds

localities. The lower channel fill feature gives consistent cross-bedding based flow direction to the southeast (n = 12). At two locations within the sequence there are accumulations of calcified plant debris and occasional insect remains between the channel fill features and just above Tuff 3.

Sedimentary Litho-Facies and Depositional Environments

Hay (1987) recognized three main litho-facies within the Upper Laetolil Beds these were 'Aeolian Tuff', 'Water-worked Tuff', and 'Air-Fall Tuff'. He also described a sub-facies of

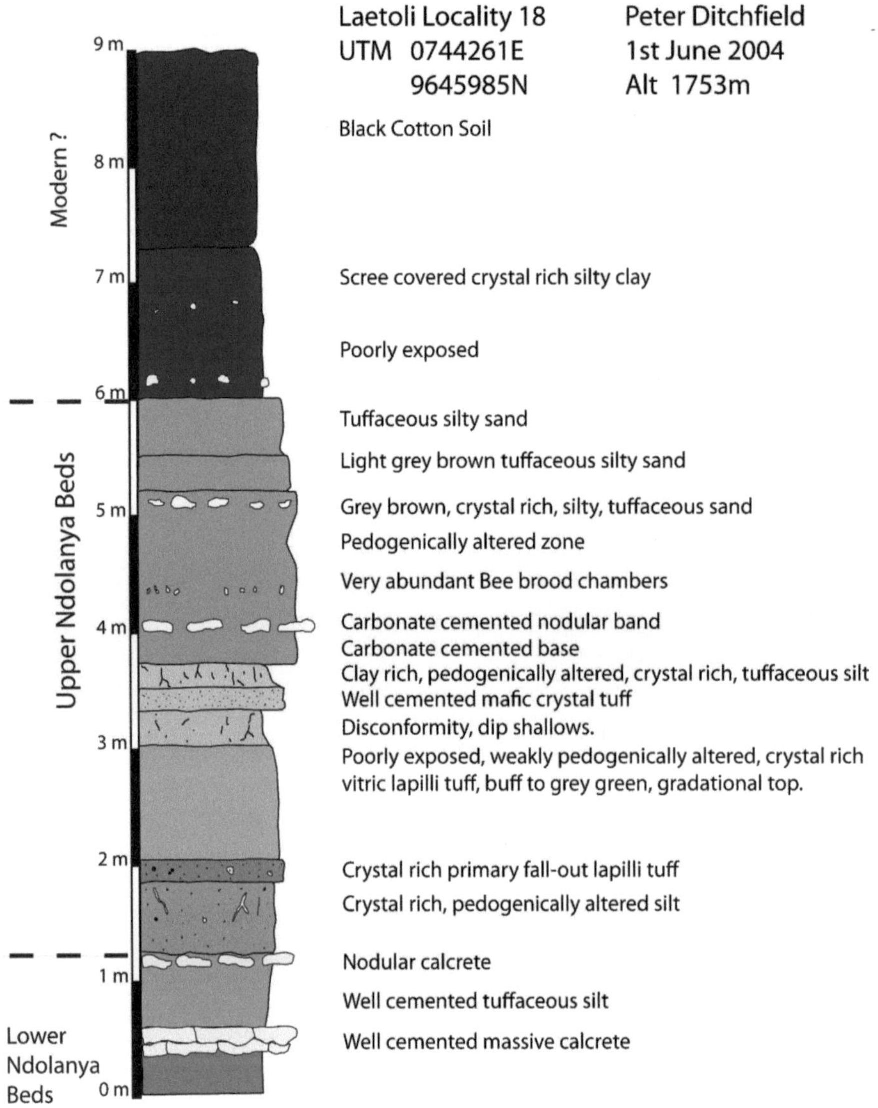

Fig. 3.17 Representative lithological sections recorded from Loc. 18 at UTM 744261E, 9645985N, showing the upper part of the Lower Ndolanya Beds and the highly fossiliferous sequence of the Upper Ndolanya Beds

Fig. 3.18 Lithological section recorded at Loc. 19 at UTM 36, 746007E, 9642798N, showing Upper Laetolil beds probably from below Tuff 6

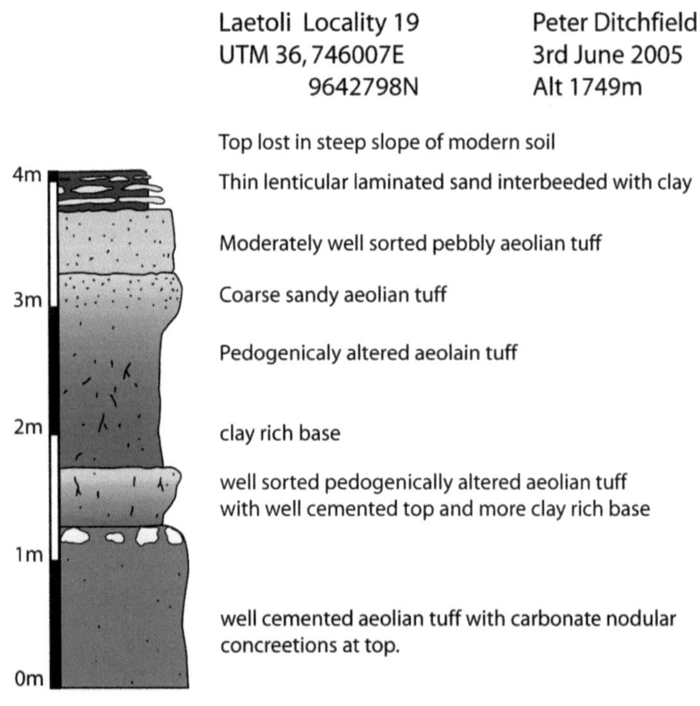

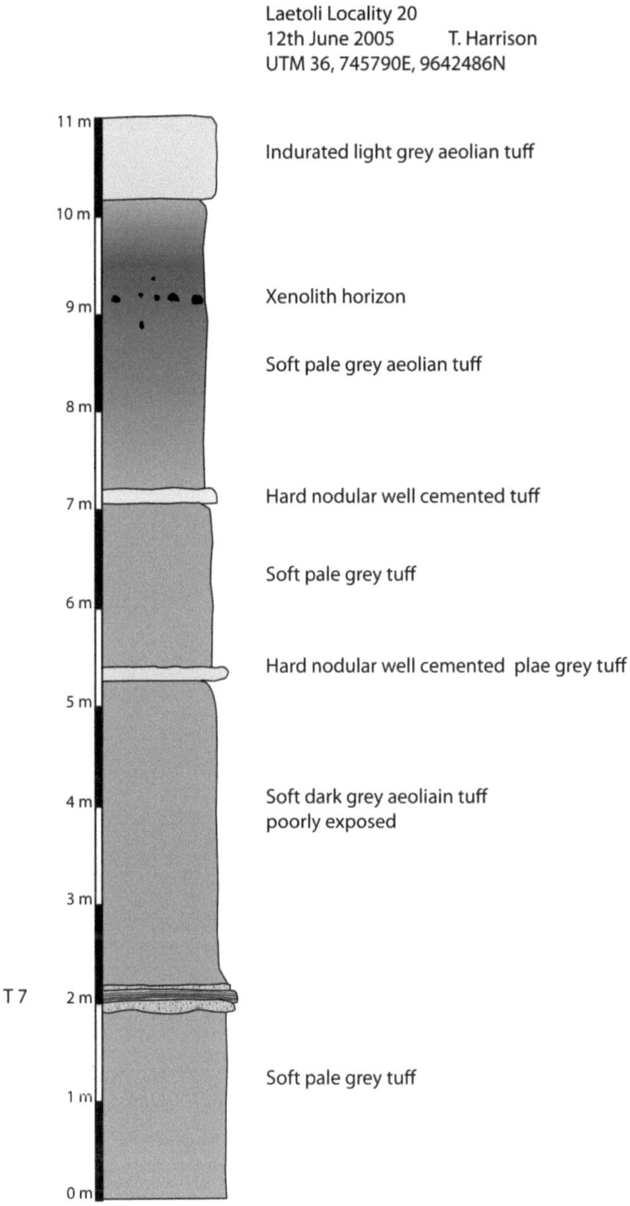

Fig. 3.19 Lithological section recorded at Loc. 20 at UTM 36, 745790E, 9642486N, showing Upper Laetolil beds above Tuff 7

the Aeolian Tuff with 'Xenolith Horizons'. Of these litho-facies the Aeolian Tuff is volumetrically most significant making up over 80% of the sequence based on the sections recorded for this study.

Aeolian Tuff

The majority of the sediments of the Upper Laetolil Beds consist of aeolian tuff as defined by Hay (1978). Yet it is a rather enigmatic litho-facies, as for the most part it is devoid of any sedimentary structures that would help constrain its depositional mode. Hay (1987) implies that this is due to the destructive effects of bioturbation and pedogenesis. Whilst both are locally important in the Laetoli Beds it is unlikely to have completely homogenized the sedimentary structures, while leaving the bed bounding surfaces intact. Pedogenesis also seems unlikely to have removed all sedimentary traces related to aeolian deposition given the lack of any well-developed paleosol horizons within the beds (Kingston and Harrison 2007).

Hay's views on the Aeolian tuff in the Laetoli area were largely influenced by his work at Olduvai Gorge and particularly of his study of the Holocene sediments on the surrounding plain (Hay 1976). Here he described blanket like deposits of largely tuffaceous material interspersed with occasional relict or in some cases active aeolian dune fields. These Holocene sediments do seem to be at least partially analogous to the Aeolian Tuff of the Upper Laetolil Beds. However, there are some important sedimentological and environmental differences that should be considered. The Holocene sediments from the plain around Olduvai Gorge are much more varied in grain size and constituent clasts compared to those in the Upper Laetolil Beds and there appears to have been more mixing of various sediment sources. Secondly the sediments associated with the Holocene dune systems on the Olduvai Plain show marked concentration of the heavy mineral component with the lighter mineral fraction being winnowed out of the dune sequences during dune migration (Hay 1978). This does not seem to be the case in the Upper Laetolil beds. In the rare instances where relics of probably aeolian dune cross-bedding can be found (such as Loc.10 between Tuffs 4 and 6) no such concentration of the heavier mineral fraction has been observed. These differences may be a function of sedimentation rate and availability. Thus, the uniformity of the aeolian tuff sediments of the Upper Laetolil Beds and the lack of mixing of different components or differential sorting of mineralogies by aeolian transport processes may point to higher sedimentation rates than those seen in the Holocene sediments around Olduvai Gorge.

The environmental/geographic differences between the Olduvai Plain and the paleogeography of the Upper Laetolil Beds as proposed by Hay (1987) also need to be considered. The Holocene sediments of the Olduvai Plain are quite distal as well as geographically and topographically separated from the potential sources of their tuffaceous components. Thus, they are only likely to have been supplied primarily with such sediment via aeolian or primary fall-out processes. However, the paleogeography proposed for the Upper Laetolil Beds has them as much more proximal to potential volcanic sediment sources, possibly as a contiguous plain with a gentle slope away from the contemporary volcanic highlands in the east to a depositional low to the west. This opens up the possibility of other volcano-sedimentary transport processes as being important within the Upper Laetolil Beds.

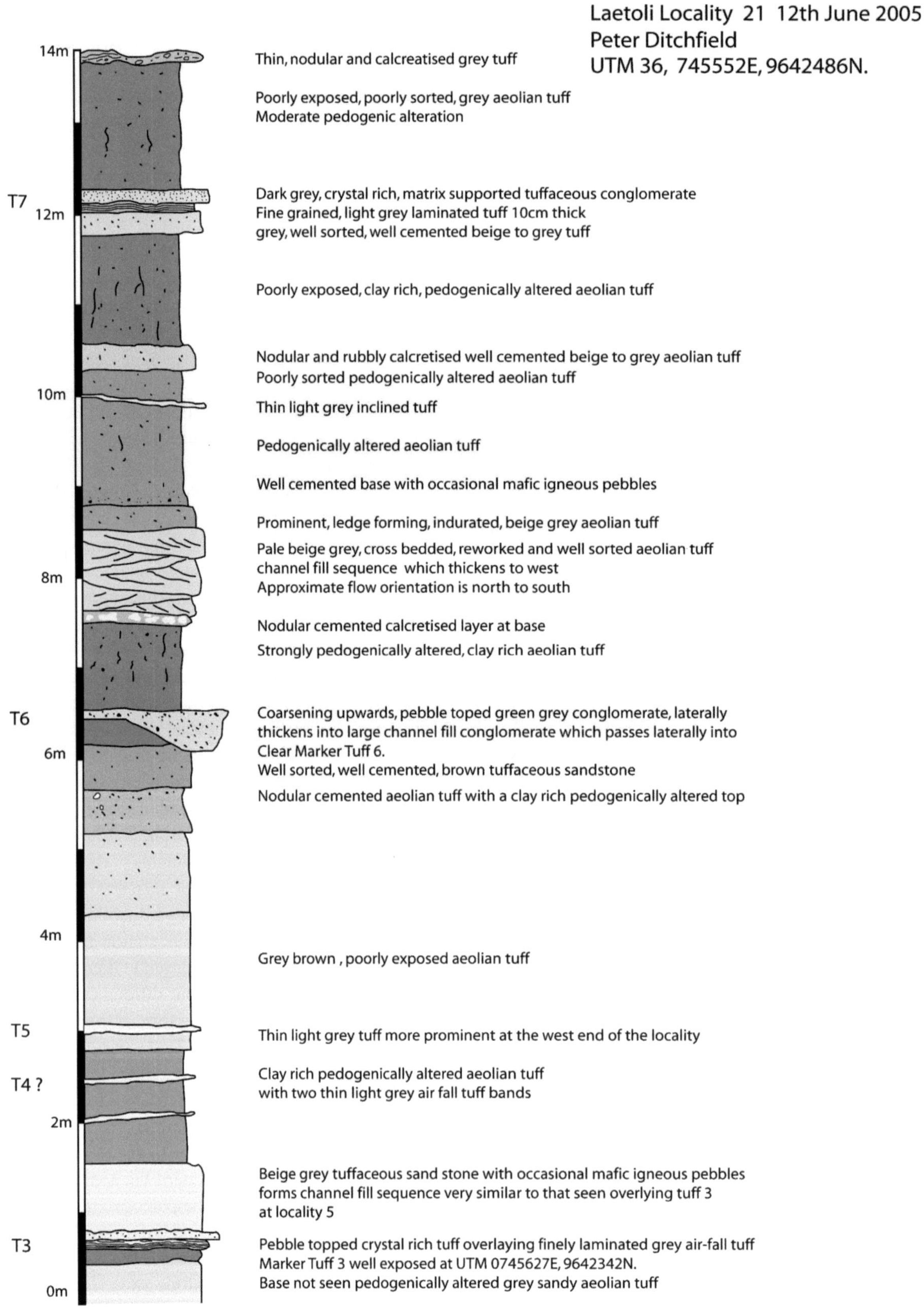

Fig. 3.20 Lithological section recorded at Loc. 21 at UTM 36, 745552E, 9642486N, showing a sequence from below Tuff 3 to above Tuff 7

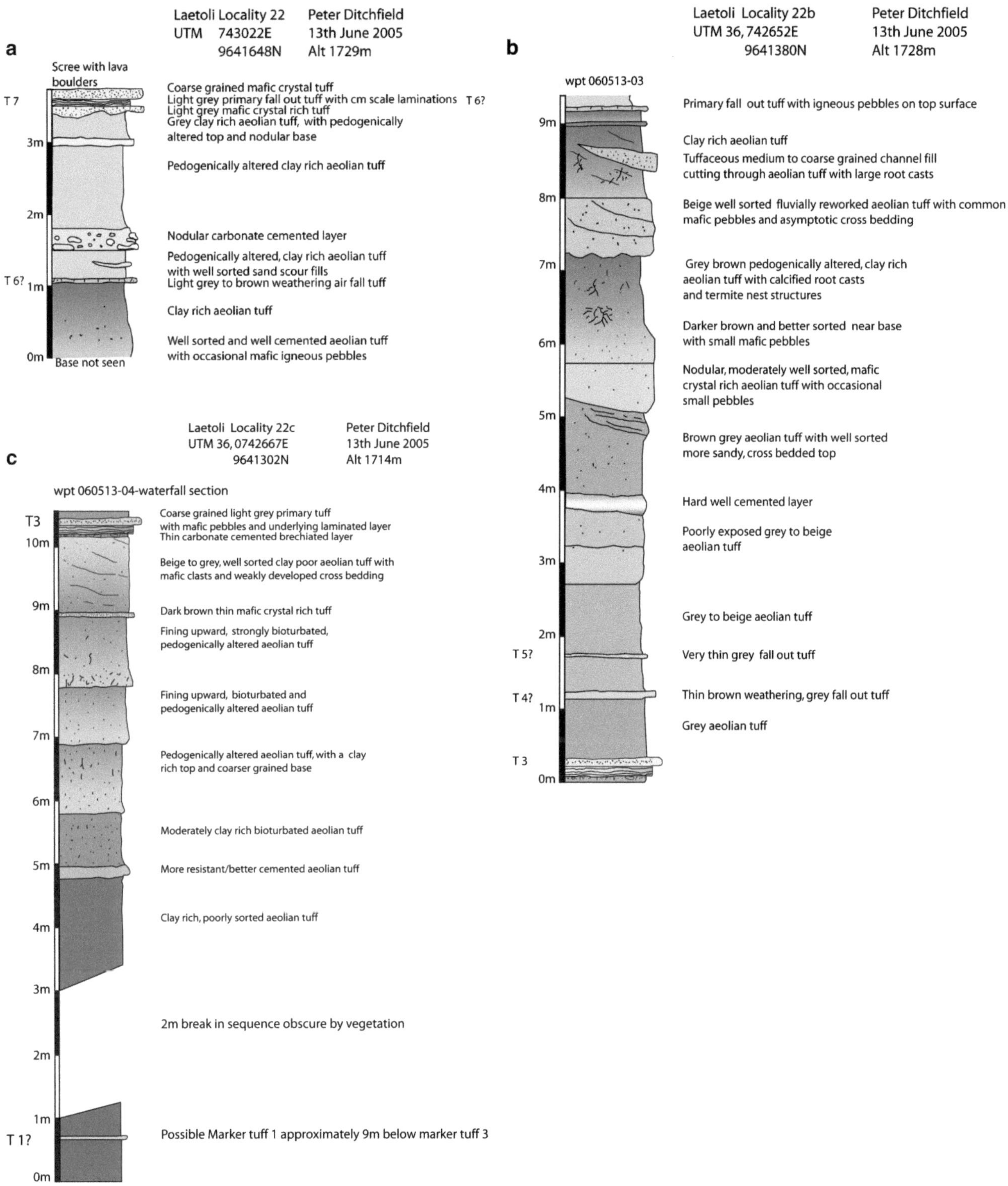

Fig. 3.21 (**a**) Lithological sequence recorded from Loc. 22 at UTM 36, 743022E, 9641648N, Altitude 1,729 m, showing a relatively condensed section within the Upper Laetolil Beds between Tuffs 6 and 7. (**b**) Lithological sequence recorded from Loc. 22 at UTM 36, 742652E, 9641380N, Altitude 1,728 m, showing Upper Laetolil Beds between Tuffs 3 and 6. (**c**) Lithological section recorded from the prominent waterfall in Loc. 22 at UTM 36, 742667E, 9641302N, Alt 1,714 m, showing Upper Laetolil Beds from Tuff 3 down to a possible outcrop of Tuff 1

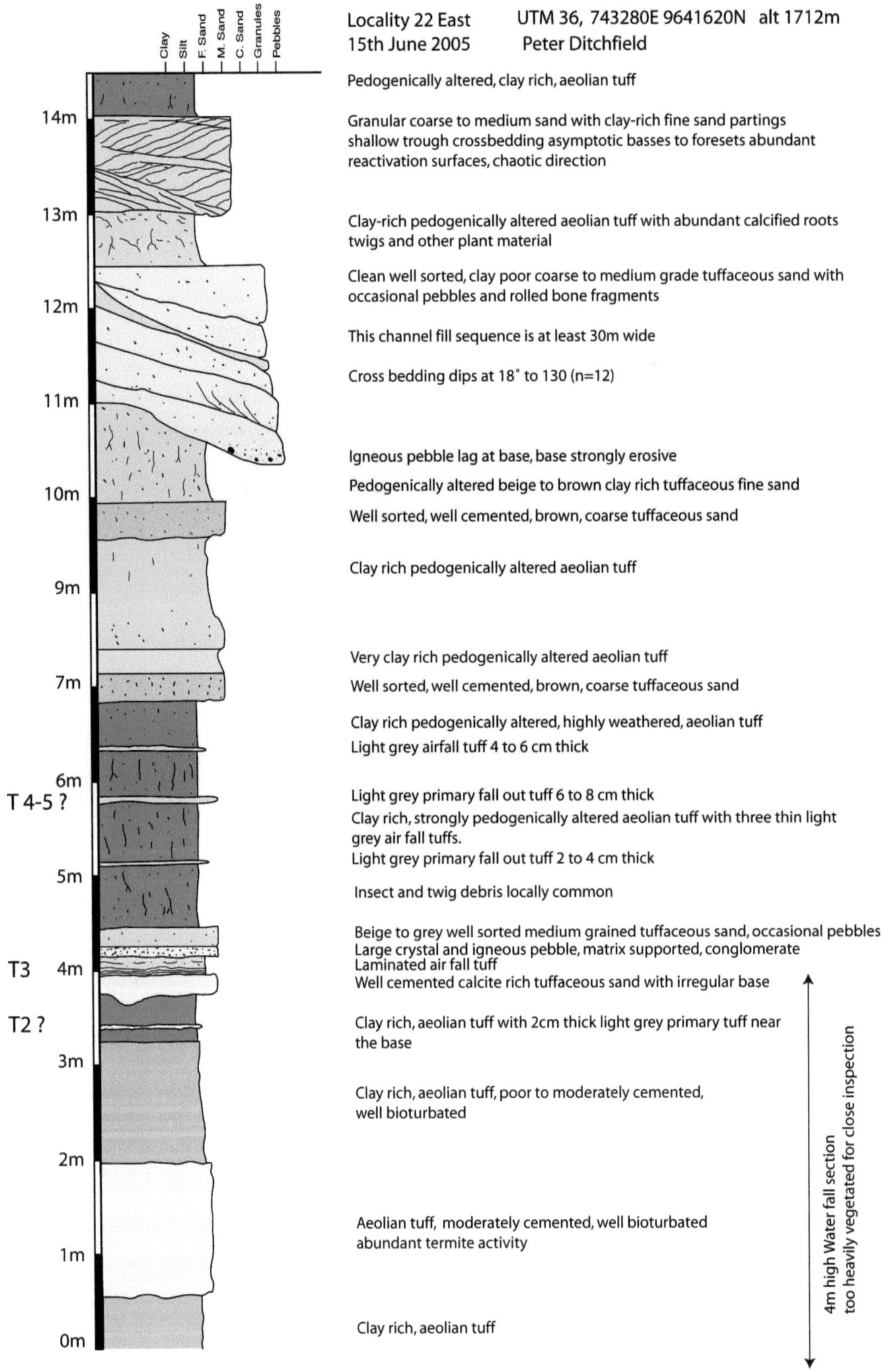

Fig. 3.22 Lithological section recorded at Loc. 22E, at UTM 36, 743280E, 9641620N, Altitude 1,712 m, showing a section through the Upper Laetolil Beds from below Tuff 3 to a channel complex roughly at the level of Tuff 6. Locally, Tuff 7 is exposed some 4.5 m above the top of this logged sequence

In the light of subsequent work on sedimentary processes in volcanic environments (e.g., Smith and Lowe 1991) other depositional processes (such as distal lahars and hyper-concentrated flows) may be implicated in the deposition of some of these units. This may be particularly applicable to the zones of "aeolian tuff with xenoliths" as described by Hay (1987), which characterize numerous horizons within the Laetolil Beds. Hay (1987) ascribed these lithic clasts to air-fall deposits from "highly explosive eruptive events" that were then redistributed throughout the full thickness of certain beds by bioturbation. There appears to be little supporting evidence for this. Beds with significant concentrations of "xenoliths" do not contain matrix of a different mineralogical or textural composition as might be expected from a different style of eruptive event, neither do they show an increase in the amount of identifiable fine-grained primary fall-out material. These units could be reinterpreted as lahar deposits. This would have serious implications for the use of these 'xenolith' bearing beds as stratigraphic markers for correlation. Even if they were produced by relatively unconfined flows, such deposits are likely to be lobate in their geometry and restricted in their lateral continuity, therefore they should only be used for local correlation with caution and only where they can be traced laterally in contiguous outcrops.

Water Worked Tuff

Water worked tuffs as defined by Hay (1987) can be spilt into two groups; the first and most abundant are those where fluvial processes dominate deposition. The second minor group seem to have been deposited in standing water.

Fluvial Tuff

Fluvially reworked tuffaceous material can be found within minor channel fill sequences at many of the localities. Minor erosive gully features up to 2 m deep are relatively common within the Aeolian Tuff of the Laetolil Member throughout the sequence but especially between Tuffs 5 and 6. Typical examples can be seen at Loc. 7 where an erosive gully system can be traced from a narrow set of runnels to a broader minor channel system. While such features make up a relatively minor component of the sedimentary sequence, they do demonstrate that a well-established drainage system was present across the paleolandscape. Whilst there is little sedimentary evidence to support widespread re-deposition of the aeolian tuffs by water, there is evidence of substantial water reworking at several locations. In particular, the section seen in Loc. 10W contains a well-developed fluvial system with a channel belt at least 200 m wide in the stratigraphic interval between marker Tuffs 1 and 2. This fluvial system preserves abundant examples of epsilon (sigmoidal) cross-bedding representing deposition on former point bar surfaces within a moderate to high sinuosity fluvial system. Individual point bar surfaces are marked with clay drapes and can be traced for several tens of meters up original depositional dip of the point bar. They often show a clear fining upwards trend as they are traced from a low point near the thalweg to a high point near the top of the bar where internal stratification is lost due to increased bioturbation (Fig. 3.13d). The channel belt has an approximate North-East to South-West trend and ripple cross-bedding towards the base of some point bar sequences give flow directions to the Southwest.

This represents a relatively sizable fluvial system, larger than the modern-day Garusi system that currently drains the Laetoli area. Sediments within this feature are still rather poorly sorted and dominated by tuffaceous material characteristic of the rest of the Laetolil Beds. Thus, it seems likely that this fluvial system was part of the local drainage system, rather than anything of more regional significance. However, its location at the southwest edge of the main Laetoli area may be significant. The majority of paleoflow directions recorded in this study point to an overall southwesterly direction for fluvial sediment transport (although Hay [1987] suggests a more southerly gross transport direction). Smaller, more variably oriented, channels seem to be more prominent in the northeastern part of the main Laetoli area. Thus, a model with more dendritic drainage in the northeast coalescing to give a more coherent channelized fluvial system to the southwest side of the main Laetoli area is plausible. It is likely that this type of drainage system would have provided a suitable setting for a significant riparian woodland component as part of a complex vegetational mosaic (Kingston and Harrison 2007; Kingston 2011; Andrews et al. 2011; Bamford 2011a, b) and that with increasing channel size to the southwest of the area the density of this woodland would have increased. Given that fluvial development is present throughout the sequence it is likely that such a drainage system would have been re-established relatively quickly after episodes of major landscape remodelling, such as the rapid deposition of significant thicknesses of aeolian tuff.

There is evidence from the logged sections to suggest that this type of drainage development was more intense in the interval between Tuffs 6 and 5. As the sediment in this interval is not markedly different texturally from that found elsewhere in the sequence, and as there is no bedding disconformity between this interval and those above and below, it would seem unlikely that this increase in the intensity of drainage development could be explained by either a reduction in sedimentation rate or tectonic rejuvenation. Thus, it would seem more likely that this could be due to enhanced levels of precipitation in the area during the interval between Tuffs 5 and 6.

Pond Deposits

There is evidence of deposition in relatively shallow (generally less than 2 m deep) bodies of standing water at several locations. The largest of these described so far is in the northern part of Loc. 9 where the ponded area was large enough to be susceptible to wind wave development, which is shown in the sedimentary structures seen in the lower part of the section (see Fig. 3.12). However, this feature was rather short lived and is unlikely to have been a major feature of the landscape. Similarly, other pond deposits are relatively common throughout the sequence of the Upper Laetolil Beds. These tend to be rather limited in size and probably represent relatively short-lived small to medium-sized ponds sometimes but not exclusively associated with local minor fluvial development. Whilst they make up only a tiny fraction of the sedimentary sequence they are important paleontologically as these low-energy environments can give rise to suitable depositional conditions for the preservation of insects and plant fossils, which are lacking from other litho-facies. As water sources, even if temporary and ephemeral, they make up an important part of the paleoenvironmental reconstruction along with the fluvial systems.

Air Fall Tuff

The term 'air fall tuff' was used by Hay (1987) to denote litho-facies that would be called primary ash fall deposits in modern terminology. These include the Marker Tuffs used by Hay and in this study for local stratigraphic correlation as well as other usually thin volcanic ashes. The sedimentology of the 'Marker Tuffs' particularly Tuff 7 is well described in Hay (1987) and detailed geochemistry of these units is given in McHenry (2011).

Discussion and Summary

Outcrops of fossil-bearing sediments are exposed over a very large area of the Eyasi Plateau, which is located on the southern margin of the Serengeti Plains just north of Lake Eyasi. Fossils have been recovered from 60 localities that extend from Laetoli in the east to Kakesio in the west, over a distance of about 45 km. However, the vast majority of fossils are derived from the Upper Laetolil and Upper Ndolanya Beds in the Laetoli area, from localities in the Garusi, Gadjingero, Olaitole and Nompopong River valleys (see Harrison and Kweka 2011).

Previous geological investigations at Laetoli have laid an excellent foundation for understanding the stratigraphy, sedimentology and geochronology (Hay 1976, 1978, 1987; Hay and Reader 1978; Hay and Leakey 1982; Drake and Curtis 1979, 1987; Ndessokia 1990; Manega 1993), and this work has been supplemented and revised by the recent geological contributions presented in this volume (Deino 2011; McHenry 2011; Mollel et al. 2011; Adelsberger et al. 2011). A review of the stratigraphy and paleoenvironmental context of the main sedimentary units that outcrop at the paleontological collecting localities at Laetoli is presented here, with a primary focus on the Laetolil and Ndolanya Beds. The lithological sequences at most of the Laetoli paleontological localities are described and sedimentary logs presented. The litho-facies as described by Hay (1987) are reassessed in the light of recent developments in the understanding of volcano-sedimentary processes.

Hay (1987) recognized the majority of sediments in the Laetolil Beds as aeolian tuffs, but because these tend to be massive and largely devoid of any sedimentary structure the depositional history has proved difficult to interpret. Hay (1987) attributed the lack of structure to the destructive effects of bioturbation and pedogenesis. While both may have been important factors, they do not account fully for the degree of homogenization of the sediments. Another possibility, given the proximal location of Laetoli relative to the volcanic center, is that the sediments are the product of lahars or hyper-concentrated flows. In this case, such deposits would be restricted in their lateral extent, and would only permit reliable correlations between stratigraphic units on a local scale.

Gullies and minor channels are relatively common in the Upper Laetolil Beds, indicating that a well-developed drainage system, larger than the present-day Garusi River system, was in place during the Pliocene. This implies an abundant availability of water during certain times of the year, although the streams and ponds were probably seasonal and ephemeral. The general direction of the paleoflow to the southwest suggests that run-off originated as precipitation in the elevated areas towards the volcanic highlands in the east and that it flowed across the Laetoli area towards the Eyasi basin. The drainage system in the northeastern part of Laetoli, proximal to the volcanic highlands, appears to have been dendritic in nature, while a more coherent channelized fluvial system developed more distally in the southwestern part of Laetoli. This type of drainage system would have supported a complex vegetational mosaic, including woodland along the river courses, probably with an increased density of woodlands towards the southwest. Gully and channel features have their greatest intensity in the Upper Laetolil Beds in the interval between Tuffs 5 and 6, and this may imply wetter condition at 3.7–3.8 Ma, at least locally.

There is also evidence that ponds dotted the landscape during the Pliocene, but these were temporary bodies of water during the rainy season and dry for much of the year.

The absence of fossil fishes and turtles in the Upper Laetolil Beds, as well as larger aquatic and hydrophilic vertebrates, such as crocodiles and hippopotamuses, indicates that the ponds and rivers were present seasonally and that there was no communication between the local drainage system and more regional hydrological systems that would allow sporadic influxes of aquatic taxa. Rage and Bailon (2011) report the occurrence of fossil anurans in the lower part of the Upper Laetolil Beds at Locs. 5 and Loc 10W, but such taxa are able to survive and reproduce in temporary pools formed during the rainy season. A small pond, with a minimum diameter of 40 m, is represented at Loc. 3 by a series of finely laminated clays above Tuff 7, and is notable for the preservation of remarkably preserved fossil insects, plant remains, and ruminant dung (Harrison and Kweka 2011).

Primary ash fall deposits periodically blanketed the Laetoli area, forming distinctive marker tuffs. Most of these are a few centimetres thick, but some formed substantial and widespread calcretes. The thickest and most distinctive marker tuffs is Tuff 7 (the Footprint Tuff). Such a massive inundation of natrocarbonatitic ash, in the form of a series of eruptive events that followed in quick succession, would have had a significantly adverse effect on the local ecosystem. Initially, the ash falls would have buried the vegetation and caused standing bodies of water to become highly alkaline and toxic for mammals to drink from (Peters et al. 2008). As the tuffs became consolidated and cemented, buried trees would have died and the roots of germinating seedlings would not have been able to penetrate the hard and impervious calcretes, thereby preventing regeneration of the woody vegetation. During these periods, the landscape would have been dominated by grasslands, with open woodland occurring along watercourses or in areas of higher elevation where the calcretes were thinner or absent (Andrews 1989; Peters et al. 2008; Andrews et al. 2011). Since there are no indications of specialized grassland adapted faunal communities in any of the assemblages from the Upper Laetolil Beds (Kingston and Harrison 2007; Su and Harrison 2007, 2008; Su 2011; Kovarovic and Andrews 2011; Kaiser 2011; Kingston 2011), we can presume that these periods of disruption in the climax ecosystem were relatively short-lived, with grasslands being quickly replaced by grassy woodlands and bushland as soils developed on top of the calcretes.

References

Adelsberger, K. A., Wirth, K. R., Mabulla, A. Z. P., & Bowman, D. C. (2011). Geochemical and mineralogical characterization of Middle Stone Age tools of Laetoli, Tanzania and comparisons with possible source materials. In T. Harrison (Ed.), *Paleontology and geology of Laetoli: Human evolution in context*, (Geology, geochronology, paleoecology and paleoenvironment, vol. 1, pp. 143–165). Dordrecht: Springer.

Andrews, P. J. (1989). Palaeoecology of Laetoli. *Journal of Human Evolution, 18*, 173–181.

Andrews, P., Bamford, M. K., Njau, E. F., & Leliyo, F. (2011). The ecology and biogeography of the Endulen-Laetoli area in northern Tanzania. In T. Harrison (Ed.), *Paleontology and geology of Laetoli: Human evolution in context,* (Geology, geochronology, paleoecology and paleoenvironment, vol. 1, pp. 167–200). Dordrecht: Springer.

Bamford, M. K. (2011a). Fossil wood. In T. Harrison (Ed.), *Paleontology and geology of Laetoli: Human evolution in context,* (Geology, geochronology, paleoecology and paleoenvironment, vol. 1, pp. 217–233). Dordrecht: Springer.

Bamford, M. (2011b). Fossil leaves, fruits and seeds. In T. Harrison (Ed.), *Paleontology and geology of Laetoli: Human evolution in context*, (Geology, geochronology, paleoecology, and paleoenvironment vol. 1, pp. 235–252). Dordrecht: Springer.

Deino, A. L. (2011). ^{40}Ar/^{39}Ar dating of Laetoli, Tanzania. In T. Harrison (Ed.), *Paleontology and geology of Laetoli: Human evolution in context*, (Geology, geochronology, paleoecology and paleoenvironment, vol. 1, pp. 77–97). Dordrecht: Springer.

Dietrich, W. O. (1942). Altesquartäre Säugetiere aus der südlichen Serengeti, Deutsch-Ostafrika. *Paläontographica, 94*, 44–75.

Drake, R., & Curtis, G. H. (1979). Radioisotope date of the Laetoli Beds, the Hadar Formation and the Koobi Fora-Shungura formation. *American Journal of Physical Anthropology*, 50, 433–434.

Drake, R., & Curtis, G. (1987). K–Ar geochronology of the Laetoli fossil localities. In M. D. Leakey & J. M. Harris (Eds.), *Laetoli: A Pliocene site in northern Tanzania* (pp. 48–52). Oxford: Clarendon.

Harrison, T. (2011a). Hominins from the Upper Laetolil and Upper Ndolanya Beds, Laetoli. In T. Harrison (Ed.), *Paleontology and geology of Laetoli: Human evolution in context*, (Fossil Hominins and the associated fauna, vol. 2, pp. 141–188). Dordrecht: Springer.

Harrison, T. (2011b). Coprolites: Taphonomic and paleoecological implications. In T. Harrison (Ed.), *Paleontology and geology of Laetoli: Human evolution in context,* (Geology, geochronology, paleoecology, and paleoenvironment, vol. 1, pp. 279–292). Dordrecht: Springer.

Harrison, T., & Kweka, A. (2011). Paleontological localities on the Eyasi Plateau, including Laetoli. In T. Harrison (Ed.), (*Paleontology and geology of Laetoli: Human evolution in context,* (Geology, geochronology, paleoecology, and paleoenvironment, vol. 1, pp. 17–45). Dordrecht: Springer.

Hay, R. L. (1976). *Geology of the Olduvai Gorge*. Berkeley: University of California Press.

Hay, R. L. (1978). Melilitite–carbonatite tuffs in the Laetoli Beds of Tanzania. *Contributions to Mineralogy and Petrology, 17*, 255–274.

Hay, R. L. (1987). Geology of the Laetoli area. In M. D. Leakey & J. M. Harris (Eds.), *Laetoli: A Pliocene site in northern Tanzania* (pp. 23–47). Oxford: Clarendon.

Hay, R. L., & Leakey, M. D. (1982). The fossil footprints of Laetoli. *Scientific American, 246*, 50–57.

Hay, R. L., & Reader, R. J. (1978). Calcretes of Olduvai Gorge and the Ndolanya Beds of northern Tanzania. *Sedimentology, 25*, 649–673.

Hedberg, H. D. (Ed.). (1976). *International stratigraphic guide: A guide to stratigraphic classification, terminology, and procedure by The International Subcommission on Stratigraphic Classification of IUGS Commission on Stratigraphy*. New York: Wiley.

Kaiser, T. M. (2011). Feeding ecology and niche partitioning of the Laetoli ungulate faunas. In T. Harrison (Ed.), *Paleontology and geology of Laetoli: Human evolution in context*, (Geology, geochronology, paleoecology and paleoenvironment, vol. 1, pp. 329–354). Dordrecht: Springer.

Kent, P. E. (1941). The recent history and Pleistocene deposits of the plateau north of Lake Eyasi, Tanganyika. *Geological Magazine, 78*, 173–184.

Kingston, J. D. (2011). Stable isotope analyses of Laetoli fossil herbivores. In T. Harrison (Ed.), *Paleontology and geology of Laetoli: Human evolution in context*, (Geology, geochronology, paleoecology and paleoenvironment, vol. 1, pp. 293–328). Dordrecht: Springer.

Kingston, J. D., & Harrison, T. (2007). Isotopic dietary reconstructions of Pliocene herbivores at Laetoli: Implications for early hominin paleoecology. *Palaeogeography, Palaeoclimatology, Palaeoecology, 243*, 272–306.

Kovarovic, K., & Andrews, P. (2011). Environmental change within the Laetoli fossiliferous sequence: Vegetation catenas and bovid ecomorphology. In T. Harrison (Ed.), *Paleontology and geology of Laetoli: Human evolution in context*, (Geology, geochronology, paleoecology and paleoenvironment, vol. 1, pp. 367–380). Dordrecht: Springer.

Krell, F.-T., & Schawaller, W. (2011). Beetles (Insecta: Coleoptera). In T. Harrison (Ed.), *Paleontology and geology of Laetoli: Human evolution in context,* Fossil hominins and the associated fauna, vol. 2, pp. 535–548). Dordrecht: Springer.

Leakey, M. D. (1987a). Introduction. In M. D. Leakey & J. M. Harris (Eds.), *Laetoli: A Pliocene site in northern Tanzania* (pp. 1–22). Oxford: Clarendon.

Leakey, M. D. (1987b). The Laetoli hominid remains. In M. D. Leakey & J. M. Harris (Eds.), *Laetoli: A Pliocene site in northern Tanzania* (pp. 108–117). Oxford: Clarendon.

Leakey, M. D. (1987c). Animal prints and trails. In M. D. Leakey & J. M. Harris (Eds.), *Laetoli: A Pliocene site in northern Tanzania* (pp. 451–469). Oxford: Clarendon.

Magori, C. C., & Day, M. H. (1983). Laetoli Hominid 18: An early *Homo sapiens* skull. *Journal of Human Evolution, 12*, 747–753.

Manega, P. C. (1993). Geochronology, geochemistry and isotopic study of the Plio-Pleistocene hominid sites and the Ngorongoro Volcanic Highlands in northern Tanzania. Ph.D. dissertation, University of Colorado, Boulder.

McHenry, L. J. (2011). Geochemistry and mineralogy of the Laetoli area tuffs: Lower Laetolil through Naibadad Beds. In T. Harrison (Ed.), *Paleontology and geology of Laetoli: Human evolution in context,* (Geology, geochronology, paleoecology and paleoenvironment, vol. 1, pp. 121–141). Dordrecht: Springer.

Mollel, G. F., Swisher, C. C., III, Feigenson, M. D., & Carr, J. D. (2011). Petrology, geochemistry and age of Satiman, Lemagurut and Oldeani: Sources of the volcanic deposits of the Laetoli area. In T. Harrison (Ed.), *Paleontology and geology of Laetoli: Human evolution in context,* (Geology, geochronology, paleoecology and paleoenvironment, vol. 1, pp. 99–119). Dordrecht: Springer.

Murphy, M. A., & Salvador, A. (1999). International stratigraphic guide – An abridged version. *Episodes, 22*, 255–271.

Ndessokia, P.N.S. (1990). The mammalian fauna and archaeology of the Ndolanya and Olpiro Beds, Laetoli, Tanzania. Ph.D. dissertation, University of California, Berkeley.

Peters, C. R., Blumenschine, R. J., Hay, R. L., Livingstone, D. A., Marean, C. W., Harrison, T., Armour-Chelu, M., Andrews, P., Bernor, R. L., Bonnefille, R., & Werdelin, L. (2008). Paleoecology of the Serengeti-Mara ecosystem. In A. R. E. Sinclair, C. Packer, S. A. R. Mduma, & J. M. Fryxell (Eds.), *Serengeti III: Human impacts on ecosystem dynamics* (pp. 47–94). Chicago: University of Chicago Press.

Pickering, R. (1964). Endulen. Quarter Degree Sheet, 52. Dodoma: Geological Survey of Tanzania.

Rage, J. C., & Bailon, S. (2011). Amphibia and Squamata. In T. Harrison (Ed.), *Paleontology and geology of Laetoli: Human evolution in context,* (Fossil hominins and the associated fauna, vol. 2, pp. 467–478). Dordrecht: Springer.

Smith, G., & Lowe, D. R. (1991). Lahars: Volcano-hydrologic events and deposition in the debris flow-hyperconcentrated flow continuum. In R. V. Fisher & G. A. Smith (Eds.), *Sedimentation in volcanic settings*. (pp. 59–74) *SEPM Special Publication No. 45*.

Su, D. F. (2011). Large Mammal Evidence for the Paleoenvironment of the Upper Laetolil an Upper Ndolanya Beds of Laetoli, Tanzania. In T. Harrison (Ed.), *Paleontology and geology of Laetoli: Human evolution in context,* (Geology, geochronology, paleoecology and paleoenvironment, vol. 1, pp. 381–392). Dordrecht: Springer.

Su, D. F., & Harrison, T. (2007). The paleoecology of the Upper Laetolil Beds at Laetoli: A reconsideration of the large mammal evidence. In R. Bobe, Z. Alemseged, & A. K. Behrensmeyer (Eds.), *Hominin environments in the East African Pliocene: An assessment of the faunal evidence* (pp. 279–313). Dordrecht: Springer.

Su, D., & Harrison, T. (2008). Ecological implications of the relative rarity of fossil hominins at Laetoli. *Journal of Human Evolution, 55*, 672–681.

Whittaker, A., Cope, J. C. W., Cowie, J. W., Gibbons, W., Hailwood, E. A., House, M. A., Jenkins, D. G., Rawson, P. F., Rushton, A. A. W., Smith, D. G., Thomas, A. T., & Wimbledon, W. A. (1991). A guide to stratigraphic procedure. *Journal of the Geological Society, 148*, 813–824.

Chapter 4
$^{40}Ar/^{39}Ar$ Dating of Laetoli, Tanzania

Alan L. Deino

Abstract $^{40}Ar/^{39}Ar$ dating of Pliocene tuffs from Laetoli, northern Tanzania, has refined the geochronological framework of the Laetolil Beds and overlying strata. Dated units include the Lower and Upper Laetolil Beds (4.36–3.85 Ma and 3.85–3.63 Ma, respectively), the Lower and Upper Ndolanya Beds (3.58 and 2.66 Ma, respectively), the Naibadad Beds (2.155–2.057 Ma), and the Olpiro Beds (<2.057 Ma). Sedimentation rates of ~50 cm/ka are obtained for the lower half of the Upper Laetolil Beds through Tuff 2, compared to a rate of ~15 cm/ka for the upper half (Tuff 4 to Tuff 8). Accumulation of the Upper Laetolil Beds required less than 220 ka, corresponding to a minimum overall sedimentation rate of 22 cm/ka. The new dates provide a more precise assessment of the age of the fossil hominins at Laetoli. *Australopithecus afarensis* specimens from the Upper Laetolil Beds are constrained to 3.85–3.63 Ma, and a new interpolated age for the Footprint Tuff (Tuff 7) is 3.66 Ma. The *Paranthropus aethiopicus* maxilla (Silal Artum) from the Upper Ndolanya Beds at 2.66 Ma is amongst the oldest known specimen attributable to this genus.

Keywords $^{40}Ar/^{39}Ar$ dating • Laetoli • Hominin • Pliocene • Tanzania • Laetolil Beds • Ndolanya Beds • Naibadad Beds • Olpiro Beds

Introduction

Laetoli is a world-renowned paleoanthropological research area in northern Tanzania, situated in the Eastern Rift Valley on the Eyasi Plateau, between Lake Eyasi on the southwest and Olduvai Gorge on the northeast (Fig. 4.1). At the north-eastern end of the plateau lie the Pliocene volcanoes Lemagurut, Satiman, and Ngorongoro. Laetoli provides an archive of sedimentary and volcanic strata spanning the interval 4.3–2.0 Ma, and contains numerous vertebrate fossil localities including more than 30 fossil hominid specimens (White 1977, 1980; Leakey 1987a; Harrison 2011). Laetoli is also the site of the famous hominid fossil footprint tracks imprinted in carbonatite ash, discovered in 1978 by members of Mary Leakey's team (Leakey 1987b; Robbins 1987; Tuttle 1987; Hay 1987; Hay and Leakey 1982). This work represents new geochronological efforts at Laetoli, begun in 2001 and carried out in conjunction with the Eyasi Plateau Paleontological Expedition (EPPE), under the direction of Terry Harrison.

The Eyasi Plateau consists of Precambrian basement rocks overlain by Plio-Pleistocene waterlain, air-fall, and aeolian tuffs, with interbedded mafic lavas and medium- to fine-grained epiclastic strata. The deposits are typically preserved in thin (5–120 m), discontinuous exposures. They have been subdivided into eight lithostratigraphic units (Fig. 4.2), from oldest to youngest: the Lower Laetolil Beds (LLB, >64 m), the Upper Laetolil Beds (ULB, 44–59 m), Lower Ndolanya Beds (4.5–11 m), Upper Ndolanya Beds (8–16 m), Ogol Lavas (10–20 m), Naibadad Beds (10–15 m), Olpiro Beds (≤5 m), and Ngaloba Beds (2–6 m) (Hay 1987). This study provides new $^{40}Ar/^{39}Ar$ ages for most of these units, with the exception of the Ogol Lavas and Ngaloba Beds. The aim is to provide improved chronostratigraphic control of the Pliocene geological record and fossil occurrences. Particular focus is placed on the sparsely dated, highly fossiliferous ULB.

Previous Research

Geochronological studies of the Laetoli strata were initiated in 1974–1975 employing conventional K-Ar dating (and one biotite $^{40}Ar/^{39}Ar$ age incremental heating experiment) on tuffs and lavas (Drake and Curtis 1987; Leakey et al. 1976). This pioneering effort has proved remarkably robust, and subsequent studies have served to refine various aspects of this early chronostratigraphic framework. Extensive $^{40}Ar/^{39}Ar$ dating accompanied the Ph.D. thesis studies of two Tanzanian students

A.L. Deino (✉)
Berkeley Geochronology Center, 2455 Ridge Road,
Berkeley, CA 94709, USA
e-mail: al@bgc.org

T. Harrison (ed.), *Paleontology and Geology of Laetoli: Human Evolution in Context. Volume 1: Geology, Geochronology, Paleoecology and Paleoenvironment*, Vertebrate Paleobiology and Paleoanthropology, DOI 10.1007/978-90-481-9956-3_4,

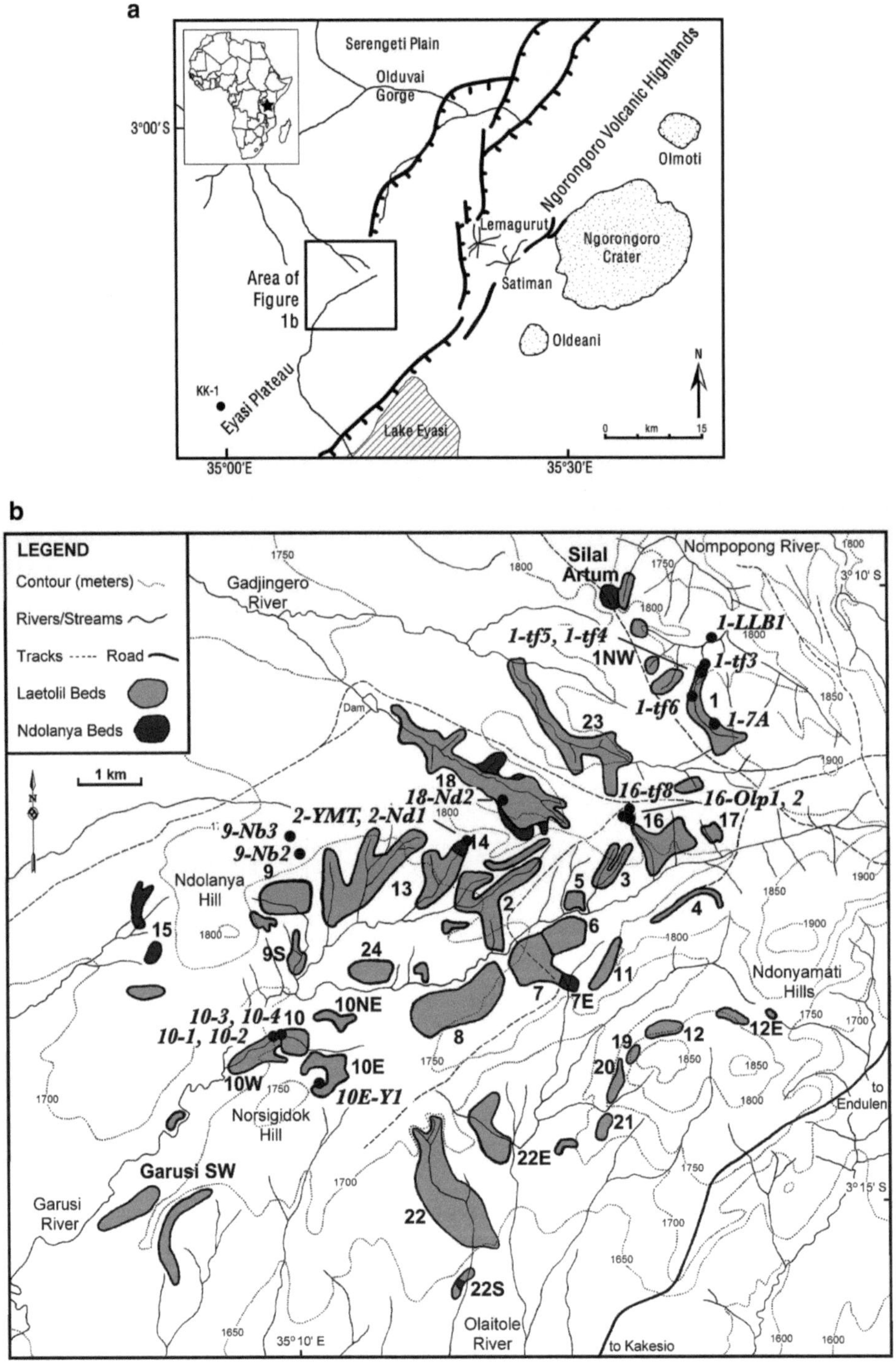

Fig. 4.1 (**a**) Regional map showing location of Laetoli field collection area (Fig. 4.1b). 'KK-1' shows the location of a dating sample from the vicinity of Kakesio, outside the area of Fig. 4.1b. (**b**) Laetoli field collection area, showing dating sample locations (italicized) and site designations (non-italicized). Outcrops of the Laetolil Beds and Ndolanya Beds are shown by the light and dark shaded areas, respectively

in the early 1990s (Ndessokia 1990; Manega 1993). Their coverage of the Pliocene sequence was necessarily incomplete, given the focus and constraints of their individual projects. These data were also never formally published. Prior to the current study, details of the chronostratigraphy of the Laetolil Beds have remained poorly known. It might be noted that all conventional K-Ar and $^{40}Ar/^{39}Ar$ dating at Laetoli, including the new work presented herein, has been carried out in the same laboratory, though its location, name, and affiliation have changed through the years (pre-1985, the K-Ar Laboratory of the Department of Geology, University of California, Berkeley; 1985–1994, the Geochronology Center of the Institute of Human Origins, 1994–present, the Berkeley Geochronology Center, 'BGC'). The fact that the new dating results are generally in good agreement with the prior ages points to the consistency of the overall dating effort across multiple techniques and researchers, encompassing a period of more than three decades.

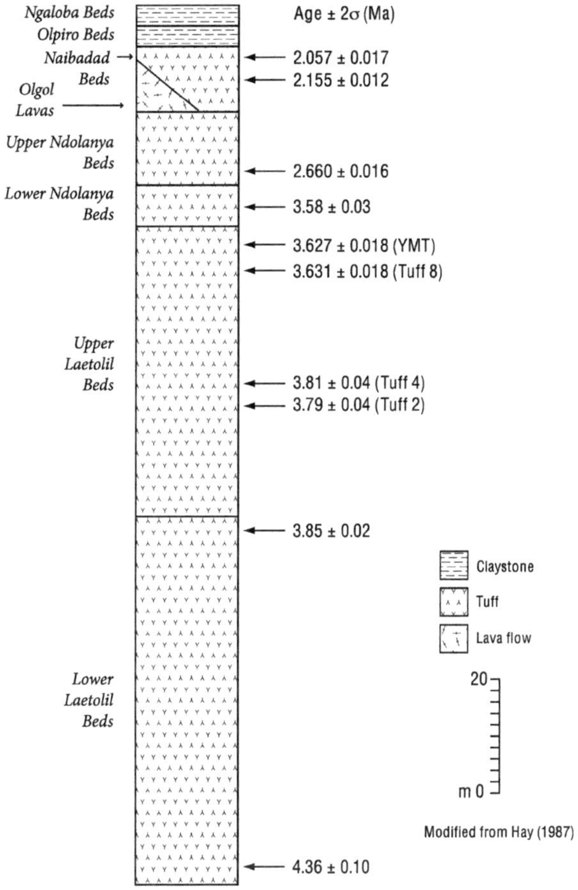

Fig. 4.2 Schematic stratigraphic section (Hay 1987) of the Plio-Pleistocene strata of the Laetoli area, with new $^{40}Ar/^{39}Ar$ age determinations

Methodology

The principal means for establishing chronostratigraphic control of the Laetoli Pliocene sedimentary strata in this study is $^{40}Ar/^{39}Ar$ dating of phenocrystic K-feldspar of volcanic origin in tuffaceous units. There are numerous primary and reworked tuffs in the Laetoli stratigraphic sequence, some of which contain excellent primary K-feldspar suitable for precise and accurate $^{40}Ar/^{39}Ar$ dating. However, many tuff units either do not contain K-feldspar or contain only rare or small grains, or are dominated by anomalously old material (i.e., xenocrysts). Biotite, also present as phenocrysts in many of the tuffs, has been extensively dated in this study to supplement the K-feldspar geochronologic record. However, results from biotite must be viewed cautiously, as its phyllosilicate structure is more susceptible to diagenetic alteration and open-system isotopic behavior than K-feldspar. Phenocrystic hornblende has also been analyzed from a number of tuff units in the Laetolil Beds. Attempts to obtain useful chronostratigraphic information from hornblende have been mostly thwarted, however, by its low potassium and high calcium content, and isotopic complexities as revealed by incremental-release argon dating experiments.

All $^{40}Ar/^{39}Ar$ analytical work was performed at the BGC. Samples were gently disaggregated using a ceramic mortar and pestle, and sieved to retain the >0.35 mm size fraction. Phenocrysts were extracted from pumice lapilli where possible, but often lapilli are rare or small, and in such cases bulk tuff was used. K-feldspar, biotite, and hornblende were concentrated from the same source material using magnetic and heavy-liquid separation techniques. Typically, K-feldspar and hornblende could be adequately concentrated using magnetic separation. Biotite, however, after magnetic separation was further split into density fractions using pure Li-metatungstate liquid (density ~3.1 g/cc) to concentrate the heavier, typically less altered material. Feldspars were treated with dilute HF to remove adhering matrix. All mineral separates were washed in distilled water in an ultrasonic bath, and hand-picked to remove impurities and grains with visible glass or mineral inclusions, or adhering matrix.

The crystal concentrates were irradiated in seven separate batches for 0.33–7 h in the Cd-lined, in-core CLICIT facility of the Oregon State University TRIGA reactor. Sanidine from the Fish Canyon Tuff of Colorado was used as a mineral standard, with an astronomically calibrated reference age of 28.201 ± 0.046 Ma (Kuiper et al. 2008). Standards and unknowns were placed in 2 mm deep wells in various semicircular configurations on 18 mm diameter aluminum disks, with standards placed strategically so that the lateral neutron flux gradients across the disk could be evaluated. Planar regressions were fit to the standard data, and the $^{40}Ar/^{39}Ar$ neutron fluence parameter, J, interpolated for the unknowns. Uncertainties in J are conservatively estimated at 0.2–0.3% (1σ), based on Monte Carlo error analysis of the planar regressions (Deino et al. 2006).

Gas extractions were performed under ultra-high vacuum, using either a focused CO_2 laser to quickly fuse individual feldspar crystals (the 'single-crystal laser-fusion' method, or 'SCLF'), or using a faceted integrator lens that yields a square 6 × 6 mm, uniform-energy beam profile for incremental heating experiments on biotite and hornblende. Typical sample weights for the bulk incremental heating experiments were ~10–50 mg. Gasses were exposed for several minutes to an approximately −130°C cryosurface to trap H_2O, and to SAES getters to remove reactive compounds (CO, CO_2, N_2, O_2, and H_2). The cleanup procedure was followed immediately by measurement of five argon isotopes on a MAP 215-50 mass spectrometer for approximately 30 minutes. Further details of the dating methodology are provided elsewhere (Deino and Potts 1990; Deino et al. 1998, 2006).

It should be noted that 'inverse isochrons,' (isotope correlation diagrams in which $^{40}Ar/^{36}Ar$ is plotted against $^{39}Ar/^{36}Ar$ for a series of runs), are not employed in the final data reduction of either the feldspar single-crystal analyses, or the biotite incremental-heating experiments. For K-feldspar of this age, isochrons are inappropriate, as the material yields such a limited spread in the relative abundance of radiogenic argon ($^{40}Ar^*$) relative to atmospheric argon (^{40}Ar and ^{36}Ar in a ratio of 295.5; Steiger and Jäger 1977), that there is insufficient variation in $^{39}Ar/^{36}Ar$ to form a useful isochron. Biotite incremental heating experiments, on the other hand, often exhibit a significant spread of the percent of $^{40}Ar^*$ of total ^{40}Ar, hence also in $^{39}Ar/^{36}Ar$, but observation of the 'trapped' $^{40}Ar/^{36}Ar$ composition of the mineral is masked by a strong atmospheric component introduced during diagenetic alteration. The $^{40}Ar/^{36}Ar$ composition inferred from the $^{36}Ar/^{40}Ar$ axis of the isochron is atmospheric, within error: the weighted mean of $^{40}Ar/^{36}Ar_{trapped}$ for all the multi-grain biotite step-heating experiments is essentially air composition (294 ± 2), and none of the experiments, individually, fall outside of air composition within 2σ. Errors in the ages derived from isochrons are about a factor of two higher than those assigned to apparent age plateaus, and are indeed quite exaggerated in the case of the best biotite experiments, those yielding high $^{40}Ar^*$ throughout. In these situations the errors in the isochron age are highly exaggerated, about a factor of five greater than the plateau apparent ages, because there is little spread in the data along the isochron leading to huge error magnification at the $^{39}Ar/^{36}Ar$ axis intercept, where the age is calculated. In lieu of isochron ages, the final age results are reported as a weighted mean of the individual ages of the 'primary' population in the case of the single-crystal feldspar data, and as the inverse-variance weighted mean of the plateau step ages for the incremental-heating experiments. For both, the reported error is the 2σ standard error of the weighted mean, expanded by root MSWD (mean square weighted deviation) when excess scatter may be present (MSWD > 1). Supplementary data are available online at http://www.nyu.edu/gsas/dept/anthro/programs/csho/pmwiki.php/Home/SupplementaryInformation.

Previously published K–Ar and $^{40}Ar/^{39}Ar$ dates quoted in this manuscript have been adjusted upward for consistency with a reference age of 28.201 Ma for the Fish Canyon sanidine (Kuiper et al. 2008), the irradiation monitor mineral used in this study. For the conventional K–Ar ages of Drake and Curtis (1987) and Leakey et al. (1976), this conversion results in an increase of ~0.6% from the published ages. For the previously published $^{40}Ar/^{39}Ar$ ages, the following conversions apply: Manega (1993), 1.26%; Ndessokia (1990), 1.26%; Mollel et al. (2011), 0.65%; McDougall and Brown (2008), 0.36%. Also, prior results are herein quoted at 2σ analytical error.

Dating Results Overview

K-Feldspar

A total of 325 single-crystal, total-fusion $^{40}Ar/^{39}Ar$ age determinations were completed on K-feldspar phenocrysts (Ca/K <1) from 18 samples, of 16 different tuff units (Table 4.1). While many of these analyses are representative of the eruptive age of the tuff, others are clearly anomalous, and it is therefore necessary to establish criterion to filter these runs. First, while most grains yield a high percentage of radiogenic ^{40}Ar ('% $^{40}Ar^*$') to total ^{40}Ar (atmospheric plus radiogenic) as expected for Pliocene and older K-feldspars, a significant number demonstrate a lower than expected radiogenic proportion. These anomalous grains may have slight alteration, adhering glass, trapped air pockets along fractures, or glass, vapor or fluid inclusions. A cutoff of 75% $^{40}Ar^*$ was established at a gap in the frequency distribution of radiogenic content to exclude such grains from further data analysis, eliminating 23 (7%) of the feldspars analyzed.

Of the remaining 302 analyses, many are simply too old to reflect a true eruptive age. These are most likely xenocrysts, acquired immediately prior to or during the eruptive process, or during transport to the site of deposition. At Laetoli, where the current dating effort, as well as prior work, indicates that the oldest strata of the Laetoli Beds are ~4.2–4.4 Ma, a cutoff of 4.7 Ma is used to classify 119 grains as 'obvious xenocrysts.' The most precise of these (< 3% error in age) are up to 600 Ma, with prominent modes at ~7.6, 6.5, and 5.5 Ma.

Also present are more subtle components that are only slightly older than an inferred eruption age. These may be recognizable as skewed tails to a Gaussian-like primary age distribution, or grains that exhibit ages slightly older than reliable results on stratigraphically lower units. These subtly older materials may be partially or undegassed xenocrysts acquired from eruptive sources (vent material, tephra mantling volcanic landscapes), derived from intraformational erosion of recently deposited tuff units, subtly altered material, or grains containing excess ^{40}Ar. The particular characteristics of the age distribution for each tuff or group of tuffs are discussed below.

Biotite

Biotite was analyzed as multi-grain separates (~10–30 mg; 13 step-heating experiments on six samples; Table 4.2) and as individual grains (~1–3 mm phenocrysts; 65 step-heating experiments on seven samples, Table 4.3) using the laser incremental-heating approach.

Most of the multi-grain experiments yielded broad apparent-age 'plateaus' encompassing >90% of the total ^{39}Ar release (Fig. 4.3). Replicate experiments on aliquots of irradiated mineral separates demonstrate reproducibility within the stated errors, with an overall pooled standard deviation of about 0.04 Ma (1%). The mean sample ages appear geologically reasonable based on prior dating, which place the ULB at about 3.8–3.5 Ma (Drake and Curtis 1987). However, the ULB biotite sample ages exhibit repeated age reversals in

Table 4.1 Summary of SCLF K-feldspar analyses, listed in stratigraphic order starting with youngest at the top

Sample	Lab ID#	Locality	UTM (WGS'84 Grid)	Ca/K ± 2σ		Age (Ma) ± 2σ MSE		MSWD	Prob.	n/n_{total}
Olpiro Beds										
16-OLP1	25171	16	35745952/9645948	0.0028		(2.070	±0.020)			1/21
16-OLP2	25172	16	36746029/9645933	0.0035		(2.057	±0.015)			1/19
Naibadad Beds										
9-NB3	25168	9	36740918/9645689	0.0210	±0.0004	2.057	±0.017	6.6	0.00	12/21
9-NB2	25169, 25170	9	36741062/9645436	0.0363	±0.0003	2.155	±0.012	3.6	0.00	15/21
Ndolanya Beds										
18-Nd2	25187	18	36744177/9646250	0.0078	±0.0010	2.660	±0.016	0.9	0.52	12/12
2-Nd1	25181	2	36743682/9645729	0.0223	±0.0012	3.58	±0.03	1.2	0.32	6/21
Upper Laetolil Beds										
10E-Y1	25182	10E	36741307/9642111	0.0038	±0.0006	3.64	±0.02	1.5	0.11	14/20
2-YMT	25188	2	36743682/9645729	0.0063	±0.0011	3.61	±0.03	1.4	0.16	11/21
16-tf8	25189	16	36746030/9646062	0.0092	±0.0005	3.631	±0.018	1.7	0.07	13/22
1-tf6	24700	1	36747003/9647710	0.0043	±0.0009	(3.73	±0.09)	14.6	0.00	3/7
1-tf5	24703, 24701	1	36747140/9648057	0.0206	±0.0034	(3.76	±0.08)	3.5	0.06	2/40
10-4	24524, 22532	10	36740768/9642767	0.0309	±0.0177	3.81	±0.04	1.9	0.09	6/56
1-tf4	24697	1	36747152/9648060	0.1141	±0.0051	(4.04	±0.08)	1.7	0.18	3/18
1-tf3	24698	1	36747189/9648170	0.0153	±0.0011	(3.86	±0.05)	2.6	0.07	3/7
10-2	22602	10	36740624/9642736	0.0134	±0.2105	3.79	±0.04	0.6	0.73	8/13
10-1	22535	10	36740619/9642735	0.0018	±0.0002	(3.93	±0.06)	1.5	0.17	6/9
Upper part of Lower Laetolil Beds										
1-LLB1	24706	1	36747305/9648569	0.0073	±0.0010	3.85	±0.02	2.0	0.01	14/22
Lower part of Lower Laetolil Beds										
KK-1	22527	Kekesio	36721999/9629501	0.0047	±0.0006	4.36	±0.10	0.8	0.49	4/5

Table 4.2 Summary of incremental heating biotite multi-grain experiments

		Integrated age			Apparent age plateau						Inverse isochron				
Sample	Lab ID	(Ma ±2σ)		%^{40}Ar*	Age (Ma ± 2σ)[a]		MSWD	Steps	n/n_{tot}	%^{39}Ar	Age (Ma ± 2σ)[a]		(^{40}Ar/^{36}Ar)$_{tr}$	MSE	MSWD
Tuff 7															
1-7A	22536-01	3.64	±0.06	55.8	3.65	±0.03	0.5	D-M	10/10	100.0	3.66	±0.04	293	±4	0.3
”	22536C-01	3.75	±0.12	40.5	3.74	±0.07	1.7	D-L	9/10	92.4	3.78	±0.21	293	±11	1.9
”	24523-02	3.61	±0.10	63.6	3.62	±0.07	1.5	C-M	11/11	100.0	3.66	±0.14	288	±20	1.5
”	24523-03	3.66	±0.10	84.9	3.62	±0.09	1.5	F-M	8/10	88.9	3.71	±0.32	252	±176	1.7
”	24523-04	3.65	±0.07	78.5	3.66	±0.06	1.7	B-L	11/11	100.0	3.72	±0.27	280	±78	1.8
				Wtd. Mean =	*3.66*	*±0.02*									
				Median=	*3.65*										
Tuff 4															
10-4	22534-01	3.87	±0.07	48.8	3.88	±0.03	2.0	E-L	8/11	70.6	3.87	±0.06	297	±5	2.2
”	22534C-01	3.81	±0.08	55.5	3.81	±0.04	1.0	E-O	11/11	100.0	3.83	±0.06	293	±6	1.0
				Wtd. Mean =	*3.86*	*±0.02*									
				Median=	*3.845*										
Tuff 3															
10-3	22603-02	3.70	±0.05	93.5	3.72	±0.05	0.9	H-N	7/8	96.7	3.78	±0.1	216	±108	0.8
”	22603-03	3.76	±0.09	82.5	3.71	±0.07	1.3	I-N	6/6	100.0	3.69	±0.09	315	±34	1.2
				Wtd. Mean =	*3.72*	*±0.04*									
				Median=	*3.715*										
Tuff 2															
10-2	22538-01	3.92	±0.06	61.4	3.86	±0.03	1.5	E-N	10/11	95.0	3.86	±0.04	298	±4	1.5
Tuff 1															
10-1	22529-01	3.76	±0.03	85.8	3.75	±0.03	1.7	G-O	9/11	93.5	3.75	±0.03	311	±13	0.9
Lower part of Lower Laetolil Beds															
KK-1	22531-01	4.24	±0.13	39.4	4.33	±0.07	1.6	E-M	9/10	96.1	4.42	±0.09	287	±6	0.8
”	22531C-01	4.33	±0.18	30.6	4.28	±0.07	1.3	C-M	11/11	100.0	4.25	±0.12	297	±5	1.4
				Wtd. Mean =	*4.3*	*±0.05*									
				Median=	*4.33*										

10-3 was sampled at UTM 36740741/9642757, and 1-7A at UTM 36747333/9647290. All other UTMs are listed in Table 4.1

[a]Incorporates error in *J*, the neutron fluence parameter

Table 4.3 Summary of incremental heating biotite single-grain experiments

		Integrated age			Apparent age plateau					
					Age					
Sample	Run ID	(Ma ± 2σ)		% $^{40}Ar^*$	(Ma ± 2σ)[a]		MSWD	Steps	n/n_{tot}	% ^{39}Ar
Tuff 7										
1-7A	24523-05	3.7	±0.3	67.8	3.7	±0.2	0.8	I-P	4/4	100
Tuff 6										
1-tf6	24708-05	3.3	±0.3	43.7	3.41	±0.18	1.0	A-F	6/6	100
"	24709-01	4.03	±0.12	71.8	3.99	±0.10	2.2	B-E	4/4	100
"	24709-02	3.63	±0.12	68.5	3.70	±0.09	0.6	B-E	4/5	97.4
"	24709-03	3.64	±0.15	82.3	3.80	±0.11	1.7	A-C	3/4	97.2
"			*Weighted mean =*		*3.78*	*±0.05*				
"			*Median=*		*3.75*					
Tuff 5										
1-tf5	24714-04	3.5	±0.3	22.9	3.62	±0.19	1.7	A-E	5/7	94.8
Between Tuffs 4 & 5										
1-tf4	24711-01	4.0	±0.2	55.6	3.95	±0.16	1.8	A-D	4/7	86.1
"	24711-05	3.62	±0.19	87.5	3.64	±0.16	1.3	B-G	6/6	100
			Weighted mean =		*3.79*	*±0.11*				
			Median=		*3.80*					
Tuff 4										
10-4	22534B-01	4.1	±0.3	57.8	4.0	± 0.2	1.3	B-F	5/5	100
"	22534C-03	3.9	±0.2	99.8	3.95	±0.17	1.1	C-J	8/8	100
			Weighted mean =		*3.97*	*±0.13*				
			Median=		*3.98*					
Tuff 1										
10-1	22529C-01	3.74	±0.04	77.5	3.73	±0.03	1.0	G-M	7/9	89.5
"	22529C-02	3.73	±0.06	95.6	3.75	±0.05	1.2	N-K	9/9	100
"	22529C-03	3.83	±0.15	79.0	3.84	±0.12	0.7	C-K	9/9	100
"	22529C-04	3.9	±0.3	98.0	3.80	± 0.2	0.8	E-M	8/8	100
"	22529C-05	3.95	±0.17	98.0	3.92	±0.14	0.3	C-J	8/8	100
			Weighted mean =		*3.75*	*±0.02*				
			Median=		*3.80*					
Upper part of Lower Laetolil Beds										
1-LLB1	24707-03	4.0	±0.3	24.6	4.1	± 0.2	2.0	A-C	3/4	96.3
Lower part of Lower Laetolil Beds										
KK-1	22531C-03	4.3	±0.2	48.3	4.3	± 0.2	1.5	C-K	7/7	100.0

[a] Incorporates error in *J*, the neutron fluence parameter

stratigraphic context that are well beyond analytical uncertainty (e.g., Tuffs 1, 2, 3 and 4 in Fig. 4.3). Clearly, the multi-grain biotite plateau results, while seemingly analytically sound, are nevertheless inaccurate to an extent beyond that reflected in the calculated analytical sample errors. The most likely explanations include open-system isotopic behavior accompanying diagenesis, contamination of the multi-grain samples with xenocrystic biotite, or possibly excess ^{40}Ar entrapped in the biotite. Excess ^{40}Ar appears unlikely, as the high K content of biotite and the Pliocene age of the samples necessitate incorporation of a large initial trapped excess $^{40}Ar/^{36}Ar$ ratio to influence the apparent age, whereas the trapped $^{40}Ar/^{36}Ar$ ratio of the biotite suggested by isotope $^{36}Ar/^{40}Ar$ vs. $^{39}Ar/^{40}Ar$ correlation diagrams ('inverse isochrons') are almost entirely within 2σ error of atmospheric composition (295.5). The potential for xenocrystic biotite remains a possibility, partially addressed by single-crystal biotite dating.

Biotite single-crystal incremental heating results (Fig. 4.4; Table 4.3) are about 2–10 times less precise than the multi-grain experiments, due to the relatively small gas yields, leading to greater error due to counting statistics, and the greater influence of errors in blank measurement. The best single-grain biotite runs are summarized in Table 4.3, comprising incremental-heating experiments with more than three significant steps (steps with >2% of total ^{39}Ar), plateaus encompassing >85% of total ^{39}Ar, and a relatively precise apparent age plateau (2σ errors < ±0.3 Ma). Six tuffs from the ULB and two from the LLB are represented. The pooled standard deviation of replicate plateau ages from these selected runs is 0.17 Ma (~4%). Tuff 1 is the most reproducible, with a grain-to-grain standard deviation about half that

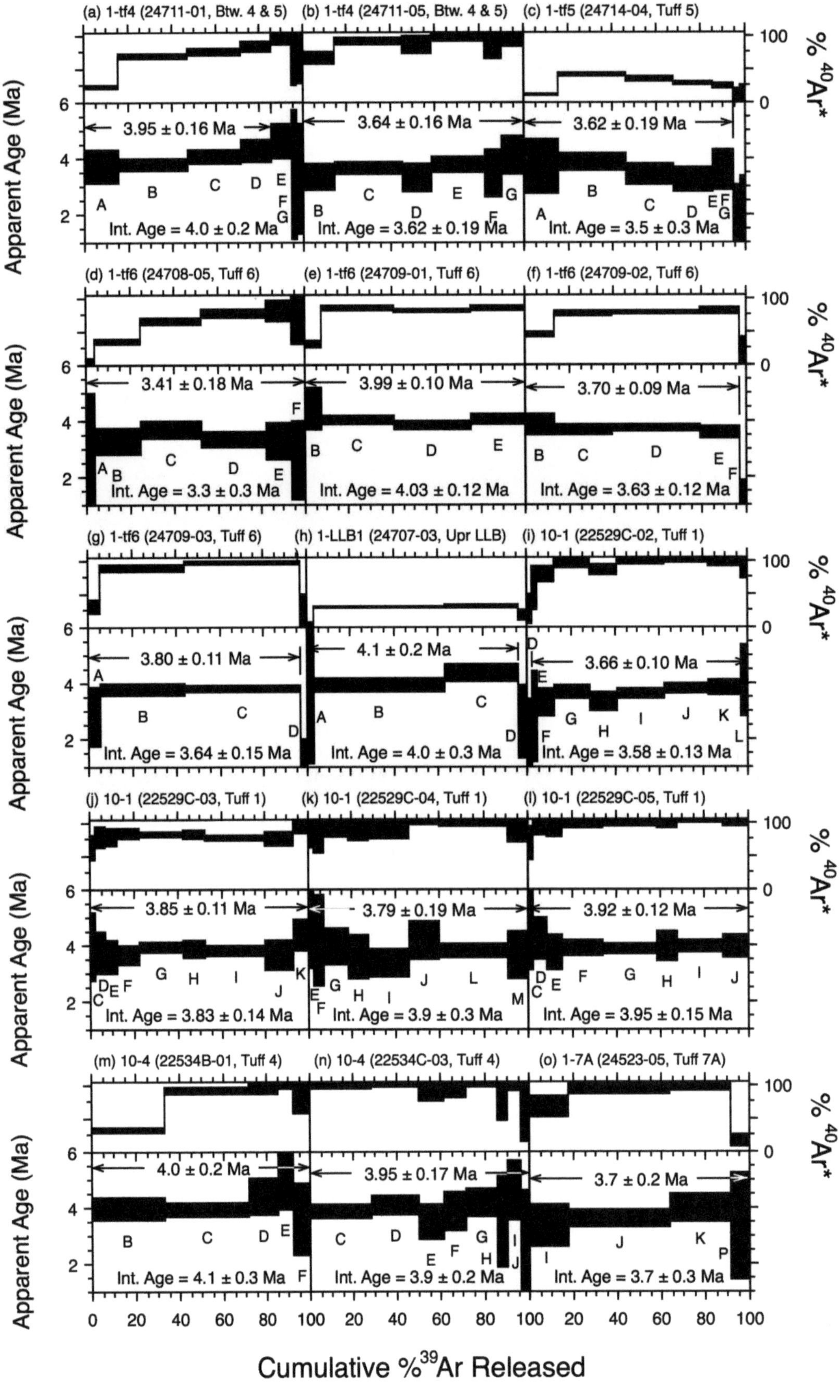

Fig. 4.3 Multi-grain biotite $^{40}Ar/^{39}Ar$ incremental heating spectra for the LLB and ULB. Labels (**a**)–(**o**) give the sample name (e.g., KK-1), the dated aliquot designation (e.g., 22531-01), and tuff attribution or stratigraphic level (e.g., Lwr. LLB). Capital letters below each apparent-age steps identify the individual run and are keyed to the on-line table of full analytical results. Apparent-age plateau errors and integrated-age errors are quoted at 2σ, and include error in *J*, the neutron fluence parameter. 'Int. Age' refers to the integrated age of the overall incremental heating experiment, obtained by mathematical recombination of the step-wise isotopic yields. '% $^{40}Ar^{*}$' refers to the percent radiogenic ^{40}Ar content compared to overall ^{40}Ar (radiogenic plus atmospheric) of the recombined sample gas

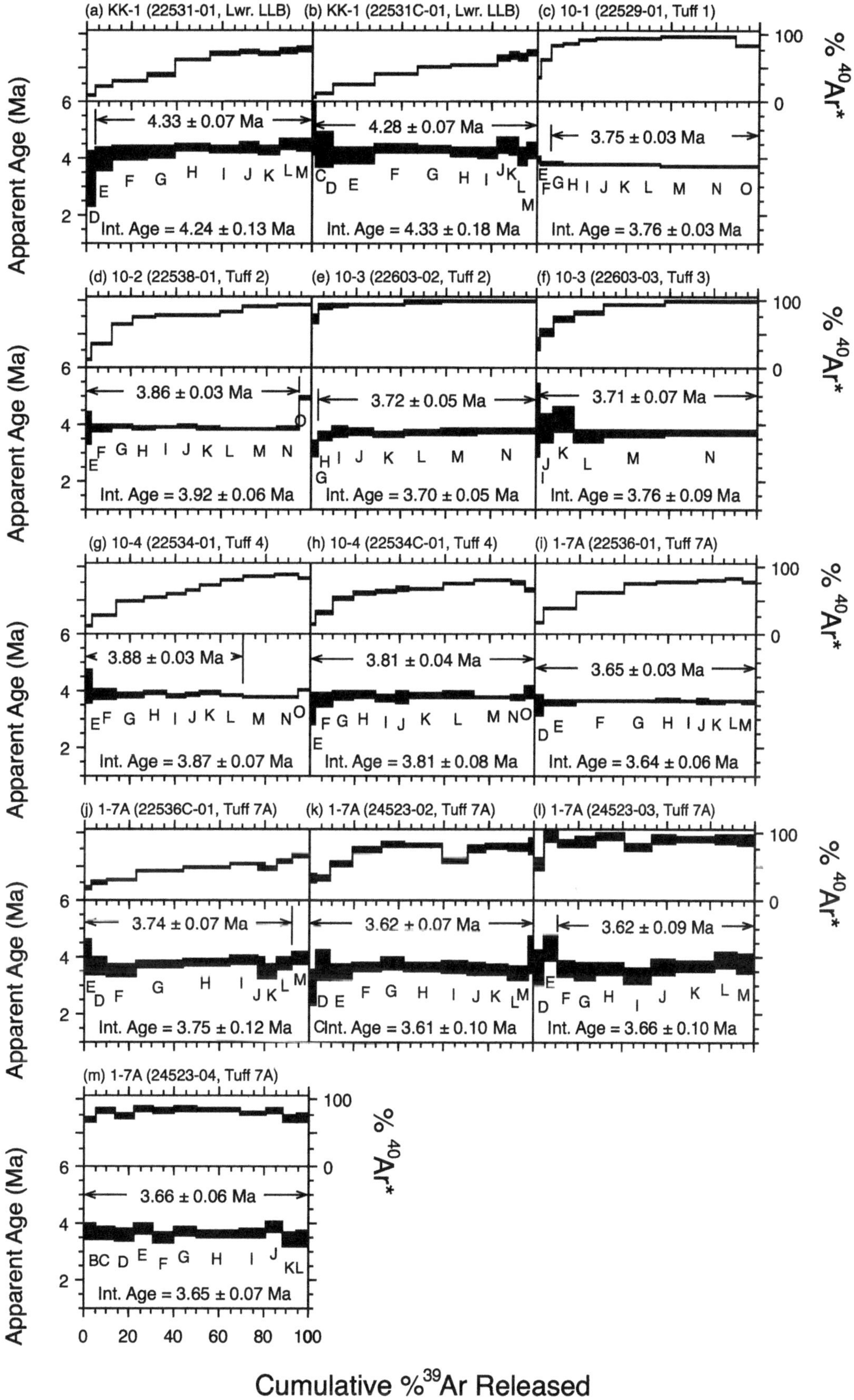

Fig. 4.4 Single-grain biotite $^{40}Ar/^{39}Ar$ incremental heating spectra

of the pooled value. Median tuff ages for the best biotite single-crystal experiments correspond modestly well to stratigraphic position, with some reversals, and numerically support a considerably older age for the lowermost strata of the LLB at Kakesio. Although the single-crystal biotite incremental heating experiments are relatively imprecise, they nevertheless provide an indicator of age, and reveal details of the grain-to-grain isotopic systematics of these tephra. Obviously old xenocrystic biotite was not observed, and may be uncommon or absent.

The preferred remaining explanation for the inconsistent biotite chronostratigraphy is post-depositional alteration. Despite efforts in the field to collect the freshest appearing biotite, and in the laboratory to grade mineral separates by density to avoid altered (light) material, alteration effects were impossible to eliminate. Evidence for variable alteration of biotite samples is reflected in the large between-sample spread in integrated-age calculation of the % $^{40}Ar^*$ of the multi-grain experiments (31–94% $^{40}Ar^*$; Table 4.2), and in the case of sample 1–7A where five replicate analyses are available, the large within-sample spread (40–85% $^{40}Ar^*$).

Evidence for correlation of age and alteration of biotite, however, as measured by the integrated yield of $^{40}Ar^*$, is equivocal. For example, the single-grain biotite incremental heating experiments yielded six grains with $^{40}Ar^*$ >80%, from three different samples (Fig. 4.4; Table 4.3). Comparing each of these runs to its median sample plateau age gives two above, one equal, and three below the sample median. In regards to the multi-grain biotite experiments, two of the five aliquots of sample 1–7A tested had high radiogenic yields, one of which falls above the median and one below. Therefore, the within-sample variability indicates the absence of a strong correlation of age and alteration, at least as measured by the percent radiogenic ^{40}Ar proxy. However, there is a suggestion of susceptibility of some tuff units to alteration and perturbation of biotite ages. Near the base of ULB, the multi-grain biotite experiments for Tuffs 1 and 3, which yielded high% $^{40}Ar^*$, are considerably younger (3.75, 3.72 Ma) than Tuffs 2 and 4 (both 3.86 Ma), which demonstrated much lower radiogenic yields and are presumably more altered. The application of these chronostratigraphic relationships to the question of age trends with alteration is uncertain, however, as both the highly radiogenic and low radiogenic tuff ages depart about equally from the age–trend line established by the feldspar data.

Hornblende

Hornblende was analyzed by a modified SCLF method (a degas followed by a fusion step), and as multi-grain incremental heating experiments (Table 4.4). Overall, these efforts failed to indicate accurate eruptive ages, with the exception of Tuff 7A, discussed below. Difficulties in dating hornblende in this context includes low radiogenic yield due to low K content, complex incremental heating isotopic release systematics (irregular or saddle-shaped cumulative ^{39}Ar release spectra), and anomalously old total-fusion and step apparent ages suggestive of excess argon or mixed age populations.

Chronostratigraphy

Lower Laetolil Beds

The oldest tuff dated in this study was sampled from remote outcrops of the LLB at Kakesio, ~23 km southwest of the main Laetoli area, where ~15 m of LLB unconformably overlie Precambrian granitic basement. The ULB is not present at this locality. Sample KK-1 is from a ~10–15 cm brown, lithic tuff resting on conglomeratic sediments 1.3 m above basement. The tuff contains rare, small phenocrysts of biotite and anorthoclase. It yielded only a few clear K-feldspars, near the lower end of the size range (300–350 μ) for acceptably precise dating in this time range. Out of the five phenocrysts analyzed by the SCLF method, four exhibited $^{40}Ar^*$ above the cutoff of 75%, comprising a unimodal age distribution with a weighted-mean of 4.36 ± 0.10 Ma (2σ; MSWD = 1.5; Fig. 4.5; Table 4.1). Two multi-grain and one single-grain biotite incremental heating experiments on aliquots from a single irradiated separate were also performed. These yielded apparent-age plateaus from 4.3–4.25 Ma (Tables 4.2 and 4.3; Figs. 4.3 and 4.4), with an overall weighted mean of 4.30 ± 0.05 Ma, consistent with the K-feldspar mean. The total gas % $^{40}Ar^*$ yield of these experiments was quite low (39–48%), possibly reflecting significant alteration. Although the biotite mean is more precise than the K-feldspar age, uncertainties regarding the accuracy of biotite ages at Laetoli dictates that the K-feldspar result be taken as the reference age for this tuff, representing the oldest known outcrops of the LLB.

The youngest strata of the LLB were calibrated with a tuff sampled from the uppermost part of the unit, just below the contact with the ULB at Locality 1. The sample was obtained from a yellow, massive eolian tuff ~9 m thick, with ca. 5% phenocrysts of augite, biotite and anorthoclase (sample 1-LLB1). SCLF dating of K-feldspar yielded a well-defined primary mode interpreted as an eruption age, with a weighted mean of 3.85 ± 0.02 Ma.

Previous conventional K-Ar dating results from strata attributed to the lowermost Laetolil Beds gave 4.35 ± 0.12 Ma (2σ) on anorthoclase from a tuff 15 km southeast of Laetoli near Olpiro (Drake and Curtis 1987). This determination is in good agreement with the new $^{40}Ar/^{39}Ar$ age at Kakesio for the lowermost Laetolil Beds. The LLB below

Table 4.4 Summary of incremental heating hornblende multi-grain experiments

	Integrated age			Apparent Age Plateau					
Sample /Lab ID	(Ma ± 2σ)		% $^{40}Ar^*$	Age (Ma ± 2σ)[a]		MSWD	Steps	n/n_{tot}	% ^{39}Ar
1-7A	*Tuff 7, ULB*								
22506-01	3.66	±0.05	85.4	No plateau					
22506-02	3.67	±0.04	86.8	3.55	±0.03	0.8	D-F	3/5	93.3
22519-01	9.3	±0.7	20.2	4.51	±0.09	1.6	D-F	3/7	84.9
22621-01	9.1	±0.7	34.3	No plateaus					
22621-02	11.3	±0.5	78.6	No plateaus					
22519B-01	12.9	±0.7	27.1	No plateaus					
22622-19	7.6	±0.2	30.4	No plateaus					
1-tf6	*Tuff 6, ULB*								
24768-01	500	±70.	5.7	230	±40	1.9	A-C	3/9	65.0
24768-02	710	±150.	4.6	No plateaus					
24768-03	1060	±110.	11.3	800	±70	2.7	A-C	3/9	62.4
1-tf5	*Tuff 5, ULB*								
24769-01	1500	±300	7.7	1500	±200	0.9	A-E	5/9	74.2
24769-02	1840	±170	14.2	No plateaus					
24769-03	870	±50	24.4	No plateaus					
10-5	*Tuff 5, ULB*								
22522-01	30	±4	10.7	No plateaus					
1-tf4	*Betw. Tuffs 4&5, ULB*								
24765-01	410	±50	58.7	No plateaus					
24765-02	541	±11	39.1	No plateaus					
24765-03	410	±20	20.9	No plateaus					
10-4	*Tuff 4, ULB*								
22507-01	12	±2	7.8	No plateaus					
22508-01	40	±8	9.7	No plateaus					
1-tf3	*Tuff 3, ULB*								
24767-01	630	±100	8.7	470	±60	1.8	B-H	7/9	78.5
24767-02	1070	±130	23.7	No plateaus					
24767-03	780	±180	20.4	570	±160	1.7	D-J	7/10	61.5
10-3	*Tuff 3, ULB*								
22520-01	33	±8	8.1	32	±7	0.3	D-G	4/10	50.5
22520B 01	23.8	±1.9	17.7	No plateaus					
10-2	*Tuff 2, ULB*								
22620-01	35	±20	9.2	4	±6	0.6	B-Q	12/12	100.0
22620-02	36	±14	9.6	17	±10	1.9	D-L	7/10	78.2
22523-01	18	±4	6.3	No plateaus					
10-1	*Tuff 1, ULB*								
22510-01	11	±13	1.5	No plateaus					
22511-01	35	±10	7.8	24	±6	2.2	A-D	4/10	57.0
1-LLB1	*Uppermost LLB*								
24766-01	690	±100	44.0	No plateaus					
24766-02	1580	±130	38.3	No plateaus					
24766-03	1400	±200	46.2	1170	±130	1.6	B-F	5/8	64.8
KK-1	*Lowermost LLB*								
22509-01	5.88	±0.14	48.5	No plateaus					
22509B-01	10.7	±0.4	36.0	No plateaus					

10-3 was sampled at UTM 36740741/9642757, and 1-7A at UTM 36747333/9647290. All other UTM's are listed in Table 4.2

[a] Incorporates error in *J*, the neutron fluence parameter

the ULB in the main Laetoli exposures has not been previously dated, although biotite from a crystal-lithic tuff at the ULB/LLB contact north of Locality 1 yielded 3.78 ± 0.06 Ma by the conventional K-Ar method (Drake and Curtis 1987), consistent with the new date on the uppermost ULB.

The age range of the LLB (4.4 – 3.8 Ma) partially overlaps with the known activity of the silica-undersaturated Satiman volcano, 24 km to the east. Mollel et al. (2011) identify a range of ~4.6 – 4.0 Ma for lava flows on this edifice. They postulate that Satiman is the probable source of the lower LLB, but note that Satiman lacks younger rocks that would correspond to the upper LLB and ULB deposits. If Satiman was the source of the younger Laetolil Beds, the record of this activity has been removed by subsequent erosion. Alternatively, the source is another vent now buried beneath the younger Lemagurut volcano, 6 km to the northwest, or a vent destroyed during development of the ~2 Ma Ngorongoro caldera, ~10 km to the northeast of Satiman.

Upper Laetolil Beds

The ULB contains nine named marker tuffs (Hay 1987), and, with the exception of Tuff 7, attempts were made to date all of them by the K-feldspar SCLF method. Tuffs 1–7 were also analyzed by incremental heating of biotite. Further, hornblende from Tuffs 1–7 were analyzed by single-crystal and multi-grain incremental heating techniques; of these, only Tuff 7A yielded useful geochronological information. The sample names, material dated, and tuff attribution are summarized in order of locality in Table 4.5.

ULB K-feldspar separates from Tuffs 1, 2, 4, 8, and Yellow Marker Tuff (YMT) yield distinct youngest age modes that in most cases can be confidently interpreted as eruptive ages. The youngest mode typically also dominates the age-probability density spectrum, with the exception of Tuff 4 (10 – 4), were a distinctly older xenocrystic mode at ~4.2 Ma is prominent (Figs. 4.5 and 4.6).

The primary youngest mode for Tuff 1, however, is significantly older (80 ka) at the 95% confidence level than the excellent K-feldspar result obtained for the stratigraphically lower tuff at the top of the LLB of 3.825 ± 0.017 Ma. The feldspar population in Tuff 1 may be entirely xenocrystic.

The best K-feldspar dating result from the ULB was obtained from Tuff 8 (16-tf8), situated ~1 m below the YMT at the top of the ULB. Tuff 8 exhibited a sharply resolved Gaussian-like primary age mode, yielding an inferred eruption age of 3.631 ± 0.018 Ma ($n = 13$).

The top of the ULB is defined as the top of the YMT. Two samples of the YMT, from Locs. 2 and 10E, gave statistically indistinguishable K-feldspar ages of 3.61 ± 0.03 Ma ($n = 11$) and 3.64 ± 0.02 Ma ($n = 14$), respectively, after exclusion of distinct submodes ~200 ka older than the youngest, dominant primary modes. The weighted mean of these samples, 3.627 ± 0.018 Ma, provides a firm chronostratigraphic tie point for the top of the ULB.

Figure 4.7 shows detailed sections for Locs. 1, 10, and 16 (modified from Hay 1987) illustrating the stratigraphic position and preferred ages of the K-feldspar geochronology for the ULB and upper LLB. Fig. 4.8 shows preferred tuff ages against stratigraphic height of Locality 1 for K-feldspar dating of the LLB, ULB and Lower Ndolanya Beds, discussed below.

Biotite multi- and single-grain $^{40}Ar/^{39}Ar$ incremental heating analyses were performed on many of the ULB tuffs. The multi-grain samples show a fair degree of internal precision and reproducibility, but show repeated age reversals when taken in stratigraphic context (Tables 4.2 and 4.3; Fig. 4.8). The biotite multi-grain plateau apparent ages scatter about the K-feldspar ages; the biotite age for Tuff 7A near the top (3.66 ± 0.02 Ma) is the same as an age for this tuff interpolated from the K-feldspar ages (see below), while the mean of the three mid-ULB tuffs (2, 3, 4) biotite ages (3.84 Ma) is about the same as the Tuff 2 and Tuff 4 K-feldspar ages (3.79, 3.81 Ma). Although the biotite results show considerable between-sample age variability, overall they scatter about the age trend defined by the K-feldspar data. Four tuffs from the ULB at Loc. 1 were also analyzed by single-crystal incremental heating experiments on biotite (Table 4.3). The plateau apparent-age results are scattered over a broad range from ~4.0 – 3.4 Ma, and fail to indicate an age progression with stratigraphic height. Nevertheless, the mean age of this group of 3.75 Ma is supportive of the eruptive age obtained from the feldspar data. Overall, the biotite results often show good internal reproducibility, but display considerable scatter and age reversals in stratigraphic context. Although the biotite ages must be considered potentially less geologically accurate than the feldspar ages, on aggregate the biotites give reasonable approximate ages for ULB strata.

Hornblende is a common phenocryst in many of the Laetoli Beds tuff units, and a concerted effort was undertaken

Table 4.5 Dating samples of the Upper Laetolil Beds (K-fsp = K-feldspar, hb = hornblende, bio = biotite, YMT = Yellow Marker Tuff)

Locality	Sample	Tuff	Mineral
1	1-tf3	3	K-fsp, hb
1	1-tf4	Between 4 & 5	K-fsp, bio, hb
1	1-tf5	5	K-fsp, bio, hb
1	1-tf6	6	K-fsp, bio, hb
1	1-7A	7A	Bio, hb
2	2-YMT	YMT	K-fsp
10	10-1	1	K-fsp, bio, hb
10	10-2	2	K-fsp, bio, hb
10	10-3	3	Bio, hb
10	10-4	4	K-fsp, bio, hb
10	10-5	5	Hb
10E	10E-Y1	YMT	K-fsp
16	16-tf8	8	K-fsp

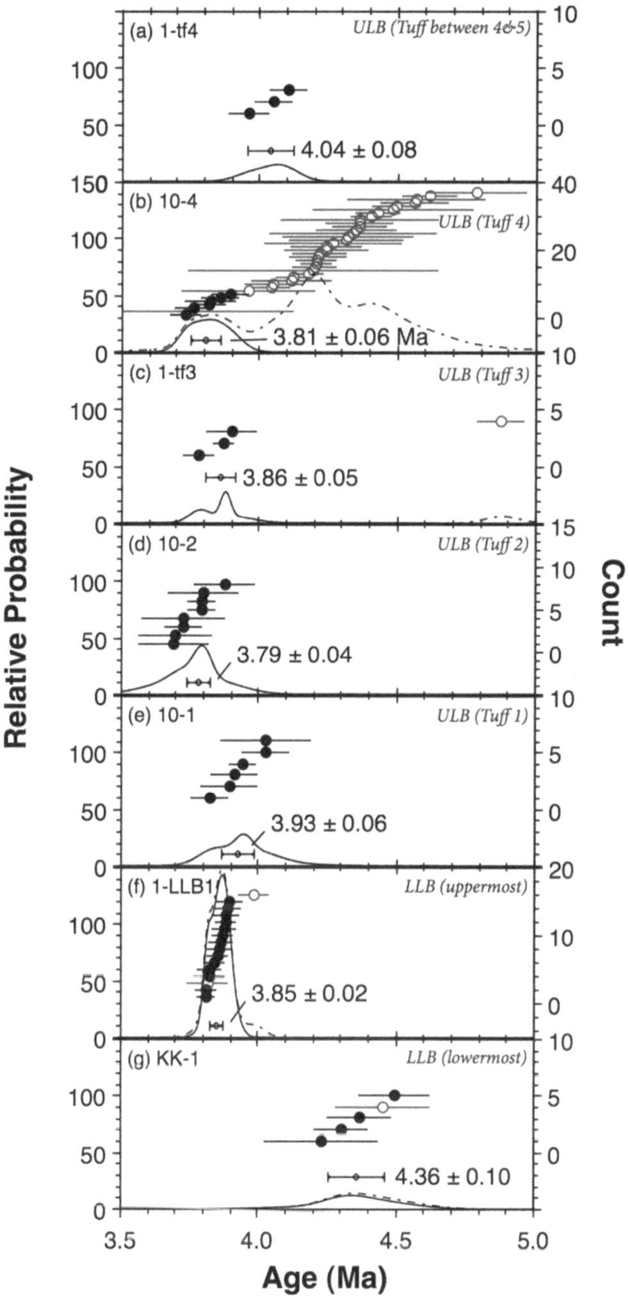

Fig. 4.5 Age-probability density spectra of K-feldspar $^{40}Ar/^{39}Ar$ dating results for the stratigraphically oldest seven samples (LLB up through a tuff between ULB Tuffs 4 and 5). Individual analyses with 1σ analytical uncertainty are shown by circles with horizontal error bars, ordered vertically by increasing age. Open circles represent excluded analyses. The weighted mean age and 2σ analytical uncertainty (including error in *J*, the neutron fluence parameter) are shown numerically (in Ma) and by the diamond symbol with error bars. Solid curves represent the age-probability spectrum for included analyses only; the dashed curve includes all analyses. Additional older analyses may be off-scale to the right and now shown. Unit names and stratigraphic level (by tuff designation or relative position) are shown to the upper right of each panel

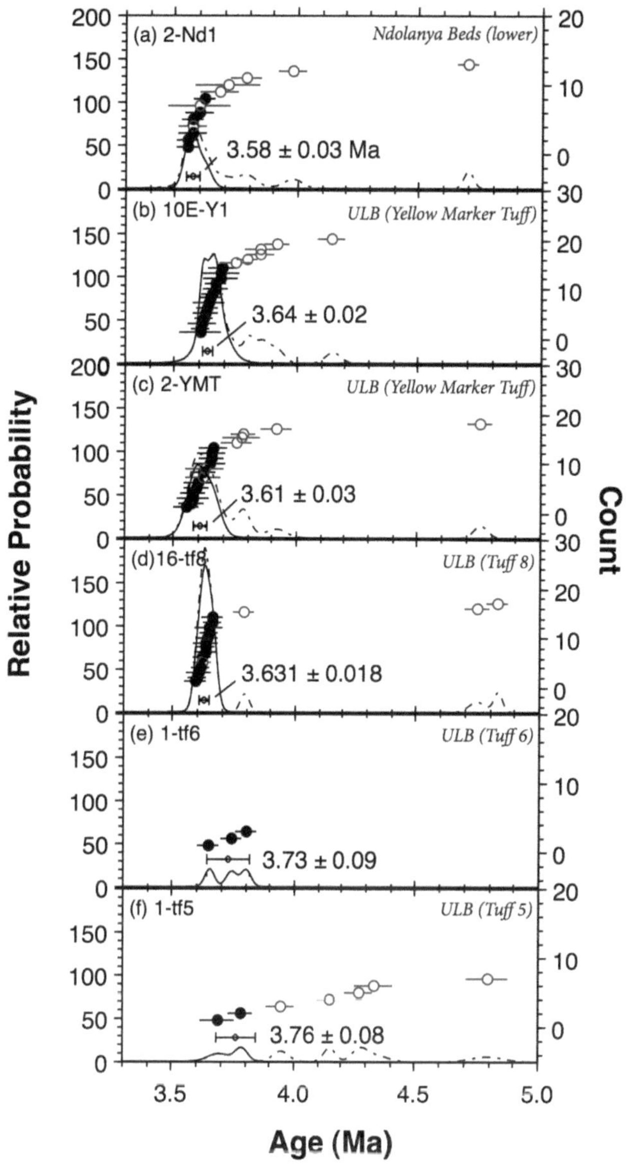

Fig. 4.6 Age-probability density spectra of K-feldspar $^{40}Ar/^{39}Ar$ dating results, ULB Tuff 5 through Lower Ndolanya Beds

to examine the potential for this low-K phase to supplement the K-feldspar and biotite geochronology (Table 4.4; Figs. 4.9 and 4.10). In one type of experiment, designed to examine the grain-to-grain heterogeneity of hornblende populations, a modified total-fusion technique (a low-power degas step followed by a total-fusion step) was applied to single crystals of hornblende from Tuff 7A. The resulting age-probability density spectra (Fig. 4.9) is characterized by a skewed distribution toward older ages, with a dominant mode at 4.6 Ma, a lesser mode at 11.2 Ma, and ages tailing up to 55 Ma with one extreme age at 150 Ma. Both the 4.6 and 11.2 Ma modes are much older than even the oldest LLB strata at Kakesio (~4.4 Ma). This type of near-continuous tailing is attributed to various degrees of excess ^{40}Ar contamination of the hornblendes. No valid geological age can be inferred from these data. Three analyses, however, separated from the others in the second age-probability spectrum shown in Fig. 4.9, have much higher ^{39}Ar yields, are at

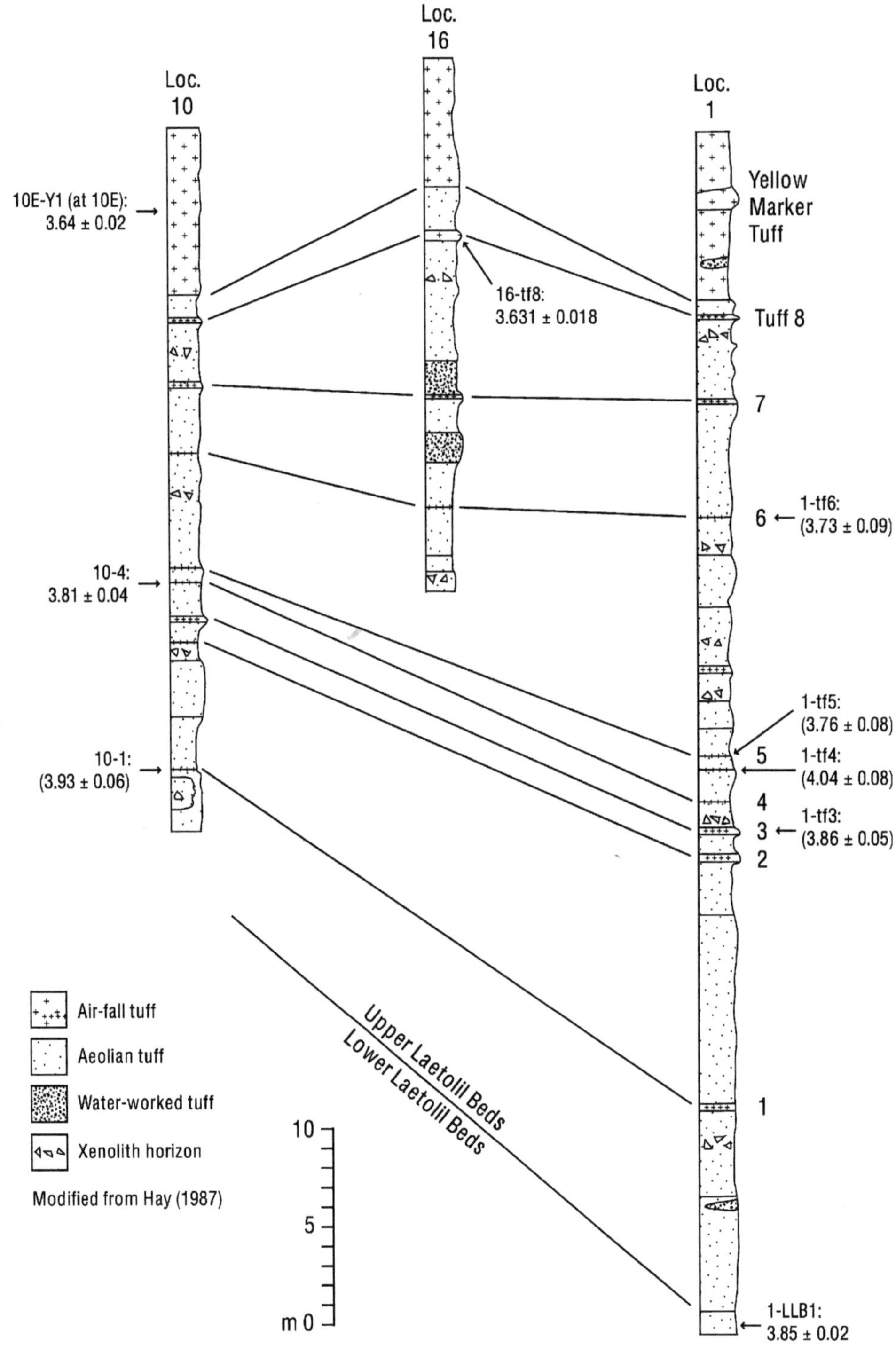

Fig. 4.7 Stratigraphic sections of the ULB, identifying location of $^{40}Ar/^{39}Ar$ dating samples and reference ages (in Ma, uncertainties are 2σ). Sections are modified from Hay (1987). Sample 2-YMT from Loc. 2, not shown, gave 3.61 ± 0.03 Ma

the high range of% $^{40}Ar^*$ yield of the hornblendes (~30–60%), have low Ca/K ratios (~30 compared to median value of 250), and much higher age precision (3–4%) compared to the median hornblende age uncertainty (20%). These crystals probably contain inclusions of biotite, a common phenocrystic phase in Tuff 7A. Two of these analyses are similar and give a geologically reasonable age of 3.7 ± 0.2 Ma (2σ).

Tuff 7A is also the only tuff that provided useful multi-grain incremental-heating age information from hornblende (Table 4.4). In particular, experiment number 22506–02 exhibits an apparent-age plateau of 3.55 ± 0.03

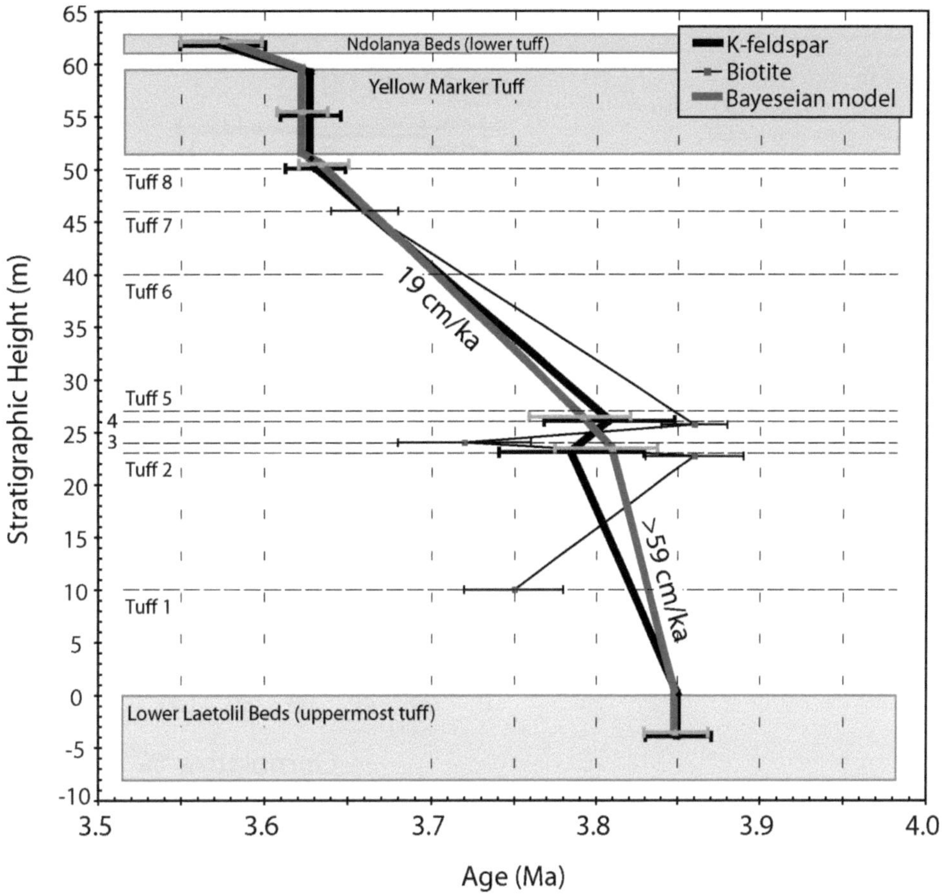

Fig. 4.8 Sedimentation history of the ULB and bracketing strata. The heavy black line connects K-feldspar SCLF control points, while the fine *black* line connects biotite multi-grain plateau age points. The heavy *grey* line represents the K-feldspar SCLF results smoothed via a Bayesian stacked age model. All control points are show with 2σ error bars

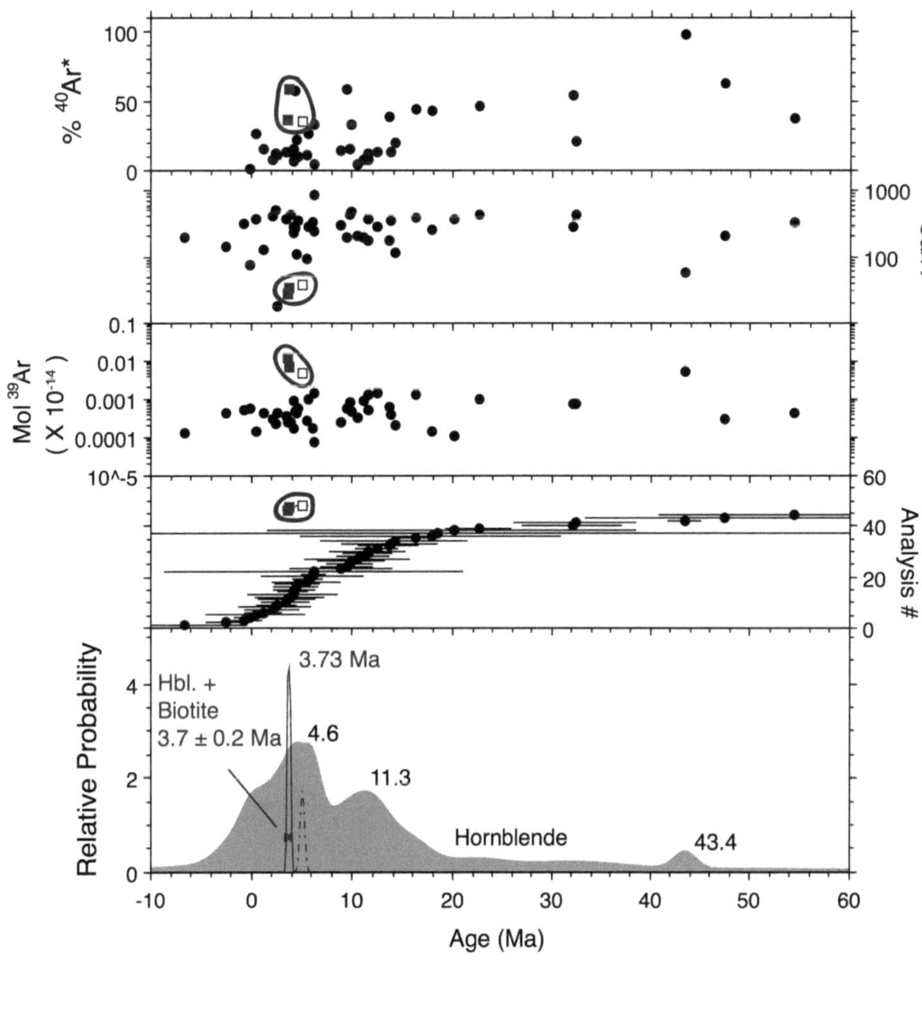

Fig. 4.9 Age-probability density spectra of modified SCLF dating of hornblende phenocrysts from Tuff 7A (sample 1–7A). Represented are the fusion steps of individual hornblende phenocrysts of a 'two-step' incremental heating experiment (a low-power degas step, followed by a fusion step). The density spectrum of 'pure' hornblende analyses is shown in *grey*, with anomalously old modes at 4.6, 11.3, and 43.4 Ma. The density spectrum of hornblende phenocrysts with inferred biotite inclusions is labeled 'Hbl. + Biotite' and shown as unfilled peaks. The circled data points in the auxiliary plots above the density spectra panel identify the three hornblende grains with possible biotite inclusions

Fig. 4.10 Two $^{40}Ar/^{39}Ar$ incremental heating experiments on multi-grain hornblende separates from Tuff 7A (sample 1-7A). (**a**) Aliquot 22506-02, illustrating results that are inferred to incorporate degassing of biotite inclusions. This aliquot yields reasonably geologically accurate plateau and integrated ages. (**b**) Aliquot 22519-01, with lesser or no influence of biotite inclusions, showing a more typical pattern of elevated initial and final apparent ages, forming a 'saddle-shaped' spectrum believed indicative of excess argon. Both the plateau and integrated apparent ages for this aliquot are much older than geochronological constraints permit

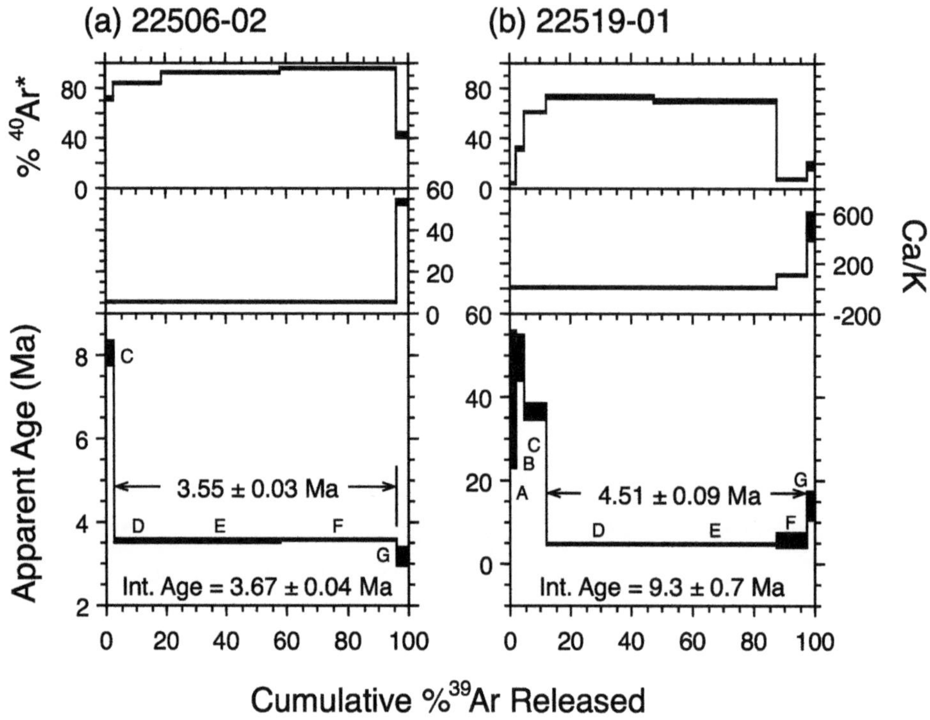

Ma and an integrated age of 3.67 ± 0.04 Ma (2σ) (Fig. 4.10). The plateau age is slightly out of sequence with respect to the K-feldspar age on overlying Tuff 8 (3.631 ± 0.018 Ma); nevertheless, the result suggests that hornblende multi-grain incremental heating can occasionally yield useful age information. However, as with the single-grain hornblende dating of this tuff, biotite inclusions are likely responsible for this good result. It is apparent on the basis of the relatively low Ca/K ratio (5.5–5.7) of the plateau steps that entrapped biotite in the hornblende dominates the ^{39}Ar and $^{40}Ar^*$ budget of this part of the experiment, as compared, for example, to another Tuff 7 hornblende incremental experiment shown in Fig. 4.10 (plateau Ca/K = 6–101), and to the median Ca/K ratio (250) obtained from the single-grain hornblende fusions. Thus, the dated multi-grain separates are not pure hornblende, though potentially the biotite inclusions are more pristine than their counterparts found as individual phenocrysts in the tuff matrix.

Previous K-Ar dates for the ULB were reported in Drake and Curtis (1987). K-Ar ages on biotite yielded 3.78 ± 0.6 Ma for the base of the ULB, 3.58 ± 0.04 Ma for the xenolithic horizon between Tuffs 7 and 8, and 3.48 ± 0.24 Ma for Tuff 8. These biotite ages are younger than the corresponding $^{40}Ar/^{39}Ar$ K-feldspar ages (measured or interpolated) reported herein by ~50–150 ka. Manega (1993) attempted to date K-feldspars from the upper part of the ULB (Tuff 6?) by the SCLF $^{40}Ar/^{39}Ar$ method, but obtained an old age from a few grains that he discounted as contamination (3.83 ± 0.06 Ma).

Ndolanya Beds

Two tuff samples from the Ndolanya Beds were found suitable for SCLF K-feldspar $^{40}Ar/^{39}Ar$ dating, one each from the upper and lower units. The lower tuff (sample 2-Nd1), similar in appearance to the YMT, was sampled at Loc. 2 where it lies above a meter-thick irregular limestone defining the base of the Ndolanya Beds. The limestone, in turn, lies about 1 m above the YMT (dated sample 2-YMT-1). The yellow, compact, devitrified lower Ndolanya tuff contains rare pumice to 1–2 cm and rare phenocrysts of K-feldspar, plagioclase and biotite. This tuff yielded a clearly defined young primary mode, skewed toward older ages (Fig. 4.6). Several distinctly older contaminants were omitted from the calculation of a mean age, yielding 3.58 ± 0.02 Ma (n = 6, MSWD = 1.2), only ~60 ka younger than the underlying YMT.

A sample of light gray, massive, sparsely pumiceous (up to 2–3 mm) vitric tuff (10–20 cm thick) ~1.5 m above the exposed base of the Upper Ndolanya Beds at Loc. 18 also exhibited a distinct youngest mode with a minor tail toward older ages. Overall, this population yields a weighted-mean age of 2.660 ± 0.014 Ma (n = 12, MSWD = 0.9, sample 18-Nd2; Fig. 4.11). This result is much younger than ages obtained for the Lower Ndolanya Beds, confirming the presence of a major unconformity between the Upper and Lower Ndolanya Beds (Hay 1987).

Previous attempts to date the Ndolanya Beds yielded 2.69 ± 0.046 Ma (2σ) by SCLF $^{40}Ar/^{39}Ar$ dating of a pumice tuff in the upper unit at Loc. 14 (Ndessokia 1990). This result is statistically indistinguishable from the age obtained here for possibly

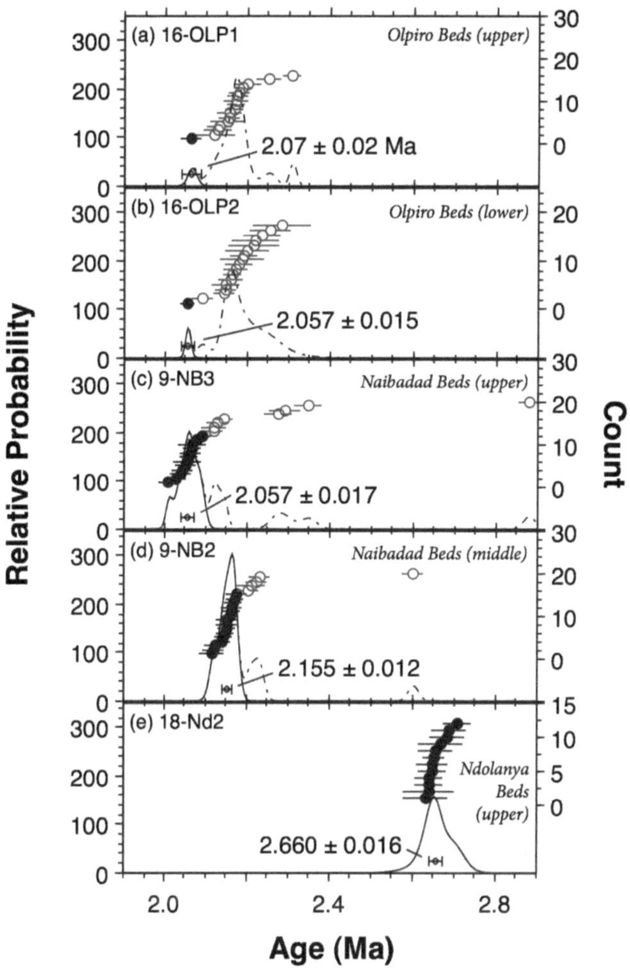

Fig. 4.11 Age-probability density spectra of SCLF K-feldspar $^{40}Ar/^{39}Ar$ dating results, Upper Ndolanya Beds through Olpiro Beds

the same pumice tuff. These dates are likewise consistent with a K-Ar date on the overlying Ogol Lava of 2.42 ± 0.024 Ma (Drake and Curtis 1987), and newer $^{40}Ar/^{39}Ar$ ages of 2.32 ± 0.02–2.28 ± 0.02 Ma on the Ogol Lava (Mollel et al. 2011).

The source of the Lower Ndolanya Beds is probably the same as the Laetolil Beds, based on close age relationships and lithologic similarities. These pyroclastic eruptions are believed to have originated, at least in part, from Satiman volcano, or another now buried or destroyed vent (Mollel et al. 2011). The upper Ndolanya Beds, however, were deposited during a gap in the known record of volcanic activity within the Ngorongoro Volcanic Highlands lasting from 4.0 – 2.4 Ma (Mollel et al. 2011). Lemagurut volcano is the closest in age to the Upper Ndolanya Beds, with the earliest known lavas from this edifice yielding dates of 2.4 Ma. It is possible the dated tuff in the Upper Ndolanya Beds (18-Nd2) has a more distal origin; silicic tuffs of similar age are found in Kenya in the Chemeron Formation in the Tugen Hills near Lake Baringo (three tuffs 2.61–2.60 Ma and an older tuff at 2.74 Ma; Deino et al. 2006), and in the Omo Group in the Turkana Basin (the Burgi Tuff, 2.63 Ma; McDougall and Brown 2008).

Naibadad Beds

Two tuffs from the Naibadad Beds were dated by the K-feldspar SCLF technique. Sample 9-Nb2 was obtained from brown-red 0.5 m lapillistone horizon ~7.5 m above a calcrete at the top of the Ndolanya Beds at Loc. 9, bearing monolithologic rounded, reworked trachytic pumice to 2 cm containing 3–4 mm K-feldspar phenocrysts (corresponds to level of 9.02 in the stratigraphic section of Fig. 4.3 in McHenry 2011). Sample 9-Nb3 was obtained from a 10–20 cm pumice tuff ~3.5 m higher up in the section, containing ~20–30% stretched, glassy, crystal-poor pumice generally less than 1 cm in size (corresponding to level 9.03 or 9.04 of Fig. 6.3 in McHenry 2011). While dating results for both samples show dominant, youngest primary modes, each has a secondary mode about 60 ka older than the primary mode (Fig. 4.11). After eliminating these distinct older components, the resulting weighted mean ages are 2.155 ± 0.009 Ma (9-Nb2), and 2.057 ± 0.016 Ma (9-Nb3) for the lower and upper tuffs, respectively. All K-feldspars in the primary modes exhibit Ca/K between 0.1–0.01, corresponding to the "dominant higher-K population" identified in the feldspar phenocryst populations of the Naibadad tuffs by McHenry (2011).

The Naibadad Beds was the most intensively dated stratigraphic unit prior to this study. Drake and Curtis (1987) report an average K-Ar age of 2.27 ± 0.12 Ma for anorthoclase from 'tuffaceous samples' of the Naibadad Beds. They noted that several other K-Ar ages were anomalously old, and suspected detrital contamination. Ndessokia (1990) dated a tuff in the Naibadad Beds by the SCLF $^{40}Ar/^{39}Ar$ method and obtained 2.18 ± 0.044 Ma from an isolated pumice tuff at Loc. 14 immediately above the Ndolanya Beds. Both these results are indistinguishable from the new age of 2.155 ± 0.009 Ma obtained from a tuff from the middle of the sequence at Loc. 9 (9-Nb2). Manega (1993) dated three tuffs in the Naibadad Beds by the SCLF $^{40}Ar/^{39}Ar$ method, and obtained ages, from bottom to top, of 2.29 ± 0.06, 2.16 ± 0.04, and 2.12 ± 0.02 Ma. While Manega's (1993) age for the middle tuff is in close agreement with that for 9-Nb2, our results for the upper part of the Naibadad Beds differ significantly. It is possible that the new $^{40}Ar/^{39}Ar$ dating results reported herein were derived from a tuff higher in the section than that previously dated by Manega (1993). The distinction is significant, as the younger Naibadad Beds tuff provides an important chronostratigraphic constraint for the age of the overlying Olpiro Beds, as discussed below. The ages cited above for the Naibadad Beds are in general agreement with the constraints imposed by previous K-Ar (2.42 ± 0.024 Ma; Drake and Curtis 1987) and $^{40}Ar/^{39}Ar$ dating (2.32 ± 0.02 – 2.28 ± 0.02 Ma; Mollel et al. 2011) of the underlying Ogol Lava.

Mollel et al. (2011) presented new $^{40}Ar/^{39}Ar$ ages for Lemagurut volcano of 2.4 – 2.2 Ma. While this range is slightly younger than the two dates obtained here for the Naibadad Beds, Lemagurut remains chronologically a potential source.

However, the Naibadad Beds tuffs contain phenocrystic quartz, which are not present in Lemagurut lavas (Mollel et al. 2011). New $^{40}Ar/^{39}Ar$ ages on Ngorongoro lavas (~2.3–2.0 Ma; Mollel, in preparation) and similar composition (Hay 1987) suggest that this large silicic center is a more likely source for the Naibadad Beds.

The age obtained in this study for the upper Naibadad Beds tuff (2.057 ± 0.016 Ma) is indistinguishable from that of the Coarse Feldspar Crystal Tuff ('CFCT'; 2.05 ± 0.04 Ma; Walter et al. 1992) and the Naabi Ignimbrite at Olduvai Gorge (2.06 ± 0.03 Ma). However, there are sufficient mineralogical differences between these units to indicate that a direct correlation is unlikely (McHenry 2011).

Olpiro Beds

Lastly, two tuffs of the Olpiro Beds were dated by the SCLF technique on K-feldspar, where this unit unconformably overlies the Laetoli Beds at Loc.16. Here the Olpiro Beds consist mainly of laminated mudstones enclosing laterally persistent exposures of the 30–50 cm thick 'Olpiro Marker Tuff' (sample 16-Olp1), and a white, fine-grained tuff (15 cm thick; sample 16-Olp2) about 1 m below the marker tuff. The marker tuff is a reworked laminated to crudely bedded, well-sorted, crystal to crystal-vitric tuff with K-feldspar phenocrysts to 2 mm. The base of the tuff is almost entirely free of glass, but toward the top ~10–20% shard pseudomorphs attest to the volcanic origin of this unit.

The age-probability density spectra of SCLF K-feldspar results for the two Olpiro Beds tuffs are similar (Fig. 4.11). They are each characterized by a dominant mode at ~2.16 Ma with slight skewness toward older ages, and one or two high-quality analyses at ~2.06–2.09 Ma. This dominant mode may be xenocrystic, as it is ~100 ka older that the new $^{40}Ar/^{39}Ar$ age for the upper Naibadad Beds (sample 9-Nb3), but about the same age as the middle Naibadad Beds (sample 9-Nb2). The youngest grain in each of the two Olpiro Beds tuff samples may actually hint at the true primary feldspar population. These sanidine crystals have a Ca/K = 0.003 are distinct from all other Olpiro and Naibadad tuff K-feldspars analyzed (Ca/K = 0.01–0.26), but are consistent in age (2.06 ± 0.03 Ma) and similar in Ca/K ratio (0.001) to the Naabi Ignimbrite exposed at nearby Olduvai Gorge (Walter et al. 1991, 1992). Although additional work is clearly needed on the detailed $^{40}Ar/^{39}Ar$ age systematics of the Olpiro Beds, at this stage it can be concluded that the unit is probably younger than ~2.06 Ma (the age of the upper Naibadad Bed tuff and of the younger grains in the Olpiro tuff samples), and is probably older than ~2.0 Ma, because no feldspars younger than this were encountered in the current study, despite extensive volcanism throughout the 2.1 – 1.7 Ma interval recorded at Olduvai Gorge.

Previous dating efforts directed toward the Olpiro Beds includes Ndessokia (1990) on the Olpiro Marker Tuff at Loc. 16 of 2.17 ± 0.036 Ma, and Manega (1993), who dated five samples of the tuff from three localities. Manega concluded that many feldspars encountered in his analyses were older contaminants, and that the best age was derived from the youngest feldspars, with an age of 2.02 ± 0.02 Ma, within the age range inferred above.

Eruptive activity at both Ngorongoro (~2.4 – 2.0 Ma) and Olmoti Crater (~2.0 – 1.8 Ma; 54 km northeast of Laetoli) are contemporaneous with the inferred depositional age of the Olpiro Beds. The phonolitic composition of the tuffs in the Olpiro Beds may be a better match to Olmoti lavas (Mollel et al. 2011).

Sedimentation History of the Upper Laetolil Beds

Sufficient age and stratigraphic information is available to examine the sedimentation history of the ULB portion of the Laetoli volcaniclastic sequence. Using published information for Loc. 1 (Hay 1987), where the section is complete and the ULB is relatively thick (Fig. 4.7), the age estimates for dated ULB tuffs and immediately adjacent tuff units (top of the LLB, and Lower Ndolanya Beds) are plotted against stratigraphic height in Fig. 4.8. The K-feldspar tuff ages generally conform to the constraints of stratigraphic superposition within error. Although there is a slight reversal in absolute age between Tuffs 2 and 4, the errors broadly overlap and the means are indistinguishable at the 95% confidence level.

The age progression can be further interpreted using a Bayesian approach, employing prior knowledge of the stratigraphic succession to refine the tuff ages and their uncertainties (Fig. 4.8; Bernardo and Smith 1994; Ludwig 2003). Bayesian estimates of ages have the property that they will always fall in the correct stratigraphic order even if the measured ages are identical or in reversed order. Using the smoothed Bayesian age progression, it is straightforward to calculate sedimentation rates for the ULB sequence. Accumulation of strata was considerably more rapid in the bottom half of the ULB (>59 cm/ka; base of ULB to Tuff 2) than the upper half (19 cm/ka; Tuff 2 through YMT). Overall, the accumulation interval of the ULB is greater than 160 ka (interval between the YMT and the Bayesian age for Tuff 2), but probably closer to 230 ka (interval of YMT to the uppermost LLB tuff). The latter result corresponds to an overall sedimentation rate of >26 cm/ka. For comparison, this is much faster than the average Bed I sedimentation rate between ~1.85 – 1.78 Ma at Olduvai Gorge for the central part of the basin of ~6.0 cm/ka, but about equal to that of the eastern pyroclastic fan facies of ~24 cm/ka (McHenry 2005;

Deino, in preparation). This part of the Olduvai basin is not only closer to volcanic source areas in the Ngorongoro Volcanic Highlands, but strata may be thickening eastward inside a half-graben structure, with fault-controlled subsidence along the eastern margin of the basin (Hay 1976). In contrast, the Laetolil Beds were deposited on a peneplain on the drainage divide between the Eyasi and Olduvai basins (Hay 1987), so that accumulation rates may have been dominated by the frequency and volume of material supplied by repeated eruptive activity of the Ngorongoro Volcanic Highlands beginning 10–15 km to the east.

It is also more straightforward, using the Bayesian age model, to interpolate the ages of undated units in the ULB, assuming linear sedimentation rates between chronostratigraphic tie points (Tables 4.5 and 4.6). Errors for the directly dated tuffs are calculated as part of the Bayesian analysis, while errors in the interpolated ages must be at least as large as the errors of the dated tuffs (~1%). Tuff 7, the 'Footprint Tuff,' is an especially noteworthy horizon of the ULB. In Loc. 1, this tuff lies only 4 m below dated Tuff 8, and has an interpolated age of 3.66 Ma.

Conclusions

New $^{40}Ar/^{39}Ar$ ages on K-feldspar, biotite, and hornblende have been acquired from most of the Pliocene stratigraphic units of the Laetoli area. K-feldspar (anorthoclase and sanidine) provides the most definitive chronostratigraphy. Dated units include the Lower and Upper Laetolil Beds (4.36–3.85 Ma and 3.85–3.63 Ma, respectively), the Lower and Upper Ndolanya Beds (3.58 and 2.66 Ma, respectively), the Naibadad Beds (2.155–2.057 Ma), and the Olpiro Beds (<2.057 Ma). Biotite ages show much greater dispersion than those for K-feldspar, but on average give nearly the same result. Hornblende yielded little useful geochronological information except in rare cases where biotite inclusions were likely present. The best of the geochronological results for the different phenocrystic mineral phases are summarized in Table 4.6.

Using a smoothed Bayesian age progression through the ULB, sedimentation rates of ~50 cm/ka are obtained for the lower half of the sequence through Tuff 2, compared to a rate of ~15 cm/ka for the upper half (Tuff 4 to Tuff 8).

Table 4.6 Preferred $^{40}Ar/^{39}Ar$ ages for K-feldspar, biotite, and hornblende phenocrysts from the Laetoli sequence. Feldspar results based on three or fewer analyses are not shown. K-feldspar ages in parenthesis denote potentially lesser geological reliability. The best radiometric age estimate for a given tuff, all from SCLF K-feldspar experiments, is shown in bold. Biotite ages are the weighted mean of single-crystal and multi-grain means where both dating approaches were used on the same tuff. Also shown are Bayesian Monte Carlo age estimates for the older part of the sequence, employing known stratigraphic superposition as the prior-knowledge constraint. Ages listed in the Bayesian column in italics are interpolation estimates assuming linear sedimentation rates between Bayesian-modeled K-feldspar tuff ages

Unit dated	Bayesian Age (Ma) [+/– 95% Conf. Int.]		K Feldspar Age (Ma) ± 2σ MSE		Biotite Age (Ma) ± 2σ MSE		Hornblende Age (Ma) ± 2σ MSE	
Olpiro Beds								
Olpiro Marker Tuff			(2.070	±0.020)				
Tuff below marker tuff			(2.057	±0.015)				
Naibadad Beds								
Upper			**2.057**	**±0.017**				
Middle			**2.155**	**±0.012**				
Upper Ndolanya Beds			**2.660**	**±0.016**				
Lower Ndolanya Beds	3.58	[+0.023, –0.026]	**3.58**	**±0.03**				
Upper Laetolil Beds								
YMT	3.623	[+0.015, –0.015]	**3.627**	**±0.018**				
Tuff 8	3.634	[+0.017, –0.013]	**3.631**	**±0.018**				
Tuff 7A	*3.66*				3.66	±0.02	3.55	±0.03
Tuff 6	*3.70*				3.78	±0.05		
Tuff 5	*3.79*				3.62	±0.19		
Tuff between 4 & 5	*3.79*				3.79	±0.11		
Tuff 4	3.79	[+0.029, –0.033]	**3.81**	**±0.04**	3.86	±0.02		
Tuff 3	*3.80*				3.72	±0.04		
Tuff 2	3.81	[+0.028, –0.035]	**3.79**	**±0.04**	3.86	±0.03		
Tuff 1	*3.83*		(3.93	±0.06)	3.75	±0.02		
Lower Laetolil Beds								
Uppermost part	3.85	[+0.021, –0.017]	**3.85**	**±0.02**	4.1	±0.2		
Lowermost part			**4.36**	**±0.10**	4.30	±0.04		

Accumulation of the entire ULB required less than 220 ka from emplacement of the uppermost aeolian tuff of the LLB, through the ULB inclusive of the Yellow Marker Tuff, corresponding to a minimum sedimentation rate of 22 cm/ka.

The new dates provide a more refined assessment of the age of the fossil hominins at Laetoli. *Australopithecus afarensis* specimens have been recovered from horizons throughout the ULB, ranging from between Tuffs 1 and 2 (L.H. 11) to 30 cm above Tuff 8 (EP 2400/00), and are constrained to 3.85–3.63 Ma (Leakey 1987a; Harrison 2011). A new interpolated age for the Footprint Tuff (Tuff 7) is 3.66 Ma. The *Paranthropus aethiopicus* maxilla (from Silal Artum) and the proximal tibia (from Loc. 22 S) from the Upper Ndolanya Beds are dated to 2.66 Ma, which makes them among the oldest known specimens attributable to *Paranthropus* (Harrison 2011).

The ULB documents an otherwise sparsely represented interval in the fossil, and especially hominid, record of East Africa between ca. 3.8 – 3.6 Ma. The only other well-dated hominid-bearing strata of this age occur at the opposite geographic extreme of the East African Rift system, in the Woranso-Mille area in the west-central Afar region of Ethiopia (Haile-Selassie et al. 2007; Deino et al. 2010). Dikika in the lower Awash River Valley of Ethiopia contains roughly contemporaneous hominid-bearing strata but these have not yet been precisely dated (Alemseged et al. 2005; Wynn et al. 2008). Somewhat older sites are known in Baringo, Kenya (6.6–3.8 Ma; Deino et al. 2002), Kanapoi, Kenya (4.2–4.1 Ma; Leakey et al. 1995), Allia Bay, Kenya (3.9 Ma; Leakey et al. 1995), and Asa Issie, Middle Awash, Ethiopia (4.1–3.8 Ma; White et al. 2006), while slightly younger sites are found at Hadar, Ethiopia (3.4–2.9 Ma; Walter and Aronson 1993; Alemseged et al. 2006) and also at Baringo (3.0–1.6 Ma, and older; Deino and Hill 2002; Deino et al. 2006; Kingston et al. 2007; and work in progress).

Acknowledgments I would like to thank the Tanzanian Commission for Science and Technology and the Department of Antiquities in Dar es Salaam for permission to conduct research in Tanzania. Lindsay McHenry and Peter Ditchfield provided valuable insights into Laetoli stratigraphy in the field, and during preparation of this manuscript. Operation of the $^{40}Ar/^{39}Ar$ laboratory at BGC at the time the analyses were performed was partially supported through NSF grant BCS-0211172. Additional support was provided by NSF grant BCS-0309513 (to Terry Harrison). The manuscript benefited from constructive comments by John Kingston and an anonymous reviewer.

References

Alemseged, Z., Wynn, J. G., Kimbel, W. H., Reed, D., Geraads, D., & Bobe, R. (2005). A new hominin from the basal members of the Hadar Formation, Dikika, Ethiopia, and its geological context. *Journal of Human Evolution, 49*, 499–514.

Alemseged, Z., Spoor, F., Kimbel, W. H., Bobe, R., Geraads, D., Reed, D., & Wynn, J. G. (2006). A juvenile early hominin skeleton from Dikika, Ethiopia. *Nature, 443*, 296–301.

Bernardo, J. M., & Smith, A. F. (1994). *Bayesian theory*. New York: Wiley.

Deino, A. L., $^{40}Ar/^{39}Ar$ dating of Bed I, Olduvai Gorge, Tanzania. *Journal of Human Evolution* (in prep.)

Deino, A. L., & Potts, R. (1990). Single-crystal $^{40}Ar/^{39}Ar$ dating of the Olorgesailie Formation, southern Kenya Rift. *Journal of Geophysical Research, B, Solid Earth and Planets, 95*, 8453–8470.

Deino, A. L., & Hill, A. (2002). $^{40}Ar/^{39}Ar$ the Chemeron Formation strata encompassing the site of hominid KNM-BC 1, Tugen Hills, Kenya. *Journal of Human Evolution, 42*, 141–151.

Deino, A. L., Renne, P. R., & Swisher, C. C. (1998). $^{40}Ar/^{39}Ar$ dating in paleoanthropology and archaeology. *Evolutionary Anthropology, 6*, 63–75.

Deino, A. L., Tauxe, L., Monaghan, M., & Hill, A. (2002). $^{40}Ar/^{39}Ar$ geochronology and paleomagnetic stratigraphy of the Lukeino and Lower Chemeron Formations at Tabarin and Kapcheberek. Tugen Hills, Kenya. *Journal of Human Evolution, 42*, 117–140.

Deino, A. L., Kingston, J., Glen, J. M., Edgar, R. K., & Hill, A. (2006). Precessional forcing of lacustrine sedimentation in the late Cenozoic Chemeron basin, Central Kenya Rift, and calibration of the Gauss/Matuyama boundary. *Earth and Planetary Science Letters, 247*, 41–60.

Deino, A. L., Scott, G. R., Saylor, B., Alene, M., Angelini, J. D., & Haile-Selassie, Y. (2010). $^{40}Ar/^{39}Ar$ dating, paleomagnetism, and tephrochemistry of Pliocene strata of the hominid-bearing Woranso-Mille area, West-Central Afar Rift, Ethiopia. *Journal of Human Evolution, 58*, 111–126.

Drake, R., & Curtis, G. H. (1987). K-Ar geochronology of the Laetoli fossil localities. In M. D. Leakey & J. M. Harris (Eds.), *Laetoli: A Pliocene site in northern Tanzania* (pp. 48–52). Oxford: Clarendon.

Haile-Selassie, Y., Deino, A., Saylor, B., Umer, M., & Latimer, B. (2007). Preliminary geology and paleontology of new hominid-bearing Pliocene localities in the central Afar region of Ethiopia. *Anthropological Science, 115*, 215–222.

Harrison, T. (2011). Hominins from the Upper Laetolil and Upper Ndolanya Beds, Laetoli. In T. Harrison (Ed.), *Paleontology and geology of Laetoli: Human evolution in context* (Fossil hominins and the associated fauna, vol. 2, pp. 141–188). Dordrecht: Springer.

Hay, R. L. (1976). *Geology of the Olduvai Gorge*. Berkeley: University of California Press.

Hay, R. L. (1987). Geology of the Laetoli area. In M. D. Leakey & J. M. Harris (Eds.), *Laetoli: A Pliocene site in northern Tanzania* (pp. 23–47). Oxford: Clarendon.

Hay, R. L., & Leakey, M. D. (1982). The fossil footprints of Laetoli. *Scientific American, 246*, 50–57.

Kingston, J. D., Deino, A., Hill, A., & Edgar, R. (2007). Astronomically forced climate change in the Kenyan Rift Valley 2.7–2.55 Ma: Implications for the evolution of early hominin ecosystems. *Journal of Human Evolution, 53*, 487–503.

Kuiper, K. F., Deino, A., Hilgen, F. J., Krijgsman, W., Renne, P. R., & Wijbrans, J. R. (2008). Synchronizing rock clocks of Earth history. *Science, 320*, 500–504.

Leakey, M. D. (1987a). The Laetoli hominid remains. In M. D. Leakey & J. M. Harris (Eds.), *Laetoli: A Pliocene site in northern Tanzania* (pp. 108–117). Oxford: Clarendon.

Leakey, M. D. (1987b). The hominid footprints. Introduction. In M. D. Leakey & J. M. Harris (Eds.), *Laetoli: A Pliocene site in northern Tanzania* (pp. 490–496). Oxford: Clarendon.

Leakey, M. D., Hay, R. L., Curtis, C. H., Drake, R. E., Jackes, M. K., & White, T. D. (1976). Fossil hominids from the Laetoli Beds, Tanzania. *Nature, 262*, 461–466.

Leakey, M. D., Feibel, C. S., McDougall, I., & Walker, A. (1995). New four million-year-old hominid species from Kanapoi and Allia Bay, Kenya. *Nature, 376*, 565–571.

Ludwig, K. R. (2003). User's Manual for Isoplot 3.0: A Geochronological Toolkit for Microsoft Excel. *Berkeley Geochronology Center Special Publication No. 4*, Berkeley.

Manega, P. C. (1993). Geochronology, geochemistry and isotopic study of the Plio-Pleistocene hominid sites and the Ngorongoro volcanic highland in northern Tanzania. Ph.D. dissertation, University of Colorado, Boulder.

McDougall, I., & Brown, F. H. (2008). Geochronology of the pre-KBS Tuff, Omo Group, Turkana Basin. *Journal of the Geological Society of London, 165*, 549–562.

McHenry, L. J. (2005). Phenocryst composition as a tool for correlating fresh and altered tephra, Bed I, Olduvai Gorge, Tanzania. *Stratigraphy, 2*, 101–115.

McHenry, L. J. (2011). Geochemistry and mineralogy of Laetoli area tuffs: Lower Laetolil through Naibadad Beds. In T. Harrison (Ed.), *Paleontology and geology of Laetoli: Human evolution in context* (Geology, geochronology, paleoecology and paleoenvironment, vol. 1, pp. 121–141). Dordrecht: Springer.

Mollel, G. F., Swisher, C. C., Feigenson, M. D., & Carr, M. J. (2011). Petrology, geochemistry and age of Satiman, Lemagurut and Oldeani: sources of the volcanic deposits of the Laetoli area. In T. Harrison (Ed.), *Paleontology and geology of Laetoli: Human evolution in context* (Geology, geochronology, paleoecology and paleoenvironment, vol. 1, pp. 99–119). Dordrecht: Springer.

Ndessokia, P. N. S. (1990). The mammalian fauna and archaeology of the Ndolanya and Olpiro Beds, Laetoli. Tanzania. Ph.D. dissertation, University of California, Berkeley.

Robbins, L. M. (1987). Hominid footprints from Site G. In M. D. Leakey & J. M. Harris (Eds.), *Laetoli: A Pliocene site in northern Tanzania* (pp. 497–502). Oxford: Clarendon.

Steiger, R. H., & Jäger, E. (1977). Subcommission on Geochronology: Conventions on the use of decay constants in geo- and cosmochronology. *Earth and Planetary Science Letters, 36*, 359–362.

Tuttle, R. H. (1987). Kinesiological inferences and evolutionary implications from Laetoli bipedal trails G-1, G-2/3, and A. In M. D. Leakey & J. M. Harris (Eds.), *Laetoli: A Pliocene site in northern Tanzania* (pp. 503–523). Oxford: Clarendon.

Walter, R. C., & Aronson, J. L. (1993). Age and source of the Sidi Hakoma Tuff, Hadar Formation, Ethiopia. *Journal of Human Evolution, 25*, 229–240.

Walter, R. C., Manega, P. C., Hay, R. L., Drake, R. E., & Curtis, C. H. (1991). Laser-fusion $^{40}Ar/^{39}Ar$ dating of Bed I, Olduvai Gorge, Tanzania. *Nature, 354*, 145–149.

Walter, R. C., Manega, P. C., & Hay, R. L. (1992). Tephrochronology of Bed I, Olduvai Gorge; an application of laser-fusion $^{40}Ar/^{39}Ar$ dating to calibrating biological and climatic change. *Quaternary International, 13–14*, 37–46.

White, T. D. (1977). New fossil hominids from Laetoli, Tanzania. *American Journal of Physical Anthropology, 46*, 197–230.

White, T. D. (1980). Additional fossil hominids from Laetoli, Tanzania: 1976–1979 specimens. *American Journal of Physical Anthropology, 53*, 487–504.

White, T. D., WoldeGabriel, G., Asfaw, B., Ambrose, S., Beyene, Y., Bernor, R., Boisserie, J. R., Currie, B., Gilbert, H., Haile-Selassie, Y., Hart, W. K., Hlusko, L. J., Howell, F. C., Kono, R. T., Lehmann, T., Louchart, A., Lovejoy, C. O., Renne, P. R., Saegusa, H., Verba, E. S., Wesselman, H., & Suwa, G. (2006). Asa Issie, Aramis and the origin of *Australopithecus*. *Nature, 440*, 883–889.

Wynn, J. G., Roman, D. C., Alemseged, Z., Reed, D., Geraads, D., & Munro, S. (2008). Stratigraphy, depositional environments, and basin structure of the Hadar and Busidima Formations at Dikika, Ethiopia. In J. Quade & J. G. Wynn (Eds.), *The geology of early humans in the horn of Africa* (pp. 1–32). Geological Society of America Special Paper 446.

Chapter 5
Petrology, Geochemistry and Age of Satiman, Lemagurut and Oldeani: Sources of the Volcanic Deposits of the Laetoli Area

Godwin F. Mollel, Carl C. Swisher III, Mark D. Feigenson, and Michael J. Carr

Abstract We report on the petrochemistry and geochronology of lavas from Satiman, Lemagurut and Oldeani volcanoes, as well as from isolated Ogol flows. Comparisons of their age and mineral compositions are made with those of the Laetolil Beds to confirm the sources for these latter volcaniclastic deposits. A comparison with other volcanic centers in the region and an attempt to constrain the depth at which melting occurs is also made. Lava compositions vary between centers, with Satiman composed of foidite and phonolite. Lemagurut ranges from basalt through benmorite, whereas Oldeani is basalt and trachyandesite, and Ogol is basaltic. The Sr-Nd isotopes of these lavas are characterized by relative unradiogenic Nd ($^{143}Nd/^{144}Nd$ = 0.512056 – 0.512610), but with radiogenic Sr ($^{87}Sr/^{86}Sr$ = 0.70371 – 0.70609), suggesting a source explained by mixing between high-μ and enriched mantle reservoirs. These isotope ratios, together with trace elements patterns and ratios (e.g., Tb/Yb), indicate melting in the lithospheric mantle consistent with other studies in the region. $^{40}Ar/^{39}Ar$ dating indicates an age range for Satiman of 4.63 ± 0.05 to 4.02 ± 0.02 Ma, Lemagurut and Ogol from 2.40 ± 0.01 to 2.22 ± 0.10 Ma and Oldeani from 1.61 ± 0.01 to 1.52 ± 0.02 Ma. The age of Satiman and its silica undersaturated composition make it the likely source of the > 4.3 – 3.5 Ma Laetolil Beds. The age and composition of Naibadad Beds that overlie the Laetoli Beds are consistent with the composition and revised ages of 2.28 ± 0.02 – 2.04 ± 0.01 Ma for Ngorongoro. The Olpiro Beds, which are fine-grained and of phonolitic composition, may correlate best with the more distal Olmoti Crater.

Keywords $^{40}Ar/^{39}Ar$ dating • Petrology • Geochemistry • Ogol • Tanzania

G.F. Mollel (✉)
Department of Earth and Planetary Sciences, Rutgers University, 610 Taylor Road, Piscataway, NJ 08854, USA
and
Department of Earth and Atmospheric Sciences, University of Alberta, B-14 Earth Sciences Building, Edmonton, Alberta, T6G 2E3, Canada
e-mail: gmollel@gmail.com

C.C. Swisher III, M.D. Feigenson, and M.J. Carr
Department of Earth and Planetary Sciences, Rutgers University, 610 Taylor Road, Piscataway, NJ 08854, USA
e-mail: cswish@rci.rutgers.edu; feigy@rci.rutgers.edu; carr@rci.rutgers.edu

Introduction

The Pliocene Laetolil Beds of northern Tanzania, as described by Hay (1987), consist primarily of coarse to fine grained volcaniclastic and sedimentary deposits, rich in invertebrate and vertebrate fossil remains, including hominins. Interbedded air-fall, lahar, water-worked tuff, conglomerates, and breccia are considered to be derived from eruptions from the adjacent uplands of the Ngorongoro Volcanic Highland Complex (NVHC). The interbedded volcanics and volcaniclastic deposits of the Laetolil Beds, as well as the younger overlying Naibadad and Olpiro Beds, have been correlated to source volcanoes based on their general structure, lithology, and composition. A more detailed description of the sediments is given by Hay (1987) and Ditchfield and Harrison (2011). In this chapter, we: (1) provide detailed petrochemistry and new $^{40}Ar/^{39}Ar$ geochronology for the NVHC volcanoes: Satiman, Lemagurut and Oldeani, as well as that of Ogol flows; (2) compare the genesis of these lavas to those of the other volcanics in northern Tanzania and the Gregory Rift; and (3) relate the compositions and age of these lavas to that of the Laetoli sequence as established by Hay (1987). These volcanoes are situated to the south and southwest of the Ngorongoro Caldera, the source of much of the volcanic materials in the lower part of the nearby Olduvai Beds (Hay 1976; McHenry et al. 2008).

The closest of the NVHC volcanoes to the Laetoli deposits is Lemagurut, a highly eroded and dissected volcano, today reaching a maximum elevation of 3,090 m (Fig. 5.1). Descriptions of volcanic clasts from streams draining Lemagurut indicate that the volcano is composed primarily of trachybasalt and nephelinite (Hay 1976). Hay (1976) reported a K-Ar date of 2.39 Ma on a vogesite flow collected in the Laetoli area and interpreted to have originated from Lemagurut. Later, Hay (1987) reinterpreted the vogesite to have instead originated from the Ogol lavas that occur as a series of volcanic remnants overlying the Laetolil Beds. Thus, the 2.39 Ma date may not represent the age of Lemagurut activity.

T. Harrison (ed.), *Paleontology and Geology of Laetoli: Human Evolution in Context. Volume 1: Geology, Geochronology, Paleoecology and Paleoenvironment*, Vertebrate Paleobiology and Paleoanthropology, DOI 10.1007/978-90-481-9956-3_5,

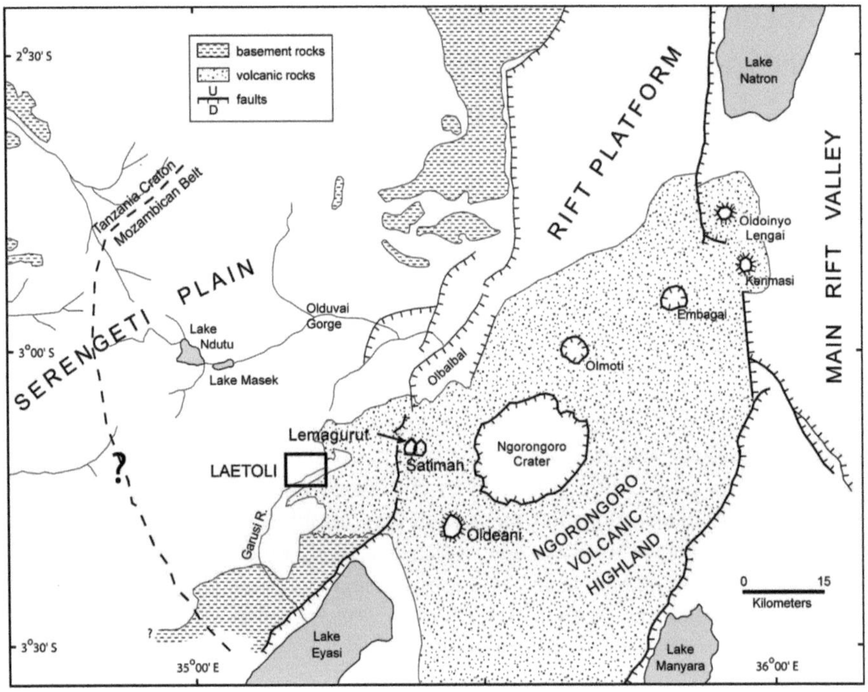

Fig. 5.1 Regional map showing major volcanoes of the Ngorongoro Volcanic Highland Complex (NVHC), Laetoli and Olduvai paleontological sites, as well as inferred boundary for the Tanzania Craton and the Mozambican Belt (modified from Hay 1976)

Bagdasaryan et al. (1973) reported a K-Ar date of 4.3 Ma on a phonolite from Lemagurut. However, based on its composition, it is more likely that the phonolite is derived from nearby Satiman, as phonolites are unknown from Lemagurut (Mollel 2002). Manega (1993) reported a $^{40}Ar/^{39}Ar$ age of 1.92 ± 0.04 Ma on an ignimbrite from the western flanks of Lemagurut. However, we now consider this ignimbrite to be a distal flow originating from Ngorongoro, rather than derived from Lemagurut.

Satiman is situated to the east of Laetoli, on the far side of Lemagurut, and it reaches a maximum elevation of 2,880 m. The volcano is highly eroded and lacks a caldera, and for the most part is heavily covered by vegetation. Outcrops of Satiman are rare and of limited extent. Satiman consists primarily of nephelinitic and phonolitic lavas (Pickering 1964; Hay 1976). K-Ar dates on nepheline suggest that Satiman was active between 4.5–3.7 Ma (Curtis and Hay 1972; Bagdasaryan et al. 1973).

The largest volcano adjacent to the Laetolil Beds is the 3,000 m high Oldeani volcano, located southeast of Satiman and southwest of Ngorongoro. Pickering (1964) described Oldeani as consisting primarily of basalt. No dates have been reported previously for Oldeani. However, based on the degree of weathering of its slopes and adjacent valleys, it was thought to fall within the range of activity of Ngorongoro and Lemagurut. Based on similarity in lava composition, Oldeani has been inferred to be the source for stone tools recovered from the hominin-bearing Upper Ngaloba Beds that overlie the Laetolil Beds (Adelsberger et al. 2002, 2011).

Hay (1987) described scoriaceous lavas that overlie the Laetolil Beds southwest of Lemagurut, associated with a series of cones or vents. The outcrops are of limited extent and were collectively grouped as the Ogol lavas. The Ogol lavas lie in an east–west orientation extending for about 25 km, and are situated 15 km from the lower limits of Lemagurut. Individual flows range from 0.5–1.5 km in length and 50–150 m above the Laetolil Beds. Most of these flows have rounded erosional tops, although Hay (1987) reported one to the southeast edge of the Laetoli area to have a well-preserved crater. According to Hay (1987), the Ogol lavas are porphyritic with phenocrysts of olivine, augite, alkali feldspar, and biotite and may include plagioclase, nepheline and opaque minerals. Hay (1987) reported reverse magnetic polarity on the Ogol lavas, based on field measurements obtained by Alan Cox of Stanford University. Drake and Curtis (1987) gave whole rock K-Ar dates on vogesite ranging from 2.52–2.21 Ma, with a mean age of 2.41 Ma, placing the Ogol lavas within the lower Matuyama Chron (Cox 1969).

Methods

Field Methods

We collected a total of 33 lava samples from Lemagurut, Satiman, and Oldeani volcanoes and Ogol flows between

2000 and 2004. Samples were taken from the tops and flanks of Satiman, Lemagurut and Oldeani to ensure unambiguous assignment to these particular volcanoes. Samples collected from lower elevations as part of previous studies are now known to be distal flows from various volcanic centers. Sample selection was based on availability of outcrops, sample freshness, and the presence of datable mineral phases. Locations of sampled sites are provided in UTM coordinates (Table 5.1).

Laboratory Methods

Samples were examined in the laboratory using binocular and petrographic microscopes. Thin sections were made and analyzed prior to all dating and chemistry sample preparations. All samples were prepared for major and minor elemental analyses by X-ray fluorescence and trace elemental analyses by LA-ICPMS (Laser Ablation Inductively Coupled Plasma Mass Spectrometry). Representative samples were prepared for $^{40}Ar/^{39}Ar$ dating, Sr and Nd by TIMS (Thermal Ionization Mass Spectrometry), and mineral phase determinations by electron microprobe.

Major Element Analyses

Splits of all samples were pulverized and powdered. At Michigan State University, the powders were fused into glass discs using high dilution fusion (HDF) technique by mixing the powder with lithium tetraborate as a flux and ammonium nitrate as an oxidizer. Major and minor elements were determined by X-ray fluorescence using a SMAX Rigaku X-ray spectrograph. Trace elements were determined using LA-ICPMS following procedures similar to those of Hannah et al. (2002). Analytical data were monitored using co-analyzed USGS rock standards. Major element oxides in weight percent (wt%) and trace elements in parts per million (ppm) are given in Table 5.1.

Sr and Nd Isotopes

Sr and Nd isotopic analyses were made on selected samples from Satiman, Lemagurut, and Oldeani at Rutgers University (RU) using a multicollector GV Isoprobe TIMS. Sample preparation followed the method of Bolge et al. (2006). Measured values for SRM-987 Sr standard and La Jolla-Nd standard were $^{87}Sr/^{86}Sr = 0.710237 \pm 5$ and $^{144}Nd/^{143}Nd = 0.511846 \pm 2$ respectively. Analyses are reported in Table 5.1.

Mineral Phase Analyses

Mineral phases were determined using a Cameca SX100 electron microprobe at the American Museum of Natural History following procedures and standards similar to those of Mandeville et al. (2002). Running conditions were 10 nA beam current, 15 KV accelerating voltage and 20 μm beam diameter. Samples were also analyzed at Rutgers University using a JEOL-8600 electron microprobe (see Mollel 2002). At RU, the same beam current was used; accelerating voltage was set at 20 KV with a beam diameter of one micrometer. Microprobe analyses are reported in Tables 5.2–5.5 as oxides in wt%.

$^{40}Ar/^{39}Ar$ Dating

Representative samples from all volcanic centers were crushed and sieved to a size of 0.425–0.25 mm and washed in ultrasonic baths of distilled water. Minerals rich in K were handpicked under a binocular microscope. Matrix (crushed whole rock) was purified by removing any altered or secondary materials. The purified samples were loaded in Al sample disks along with aliquots of the Fish Canyon sanidine or Alder Creek sanidine standards, wrapped in Al foil, sealed in quartz glass tubes, and then irradiated in the Cadmium-Lined, In-Core Irradiation Tube [CLICIT] facility of the Oregon State University Triga Research Reactor [OSTR]. Interference corrections used are $^{36}Ar/^{37}Ar_{Ca} = 2.72 \pm 0.01 (\times 10^{-4})$, $^{39}Ar/^{37}Ar_{Ca} = 7.11 \pm 0.024\ (10^{-4})$, and $^{40}Ar/^{39}Ar_K = 7 \pm 3\ (10^{-4})$ (Deino and McBrearty 2002). At the RU Ar-lab, about 50 mg of the matrix were heated in step-wise increments for 60 s each using a 40 W CO_2 laser focused though a 6 mm integrator lens. The released argon isotopes were measured on a MAP215-50 mass spectrometer, and $^{40}Ar/^{39}Ar$ ages calculated using automated software by A. Deino following procedures similar to those outlined in Renne et al. (1998). The $^{40}Ar/^{39}Ar$ data are reported as plateau, inverse isochron (simply isochron) and total fusion ages (see Tables 5.6 and 5.7). Apparent ages from individual steps were plotted against cumulative percent $^{39}Ar^*$ obtained from all incremental-heating steps. All plateau ages were calculated following typical procedures constituting at least three contiguous steps that are indistinguishable at 95% confidence level, and totaling at least 50% of the total $^{39}Ar^*$ released. Detailed evaluation of analytical data can be found in Carr et al. (2007).

Isochron ages were obtained from a slope produced by plotting and fitting the atmospheric related $^{36}Ar/^{40}Ar$ component against the radiogenic component $^{39}Ar/^{40}Ar$. Isochron ages were considered reliable if they were analytically indistinguishable from the plateau age and the total fusion age (weighted sum of individual increment ages). The Y-intercept ($^{36}Ar/^{40}Ar$) should be analytically indistinguishable from the

Table 5.1 UTM coordinate of locations, major and trace elements, and Sr, and Nd isotope analyses of lavas from volcanics associated with the Laetoli area

Sample	03-OLDI-1	03-OLDI-2	03-OLDI-3	03-OLDI-4	03-OLDI-5	03-OLDI-6	04-OLDI-1A	04-OLDI-1B	04-OLDI-2	04-OLDI-3	04-OLDI-4	01-SM-1	01-SM-2	01-Sm-3	01-Sm-4	01-Sm-5	00-SM-5	04-NT-3
UTM E	761335	762177	762318	762724	763286	763460	768361	768361	770104	769777	768331	766339	766403	767024	768029	768059	763994	747715
UTM S	9632496	9633434	9633420	9633574	9633408	9633394	9640028	9640028	9639102	9640084	9641768	9646910	9646618	9646684	9647014	9646774	9645907	9634578
SiO_2	50.1	49.5	46.4	52.0	50.1	52.2	52.4	51.6	52.2	54.1	52.8	45.3	45.7	41.9	51.0	44.4	53.0	53.6
TiO_2	3.3	3.4	3.3	3.1	3.0	3.0	3.0	2.8	2.8	2.1	2.7	1.9	1.8	2.0	0.6	1.9	0.7	1.1
Al_2O_3	14.6	14.6	14.2	15.4	14.7	15.1	16.1	15.0	15.0	16.8	15.5	17.7	17.6	13.5	18.4	15.7	19.6	19.7
FeO	12.7	13.5	13.2	11.9	12.7	11.6	12.5	11.8	11.9	9.7	11.7	9.1	8.8	13.1	8.3	10.9	6.4	6.6
MnO	0.2	0.2	0.2	0.1	0.2	0.2	0.2	0.2	0.2	0.1	0.2	0.2	0.2	0.4	0.2	0.3	0.2	0.2
MgO	3.7	3.8	3.5	3.2	4.2	3.1	1.8	3.4	3.1	2.2	2.9	1.8	1.6	2.2	0.2	1.8	0.4	0.7
CaO	7.8	8.2	10.3	7.2	7.6	6.8	5.7	6.8	6.6	6.1	6.7	7.7	7.1	9.5	2.9	8.2	2.9	2.9
Na_2O	3.8	3.5	3.3	4.1	3.7	4.4	3.8	3.9	4.0	4.6	3.6	6.9	7.0	6.9	9.8	8.8	9.7	8.9
K_2O	1.3	1.3	1.2	1.6	1.3	1.8	1.6	1.6	1.7	1.8	1.6	4.2	2.5	3.8	6.3	3.2	4.7	4.9
P_2O_5	0.5	0.4	0.4	0.6	0.5	0.6	0.6	0.6	0.6	0.5	0.5	0.3	0.3	0.5	0.1	0.3	0.1	0.2
Total	97.9	98.6	96.1	99.1	98.0	98.7	97.6	97.7	98.0	98.1	98.1	95.2	92.5	93.9	97.8	95.5	97.6	98.6
Rb	27.7	32.5	28.5	32.5	25.7	39.7	41.0	40.0	41.0	39.0	40.0	93.0	70.0	124.0	164.0	83.0	113.5	132.0
Sr	582.2	626.7	609.3	607.2	591.9	655.9	901.0	676.0	634.0	702.0	661.0	1892.0	1022.0	1700.0	1873.0	1131.0	1401.1	1412.0
87Sr/86Sr	0.70406	0.70432	0.70476		0.70427									0.70412	0.70527	0.70609		0.70630
143Nd/144Nd	0.51250	0.51248	0.51239		0.51239										0.51240	0.51206		
Zr/Nb	5.61	5.3	5.42	5.4	4.8	6.08	5.09	5.02	5.02	5.51	5.46	1.79	1.86	2.48	2.72	2.11	2.03	1.24
Zr	308.9	247.7	243.6	302.4	286.1	339.1	347.0	324.0	336.0	331.0	309.0	352.8	350.8	525.7	444.7	501.2	422.0	210.0
Ba	477.5	566.3	495.5	539.7	551.9	569.5	963.6	543.8	569.9	623.1	592.0	1434.9	1362.0	1545.8	2397.2	1278.0	1602.3	1695.7
La	51.2	47.0	43.0	53.8	55.7	55.2	78.1	59.4	67.9	59.8	58.1	138.7	115.8	91.9	115.2	130.6	155.0	118.8
Ce	93.5	83.3	80.7	94.3	100.9	100.4	101.9	101.1	108.2	100.7	98.5	213.6	167.3	103.7	149.7	194.4	241.1	179.6
Pr	12.0	11.0	10.4	12.6	13.3	13.2	16.1	13.3	15.9	13.4	13.1	22.9	17.8	10.8	14.4	21.9	21.8	18.5
Nd	47.4	43.7	41.6	50.9	53.2	53.2	65.0	53.5	65.0	53.7	53.0	79.5	60.7	36.3	44.5	78.8	66.7	61.2
Sm	9.6	8.8	8.4	10.1	10.6	10.7	12.1	10.0	12.9	9.8	9.9	12.7	10.2	7.6	7.5	13.7	10.0	8.9
Eu	2.9	2.7	2.6	3.1	3.3	3.2	3.8	2.9	3.9	3.0	2.9	3.5	3.0	2.5	2.2	3.8	2.9	2.5
Gd	9.2	8.4	7.9	9.6	10.1	9.7	13.1	10.7	13.7	10.3	10.3	12.9	11.1	8.9	8.8	13.7	9.3	11.5
Tb	1.3	1.2	1.1	1.3	1.4	1.4	1.7	1.4	1.9	1.3	1.3	1.6	1.5	1.4	1.2	1.8	1.2	1.1
Y	36.6	30.5	29.4	37.2	39.7	37.7	47.7	38.9	51.7	36.8	38.9	33.6	33.4	47.7	36.2	43.7	36.4	31.4
Dy	6.8	6.1	5.7	7.0	7.3	7.0	8.8	7.1	9.8	6.4	6.9	6.8	6.4	7.6	5.7	8.2	6.0	5.2
Ho	1.3	1.2	1.1	1.4	1.4	1.4	1.7	1.3	1.9	1.2	1.3	1.3	1.2	1.6	1.2	1.6	1.3	1.0
Er	3.4	2.8	2.7	3.3	3.5	3.3	3.9	3.1	4.5	2.9	3.0	3.1	2.9	4.1	3.2	4.1	3.4	2.6
Yb	3.4	2.7	2.6	3.2	3.3	3.1	3.9	3.0	4.5	2.7	2.9	6.2	5.5	5.6	5.7	6.7	3.2	2.9
Lu	0.4	0.3	0.3	0.4	0.4	0.4	0.6	0.4	0.7	0.4	0.4	0.4	0.4	0.5	0.4	0.5	0.5	0.4
V	241.0	258.4	179.5	194.7	129.5	250.5	137.3	141.7	131.8	109.1	423.4	133.0	99.0	238.4	54.6	138.2	32.0	40.8
Cr	14.9	3.9	3.8	6.2	2.0	47.1	4.5	3.9	3.3	4.7	16.5	2.9	1.5	1.9	1.8	4.2	17.5	3.6
Nb	55.1	46.7	44.9	56.0	59.6	55.8	68.2	64.6	67.0	60.1	56.6	197.4	189.1	212.4	163.4	237.5	208.1	169.2
Hf	8.2	6.5	6.3	8.4	8.1	8.2	9.0	8.4	8.7	8.2	7.6	8.0	7.6	10.7	8.8	11.1	7.8	5.9
Ta	3.6	3.1	3.0	3.7	3.8	3.6	4.1	3.8	4.0	3.4	3.2	9.8	8.6	4.5	3.1	10.5	5.9	7.8
Pb	3.9	3.5	1.8	4.0	3.7	3.9	5.1	3.8	4.1	6.6	4.7	9.2	8.4	14.0	22.1	10.8	41.1	14.4
Th	6.3	5.6	5.5	6.5	6.9	6.3	8.1	7.4	8.1	8.5	7.4	22.8	16.4	5.7	26.9	16.5	39.0	24.5
U	1.1	0.9	0.6	1.0	1.3	0.9	1.5	1.2	1.3	1.3	1.2	3.0	2.7	4.4	4.4	2.7	6.8	0.6

Sample	00-LM-3	00-LM-4	00-LM-5	00-LM-6	00-LM-7	00-LM-8	02-LM-1	02-LM-2	02-LM-3	02-LM-4	04-OS-1	04-OS-2	04-OS-3	04-OS-4	04-OS-5	04-OG-1	04-OG-2	04-OG-3	04-OG-4
UTM E	763806	764002	764010	765702	765714	764257	756819	759091	759091	759091	754434	754534	754996	755157	756058	741129	748119	747489	739719
UTM S	9658840	9646167	9646244	9648519	9648623	9647742	9656243	9657156	9657156	9657156	9643730	9643654	9643374	9643570	9643206	9642378	9644426	9644950	9644638
SiO_2	44.9	45.7	53.7	55.5	54.8	52.2	51.4	50.7	50.6	50.0	57.1	57.5	57.3	57.2	48.0	43.8	45.5	45.6	46.2
TiO_2	2.5	3.6	2.0	1.9	1.4	2.0	1.9	2.6	2.7	2.8	1.4	1.5	1.5	1.5	3.1	3.1	3.0	3.0	3.2
Al_2O_3	10.3	13.2	17.3	17.4	17.5	17.4	16.8	15.9	16.0	15.5	18.0	17.9	17.7	17.8	9.5	7.4	9.4	9.2	10.8
FeO	14.6	15.9	9.2	9.1	8.2	9.6	8.8	10.9	11.3	11.7	7.3	7.5	7.6	7.5	15.6	14.8	14.5	14.4	14.4
MnO	0.2	0.2	0.2	0.2	0.2	0.2	0.2	0.2	0.2	0.2	0.2	0.2	0.2	0.2	0.2	0.2	0.2	0.2	0.2
MgO	10.0	5.2	2.1	2.1	1.5	2.2	2.5	3.1	3.0	3.2	1.6	1.6	1.7	1.8	9.8	11.2	11.4	11.7	7.8
CaO	11.4	8.2	5.4	4.9	3.9	5.1	5.1	5.9	6.1	6.3	4.0	3.9	4.0	4.1	10.2	14.2	10.8	11.3	11.1
Na_2O	2.2	4.1	5.2	5.6	6.2	5.3	4.8	4.2	4.0	3.8	5.9	5.9	5.9	6.0	2.2	1.6	2.7	2.2	2.8
K_2O	0.9	1.8	2.4	2.5	3.1	2.6	2.6	2.5	2.6	2.6	3.3	3.2	3.3	3.3	1.1	0.9	1.1	1.1	1.3
P_2O_5	0.3	0.6	0.8	0.7	0.5	0.8	0.5	0.8	0.8	0.8	0.5	0.5	0.5	0.5	0.3	0.5	0.4	0.4	0.5
Total	97.2	98.6	98.4	99.8	97.3	97.2	94.6	96.7	97.2	96.8	99.1	99.7	99.6	99.7	99.8	97.6	99.1	99.0	98.2
Rb	21.6	42.3	53.4	66.8	81.3	61.9	62.0	70.0	67.0	62.0	92.0	80.0	85.0	84.0	34.0	23.0	24.0	24.0	30.0
Sr	535.3	1040.5	1210.3	1294.1	1103.2	1169.4	1293.0	1198.0	1419.0	1348.0	1688.0	1087.0	1289.0	1059.0	424.0	522.0	626.0	616.0	762.0
87Sr/86Sr		0.70472		0.70503				0.70472			0.70519		0.70517	0.70522		0.70375	0.70371	0.70379	0.70381
143Nd/144Nd		0.51246		0.51241				0.51247			0.51219		0.51238	0.51238		0.51261	0.51257		0.51255
Zr/Nb	3.44	3.6	3.92	3.52	3.76	4.05	3.32	3.68	3.53	3.58	3.54	3.55	3.62	3.63	4.82	4.04	3.58	3.53	3.79
Zr	190.5	368.5	438.8	430.2	499.7	475.3	423.0	395.0	399.0	382.0	438.0	436.0	455.0	449.0	176.0	191.0	213.0	207.0	246.0
Ba	403.6	853.6	1151.2	1104.9	1386.2	1172.2	1079.5	1051.7	986.6	931.3	1558.1	2064.0	1111.6	1301.3	358.9	593.1	475.8	468.5	567.3
La	48.0	88.6	111.3	104.9	110.8	135.1	105.8	95.0	98.4	92.7	103.2	121.2	103.1	106.1	48.1	40.1	54.9	49.7	57.6
Ce	87.6	158.9	208.8	196.8	206.1	194.7	180.5	165.0	168.0	160.7	176.4	181.9	176.0	180.2	63.0	73.6	92.2	85.6	98.9
Pr	10.0	20.2	23.3	22.6	22.5	27.5	22.3	21.3	21.4	20.4	20.9	22.0	20.9	21.3	11.0	10.0	12.0	11.4	13.0
Nd	38.9	79.1	86.7	83.4	81.0	102.4	83.0	83.5	82.7	79.2	74.0	79.8	76.9	77.5	44.1	41.0	47.8	46.6	52.1
Sm	7.7	14.0	14.8	14.8	13.5	17.3	13.6	14.3	14.2	13.6	11.9	12.4	12.6	12.2	8.1	8.5	8.8	8.4	9.2
Eu	2.2	3.9	4.4	4.2	3.9	4.7	3.9	4.0	4.0	3.8	3.4	3.5	3.5	3.4	2.3	2.6	2.6	2.4	2.7
Gd	7.1	11.8	12.7	12.8	11.4	14.8	15.0	15.4	15.0	15.3	13.6	14.2	14.3	13.8	8.1	8.9	9.2	8.8	9.7
Tb	1.0	1.5	1.7	1.7	1.5	1.9	1.7	1.7	1.7	1.7	1.5	1.6	1.5	1.5	1.1	1.1	1.1	1.0	1.1
Y	25.6	39.2	39.5	42.4	39.8	53.6	41.0	40.9	41.5	41.0	37.7	46.3	37.8	37.1	27.3	21.6	26.5	25.4	28.8
Dy	5.0	7.8	7.7	8.0	7.2	9.7	7.6	8.0	8.0	7.7	6.9	7.3	6.8	7.0	5.4	5.0	5.3	4.9	5.5
Ho	0.9	1.4	1.4	1.3	1.3	1.8	1.4	1.5	1.5	1.4	1.3	1.4	1.3	1.3	1.0	1.0	0.9	0.9	1.0
Er	2.2	3.3	3.4	3.6	3.1	4.4	3.2	3.3	3.4	3.1	3.0	3.4	3.1	3.0	2.4	2.5	2.2	1.9	2.2
Yb	2.0	2.8	3.1	3.1	3.3	3.9	3.2	3.3	3.3	3.2	3.2	3.6	3.3	3.2	2.4	2.6	2.0	1.8	2.1
Lu	0.3	0.4	0.5	0.4	0.4	0.5	0.4	0.4	0.4	0.4	0.4	0.5	0.4	0.4	0.3	0.4	0.3	0.3	0.3
V	333.4	299.0	89.0	44.9	17.9	69.0	46.7	100.5	101.1	114.6	22.2	23.5	21.3	24.8	335.5	271.6	260.5	258.5	217.7
Cr	494.9	43.0	20.7	14.7	15.2	13.1	3.4	4.8	4.9	6.1	4.3	4.5	3.9	3.6	537.5	863.2	538.3	574.5	513.9
Nb	55.4	102.4	112.0	122.2	133.0	117.4	127.5	107.4	112.9	106.8	123.9	122.7	125.8	123.7	36.5	47.3	59.5	58.6	64.9
Hf	4.5	8.7	9.3	9.6	10.7	11.1	10.3	10.0	10.3	8.9	10.3	10.6	10.9	10.3	4.6	5.6	5.4	5.3	6.1
Ta	3.0	6.0	5.9	6.7	7.3	6.4	7.1	6.2	6.4	5.6	6.7	6.5	6.8	6.6	2.2	2.9	3.5	3.2	3.6
Pb	4.4	15.1	13.7	14.3	17.2	12.7	9.9	9.3	9.5	8.8	12.8	11.5	12.8	11.6	4.2	3.0	4.0	3.7	1.7
Th	6.2	10.5	12.1	13.8	16.2	14.1	15.9	13.7	13.7	11.8	16.7	15.8	16.7	16.0	6.5	4.7	6.3	6.0	6.8
U	1.3	2.3	2.5	1.4	3.0	1.7	2.7	2.4	2.6	1.5	3.5	1.7	3.3	2.7	0.6	0.7	1.1	0.9	1.1

Table 5.2 Satiman mineral compositions (electron microprobe analyses)

	SiO_2	TiO_2	Al_2O_3	MgO	MnO	FeO	CaO	Na_2O	K_2O	P_2O_5	BaO	Total
Pyroxene												
1	49.66	1.01	1.32	6.46	0.64	18.14	20.39	3.15	0.01	0.00	0.15	100.92
2	50.13	1.05	1.11	5.73	0.56	19.61	18.47	3.82	0.03	0.01	0.00	100.51
3	50.31	0.96	1.11	5.93	0.65	19.58	18.64	3.86	0.01	0.00	0.07	101.11
4	51.21	1.29	0.88	5.60	0.62	19.67	17.04	4.57	0.00	0.02	0.00	100.90
5	49.00	1.54	3.20	11.00	0.31	11.00	23.00	1.00	0.00	0.00	0.07	100.12
Andradite												
1	28.00	15.00	0.45	0.56	0.36	20.00	32.00	0.28	0.00	0.00	0.25	97.50
2	29.00	15.00	0.23	0.73	0.32	21.00	31.00	0.39	0.00	0.00	0.21	97.89
3	27.00	18.00	0.45	0.70	0.33	19.00	32.00	0.37	0.00	0.00	0.20	98.21
Nepheline												
1	46.86	0.10	28.71	0.32	0.09	3.58	1.25	17.00	3.21	0.02	0.00	101.13
2	46.57	0.00	31.42	0.06	0.01	1.94	0.00	17.39	3.77	0.00	0.15	101.30
3	42.00	0.00	32.00	0.08	0.00	1.35	0.03	17.00	6.70	0.00	0.01	99.91
4	42.00	0.04	34.00	0.00	0.00	1.51	0.32	16.00	6.10	0.00	0.04	100.12
5	41.00	0.06	31.00	0.00	0.02	3.60	0.06	16.00	8.30	0.00	0.00	100.63
Apatite												
1	0.83	0.00	0.00	0.01	0.02	0.12	57.75	0.06	0.04	39.21	0.09	98.12
Tmagnetite												
1	0.03	14.62	0.00	0.06	1.56	78.90	0.00	0.01	0.01	0.00	0.00	95.19
2	0.50	14.38	0.06	0.03	1.41	76.07	1.24	0.11	0.02	0.79	0.00	94.58
3	0.07	14.49	0.05	0.03	1.52	78.39	0.02	0.03	0.01	0.01	0.04	94.65
4	0.02	14.03	0.05	0.04	1.90	78.51	0.00	0.02	0.01	0.03	0.05	94.64
5	0.47	14.18	0.16	0.04	1.29	77.79	0.06	0.04	0.04	0.02	0.00	94.10
Feldspar												
1	65.32	0.18	20.26	0.01	0.00	0.26	0.23	7.47	5.88	0.01	0.47	100.07
2	65.61	0.07	19.72	0.00	0.02	0.51	0.26	6.83	6.32	0.00	0.98	100.31
3	64.86	0.11	18.94	0.00	0.00	0.42	0.14	5.72	8.58	0.00	0.17	98.94
Titanite												
1	29.95	37.55	1.54	0.02	0.02	1.28	30.14	0.03	0.00	0.06	0.11	100.69
2	32.55	34.09	3.52	0.01	0.05	1.68	26.52	1.78	0.50	0.02	0.00	100.74
3	30.00	37.00	0.53	0.00	0.04	1.95	27.00	0.22	0.10	0.00	1.82	98.25

Table 5.3 Lemagurut mineral compositions (electron microprobe analyses)

	SiO_2	TiO_2	Al_2O_3	MgO	MnO	FeO	CaO	Na_2O	K_2O	P_2O_5	F	Total
Olivine												
1	37.42	0.12	0.00	35.03	0.30	27.90	0.25	0.00	0.00	0.05	0.05	101.11
2	37.73	0.05	0.01	35.49	0.36	27.10	0.24	0.04	0.01	0.00	0.00	101.04
3	34.82	0.12	0.00	23.66	0.64	41.61	0.34	0.01	0.02	0.02	0.06	101.28
4	35.26	0.08	0.00	25.52	0.52	39.12	0.35	0.00	0.00	0.00	0.10	100.95
Pyroxene												
1	51.09	1.03	1.49	11.57	0.63	13.35	20.81	0.74	0.05	0.00	0.07	100.83
2	51.74	0.88	1.03	11.80	0.68	12.91	21.04	0.68	0.02	0.00	0.08	100.87
3	49.00	2.52	3.63	14.00	0.24	10.00	21.00	0.62	0.01	0.00	0.00	100.50
4	50.92	0.91	1.32	11.20	0.53	12.89	21.07	0.86	0.03	0.01	0.09	99.84
5	48.00	3.01	4.66	14.00	0.28	10.00	21.00	0.63	0.02	0.00	0.00	100.94
6	54.26	0.54	1.41	17.63	0.09	5.29	20.75	0.36	0.00	0.01	0.00	100.33
Feldspar												
1	65.34	0.04	21.96	0.00	0.00	0.55	1.93	8.65	2.38	0.05	0.00	100.89
2	66.12	0.10	19.98	0.04	0.00	0.58	0.24	5.38	8.26	0.01	0.04	100.74
3	66.20	0.12	21.18	0.00	0.00	0.37	1.12	7.81	4.24	0.00	0.01	101.04
4	66.09	0.14	20.73	0.00	0.00	0.38	0.89	7.27	5.37	0.00	0.00	100.86
5	64.46	0.10	22.63	0.04	0.00	0.48	2.61	8.36	2.29	0.00	0.00	100.96
6	65.92	0.06	22.80	0.04	0.00	0.53	2.68	4.95	2.28	0.05	0.00	99.31
7	66.59	0.12	20.91	0.00	0.02	0.36	0.87	7.73	4.73	0.00	0.01	101.34
8	66.00	0.06	20.87	0.01	0.00	0.30	0.85	6.80	6.05	0.04	0.00	100.99
9	65.94	0.10	20.44	0.02	0.01	0.54	0.63	5.93	7.11	0.00	0.00	100.73
10	66.25	0.14	20.87	0.00	0.00	0.36	0.78	7.38	5.46	0.02	0.00	101.26

(continued)

Table 5.3 (continued)

	SiO_2	TiO_2	Al_2O_3	MgO	MnO	FeO	CaO	Na_2O	K_2O	P_2O_5	F	Total
Tmagnetite												
1	0.11	24.12	1.21	0.65	1.26	68.83	0.01	0.08	0.01	0.00	0.12	96.38
2	0.04	23.68	1.18	0.71	1.19	72.22	0.00	0.03	0.01	0.00	0.13	99.19
3	0.10	24.76	1.25	0.75	1.31	67.60	0.01	0.10	0.00	0.00	0.15	96.02
4	0.19	25.03	1.24	0.67	1.27	68.34	0.00	0.01	0.00	0.00	0.18	96.93
5	0.10	23.73	0.66	0.53	1.29	69.06	0.24	0.00	0.03	0.07	0.14	95.85
6	0.10	25.19	1.07	1.84	0.52	66.62	0.26	0.10	0.03	0.00	0.13	95.86
7	0.04	23.67	1.44	1.50	0.52	69.01	0.21	0.00	0.03	0.00	0.13	96.55
8	0.14	25.23	1.06	1.76	0.52	67.83	0.05	0.22	0.04	0.00	0.14	96.99
9	0.07	24.93	1.10	1.77	0.51	67.72	0.28	0.08	0.03	0.00	0.17	96.65
10	0.17	23.98	1.24	1.77	0.40	67.46	0.40	0.01	0.03	0.00	0.14	95.61
Amphibole												
1	41.81	5.07	11.30	12.03	0.36	14.03	11.45	3.16	0.78	0.05	0.82	100.86
2	41.10	5.17	11.38	12.09	0.32	14.64	11.26	3.24	0.74	0.03	1.28	101.25
3	40.66	5.75	12.43	12.33	0.18	12.95	11.51	2.76	0.74	0.03	0.72	100.06
4	41.42	5.03	11.44	12.10	0.26	13.67	11.43	2.99	0.79	0.04	0.70	99.88
5	45.42	1.87	7.18	13.52	0.37	13.87	10.50	3.43	1.23	0.01	3.84	101.24
6	53.03	0.87	7.79	8.04	0.14	7.81	16.10	2.49	1.76	1.62	0.20	99.84
7	57.63	0.85	14.34	5.28	0.09	5.52	10.11	4.91	1.79	0.02	0.04	100.57
Apatite												
1	0.16	0.00	0.00	0.20	0.00	0.42	54.75	0.07	0.01	41.04	5.86	102.50
2	0.24	0.03	0.00	0.20	0.11	0.64	55.23	0.08	0.01	40.97	4.94	102.46

Table 5.4 Oldeani mineral compositions (electron microprobe analyses)

	SiO_2	TiO_2	Al_2O_3	MgO	MnO	FeO	CaO	Na_2O	K_2O	P_2O_5	BaO	Total
Olivine												
1	33.90	0.16	0.00	17.73	0.71	46.47	0.29	0.03	0.01	0.21	0.07	99.58
2	36.32	1.88	1.88	19.19	0.56	38.88	0.30	0.53	0.91	0.03	0.00	100.46
3	31.43	0.14	0.04	23.29	0.76	43.52	0.35	0.04	0.02	0.15	0.03	99.76
4	33.70	0.10	0.00	19.81	0.81	45.63	0.26	0.02	0.01	0.10	0.00	100.45
5	34.88	0.09	0.00	20.43	0.60	44.13	0.28	0.03	0.01	0.09	0.05	100.59
Pyroxene												
1	51.02	1.39	2.83	14.57	0.14	8.42	22.36	0.44	0.02	0.07	0.06	101.32
2	50.55	1.70	2.70	14.32	0.24	11.22	20.35	0.33	0.00	0.05	0.00	101.46
3	49.93	1.79	2.54	13.21	0.36	12.48	20.17	0.53	0.06	0.08	0.14	101.28
4	50.34	1.50	1.80	14.07	0.32	13.25	19.32	0.38	0.03	0.21	0.06	101.28
5	52.07	1.11	3.15	12.40	0.39	12.74	17.60	0.80	0.58	0.02	0.03	100.86
Feldspar												
	54.55	0.16	28.10	0.09	0.00	0.80	11.41	5.30	0.37	0.00	0.01	100.77
	58.20	0.26	25.87	0.11	0.01	1.33	9.11	6.44	0.63	0.05	0.14	102.15
	60.97	0.21	24.23	0.04	0.00	0.78	7.06	6.82	1.87	0.05	0.09	102.12
Tmagnetite												
1	0.04	19.15	3.79	3.53	0.34	69.11	0.00	0.03	0.00	0.01	0.05	96.04
2	0.38	18.55	4.16	3.84	0.25	67.05	0.00	0.07	0.02	0.01	0.07	94.41
3	0.07	18.50	3.78	4.11	0.29	68.23	0.02	0.01	0.01	0.00	0.12	95.15
4	0.08	19.03	3.81	3.75	0.32	70.09	0.00	0.04	0.00	0.00	0.00	97.12

atmospheric ratio of 295 ± 1. The quality of the fit obtained by the mixing line between $^{36}Ar/^{40}Ar$ and $^{39}Ar/^{40}Ar$ determines the slope and thus the age. The goodness of the fit is defined as the mean sum of the weighted deviates (MSWD) of individual temperature-increment ages. MSWD is a reduced χ^2 variable that determines the scatter of data about a best-fit linear regression (Bevington 1969). Results with MSWD > 1 indicate poor fitting and large uncertainty. Data with MSWD > 2.5 were considered unreliable (see Turrin et al. 2007; Carr et al. 2007).

Results

Lavas from the NVHC volcanoes associated with the Laetoli deposits are defined as silica undersaturated to silica saturated compositions based on the alkali-silica classification of Cox et al. (1979). The silica undersaturated lavas (Table 5.1) are restricted to Satiman and include foidite and phonolite (Fig. 5.2). The silica saturated lavas are from Lemagurut, Oldeani, and Ogol. Lemagurut ranges in composition from basalt through benmorite, and includes

Table 5.5 Ogol mineral compositions (electron microprobe analyses)

	SiO_2	TiO_2	Al2O_3	MgO	MnO	FeO	CaO	Na_2O	K_2O	P_2O_5	BaO	Total
Olivine												
1	37.93	0.04	0.00	37.35	0.37	24.01	0.28	0.03	0.00	0.00	0.00	100.01
2	39.23	0.01	0.00	39.11	0.31	21.05	0.31	0.00	0.00	0.00	0.09	100.10
3	39.03	0.00	0.00	40.50	0.25	20.58	0.19	0.01	0.00	0.05	0.01	100.60
4	39.50	0.03	0.00	40.91	0.25	19.74	0.21	0.00	0.00	0.03	0.00	100.67
5	39.22	0.05	0.00	41.23	0.30	19.94	0.18	0.00	0.00	0.00	0.02	100.95
6	37.88	0.00	0.00	37.63	0.36	24.50	0.35	0.00	0.01	0.00	0.01	100.73
Pyroxene												
1	53.10	0.54	1.60	17.50	0.13	4.72	22.27	0.73	0.01	0.01	0.03	100.63
2	52.48	0.72	2.73	16.34	0.13	6.82	21.72	0.79	0.00	0.02	0.02	101.76
3	49.38	2.29	3.88	13.82	0.18	7.48	24.00	0.50	0.00	0.02	0.00	101.53
4	53.48	0.59	2.45	16.63	0.18	6.17	19.71	0.74	0.04	0.01	0.00	100.00
5	47.88	2.77	4.22	13.33	0.19	9.12	23.12	0.56	0.06	0.11	0.00	101.36
6	47.27	2.61	4.64	12.69	0.20	9.07	23.53	0.89	0.03	0.07	0.00	101.00
Feldspar												
1	60.63	0.13	23.23	0.24	0.00	1.02	5.28	7.85	1.42	0.02	0.42	100.25
Tmagnetite												
1	0.40	23.89	0.66	1.63	0.69	67.58	0.55	0.06	0.02	0.10	0.11	95.69
2	0.02	24.31	0.27	1.46	0.68	68.33	0.24	0.06	0.02	0.00	0.05	95.44
3	0.07	24.05	0.35	1.43	0.69	68.15	0.36	0.06	0.04	0.01	0.05	95.27
4	0.17	24.51	0.41	1.42	0.71	67.11	0.37	0.04	0.04	0.00	0.15	94.93
5	0.53	23.86	0.42	1.72	0.73	68.10	0.47	0.04	0.01	0.00	0.11	95.98

Table 5.6 New $^{40}Ar/^{39}Ar$ ages for volcanics associated with the Laetoli area

Sample Id	Material	Total Fusion Age (Ma)	Plateau Age (Ma)	Isochron Age (Ma)	$^{40}Ar/^{36}Ar$ Intercept	MSWD	% $^{39}Ar^*$ on Plateau
03-OLDI-1	Matrix	1.42 +/− 0.04	**1.52±0.02**	1.53±0.05	294.7±1.6	2.9	86.10
03-OLDI-2	Matrix	1.63 +/− 0.05	**1.58±0.02**	1.55±0.03	296.8±1.2	2.6	63.7
03-OLDI-4	Matrix	1.57±0.03	**1.61±0.01**	1.59±0.03	295.1±1.1	2.0	75.8
03-OLDI-5	Matrix	1.50±0.04	**155±0.02**	1.56±0.03	294.4±1.5	2.1	98.5
04-OLDI-3	Matrix	1.73±0.01		**1.57±0.03**	296±2	4.3	
04-OLDI-4	Matrix	1.61±0.01		**1.61±0.03**	292±3	3.4	
00-LM-6	Matrix	2.25±0.01	**2.25±0.01**	2.25±0.01	295±8	0.92	100
00-LM-7	Matrix	2.27±0.004	**2.26±0.005**	2.24±0.02	303±9	4.3	75.1
02-LM-4	Matrix	2.23±0.01	**2.23±0.01**	2.22±0.02	297±3	1.0	100
04-OS-1	Anorthoclase	2.20±0.10	**2.22±0.10**	2.23±0.1	293±3	0.72	100
04-OS-3	Matrix	2.55±0.01	**2.40±0.01**	2.38±0.02	294.7±1.4	2.9	53.3
04-OS-4	Matrix	2.34±0.004	**2.29±0.004**	2.30±0.02	297±6	14	56
04-OG-1	Matrix	2.41±0.02	**2.31±0.01**	2.29±0.01	299.8±1	1.0	55
04-OG-2	Matrix	2.3±0.05	**2.27±0.05**	2.30±0.12	296±3	1.5	75.6
04-OG-4	Matrix	2.52±0.01		2.51±0.05	316±6	9.9	
01-SM-3	Matrix	6.0±3.00	**4.63±0.05**	5.7±0.3	271±4	12	53.2
01-SM-3	Nepheline	6.29±0.04		5.7±0.3	290±20	6.4	
01-SM-5	Nepheline	4.99±0.03		4.51±0.02	291±6	7.1	
01-SM-5	Matrix	4.35±0.01		**4.19±0.02**	297.3±1.1	2.9	
04-NT-3	Matrix	4.11±0.01		**4.02±0.02**	301±4	5.1	

MSWD Mean sum of the weighted deviates, Bold ages are preferred.

Table 5.7 Standard used, irradiation time and irradiation parameter (J)

Sample Id	Standard used	Time	J value
03-OLDI-1	AC	40 min	$1.791 \times 10^{-4} \pm 7.7782 \times 10^{-7}$
03-OLDI-2, 4, 5	AC	40 min	$1.785 \times 10^{-4} \pm 2.1524 \times 10^{-6}$
04-OLDI-3, 4	AC	40 min	$1.8039 \times 10^{-4} \pm 3.287 \times 10^{-6}$
00-LM-6	FC2	60 min	$2.5849 \times 10^{-4} \pm 2.8788 \times 10^{-7}$
00-LM-7	AC	40 min	$1.734 \times 10^{-4} \pm 2.9112 \times 10^{-7}$
02-LM-4	FC2	60 min	$2.5891 \times 10^{-4} \pm 3.3285 \times 10^{-7}$
04-OS-1	FC2	6 h	$1.58 \times 10^{-3} \pm 1.0 \times 10^{-5}$
04-OS-3, 4	AC	40 min	$1.8139 \times 10^{-4} \pm 2.653 \times 10^{-7}$
04-OG-1, 2, 4	AC	40 min	$1.8009 \times 10^{-4} \pm 2.7542 \times 10^{-7}$
01-SM-3, 5	AC	15 min	$7.16 \times 10^{-5} \pm 1.3781 \times 10^{-7}$
04-NT-3	AC	40 min	$1.8009 \times 10^{-4} \pm 2.7542 \times 10^{-7}$

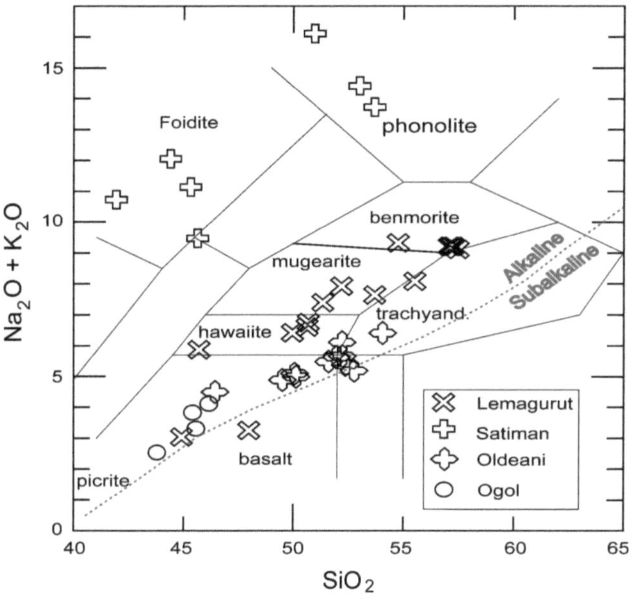

Fig. 5.2 Total alkali versus silica classification of Cox et al. (1979) for NVHC lavas associated with the Laetolil Beds

hawaiite and mugearite. Oldeani lavas range from basalt to trachyandesite, while Ogol flows are primarily basaltic in composition.

The silica undersaturated (i.e., Satiman) and saturated (i.e., Lemagurut, Oldeani, and Ogol) lavas display distinct textural features and mineral assemblages. Foidite and phonolite from Satiman are primarily aphanitic, porphyritic with nepheline and augite phenocrysts, and occasionally andradite sitting in a groundmass of augite and nepheline (Fig. 5.3a; Table 5.2). The mineral assemblage also includes titanomagnetite ± analcite, apatite, olivine, sphene and sanidine. The augite is commonly zoned and embayed, and its composition ranges from $En_{32}Fs_{18}Wo_{50}$ to $En_{10}Fs_{48}Wo_{42}$, concentrations of TiO_2 and Na_2O range up to 1.5 and 4.6 wt% respectively. Nepheline compositions from different crystals range from $Na_{5.41}K_{0.82}Al_{5.4}Si_{6.3}O_{24}$ to $Na_{2.94}K_{0.84}Al_{5.5}Si_{6.8}O_{24}$. Andradite is cubic shaped, pink to dark brown and its average composition is $Ca_{6.1}(Fe_{1.8}Ti_{2.4})(Fe_{1.1}Si_{4.9})O_{24}$. The Satiman andradite is schorlomitic (TiO_2 >15% wt; Mollel 2002), and is different from pyrope garnet found in Lashaine and Labait cones in northern Tanzania that are characterized by TiO_2 content < 8.5% wt (Jones et al. 1983; Lee and Rudnick 1999). One sample contains sanidine laths ranging in composition from $Ab_{51}Or_{49}$ to $Ab_{47}Or_{53}$ (Mollel 2002). Titanomagnetite is ophitic with augite and nepheline, low in TiO_2 (14.33 wt%) and rich in MnO (1.57 wt%).

Lavas from Lemagurut are phaneritic, mostly fine to medium grained and occasionally porphyritic, ranging in composition from basalt through hawaiite, mugearite and benmorite. The Lemagurut mineral assemblage includes augite, plagioclase, anorthoclase, and titanomagnetite, sodic and titanian amphibole ± olivine and fluoroapatite (Fig. 5.3b; Table 5.3). The phenocrysts are of plagioclase, augite, olivine and occasionally amphibole in a groundmass of plagioclase, anorthoclase and augite. Feldspar composition ranges from $An_{14}Ab_{64}Or_{22}$ to $An_{27}Ab_{64}Or_9$. Augite is occasionally zoned, its composition ranging between $En_{41}Fs_{15}Wo_{44}$ and $En_{32}Fs_{22}Wo_{46}$ and containing up to 3 wt% TiO_2. Olivine phenocrysts range in composition between forsteritic ($Fo_{65}Fa_{35}$) in the most basaltic to fayalitic ($Fo_{30}Fa_{70}$) in the most evolved lava. Titanomagnetite is ophitic with augite and olivine and its TiO_2 content ranges from ~23–25 wt%. The titanomagnetite is Al_2O_3 and MnO rich, with concentrations averaging at 1 and 1.28 wt% respectively. One sample contains rare phenocrysts of kaersutite and fluororichterite.

Basalt and trachyandesite from Oldeani are mostly fine-grained phaneritic to aphanitic and occasionally porphyritic and scoriaceous. Oldeani mineral assemblages include plagioclase, augite, olivine and titanomagnetite (Fig. 5.3c; Table 5.4). Basalt phenocrysts are dominated by olivine and augite whereas those of trachyandesite are mostly composed of augite and plagioclase. The groundmass is composed of plagioclase and augite. Amphibole and biotite are common in the trachyandesite and are rare in the basalts. Augite compositions average is $En_{37.23}Fs_{23.41}Wo_{39.36}$ and its TiO_2 content range from 1.79–1.11wt%. Feldspar composition varies from alkalic ($An_{32}Ab_{57}Or_{11}$) to plagioclase ($An_{53}Ab_{45}Or_2$). The olivine is fayalite ranging from Fa_{72} to Fa_{65}. Titanomagnetite is ophitic with augite and olivine. The TiO_2 content in the titanomagnetite ranges from 18.5–19.0 wt% and its Al_2O_3 and MgO content ranges from 4.2–3.8 wt% and 4.1–3.5 wt% respectively.

The Ogol lavas are phaneritic coarse-grained to porphyritic basalt; phenocrysts are mostly olivine and augite within a feldspar and augite groundmass (Fig. 5.3d; Table 5.5). The olivines range in composition from Fo_{78} to Fo_{73}. Augite is mostly lath-shaped and its composition ranges from $En_{48}Fs_8Wo_{44}$ to $En_{35}Fs_{16}Wo_{49}$ with TiO_2 and Na_2O contents up to 2.8 and 0.9 wt% respectively. The plagioclase is andesine (An_{23}). Titanomagnetite is ophitic with olivine, augite, and plagioclase and its TiO_2 content ranges from 24.5–23.9 wt%.

Harker diagrams (Fig. 5.4) show a continuous decrease in TiO_2, FeO_t, MgO, CaO, and P_2O_5, and an increase in Al_2O_3, Na_2O and K_2O with increasing silica content. The Mg number (Mg#) [100*(Mg/Mg + Fe)] is lowest in Satiman lavas (5–27) and highest in Ogol flows (56–59). For Lemagurut and Oldeani lavas, the Mg# varies from 18–39. Two Lemagurut lavas (00-LM-3 and OS-1) have higher values of 56 and 54 respectively. MnO is generally low for all samples 0.15–0.2 wt% (Table 5.1), although some Satiman lavas display values as high as 0.37 wt%. The Na/K ratio is higher in Oldeani (>2) compared to Satiman, Lemagurut and Ogol flows (<2). Two samples (01-SM-1 and 01-SM-5) from

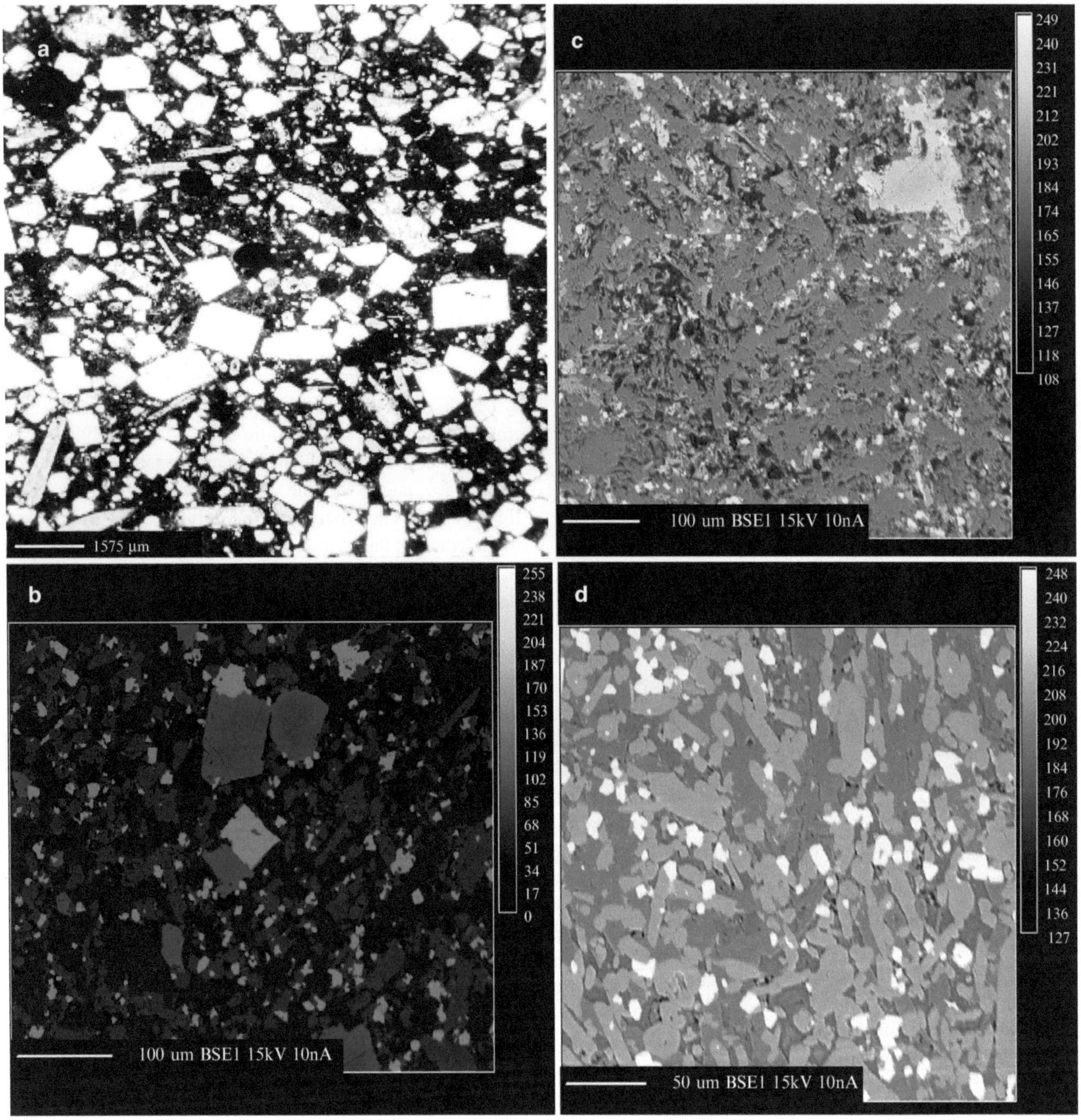

Fig. 5.3 (**a**) Microphotograph of porphyritic foidite from Satiman volcano showing abundant titanian andradite (black "cubes") sitting on nepheline (white tabular grains) and augite (white lath-shaped grains). (**b**) Back-Scatter Electron microprobe (BSE) image of Lemagurut basalt showing abundant feldspar (dark gray), augite (light gray), large olivine (at the *middle* of the image) and titanomagnetite (white specks). (**c**) Back-Scatter Electron (BSE) image of fine-grained Oldeani basalt showing abundant feldspar and a large olivine grain on *top right*. The white specks are of titanomagnetite. (**d**) Back-Scatter Electron (BSE) image of Ogol basalt showing near equigranular texture. Abundant pyroxene and olivine (200–220 on the gray scale) sitting on plagioclase dark colored. The white specks are of titanomagnetite

Satiman and (00-LM-3) from Lemagurut have values >2 (not illustrated). In general, FeO is highest in Ogol (>12 wt%) followed by Oldeani (~9.5–12 wt%), and is lowest in Lemagurut and Satiman (~5–10 wt%; Table 5.1). Samples 00-LM-3, 00-LM-4, and 04-OS-5 from Lemagurut and 01-SM-3 from Satiman are outliers.

Compatible trace element concentrations (e.g., Cr) are generally low in Satiman, Lemagurut, and Oldeani ranging from 1.5–47 ppm (Table 5.1), but significantly higher in Ogol lavas ranging from 514–863 ppm. Two Lemagurut lavas (00-LM-3 and 04-OS-5) show higher values of 495 ppm and 537 ppm respectively. Compared to ordinary

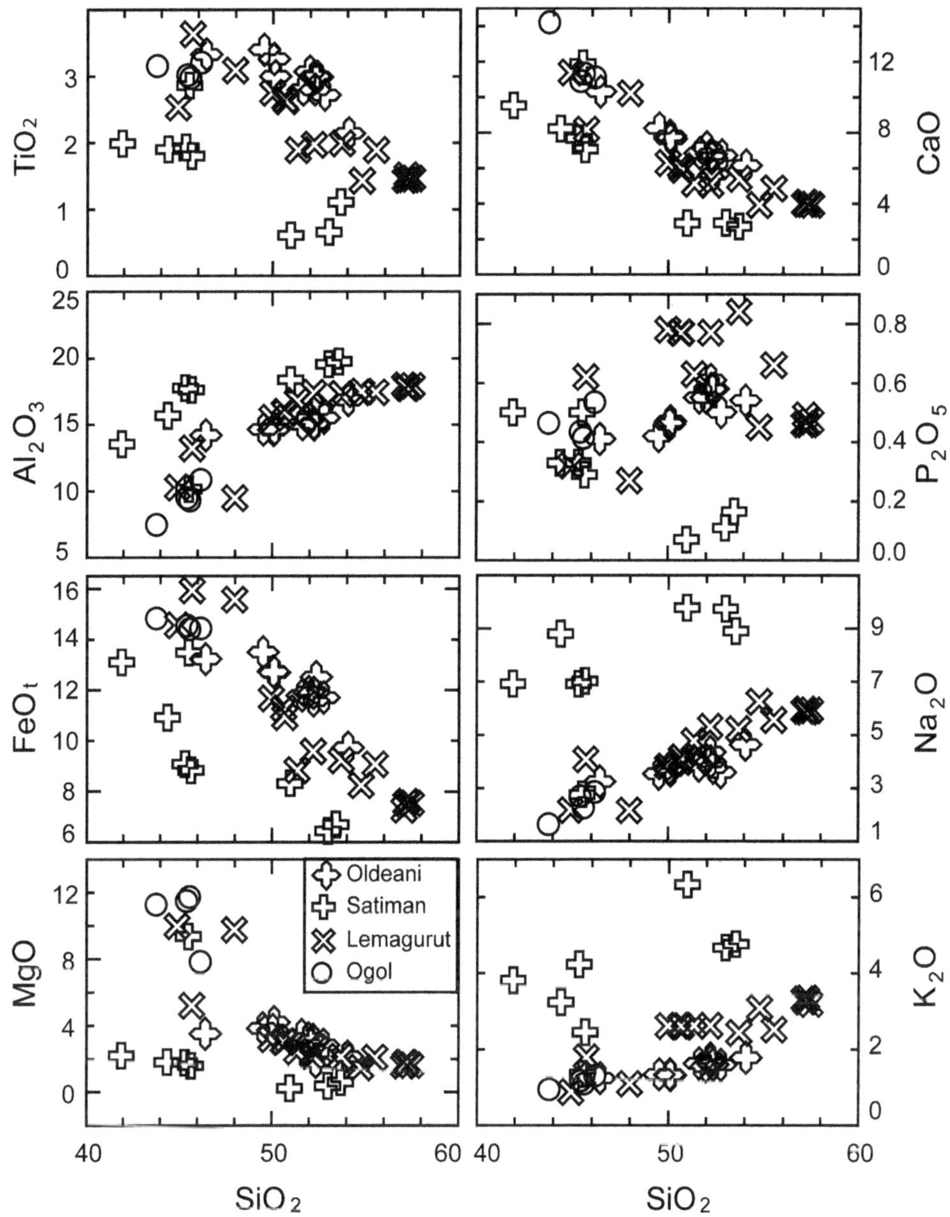

Fig. 5.4 Major elements versus silica Harker diagrams. Decrease in MgO, FeOt, CaO and TiO_2 with increasing silica content is consistent with crystal-liquid fractionation. The fractionating phases are of olivine, augite, and titanomagnetite for the silicic lavas. The decrease of TiO_2 and P_2O_5 also indicates the crystallization of titanite and apatite in the silica undersaturated lavas

chondrites, all the lavas associated with Laetoli deposits are enriched in rare earth elements (REEs, Fig. 5.5). Compared to primitive mantle, the Satiman lavas show negative Ti and P anomalies (Fig. 5.6a). Satiman lavas are also more enriched in Large Ion Lithophile Elements (LILE) compared to the other volcanoes. In all lavas REEs increase with decreasing MgO (not illustrated).

Abundances of most of the trace elements and REEs are positively correlated with La in all silica-saturated lavas, but show almost no change in foidite and phonolite (e.g., the Ba, Pb, Zr, and Nb data for Satiman in Fig. 5.7).

Zr/Nb and La/Nb are not affected by crystal-liquid fractionation (MgO value in Fig. 5.8) and clearly separate into three groups, Satiman, Lemagurut + Ogol, and Oldeani. There is an apparent temporal progression from low Zr/Nb and La/Nb at Satiman to high values of these ratios at Oldeani (Fig. 5.8). Ratios of other incompatible elements (e.g. Ba/La, La/Yb, U/Th, Zr/Hf) are heterogeneous among all the volcanoes (not shown).

Isotopic ratios of Sr and Nd (Table 5.1) display an overall negative correlation, typical of most volcanic areas. Ogol has the lowest Sr isotope ratios (0.70371–0.70381), highest

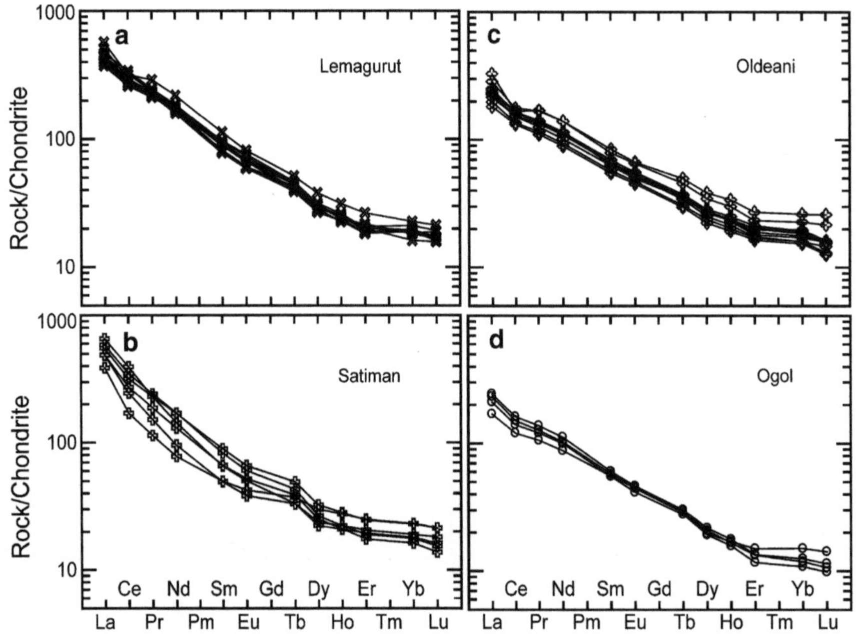

Fig. 5.5 Chondritic normalized rare earth elements (REEs) diagram indicate that volcanics associated with the deposition of Laetolil Beds are enriched in REEs abundances. Normalizing values of Sun and McDonough (1989)

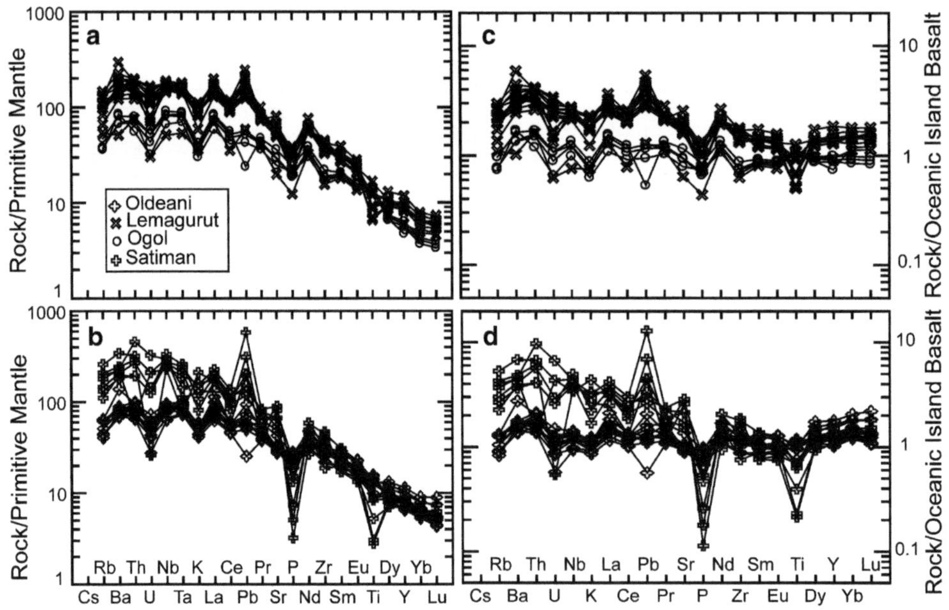

Fig. 5.6 (**a**, **b**) Primitive mantle normalized incompatible elements (REEs) diagram indicate that the lavas associated with the Laetolil Beds are enriched in incompatible elements. Troughs of P and Ti indicate fractionation of apatite and titanite while the positive Pb anomaly indicates more crustal contamination of Satiman lavas compared to others. (**c**, **d**) Oceanic Island Basalt (OIB) normalized trace element ratios for volcanics associated with Laetolil Beds are close to one suggesting an OIB source. (Normalizing values of Sun and McDonough 1989)

Nd isotope ratios (0.512554 – 0.512610), next is Oldeani ($^{87}Sr/^{86}Sr = 0.70406 - 0.70476$, $^{143}Nd/^{144}Nd = 0.512390 - 0.512482$), then Lemagurut ($^{87}Sr/^{86}Sr = 0.70472 - 0.70522$, $^{143}Nd/^{144}Nd = 0.512190 - 0.512470$), and finally Satiman, which has the highest Sr isotope ($^{87}Sr/^{86}Sr = 0.70412 - 0.70609$, and the lowest Nd isotope ($^{143}Nd/^{144}Nd = 0.512056 - 0.512402$). Satiman has a wide dispersion in Sr isotope ratios, suggestive of variable degree of contamination, supported by positive correlation between Sr isotope ratio and SiO_2 content (not illustrated). There is no simple correlation between Sr isotope ratios and SiO_2 content in Ogol and Oldeani lavas perhaps suggesting that these lavas experience limited contamination. Sr isotope ratios are negatively correlated to Mg# in all lavas (Fig. 5.9).

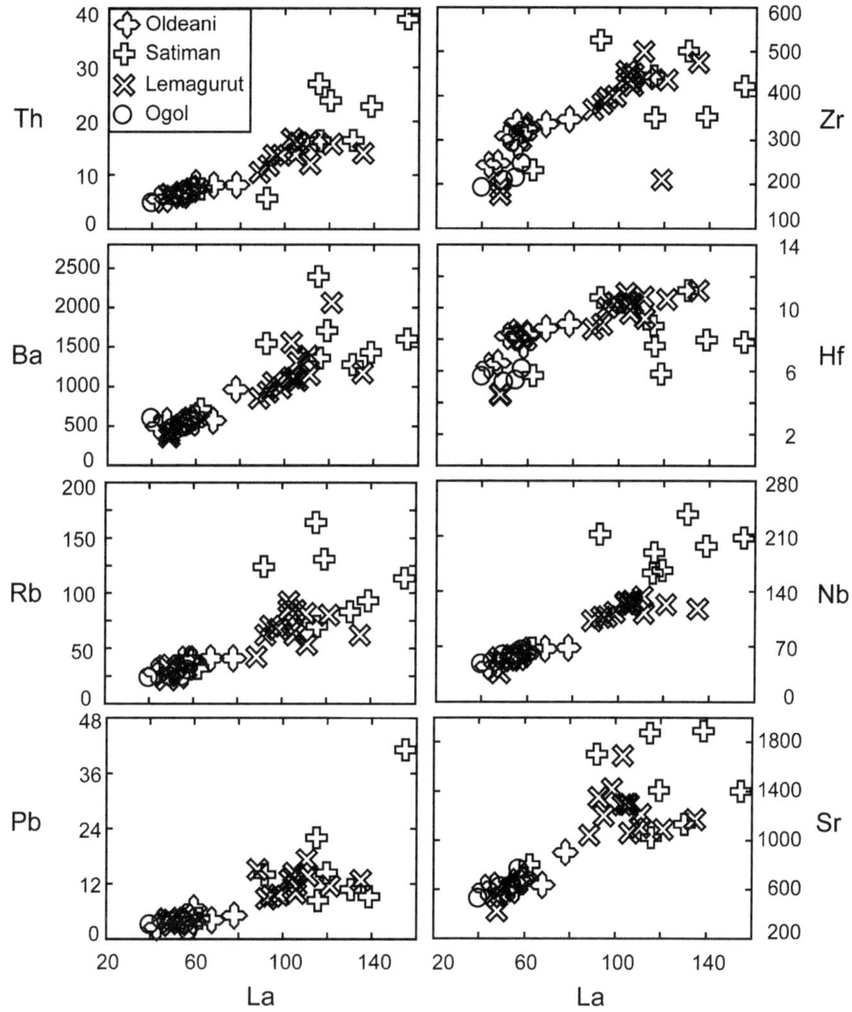

Fig. 5.7 Large ion lithophile elements (LILE) as well as rare earth elements (REEs) correlate positively with La consistent with fractional crystallization for all of the lavas associated with Laetolil Beds. Satiman appears to be higher in most of these elements suggesting an origin from a more enriched liquid or from lesser degree partial melting of the source compared to the others

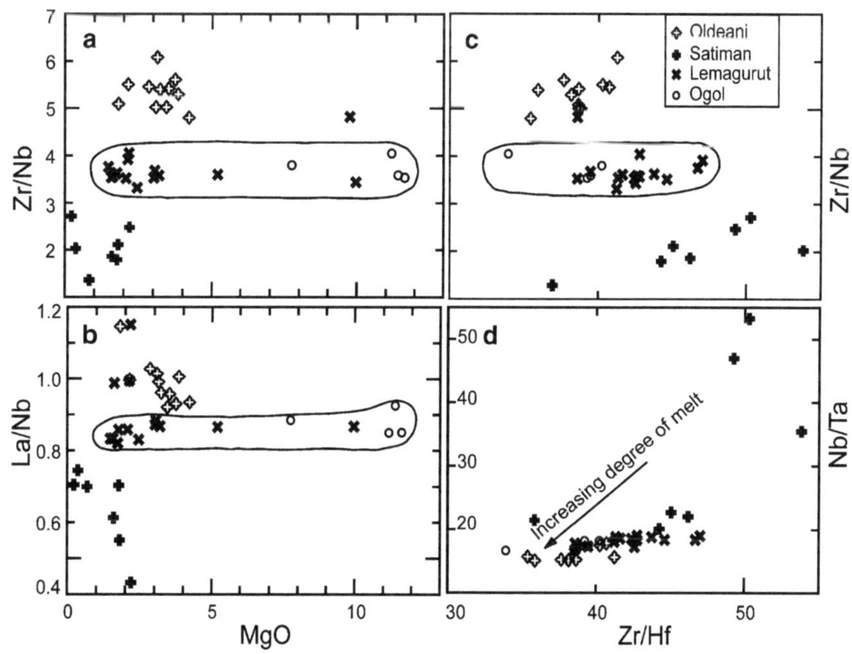

Fig. 5.8 High field strength elements (HFSE) ratios are similar for Lemagurut and Ogol flows and different from that of Oldeani and Satiman. Zr/Hf and Nb/Ta show that Satiman was derived by small degree partial melt compared to the other silicic lavas

Discussion

Evaluation of major elements for the basalt, hawaiite, mugearite, and benmorite from Lemagurut and Ogol, and for basalt and trachyandasite from Oldeani (Fig. 5.4), shows that fractional crystallization was a key process in creating the observed variation. Although the same can be said for the foidite and phonolite from Satiman, the process appears slightly different. The foidite and phonolite, as well as the silica-saturated lavas, indicate similar fractionation trends of major elements where there is a decrease in TiO_2, FeO_t, MgO, MnO and CaO with increasing silica content. Incompatible elements Na_2O and K_2O increase with increasing silica content consistent with fractionation. P_2O_5 is incompatible in the silica-saturated lavas and decreases with increasing silica content in the foidite and phonolite, indicating crystallization of apatite from the melt. The decrease in FeO_t, MgO and CaO accounts for fractional crystallization of olivine and augite. Decrease in TiO_2 indicates the crystallization of titanomagnetite in the silica saturated lavas and sphene in the foidite and phonolite. The foidite and phonolite are consistently lower in TiO_2, FeO_t, MgO, CaO, and P_2O_5 (Fig. 5.4) compared to the silica-saturated lavas, but are higher in incompatible elements (i.e., Na_2O, and K_2O), perhaps indicating that they are derived from a more enriched melt. P_2O_5 is, however low, probably suggesting that the source is not enriched in P. The silica-undersaturated lavas are also higher in LILE (e.g., Rb, Sr, Ba and Th in Figs. 5.5–5.7) compared to the saturated ones. While the variations of trace elements with incompatible element La supports fractional crystallization process (positively correlated) for the silica saturated lavas (Fig. 5.7), High Field Strength Elements (HFSE), such as Nb, Zr, Y and Hf, are almost constant for the silica undersaturated lavas, perhaps indicating that fractional crystallization was not as important in generating the Satiman lavas. The higher concentrations of LILE in Satiman lavas suggest an origin from a more enriched source or by a smaller degree of partial melt (e.g., Mertz et al. 2001; Johnson et al. 2005). HFSE ratios (e.g., Zr/Nb, Hf/Nb and La/Nb) are distinct for the different volcanic centers, perhaps indicating that they are derived as a result of different degrees of partial melting. The degree of partial melting is estimated from the ratios of Zr/Hf and Nb/Ta (Fig. 5.8d). These elements are used because only a very small degree of partial melt can fractionate Zr from Hf and Nb from Ta (Vukadinovic 1994; Bolge et al. 2006). Because Zr and Nb are slightly more incompatible than Hf and Ta (Sun and McDonough 1989), lower Zr/Hf and Nb/Ta indicate a higher degree of melting. Three Satiman lavas have unusually high Nb/Ta ratios, but even discounting these, Fig. 5.8d suggests that Satiman lavas were derived from lower degrees of partial melting than the lavas of Lemagurut, Ogol and Oldeani.

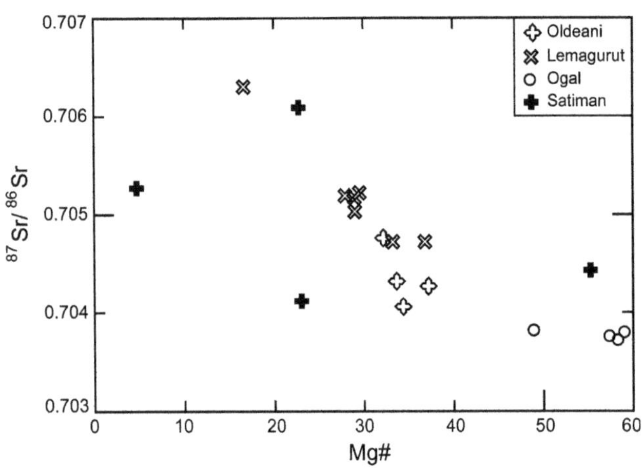

Fig. 5.9 $^{87}Sr/^{86}Sr$ versus Mg# for lavas associated with Laetoli sequence. Increasing $^{87}Sr/^{86}Sr$ with decreasing Mg# suggest increasing crust-magma interaction concomitant with fractionation

Source and Crustal Influence

There is a general consensus that an Oceanic Island Basalt (OIB) source is responsible for lavas in northern Tanzania and in other parts of the East African Rift (Paslick et al. 1995; Späth et al. 2001; Furman et al. 2004, 2006). OIB-normalized trace element ratios close to one for Oldeani and Ogol lavas and higher for Lemagurut and Satiman (Fig. 5.6b), possibly indicate that Lemagurut and Satiman are more fractionated or derived by lower degrees of partial melt. Although there is clear evidence for crustal-magma interaction (Fig. 5.9), the $^{87}Sr/^{86}Sr$ versus $^{144}Nd/^{143}Nd$ plot for most of the lavas under study, fall within or near the field defined by lavas from northern Tanzania that include lavas from Oldoinyo Lengai (Fig. 5.10a; Paslick et al. 1995; Bell and Tilton 2001). Sources for lavas from Oldoinyo Lengai and from other volcanic centers in northern Tanzania have been described to be a result of mixing between a high-μ (HIMU) and an enriched mantle (EMI) reservoir (Paslick et al. 1995; Bell and Tilton 2001; Bizimis et al. 2003), suggesting similar source for the lavas associated with the Laetolil Beds. Moderate, chondritic normalized $(Tb/Yb)_n$ ~2.6–1.0 for these lavas [normalizing values of Sun and McDonough (1989)], are within the range of values found in Turkana lavas that have been implied to indicate melting depth in the lithospheric mantle (Furman et al. 2006).

Compared to the lower section of the Ngorongoro Caldera that yielded a wide range of $^{87}Sr/^{86}Sr$ (0.70405–0.70801; Mollel et al. 2008), the Oldeani and Ogol lavas are less radiogenic, perhaps indicating that they are less contaminated by crustal rocks. Ce/Pb is 25±5 for mantle derived fluids (Hofmann et al. 1986; Sun and McDonough 1989) and is <20 for crustal contaminated melts (Furman 2007). Compared to Naivasha in southern Kenya Rift where crust-magma interaction has been documented (Davies and MacDonald 1987), most of the Oldeani and Ogol samples plot within the range of mantle values, whereas most of Lemagurut and Satiman are below the mantle range (Fig. 5.10b). Augite from Satiman are zoned and embayed, a feature that was also identified by Paslick and Halliday (1996) in a study of the chemistry of

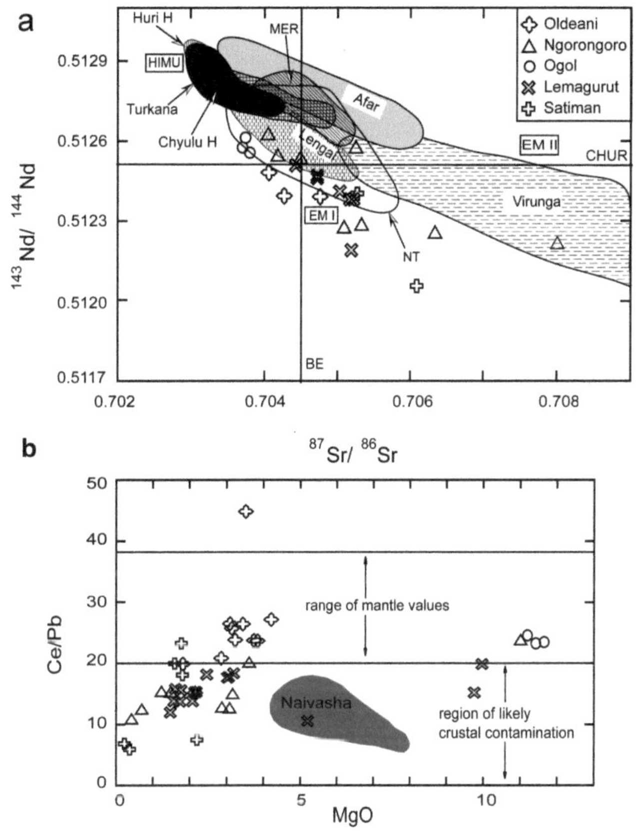

Fig. 5.10 (**a**) $^{143}Nd/^{144}Nd$ versus $^{87}Sr/^{86}Sr$ variation for lavas from volcanics associated with Laetolil Beds compared to other lavas from the East African Rift System, as well as hypothetic mantle endmember compositions. BE, Bulk Earth; CHUR, Chondritic Uniform Reservoir; Huri H., Huri Hills; Chyulu H., Chyulu Hills; MER, Main Ethiopian Rift; NT, Northern Tanzania; Lengai, Oldoinyo Lengai. Source of data: Class et al. 1994; Deniel et al. 1994; Paslick et al. 1995; Bell and Simonetti 1996; Kalt et al. 1997; Rogers et al. 1998; Trua et al. 1999; Bell and Tilton 2001; Späth et al. 2001; Furman et al. 2004, 2006; Keller et al. 2006; Mollel et al. 2008, 2009. HIMU, EMI and EM II fields from Hart et al. 1992). Most of the representative lavas from Lemagurut, Oldeani and Ogol flow plot within or close to a field defined by lavas from northern Tanzania and Oldoinyo Lengai suggesting a source defined by mixing EMI and HIMU mantle endmembers. Ngorongoro, Olmoti, Embagai and Engelosin from Mollel (2007). (**b**) Variation of Ce/Pb sensitive to crustal contamination as inferred from Naivasha (*shaded*) area (Davies and MacDonald 1987). The majority of Satiman and Lemagurut lavas plot with the region of crustal contamination similar to lavas from lower Ngorongoro section (Mollel et al. 2008). Most of Oldeani and Ogol lavas are less contaminated and fall within the range of mantle values (Hofmann et al. 1986; Sun and McDonough 1989)

minerals from alkali basalts and nephelinites from northern Tanzania. The latter study concluded that there is a chemical and isotopic disequilibrium between the minerals and their host rocks. They argued that the disequilibrium resulted from up to 10% assimilation of upper crust. The NVHC is erupting through Precambrian basement rocks, the Tanzanian Craton (Cahen et al. 1984), on the west of the NVHC, and the Mozambican mobile belt (Key et al. 1989; Möller et al. 1998) to the east. Assimilation-fractional crystallization (AFC) modeling on Pb isotopic ratios (Paslick et al. 1995) is in agreement with a lower crust contaminant. High $^{87}Sr/^{86}Sr$ ratios measured on some mantle xenoliths collected from Lashaine e.g., from clinopyroxene 0.8360, garnet 0.8345, and phlogopite 0.8196, separated from a garnet lherzolite (BD-738, Cohen et al. 1984), are high enough to make lower crust a contaminant.

A model for the NVHC evolution thus follows that of Späth et al. (2001) for the Chyulu Hills, whereby fluids/melts originating from an ascending mantle plume (Nyblade et al. 2000; Weeraratne et al. 2003) metasomatize and partially melt the lithospheric mantle beneath the NVHC. Early low-degree partial melts reflects the geochemistry of the plume itself, while later higher-degree partial melts are mixtures of the metasomatic component (plume) and the lithospheric wall-rock. The NVHC geochemistry therefore reflects a binary mixture of these two components. It is clear, however, from isotope and trace element data that the lavas under study have undergone variable degrees of crustal contamination. A third component for the NVHC chemistry therefore comes from the crust. This last component most certainly accounts for the wider variability in the isotope geochemistry displayed by these lavas.

Geochronology

Oldeani

Matrix from four lava samples collected from the eastern (03-OLDI-1 and 03-OLDI-2) and western (04-OLDI-4 and 04-OLDI-5) flanks of Oldeani were dated using $^{40}Ar/^{39}Ar$ laser incremental-heating techniques. The analyses resulted in well-behaved Ar release patterns forming plateau ages ranging from 1.61±0.01 Ma to 1.52±0.02 Ma (Fig. 5.11a; Table 5.6). Integrated ages are comparable to the plateau ages except for 03-OLDI-1, which produced a much younger age, likely due to low temperature alteration of the sample. A combined isochron age for the four analyses yielded an age of 1.57±0.02 Ma that is indistinguishable from the mean plateau age. Two other samples (04-OLDI-3 and 04-OLDI-4) from the western flanks did not yield plateau ages, but gave comparable integrated ages of 1.73 ± 0.01 Ma and 1.61 ± 0.01 Ma respectively. These two samples also yielded indistinguishable isochron ages of 1.57 ± 0.03 Ma and 1.61 ± 0.03 Ma respectively (Table 5.6). These ages suggest that Oldeani activity was of very short duration, lasting for about 90 ka between 1.61 ± 0.01 Ma and 1.52 ± 0.02 Ma.

Satiman

Matrix and nepheline from three lavas (01-SM-3, 01-SM-5 and 04-NT-3) were analyzed by the incremental-heating technique. 01-SM-3 and 01-SM-5 was collected from the present-day summit of Satiman volcano, whereas 04-NT-3 is from the Noiti area, a lowland south of Satiman from which Satiman lahar has been deposited. Analysis of nepheline from lava 01-SM-3 produced an integrated age of 6.29 ± 0.04 Ma (Table 5.6) that is older than its isochron age of 5.7 ± 0.3 Ma (Table 5.6). Although the $^{40}Ar/^{36}Ar$ intercept of the isochron is within atmospheric ratio (295.5), the mean sum of the weighted deviates (MSWD) is large (6.4).

Analysis of matrix from the same sample produced an integrated age of 6 ± 3 Ma that is indistinguishable from the nepheline isochron age as well as its plateau age of 4.63 ± 0.05 Ma (Table 5.6). Its isochron age of all increments is 5.7 ± 0.3 Ma, analytically indistinguishable from that obtained from the nepheline separate. The $^{40}Ar/^{36}Ar$ intercept is, however, not atmospheric (271 ± 4) and the MSWD is large (12). The isochron of only the plateau increments yielded an indistinguishable age of 4.8 ± 0.4 Ma (MSWD = 2.1, and $^{40}Ar/^{36}Ar$ intercept of 292 ± 7) suggesting that the plateau age is a better age estimate for this sample. The higher increments may contain excess ^{40}Ar. Analysis of nepheline from 01-SM-5 gave an integrated age of 4.99 ± 0.03 Ma that is older than its isochron age of 4.51 ± 0.02 Ma (Table 5.6), again suggesting slight excess ^{40}Ar in the higher temperature increments. Analysis of matrix from the same sample gave an integrated age of 4.35 ± 0.01 Ma and an isochron age of 4.19 ± 0.02 Ma (Table 5.6). The $^{40}Ar/^{36}Ar$ intercept for 01-SM-5 is within atmospheric ratio and the MSWD is 2.9 suggesting that the isochron age is a better age estimate for this sample.

Analysis of matrix from a phonolite 04-NT-3, gave an integrated age of 4.11 ± 0.01 Ma (Fig. 5.11b; Table 5.6) and a slightly younger (by 2.2%) isochron age of 4.02 ± 0.02 Ma. The $^{40}Ar/^{36}Ar$ intercept is within atmospheric ratio but the MSWD of the isochron is large (5.1).

Taken together, the $^{40}Ar/^{39}Ar$ dates for Satiman volcano range between 4.63 ± 0.05 Ma and 4.02 ± 0.02 Ma and can be as old as 4.8 ± 0.4 Ma. The data also indicate that excess ^{40}Ar in nepheline may explain some of the erroneously old K-Ar ages for Satiman. The new ages and mineral assemblage support a conclusion by Mutakyahwa (1997) who suggested a carbonatitic volcanic source for the Wembere-Manonga sediments. Biostratigraphic data on fauna collected from the Manonga Valley sediments (Harrison et al. 1993; Harrison and Baker 1997) suggested a late Miocene to early Pliocene (5.5–4.0 Ma) age for the Manonga-Wembere Formation.

Lemagurut

Matrix from three representative lava samples 00-LM-6, and 00-LM-7, from the south and 02-LM-4 from the northwestern flanks of Lemagurut were analyzed. Matrix from 02-LM-4 yielded an integrated and plateau age of 2.23 ± 0.01 Ma and an isochron age of 2.22 ± 0.02 Ma (Fig. 5.11c; Table 5.6). Matrix from 00-LM-6 gave the same integrated, plateau and isochron age of 2.25 ± 0.01 (Table 5.6).

Matrix from 00-LM-7 gave a plateau age of 2.26 ± 0.005 Ma that is slightly younger than its integrated age of 2.27 ± 0.004 Ma (Table 5.6). Its isochron age of 2.24 ± 0.2 Ma is indistinguishable from the plateau age, but has a large MSWD (4.3). Three samples (04-OS-1, 04-OS-3 and 04-OS-4) collected from the southeast flanks of Lemagurut in Osilale village were also dated. Anorthoclase from 04-OS-1 gave a plateau age of 2.22 ± 0.10 Ma that is indistinguishable from its integrated and isochron age of 2.20 ± 0.10 Ma and 2.23 ± 0.10 respectively (Table 5.6). Matrix from 04-OS-4 yielded a plateau age of 2.29 ± 0.004 Ma that is slightly younger than its integrated age of 2.34 ± 0.004. This plateau age is indistinguishable from that obtained from 04-OS-1 anorthoclase. A combined isochron plot of the two analyses gave an age of 2.30 ± 0.02 Ma that is indistinguishable from both plateau ages. Matrix from the third sample (04-OS-3) yielded a plateau age of 2.40 ± 0.01 Ma that is much younger than its integrated age of 2.55 ± 0.01 Ma (Fig. 5.11c; Table 5.6). The isochron age of 2.38 ± 0.02 Ma is analytically indistinguishable from the plateau age. These age data suggest that Lemagurut activity was short in duration lasting only about 180 ka between 2.40 ± 0.01 Ma and 2.22 ± 0.10 Ma.

Ogol

Three lavas (04-OG-1, 04-OG-2 and 04-OG-4) were analyzed. 04-OG-1 was collected from a ridge near fossil Loc. 10E, 04-OG-2 from Oldonyo Mati (Ndonyamati) and 04-OG-4 from a ridge near Ndolanya Hill. Whole rock analysis of 04-OG-1 and 04-OG-2 yielded indistinguishable plateau ages of 2.31 ± 0.01 Ma and 2.27 ± 0.05 Ma respectively.

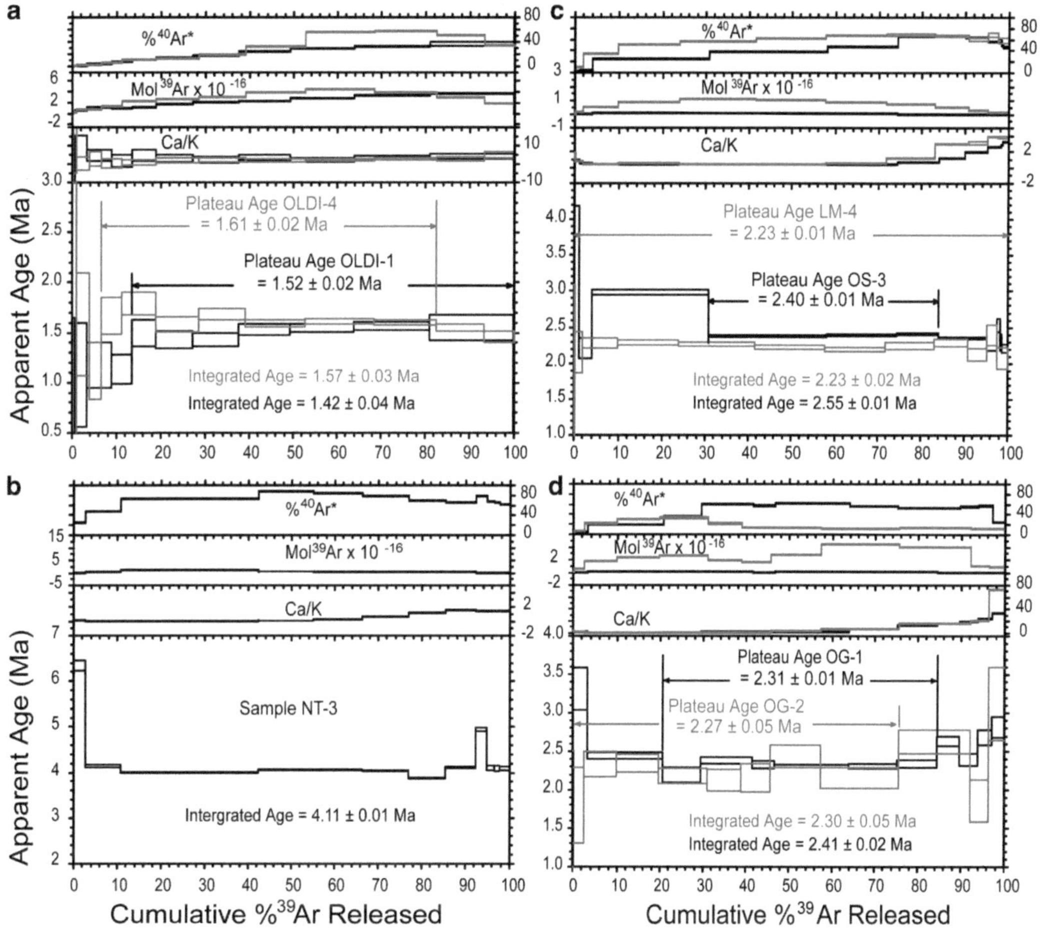

Fig. 5.11 $^{40}Ar/^{39}Ar$ Incremental heating spectra for lavas associated with Laetolil Beds. (**a**) Seven-step plateau ages for matrix from Oldeani basalts. The integrated age is indistinguishable from its plateau age for the lower boundary, whereas that of the upper boundary is younger than its plateau age likely due to K loss during low temperature alteration of the lava. Cumulative% $^{39}Ar^*$ released on the plateau steps are 86.10 and 75.80 for the upper and lower boundary respectively. Plateau temperatures are about 850°C and 1,100°C. (**b**) Total fusion age spectra for Satiman phonolite collected from Noiti 3 area. The total fusion age is older than its isochron age by 2.2%. (**c**) Three and twelve-step plateau ages for matrix from Lemagurut lavas. The integrated age is indistinguishable from its plateau age for the upper boundary, but that of the lower boundary is much older (2.55±0.01 Ma) likely due to excess argon resulted from low-temperature alteration of the lava (lower temperature ages are much older). Cumulative%$^{39}Ar^*$ released on the plateaus are 100 and 53.3 for the upper and lower boundaries respectively. Plateau temperatures are between 500°C and 1,200°C for the upper boundary and 850°C and 950 °C for the lower boundary. (**d**) Plateau age for matrix from Ogol basalt is indistinguishable from its integrated age for the upper boundary while that of the lower boundary is much younger than its integrated age likely due to excess argon resulted from low-temperature alteration of the basalt. Cumulative%$^{39}Ar^*$ for the plateau steps are 75.6 and 55 for the upper and lower boundary, obtained from a nine and six step plateau produced at temperatures between 550–1,000°C and 570–900°C respectively

Integrated ages for both samples are comparable to their plateau ages (Fig. 5.11d; Table 5.6). Combined isochron ages of the two analyses yielded an age of 2.29±0.02 Ma indistinguishable from the plateau ages. The third sample (04-OG-4) did not produce a plateau age, but yielded an integrated age of 2.52±0.01 Ma that is indistinguishable from an isochron age of 2.51±0.05 Ma (Table 5.6). The $^{40}Ar/^{36}Ar$ intercept for the isochron is 316±6 indicating that the lava contains excess argon (McDougall and Harrison 1999) and its age may therefore be unreliable. These data suggest that Ogol flows erupted between 2.31±0.01 Ma and 2.27±0.05 Ma.

Implications for Laetoli Deposits and NVHC

The new $^{40}Ar/^{39}Ar$ ages for Satiman, Lemagurut, Oldeani, and Ogol lavas indicate that Satiman volcano was active between 4.63±0.05 Ma and 4.02±0.02 Ma. These ages are comparable to those previously reported by Bagdasaryan et al. (1973). Dated samples range from the flanks of Satiman to core samples near its summit. The ages do not cover the entire range of ages reported for the duration of the Laetolil Beds from >4.30 Ma to 3.49±0.12 Ma (Drake and Curtis 1987; Hay 1987; Deino 2011). Moreover, no coarse-grained xenocrystic flows comparable to those interbedded within the Laetoli Beds were recovered on Satiman. There are three likely explanations for this discrepancy: (1) eruptions associated with the deposition of Laetolil Beds were primarily tuffaceous with no lava flows. Although Plinian and Vulcanian eruptions are typical in silicic lavas (e.g., the Vesuvius 79 AD eruption), they can also occur in silica-undersaturated volcanism (e.g., Oldoinyo Lengai, Dawson et al. 1968; Nielsen and Veksler 2002); (2) younger eruptive events may now be entirely eroded and missing from Satiman itself, and (3) a slightly younger vent of similar composition was the source of the Upper Laetolil ashes, perhaps now buried by the thick flows of the younger Lemagurut volcano. For these reasons we also failed to locate the source for the Ndolanya Beds (3.5–2.6 Ma; Hay 1987; Ndessokia 1990). Our data also supports a Satiman source for the Wembere-Manonga Formation (Mutakyahwa 1997; Verniers 1997).

The composition and age of analyzed samples from Lemagurut and the Ogol lavas are similar, ranging in age from 2.40 ± 0.01–2.22 ± 0.10 Ma. Previously older and younger ages (Bagdasaryan et al. 1973; Manega 1993) for Lemagurut are attributed to analyses made of distal flows of adjacent volcanoes, not Lemagurut itself. Ages for Ogol lavas from Loc. 10E and that from Oldoinyo Mati are indistinguishable, suggesting that the lavas were likely from the same event. It is possible that they are different lava flows, given the distance (~6 km) between them, but that cannot be resolved with the current data. The new plateau age of 2.31±0.01 Ma is younger than the 2.52±0.07 Ma obtained by K-Ar technique from lava collected from Loc. 10E (Drake and Curtis 1987). Age data of lava from the Ndolanya Hill contains excess argon, indicating the age may be less reliable. The presence of excess argon was likely the cause for older K-Ar ages.

Overall, these dates indicate that the Ogol lava erupted between 2.31±0.01 Ma and 2.27±0.05 Ma. These ages fall within the range of Lemagurut activity (from 2.40±0.01 Ma to 2.22±0.10 Ma) suggesting overlapping activity between Lemagurut and Ogol. Although Hay (1987) suggested that Ogol lavas erupted from a series of individual vents, three lines of evidence from our study suggest that the Ogol flows likely erupted from Lemagurut. (1) Absence of volcanic vents: Three lava ridges were sampled during this study and no intrusive or obvious vent was seen in the vicinity of these ridges. Most of the lavas are massive and only a few outcrops of scoria were seen on one of the ridges. (2) Petrography: Although Ogol lavas are mostly porphyritic compared to most of Lemagurut, which are generally medium-grained, our work has shown that their chemistry and mineral assemblages are similar. (3) Geochronology: Age and duration of volcanic activity for Ogol and Lemagurut from this study are indistinguishable within experimental error suggesting overlapping activity. We suggest that the Ogol lavas may be genetically related to nearby Lemagurut volcanism and may be distal flows that followed paleodrainages. The limited expanse of the Ogol flows are now the result of inverted topography and differential erosion of the adjacent sediments compared with the much resistant lavas. Additional fieldwork is required to verify this hypothesis.

The age of Lemagurut is consistent with volcanic deposits of the Naibadad Beds overlying the Laetolil Beds. However, the Naibadad mineral assemblage includes quartz (Hay 1987), a composition not compatible with Lemagurut. Revised $^{40}Ar/^{39}Ar$ age for Ngorongoro (2.28 ± 0.02 – 2.04 ± 0.01 Ma; Mollel et al. 2008) suggest overlapping activity with Lemagurut, and indicate that Ngorongoro is a better source for Naibadad deposits in terms of age and composition, as previously inferred by Hay (1987). The Ngorongoro age is also consistent with the deposition of the Olpiro Beds (2.00±0.01 Ma; Ndessokia 1990; Manega 1993) that overlie the Naibadad Beds. The phonolitic composition of Olpiro is, however, not compatible with that of Ngorongoro. Olmoti Crater, whose activity overlaps in age (2.01±0.03 Ma to 1.80±0.01 Ma; Mollel et al. 2009) with Ngorongoro, could be a better source. Correlation of the Olpiro deposits with Olduvai Beds has been suggested by Hay (1987) and supported by age data from Ndessokia (1990) and Manega (1993). The Olduvai Coarse Feldspar Crystal Tuff (CFCT), which has been correlated with Olpiro by Ndessokia (1990) and Manega (1993), has been shown to consist of three different compositions: basaltic, trachytic and rhyolitic (McHenry 2004). Olmoti, which has been inferred as the source for most of the Olduvai Bed I (Hay 1976; McHenry et al. 2008), is basaltic to trachytic in composition (Mollel et al. 2009). The trachyte from Olmoti contains trace amounts of silica undersaturated minerals, e.g., sodalite, and may in fact be a better source for the Olpiro Beds. The fine-grained nature of the Olpiro Beds (Hay 1987) may also support a distal source.

Oldeani is the youngest basaltic center in the NVHC. Oldeani activity was apparently short in duration lasting for 90 ka between 1.61±0.01 and 1.52±0.02 Ma. The age data and the geographic location of Oldeani are inconsistent with the northward younging direction previously proposed by Manega (1993). Our ages for Oldeani are inconsistent with this general trend and may relate to satellite vents of similar age that followed the collapse of the 2.28–2.04 Ma Ngorongoro (Mollel et al. 2008).

Plotting of the La/Nb ratio (an index of relative degree of partial melting, discussed in the text) versus age for the entire NVHC (Fig. 5.12) suggests the dominance of silica-undersaturated lavas (phonolite, foidite and nephelinite) from Satiman, Engelosin, Embagai, and Oldoinyo Lengai over time. The silica poor lavas are derived by lower degree partial melting of the source relative to the silica-saturated lavas from Lemagurut, Ngorongoro, Olmoti and Oldeani. From about 5 Ma to present, the NVHC eruptions were predominantly

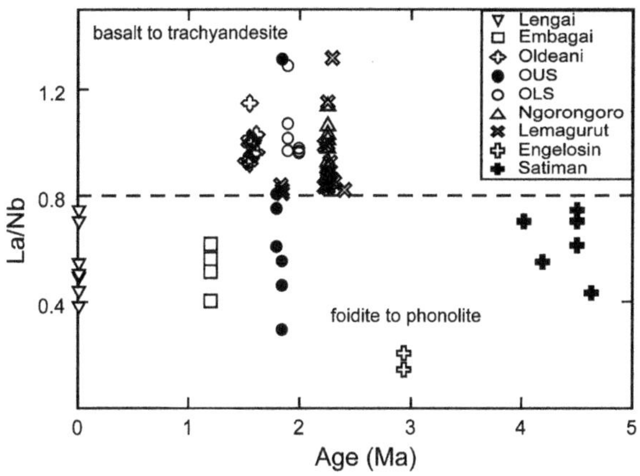

Fig. 5.12 La/Nb versus age for lavas from the Ngorongoro Volcanic Highland Complex (NVHC) indicate that over time the NVHC is dominated by silica undersaturated lavas, derived from lower degree partial melting of the source relative to the silica saturated. The eruption of silicic lavas between 2.4 and 1.6 Ma may reflect perturbed mantle potential temperatures or melting from a deeper source. Data source: Oldoinyo Lengai from Klaudius and Keller (2006); Ngorongoro, Olmoti, Embagai and Engelosin from Mollel (2007); Mollel et al. (2008, 2009). OLS, Olmoti Lower Section; OUS, Olmoti Upper Section

silica-poor with silica-saturated lavas erupting between 2.4 and 1.6 Ma. The eruption of silica-saturated lavas may reflect perturbed mantle potential temperatures.

Conclusions

Major and trace elements from volcanoes associated with the formation of the Laetolil Beds show that fractional crystallization was a key process in producing Oldeani, Lemagurut and Ogol lavas. Fractionating phases include olivine, augite, plagioclase and titanomagnetite. Apatite, amphibole and biotite are rare. Satiman appears to have been derived by small degrees of partial melting of the source that led to the enrichment of HFSE, and LILE. The Satiman mineral assemblage includes nepheline, augite ± titanomagnetite, andradite, analcite, apatite, olivine, sphene and sanidine.

Similarity in HFSE ratios for Lemagurut and Ogol (Zr/Nb = 3.75) and La/Nb suggests a common parent lineage. Satiman and Oldeani have unique HFSE ratios, e.g., Zr/Nb = 2.16 and 5.54 respectively. Sr and Nd isotope data from this work are in agreement with previous workers (Paslick et al. 1995; Bell and Tilton 2001), who suggested an OIB source for the volcanics in northern Tanzania. In a plot of $^{87}Sr/^{86}Sr$ verses $^{144}Nd/^{143}Nd$, most of the samples fall within the fields defined by lavas from northern Tanzania and other volcanic centers in the EARS whose source is explained by mixing of the HIMU and EMI mantle endmembers. However, the wider dispersion of Sr and Nd isotope in these lavas indicates variable degrees of contamination by crustal rocks.

Our work in the NHVC has so far failed to recover lavas dating between 4.0–2.5 Ma, an age bracketing a large part of the Laetoli section (Upper Laetolil and Ndolanya Beds). Satiman is still the only plausible source for Upper Laetolil and Ndolanya Beds based on its composition and geographical location. This suggests that tuffs and volcaniclastic material forming the Laetolil and Ndolanya Beds were deposited by Plinian eruptions from Satiman or a satellite vent from Satiman now covered by younger eruptions from Lemagurut.

Lavas from the ancestral Satiman volcano are dated between 4.63 ± 0.05 Ma to 4.02 ± 0.02 Ma. Lemagurut and Ogol flows are coeval and erupted between 2.40 ± 0.01 Ma and 2.22 ± 0.10 M. Lemagurut ages indicate overlapping activity with Ngorongoro and age data from both centers are consistent with the deposition of Naibadad Beds. The Naibadad composition is, however, compatible with Ngorongoro as a source. Olpiro Beds may have originated from a more distal Olmoti Crater.

Oldeani is the youngest (1.61 ± 0.01 and 1.52 ± 0.02 Ma) basaltic center in the NVHC. Its age and geographic location are inconsistent with the northward younging direction previously proposed by Manega (1993). This may relate to satellite vent eruptions following the collapse of the adjacent 2.28–2.04 Ma Ngorongoro.

Acknowledgements We would like to thank the Tanzania Commission for Science and Technology and the Tanzania Antiquities Department for granting permission to conduct research at Laetoli, Olduvai, and Ngorongoro volcanic highlands. The Laetoli Project and Olduvai Landscape Paleoanthropology Project (OLAPP) provided field support. Terry Harrison and Gail Ashley provided plane tickets to the field, Robert Blumenschine and Fidelis Masao provided field gear, and Craig Feibel and Jerry Delaney provided support during data collection. Funding for this research was partly provided by the L.S.B. Leakey Foundation to Mollel, Sigma Xi, NSF (Harrison: BCS-0309513; Ashley: EAR 9903258; Swisher: BCS-0109027; Feibel: BCS 0218511) and the Department of Earth and Planetary Sciences, Rutgers University. Comments on an earlier version of this manuscript by anonymous reviewers were of great value.

References

Adelsberger, K. A, Wirth, K. R., & Mabulla, A. Z. P. (2002). Provenance of middle stone age tools of the Laetoli archaeological site. Tanzania. *Geological Society of America (GSA), Abstract with Programs* 34, 110.

Adelsberger, K. A., Wirth, K. R., Mabulla, A. Z. P., & Bowman, D. C. (2011). Geochemical and mineralogical characterization of middle stone age tools of Laetoli, Tanzania and comparisons with possible source materials. In T. Harrison (Ed.), *Paleontology and geology of Laetoli: Human evolution in context* (Geology, geochronology, paleoecology and paleoenvironment, vol. 1, pp. 143–165). Dordrecht: Springer.

Bagdasaryan, G. P., Gerasimovskiy, V., Polyakov, A. I., & Gukasyan, R Kh. (1973). Age of volcanic rocks in the rift zones of East Africa. *Geochemistry International, 10*, 66–71.

Bell, K., & Simonetti, A. (1996). Carbonatite magmatism and plume activity: Implications from the Nd, Pb and Sr isotope systematics of Oldoinyo Lengai. *Journal of Petrology, 37*, 1321–1339.

Bell, K., & Tilton, G. (2001). Nd, Pb and Sr isotopic compositions of East African carbonatites: Evidence for mantle mixing and plume inhomogeneity. *Journal of Petrology, 42*, 1927–1945.

Bevington, P. R. (1969). *Data reduction and error analysis for physical sciences*. New York: McGraw-Hill.

Bizimis, M., Salters, V. J. M., & Dawson, B. D. (2003). The brevity of carbonatite source in the mantle: Evidence from Hf isotope. *Contributions to Mineralogy and Petrology, 145*, 281–300.

Bolge, L., Carr, M. J., Feigenson, M., & Alvarado, G. (2006). Geochemical stratigraphy and magmatic evolution at Arenal volcano, Costa Rica. *Journal of Volcanology and Geothermal Research, 157*, 34–48.

Cahen, L., Snelling, J. J., Delhal, J., & Vail, J. R. (1984). *The geochronology and evolution of Africa*. Oxford: Clarendon.

Carr, M. J., Saginor, I., Alvarado, G. E., Bolge, L. L., Lindsay, F. N., Milidakis, K., Turrin, B., Feigenson, M. D., & Swisher, C. C., III. (2007). Element fluxes from the volcanic front of Nicaragua and Costa Rica. *Geochemistry, Geophyics, Geosystems, 8*, Q06001. doi:10.1029/2006GC001396.

Class, C., Altherr, R., Volker, F., Eberz, G., & McCulloch, M. (1994). Geochemistry of Pliocene to Quaternary alkali basalts from the Huri Hills, northern Kenya. *Chemical Geology, 113*, 1–22.

Cohen, R. S., O'Nions, R. K., & Dawson, J. B. (1984). Isotope geochemistry of xenoliths from East Africa: Implications for development of mantle reservoirs and the interaction. *Earth and Planetary Science Letters, 68*, 209–220.

Cox, A. (1969). Geomagnetic reversals. *Science, 163*, 237–245.

Cox, K. G., Bell, J. D., & Pankhurst, R. J. (1979). *The interpretation of the igneous rocks*. London: Allen and Unwin.

Curtis, G. H., & Hay, R. L. (1972). Further geological studies and potassium-argon dating at Olduvai Gorge and Ngorongoro Crater. In W. W. Bishop & J. A. Miller (Eds.), *Calibration of hominoid evolution* (pp. 289–301). Edinburgh: Scottish Academic.

Davies, G. R., & MacDonald, R. (1987). Crustal influence in the petrogenesis of the naivasha basalt–comendite complex: Combined trace element and Sr–Nd–Pb isotope constraints. *Journal of Petrology, 28*, 1009–1031.

Dawson, J. B., Bowden, P., & Clark, G. C. (1968). Activity of the carbonatite volcano Oldoinyo Lengai, 1966. *International Journal of Earth Sciences, 57*, 865–879.

Deino, A. L. (2011). $^{40}Ar/^{39}Ar$ dating of Laetoli, Tanzania. In T. Harrison (Ed.), *Paleontology and geology of Laetoli: Human evolution in context* (Geology, geochronology, paleoecology and paleoenvironment, vol. 1, pp. 77–97). Dordrecht: Springer.

Deino, A. L., & McBrearty, S. (2002). $^{40}Ar/^{39}Ar$ dating of the Kapthurin Formation, Baringo, Kenya. *Journal of Human Evolution, 42*, 185–210.

Deniel, C., Vidal, P., Coulon, C., Vellutini, P., & Piguet, P. (1994). Temporal evolution of mantle sources during continental rifting: The volcanism of Djibouti (Afar). *Journal of Geophysical Research, 99*, 2853–2869.

Ditchfield, P., & Harrison, T. (2011). Sedimentology, lithostratigraphy and depositional history of the Laetoli area. In T. Harrison (Ed.), *Paleontology and geology of Laetoli: Human evolution in context* (Geology, geochronology, paleoecology, and paleoenvironment, vol. 1, pp. 47–76), Dordrecht: Springer.

Drake, R., & Curtis, G. (1987). K-Ar geochronology of the Laetoli fossil localities. In M. D. Leakey & J. M. Harris (Eds.), *Laetoli: A Pliocene site in northern Tanzania* (pp. 48–52). Oxford: Clarendon.

Furman, T. (2007). Geochemistry of East African rift basalts: An overview. *Journal of African Earth Sciences, 48*, vol. 1, pp.147–160.

Furman, T., Bryce, J., Karson, J., & Iotti, A. (2004). East African Rift System (EARS) plume structure: Insights from Quaternary mafic lavas from Turkana, Kenya. *Journal of Petrology, 45*, vol. 1, pp. 1069–1088.

Furman, T., Kaleta, K. M., Bryce, J. G., & Hanan, B. B. (2006). Tertiary mafic lavas of Turkana, Kenya: Constraints on the East African plume structure and the occurence of high-μ volcanism in Africa. *Journal of Petrology, 47*, 1221–1244.

Hannah, R. S., Vogel, T. A., Patino, L. C., Alvarado, G. E., Pérez, W., & Smith, D. R. (2002). Origin of silicic volcanic rocks in Central Costa Rica: A study of a chemically variable ash-flow sheet in the Tiribí Tuff. *Bulletin of Volcanology, 64*, 117–133.

Harrison, T., & Baker, E. (1997). Paleontology and biochronology of fossil localities in the Manonga Valley, Tanzania. In T. Harrison (Ed.), *Neogene paleontology of the Manonga Valley, Tanzania* (pp. 361–393). New York: Plenum Press.

Harrison, T., Verniers, J., Mbago, M. L., & Krigbaum, J. (1993). Stratigraphy and mammalian paleontology of Neogene sites in the Manonga valley, northern Tanzania. *Discovery and Innovation, 5*, 269–275.

Hart, S. R., Hauri, E. H., Oschmann, L. A., & Whitehead, J. A. (1992). Mantle plumes and entrainment: Isotopic evidence. *Science, 256*, 517–520.

Hay, R. (1976). *Geology of the Olduvai Gorge: A study of sedimentation in a semi-arid basin*. Berkeley: University of California Press.

Hay, R. L. (1987). Geology of the Laetoli area. In M. D. Leakey & J. M. Harris (Eds.), *Laetoli: A Pliocene site in northern Tanzania* (pp. 23–47). Oxford: Clarendon.

Hofmann, A. W., Jochum, K. P., Seufert, M., & White, W. M. (1986). Nb and Pb in oceanic basalts: New constraints on mantle evolution. *Earth and Planetary Science Letters, 79*, 33–45.

Johnson, J. S., Gibson, S. A., Thompson, R. N., & Nowell, G. M. (2005). Volcanism in the Vitim volcanic field, Siberia: Geochemical evidence for a mantle plume beneath the Baikal rift zone. *Journal of Petrology, 46*, 1309–1344.

Jones, A. P., Smith, J. V., Hansen, E. C., & Dawson, B. D. (1983). Metamorphism, partial melting and K-metasomatism of garnet-scapolite-kyanite granulite xenoliths from Lashaine, Tanzania. *Journal of Geology, 91*, 143–165.

Kalt, A., Hegner, E., & Satir, M. (1997). Nd, Sr, and Pb isotopic evidence for diverse lithospheric mantle sources of East African rift carbonatites. *Tectonics, 278*, 31–45.

Keller, J., Zaitsev, A. N., & Wiedenmann, D. (2006). Primary magmas at Oldoinyo Lengai: The role of olivine melilitites. *Lithos, 91*, 150–172.

Key, R., Charsley, T., Hackman, B., Wilkinson, A., & Rundle, C. (1989). Superimposed upper Proterozoic collision-controlled orogenies in the Mozambican orogenic belt of Kenya. *Precambrian Research, 44*, 197–225.

Klaudius, J., & Keller, J. (2006). Peralkaline silicate lavas at Oldoinyo Lengai, Tanzania. *Lithos, 91*, 173–190.

Lee, C.-T., & Rudnick, R. L. (1999). Compositionally stratified cratonic lithosphere: Petrology and geochemistry of peridotite xenoliths from the Labait tuff cone, Tanzania. In J. J. Gurney, J. L. Gurney, M. D. Pascoe, & S. H. Richardson (Eds.), *The Nixon Volume. Proceedings of the seventh international kimberlite conference* (pp. 503–521). Cape Town: University of Cape Town.

Mandeville, C. W., Webster, J. D., Rutherford, M. J., Taylor, B. E., Timbal, A., & Faure, K. (2002). Determination of molar absorptivities for infrared absorption bands of H_2O in andesitic glasses. *American Mineralogist, 87*, 813–821.

Manega, P. C. (1993). Geochronology, geochemistry and isotopic study of the Plio- Pleistocene hominid sites and the Ngorongoro volcanic highlands in northern Tanzania. Ph.D. dissertation. University of Colorado, Boulder.

McDougall, I., & Harrison, M. (1999). *Geochronology and thermochronology by the $^{40}Ar/^{39}Ar$ method*. New York: Oxford University Press.

McHenry, L. J. (2004). Characterization and correlation of altered Plio-Pleistocene tephra using a "multiple technique" approach: Case study at Olduvai Gorge Tanzania. Ph.D. dissertation, Rutgers University, New Brunswick.

McHenry, L. J., Mollel, G. M., & Swisher, C. C., III. (2008). Compositional and textural correlations between Olduvai Gorge Bed I tephra and volcanic sources in the Ngorongoro volcanic highlands, Tanzania. *Quaternary International, 178*, 306–319.

Mertz, D. F., Weinrich, A. J., Sharp, W. D., & Renne, P. R. (2001). Alkaline intrusions in a near-trench setting, Franciscan complex, California: Constrains from geochemistry, petrology and $^{40}Ar/^{39}Ar$ chronology. *American Journal of Science, 301*, 877–911.

Mollel, G. (2002). Petrology and geochemistry of the southeastern Ngorongoro Volcanic Highland (NVH); and contribution to "sourcing" of stone tools at Olduvai Gorge, Tanzania. Masters thesis, Rutgers University, Piscataway.

Mollel, G. F. (2007). Petrochemistry and geochronology of the Ngorongoro Volcanic Highland Complex (NVHC) and its relationship to Laetoli and Olduvai Gorge, Tanzania. Ph.D. dissertation, Rutgers University.

Mollel, G. F., Swisher, C. C., Feigenson, M. D., & Carr, M. J. (2008). Geochemical evolution of lavas from Ngorongoro Caldera, Tanzania: Implications for crust-magma interaction. *Earth and Planetary Science Letters, 271*, 337–347.

Mollel, G. F., Swisher, C. C., McHenry, L. J., Feigenson, M. D., & Carr, M. J. (2009). Petrogenesis of basalt-trachyte lavas from Olmoti crater, Tanzania. *Journal of African Earth Sciences, 54*, 127–143.

Möller, A., Mezger, K., & Schenk, V. (1998). Crustal age domains and the evolution of continental crust in Mozambique belt of Tanzania: Combined Sm–Nd, Rb–Sr, and Pb–Pb isotopic evidence. *Journal of Petrology, 39*, 749–783.

Mutakyahwa, M. (1997). Mineralogy of the Wembere-Manonga formation, Manonga Valley, Tanzania, and possible provenance of the sediments. In T. Harrison (Ed.), *Neogene paleontology of the Manonga Valley. Tanzania* (vol. 1, pp. 67–78). New York: Plenum.

Ndessokia, P. (1990). The mammalian fauna and archaeology of the Ndolanya and Olpiro Beds, Laetoli, Tanzania. Ph.D. dissertation. University of California, Berkeley.

Nielsen, T. F. D., & Veksler, I. V. (2002). Is natrocarbonatite a cognate fluid condensate? *Contributions to Mineralogy and Petrology, 142*, 425–435.

Nyblade, A. A., Owens, J. T., Gurrola, H., Retsema, J., & Langston, C. A. (2000). Seismic evidence for a deep upper mantle thermal anomaly beneath East Africa. *Geology, 28*, 599–602.

Paslick, C., & Halliday, A. N. (1996). Indirect crustal contamination: Evidence from isotopic and chemical disequilibria in minerals from alkali basalts and nephelines from northern Tanzania. *Contributions to Mineralogy and Petrology, 125*, 277–292.

Paslick, C., Halliday, A. N., James, D., & Dawson, J. B. (1995). Enrichment of continental lithosphere by OIB melts: Isotopic evidence from the volcanic province of northern Tanzania. *Earth and Planetary Science Letters, 130*, 109–126.

Pickering, R. (1964). Endulen. Quarter degree sheet, 52. Geological Survey of Tanzania.

Renne, P., Swisher, C., III, Deino, A., Kamer, D., Owens, T., & DePaolo, D. (1998). Intercalibration of standards, absolute ages, and uncertainties in $^{40}Ar/^{39}Ar$ dating. *Chemical Geology, 145*, 117–152.

Rogers, W. M., James, D., Kelley, S. P., & De Mulder, M. (1998). The generation of potassic lavas from eastern Virunga province, Rwanda. *Journal of Petrology, 39*, 1223–1247.

Späth, A., Le Roex, A., & Opiyo-Akech, N. (2001). Plume-lithosphere interaction and the origin of continental rift-related alkaline volcanism-the Chyulu Hills volcanic province, southern Kenya. *Journal of Petrology, 42*, 765–787.

Sun, S., & McDonough, W. F. (1989). Chemical and isotopic systematics of oceanic basalts: Implications for mantle composition and processes. In A. D. Sounders & M. J. Norry (Eds.), *Magmatism in the ocean basins* (Vol. 42, pp. 313–345). London: Geological Society Special Publication.

Trua, T., Deniel, C., & Mazzuoli, R. (1999). Crustal control in the genesis of Plio-quaternary biomodal magmatism of the Main Ethiopian Rift (MER): Geochemical and isotopic (Sr, Nd, Pb) evidence. *Chemical Geology, 155*, 201–231.

Turrin, B. D., Muffler, L. J. P., Clynne, M. A., & Champion, D. E. (2007). Robust 24±6 ka $^{40}Ar/^{39}Ar$ age of a low-potassium tholeiitic basalt in the Lassen region of NE California. *Quaternary Research, 68*, 96–110.

Verniers, J. (1997). Detailed stratigraphy of the Neogene sediments at Tinde and other localities in the central Manonga basin. In T. Harrison (Ed.), *Neogene paleontology of the Manonga Valley, Tanzania* (pp. 67–78). New York: Plenum.

Vukadinovic, D. (1994). High-field-strength elements in Javanese arc basalts and chemical layering in the mantle wedge. *Journal of Mineralogy and Petrology, 55*, 293–308.

Weeraratne, D. S., Forsyth, D. W., Fischer, K. M., & Nyblade, A. A. (2003). Evidence for an upper mantle plume beneath the Tanzania craton from rayleigh wave tomography. *Journal of Geophysical Research, 108*(B9), 2427.

Chapter 6
Geochemistry and Mineralogy of Laetoli Area Tuffs: Lower Laetolil through Naibadad Beds

Lindsay J. McHenry

Abstract Tuffs from the Lower Laetolil, Upper Laetolil, Ndolanya, and Naibadad Beds of the Laetoli and Kakesio areas of Tanzania are characterized and compared to serve as correlation tools for this important paleoanthropological region. These beds contain variably-altered tuffs of original carbonatitic, nephelinitic, phonolitic, trachytic, and rhyolitic composition, tracking changes in volcanic source area and local diagenetic conditions. Previous studies (e.g., Hay 1987) show compositional differences between tuffs from different Laetoli area beds, but do not provide the tuff-specific information needed to create a high-resolution tephrostratigraphic framework. This study details differences in mineral composition between individual tuffs (where distinguishable) within each bed. The Lower Laetolil tuffs have similar compositions but can generally be distinguished by their andradite compositions. Titanomagnetite and perovskite compositions help distinguish individual Upper Laetolil tuffs, and divide the bed compositionally into lower (UL 1–4) and upper (UL 5–8) units. Augite compositions distinguish the individual Ndolanya tuffs analyzed in this study, whereas glass compositions distinguish individual Naibadad tuffs. The variable degree of alteration observed within and between sites at Laetoli makes the comparison of mineral assemblages an unreliable means for identifying tuffs. Glass, nepheline, and melilite were likely significant components of the original Laetolil Beds tuff mineral assemblages, but are now recognized in few (if any) samples except by the presence of likely alteration products. Feldspar and biotite also show signs of degradation. A comparison of glass and phenocryst compositions between the Naibadad tuffs and the tuffs of lower Bed I at Olduvai Gorge does not support a direct correlation between the two sites. The presence of quartz and rhyolitic and trachytic glass in both does suggest a common source, likely Ngorongoro Crater.

L.J. McHenry (✉)
Department of Geosciences, University of Wisconsin-Milwaukee, 3209 N. Maryland Avenue, Milwaukee, WI, 53211, USA
e-mail: lmchenry@uwm.edu

Keywords Carbonatite • Nephelinite • Phonolite • Trachyte • Rhyolite • Tephrostratigraphy • Ndolanya Beds • Laetolil Beds • Pliocene • Tanzania

Introduction

The Laetoli area contains deposits of Pliocene through Holocene sedimentary and volcanic rocks, overlying Precambrian metamorphic basement rocks. These deposits are divided into a series of beds distinguished by lithological and paleontological character, often separated by erosive disconformities. These beds are (from oldest to youngest) the Lower and Upper Laetolil (LL and UL), Ndolanya (ND), Naibadad (NB), Olpiro, and Ngaloba Beds (Hay 1987). This chapter describes tuff compositions from the Lower Laetolil through Naibadad beds (Fig. 6.1), supplementing Hay's (1987) mineral assemblage and general descriptions. Mineral composition helps distinguish between otherwise similar tuffs, and shows the general compositional trends between individual tuffs within a single bed. Changes in tuff composition follow changes in the volcanic source area (the Ngorongoro Volcanic Highlands), and, together with the tuff record of nearby Olduvai Gorge, trace the history of evolution of this volcanic province from older than 4.4 Ma (Drake and Curtis 1987; Deino 2011) to Recent (Drake and Curtis 1987; Ndessokia 1990; Manega 1993; Deino 2011).

Descriptions of tuffs from the different Laetoli area beds, together with their general mineral content, can be found elsewhere (see Hay 1978, 1981, 1987; Hay and Reeder 1978; Hay and O'Neil 1983). These papers provide generalized descriptions detailing the characteristics of tuffs within a single bed. Detailed discussions of authigenic mineralogy, inferred original compositions, and the inferred original carbonatite component of the Upper Laetolil tuffs can be found in previous works by Hay (Hay and O'Neil 1983; Hay 1978, 1986). The present study focuses on the original silicate pyroclastic components remaining after diagenesis.

T. Harrison (ed.), *Paleontology and Geology of Laetoli: Human Evolution in Context. Volume 1: Geology, Geochronology, Paleoecology and Paleoenvironment*, Vertebrate Paleobiology and Paleoanthropology, DOI 10.1007/978-90-481-9956-3_6,

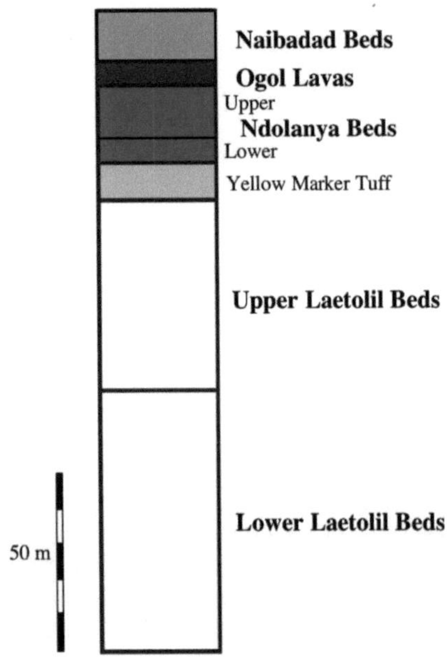

Fig. 6.1 General stratigraphy of the Lower Laetolil through Naibadad Beds at Laetoli, after Hay (1987) and Harrison and Msuya (2005)

These tuffs are exposed at sites of paleoanthropological interest and form the stratigraphic framework used to correlate temporally between geographically separated sites within and near Laetoli (e.g., Leakey 1987a; Ndessokia 1990). Therefore, accurate geochemical or mineralogical fingerprints for these tuffs are important for establishing the relative ages of the various archaeological sites, including the famous hominin trackway preserved in Upper Laetolil Tuff 7 at Loc. 8 (Leakey and Hay 1979) and localities that have produced *Australopithecus afarensis* fossils (Leakey 1987b).

The objectives of the current study are to (1) describe differences in mineral composition between and within individual beds to help create a higher resolution tephrostratigraphic framework for Laetoli, (2) explore the difficulties of characterizing and correlating highly altered tuffs, and (3) compare Laetoli and Olduvai Gorge Bed I tuffs for possible correlation.

Background

The Lower Laetolil Beds, which are best exposed at Kakesio (southwest of Laetoli), directly overlie Precambrian metamorphic rocks and consist largely of eolian or fluvially-reworked tuffs of nephelinite or phonolite composition (Hay 1987). The Upper Laetolil Beds are more fully eolian, and layers of xenoliths are common. Highly weathered melilitite and carbonatite airfall tuffs are more common in the Upper Laetolil Beds, separated by thick deposits of massive bedded eolian tuff (Hay 1987).

The Yellow Marker Tuff (UL YMT) at the top of the Upper Laetolil Beds is compositionally similar to nepheline phonolite tuffs in the disconformably overlying Ndolanya Beds (Hay and Reeder 1978; Hay 1987). The similar tuff compositions suggest that the disconformity may represent only a short hiatus between the Upper Laetolil and Ndolanya Beds (but see Deino 2011). The Lower Ndolanya Beds are clay-rich with some tuff and limestone, whereas the Upper Ndolanya Beds consist predominantly of eolian tuff (Hay 1987).

The mineralogically distinctive vogesite Ogol Lavas lie between the Ndolanya Beds and the younger Naibadad Beds (Fig. 6.1). The Naibadad Beds consist mostly of massive, crudely bedded tephra, interspersed with a few layers of better-stratified tuff. The composition diverges from the underlying Laetoli area tuffs by displaying a trachytic/rhyolitic composition similar to those of Olduvai Gorge Bed I and Ngorongoro Crater in the Ngorongoro Volcanic Highlands (NVH). These general differences in tuff composition distinguish the various beds. A more detailed examination of the mineral compositions may help to distinguish between tuffs within the same bed, and show general evolutionary trends in volcanic composition within individual beds.

Samples and Sample Sites

Samples were collected from sites within the Laetoli area and from Kakesio (Fig. 6.2). Stratigraphic columns for the sections sampled are presented in Fig. 6.3. Four sites (Kakesio, Locs. 1, 18, and 9) were chosen because they individually represent the longest, best preserved, or most complete sections of their respective beds. Samples of the UL Yellow Marker Tuff were also collected at Locs. 10E and 16. Table 6.1 reports GPS coordinates for each measured section and for most individual samples collected.

Lower Laetolil (LL) tuffs were sampled at Kakesio, to the southwest of Laetoli. Three layers were collected: the oldest (KAK-01, 9.4 m above the Precambrian basement) consists of "tuff balls" preserved in a bioturbated, calcite-rich matrix. Above it lies a massive, brown tuffaceous unit resembling the underlying tuff balls without the matrix (KAK-02), and this is overlain in turn by a grey, pisolitic tuff (KAK-03) of possible original carbonatite composition.

The Upper Laetolil sequence is well exposed at Loc. 1. There, the Upper Laetolil Beds directly overlie the Lower Laetolil Beds. One Lower Laetolil tuff (sample LA-1.19) was sampled just below the contact between the two beds (10 m below the lowest Upper Laetolil tuff sampled). The named Upper Laetolil tuffs (UL 1–8 and Yellow Marker Tuff) were all sampled using Hay's (1987) Loc. 1 section for proper identification. All are exposed in a stream drainage that was traced from the Lower Laetolil Beds and UL Tuff 1

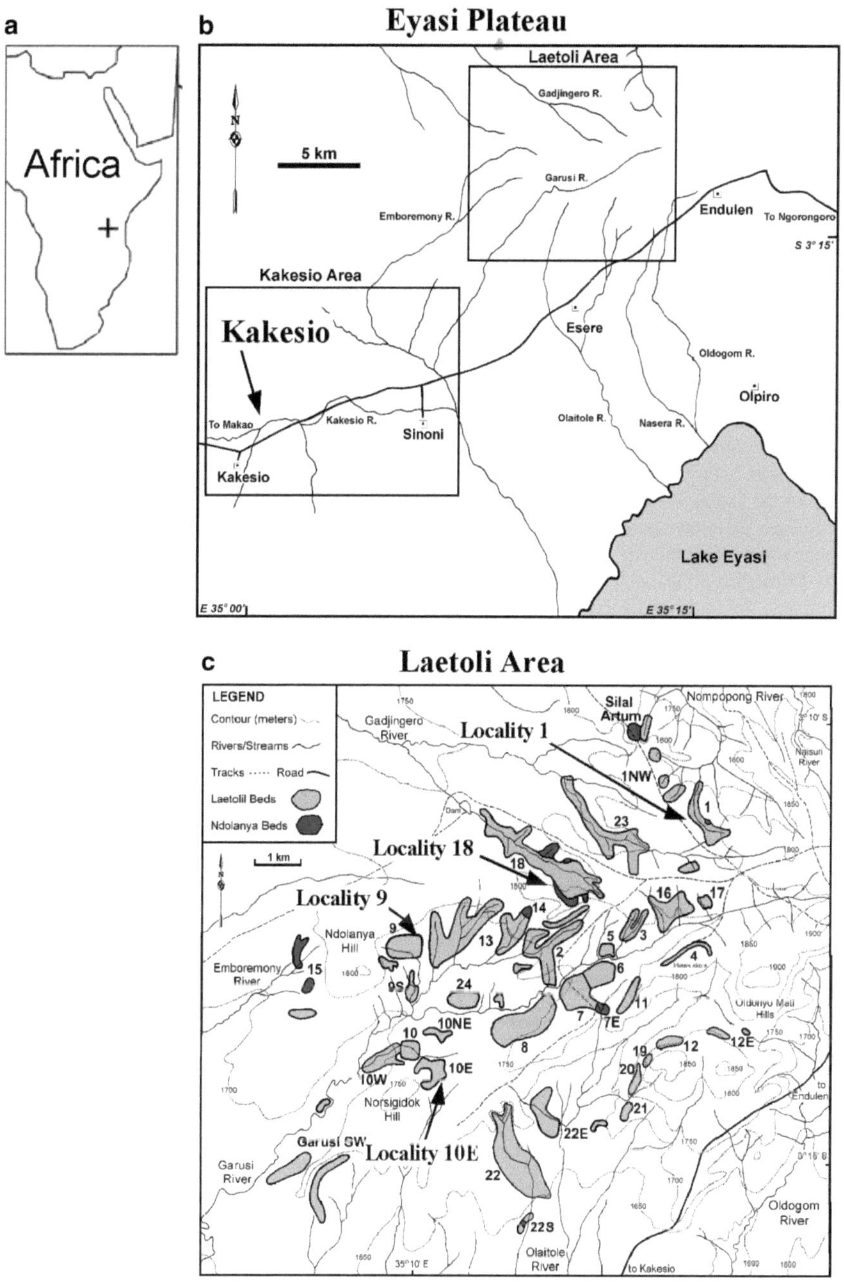

Fig. 6.2 Maps of the Eyasi Plateau and Laetoli areas, showing sample localities. (**a**) Map of Africa. (**b**) Map of the Eyasi Plateau, showing the relative positions of Kakesio (Lower Laetolil Bed sampling) and Laetoli (Upper Laetolil, Ndolanya, and Naibadad Beds sampling) areas (after Harrison and Msuya 2005). (**c**) Detailed map of the Laetoli area. Gray shaded areas show exposures of the Upper Laetolil Beds, black shaded areas show exposures of the Ndolanya Beds. White numbers represent Mary Leakey's collecting localities (Leakey 1987b), the black numbers and arrows show the sample localities reported in the current study (after Harrison and Msuya 2005)

(to the north) all the way to UL Tuff 6 (to the south), where Upper Laetolil Tuffs 6, 7, 8, and the Yellow Marker Tuff are exposed in "badlands" topography near the discovery site of Laetoli hominid 1 (Leakey 1987b). Strata between the airfall tuffs consist largely of eolian-reworked tephra.

The tuffs of the Ndolanya Beds were sampled at Loc. 18 where four of them are exposed in a steep outcrop. This is the same section where Ndessokia (1990) excavated and analyzed the Upper Ndolanya Bed faunal remains. Kovarovic et al. (2002) used Ndessokia's data to reconstruct the paleoecology of the Ndolanya Beds. All four tuffs sampled lie above Ndessokia's lower calcrete marker bed, thus placing them all within the fossiliferous Upper Ndolanya Beds (Fig. 6.1). The fourth tuff sampled (sample 18.04) corresponds to Ndessokia's (1990) augitic sandstone. The phonolitic Lower Ndolanya Beds (Hay and Reeder 1978) are not well exposed at Loc. 18 and were not sampled in this study.

Four tuffs were sampled in the Naibadad Beds at Loc. 9, where they directly overlie the Ndolanya Beds and form part of the face of a steep hill. Most of the Naibadad Beds at this locality consist of weathered tuffaceous material, thus sampling was limited to the thinner, less-weathered gray airfall or fluvially reworked tuffs. All Naibadad tuffs sampled contain unweathered volcanic glass. The fourth sample (LA-9.04) is from a pumice-rich layer directly overlying the third tuff (sample LA-9.03).

Fig. 6.3 Stratigraphic columns for sampled sections at Kakesio and Laetoli Locs. 1, 9, and 18. The columns for Kakesio, Loc. 9, and Loc. 18 are on the same scale; the Loc. 1 section is thicker and has its own scale. (**a**) The tuff at the base of the section marks the top of the Lower Laetolil Bed, with the Upper Laetolil tuffs above. Tuffs in this section were identified using the stratigraphic positions reported in Hay (1987). (**b**) The base of the Kakesio section lies at the contact between the Precambrian basement rock and the Lower Laetolil Beds. (**c**) The top of the Loc. 18 section corresponds to the uppermost pedogenic calcrete layer reported by Ndessokia (1990) at the same site, and the fourth tuff (18.04) corresponds to his augitic sandstone. (**d**) The base of the Loc. 9 section lies at the contact between the Ndolanya and Naibadad Beds (marked by a calcrete). Sample 9.04 is fresh pumice directly above the ash sampled in 9.03

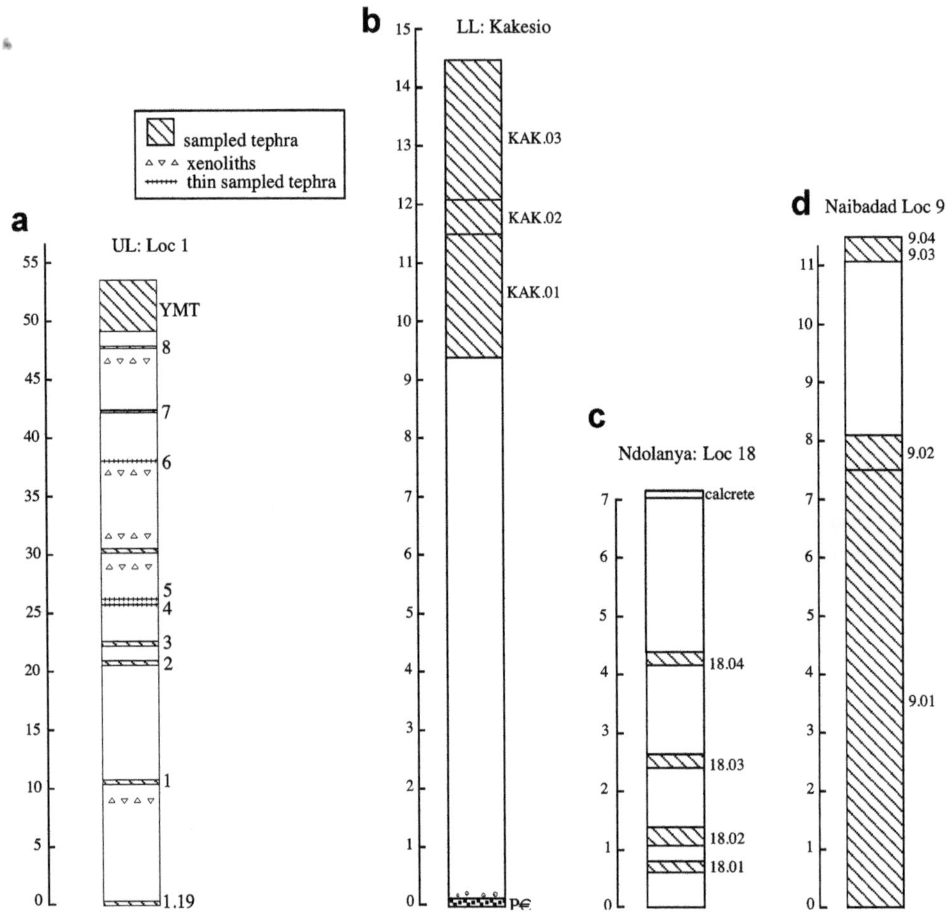

Table 6.1 Site and sample locations (GPS), listed in stratigraphic order

Locality	Latitude	Longitude	Sample ID	Tuff
9			LA-9.04	Top of LA-9.03
9	S 03.20565°	E 35.16868°	LA-9.03	Highest Naibadad tuff
9			LA-9.02	Middle Naibadad tuff
9	S 03.20542°	E 35.16922°	LA-9.01	Lowest Naibadad tuff
18	S 03.20026°	E 35.19733°	LA-18.04	Highest ND tuff (dark, crystal-rich)
18			LA-18.03	3rd ND tuff (well-cemented, crystal-rich)
18			LA-18.02	2nd ND tuff (clayey, with altered pumice)
18	S 03.20048°	E 35.19791°	LA-18.01	Lowest ND tuff (light color)
10 E	S 03.23562°	E 35.17224°	LA-10E.04	UL Yellow Marker Tuff
16	S 03.20137°	E 35.21369°	LA-16.03	UL Yellow Marker Tuff
1	S 03.18902°	E 35.22508°	LA-1.18	UL Yellow Marker Tuff
1	S 03.18845°	E 35.22513°	LA-1.16	UL Tuff 8
1	S 03.18593°	E 35.22366°	LA-1.10	UL Tuff 7 (upper coarse)
1			LA-1.09	UL Tuff 7 (lower fine)
1	S 03.18626°	E 35.22358°	LA-1.06	UL Tuff 6
1			LA-1.24	UL Tuff 5
1	S 0318148°	E 35.22391°	LA-1.23	UL Tuff 4
1			LA-1.22	UL Tuff 3
1	S 03.18065°	E 35.22444°	LA-1.21	UL Tuff 2
1	S 03.18004°	E 35.22416°	LA-1.20	UL Tuff 1
1	S 03.17704°	E 35.22555°	LA-1.19	LL tuff
Kakesio	S 03.34989°	E 34.99804°	LA-KAK.03	LL carbonatite tuff
Kakesio			LA-KAK.02	LL tuff above KAK.01
Kakesio	S 03.34997°	E 34.99807°	LA-KAK.01	LL tuff balls

ND Ndolanya, *UL* Upper Laetolil, *LL* Lower Laetolil, *Datum* WGS 84

Methods

Thin sections of most Laetoli area and Kakesio tuffs were prepared, polished, and carbon coated for electron microprobe analysis without the use of acid. Phenocrysts from three samples (LA-1.10, LA-16.03, LA-10E.04) were instead prepared as mounts for electron microprobe analysis using the methodology described in McHenry (2005). The preparation procedure for the mounts did involve a one-minute rinse in 4% HF, a step that could damage nepheline, if present. Most samples were analyzed by electron microprobe at Rutgers University (JEOL JXA-8600 SuperProbe), running at 15 kV and 15 nA with a rastering beam (count times, standards, and data analysis methods reported in McHenry [2005]). A volatile correction routine was applied to all glass samples to minimize Na-loss during analysis (Donovan 2000). Additional analyses were conducted by electron microprobe at the University of Wisconsin-Madison (Cameca SX50) running at 15 kV and 20 nA with a focused electron beam (for augite, andradite, titanomagnetite, perovskite, and sphene), 15 kV and 10 nA (for feldspar, nepheline, and hornblende), or 15 kV and 6 nA with a defocused beam (for glass). Individual analyses with particularly low totals (glass, titanomagnetite <90%, perovskite <93%, feldspar, andradite <94%, augite <96%) were excluded, as were those with poor stoichiometry, particularly for feldspars with significant Na-loss during analysis. Where mineral compositions varied within an individual sample, analyses of similar phenocrysts were grouped together before calculating a mean.

Results

The mineral assemblages analyzed for each sample are reported in Table 6.2. Data in this table represent the phases remaining after alteration: thus nepheline, melilite, and glass are only present in the freshest samples. Mineral identification in this study is based on compositional analysis, whereas Hay's (1987) study relied more heavily on pseudomorphs, alteration products, and other indicators of original mineralogy. The absence of an easily weathered mineral phase in the current study does not imply its absence in the original assemblage, as it may have altered. The likely effects of alteration are addressed in the discussion.

Compositional results are presented in Tables 6.3–6.8. Feldspar compositions are reported for all mineral populations represented by two or more analyses; for all other minerals a population of three or more is required. Where a single sample contains multiple compositions of a mineral, each is treated separately and a number (1–4) is appended to the sample ID in the data table for each (for example, LA-9.03–2 in the glass Table 6.8 represents the second most abundant glass population within sample LA-9.03).

The Lower Laetolil tuffs (KAK.01-KAK.03 and 1.19) are dominated by augite and titanomagnetite, with smaller amounts of andradite, sphene, and perovskite (Table 6.2). Feldspar is present in the three Kakesio samples, but not in the Loc. 1 tuff.

The Upper Laetolil tuffs contain abundant augite and titanomagnetite except for the lower, fine part of UL 7 (sample 1.09) and the YMT at Loc. 1. The Upper Laetolil tuffs contain variable amounts of other minerals (nepheline, andradite, sphene, perovskite) and are generally devoid of feldspar until the YMT.

The Ndolanya tuffs are sodic augite (aegerine to aegerine-augite) and feldspar rich, with variable amounts of andradite, sphene, and titanomagnetite. The Naibadad tuffs contain abundant fresh or only partly altered volcanic glass, along with feldspar, augite, titanomagnetite, and minor quartz, andradite, and ilmenite.

Discussion

Lower Laetolil Tuffs

The three Lower Laetolil tuffs from Kakesio have similar mineralogical compositions to each other, but that are distinct from the other tuffs. Unlike the Upper Laetolil tuffs, sphene and feldspar are common. The presence of andradite (melanite) and perovskite and apparent absence of hornblende indicate that these three tuffs are compositionally more similar to the Lower Laetolil aeolian tuffs reported in Hay (1987) than to his phonolitic air-fall tuffs.

Titanomagnetites from the Lower Laetolil tuffs have slightly higher TiO_2 and lower FeO contents than those of Upper Laetolil tuffs: the main titanomagnetite population of the Lower Laetolil tuffs varies from 11.7 to 14.8 wt.% TiO_2, whereas Upper Laetolil Tuffs 1–7 range from 8.9 to 11.8% (Fig. 6.4a). The compositional difference between the Lower and Upper Laetolil Bed tuff augite is subtle, but can be seen on a plot of MgO vs. FeO (Figs. 6.5 and 6.6). The Lower Laetolil augites follow a parallel trend, with slightly higher MgO concentrations for a given FeO value. Minor augite populations overlap between the various beds, perhaps because they are derived from a common contaminant source. Andradite (melanite) phenocrysts also serve to distinguish the Lower Laetolil from the Upper Laetolil tuffs. Andradites from Lower Laetolil tuffs have higher TiO_2 and lower FeO concentrations than those of the Upper Laetolil tuffs (Fig. 6.7).

Table 6.2 Mineral assemblage

Sample	Locality	Tuff	Augite	Hornblende	Biotite	Nepheline	Melilite	Feldspar	Andradite	Sphene	Perovskite	Titanomagnetite	Ilmenite	Quartz	Glass
LA-9.04	9	NB 4	X	+	–	–	–	XXX	–	–	–	XX	+	+	XXX
LA-9.03	9	NB 3	XX	–	–	–	–	XX	+	–	–	X	X	+	XXX
LA-9.02	9	NB 2	X	+	–	–	–	XXX	–	–	–	XX	X	+	XXX
LA-9.01	9	NB 1	XXX	–	–	–	–	XXX	X	+	–	XX	X	+	XXX
LA-18.04	18	ND 4	XXX	–	–	XXX	–	–	XXX	–	–	XX	–	–	–
LA-18.03	18	ND 3	XXX	–	–	–	–	XX	XXX	–	–	X	–	–	–
LA-18.02	18	ND 2	XXX	–	–	–	–	XXX	X	XXX	–	+	–	–	–
LA-18.01	18	ND 1	XXX	–	–	–	–	XXX	+	X	–	+	–	–	–
LA-16.03	16	UL YMT	XXX	–	–	–	–	XX	X	+	–	+	–	–	–
LA-10E.04	10E	UL YMT	XXX	–	–	–	–	XX	X	+	X	XXX	–	–	–
LA-1.18	1	UL YMT	+	–	–	–	–	XXX	XX	XX	X	+	–	–	–
LA-1.16	1	UL 8	XXX	–	X	–	–	+	XXX	+	XX	XXX	+	–	–
LA-1.10	1	UL 7	XXX	–	+	XXX	–	–	XXX	–	XX	XX	–	–	–
LA-1.09	1	UL 7	+	–	+	XX	XX	–	–	–	XX	XX	–	–	–
LA-1.06	1	UL 6	XXX	–	X	+	–	+	XXX	XX	X	XXX	–	–	–
LA-1.24	1	UL 5	XXX	–	–	–	–	–	XX	–	+	XX	–	–	–
LA-1.23	1	UL 4	XXX	–	–	–	–	–	–	–	–	XXX	–	–	–
LA-1.22	1	UL 3	XXX	–	–	–	–	–	+	–	XX	XXX	–	–	–
LA-1.21	1	UL 2	XX	–	–	–	–	–	–	–	X	XXX	–	–	–
LA-1.20	1	UL 1	XXX	–	+	–	–	–	X	–	XX	XXX	–	–	–
LA-1.19	1	LL	XXX	–	X	–	–	–	XX	X	+	XXX	–	–	–
LA-KAK.03	Kakesio	LL	XXX	–	+	–	–	X	X	X	X	XXX	–	–	–
LA-KAK.02	Kakesio	LL	XXX	–	–	–	–	XXX	X	X	X	XXX	–	–	–
LA-KAK.01	Kakesio	LL	XXX	–	+	–	–	XX	XXX	X	X	XX	+	–	–

NB Naibadad, *ND* Ndolanya, *UL* Upper Laetolil, *LL* Lower Laetolil
XXX = abundant; XX = common; X = between common and rare; + = rare; – = absent

Table 6.3 Andradite compositions, in weight% oxide

Sample	Tuff	n	SiO_2	TiO_2	Al_2O_3	FeO	MnO	MgO	CaO	Na_2O	SUM
LA-9.01	NB 1	3	29.06	14.64	1.27	19.50	0.23	0.84	32.06	0.27	98.01
StDev			1.56	3.81	0.27	2.37	0.09	0.34	0.14	0.07	0.87
LA-18.04-1	ND 4	12	32.89	7.79	1.11	23.20	0.37	0.33	31.22	0.19	97.14
StDev			0.65	1.09	0.48	0.81	0.11	0.11	0.57	0.07	0.70
LA-18.04-2	ND 4	4	27.97	15.86	1.30	18.80	0.37	1.03	30.25	0.27	95.91
StDev			0.30	0.70	0.47	0.50	0.04	0.12	0.74	0.06	0.65
LA-18.04-3	ND 4	4	31.24	12.00	1.03	21.23	0.42	0.56	30.39	0.32	97.25
StDev			0.61	0.31	0.53	0.76	0.07	0.18	0.80	0.15	1.29
LA-18.03	ND 3	18	32.68	8.76	1.04	22.63	0.41	0.31	31.03	0.25	97.15
StDev			1.47	2.31	0.63	1.52	0.12	0.11	1.00	0.10	1.12
LA-18.02	ND 2	5	34.51	8.01	1.28	23.04	0.32	0.24	32.27	0.27	99.95
StDev			0.76	1.34	0.77	1.19	0.05	0.19	1.31	0.11	1.43
LA-18.01	ND 1	2	29.77	12.78	1.78	20.19	0.32	0.93	33.04	0.23	99.07
StDev			0.79	1.63	1.39	1.82	0.01	0.23	0.16	0.09	0.32
LA-1.18	UL YMT	7	27.90	15.76	0.85	19.78	0.27	0.79	30.54	0.31	96.30
StDev			1.27	1.73	0.38	1.10	0.12	0.21	1.42	0.13	1.25
LA-1.16	UL 8	15	29.95	11.03	1.02	21.02	0.39	0.62	31.28	0.24	95.62
StDev			1.81	3.46	0.53	1.50	0.20	0.32	0.94	0.21	1.09
LA-1.10	UL 7	5	28.65	13.82	1.59	19.49	0.26	0.97	31.73	0.18	96.73
StDev			1.05	1.91	0.68	1.12	0.04	0.15	0.37	0.03	0.48
LA-1.06	UL 6	22	28.55	12.88	0.70	20.76	0.32	0.76	32.53	0.23	96.87
StDev			2.00	1.98	0.19	1.16	0.17	0.22	0.84	0.09	1.11
LA-1.24	UL 5	9	31.68	9.67	0.55	21.40	0.40	0.46	31.59	0.33	96.10
StDev			1.90	2.81	0.22	1.04	0.14	0.29	0.27	0.12	0.63
LA-1.22	UL 3	2	28.83	12.32	0.56	20.48	0.25	0.73	32.28	0.22	95.71
StDev			0.52	1.40	0.05	1.01	0.04	0.11	0.19	0.03	0.01
LA-1.20	UL 1	4	29.70	12.67	0.25	21.18	0.37	0.63	31.61	0.46	96.94
StDev			1.67	3.00	0.09	1.65	0.09	0.32	0.19	0.08	0.86
LA-1.19	LL	9	26.61	16.56	0.97	19.72	0.35	0.85	31.08	0.29	96.52
StDev			1.41	0.71	0.45	0.67	0.12	0.13	1.32	0.15	1.10
KAK.03	Kakesio 3	6	25.90	13.29	0.88	20.76	0.33	0.69	31.84	0.40	94.19
StDev			2.20	3.00	0.85	1.48	0.12	0.44	0.88	0.25	0.99
KAK.02	Kakesio 2	4	26.80	13.62	0.89	20.40	0.37	0.72	31.45	0.28	94.61
StDev			0.91	1.70	0.33	1.54	0.11	0.30	0.65	0.12	0.60
KAK.01	Kakesio 1	15	28.93	13.21	1.10	21.05	0.34	0.92	32.13	0.26	98.00
StDev			1.51	2.44	0.44	1.19	0.12	0.30	0.79	0.10	1.18

NB Naibadad, *ND* Ndolanya, *UL* Upper Laetolil, *LL* Lower Laetolil

Table 6.4 Augite compositions, in weight% oxide

Sample	Tuff	n	SiO_2	TiO_2	Al_2O_3	FeO	MnO	MgO	CaO	Na_2O	SUM
LA-9.04	NB 4	2	49.41	0.35	0.36	24.90	1.28	3.26	18.40	0.63	98.62
StDev			0.40	0.05	0.02	1.15	0.07	0.81	0.09	0.05	0.09
LA-9.03	NB 3	6	48.49	0.38	0.41	26.15	1.35	3.05	18.48	0.60	98.96
StDev			0.67	0.10	0.09	1.41	0.16	0.91	0.51	0.09	0.58
LA-9.01-1	NB 1	5	49.14	0.47	0.45	23.46	1.40	3.86	18.68	0.74	98.23
StDev			0.87	0.05	0.09	1.74	0.14	1.28	0.47	0.13	1.69
LA-9.01-2	NB 1	4	50.90	0.37	0.47	23.74	0.49	2.41	11.05	6.93	96.42
StDev			1.24	0.18	0.30	1.12	0.19	0.77	2.70	1.81	1.28
LA-18.04-1	ND 4	10	51.03	0.40	0.76	19.19	0.76	5.79	18.71	2.50	99.17
StDev			1.12	0.11	0.19	1.29	0.27	0.91	0.66	0.31	1.00
LA-18.04-2	ND 4	7	51.15	0.45	0.69	22.13	0.81	3.87	15.32	4.21	98.64
StDev			0.37	0.17	0.22	0.65	0.18	0.45	1.28	0.62	1.02
LA-18.04-3	ND 4	4	52.45	0.43	0.90	13.45	0.43	9.48	20.95	1.60	99.69
StDev			0.85	0.06	0.33	0.71	0.13	0.46	0.34	0.25	0.93
LA-18.04-4	ND 4	3	51.19	1.09	2.59	9.94	0.28	11.51	21.86	0.99	99.47
StDev			1.19	0.41	0.72	1.44	0.05	0.95	0.64	0.04	0.93
LA-18.03	ND 3	18	50.81	0.48	0.66	21.21	0.79	4.49	17.65	3.06	99.17
StDev			0.77	0.31	0.34	3.05	0.25	2.00	2.24	1.11	0.76
LA-18.02-1	ND 2	10	50.24	0.38	1.02	21.25	0.80	3.67	16.55	3.87	97.81
StDev			0.49	0.09	0.33	2.22	0.24	1.43	2.33	1.29	0.51
LA-18.02-2	ND 2	3	51.41	0.59	1.08	24.59	0.48	1.67	7.60	9.02	96.46
StDev			0.60	0.17	0.12	0.48	0.26	0.29	3.48	2.16	0.77
LA-18.01	ND 1	10	52.07	0.40	0.82	24.50	0.64	2.03	11.66	6.58	98.71
StDev			0.55	0.11	0.20	1.22	0.26	0.66	2.09	1.35	1.01
LA-1.18	UL YMT	1	49.24	1.54	3.51	9.16	0.06	11.56	23.45	0.67	99.21
LA-1.16-1	UL 8	16	48.33	1.73	3.73	6.63	0.15	13.30	23.29	0.62	97.81
StDev			1.74	0.33	0.96	1.39	0.13	1.06	0.75	0.15	0.83
LA-1.16-2	UL 8	3	51.17	0.38	1.17	10.40	0.34	11.23	22.54	0.97	98.19
StDev			1.35	0.20	0.57	1.45	0.05	1.21	1.05	0.30	0.48
LA-1.10-1	UL 7	9	52.26	0.75	1.28	9.40	0.30	11.84	22.86	0.98	99.68
StDev			0.74	0.29	0.53	1.63	0.07	0.99	0.53	0.21	0.33
LA-1.10-2	UL 7	6	46.82	2.16	5.74	8.31	0.12	11.48	23.63	0.63	98.92
StDev			1.72	0.60	1.40	0.90	0.05	0.54	0.28	0.09	0.50
LA-1.06-1	UL 6	8	50.03	0.64	1.34	9.97	0.28	11.81	23.10	0.93	98.16
StDev			1.44	0.42	1.19	2.30	0.15	1.33	0.84	0.39	0.84
LA-1.06-2	UL 6	7	49.95	0.39	0.59	17.62	0.72	6.87	20.22	2.09	98.53
StDev			1.10	0.11	0.33	1.42	0.23	0.99	0.91	0.42	1.26
LA-1.06-3	UL 6	3	46.78	2.03	5.65	8.12	0.16	12.30	23.77	0.57	99.48
StDev			1.56	0.12	0.67	1.10	0.06	1.02	0.67	0.16	1.14
LA-1.24-1	UL 5	13	51.72	0.82	1.69	7.75	0.24	12.61	23.22	0.78	98.86
StDev			1.62	0.44	1.27	1.97	0.11	1.27	0.78	0.34	0.64

LA-1.24-2	UL 5	6	51.72	0.46	0.64	17.76	0.75	6.30	16.62	4.02	98.30
StDev			1.57	0.31	0.33	3.76	0.19	2.45	2.47	1.22	0.59
LA-1.23-1	UL 4	6	48.28	1.60	4.48	6.73	0.19	12.90	23.69	0.52	98.41
StDev			1.38	0.48	1.75	0.65	0.05	0.75	0.39	0.04	0.79
LA-1.23-2	UL 4	5	52.33	0.46	0.62	8.16	0.31	12.85	22.87	1.01	98.63
StDev			1.38	0.11	0.46	1.79	0.09	1.09	0.81	0.23	0.64
LA-1.23-3	UL 4	4	44.09	2.51	8.24	8.29	0.15	10.82	23.72	0.49	98.35
StDev			0.92	0.15	0.62	0.41	0.05	0.43	0.50	0.02	1.04
LA-1.23-4	UL 4	3	49.50	0.37	0.84	19.83	0.71	5.17	19.27	2.35	98.05
StDev			0.67	0.10	0.17	1.26	0.03	0.97	0.88	0.58	0.82
LA-1.22-1	UL 3	8	51.55	0.99	0.64	7.56	0.25	13.25	22.89	1.25	98.41
StDev			1.12	0.57	0.48	1.24	0.04	0.90	0.59	0.39	0.68
LA-1.22-2	UL 3	5	49.96	1.51	2.86	6.40	0.11	13.62	23.32	0.49	98.26
StDev			0.77	0.15	0.48	1.35	0.05	0.88	0.75	0.15	0.51
LA-1.21	UL 2	9	51.90	1.11	1.94	4.95	0.12	14.92	23.75	0.45	99.15
StDev			1.00	0.38	0.55	0.53	0.06	0.42	0.63	0.09	0.59
LA-1.20-1	UL 1	10	50.42	1.51	2.90	5.89	0.11	13.97	23.54	0.49	98.86
StDev			1.80	0.41	0.86	1.07	0.07	0.82	0.67	0.14	0.75
LA-1.20-2	UL 1	6	51.84	0.47	0.49	11.32	0.53	11.02	22.15	1.44	99.28
StDev			0.73	0.26	0.26	2.56	0.22	1.46	0.95	0.53	0.98
LA-1.20-3	UL 1	3	49.44	0.59	0.35	20.71	0.89	4.59	15.80	4.62	97.00
StDev			1.60	0.43	0.12	1.05	0.43	0.88	2.16	1.42	0.54
LA-1.19	LL	18	49.27	1.48	2.75	8.04	0.19	12.71	23.11	0.70	98.28
StDev			1.48	0.36	0.79	2.82	0.15	2.02	0.97	0.28	0.89
KAK.03	Kakesio 3	16	44.67	1.67	3.57	7.40	0.18	15.25	23.70	0.58	97.03
StDev			1.52	0.46	1.33	1.80	0.11	1.53	0.91	0.21	0.77
KAK.02-1	Kakesio 2	9	48.98	1.36	3.00	10.96	0.22	12.98	21.62	1.18	100.37
StDev			0.70	0.30	0.88	3.25	0.16	2.10	1.37	0.61	1.42
KAK.02-2	Kakesio 2	4	48.09	0.52	1.12	21.70	0.71	5.13	17.98	3.17	98.41
StDev			1.19	0.13	0.25	1.43	0.18	1.03	1.33	1.05	0.99
KAK.01-1	Kakesio 1	11	49.98	1.55	2.94	7.18	0.15	13.38	24.77	0.74	100.73
StDev			1.81	0.50	1.02	1.54	0.07	1.27	0.72	0.23	0.92
KAK.01-2	Kakesio 1	5	48.43	0.58	0.72	21.84	0.79	3.75	19.05	3.64	98.81
StDev			2.29	0.16	0.47	2.18	0.15	1.72	1.96	1.32	2.97

NB Naibadad, *ND* Ndolanya, *UL* Upper Laetolil, *LL* Lower Laetolil

Table 6.5 Perovskite compositions, in weight% oxide

Sample	Tuff	n	SiO_2	TiO_2	Al_2O_3	FeO	MnO	MgO	CaO	Na_2O	SUM
LA-1.18	UL YMT	2	0.00	57.80	0.07	0.86	0.05	0.02	40.14	0.37	99.70
StDev			0.00	0.19	0.02	0.03	0.06	0.03	0.28	0.06	0.03
LA-1.16	UL 8	9	0.15	57.50	0.16	1.01	0.07	0.02	37.80	0.45	97.57
StDev			0.44	0.93	0.14	0.40	0.07	0.03	2.00	0.16	1.74
LA-1.10	UL 7	10	0.01	56.77	0.18	0.81	0.02	0.01	39.57	N.A.	97.40
StDev			0.01	0.26	0.03	0.06	0.03	0.02	0.21		0.30
LA-1.09	UL 7	15	0.11	54.86	0.18	1.45	0.04	0.02	37.51	0.48	95.19
StDev			0.27	0.94	0.13	0.26	0.05	0.02	0.94	0.08	1.19
LA-1.06	UL 6	4	0.00	56.52	0.06	0.88	0.00	0.01	41.36	0.24	99.42
StDev			0.00	0.43	0.08	0.35	0.00	0.01	2.19	0.08	2.71
LA-1.24	UL 5	3	0.07	56.14	0.19	0.98	0.01	0.03	39.13	N.A.	96.56
StDev			0.04	0.21	0.02	0.05	0.02	0.00	0.67		0.41
LA-1.22	UL 3	10	0.00	54.11	0.08	0.88	0.06	0.00	38.64	0.47	94.65
StDev			0.00	0.55	0.07	0.27	0.04	0.01	0.82	0.30	0.87
LA-1.21	UL 2	9	0.03	53.65	0.16	1.40	0.05	0.02	38.79	0.30	94.82
StDev			0.07	0.67	0.06	0.33	0.06	0.02	0.77	0.07	0.73
LA-1.20	UL 1	10	0.03	54.09	0.14	1.23	0.05	0.01	38.25	0.44	94.62
StDev			0.06	0.53	0.03	0.41	0.06	0.01	0.80	0.15	0.89
LA-1.19	LL	3	0.34	55.55	0.14	0.91	0.00	0.08	38.40	0.66	96.50
StDev			0.53	1.37	0.08	0.14	0.00	0.13	1.53	0.19	1.35
KAK.03	Kakesio 3	8	0.01	54.05	0.10	0.87	0.10	0.01	39.87	0.40	95.75
StDev			0.02	0.70	0.04	0.12	0.08	0.02	0.68	0.12	0.68
KAK.02	Kakesio 2	4	0.03	53.56	0.13	0.90	0.07	0.01	38.10	0.42	93.68
StDev			0.05	0.59	0.08	0.25	0.04	0.01	0.50	0.09	0.87
KAK.01	Kakesio 1	8	0.46	53.92	0.27	1.16	0.03	0.06	38.38	0.52	95.24
StDev			0.78	1.57	0.32	0.43	0.04	0.10	0.96	0.32	1.25

NB Naibadad, *ND* Ndolanya, *UL* Upper Laetolil, *LL* Lower Laetolil

Table 6.6 Feldspar compositions, in weight% oxide

Sample	Tuff	n	SiO_2	Al_2O_3	FeO	CaO	Na_2O	K_2O	BaO	SUM
LA-9.04-1	NB 4	7	66.90	19.46	0.39	0.26	7.40	5.61	0.15	100.18
StDev			0.62	0.23	0.05	0.24	0.28	0.68	0.15	0.57
LA-9.04-2	NB 4	7	64.22	21.86	0.28	2.28	8.28	2.61	0.24	99.77
StDev			0.85	0.93	0.06	1.00	0.21	0.66	0.10	0.38
LA-9.03-1	NB 3	6	62.83	21.71	0.35	3.02	8.35	2.07	0.31	98.73
StDev			0.46	0.32	0.05	0.34	0.15	0.21	0.21	0.60
LA-9.03-2	NB 3	4	58.39	24.88	0.38	6.73	7.10	0.81	0.09	98.55
StDev			1.27	0.75	0.06	0.98	0.57	0.18	0.04	0.22
LA-9.03-3	NB 3	4	66.18	18.88	0.40	0.20	7.53	5.85	0.03	99.12
StDev			0.47	0.19	0.10	0.08	0.10	0.24	0.04	0.48
LA-9.02-1	NB 2	11	66.00	19.02	0.42	0.29	7.81	5.36	0.14	99.10
StDev			0.56	0.40	0.19	0.20	0.28	0.42	0.12	0.67
LA-9.02-2	NB 2	3	61.89	22.59	0.35	3.95	8.16	1.66	0.49	99.25
StDev			2.07	1.06	0.10	1.76	0.48	0.77	0.13	0.54
LA-9.01-1	NB 1	11	66.52	19.45	0.31	0.57	7.94	4.84	0.31	100.02
StDev			0.69	0.65	0.11	0.37	0.47	0.90	0.23	0.61
LA-9.01-2	NB 1	2	59.43	25.01	0.45	6.66	7.06	0.78	0.08	99.60
LA-18.02-1	ND 2	4	65.33	19.28	0.35	0.00	2.29	13.01	0.66	101.03
StDev			0.67	0.19	0.05	0.00	0.87	1.55	0.31	0.66
LA-18.02-2	ND 2	2	66.26	19.83	0.28	0.05	7.13	5.67	0.46	99.82
LA-18.02-3	ND 2	2	66.26	19.83	0.28	0.05	7.13	5.67	0.46	99.82
LA-18.01-1	ND 1	5	64.64	18.54	0.43	0.08	2.69	12.68	0.57	99.77
StDev			1.20	0.56	0.50	0.12	0.56	0.96	0.41	1.68
LA-18.01-2	ND 1	3	66.98	19.29	0.28	0.02	4.84	9.51	0.43	101.39
StDev			0.44	0.20	0.06	0.01	0.37	0.75	0.08	0.77
LA-1.18	UL YMT	12	63.92	19.73	0.75	0.06	5.03	8.36	1.54	99.56
StDev			1.56	0.49	0.20	0.04	0.36	0.46	0.65	1.78
LA-1.06-1	UL 6	1	62.38	19.21	0.66	0.06	3.95	10.96	0.15	97.39
KAK.03-1	Kakesio 3	2	55.13	24.58	0.37	5.30	7.74	0.71	0.01	94.08
KAK.03-2	Kakesio 3	2	54.48	27.50	0.07	7.49	7.03	0.15	0.02	96.77
KAK.02	Kakesio 2	5	60.26	19.60	0.88	0.41	5.39	7.92	0.88	95.63
StDev			1.65	0.65	0.29	0.45	1.15	1.81	0.14	1.79
KAK.01-1	Kakesio 1	2	64.59	19.97	0.58	0.20	5.14	8.28	1.20	100.04
KAK.01-2	Kakesio 1	2	63.05	23.75	0.13	4.62	8.81	0.22	0.05	100.70

NB Naibadad, *ND* Ndolanya, *UL* Upper Laetolil, *LL* Lower Laetolil

Table 6.7 Titanomagnetite compositions, in weight% oxide

Sample	Tuff	n	SiO_2	TiO_2	Al_2O_3	FeO	MnO	MgO	CaO	SUM
LA-9.04	NB 4	4	0.16	23.76	1.12	65.74	1.35	0.68	0.04	92.86
StDev			0.02	0.83	0.11	0.76	0.06	0.17	0.03	1.08
LA-9.03	NB 3	6	0.22	23.06	1.39	68.00	1.05	1.21	0.03	95.24
StDev			0.16	2.41	0.86	2.86	0.14	1.06	0.02	2.01
LA-9.02	NB 2	8	0.12	26.02	0.75	64.30	1.53	0.57	0.03	93.32
StDev			0.04	0.99	0.27	2.11	0.09	0.27	0.03	1.95
LA-9.01	NB 1	7	0.30	24.77	1.06	65.12	1.21	0.77	0.05	93.35
StDev			0.21	1.32	0.49	1.94	0.27	0.50	0.03	1.46
LA-18.04	ND 4	9	0.09	10.89	1.08	79.55	0.99	0.94	0.05	93.65
StDev			0.15	1.35	0.39	1.47	0.15	0.50	0.06	0.76
LA-18.03	ND 3	3	0.34	5.74	0.09	84.47	1.06	0.14	0.10	92.06
StDev			0.52	2.00	0.11	2.56	0.38	0.13	0.16	1.19
LA-18.02	ND 2	2	0.06	9.37	3.49	76.40	0.60	3.84	0.01	93.97
StDev			0.08	0.86	0.68	0.86	0.11	0.41	0.02	0.79
LA-18.01	ND 1	1	0.01	10.17	2.21	78.74	0.91	2.06	0.00	94.38
LA-1.16-1	UL 8	9	0.16	11.78	2.33	75.63	0.70	3.15	0.10	94.01
StDev			0.32	1.09	0.50	1.24	0.13	0.67	0.14	1.11
LA-1.16-2	UL 8	5	0.16	11.46	0.72	76.03	1.59	1.83	0.22	92.18
StDev			0.16	1.52	0.36	1.42	0.40	0.63	0.16	1.51
LA-1.10	UL 7	12	0.05	10.32	2.52	77.99	0.72	2.69	0.03	94.33
StDev			0.03	0.74	0.60	2.05	0.08	0.87	0.03	0.73
LA-1.09	UL 7	14	0.05	9.50	2.31	79.70	0.83	2.73	0.07	95.28
StDev			0.08	0.35	0.26	1.45	0.19	0.22	0.06	1.77
LA-1.06-1	UL 6	8	0.02	9.13	2.39	79.43	0.77	2.71	0.06	94.61
StDev			0.03	0.19	0.69	1.03	0.25	0.55	0.04	1.21
LA-1.06-2	UL 6	4	0.02	8.22	0.46	82.87	0.94	0.72	0.18	93.45
StDev			0.04	2.24	0.36	2.29	0.24	0.78	0.23	1.44
LA-1.24	UL 5	16	0.05	9.10	2.80	77.97	0.72	2.63	0.09	93.38

StDev			0.02	0.39	0.40	1.01	0.08	0.23	0.04	0.76
LA-1.23-1	UL 4	13	0.04	9.31	3.12	76.77	0.70	3.07	0.05	93.25
StDev			0.04	0.32	0.36	1.62	0.10	0.45	0.08	1.47
LA-1.23-2	UL 4	4	0.00	9.14	1.91	79.28	0.84	1.83	0.00	93.26
StDev			0.00	0.20	0.12	1.01	0.04	0.11	0.00	1.27
LA-1.23-3	UL 4	3	0.01	6.14	0.42	86.10	1.00	0.34	0.00	94.31
StDev			0.02	1.12	0.15	1.50	0.12	0.04	0.00	1.95
LA-1.22	UL 3	19	0.02	9.42	3.51	74.34	0.67	4.82	0.04	92.97
StDev			0.03	0.29	0.12	0.80	0.09	0.11	0.03	0.71
LA-1.21	UL 2	19	0.01	9.19	3.43	74.98	0.60	4.94	0.09	93.45
StDev			0.02	0.35	0.13	0.98	0.10	0.17	0.07	0.98
LA-1.20	UL 1	19	0.01	9.72	1.84	77.65	0.79	2.80	0.06	93.07
StDev			0.02	0.72	0.59	1.32	0.07	0.49	0.04	0.97
LA-1.19	LL	11	0.07	14.88	1.89	73.43	0.93	3.13	0.08	94.57
StDev			0.06	1.06	0.48	2.37	0.32	1.02	0.15	1.63
KAK.03-1	Kakesio 3	14	0.16	11.67	2.47	76.22	0.75	5.63	0.10	97.16
StDev			0.29	1.42	0.72	2.65	0.25	1.18	0.11	1.89
KAK.03-2	Kakesio 3	5	0.01	9.40	0.68	83.08	0.92	2.54	0.04	96.99
StDev			0.02	1.46	0.42	1.15	0.22	1.44	0.07	1.75
KAK.02-1	Kakesio 2	6	0.23	14.01	2.60	73.14	0.55	6.01	0.07	96.75
StDev			0.34	1.57	0.29	1.81	0.16	1.26	0.09	1.12
KAK.02-2	Kakesio 2	5	0.09	11.17	0.69	80.07	1.19	0.72	0.01	94.16
StDev			0.10	2.95	0.32	4.47	0.22	0.29	0.02	1.57
KAK.02-3	Kakesio 2	3	0.35	13.71	1.66	74.29	0.72	2.84	0.03	93.72
StDev			0.47	3.17	0.19	3.23	0.12	0.33	0.01	1.85
KAK.01	Kakesio 1	6	0.03	12.19	2.03	78.11	0.79	3.86	0.08	97.24
StDev			0.03	1.83	0.33	3.10	0.16	1.75	0.07	1.39

NB Naibadad, *ND* Ndolanya, *UL* Upper Laetolil, *LL* Lower Laetolil

Table 6.8 Other mineral compositions

Sample	Tuff	n	SiO_2	TiO_2	Al_2O_3	FeO	MnO	MgO	CaO	Na_2O	K_2O	BaO	SUM
Melilite													
LA-1.09	UL 7	7	44.36	0.01	8.16	3.97	0.10	6.40	32.64	4.36	0.04	N.A.	100.06
StDev			1.62	0.01	0.54	0.13	0.06	0.16	1.00	0.17	0.03		2.37
Glass													
LA-9.04	NB 4	16	70.61	0.20	10.77	2.90	0.06	0.04	0.29	3.33	5.12	0.07	93.65
StDev			0.79	0.06	0.16	0.22	0.05	0.04	0.07	0.58	0.54	0.10	1.03
LA-9.03-1	NB 3	10	61.36	0.75	13.75	5.54	0.21	0.53	1.62	4.49	4.03	0.17	92.45
StDev			1.69	0.13	0.77	0.68	0.08	0.16	0.37	0.38	0.21	0.13	0.75
LA-9.03-2	NB 3	5	68.97	0.22	10.75	3.04	0.10	0.04	0.29	3.11	5.38	0.01	91.93
StDev			1.34	0.06	0.49	0.33	0.06	0.02	0.10	0.39	0.80	0.03	0.61
LA-9.02	NB 2	13	64.37	0.59	12.30	6.36	0.23	0.16	0.87	2.12	4.02	0.06	91.07
StDev			1.74	0.08	0.65	1.07	0.06	0.04	0.12	0.73	0.99	0.08	0.46
LA-9.01	NB 1	7	62.55	0.43	12.96	4.95	0.18	0.13	0.81	3.89	5.50	0.12	91.51
StDev			0.91	0.06	0.62	0.26	0.05	0.04	0.14	0.73	0.55	0.23	0.58
Nepheline													
LA-1.10	UL 7	13	41.00	N.A.	35.55	0.76	N.A.	N.A.	0.82	16.29	5.98	N.A.	100.44
StDev			0.48		0.58	0.27			0.35	0.26	0.50		0.93
LA-1.09	UL 7	10	42.07	N.A.	34.27	0.67	N.A.	N.A.	1.00	14.82	6.29	N.A.	99.32
StDev			0.75		0.42	0.10			0.08	0.86	0.45		1.50
LA-1.06	UL 6	3	40.63	N.A.	33.56	1.37	N.A.	N.A.	0.74	14.07	8.61	N.A.	99.18
StDev			1.50		0.92	0.33			0.68	0.35	0.70		2.66
Sphene													
LA-18.02	ND 2	17	31.89	35.96	0.66	1.87	N.A.	N.A.	28.35	0.22	N.A.	N.A.	99.23
StDev			0.66	1.36	0.18	0.49			0.72	0.13			1.49
LA-18.01	ND 1	6	31.61	35.09	0.55	1.89	N.A.	N.A.	29.09	0.26	N.A.	N.A.	98.70
StDev			0.90	0.89	0.23	0.31			0.37	0.16			1.33
LA-1.18	UL YMT	11	29.82	36.88	0.75	1.84	N.A.	N.A.	26.72	0.18	N.A.	N.A.	96.52
StDev			0.75	1.29	0.29	0.24			1.00	0.12			1.31
LA-1.16	UL 8	1	29.21	36.81	0.27	1.48	N.A.	N.A.	27.59	0.17	N.A.	N.A.	95.65
LA-1.19	LL	9	28.91	36.29	1.07	1.99	N.A.	N.A.	27.69	0.11	N.A.	N.A.	96.47
StDev			1.32	1.23	0.54	0.29			1.19	0.05			1.42
KAK.03	Kakesio3	3	27.31	36.26	0.57	1.48	N.A.	N.A.	27.98	0.17	N.A.	N.A.	94.12
StDev			0.63	0.34	0.37	0.47			0.31	0.15			0.53
KAK.02	Kakesio2	8	29.11	34.97	1.07	1.75	N.A.	N.A.	28.63	0.11	N.A.	N.A.	96.07
StDev			1.08	1.18	0.43	0.35			0.50	0.05			1.74
KAK.01	Kakesio1	6	29.50	35.40	0.55	1.91	N.A.	N.A.	28.21	0.14	N.A.	N.A.	95.98
StDev			0.64	0.61	0.22	0.16			0.33	0.06			0.62

NB Naibadad, *ND* Ndolanya, *UL* Upper Laetolil, *LL* Lower Laetolil

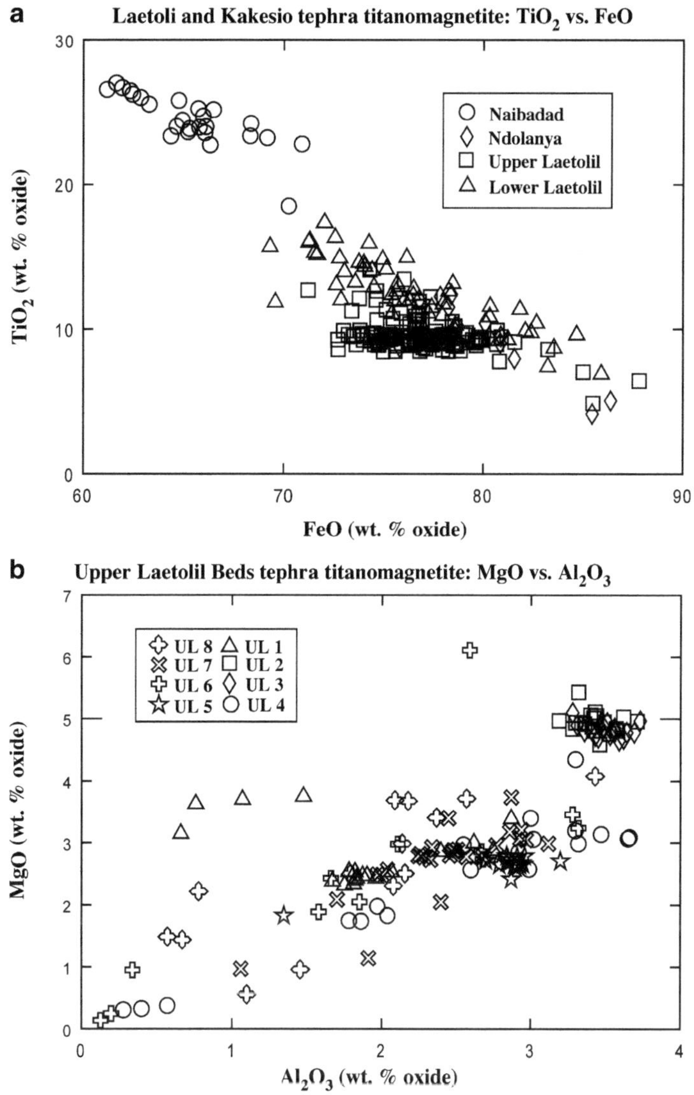

Fig. 6.4 Titanomagnetite compositions. (**a**) Lower and Upper Laetolil, Ndolanya, and Naibadad tuff titanomagnetite compositions, TiO_2 vs. FeO. The Naibadad tuffs have distinctively higher TiO_2 and lower FeO titanomagnetites than the other Beds, and the Lower Laetolil titanomagnetites are slightly more TiO_2 rich than those from the Upper Laetolil tuffs. (**b**) Upper Laetolil titanomagnetites, MgO vs. Al_2O_3. The Upper Laetolil titanomagnetites form compositional clusters: UL 2 and 3, UL 5 and 7, and UL 1. UL 4, 6, and 8 have more variable compositions

Individual Lower Laetolil tuffs can be distinguished using their andradite compositions: the first (KAK.01) has higher SiO_2 (28.9%) and lower TiO_2 (13.2%), and the youngest (LA-1.19) has a high TiO_2 concentration (16.6%, Fig. 6.7). The third Kakesio sample, KAK.03, has a unique composition: its augite has higher MgO (15.2 wt.%) than those from any other tuff analyzed in this study (Fig. 6.5b), and its perovskite has slightly higher CaO (39.6 wt.%) content than found in the surrounding tuffs (Fig. 6.8). The Lower Laetolil tuff at the base of the Loc. 1 section (sample LA-1.19) has mineral compositions intermediate between the Kakesio Lower Laetolil tuffs and the Upper Laetolil tuffs above. Like the other Lower Laetolil tuffs it contains abundant sphene; however it lacks feldspar. Its titanomagnetites show the same high TiO_2 concentrations (14.8 wt.%) as the other Lower Laetolil tuffs, however its other minerals (particularly augite) are more similar to those of the Upper Laetolil tuffs (Table 6.6).

Upper Laetolil Tuffs

The Upper Laetolil tuffs can be distinguished from the Lower Laetolil tuffs based on the near absence of feldspar and sphene and the presence of abundant slightly lower-TiO_2 titanomagnetite (at least for units below UL 8, Fig. 6.4a). Andradite compositions are intermediate between Lower Laetolil and Ndolanya compositions in TiO_2 (Fig. 6.7). Only UL 6 and 7 contain fresh nepheline (Table 6.8), and only the base of UL 7 (sample LA-1.09) contains melilite. The Upper Laetolil titanomagnetites vary in composition, but do not show a consistent trend with age. UL 2 and UL 3 titanomagnetites group together, with high MgO and Al_2O_3 (Fig. 6.4b). UL 5 and 7 also group together with intermediate Al_2O_3 concentrations, whereas UL 1 has lower concentrations. UL 4, 6, and 8 have more variable titanomagnetite compositions, including populations with distinctively low Al_2O_3.

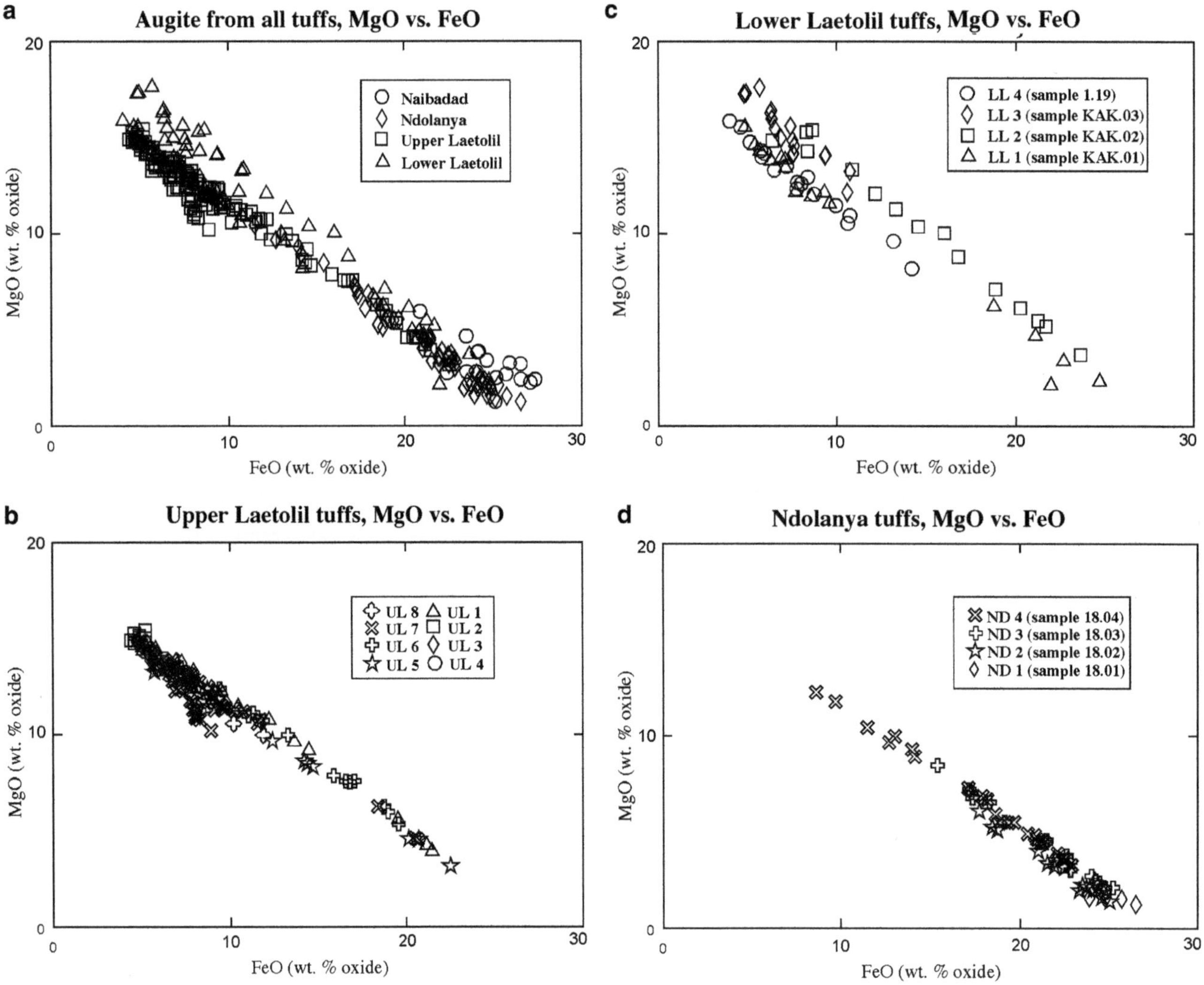

Fig. 6.5 Augite compositions, MgO vs. FeO. (**a**) All Beds, plotted together. The Lower Laetolil tuffs have slightly higher MgO concentrations for a given FeO concentration compared to the Upper Laetolil tuffs. (**b**) Upper Laetolil tuffs. (**c**) Lower Laetolil tuffs. (**d**) The Ndolanya tuffs are generally more FeO-rich than the Lower and Upper Laetolil

Among the tuffs studied, perovskite is only observed in the Lower Laetolil and Upper Laetolil Beds (Table 6.2). Whereas perovskite compositions from the Lower Laetolil tuffs and those from the lowermost tuffs within the Upper Laetolil Beds (UL 1–4) are similar, perovskites from the uppermost Upper Laetolil tuffs (5–8 and YMT) are generally more TiO_2 rich (Fig. 6.8). Perovskites from the lower, finer-grained part of UL 7 (sample LA-1.09) are intermediate in composition, whereas those from the upper, coarser part (sample LA-1.10) are more TiO_2 rich. UL 8 and the YMT, the two youngest Upper Laetolil tuffs, show the highest TiO_2 concentrations in their perovskite. This trend could serve as a "check" for stratigraphic position within the Upper Laetolil Beds.

Upper Laetolil Yellow Marker Tuff and the Ndolanya Beds

The Ndolanya tuffs from Loc. 18 are compositionally distinct from the Lower Laetolil and Upper Laetolil tuffs, yet show some similarities to the Yellow Marker Tuff (YMT, sample LA-1.18) at the top of the Upper Laetolil Beds. The YMT appears to be transitional between the Upper Laetolil and Ndolanya mineral assemblages. The YMT contains sanidine, with lower K-content than the K-rich sanidine found in some Ndolanya Tuffs. The YMT sanidine phenocrysts are also particularly Ba-rich (up to 2.5 wt.%), distinctly higher than any other Laetoli samples analyzed (Table 6.6). Sample 18.03 contains small authigenic K-feldspar: these (14 wt.%

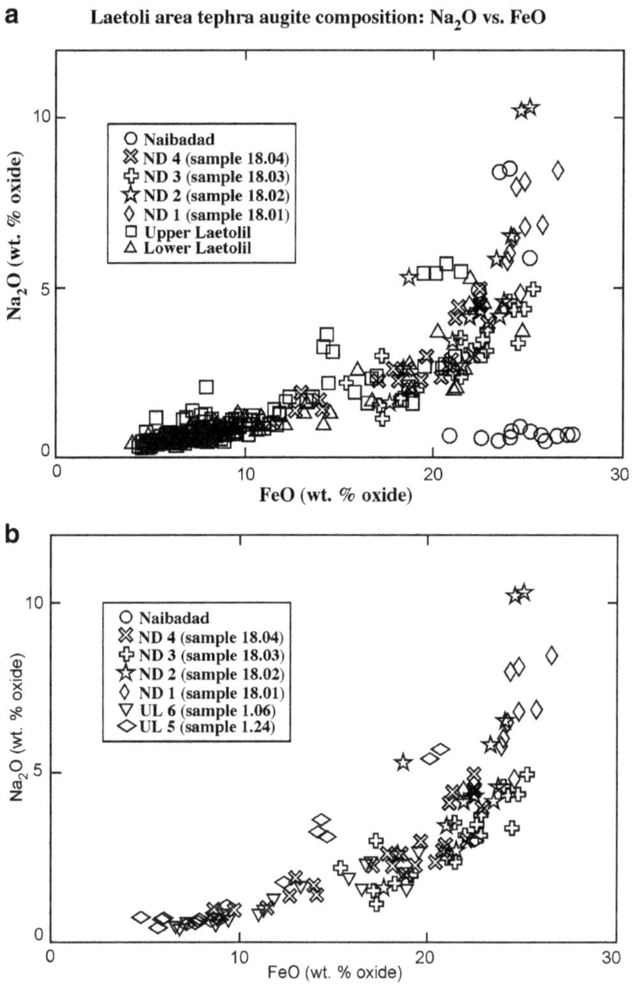

Fig. 6.6 Augite compositions, Na_2O vs. FeO. (**a**) The Ndolanya tuffs are plotted individually; the tuffs from all other Beds are plotted as groups. The Lower and Upper Laetolil tuffs are low in Na_2O and FeO, the Naibadad tuffs are FeO-rich but Na_2O poor, and the Ndolanya tuffs have high Na_2O and FeO. Except for two grains of Ndolanya sample 18.04, only the Lower and Upper Laetolil tuffs have less than 10% FeO. (**b**) Upper Laetolil Tuffs 5 and 6, with the Ndolanya tuffs. The first Ndolanya tuff (sample 18.01) has distinctively high Na_2O

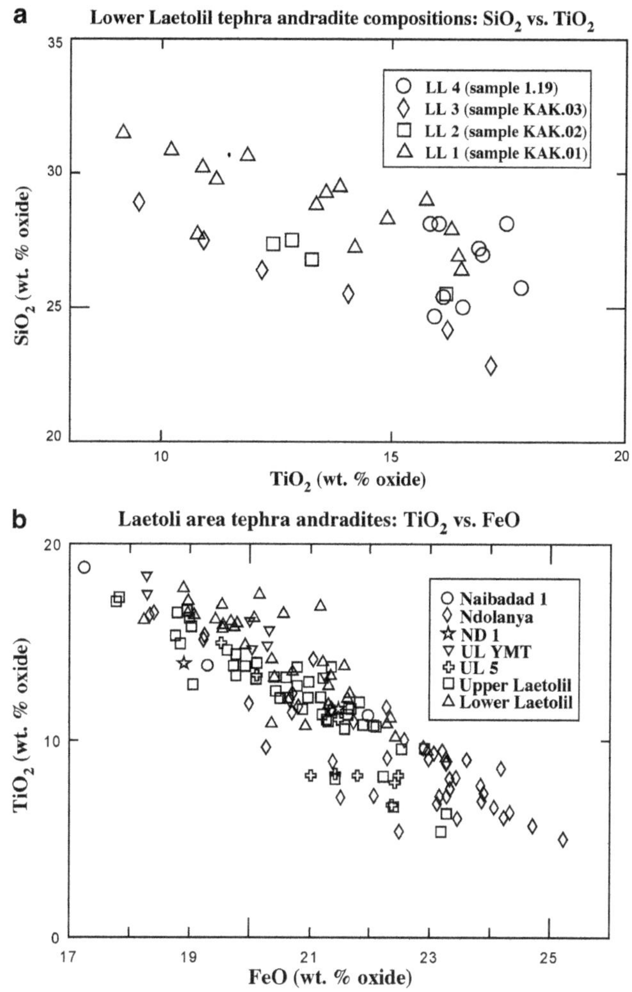

Fig. 6.7 Andradite compositions. (**a**) SiO_2 vs. TiO_2 for the Lower Laetolil tuffs. In general, the uppermost Lower Laetolil tuff (sample 1.19) has higher TiO_2 andradites, and the lowest tuff (sample KAK.01) andradites have the lowest TiO_2 concentrations. (**b**) TiO_2 vs. FeO for all Beds. For simplicity, tuffs from individual Beds with similar andradite compositions are plotted with the same symbol. The Lower Laetolil tuffs here include KAK.01, KAK.02, KAK.03, and 1.18. The Upper Laetolil tuffs include UL 1, 3, 7, and 8. The Ndolanya tuffs include sample 18.02, 18.03, and 18.04. UL 5, UL YMT, and ND 1 (sample 18.01), and are plotted separately because their andradite phenocrysts differ in composition from the others within their Beds. The Ndolanya tuff andradites have the highest FeO content, and the Lower Laetolil tuffs have the lowest

K_2O and higher) are not reported in the data tables because they are not primary volcanic minerals.

The Ndolanya tuffs lack perovskite and contain few titanomagnetite phenocrysts (Table 6.2), but Mn-rich hematite is an abundant alteration product. The titanomagnetite overlaps in composition with titanomagnetites in the UL tuffs (Fig. 6.4a). Garnets in the Upper Laetolil and Ndolanya tuffs (with the exception of UL 5) also differ in andradite composition, as the Ndolanya tuff andradites contain more TiO_2 (Fig. 6.7b) and SiO_2 and less MgO than the Upper Laetolil garnets.

Augites from the Ndolanya tuffs are much richer in Na_2O and FeO than are those in the Upper Laetolil tuffs, with Na_2O concentrations of up to 10 wt.% (Fig. 6.6), putting them in the aegerine or aegerine/augite compositional range. Of the Upper Laetolil tuffs only UL 5 and 6 contain a significant population of high-Na augites, though the dominant augite composition for both tuffs is Na-poor (Fig. 6.6b). Within the Ndolanya Beds, the first (18.01) is distinguishable from the others by its higher Na_2O and lower MgO augite (Figs. 6.5d, 6.6). The fourth (18.04) has augite of higher CaO and MgO, but the second and third (18.02 and 18.03) are indistinguishable in augite composition.

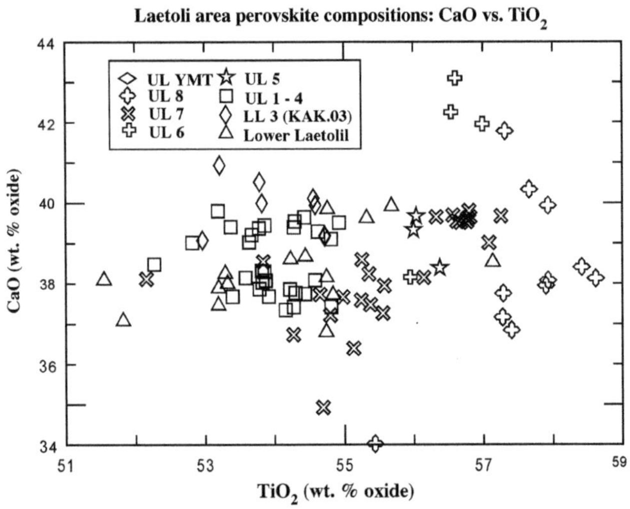

Fig. 6.8 Perovskite CaO vs. TiO_2. Among the tuffs reported in this study, only the Lower and Upper Laetolil tuffs contained perovskite. The Lower and lower Upper Laetolil tuffs (UL 1–4) contain lower-TiO_2 perovskite than the upper Upper Laetolil tuffs (5-YMT)

Naibadad Beds

Compositionally, the Naibadad tuffs depart significantly from all of the underlying tuffs. Whereas the Lower Laetolil, Upper Laetolil, and Ndolanya Beds also contain silica-undersaturated, often feldspathoid-bearing tuffs, the Naibadad tuffs are dacitic, trachytic, or rhyolitic and contain minor quartz (Fig. 6.9). Unfortunately, all of the Naibadad tuff samples contain some detrital mineral contaminants, including andradite and aegerine/augite. As they also contain glass, some primary phenocrysts could be identified because they were in direct contact with the glass shards. This was not possible for all grains analyzed, and the andradite in the lowest tuff (9.01) is most likely derived from reworked Ndolanya or Laetolil Bed material. Compositionally, the andradite more closely resembles that derived from the Upper Laetolil Beds (lower in SiO_2 and MgO, higher in TiO_2 than the Ndolanya).

Feldspar from the Naibadad tuffs varies from plagioclase to anorthoclase, even within an individual tuff. The lower three samples each contain at least two grains of plagioclase, samples 9.02, 9.03, and 9.04 contain a feldspar population intermediate between plagioclase and anorthoclase (equal parts K_2O and CaO), and all four contain a dominant higher-K anorthoclase population (Fig. 6.10). The first tuff (9.01) contains anorthoclase slightly lower in K_2O than anorthoclase from other Naibadad tuffs.

All four Naibadad tuff samples contain fresh glass. The 9.01 sample glass is trachytic, whereas the 9.02 glass is trachydacitic/dacitic, and 9.03 consists of two populations, one rhyolitic and one trachyandesitic/trachydacitic (Fig. 6.9, after Le Bas et al. 1986). Sample 9.04 contains only rhyolitic glass. The four tuffs can be distinguished based on FeO (Fig. 6.11) and CaO content in the glass, though a few shards of 9.02 composition are identified in 9.01 and a few shards of 9.03 are present in the directly overlying 9.04. These results differ from those reported by Manega (1993) because his results were normalized to 100%, and are lower in Na_2O most likely because of volatilization during electron microprobe analysis.

Augite is rare in the Naibadad tuffs, but where present it is distinct from that in all other Laetoli tuffs. It is both high in FeO and low in Na_2O (Fig. 6.6), with a few grains in the aegerine compositional range, perhaps reworked from older eruptions. Titanomagnetite phenocrysts from Naibadad tuffs are more Ti-rich (23–27 wt.%) and Fe-poor than are the titanomagnetites from the other beds (Fig. 6.4a). In general, there is little danger of mineralogically misidentifying a Naibadad tuff for a tuff from any other bed at Laetoli, though some care must be taken to avoid contaminant grains from older deposits.

Tuff Alteration

The current study found site-to-site differences in the mineral assemblage of the Yellow Marker Tuff. Of three samples analyzed (from Locs. 1, 10E, and 16: contrasting mineral assemblages reported in Table 6.2), all contained minor andradite (melanite), a phase reported as absent in Hay (1987). The Loc. 1 sample (1.18) lacks significant augite, whereas both Locs. 10E and 16 contain abundant diopside and common sodic pyroxene (Table 6.2), consistent with Hay (1987). These differences could result from local contamination (likely at Loc. 16 where the YMT is largely reworked) or they may reflect primary compositional differences between different parts of the tuff. The YMT is the product of multiple eruptions (Hay 1987), and a different part of the tuff was sampled at each site. The sample from Loc. 1 was collected 41 cm above the base of the tuff, which is 3.66 m thick; the Loc. 10E sample was collected 89 cm above the base of the tuff, which is 1.68 m thick, likely representing two different eruptive events. More detailed sampling and analysis of individual eruptive packages within the YMT is needed to resolve this inconsistency.

Hay (1987) lists melilite and nepheline as abundant or common minerals in tuffs of the Lower Laetolil, Upper Laetolil, and Ndolanya Beds. In this study, nepheline was only observed in UL 6 and 7 and ND 4, whereas melilite was only observed in the lower part of UL 7. Acid was not used to prepare the thin sections for this study, so nepheline should not have been destroyed during sample preparation. Only unaltered phenocrysts are reported in the current study, whereas Hay (1987) used pseudomorphs and other indirect means to deduce the original mineralogy. Thus, the absence of easily weathered phases in this study should not

Fig. 6.9 Naibadad glass alkali/silica diagram after LeBas et al. (1986). All of the Naibadad tuff glasses are silicic, ranging from dacitic (Naibadad 2) to rhyolitic (Naibadad 3 and 4) or trachytic (Naibadad 1). Naibadad 3 also contains a trachyandesitic glass population

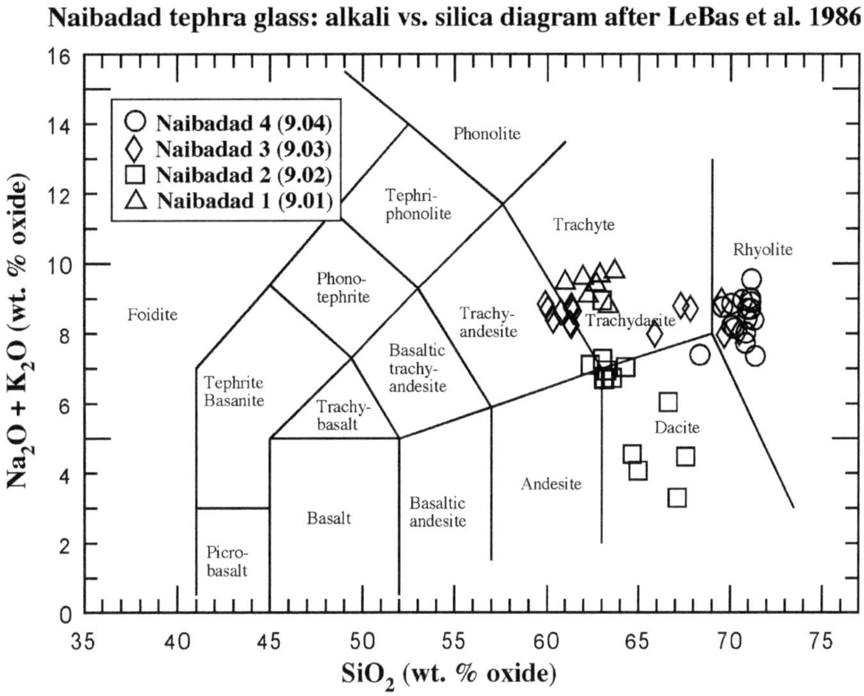

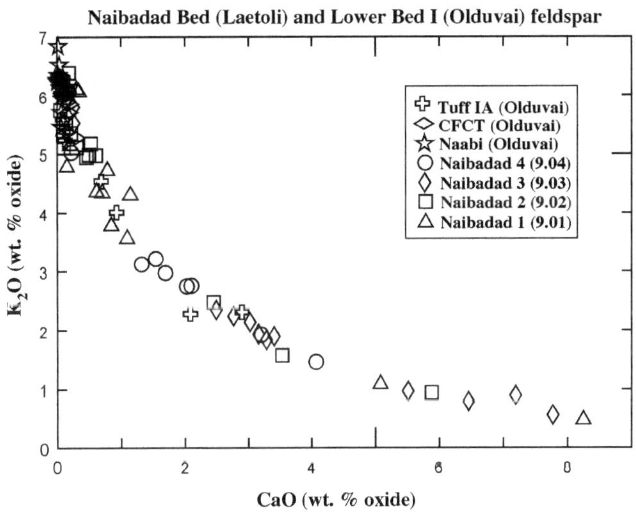

Fig. 6.10 Naibadad vs. Olduvai feldspar compositions: K_2O vs. CaO. The dominant feldspar from all Naibadad and lower Bed I Olduvai tuffs reported is high-K_2O anorthoclase, however all of the Naibadad tuffs also contain second, minor low-K_2O anorthoclase or plagioclase populations not present in the Olduvai tuffs

be interpreted to reflect the original mineral assemblage. Phillipsite, clay, and calcite are typical weathering products of nepheline and melilite (Hay and O'Neil 1983), and all three are frequently observed in the Upper Laetolil tuffs. It is possible that all Lower Laetolil, Upper Laetolil, and Ndolanya tuffs originally contained abundant nepheline, but that post-depositional alteration has since destroyed it. Some matrix calcite is likely derived from the alteration of the original carbonatite component of the UL tuffs (Hay 1981; Hay and O'Neil 1983). The high degree of alteration makes it difficult to characterize and reconstruct the original mineral assemblages of Lower and Upper Laetolil and Ndolanya Bed tuffs.

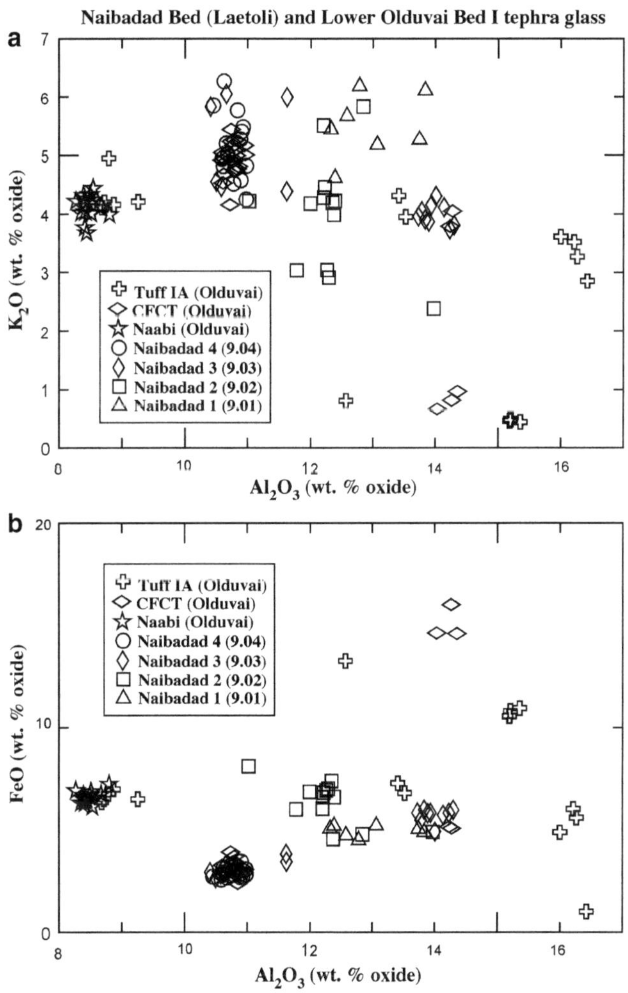

Fig. 6.11 Naibadad vs. Olduvai glass. The Naibadad tuffs are similar to tuffs from lower Bed I at Olduvai Gorge nearby. In particular, Naibadad 3 and the Olduvai coarse feldspar crystal tuff (*CFCT*) share a bimodal glass composition in (**a**) K_2O vs. Al_2O_3 and (**b**) FeO vs. Al_2O_3

The Naibadad tuffs are much less weathered, and each contains glass and feldspar, augite, and titanomagnetite phenocrysts with little evidence for alteration. The same condition persists through the Olpiro beds, which are not described in this paper. This contrast in preservation between the various beds is likely due in part to the difference in composition. The Laetolil and Ndolanya tuffs were originally carbonatite or highly silica-undersaturated silicate tephra, which degrade rapidly under surface conditions. The trachytes and rhyolites of the Naibadad Beds are more resistant to weathering, even under similar diagenetic conditions. Changes in diagenetic environment are likely partly responsible, as observed by Hay (1986).

Regional Correlations

Mollel et al. (2011) speculate about potential volcanic sources for the tuffs at Laetoli. The silica-undersaturated tuffs of the Lower and Upper Laetolil Beds may be consistent with eruption from Satiman volcano. This old (4.5–3.3 Ma; Curtis and Hay 1972; Bagdasaryan et al. 1973; Hay 1987; Manega 1993), eroded volcano in the southern NVH currently exposes phonolitic and nephelinitic lavas. So far, no evidence for carbonatite activity has been identified in the southern NVH, though it is possible that deposits of this type have been eroded away or obscured by more recent volcanism (perhaps the development of Lemagurut volcano). Hay's reconstructed stream flow directions for the Laetolil Beds are consistent with radial drainage from a point five to eight km west/southwest of Satiman volcano (Hay 1987). The Naibadad tuffs must have originated from a younger, silica-saturated volcanic source, perhaps the same source that produced the trachytes and rhyolites of lower Bed I at Olduvai Gorge.

Single Crystal Laser Fusion (SCLF) $^{40}Ar/^{39}Ar$ age dating has shown potential temporal overlap between the Naibadad (2.23–1.93 ± 0.02; Manega 1993) at Laetoli with the Naabi Ignimbrite (2.029 ± 0.005 Ma, Walter et al. 1992) and lower Bed I at Olduvai Gorge. Both the Naibadad Bed and lower Bed I also contain fresh quartz-bearing tuffs, which can be used for geochemical characterization and correlation. Three lower Bed I tuffs (Naabi Ignimbrite, coarse feldspar crystal tuff [CFCT], and Tuff IA) were analyzed by electron microprobe for phenocryst and glass composition (McHenry et al. 2008) and compared to the Naibadad tuffs to test for possible correlations. Certain mineralogical features make such a direct correlation unlikely. None of the lower Bed I tuffs analyzed contain significant plagioclase, which is present as a minor constituent in all Naibadad tuffs analyzed except 9.04. In glass composition, the Olduvai lower Bed I CFCT is similar to Naibadad 9.03 (Fig. 6.11). Both are bimodal with a dominant population with low FeO and Al_2O_3 concentrations and a second population with higher Al_2O_3. A third, minor, mafic composition present in the CFCT was not observed in Naibadad 3. A major difference between the CFCT and 9.03 is in the feldspar: whereas the anorthoclase/sanidine population of 9.03 overlaps the CFCT composition, most feldspar crystals from sample 9.03 are in the plagioclase/anorthoclase range, but feldspars in this range are absent in the CFCT (Fig. 6.10). The CFCT also contains abundant hornblende, whereas none was observed in this Naibadad tuff. These mineralogical differences highlight the need for mineralogical in addition to glass geochemical data for tuff correlation for these deposits.

Ngorongoro crater is the likely source for the Naibadad and Olduvai lower Bed I tuffs because it is the only known source of quartz-bearing tuffs within the NVH (Mollel 2002). The Naabi Ignimbrite of lower Bed I Olduvai has been correlated directly (geochemically) to an ignimbrite proximal to Ngorongoro, strengthening this correlation (McHenry et al. 2008). Since there appears to be no direct tuff correlation between Laetoli and Olduvai, it is likely that the Naibadad tuffs represent the early eruptive history of Ngorongoro, whereas Olduvai preserves the later history. This is consistent with the available dates for the two sites. It is also possible that the Naibadad tuffs were erupted from an older, as yet unidentified NVH volcano, perhaps one that has been covered by more recent volcanic deposits.

Fine-grained volcanic-derived sediments of similar composition are found in the Wembere-Manonga Formation of the Manonga Valley, ~100 km southwest of Laetoli (Mutakyahwa 1997). These highly altered sediments contain minor feldspar, aegerine-augite, sphene, pseudomorphs after nepheline and melilite, and textural and compositional evidence for a carbonatitic origin.

Conclusions

The Lower Laetolil, Upper Laetolil, and Ndolanya Bed tuffs are highly altered and difficult to characterize using mineral assemblage and geochemistry. The gradual compositional evolution of the most resistant phenocrysts (i.e., augite, andradite, sphene, perovskite, and titanomagnetite) within and between the different Laetoli area beds can be used for general tuff fingerprinting. Andradite composition serves to distinguish individual Lower Laetolil tuffs. Perovskite composition distinguishes between the oldest and youngest Upper Laetolil tuffs (i.e., UL 1–4, and UL 5–8 and YMT, respectively), and titanomagnetite compositions further subdivide the Upper Laetolil Beds. The uppermost tuff in the Lower Laetolil Beds (sample 1.19 at Loc. 1) is intermediate in composition between the Lower Laetolil and Upper Laetolil tuffs, and the Yellow Marker Tuff (YMT) at the top of the Upper Laetolil Beds is intermediate between the underlying Upper Laetolil and overlying Ndolanya tuff compositions. The Ndolanya tuffs can be distinguished from other Laetoli area tuffs based on their high-Na_2O augites, and from each other using differences in augite composition. The Naibadad Bed tuffs can be individually identified using their glass compositions. They are similar in composition to the tuffs of lower Bed I at Olduvai, and likely also derived from Ngorongoro crater.

Acknowledgements I thank the Tanzania Commission for Science and Technology and the Tanzania Antiquities Department for granting me permission to conduct research at Laetoli and Olduvai Gorge. Terry Harrison, Carl Swisher, Godwin Mollel, Peter Ditchfield, Alan Deino, Jeremy Delaney, and John Fournelle provided invaluable support in both the field and analytical portions of this project. Funding was provided in part by the University of Wisconsin- Milwaukee and by grants to Terry Harrison (NSF grant BCS-0309513).

References

Bagdasaryan, G. P., Gerasimovskiy, V. I., Polyakov, A. I., Gukasyan, R. K., & Vernadskiy, V. I. (1973). Age of volcanic rocks in the rift zones of East Africa. *Geochemistry International, 10*, 66–71.

Curtis, G., & Hay, R. (1972). Further geological studies and K-Ar dating at Olduvai Gorge and Ngorongoro crater. In W. W. Bishop & J. A. Miller (Eds.), *Calibration of hominoid evolution* (pp. 289–301). Edinburgh: Scottish Academic.

Deino, A. L. (2011). $^{40}Ar/^{39}Ar$ dating of Laetoli, Tanzania. In T. Harrison (Ed.), *Paleontology and geology of Laetoli: Human evolution in context* (Geology, geochronology, paleoecology and paleoenvironment, vol. 1, pp. 77–97). Dordrecht: Springer

Donovan, J. (2000). Probe for Windows: Analysis and Automation for EPMA, version 5.11. Software package distributed by Advanced Microbeam, Vienna, OH.

Drake, R., & Curtis, G. (1987). K–Ar geochronology of the Laetoli fossil localities. In M. D. Leakey & J. M. Harris (Eds.), *Laetoli: A Pliocene site in northern Tanzania* (pp. 48–52). Oxford: Clarendon.

Harrison, T., & Msuya, C. P. (2005). Fossil struthionid eggshells from Laetoli, Tanzania: Taxonomic and biostratigraphic significance. *Journal of African Earth Sciences, 41*, 303–315.

Hay, R. L. (1978). Melilitite-carbonatite tuffs in the Laetoli Beds of Tanzania. *Contributions to Mineralogy and Petrology, 67*, 357–367.

Hay, R. L. (1981). Paleoenvironments of the Laetoli Beds, northern Tanzania. In G. Rapp Jr. & C. F. Vondra (Eds.), *Hominid sites: Their geologic settings* (pp. 7–24). Boulder: Westview.

Hay, R. L. (1986). Role of tephra in the preservation of fossils in cenozoic deposits of east Africa. In L. E. Frostick, R. W. Renaut, I. Reid, & J. J. Tiercelin (Eds.), *Sedimentation in the African rifts*. Geological Society Special Publications No. 25 (pp. 339–344). Oxford: Blackwell.

Hay, R. L. (1987). Geology of the Laetoli area. In M. D. Leakey & J. M. Harris (Eds.), *Laetoli: A Pliocene site in northern Tanzania* (pp. 23–47). Oxford: Clarendon.

Hay, R. L., & O'Neil, J. R. (1983). Carbonatite tuffs in the Laetolil Beds of Tanzania and the Kaiserstuhl in Germany. *Contributions to Mineralogy and Petrology, 82*, 403–406.

Hay, R. L., & Reeder, R. J. (1978). Calcretes of Olduvai Gorge and the Ndolanya Beds of northern Tanzania. *Sedimentology, 25*, 649–673.

Kovarovic, K., Andrews, P., & Aiello, L. (2002). The palaeoecology of the Upper Beds at Laetoli, Tanzania. *Journal of Human Evolution, 43*, 395–418.

Leakey, M. D. (1987a). Introduction. In M. D. Leakey & J. M. Harris (Eds.), *Laetoli: A Pliocene site in northern Tanzania* (pp. 1–22). Oxford: Clarendon.

Leakey, M. D. (1987b). The Laetoli hominid remains. In M. D. Leakey & J. M. Harris (Eds.), *Laetoli: A Pliocene site in northern Tanzania* (pp. 108–117). Oxford: Clarendon.

Leakey, M. D., & Hay, R. L. (1979). Pliocene footprints in the Laetolil Beds at Laetoli, northern Tanzania. *Nature, 278*, 317–323.

LeBas, M. J., LeMaitre, R. W., Streckeisen, A., & Zanettin, B. (1986). A chemical classification of volcanic rocks based on the total alkali silica diagram. *Journal of Petrology, 27*, 745–750.

Manega, P. (1993). Geochronology, geochemistry and isotopic study of the Plio-Pleistocene hominid sites and the Ngorongoro volcanic highland in northern Tanzania. Ph.D. dissertation University of Colorado, Boulder.

McHenry, L. J. (2005). Phenocryst composition as a tool for correlating fresh and altered tephra, Bed I, Olduvai Gorge, Tanzania. *Stratigraphy, 2*, 101–115.

McHenry, L. J., Mollel, G. F., & Swisher, C. C. (2008). Compositional and textural correlations between Olduvai Gorge Bed I tephra and volcanic sources in the Ngorongoro volcanic highlands, Tanzania. *Quaternary International, 178*, 306–319.

Mollel, G. F. (2002). Petrology and geochemistry of southeastern Ngorongoro Volcanic Highland; and contribution to "sourcing" of stone tools at Olduvai Gorge, Tanzania. M.S. thesis, Rutgers University, New Brunswick.

Mollel, G. F., Swisher, C. C., III, Feigenson, M. D., & Carr, J. D. (2011). Petrology, geochemistry and age of Satiman, Lemagurut and Oldeani: Sources of the volcanic deposits of the Laetoli area. In T. Harrison (Ed.), *Paleontology and geology of Laetoli: Human evolution in context* (Geology, geochronology, paleoecology and paleoenvironment, vol. 1, pp. 99–119). Dordrecht: Springer.

Mutakyahwa, M. (1997). Mineralogy of the Wembere-Manonga Formation, Manonga Valley, Tanzania, and the possible provenance of the sediments. In T. Harrison (Ed.), *Neogene paleontology of the Manonga Valley, Tanzania: A window into the evolutionary history of East Africa* (pp. 67–79). New York: Plenum.

Ndessokia, P. N. S. (1990). The Mammalian Fauna and Archaeology of the Ndolanya and Olpiro Beds, Laetoli, Tanzania. Ph.D. dissertation, University of California, Berkeley.

Walter, R. C., Manega, P. C., & Hay, R. L. (1992). Tephrochronology of Bed I, Olduvai Gorge: An application of the laser-fusion 40Ar/39Ar dating to calibrating biological and climatic change. *Quaternary International, 13–14*, 37–46.

Chapter 7
Geochemical and Mineralogic Characterization of Middle Stone Age Tools of Laetoli, Tanzania, and Comparisons with Possible Source Materials

Katherine A. Adelsberger, Karl R. Wirth, Audax Z.P. Mabulla, and Daniel C. Bowman

Abstract Laetoli preserves a rich Middle Stone Age (MSA) artifact assemblage. We examine basaltic MSA materials collected through surface survey at a number of Laetoli localities, as well as representative samples of nearby volcanic rock exposures (i.e., Ogol Lavas, Lemagurut), using petrographic (mineralogic, textural) and geochemical (major and trace element composition) techniques. Petrographic analysis allows for the informal discrimination of three groups of artifacts based on lithologic criteria (Types A, B, and C). The petrographic basis for these groupings is supported by whole-rock geochemical characteristics. Type C artifacts are readily distinguished from Type A and B artifacts, as well as from local volcanic sources, using both major and trace element data. Type A and B artifacts are more similar in character, and geochemical data suggest that Type A artifacts were likely manufactured from the Ogol Lavas. Type B artifacts share some petrographic and geochemical characteristics with volcanic lavas from Ogol, Lemagurut, and Olmoti, but are not a straightforward match for any single source location. The compositions of Type C artifacts do not match those of any nearby volcanic sources used for comparison in this study, nor do they match published data from other volcanic centers in the region. These data demonstrate that multiple raw material sources were utilized in the manufacture of basaltic MSA artifacts at Laetoli. In addition, the dissimilar petrographic and geochemical features of the MSA artifacts and nearby bedrock exposures indicate more distal sources for the artifacts, despite the local availability of basalt. Raw materials for tool-making might have been transported to the site by streams or by Middle Stone Age hominins.

Keywords Geochemical fingerprinting • Basalt • Archaic *Homo sapiens* • Middle Stone Age • Provenance

Introduction

Laetoli is situated on the Eyasi Plateau of northern Tanzania, part of the southern Gregory Rift Zone of the East African Rift System (EARS). The site itself lies south of Olduvai Gorge and west of the Plio-Pleistocene volcanic centers that make up the Ngorongoro Highland (Hay 1987; Dawson 1992) (Fig. 7.1). Satiman and Lemagurut are the westernmost of these volcanic centers and the closest volcanic sources to Laetoli. The Laetoli region provides a record of volcanic processes of the past several million years, including tephra layers interbedded with sediments as well as the olivine nephelinite flows known as the Ogol Lavas (Hay 1987). The Ogol Lavas, which date to 2.41 ± 0.12 Ma (Drake and Curtis 1987), erupted from several small cones oriented along a 25 km east-west zone to the south and southwest of Lemagurut volcano (Hay 1987). Ogol Lava outcrops at Laetoli occur as flow remnants and agglomerate cones (Hay 1987) in the hills surrounding the Laetoli paleontological localities (Fig. 7.2) (Leakey et al. 1976; Leakey 1981; Leakey and Harris 1987; Harrison and Kweka 2011).

Middle Stone Age artifacts, as well as the remains of archaic *Homo sapiens*, have been recovered from the Ngaloba Beds at Laetoli, which make up the uppermost Pleistocene stratigraphic unit (Day et al. 1980; Leakey 1981; Magori and Day 1983). These beds date to 200 ka, as determined by amino acid racemization of ostrich eggshells (Manega 1993). MSA materials continue to erode out of the Ngaloba Beds and can be collected through surface survey at many Laetoli localities (Fig. 7.2). Although many of these artifacts could

K.A. Adelsberger (✉)
Department of Environmental Studies, Knox College,
2 East South Sreet, Galesburg, IL 61401, USA
e-mail: kadelsbe@knox.edu

K.R. Wirth and D.C. Bowman
Department of Geology, Macalester College,
St. Paul, MN 55105, USA
e-mail: wirth@macalester.edu; dbowman@macalester.edu

A.Z.P. Mabulla
Archaeology Unit, University of Dar es Salaam,
Dar es Salaam, Tanzania
e-mail: aumab@udsm.ac.tz

T. Harrison (ed.), *Paleontology and Geology of Laetoli: Human Evolution in Context. Volume 1: Geology, Geochronology, Paleoecology and Paleoenvironment*, Vertebrate Paleobiology and Paleoanthropology, DOI 10.1007/978-90-481-9956-3_7,

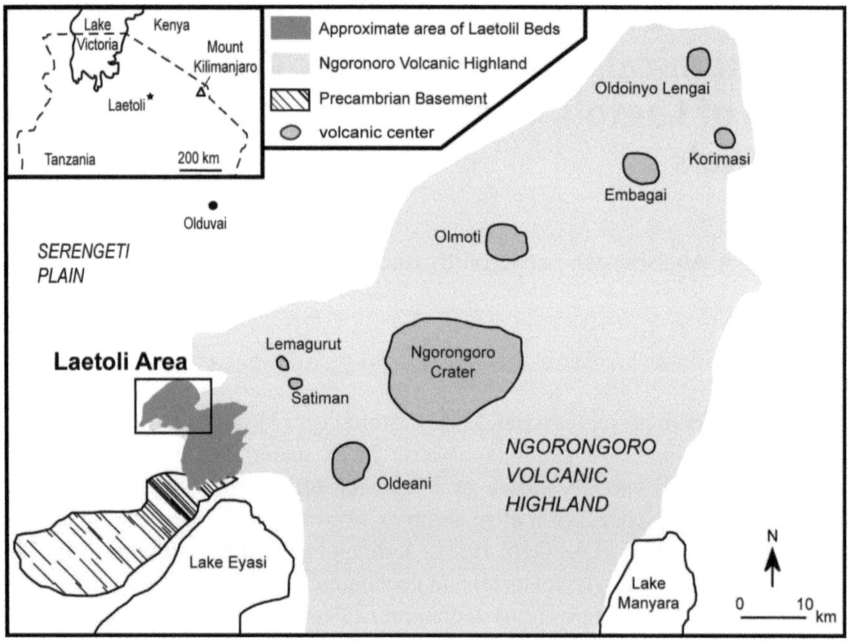

Fig. 7.1 Map of the Laetoli area and its regional context, illustrating the location of Laetoli and nearby volcanic centers associated with the Ngorongoro Volcanic Highland. Modified from Hay (1987). Boxed area illustrated in Fig. 7.2. See also Fig. 5.1 in Mollel et al. (2011)

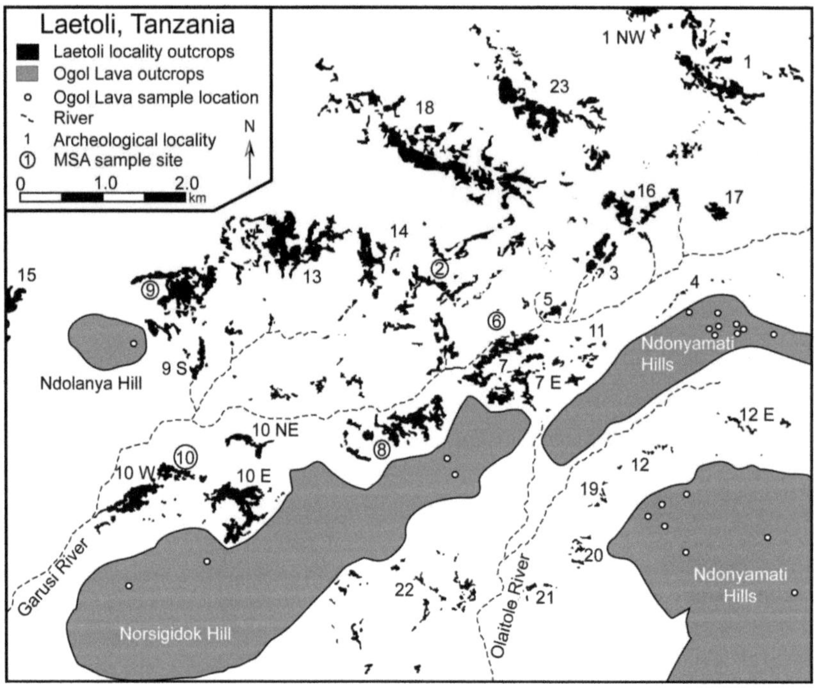

Fig. 7.2 Map of Laetoli and its numbered localities, as well as the approximate sampling area for Ogol Lava samples obtained during this study (modified after Musiba 1999). Area illustrated here is shown as the boxed region in Figs. 7.1 and 7.2. Localities where artifacts were collected are designated by circled locality numbers

not have been manufactured from local raw materials, a large proportion of Laetoli's MSA assemblage consists of basaltic artifacts that could have been sourced in the local Ogol Lavas. The potential for local tool manufacture, or alternatively for preferential transport of more suitable raw materials from elsewhere despite a locally available basalt, may provide insights into the behavioral patterns of hominins at Laetoli.

Studies of artifact source-material compositions can be used to address questions regarding raw material selection and transport (e.g., Merrick et al. 1994; Greenough et al. 2001). Major- and trace-element analyses are commonly used to distinguish between volcanic rock types and to help identify the tectonic or regional setting in which particular samples may have originated (e.g., Rollinson 1993). Previous geochemical investigations in the EARS sought the sources of Oldowan artifacts at Olduvai Gorge, Tanzania (Callow 1994; Jones 1994; Mollel et al. 2001; Mollel 2002). These studies identified the primary rock types used to make Oldowan artifacts, as well as possible source materials, including local quartzite, phonolite, basalt, trachyandesite, and nephelinite. Additional investigations examined the geochemistry of the various volcanic centers of the Ngorongoro Highlands as potential source materials for artifacts found within the EARS (Manega 1993; Mollel et al. 2001, 2011; Mollel 2002). However, these techniques have not been extensively utilized on MSA artifacts at Laetoli.

Most MSA stone artifacts at Laetoli were manufactured from volcanic rock that could have been derived from a variety of sources. Exposures of dense volcanic flows (Ogol Lavas) occur along the southern margin of the Laetoli area and may have been utilized as sources for MSA materials. The valley along the modern Garusi River (Fig. 7.2) contains abundant pebbles, cobbles, and boulders of volcanic rock transported from the nearby Ngorongoro Highlands (Fig. 7.1), which exhibit diverse compositions and textures. These too may have been utilized by Middle Stone Age hominins as a source of raw material for tool manufacture. Alternatively, tools or the raw materials necessary for their manufacture may have been transported by Laetoli hominins from more distal volcanic sources. The objectives of this study are to characterize the mineralogic and chemical "fingerprints" of MSA artifacts at Laetoli, to determine if nearby volcanic rocks have similar characteristics, and to utilize these geochemical data to address questions of raw material selection and transport during the Middle Stone Age.

Methods

A total of 358 stone tools were collected from Laetoli through non-systematic surface survey at Locs. 2, 6, 7, 8, 9, and 10 in 2000 (Adelsberger et al. 2002) (Fig. 7.2). Artifacts were identified as MSA prior to analysis; materials suspected to be Later Stone Age were not utilized for this study. MSA materials were sketched, measured, photographed, and categorized by artifact type and rock type prior to analysis (Fig. 7.3). Data in this study are limited to basaltic materials, and artifacts composed of quartzite, phonolite or chert are not be considered further in this study. Ogol Lava outcrops were sampled throughout the Ndonyamati Hills, at Ndolanya Hill, and in other areas where the Ogol Lavas are present (Wirth and Adelsberger 2002) (Fig. 7.2). In most cases, samples of volcanic materials can be associated with specific volcanic centers on the basis of volcanic landforms and geology. However, the affinities of some volcanic rocks in the eastern Laetoli region are more unambiguous in the field. In these cases, volcanic flows were attributed to either Ogol or Lemagurut volcanism on the basis of mineralogy, major and trace element chemistry and isotopic composition. Volcanics were described in terms of rock type, as well as the proportion and composition of phenocrysts, the proportion of vesicles, and the color of weathered and fresh rock surfaces.

Fig. 7.3 Photograph of representative artifacts analyzed as part of this study: (**a**, **b**) weathered flakes and (**c**) core

Examination of thin sections distinguished several petrographic types prior to geochemical analyses, which were used to compare artifacts and lava samples.

A subset of 71 basaltic samples were analyzed by X-ray fluorescence (XRF) spectrometry to determine major (SiO_2, TiO_2, Fe_2O_3, MnO, MgO, CaO, Na_2O, K_2O, P_2O_5) and trace (Sc, V, Cr, Ba, Rb, Sr, Nb, Zr, Y, Co, Ni, La, Ce, Th, U, and Pb) element compositions using a Philips PW2400 X-ray fluorescence spectrometer at Macalester College. Whole-rock samples were crushed, sifted, and powdered for major elemental analysis using a SPEX 8510 Shatterbox with a tungsten carbide (WC) vessel. Separate aliquots of most samples were powdered in a steel vessel for trace element analysis to minimize tungsten and tantalum contamination. Artifacts that were too small to be divided into two separate aliquots were powdered using only the steel vessel. Pressed powder pellets for trace element analysis were prepared by mixing sample powders with a polyvinyl alcohol binder and compressing the powder under six tons of pressure for one minute. Fused glass beads for major element analysis were prepared by first drying the sample powders to determine loss on ignition, then mixing in a 1:5 ratio with flux (a lithium tetraborate-lithium metaborate 12:22 mixture) and heating in Pt-Au crucibles. The molten sample-flux solutions were cast in Pt-Au molds and allowed to cool before analysis. In order to ensure that the results obtained from the two different grinding containers (steel and tungsten carbide) did not differ significantly, beads were created using both the steel and the WC powders for several samples and analyzed using XRF. Analytical precision is estimated to be $<1\%$ relative for the major elements (Vervoort et al. 2007).

The concentrations of the rare earth elements, Cs, Ta and Hf were determined for representative Ogol lavas and Laetoli artifacts using inductively-coupled plasma-mass spectrometry

(ICP-MS) in the GeoAnalytical Laboratories at Washington State University. Samples were prepared by acid digestion of fused beads and were analyzed using a Sciex Elan model 250 ICP-MS (Knaack et al. 1994). Analytical precision is estimated to better than 5% for most trace elements.

The artifacts from Laetoli are commonly coated with a thin layer of desert varnish (patina), and this was often included as part of the total artifact sample during whole-rock XRF analysis due to the small size of some artifacts (40 g/5 cm maximum dimension). Varnish was removed from a number of the larger artifacts and analyzed independently in order to evaluate its elemental effect on whole-rock artifact compositions. Comparison between varnish and whole-sample analyses shows that varnish is generally enriched in Fe, Mn, P, Ba, Sr, and Y relative to the fresh rock. Varnish coatings also show a general decrease in Si and Mg when compared to the bulk sample. Varnish-free samples are rare and were not analyzed as part of this study. Therefore, varnish compositions provide the only data available for determining the effect of varnish on geochemical analyses. To minimize the effects of varnish compositions on elemental analysis of the artifacts, varnished surfaces were avoided during sample preparation when possible, and comparisons were made using elements that showed little or no variation between desert varnish and whole-rock analyses.

The Nd and Sr isotopic compositions of five samples of Ogol lava and one sample of Lemagurut lava were analyzed in the GeoAnaltyical Laboratories at Washington State University using standard chemical separation techniques. Samples were dissolved for ~5 days at 160°C in steel-jacketed, high-pressure teflon dissolution vessels using a 10:1 mixture of concentrated HF and HNO_3 acids. Following oven dissolution, samples were converted to chloride complexes using H_3BO_3 and HCl acids and spiked with a mixed ^{149}Sm-^{150}Nd isotopic tracer. Sr and Rare Earth Elements were separated using ion-exchange columns with Dowex AG50W-X8 resin, eluting Sr with 2.5 M HCl and the REE with 6 M HCl. Sr was further purified in columns with Eichrom Sr Spec resin. Sm and Nd were isolated from other REE in columns using teflon powder coated with Hydrogen Di-(2-EthylHexyl)-Phosphate (HDEHP). The isotopic compositions of the resulting purified elements were analyzed using a ThermoFinnigan Neptune MC-ICPMS at Washington State University. Isotopic ratios were measured in static mode consisting of either 75, 8-second integrations (Sr and Nd) in three blocks or with one block of 30, 8-second integrations (Sm). Mass bias corrections were made using the exponential law and the following ratios: ^{86}Sr/^{88}Sr=0.1194; ^{146}Nd/^{144}Nd=0.7219; and ^{147}Sm/^{152}Sm=0.56081. Isotopic ratios of interest were corrected for the following potential interferences: Kr and Rb on Sr, Sm and Ce on Nd; and Gd and Nd on Sm. Because the elements are effectively purified during ion-exchange chemistry, these interferences are n egligible except for Ce on Nd. This interference does not, however, effect the determination of the ^{143}Nd/^{144}Nd ratio. Laboratory solution standards analyzed over the course of the study gave the following values (errors reported as two standard error): NBS 987 Sr, ^{87}Sr/^{88}Sr=0.710263±26; Ames Nd, ^{143} Nd/^{144}Nd=0.512116±22; and ^{149} Sm/^{152}Sm=0.516923±45. Final reported ^{87}Sr/^{88}Sr and ^{143}Nd/^{144}Nd values were adjusted slightly (< 100 ppm) relative to the accepted values for these standards (^{87}Sr/^{88}Sr=0.710240, ^{143}Nd/^{144}Nd=0.512138). Procedural blanks performed during this study were low (<< 1:1000) and no blank corrections were employed. Initial ε_{Nd} values were calculated assuming

$$^{143}\text{Nd}/^{144}\text{Nd}_{\text{CHUR}(0)} = 0.512638, ^{147}\text{Sm}/^{144}\text{Nd}_{\text{CHUR}(0)} = 0.1966, \text{ and } \lambda^{147}\text{Sm} = 6.54 \times 10^{-12}.$$

Results

Petrographic Characteristics

MSA artifacts collected and identified during this study are made from basalt, chert, and quartzite. The entire collection consists of 305 flakes, 52 cores, and a single hammerstone. For the purposes of this study, all artifacts were collected during surface survey but only basaltic materials (63% percent of the total collection) were examined in detail for petrographic or geochemical characteristics. To a large extent, the Laetoli artifacts can be differentiated from samples of nearby exposures of volcanic lava flows using basic mineralogy and textural characteristics that are visible in hand sample and in thin section. For example, all Ogol Lava samples collected during this study are porphyritic in texture, whereas only 28.8% of basaltic artifacts are porphyritic. This type of textural distinction serves as an initial means to distinguish between the majority of these artifacts and the lava samples. However, discrimination using the proportions of different mineral phases is much more useful as an initial comparison.

Discrimination of basaltic artifacts based on mineral assemblages and textures determined through thin-section petrography reveals three distinct artifact basalt "types" within the assemblage (Fig. 7.4). Type A artifacts (Fig. 7.4) contain large phenocrysts of clinopyroxene and olivine enclosed in a fine-grained groundmass of pyroxene, olivine, plagioclase, and ilmenite. Clinopyroxene phenocrysts are commonly zoned; olivine phenocrysts are mostly euhedral, but some grains are deeply embayed. Some groundmass plagioclase is subophitic and contains

Fig. 7.4 Photomicrographs (cross-polarized light) of representative thin sections illustrating the textures of Ogol Lava flows and artifact types described in the text

groundmass pyroxene. Elongate microcrysts of pyroxene define a trachytoid texture. In some samples, minor blue-green amphibole is also present as phenocrysts and groundmass plagioclase is less abundant. Type B artifacts (Fig. 7.4) are fine-grained basalts characterized by trachytic texture that consist of plagioclase and pyroxene in a microcrystalline groundmass of plagioclase, pyroxene, possible blue-green amphibole, and ilmenite. Some calcite is also present as an alteration mineral.

Artifacts assigned to Type C (Fig. 7.4) can be classified as medium-grained basalt consisting of plagioclase and intergranular and subophitic pyroxene, as well as ilmenite and rare olivine. These basaltic materials also contain calcite, fine-grained chlorite, and possible clay minerals, indicating considerable alteration. No comparable mineral composition is found in any of the Ogol Lava samples. Ogol Lavas (Fig. 7.4) are dominated by clinopyroxene and olivine phenocrysts accompanied by a matrix consisting of fine-grained pyroxene, ilmenite, altered glass, and minor calcite. Clinopyroxene in lava samples exhibits less pronounced zoning than that found in artifact samples; olivine is also largely subhedral and often altered to iddingsite. Few of the MSA materials exhibit mineral assemblages similar to those of the Ogol Lavas, and even those artifacts that are mineralogically similar to the Ogol Lavas have very different textural characteristics, including a lack of vesicles and smaller phenocrysts. Both of these features characterize textural changes found in the quickly cooled outer margins of lava flows.

Stone tools with different petrographic characteristics occur within single localities at Laetoli, and there is no apparent correlation between collection locality and the type defined through petrographic analysis. For the purposes of this paper, the artifacts will henceforth be referred to according to their assigned petrographic groups (Types A, B, and C), as defined through petrologic analysis. However, these assignments are intended only as informal designations for Laetoli's artifacts in order to provide a convenient method for comparing similar artifact compositions with possible source materials. Data for individual artifact and lava samples are illustrated in all cases in order to demonstrate the relationships between samples despite their designated petrographic groups.

Geochemical Discrimination

In order to better understand the nature of raw material sources for Laetoli's basaltic MSA materials, we examined geochemical data from the artifacts as well as from natural exposures of volcanic rock. The textural and geochemical properties of the three artifact types are summarized in Table 7.1. For comparison, data from the Ogol Lava flows are also included. Major and minor element compositions of the three-assigned artifact types and Ogol Lavas are listed in Table 7.2.

Table 7.1 Comparison of major textural and chemical attributes of the three artifact types and Ogol Lava basalts (see Table 7.2 for quantitative CIPW norms)

Type	Texture	Phenocrysts	Alkalinity[a]	Norm minerals[b]
Tool A	Phaneritic	Olivine & pyroxene	Alkaline	Ne+Ol
Tool B	Aphanitic	None	Alkaline	Ne+Ol
Tool C	Phaneritic	None	Subalkaline	Hy+Q or Hy+Ol
Ogol Lava	Phaneritic	Olivine & pyroxene	Alkaline	Ne+Ol

Ne normative nepheline (undersaturated), *Ol* normative olivine, *Hy* normative hypersthene (saturated), *Q* normative quartz (oversaturated)

[a] Alkalinity is based on MacDonald and Katsura (1964)

[b] CIPW Norm mineral compositions (Cross et al. 1903) are calculated from major element oxides and provide a measure of silica saturation in volcanic rocks

Major Element Compositions

Total alkalis-silica (TAS) diagrams (Le Bas et al. 1986; Le Maitre 1989) are commonly used to classify volcanic rocks and here serve to illustrate the wide compositional range of Laetoli artifacts compared to the compositions of nearby volcanic sources (Fig. 7.5). Type C artifacts range from basalt to andesite, whereas no Ogol samples fall within the andesite field. The slight differences in silica content caused by the inclusion of some desert varnish during analysis are unlikely to cause such a difference in overall relationships between these samples. Type C artifacts can be effectively distinguished from the other artifacts (Types A and B) and from the Ogol Lava samples using this diagram.

Artifact types and lava samples can also be clearly distinguished using a plot of $MgO/Fe_2O_3^*$ versus TiO_2 (Fig. 7.6), where the TiO_2 content of Type C artifacts is significantly lower than that of the other samples. This plot also separates Type B from Type A artifacts and from Ogol Lava samples because of the lower $MgO/Fe_2O_3^*$ (where $Fe_2O_3^*$ is total iron as Fe_2O_3) ratios in Type B. These major element diagrams indicate significant elemental differences between the artifact and lava samples, supporting the presence of at least two separate raw material sources within the artifact assemblage. Major elemental data alone suggest that Type C artifacts are significantly different from both Types A and B artifacts, and Type A basalts seem most similar to the local Ogol Lavas. The separation of Type C artifacts in particular is supported by the unique mineral assemblage of these samples as determined by petrographic analysis.

Trace Element Compositions

The trace element concentrations of volcanic rocks commonly show greater variation than those exhibited by the major elements (e.g., Rollinson 1993). Processes of melting and crystallization largely control trace element concentrations during the production and evolution of volcanic melts, whereas major element compositions are determined by the mineral assemblage present. As a result, two volcanic rocks with similar major element compositions can have significantly different trace element compositions. Some trace elements are also less susceptible to modification by late stage processes (e.g., alteration and weathering) and are generally regarded to be more reliable indicators of igneous processes and origins.

The rare earth elements (REE) are commonly treated as a group because they generally behave incompatibly in basaltic melts and because they exhibit progressive fractionation effects as a function of ionic radius during melting and crystallization (e.g., Rollinson 1993). REE patterns for Laetoli Type A and B artifacts have similar steep patterns, suggesting similar volcanic origins, but they differ markedly from those of Type C artifacts (Fig. 7.7a–c, Table 7.3). Analyzed Ogol and Lemagurut lavas (Figs. 7.7d, e respectively) also have similar REE patterns, but some Lemagurut lavas have higher normalized abundances and weak positive Eu-anomalies.

Trace element concentrations of the lavas and Laetoli artifacts show similar trends (Fig. 7.8a–e). Type A and B artifacts have similar trace element patterns, but some Type B artifacts exhibit higher mantle normalized concentrations and more pronounced U, P, and Ti anomalies. Type C artifacts are distinct with nearly flat mantle normalized patterns. Ogol and Lemagurut lavas have similar patterns, but Lemagurut lavas show greater variation in mantle-normalized abundances.

Nd and Sr Isotopic Compositions of Ogol and Lemagurut Lavas

Representative lavas from the Ogol flows and from Lemagurut were analyzed for their Nd and Sr isotopic compositions (Table 7.4). Initial Epsilon Nd values from the Ogol Lavas range from −1.0 to −7.9 and are inversely correlated with $^{87}Sr/^{86}Sr$ ratios (0.703502 – 0.705766). The Lemagurut lava has Nd (initial $\varepsilon_{Nd} = -3.8$) and Sr ($^{87}Sr/^{86}Sr = 0.704753$) isotopic compositions similar to the analyzed Ogol Lavas. These isotopic data are not discussed further in this paper, but the interested reader is referred to Paslick et al. (1995) and Mollel et al. (2011).

Table 7.2 Major element compositions of Laetoli MSA stone artifacts, Ogol Lavas, and nearby volcanic centers

Sample	Type	SiO_2	TiO_2	Al_2O_3	$Fe_2O_3^*$	MnO	MgO	CaO	Na_2O	K_2O	P_2O_5	LOI	Total	Mg#	Q	Ne	Ol
Laetoli MSA Stone Tools																	
TOOL 2–12	A	46.91	3.73	13.50	14.45	0.21	4.87	8.13	3.88	1.88	0.66	1.33	99.54	0.57		4.99	11.17
TOOL 2–12	A (Fe)	46.87	3.73	13.47	14.54	0.22	4.87	8.12	3.85	1.89	0.66	1.31	99.52	0.57		4.89	11.28
TOOL 2–16	A	44.85	4.12	10.91	16.68	0.21	6.58	9.85	3.33	1.37	0.53	0.78	99.21	0.61		6.38	12.56
TOOL 2–27	B	46.70	3.48	13.65	13.64	0.19	4.52	7.75	4.14	1.95	0.66	1.98	98.66	0.57		6.13	10.41
TOOL 2–28	A	46.76	3.31	12.31	14.54	0.21	5.26	8.57	3.35	1.97	0.61	2.65	99.55	0.59		3.16	11.64
TOOL 2–39	B	46.43	3.31	12.31	14.66	0.20	5.06	8.80	3.31	1.96	0.61	3.00	99.65	0.58		3.58	11.08
TOOL 2–40	B	46.55	3.71	13.49	14.45	0.20	4.62	8.04	3.85	1.96	0.67	1.72	99.26	0.56		5.20	10.85
TOOL 6–1	A	44.87	3.19	10.22	14.77	0.22	8.11	10.96	2.84	1.39	0.52	2.08	99.16	0.68		6.40	13.44
TOOL 6–11	A	45.44	3.29	10.82	14.87	0.22	7.27	11.24	2.90	1.46	0.56	1.80	99.86	0.66		6.32	11.57
TOOL 6–12	A	44.94	4.16	10.81	16.89	0.21	7.09	9.75	3.11	1.28	0.53	0.67	99.44	0.62		4.73	14.14
TOOL 6–17	B	46.45	3.60	13.53	14.03	0.22	4.53	8.18	3.88	1.96	0.66	2.03	99.07	0.56		5.67	10.26
TOOL 6–18	B	46.60	3.70	13.54	14.30	0.22	4.88	8.11	3.85	1.90	0.67	1.74	99.50	0.57		5.22	11.21
TOOL 6–20	B	46.46	3.63	13.48	14.25	0.21	4.88	8.09	3.83	1.90	0.66	1.91	99.30	0.58		5.23	11.29
TOOL 7–16	A	44.95	3.11	8.95	14.49	0.20	10.67	11.57	2.31	1.37	0.48	1.79	99.88	0.74		5.71	16.85
TOOL 7–16	A (Fe)	44.95	3.10	8.87	14.57	0.19	10.63	11.59	2.32	1.35	0.47	1.85	99.89	0.74		5.69	16.77
TOOL 7–30	A	45.03	4.15	10.90	16.64	0.22	6.96	9.53	3.27	1.32	0.52	1.00	99.54	0.62		5.24	13.86
TOOL 7–37	B	47.80	3.00	12.83	14.31	0.19	5.15	8.28	3.82	1.92	0.58	1.80	99.68	0.59		4.27	11.72
TOOL 8–12	A	45.39	3.27	10.77	14.89	0.24	7.44	10.79	3.10	1.50	0.61	1.28	99.27	0.66		7.15	12.32
TOOL 8–56	B	52.89	1.26	18.62	7.15	0.17	1.27	3.97	6.85	4.56	0.20	2.90	99.84	0.41		17.40	4.08
TOOL 8–60	C	55.80	0.86	15.18	9.74	0.15	5.51	8.08	3.02	1.29	0.16	0.23	100.03	0.69	5.97		
TOOL 8–74	A	46.73	3.68	13.57	14.27	0.21	4.75	8.13	4.15	1.93	0.66	1.14	99.21	0.57		7.06	10.44
TOOL 8–77	A	44.27	4.17	11.05	17.11	0.24	6.56	9.18	3.17	1.32	0.53	1.12	98.71	0.60		4.73	14.47
TOOL 8–82	A	44.90	3.08	8.95	14.30	0.19	10.73	11.56	2.38	1.38	0.47	1.61	99.55	0.75		6.26	16.76
TOOL 8–9	B	50.95	2.36	16.24	10.29	0.23	2.60	6.05	5.33	2.66	0.79	1.90	99.40	0.50		6.81	7.19
TOOL 9–13	A	45.95	3.09	10.22	14.13	0.20	9.31	10.96	2.78	1.32	0.49	1.28	99.72	0.72		5.19	15.18
TOOL 9–17	B	48.15	3.24	14.34	12.57	0.22	4.05	7.83	4.15	2.11	0.69	2.07	99.42	0.56		4.75	8.97
TOOL 9–20	C	59.24	1.15	14.11	9.50	0.18	3.31	5.53	3.54	1.89	0.20	1.07	99.72	0.58	12.16		
TOOL 9–28	B	47.20	3.64	13.76	14.06	0.20	4.75	8.12	4.01	1.93	0.67	1.44	99.77	0.57		5.53	10.61
TOOL 9–30	A	44.79	4.19	10.46	16.97	0.21	7.38	10.04	3.04	1.27	0.51	0.81	99.66	0.63		5.13	14.10
TOOL 9–38	C	59.67	1.13	14.00	9.56	0.14	2.97	5.88	3.65	2.15	0.19	0.00	99.34	0.55	11.29		
TOOL 10–1	C	49.59	1.55	14.75	13.07	0.19	7.08	9.60	2.24	0.98	0.34	0.99	100.38	0.68			2.23
TOOL 10–10	A	43.42	3.41	7.81	15.44	0.31	10.40	13.41	2.07	1.07	0.54	1.92	99.80	0.73		8.00	14.03
TOOL 10–12	A	43.92	3.42	7.77	15.58	0.26	11.28	12.89	1.91	0.85	0.48	1.44	99.80	0.74		5.14	16.84
TOOL 10–13	C	59.58	1.16	14.19	9.77	0.14	3.06	6.02	3.64	2.17	0.21	0.16	100.09	0.55	10.61		
TOOL 10–14	C	48.16	1.29	16.26	11.78	0.18	7.00	9.86	1.90	0.67	0.16	2.78	100.03	0.70			0.16
TOOL 10–16	C	48.90	1.69	15.32	12.42	0.17	5.36	8.51	2.89	0.69	0.26	3.50	99.71	0.63	0.21		
TOOL 10–17	C	59.50	1.14	14.07	9.74	0.14	3.17	6.13	3.62	2.08	0.20	0.00	99.79	0.56	10.74		
TOOL 10–2	C	48.28	1.23	15.80	12.43	0.23	7.10	9.73	2.18	0.93	0.20	1.75	99.86	0.69			6.61
TOOL 10–3	C	48.02	1.20	16.27	12.73	0.26	6.86	10.01	2.09	0.92	0.20	1.63	100.18	0.68			7.35
TOOL 10–39	C	59.76	1.14	14.08	9.71	0.14	3.15	6.10	3.60	2.13	0.20	0.05	100.06	0.56	10.98		

(continued)

Table 7.2 (continued)

Sample	Type	SiO_2	TiO_2	Al_2O_3	$Fe_2O_3^*$	MnO	MgO	CaO	Na_2O	K_2O	P_2O_5	LOI	Total	Mg#	Q	Ne	Ol
TOOL 10–41	C	59.59	1.14	14.00	9.59	0.14	3.06	5.98	3.66	2.11	0.20	0.03	99.49	0.56	11.04		
TOOL 10–44	C	47.54	1.22	15.67	11.33	0.17	7.11	9.62	2.32	0.31	0.15	4.11	99.56	0.71			2.59
TOOL 10–45	C	47.53	2.41	15.21	12.74	0.18	6.82	9.39	2.21	1.40	0.60	1.34	99.83	0.68			8.08
TOOL 10–56	C	49.52	1.01	17.36	10.05	0.34	4.23	11.19	3.18	0.97	0.10	1.90	99.86	0.63			9.03
TOOL 10–63	B	47.98	3.41	14.15	13.46	0.21	4.65	7.69	4.35	1.96	0.68	1.16	99.69	0.58		6.01	10.61
TOOL 10–66	B	46.76	3.57	13.61	13.82	0.20	4.51	8.41	4.03	1.92	0.67	1.79	99.28	0.56		6.46	9.48
TOOL 10–66	B (Fe)	46.89	3.56	13.58	13.88	0.20	4.52	8.41	4.07	1.92	0.67	1.83	99.53	0.56		6.57	9.46
TOOL 10–68	C	46.39	2.51	14.63	14.63	0.21	6.83	9.64	2.09	0.56	0.46	1.98	99.92	0.65			5.07
TOOL 10–71	A	46.05	3.06	9.95	14.08	0.20	9.86	10.65	2.94	1.29	0.48	0.93	99.49	0.73		5.78	16.29
TOOL 10–72	A	42.39	3.88	8.57	16.53	0.24	7.91	13.54	2.29	1.17	0.60	2.25	99.38	0.65		9.60	9.56
TOOL 10–79	A	45.75	3.16	10.48	14.45	0.22	8.22	11.10	2.85	1.37	0.52	1.46	99.58	0.69		5.60	13.26
TOOL 10–8	A	44.93	4.20	10.91	16.72	0.21	6.93	9.52	3.13	1.39	0.52	1.09	99.55	0.62		4.71	13.96
Varnish 10–1	Varnish	48.94	1.56	14.93	13.04	0.23	6.63	9.54	2.31	1.03	0.38	1.41	100.00	0.67			3.73
Varnish 10–14	Varnish	48.11	1.30	16.27	11.80	0.18	6.81	9.95	1.89	0.65	0.16	2.71	99.82	0.70	0.17		
Varnish 8–74	Varnish	46.72	3.71	13.58	14.54	0.21	4.47	7.89	3.96	1.94	0.67	1.55	99.23	0.55		5.26	10.81
Varnish 8–82	Varnish	44.24	3.16	8.97	14.97	0.20	9.34	11.98	2.39	1.39	0.49	1.92	99.04	0.71		7.00	14.09
Ogol Lavas																	
OL-01	Basanite	44.92	3.13	9.05	14.49	0.22	10.14	11.98	2.63	1.21	0.48	1.37	99.62	0.73		7.49	14.99
OL-02B																	
OL-03A1	Basalt	46.26	2.92	10.07	13.64	0.18	9.24	11.22	2.75	1.41	0.59	1.18	99.46	0.73		5.68	15.23
OL-03B	Basalt	46.41	2.95	9.92	13.75	0.18	9.47	10.91	2.39	1.37	0.60	1.25	99.20	0.73		2.18	15.69
OL-04A	Basalt	45.73	3.08	10.19	14.10	0.19	9.48	10.91	2.22	1.53	0.67	1.44	99.53	0.73		2.78	16.20
OL-06A	Basalt	45.45	3.00	9.34	14.09	0.19	10.47	11.56	2.90	1.32	0.53	0.53	99.38	0.75		8.50	15.67
OL-07	Basalt	46.08	3.05	10.23	14.12	0.19	8.85	10.82	2.46	1.52	0.63	1.52	99.47	0.71		3.15	14.91
OL-08	Basalt	45.88	3.04	10.39	14.51	0.21	9.15	9.89	2.60	1.59	0.64	1.52	99.42	0.71		3.28	17.18
OL-09B	Basanite	44.73	4.09	10.70	16.56	0.21	7.44	9.64	3.04	1.27	0.52	0.82	99.01	0.64		4.49	14.86
OL-10B	Basanite	44.83	4.14	10.33	16.87	0.21	7.61	10.20	3.21	1.34	0.49	0.41	99.64	0.64		6.93	13.82
OL-10D	Basanite	44.37	4.15	10.23	16.89	0.21	7.82	10.13	2.95	1.27	0.49	0.42	98.93	0.65		5.57	14.79
OL-12A	Basalt	45.52	3.00	9.02	14.45	0.19	11.18	11.14	2.42	1.17	0.45	0.94	99.47	0.75		4.53	18.54
OL-12B	Basalt	45.26	2.99	9.03	14.60	0.20	10.33	11.09	1.84	1.13	0.43	1.85	98.75	0.74		0.09	18.27
OL-13B	Basalt	45.49	2.89	8.81	14.17	0.19	11.17	11.47	1.99	0.98	0.43	1.83	99.42	0.76		1.48	18.64
OL-15	Basanite	44.93	3.02	9.41	14.15	0.19	9.75	11.20	2.18	1.20	0.46	2.45	98.94	0.73		2.70	16.36
OL-17	Basalt	45.46	2.98	9.27	14.23	0.19	10.91	11.09	2.88	1.23	0.46	0.43	99.12	0.75		7.50	17.43
OL-18	Basalt	45.45	3.05	9.86	14.12	0.20	9.74	11.02	2.86	1.31	0.49	1.13	99.23	0.73		6.71	15.65
OL-19D	Basalt	47.74	2.74	10.31	13.10	0.16	10.33	9.40	2.82	1.27	0.36	0.85	99.08	0.76		0.72	18.88
OL-22A	Basalt	45.17	3.04	10.03	13.79	0.19	8.91	11.11	2.90	1.10	0.49	1.98	98.70	0.72		5.83	14.03
OL-22E	Picrobasalt	40.31	3.99	9.57	26.87	0.37	3.40	9.15	2.27	0.59	2.11	0.35	98.98	0.33		0.77	19.36
OL-22F	Tephrite	42.74	2.88	9.61	13.16	0.18	8.21	13.53	2.75	1.04	0.47	4.05	98.63	0.71		11.77	8.83
OL-23B	Basanite	44.92	3.20	7.18	15.12	0.19	12.00	13.12	2.05	0.96	0.37	0.25	99.36	0.76		6.04	16.71

Endonyo empuskiyaa																	
TZ W 1	Basalt	47.03	2.81	8.84	14.41	0.17	9.56	12.26	2.45	1.13	0.33	1.24	100.23	0.72		3.38	13.69
Lemagurut																	
OL-20C	Trachyand	54.66	1.55	17.54	8.54	0.18	1.89	4.78	6.06	3.10	0.54	0.59	99.44	0.47		5.82	6.58
OL-25B	Basalt	46.98	2.78	10.47	13.67	0.19	10.16	9.70	2.61	1.17	0.44	1.53	99.70	0.75		0.55	19.19
OL-26B	Basalt	46.77	2.79	10.75	13.64	0.18	9.80	9.76	2.63	1.41	0.45	1.16	99.33	0.74		1.84	18.36
TZ Y 2	Bas. trachy	51.61	2.71	16.02	11.48	0.21	2.80	6.20	5.29	2.90	0.79	0.31	100.32	0.49		7.26	7.52
TZ Z 1	Trachybasalt	48.32	3.37	14.79	14.66	0.22	3.56	7.30	4.65	2.22	0.77	0.12	99.98	0.49		7.49	10.05
TZ AA 1	Basalt	46.59	2.90	10.19	14.51	0.20	9.94	11.96	2.95	1.09	0.45		100.78	0.73		7.12	15.11
TZ AC 1	Trachyand.	56.84	1.43	17.52	8.31	0.19	1.78	4.01	6.21	3.09	0.45	0.09	99.91	0.46		2.46	7.25
TZ AC 6	Trachyand	55.10	1.76	17.65	9.11	0.19	2.22	4.93	5.91	2.77	0.62		100.26	0.49		3.61	7.81
TZ AL 1	Trachybasalt	50.63	3.29	14.74	12.82	0.17	4.00	8.09	4.25	1.46	0.45		99.89	0.55		0.16	9.12
TZ AO 4	Trachyand	55.33	1.75	17.93	9.20	0.18	1.76	4.00	5.73	2.93	0.59	0.83	100.23	0.43		0.80	8.96
Oldeani																	
TZ AG 2	Bas. trachy	52.48	3.08	14.64	12.49	0.16	3.89	7.42	4.22	1.50	0.45		100.32	0.55			1.25
Satiman																	
TZ AK 1	Trachybasalt	52.06	2.98	14.58	12.46	0.20	3.94	7.50	4.19	1.51	0.45		99.86	0.56			2.81
Average compositions																	
Lemagarut	Bas. trachy	52.68	2.17	15.38	10.60	0.19	3.74	6.09	4.79	2.51	0.55	1.30	100.00	0.58		1.47	9.92
Ogol	Basalt	45.37	3.17	9.65	14.50	0.19	9.61	11.07	2.59	1.26	0.50	1.31	99.22	0.72		4.81	15.77
Tool Type A	Basalt	45.11	3.58	10.45	15.39	0.22	7.88	10.57	2.94	1.40	0.54	1.42	99.50	0.67		5.89	13.62
Tool Type B	Trachybasalt	47.76	3.23	14.12	13.15	0.20	4.27	7.64	4.26	2.21	0.63	1.96	99.43	0.56		6.34	9.83
Tool Type C	Basaltic andesite	52.94	1.36	15.06	11.17	0.18	5.16	8.20	2.86	1.33	0.24	1.35	99.85	0.65	3.72		

Number preceeding hypen in Laetoli MSA stone tools refers to Laetoli site number where collected. "Type" refers to stone tool petrographic type (see text for description), or in the case of lavas, rock type based on IUGS classification.$Fe_2O_3^*$ is total iron as Fe_2O_3; samples with "(Fe)" appended in sample name were ground in Fe bowl to test for contamination effects.Samples with no LOI value experienced weight gain after fusion; a result of low Loss on Ignition and/or iron oxidation. *Q*, *Ne*, and *Ol* mineral compositions are calculated following CIPW norms (Cross et al. 1903) and provide a measure of silica saturation in volcanic rocks. Mineral Abbreviations: *Ne* = normative nepheline (undersaturated), *Ol* = normative olivine, *Q* = normative quartz (oversaturated), *bd* = below detection

Discussion

Stone artifacts from Laetoli exhibit significant petrographic and chemical variations that may provide clues to raw material sources used by early Middle Stone Age toolmakers at Laetoli. In this section, we compare the characteristics of Laetoli's MSA artifacts with those of nearby lavas and regional volcanic centers.

Comparison of Laetoli Artifacts and Local Lava Sources (Ogol Lavas)

Ogol Lavas are the most proximal volcanic materials that might have served as a source for MSA materials at Laetoli.

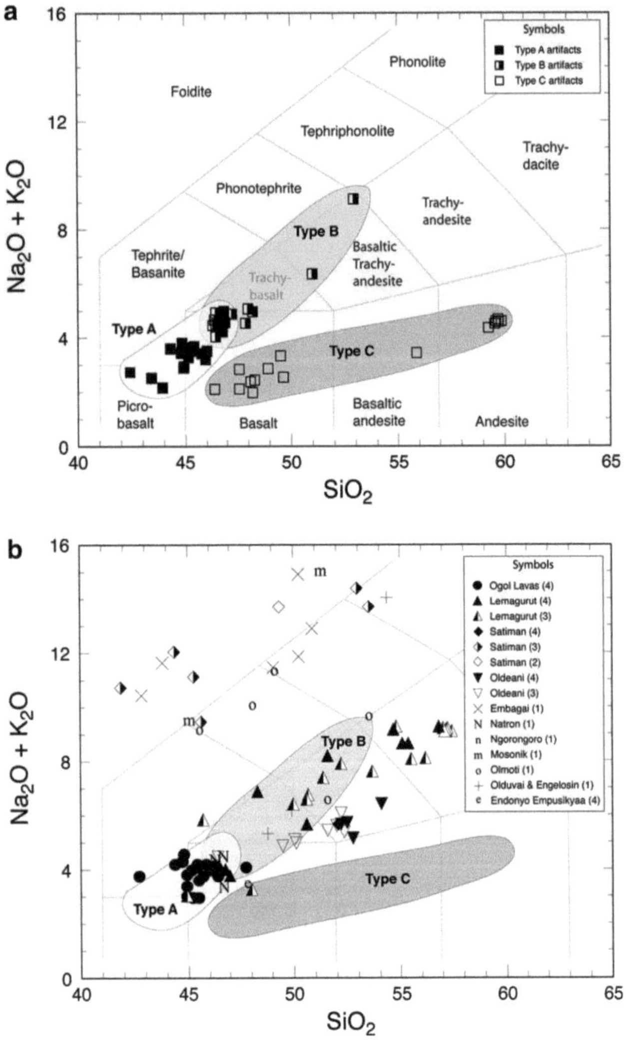

Fig. 7.5 Plot of SiO_2 versus total alkalis (Na_2O+K_2O). The compositions of Laetoli's basaltic MSA artifacts (**a**) are compared with Ogol Lavas and other nearby volcanic centers (**b**). IUGS rock classification terms and boundaries are from Le Maitre (1989). Data from (1) Manega (1993), (2) Paslick et al. (1995), (3) Mollel et al. (2011), and (4) this study

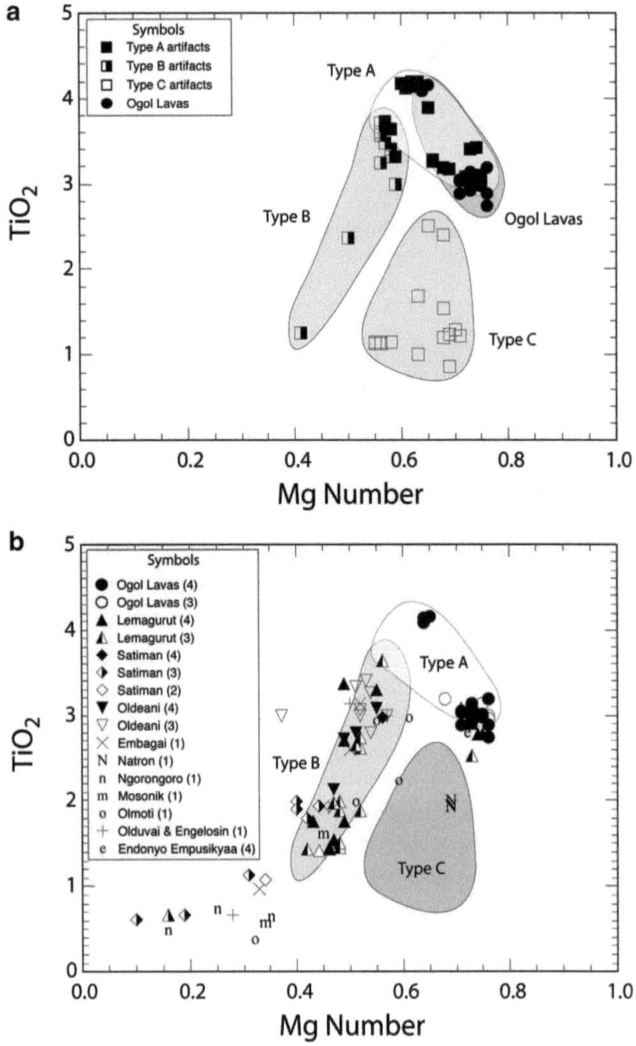

Fig. 7.6 Plot of Mg Number (Mg/(Mg+Fe)) versus TiO_2 (wt. percent) for basaltic artifacts and Ogol Lavas (**a**), as well as nearby volcanic centers (**b**). Symbols and data as in Fig. 7.5

Petrographic data indicate that the major mineralogy and textures of the Type A artifacts are most similar to the Ogol Lava samples. The lack of macroscopic olivine phenocrysts in Type B and C artifacts alone does not preclude these artifact types from being derived from Ogol Lava; they could have originated from phenocryst-poor flows, or portions of flows. However, major element data from the artifacts confirm that Type C is most dissimilar to the Ogol Lavas (Figs. 7.5a and 7.6a). On the basis of major element characteristics, it could be argued that Type B artifacts might also have derived from "more evolved" members of the Ogol Lava system. In contrast, it is highly unlikely that Type C artifacts, which have distinctive petrographic (Fig. 7.4) and major element compositions (Figs. 7.5a and 7.6a), were derived from Ogol Lavas.

Although major elemental chemistry alone does not eliminate the Ogol Lavas as a possible source material for Type A or Type B artifacts, trace elemental analysis provides an

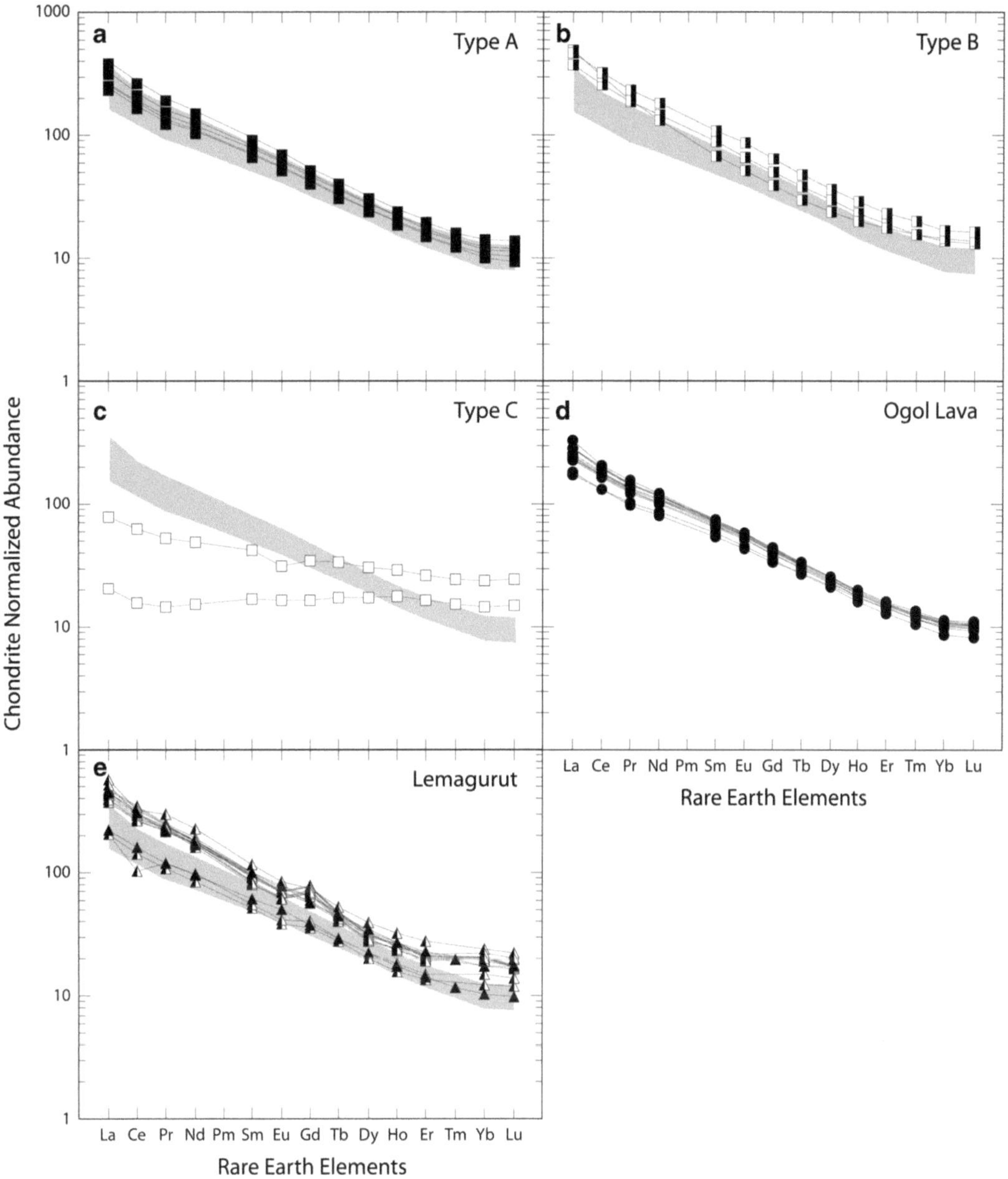

Fig. 7.7 Rare earth element plots of Laetoli MSA artifacts (**a–c**), Ogol Lavas (**d**), and Lemagurut lavas (**e**). The range of Ogol Lava compositions is shown for comparison with a shaded field in a–c and e. Concentrations are normalized to chondrite meteorite values of McDonough and Sun (1995)

additional means for differentiating these materials. The REE (Fig. 7.7) and trace element (Fig. 7.8) patterns distinguish Type C artifacts from the Ogol Lavas as well as the other artifact types, indicating that these materials most likely originated from a source outside the Laetoli area, and one that can be clearly differentiated from the sources of other artifacts found at the site.

Comparison of combined major element and trace element compositions reveals a number of differences between the MSA artifacts found at Laetoli. A plot of Nb/Zr against magnesium number (Mg# = [Mg]/[Mg + Fe^{+2}]) illustrates two distinct trends (Fig. 7.9). Ogol Lava flows and Type A and B artifacts have relatively constant and high Nb/Zr values (0.27–0.35) that span a range of Mg Numbers from high (Ogol Lava and Type A) to intermediate (Type B) values. In contrast, Type C artifacts are characterized by much lower Nb/Zr values (0.5–1.2) that decrease slightly with decreasing Mg number. The Mg number serves as an index of the degree of crystal fractionation in magma, with a lower Mg number representing greater crystallization or more evolved magma.

The two distinct trends observed on the Mg number versus Nb/Zr plot (Fig. 7.9) suggest that, whereas the artifacts with a higher Nb/Zr trend may originate from Ogol Lava sources or other melts from enriched mantle sources (e.g., Paslick et al. 1995; Mollel et al. 2011), the materials used to make Type C artifacts must come from volcanic flows that likely originated from different mantle sources. Compared with Type A and B

Table 7.3 Trace element compositions of Laetoli MSA stone artifacts, Ogol Lavas, and nearby volcanic centers

Sample	Type	Co	Cr	Ni	Sc	V	Zn	Ga	Ba	Rb	Sr	Y	Zr	Nb	Cs	Ta	Hf
Laetoli MSA Stone Tools																	
TOOL 2–12	A	50	11	19	26	307	129	23	953	44	1438	32	332	110			
TOOL 2–12	A (Fe)	42		14	25	302	127										
TOOL 2–16	A	64	278	77	30	357	132	20	2604	32	1485	29	282	85			
TOOL 2–27	B	35	16	30	23	278	133	23	1345	47	1230	33	333	111			
TOOL 2–28	A	55	22	34	25	293	120	21	1373	52	1351	31	312	108			
TOOL 2–39	B	51	24	38	26	289	124	21	1462	52	1180	31	310	108			
TOOL 2–40	B	46	11	22	23	314	129	22	1769	47	1482	34	331	110			
TOOL 6–1	A	66	504	162	33	353	112	18	1228	30	1145	27	262	81			
TOOL 6–11	A	63	455	112	33	355	109	19	981	31	1272	28	276	83			
TOOL 6–12	A	69	290	76	30	350	126	20	2149	29	1726	29	275	80			
TOOL 6–17	B	42	14	18	24	299	129	22	1733	48	1421	33	329	111			
TOOL 6–18	B	48	9	19	24	305	128	22	1917	45	1286	32	332	112			
TOOL 6–20	B	44	10	19	24	300	132	22	2211	45	1343	32	331	111			
TOOL 7–16	A	71	731	220	34	305	100	17	780	35	1123	26	238	76			
TOOL 7–16	A (Fe)	69	736	206	35	301	92		1								
TOOL 7–30	A	65	265	69	30	344	125	21	683	31	1061	28	278	83			
TOOL 7–37	B	52	40	46	26	272	120	21	907	47	1106	29	299	88			
TOOL 8–12	A	63	453	145	33	395	114	19	3122	32	1338	30	262	82	0.23	4.50	6.25
TOOL 8–56	B	18	26	17	9	67	95	26	2309	129	1474	33	290	171	1.68	8.16	7.14
TOOL 8–60	C	39	170	42	23	177	70	19	468	44	329	18	106	6			
TOOL 8–74	A	47	14	32	24	297	142	23	1580	45	1173	33	332	111	0.38	6.14	7.95
TOOL 8–77	A	67	251	74	30	372	134	20	5342	30	1558	31	290	87			
TOOL 8–82	A	73	739	234	35	304	113	17	701	35	1056	26	238	76			
TOOL 8–9	B	20	22	8	14	103	134	24	2418	64	1528	40	435	139	0.71	7.25	9.80
TOOL 9–13	A	64	609	185	32	314	106	18	903	29	1321	27	253	76			
TOOL 9–17	B	40	12	11	22	261	125	23	1115	50	1304	34	344	115			
TOOL 9–20	C	29	79	43	19	190	120	20	753	76	383	28	176	9			
TOOL 9–28	B	47	10	23	24	293	137	23	1052	46	1142	32	334	112			
TOOL 9–30	A	72	327	85	32	365	123	20	620	30	984	27	263	79			
TOOL 9–38	C	27	71	35	20	191	87	22	775	75	393	27	176	10			
TOOL 10–1	C	50	284	126	31	189	88	20	410	51	238	31	135	9			
TOOL 10–10	A	81	917	232	37	394	99	16	664	25	876	27	225	66			
TOOL 10–12	A	85	886	223	37	357	99	16	787	19	832	25	210	64	0.16	3.61	5.25
TOOL 10–13	C	28	69	31	19	201	89	22	652	74	309	27	170	11			
TOOL 10–14	C	54	122	62	31	240	76	18	463	47	353	20	90	10			
TOOL 10–16	C	51	38	93	22	231	94	21	400	40	479	22	166	17			
TOOL 10–17	C	30	65	35	20	197	87	21	655	72	324	27	167	10			
TOOL 10–2	C	56	233	137	30	226	132	17	677	96	250	25	104	6			
TOOL 10–3	C	54	223	129	30	234	210	17	217	121	182	25	106	6			
TOOL 10–39	C	29	67	32	20	192	86	21	636	76	318	27	168	10			
TOOL 10–41	C	28	69	39	21	190	94	21	668	73	330	27	170	10			

TOOL 10–44	C	52	113	69	28	221	113	16	637	23	334	18	86	9			
TOOL 10–45	C	50	228	149	30	241	96	20	452	131	189	45	262	16			
TOOL 10–56	C	72	342	180	40	300	78	19	221	56	278	28	5 8	3	0.44	0.19	1.62
TOOL 10–63	B	38	12	10	22	262	123	23	934	47	1128	33	338	114			
TOOL 10–66	B	46	8	20	24	295	128	23	1517	48	1191	32	333	113	0.45	5.94	7.74
TOOL 10–66	B (Fe)	43	11	15	23	294	125										
TOOL 10–68	C	50	257	135	30	238	96	20	1126	46	231	42	228	14	1.84	0.83	5.09
TOOL 10–71	A	64	665	195	32	294	101	18	578	29	774	26	251	76	0.26	3.99	6.01
TOOL 10–72	A	72	697	153	38	437	116	18	584	27	967	31	260	72			
TOOL 10–79	A	61	639	186	33	342	107	19	981	30	1308	28	271	82			
TOOL 10–8	A	71	255	72	30	347	133	21	679	31	818	28	276	83	0.29	4.54	6.59
Varnish 10–1	Varnish	46	280	127	31	209	89	20	344	54	250	32	149	10			
Varnish 10–14	Varnish	54	123	63	29	242	78	18	858	46	394	20	90	10			
Varnish 8–74	Varnish	47	9	19	23	309	136	23	2522	48	1317	33	336	113			
Varnish 8–82	Varnish	62	765	177	36	365	104	17	2191	39	1507	28	255	78			
Ogol Lavas																	
OL-01	Basanite	72	677	213	36	313	97	17	1300	27	1047	26	237	76			
OL-02B								19	475	30	512	23	203	51			
OL-03A1	Basalt	67	606	194	32	282	96	18	956	29	858	26	245	81	0.43	4.23	5.79
OL-03B	Basalt	66	608	199	32	282	100	18	1042	28	867	28	247	76	0.36	4.17	5.74
OL-04A	Basalt	71	580	192	31	292	107	18	866	36	878	27	248	81	0.34	4.18	5.67
OL-06A	Basalt	68	596	189	33	305	97	17	689	31	789	25	234	75	0.34	3.98	5.65
OL-07	Basalt	64	612	237	32	297	108	18	790	37	869	27	253	82	0.48	4.27	5.95
OL-08	Basalt	81	646	232	32	307	110	18	1065	38	867	28	260	83	0.38	4.28	6.10
OL-09B	Basanite	64	297	77	30	333	123	20	561	30	729	27	270	82	0.24	4.49	6.43
OL-10B	Basanite	73	338	88	30	362	124	20	569	30	741	27	260	79	0.23	4.22	6.07
OL-10D	Basanite	73	336	95	31	351	123	20	836	30	952	27	258	78			
OL-12A	Basalt	84	633	244	34	307	96	16	791	26	718	24	222	70	0.25	3.82	5.50
OL-12B	Basalt	75	663	248	34	303	87	17	766	19	738	26	227	71	0.38	3.87	5.55
OL-13B	Basalt	75	675	246	35	285	97	16	938	19	688	24	216	68			
OL-15	Basanite	74	530	220	33	304	105	17	2722	21	986	25	232	71	0.90	3.82	5.54
OL-17	Basalt	75	591	225	33	303	104	17	642	27	693	24	223	69			
OL-18	Basalt	67	496	190	33	290	102	18	1185	31	818	26	245	76			
OL-19D	Basalt	69	634	241	28	276	94	18	469	31	529	24	207	52	0.14	2.85	5.16
OL-22A	Basalt	76	594	202	31	273	111	18	709	15	799	26	242	73	0.30	4.03	6.01
OL-22E	Picrobasalt	49	32	8	42	82	161	22	213	33	234	62	115	13	1.38	0.78	3.03
OL-22F	Tephrite	76	564	188	34	258	102	17	1061	20	807	25	234	71			
OL-23B	Basanite	79	1091	232	39	335	90	15	581	23	564	21	202	56	0.15	3.13	5.08
Endonyo empuskiyaa																	
TZ W 1	Basalt	74	734	257	40	272	114	18	374	20	519	20	171	36			
Lemagurut																	
OL-20C	Trachyand	16	19	2	11	43	124	26	1420	71	1203	36	442	130	0.45	7.14	10.5

(continued)

Table 7.3 (continued)

Sample	Type	Co	Cr	Ni	Sc	V	Zn	Ga	Ba	Rb	Sr	Y	Zr	Nb	Cs	Ta	Hf
OL-25B	Basalt	66	399	163	30	279	97	18	597	26	687	24	223	65	0.15	3.58	5.48
OL-26B	Basalt	69	380	158	30	269	97	18	563	31	736	24	231	67			
TZ Y 2	Bas. trachy	17	1	7	7	109	121	23	1086	66	1358	36	409	105			
TZ Z 1	Trachybasalt	36	2	13	12	174	143	22	1025	55	1137	37	360	107			
TZ AA 1	Basalt	58	703	178	30	291	107	18	561	30	705	24	210	65			
TZ AC 1	Trachyand	4	bd	4	4	22	121	25	1297	70	1012	34	424	109			
TZ AC 6	Trachyand	6	2	10	5	37	130	25	1057	56	1245	38	402	107			
TZ AL 1	Trachybasalt	30	28	28	21	285	108	24	545	28	674	39	283	54			
TZ AO 4	Trachyand	7	0	4	6	39	132	26	1181	63	970	42	417	109			
Oldeani																	
TZ AG 2	Bas. trachy	32	35	31	22	258	125	23	668	31	620	42	306	52			
Satiman																	
TZ AK 1	Trachybasalt	34	30	31	19	231	117	23	747	29	633	42	293	49			
Average compositions																	
Lemagarut	Bas. trachy	28	100	48	13	119	118	22	1041	62	1045	38	386	102	0.30	5.88	9.18
Ogol	Basalt	73	588	198	33	303	104	18	905	27	783	26	236	72	0.35	3.95	5.73
Tool Type A	Basalt	66	450	129	32	344	117	19	1365	32	1180	29	269	83	0.26	4.56	6.41
Tool Type B	Trachybasalt	40	16	22	22	257	126	23	1591	55	1293	33	334	117	0.95	7.12	8.23
Tool Type C	Basaltic and.	44	152	84	26	216	101	20	575	69	308	27	148	10	1.14	0.51	3.36
Laetoli MSA Stone Tools																	
TOOL 2–12	A	54	135												12	12	3
TOOL 2–12	A (Fe)		129														
TOOL 2–16	A	30	114												8	7	1
TOOL 2–27	B	52	134												13	12	4
TOOL 2–28	A	56	139												13	13	3
TOOL 2–39	B	59	144												13	13	3
TOOL 2–40	B	67	165												14	12	3
TOOL 6–1	A	47	120												10	7	4
TOOL 6–11	A	43	114												11	9	8
TOOL 612	A	27	104												8	7	2
TOOL 6–17	B	54	142												12	12	3
TOOL 6–18	B	53	143												12	12	3
TOOL 6–20	B	55	148												12	12	3
TOOL 7–16	A	44	110												8	7	bd
TOOL 7–16	A (Fe)		111														
TOOL 7–30	A	20	78												9	7	2
TOOL 7–37	B	50	125												11	10	2
TOOL 8–12	A	73.4	129	14.3	56.7	11.5	3.39	9.33	1.27	6.57	1.15	2.74	0.35	1.98 0.29	13	9	3
TOOL 8–56	B	115	181	17.5	60.3	10.1	2.89	7.70	1.08	5.88	1.10	2.81	0.39	2.35 0.34	22	23	1
TOOL 8–60	C	33	65												10	5	2
TOOL 8–74	A	89.4	162	17.6	68.5	13.3	3.92	10.3	1.43	7.45	1.29	3.11	0.40	2.29 0.34	13	12	3
TOOL 8–77	A	53	174												14	9	2

TOOL 8–82	A	52	120													7	7	1
TOOL 8–9	B	109	198	21.4	83.0	16.1	4.91	12.7	1.73	8.93	1.59	3.72	0.49	2.71	0.40	15	15	3
TOOL 9–13	A	42	109													9	7	2
TOOL 9–17	B	60	143													13	13	4
TOOL 9–20	C	29	68													16	10	3
TOOL 9–28	B	53	133													13	12	3
TOOL 9–30	A	23	80													8	7	1
TOOL 9–38	C	35	74													17	9	2
TOOL 10–1	C	30	68													5	1	bd
TOOL 10–10	A	51	120													10	7	3
TOOL 10–12	A	55.3	100	11.3	46.8	9.72	2.90	8.04	1.11	5.79	1.01	2.38	0.30	1.62	0.23	7	6	2
TOOL 10–13	C	31	71													17	8	1
TOOL 10–14	C	32	66													5	2	1
TOOL 10–16	C	28	66													9	1	bd
TOOL 10–17	C	36	76													16	8	2
TOOL 10–2	C	27	62													9	2	bd
TOOL 10–3	C	32	65													12	2	
TOOL 10–39	C	30	67													17	8	1
TOOL 10–41	C	30	69													16	8	1
TOOL 10–44	C	29	66													9	2	1
TOOL 10–45	C	17	57													6	3	1
TOOL 10–56	C	4.89	9.54	1.35	6.98	2.49	0.93	3.26	0.63	4.29	0.96	2.66	0.38	2.34	0.37	5	0	1
TOOL 10–63	B	54	134													13	12	3
TOOL 10–66	B	87.6	158	17.1	66.1	13.0	3.74	10.0	1.37	7.31	1.28	3.04	0.39	2.22	0.32	12	12	2
TOOL 10–66	B (Fe)		143															
TOOL 10–68	C	18.5	38.2	4.92	22.4	6.26	1.77	6.90	1.21	7.58	1.58	4.26	0.61	3.82	0.60	8	2	1
TOOL 10–71	A	57.2	107	11.9	48.05	10.0	2.99	8.12	1.14	5.88	1.03	2.43	0.31	1.72	0.25	8	8	2
TOOL 10–72	A	43	110													10	7	2
TOOL 10–79	A	54	129													10	7	1
TOOL 10–8	A	59.0	111	12.7	52.1	11.0	3.28	8.90	1.24	6.45	1.14	2.64	0.34	1.89	0.28	8	7	2
Varnish 10–1	Varnish	32	71													7	2	1
Varnish 10–14	Varnish	35	76													5	2	bd
Varnish 8–74	Varnish	50	144													15	13	3
Varnish 8–82	Varnish	50	134													8	7	2
Ogol Lavas																		
OL-01	Basanite	49	122													8	7	1
OL-02B																7	6	1
OL-03A1	Basalt	66.4	119	13.3	52.5	10.6	3.14	8.40	1.15	5.99	1.02	2.39	0.30	1.68	0.25	9	8	1
OL-03B	Basalt	67.6	117	13.5	53.8	10.8	3.27	8.56	1.20	6.15	1.07	2.48	0.32	1.75	0.26	9	7	1
OL-04A	Basalt	66.3	119	13.4	53.2	10.7	3.22	8.55	1.17	6.06	1.04	2.42	0.31	1.70	0.25	9	8	2
OL-06A	Basalt	60.9	112	12.6	50.1	10.2	3.10	8.09	1.13	5.85	1.01	2.35	0.29	1.64	0.24	8	8	2
OL-07	Basalt	67.3	121	13.5	53.3	10.9	3.23	8.58	1.18	6.13	1.06	2.48	0.31	1.74	0.25	9	8	2

(continued)

Table 7.3 (continued)

Sample	Type	Co	Cr	Ni	Sc	V	Zn	Ga	Ba	Rb	Sr	Y	Zr	Nb	Cs	Ta	Hf
OL-08	Basalt	77.0	127	14.4	56.1	11.1	3.27	8.70	1.20	6.26	1.07	2.49	0.32	1.81 0.27	10	8	2
OL-09B	Basanite	57.3	109	12.5	50.7	10.7	3.25	8.71	1.22	6.29	1.09	2.55	0.33	1.83 0.26	7	6	2
OL-10B	Basanite	55.3	104	12.1	49.0	10.4	3.18	8.42	1.19	6.09	1.06	2.50	0.32	1.78 0.25	7	7	1
OL-10D	Basanite	21	79												7	6	2
OL-12A	Basalt	53.9	101	11.4	45.9	9.56	2.90	7.70	1.10	5.67	0.97	2.26	0.29	1.59 0.23	8	7	2
OL-12B	Basalt	57.3	102	11.8	47.4	9.83	3.01	8.10	1.12	5.92	1.02	2.36	0.30	1.69 0.25	9	7	2
OL-13B	Basalt	44	108												7	6	1
OL-15	Basanite	55	102	11.6	46.5	9.64	2.94	7.82	1.07	5.64	0.97	2.28	0.29	1.60 0.24	8	7	2
OL-17	Basalt	45	108												8	7	2
OL-18	Basalt	47	115												8	7	2
OL-19D	Basalt	42.8	79.9	9.02	36.2	7.96	2.44	6.78	0.99	5.29	0.95	2.26	0.29	1.65 0.25	9	6	1
OL-22A	Basalt	57.4	108	12.2	48.7	10.1	3.13	8.19	1.15	5.98	1.04	2.43	0.30	1.72 0.25	13	8	1
OL-22E	Picrobasalt	30.4	70.7	9.90	49.3	13.8	4.20	14.80	2.23	12.52	2.36	5.64	0.71	3.97 0.60	8	2	1
OL-22F	Tephrite	53	122												5	7	2
OL-23B	Basanite	41.2	80.7	9.41	39.2	8.62	2.58	7.04	0.97	5.13	0.87	2.06	0.26	1.40 0.20	6	5	2
Endonyo empuskiyaa																	
TZ W 1	Basalt	39	61												5	3	1
Lemagurut																	
OL-20C	Trachyand	105	189	20.2	76.7	14.7	4.41	11.23	1.59	8.47	1.49	3.63	0.48	2.78 0.42	15	14	3
OL-25B	Basalt	52.3	97.8	11.0	44.0	9.07	2.79	7.42	1.04	5.43	0.95	2.25	0.29	1.64 0.24	8	7	1
OL-26B	Basalt	41.4	99.1												9	7	1
TZ Y 2	Bas. trachy	95	182												7	2	5
TZ Z 1	Trachybasalt	97	166												6	4	4
TZ AA 1	Basalt	50	88												3	2	2
TZ AC 1	Trachyand	109	173												9	7	2
TZ AC 6	Trachyand	103	174												9	6	5
TZ AL 1	Trachybasalt	54	93												4	3	7
TZ AO 4	Trachyand	114	165												9	6	4
Oldeani																	
TZ AG 2	Bas. trachy	60	91												4	4	5
Satiman																	
TZ AK 1	Trachybasalt	62	105												5	2	4
Average compositions																	
Lemagarut	Bas. trachy	94.0	161	19.9	75.9	12.9	3.67	12.6	1.52	7.29	1.33	3.16	0.39	3.14 0.41	10.0	10.4	2.7
Ogol	Basalt	59.0	107	12.2	48.8	10.1	3.05	8.12	1.13	5.89	1.02	2.38	0.30	1.68 0.25	8.1	7.0	1.5
Tool Type A	Basalt	66.8	122	13.6	54.5	11.1	3.30	8.93	1.24	6.43	1.12	2.66	0.34	1.90 0.28	9.6	7.9	2.2
Tool Type B	Trachybasalt	104	179	18.6	69.8	13.0	3.85	10.1	1.39	7.37	1.32	3.19	0.42	2.43 0.35	13.4	13.1	2.7
Tool Type C	Basaltic and.	11.7	23.9	3.14	14.7	4.38	1.35	5.08	0.92	5.94	1.27	3.46	0.50	3.08 0.49	11.0	4.4	1.0

Trace element concentrations determined by XRF, except for samples where all REE values are reported, in which case the reported values for the REE, Cs, Ta, and Hf are by ICP-MS artifacts, the lower Nb/Zr abundances of many of Type C suggests that these artifacts were fashioned from volcanic rocks with greater contributions from the subcontinental lithosphere or crust. These relationships strengthen the argument that Type C artifacts could not have been derived from Ogol Lavas, that Type A artifacts could have been derived from Ogol Lavas, and that Type B artifacts could have been derived from volcanic rocks similar to the Ogol Lavas, but not represented in our sample collection.

The source of Type B artifacts is less clear. Although of similar composition to some Type A artifacts, there are significant differences in their petrographic and geochemical features (e.g., Figs. 7.4–7.9). Comparison of the artifact compositions with Ogol Lava flows is facilitated using normalized plots of the elements (Fig. 7.10). In this plot, the average elemental compositions of each artifact type are normalized to the average composition of Ogol Lava. It is readily apparent from this plot that Type A artifacts are most similar to the Ogol Lavas, and that Type B and C artifacts are significantly different. Furthermore, those elements in the Type A artifacts that deviate most from the average Ogol Lava composition are generally known to be sensitive to

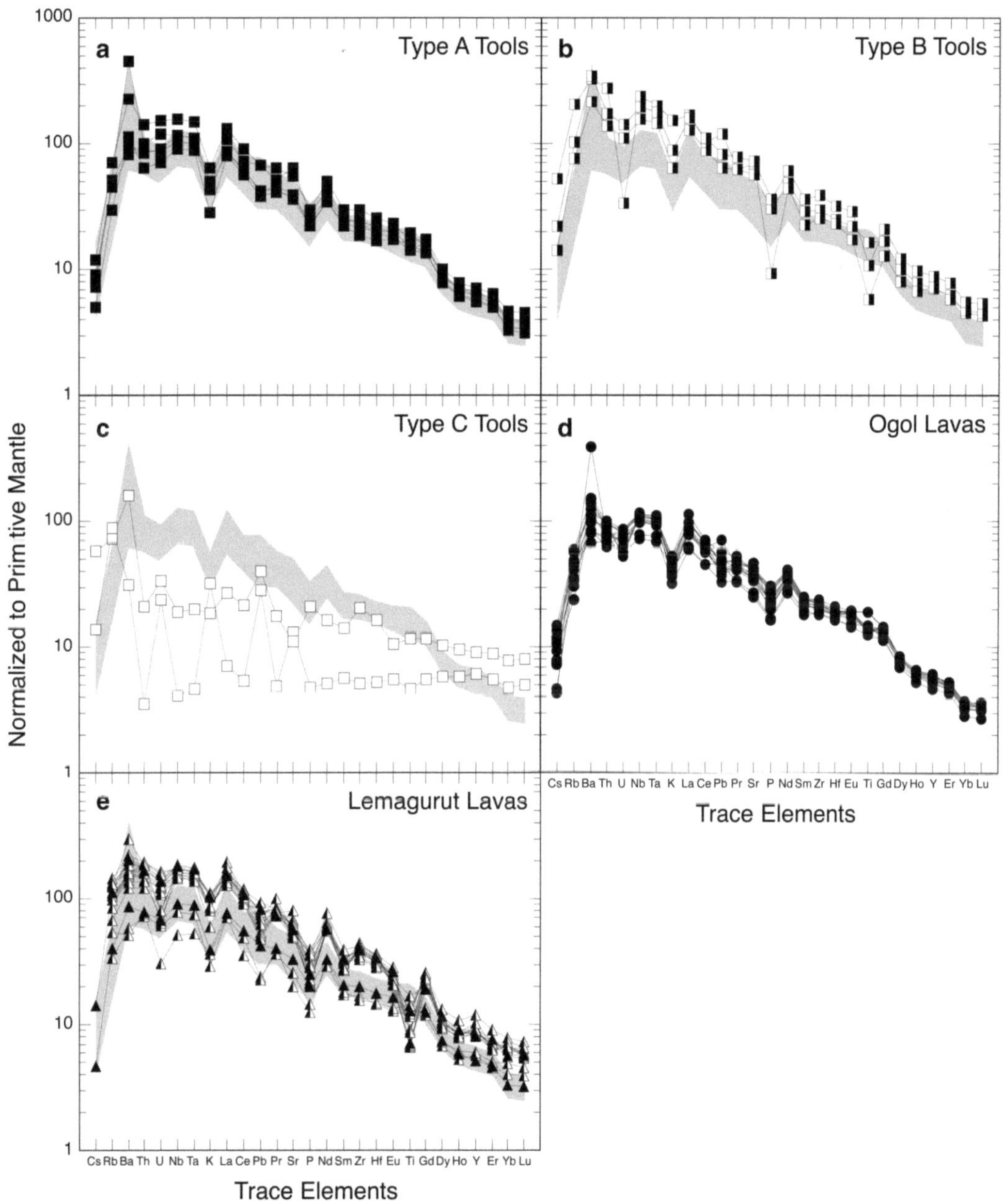

Fig. 7.8 Trace element compositions of Laetoli MSA artifacts (**a–c**), Ogol Lavas (**d**), and Lemagurut lavas (**e**). The range of Ogol Lava compositions is shown for comparison with a shaded field in a–c and e. Concentrations are normalized to primitive mantle values of McDonough and Sun (1995)

Table 7.4 Nd and Sr isotopic compositions of representative flows from the Ogol Lavas and Lemagurut

Sample#	Type	$^{143}Nd/^{144}Nd$	Initial ε_{Nd}	$^{87}Sr/^{86}Sr$
Ogol Lavas				
OL-10B	Basanite	0.512585	−1.0	0.703502
OL-12B	Basalt	0.512565	−1.4	0.703823
OL-15	Basanite	0.512560	−1.5	0.704256
OL-19D	Basalt	0.512447	−3.3	0.704850
OL-22E	Picrobasalt	0.512211	−7.9	0.705766
Lemagurut				
OL-20C	Trachyandesite	0.512445	−3.8	0.704

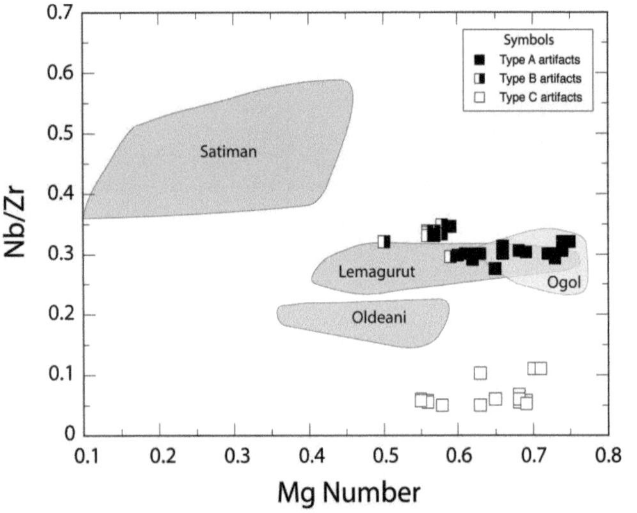

Fig. 7.9 Mg Number versus Nb/Zr for volcanic centers of the western Ngorongoro Highlands and Laetoli MSA artifacts. Note the similar Nb/Zr values of Artifact Types A and B, but the lower Mg numbers of Type B artifacts relative to Type A. Type C artifacts are distinct. Type A artifacts are most similar to Ogol Lavas, but also overlap with the compositional range of Lemagurut. Type B artifacts are more distinctly of Lemagarut affinity, but Type C artifacts have no known volcanic analog in the western Ngorongoro Highlands. Symbols and data as in Fig. 7.5

fractionation (e.g., Cr, Ni) or to alteration (e.g., Ba, Sr, Cs) effects. The Ogol Lava-normalized compositions of Type B artifacts suggest a non-Ogol source for these artifacts, although perhaps a source more closely related to the Ogol Lava Beds than that of Type C. In order to identify the source materials for all of these artifacts, we must turn to regional geochemical datasets examining several possible volcanic sources.

Comparisons with Regional Volcanic Sources

It is possible that MSA artifacts at Laetoli were prepared from more distant volcanic sources than the Ogol Lavas. Investigations by Manega (1993); Mollel et al. (2001, 2006, 2011); Mollel (2002), and D. Bowman (unpublished data) provide information about a variety of possible raw material sources in the region of northern Tanzania, including Olmoti, Ngorongoro, Satiman, Lemagurut, and Endonyo Empusikiyaa (Figs. 7.5–7.9). Incorporation of these data into the comparisons we have already made between Ogol Lava and Laetoli artifacts allows a comparison of our results with other possible source materials. Figure 7.5 illustrates the wide range of silica and total alkali compositions represented by the regional sources of volcanic raw materials. The variability within each volcanic source can then be used as a comparison with the artifacts, here still labeled according to petrologic type (Fig. 7.5b). This plot clearly illustrates that no regional volcanic material lies closer to the composition of the Type A artifacts than the Ogol Lavas, which have the lowest SiO_2 and Na_2O+K_2O compositions of any volcanic source material.

The composition of Type B artifacts overlaps with the range of samples from both Lemagurut and Oldeani (Figs. 7.5b and 7.6b), both of which are nearby volcanic centers. However, when trace element data are considered (e.g., Fig. 7.9), Oldeani seems like an unlikely source for Type B artifacts. A comparison of Type B artifacts to Lemagurut lavas (Fig. 7.11) indicates that for many elements a Lemagurut source is a good fit. Those elements that differ most significantly from a Lemagurut source composition are Cr, Ni, Sc, V, Ba, and Cs, and these differences have resulted from fractionation or alteration effects. Thus, the comparison seems to indicate that Type B artifacts were derived from Lemagurut volcanics, or a similar volcanic center.

Surprisingly, none of the available data from regional volcanic centers indicate the presence of volcanic compositions comparable to those found for Type C artifacts. One sample from a small volcanic hill (Endonyo Empusikiyaa) located on the north slope of Lemagurut (Fig. 7.12), shares some similarities with Type C artifacts (e.g., Mg number, Nb, Zr, and Rb/Sr; Bowman, unpublished data), but overall Type C is not a good geochemical match to any major volcanic center (e.g., Figs. 7.5b, 7.6b, 7.9, 7.10 and 7.11). These data support the hypothesis that Type C raw materials originated from an area farther removed from Laetoli than Type A

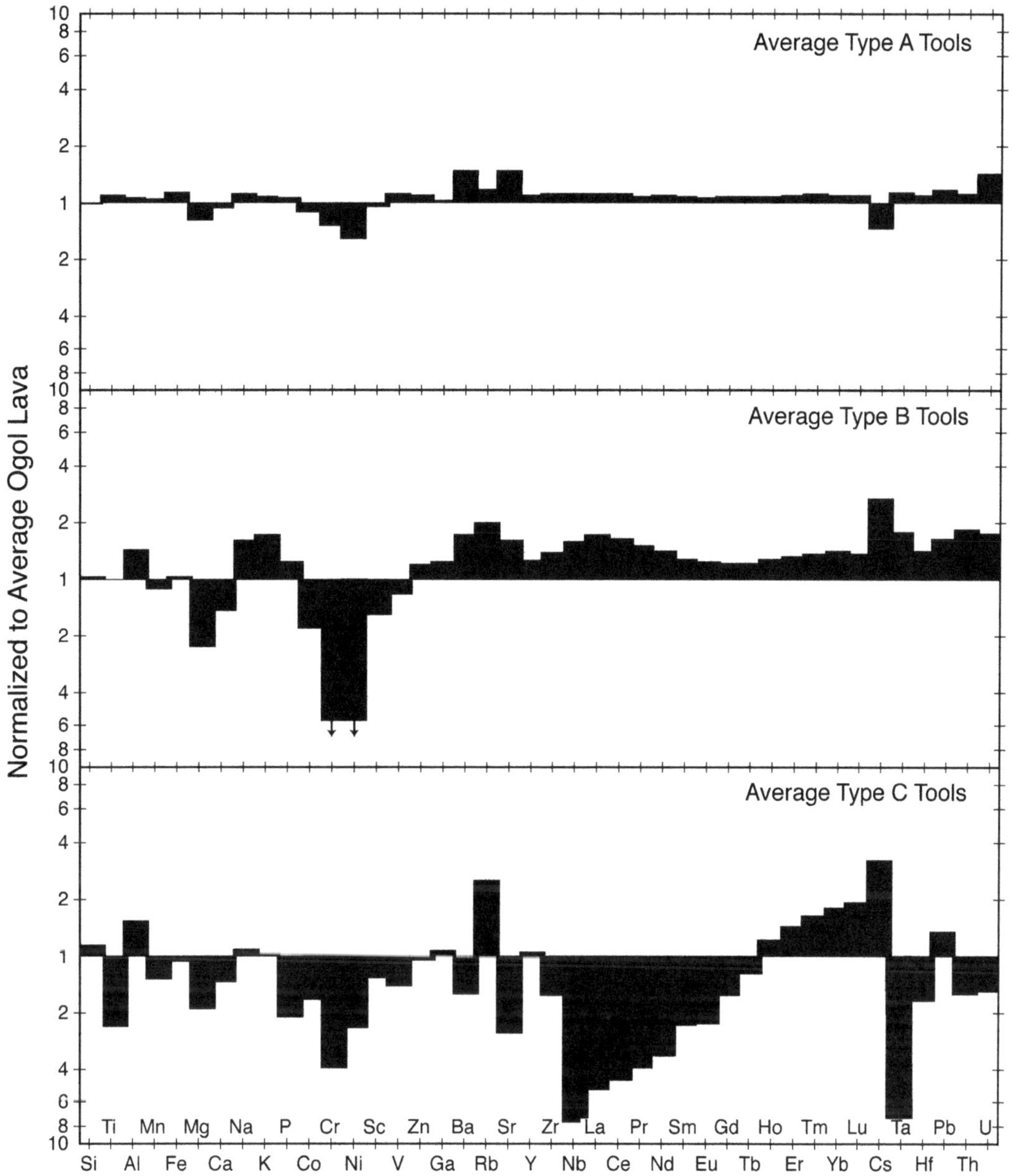

Fig. 7.10 Normalized compositions of MSA stone artifacts from Laetoli. Compositions are normalized to average Ogol Lavas. Ogol Lava normalized concentrations of all analyzed elements in Type A artifacts are near unity indicating their nearly identical compositions. In contrast, Type B and Type C artifacts have normalized compositions that deviate significantly from average Ogol Lavas

or Type B raw materials, and that of all the basaltic MSA artifacts found at Laetoli, those of Types A and B are the most likely to have originated from local outcrops (i.e., Ogol and Lemagurut, respectively). These geochemical data further illustrate the differences between the three artifact compositions and suggest that at least three separate raw materials are represented in the basaltic MSA assemblage at Laetoli, corresponding to the petrologic distinctions defined at the beginning of this study. Additional survey and geochemical analysis of the EARS volcanics will be necessary for the complete sourcing of Laetoli's basaltic artifacts.

Raw Material Acquisition

The use of multiple raw material sources for the basaltic MSA artifacts found at Laetoli should not be particularly surprising, given the presence of quartzite and phonolite artifacts in addition to basalt. Sources for these non-volcanic materials are not available within the Laetoli region, and they may have been acquired from Olduvai Gorge and areas farther north (e.g., Jones 1994). During the MSA, certain rock sources were likely favored for stone tool manufacture, presumably because of the quality

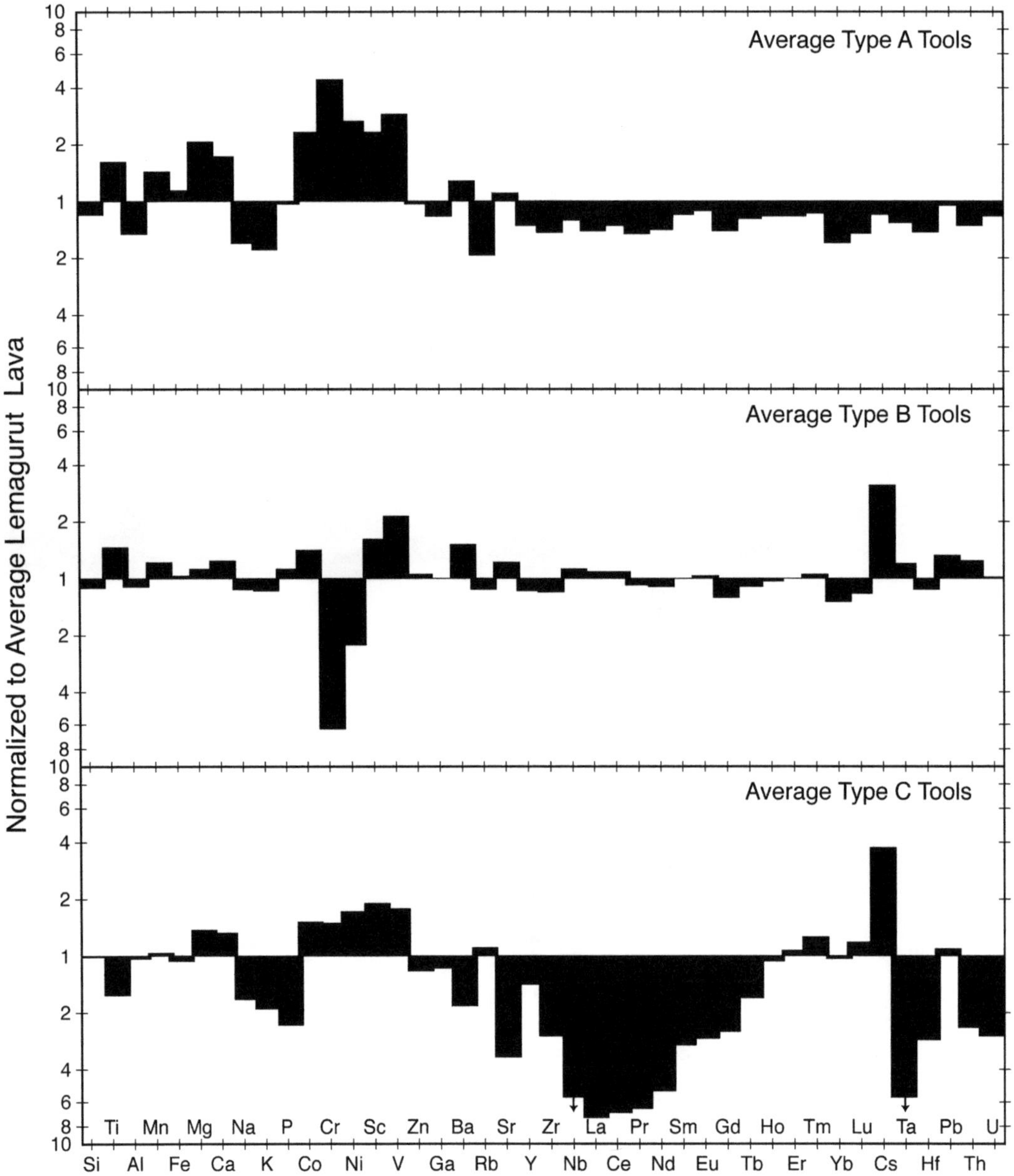

Fig. 7.11 Normalized compositions of basaltic MSA artifacts from Laetoli. Compositions are normalized to average Lemagurut lavas. Whereas the normalized compositions of Type A and C artifacts exhibit considerable variation, Type B artifacts are most similar to average Lemagurut lava

of the material for tool-making purposes or ease of rock collection and material availability (e.g., Toth 1985; Jones 1994; Kimura 1999). It seems in this case that proximity to material sources was not the primary reason for raw material use or selection, as has been illustrated for later MSA obsidian use in Ethiopia, where raw materials are sourced up to 250 km away from the archaeological sites where the relevant artifacts are found (Negash and Shackley 2006). Our data suggest that similar patterns may apply to Laetoli's assemblage, indicating the use of distant sources in this region as well.

The availability of raw materials at Laetoli may not have been limited to exposures at the Ogol Lava Beds. Laetoli lies downstream of the Ngorongoro volcanic highland today (Fig. 7.12), and evidence from the Laetolil Beds indicates that the prevailing drainage direction during hominin occupation of the region was similar to that of the present (Hay 1987). A large proportion of the cobbles in the modern Garusi River (Fig. 7.2) are basaltic, but these were not analyzed as part of this study. The basaltic cobbles may have originated from eastern volcanic sources (most likely Lemagurut, or possibly Satiman or Oldeani)

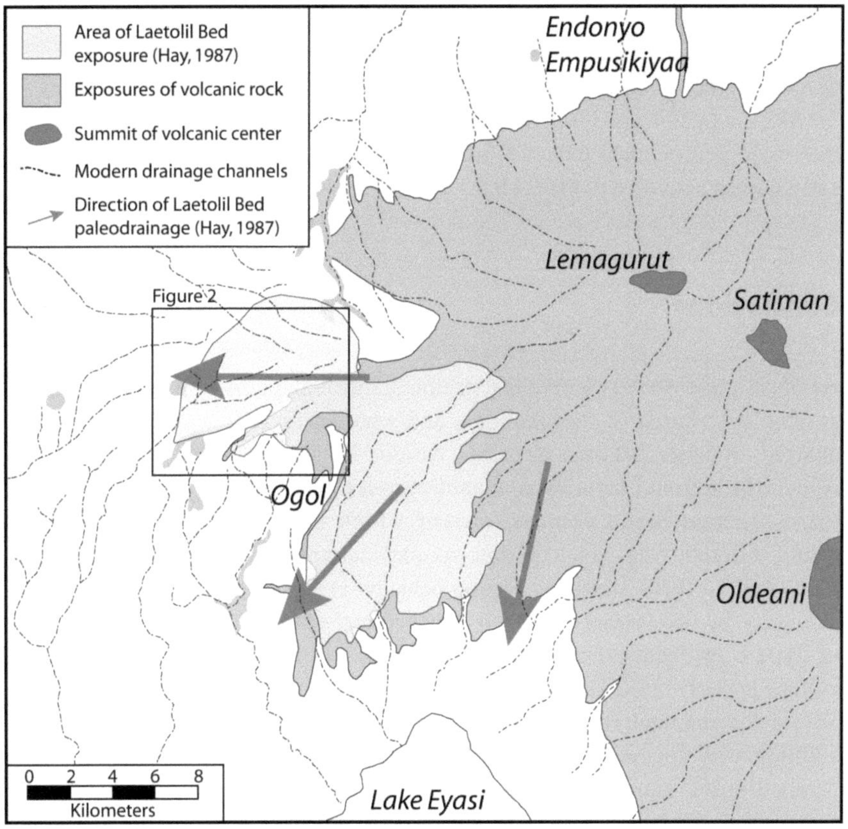

Fig. 7.12 Map of Laetolil beds showing exposures of volcanic rock, modern drainages and paleodrainages. Boxed area illustrated as Fig. 7.2. Modified from Pickering (1964) and Hay (1987)

and been washed downstream during periods of increased discharge. A similar drainage orientation in the past may have given hominin occupants of the region easy access to non-Ogol raw materials, without additional travel or raw material transport.

Despite the relative proximity of Satiman and Oldeani volcanic materials, our geochemical comparisons suggest that stone materials from these volcanic centers were not utilized as raw materials for any of the MSA artifacts we analyzed. However, these hypotheses are based on limited sampling of both regional source materials and the artifact assemblage itself. The lack of artifacts geochemically similar to Satiman may argue against a stream-transport model of raw material acquisition, as Satiman is one of the closest raw materials sources east of Laetoli. In addition, the acquisition of materials from the north or areas farther east would almost certainly have required manual transport through travel or trade. However, stream transport may partially account for the presence of several raw material types at Laetoli. Any future examination of the hominin behaviors involved in stone tool manufacture and raw material acquisition should consider both Ogol Lava basalts and stream-transported cobbles to be local rock sources, as the effort and planning required to acquire both of these materials would be similar.

Limitations and Future Work

The geochemical and mineralogical characterization of MSA artifacts presented here may be useful in later examinations of raw material acquisition, and it provides a starting point for determining the differences present among artifacts found at the various Laetoli localities based on limited and non-systematic surface survey. The timeframe covered by the archaeological samples obtained for this study comes from the identification of MSA morphologies in artifacts found during surface survey, providing a very coarse view of material use and almost no information regarding changes in raw materials over time. Although detailed knowledge of artifact sources would be ideal for this type of study and would provide a great deal of additional information relevant to hominin behavior, the current investigation aims to provide a starting point for provenance determination. Future studies may very well determine a temporal distinction between the artifact types provided here. In addition, although a more comprehensive examination of possible raw material sources was not a part of this project, the data provided here will hopefully serve as a source of information for future raw material studies to complement those completed by Manega (1993); Mollel (2002); and Mollel et al. (2001, 2006, 2011).

Continued work on the geochemical characterization of possible source materials in the EARS may reveal the sources of many of the raw materials used in tool manufacture during the MSA at Laetoli and throughout East Africa. It is our hope that these geochemical data will inform future analyses of early human activities in the EARS.

Conclusions

As illustrated by hand-sample description in the field, petrographic description of thin sections, and whole-rock XRF analyses of major and trace elements, basaltic Middle Stone Age artifacts from Laetoli were manufactured from at least three separate basalt sources, one of which is significantly different from volcanic rock sources currently sampled from northern Tanzania. This conclusion is supported primarily by the petrographic and geochemical separation of Type C artifacts from the rest of the MSA assemblage, and the lack of similar major element compositions found in any of the regional volcanic sources analyzed to date. In addition, a number of different trace elemental comparisons resulted in the same classification of artifact compositions determined through petrographic analysis, supporting the separation of these three artifact types. Analyses of mineral compositions using microbeam techniques might allow for further discrimination of Laetoli artifacts and facilitate comparisons with nearby volcanic sources.

Type C raw materials appear to have originated from an area farther removed from Laetoli than Type A or Type B raw materials, whereas Types A and B are the most likely to have originated from more local outcrops of Ogol and Lemagurut lavas (respectively). Despite the availability of comparative data, a source for the Type C artifacts could not be conclusively determined. However, the source in question may simply remain unsampled and uncharacterized; a wide range of locations for artifact raw materials exist outside the small area of Tanzania discussed in this study. This study was intended as a geochemical and mineralogical characterization of the Ogol Lava beds and the basaltic MSA artifacts of Laetoli. The lack of available geochemical data for the possible volcanic source materials from northern Tanzania and the rest of the EARS makes it difficult at present to make any definite claims about the source materials used in the manufacture of Laetoli's MSA artifacts.

Acknowledgments This research was conducted with the help and funding of the Associated Colleges of the Midwest Tanzania Program, Beloit College Department of Geology, and Macalester College Department of Geology. Chemical analyses at Macalester College were supported by National Science Foundation (NSF) grants EAR-9601475, DUE-9651385, EAR-0520870, and EAR-9903003. J. Vervoort provided invaluable expertise and guidance on isotopic analyses. In addition, we would like to thank the Tanzanian Department of Antiquities for permits to export scientific samples as well as the University of Dar es Salaam, particularly the Archaeology Unit, for facilitating the ACM program and research requirements. We also wish to thank Simon Kateyo, Said Kilindo, Carl Mendelson, Jeff Thole, Charles Saanane, Carolyn Kilberg, Cameron Davidson, Amy Hubbard, Robert Dymek, Robert Buchwaldt and Jennifer Smith for their help and support, as well as Terry Harrison, Lindsay McHenry, and two anonymous reviewers for their thoughtful comments and suggestions regarding the initial draft of this manuscript. A National Science Foundation Graduate Research Fellowship (to KAA) provided student funding for the writing stages of this project.

References

Adelsberger, K. A., Wirth, K. R., & Mabulla, A. Z. P. (2002). Provenance of Middle Stone Age tools of the Laetoli archaeological site, Tanzania. *Geological Society of America, Abstracts with Programs, 34*, 110.

Callow, P. (1994). The Olduvai bifaces: Technology and raw materials. In M. D. Leakey & D. A. Roe (Eds.), *Olduvai Gorge* (Excavations in Beds III, IV and the Masek Beds, 1968–1971, Vol. 5, pp. 235–253). Cambridge: Cambridge University Press.

Cross, W., Iddings, J. P., Pirsson, L. V., & Washington, H. S. (1903). *Quantitative classification of igneous rocks*. Chicago: University of Chicago Press.

Dawson, J. B. (1992). Neogene tectonics and volcanicity in the North Tanzania sector of the Gregory rift valley: Contrasts with the Kenya sector. *Tectonophysics, 204*, pp. 81–92.

Day, M. H., Leakey, M. D., & Magori, C. (1980). A new hominid fossil skull (L.H. 18) from the Ngaloba Beds, Laetoli, northern Tanzania. *Nature, 284*, pp. 55–56.

Drake, R., & Curtis, G. H. (1987). K–Ar geochronology of the Laetoli fossil localities. In M. D. Leakey & J. M. Harris (Eds.), *Laetoli: A Pliocene site in northern Tanzania* (pp. 48–52). Oxford: Clarendon.

Greenough, J. D., Gorton, M. P., & Mallory-Greenough, L. M. (2001). The major and trace-element whole-rock fingerprints of Egyptian basalts and the provenance of Egyptian artifacts. *Geoarchaeology: An International Journal, 16*, pp. 763–784.

Harrison, T., & Kweka, A. (2011). Paleontological localities on the Eyasi Plateau, including Laetoli. In T. Harrison (Ed.), *Paleontology and geology of Laetoli: Human evolution in context* (Geology, geochronology, paleoecology and paleoenvironment, vol. 1, pp. 99–119). Dordrecht: Springer.

Hay, R. L. (1987). Geology of the Laetoli area. In M. D. Leakey & J. M. Harris (Eds.), *Laetoli: A Pliocene site in northern Tanzania* (pp. 23–47). Oxford: Clarendon.

Jones, P. R. (1994). Results of experimental work in relation to the stone industries of Olduvai Gorge. In M. D. Leakey & D. A. Roe (Eds.), *Olduvai Gorge* (Excavations in Beds III, IV and the Masek Beds, 1968–1971, Vol. 5, pp. 254–298). Cambridge: Cambridge University Press.

Kimura, Y. (1999). Tool-using strategies by early hominids at Bed II, Olduvai Gorge, Tanzania. *Journal of Human Evolution, 37*, pp. 807–831.

Knaack, C., Cornelius, S., & Hooper, P. (1994). *Trace element analyses of rocks and minerals by ICP-MS*. Department of Geology, GeoAnalytical Laboratory, Pullman, Washington, Open-File Report, Washington State University.

Leakey, M. D. (1981). Discoveries at Laetoli in northern Tanzania. *Proceedings of the Geologists' Association, 92*, 81–86.

Leakey, M. D., & Harris, J. M. (1987). *Laetoli: A Pliocene site in northern Tanzania*. Oxford: Clarendon.

Leakey, M. D., Hay, R. L., Curtis, G. H., Drake, R. E., Jackes, M. K., & White, T. D. (1976). Fossil hominids from the Laetolil beds. *Nature, 262*, pp. 460–466.

Le Bas, M. J., Le Maitre, R. W., Streckeisen, A., & Zanettin, B. (1986). A chemical classification of volcanic rocks based on the total alkali-silica diagram. *Journal of Petrology, 27*, pp. 745–750.

Le Maitre, R. W. (Ed.). (1989). *A classification of igneous rocks and glossary of terms: Recommendations of the international union of geological sciences subcommission on the systematics of igneous rocks*. Oxford: Blackwell.

MacDonald, G. A., & Katsura, T. (1964). Chemical composition of Hawaiian lavas. *Journal of Petrography, 5*, pp. 82–133.

Magori, C. C., & Day, M. H. (1983). Laetoli Hominid 18: An early *Homo sapiens* skull. *Journal of Human Evolution, 12*, pp. 747–753.

Manega, P. C. (1993). Geochronology, geochemistry and isotopic study of the Plio-Pleistocene hominid sites and the Ngorongoro volcanic highland in northern Tanzania. Ph.D. dissertation, University of Colorado, Boulder.

McDonough, W. F., & Sun, S.-S. (1995). The composition of the earth. *Chemical Geology, 120*, pp. 223–253.

Merrick, H. V., Brown, R. H., & Nash, W. P. (1994). Use and movement of obsidian in the Early and Middle Stone Ages of Kenya and Tanzania. In T. S. Childs (Ed.), *Society, culture, and technology in Africa* (pp. 29–44). Philadelphia: Museum Applied Science Center for Archaeology.

Mollel, G. F. (2002). Petrology and geochemistry of the southeastern Ngorongoro volcanic highland; and contribution to "sourcing" of stone tools at Olduvai Gorge, Tanzania. M.Sc., Rutgers, The State University of New Jersey, Piscataway.

Mollel, G. E., Ashley, G. M., & Delaney, J. S. (2001). Bulk and mineral compositions from Plio-Pleistocene Ngorongoro Volcanic Highland (NVH), a possible source for stone tool materials at Olduvai Gorge, Tanzania. *Geological Society of America, Abstracts with Programs, 33*, 295.

Mollel, G. M., Swisher, C., III, & Feigenson, M. (2006). Petrology, geochemistry and $^{40}Ar/^{39}Ar$ dating of Oldeani volcano, northern Tanzania. *Geological Society of America, Abstracts with Programs, 38*, 485.

Mollel, G. F., Swisher, C. C., Feigenson, M. D., & Carr, M. J. (2011). Petrology, geochemistry and age of Satiman, Lemagurut and Oldeani: Source of the volcanie deposits of the Laetoli area. In T. Harrison (Ed.), *Paleontology and geology of Laetoli: Human evolution in context* (Geology, geochronology, paleoecology and paleoenvironment, vol. 1, pp. 99–119). Dordrecht: Springer.

Musiba, C.M. (1999). Laetoli Pliocene paleoecology: A reanalysis via morphological and behavioral approaches. Ph.D. dissertation, University of Chicago, Chicago.

Negash, A., & Shackley, M. S. (2006). Geochemical provenance of obsidian artifacts from the MSA site of Porc Epic, Ethiopia. *Archaeometry, 48*, pp. 1–12.

Paslick, C., Halliday, A., James, D., & Dawson, J. B. (1995). Enrichment of the continental lithosphere by OIB melts: Isotopic evidence from the volcanic province of northern Tanzania. *Earth and Planetary Science Letters, 130*, pp. 109–126.

Pickering, R (1964) Geologic map: Endulen, Tanzania. Quarter Degree Sheet 52, Scale: 1:125,000. Dodoma: Tanzania Geological Survey.

Rollinson, H. (1993). *Using geochemical data: Evaluation, presentation, and interpretation*. New York: Wiley.

Toth, N. (1985). The Oldowan reassessed: A close look at early stone artifacts. *Journal of Archaeological Science, 12*, pp. 101–120.

Vervoort, J. D., Wirth, K. R., Kennedy, B., Sandland, T., & Harpp, K. S. (2007). The magmatic evolution of the midcontinent rift: New geochronologic and geochemical evidence from felsic magmatism. *Precambrian Research, 157*, pp. 235–268.

Wirth, K. R., & Adelsberger, K. A. (2002). Mantle source and magmatic evolution of the late Pliocene Ogol Lava beds of Laetoli, Tanzania. *Geological Society of America, Abstracts with Programs, 34*, 463.

Chapter 8
The Ecology and Biogeography of the Endulen-Laetoli Area in Northern Tanzania

Peter Andrews, Marion K. Bamford, Efrem-Fred Njau, and Godson Leliyo

Abstract We have mapped the vegetational variability of the Endulen-Laetoli area of northern Tanzania based on the premise that heterogeneity at the landscape level is a function of differences in geology, soils and topography. Vegetation habitats have been described in relation to these factors, particularly soils and topography, and the effects of climate changes are estimated in relation to these other factors. We have made voucher collections of the plants in the Laetoli region, an area of about 250 km^2, and we have recorded distributions of plants by habitat, climate, soil and topography. Vegetation structure has been measured in terms of canopy cover, species richness, and canopy height. Sixteen species associations and structural types have been identified within the study area, and these have been mapped at a scale of 1:120,000. The whole area has been much disturbed by human activity, with large areas of weed growth resulting from cultivation, burning and overgrazing. Results show that areas of low relief have soils with impeded drainage or seasonal waterlogging and dense *Acacia drepanolobium* woodland, low stature when disturbed by human action, high when not. Shallow brown soils on volcanic lavas have five woodland associations, two dominated by *Acacia* species, two by *Combretum-Albizia* species, while a fifth to the east on volcanic soils has a different woodland association with *Croton-Dombeya Albizia* species. Elevated land to the east, also on volcanic soils, has two associations of montane-edge species, one with *Croton-Celtis-Lepidotrichilia*, and the other with *Acacia lahai*, grading into the eastern highlands montane forest at 2,750 m. Seasonal water channels flowing from east to west have three *Acacia* riverine woodland associations. Three deep valleys to the north of the area have dense riverine woodland with *Celtis, Albizia, Euclea, Combretum,* and *Acacia* spp. Emergence of springs at Endulen feed a perennial stream with closed gallery forest with *Ficus-Croton-Lepidotrichilia.* Finally, recent ash falls from Oldonyo Lengai have produced localized immature alkaline soils with calcrete formation and short grass vegetation. Projections of increased or decreased rainfall and how this may potentially have affected the different environments form the basis for predictions about vegetation and habitat change.

Keywords Vegetation • Soils • Topography • Geology • Climate change • Habitat variability

Introduction

One of the most striking features of East African environments is the sheer diversity of habitats, often within limited geographical areas. Descriptions of African habitats do not always do justice to this diversity, lumping vegetation associations into broad categories such as 'acacia woodland', or 'grassland'. There are good comparative data on recent habitats and ecosystems in Africa (Michelmore 1939; Burtt 1942; Pratt et al. 1966; Lind and Morrison 1974; Andrews et al. 1975; Sinclair and Norton-Griffiths 1979; White 1983; Metzger et al. 2005; Boone et al. 2006), particularly on the central and northern parts of the Serengeti ecosystem (Herlocker and Dirschl 1972; Anderson and Herlocker 1973; Anderson and Talbot 1976; de Wit 1978; Sinclair and Norton-Griffiths 1979; Jager 1982). There has been little published, however, on the southern extent of the Serengeti ecosystem (Herlocker and Dirschl 1972), and yet, because of the particular interest of this area, and the location within it of fossil sites with a rich record of human evolution, it is important to gain some insight into the factors that determine the nature of the environment.

The present paper describes the vegetation in relation to the soils and topography of the Endulen-Laetoli area (Fig. 8.1).

P. Andrews (✉)
Palaeontology Department, Natural History Museum, Cromwell Road, London, SW7 5BD, UK
e-mail: pjandrews@uwclub.net

M.K. Bamford
Bernard Price Institute for Palaeontological Research, School of Geosciences, University of the Witwatersrand, Private Bag 3, WITS 2050, Johannesburg, South Africa
e-mail: marion.bamford@wits.ac.za

E.-F. Njau and G. Leliyo
Arusha Herbarium, TPRI, PO Box 3024, Arusha, Tanzania
e-mail: efrednjau@yahoo.co.uk

T. Harrison (ed.), *Paleontology and Geology of Laetoli: Human Evolution in Context. Volume 1: Geology, Geochronology, Paleoecology and Paleoenvironment*, Vertebrate Paleobiology and Paleoanthropology, DOI 10.1007/978-90-481-9956-3_8,

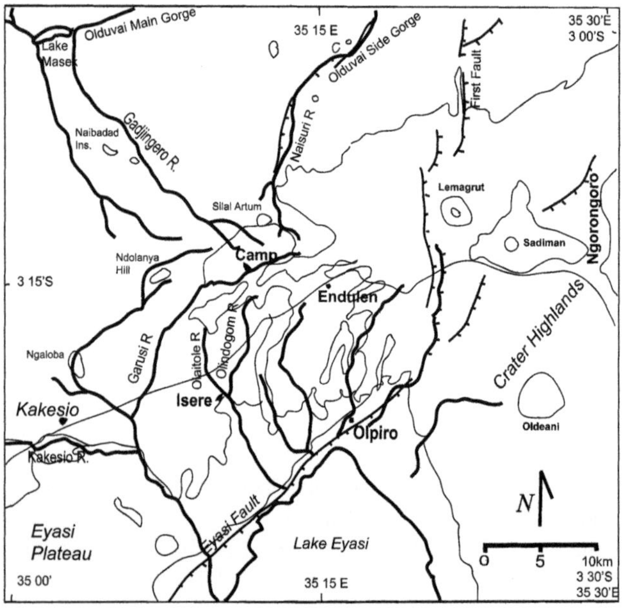

Fig. 8.1 Location map of the Endulen-Laetoli area showing the major drainage pattern and main topographical features

The fossil sites in the Olaitole and Garusi valleys are situated in this area, providing some of the earliest evidence of human evolution. The presence of footprints of the early hominin *Australopithecus afarensis* (Johanson et al. 1978) still provides the earliest unequivocal evidence of bipedal walking in the human lineage (Leakey and Hay 1979), but there is also evidence that *A. afarensis* retained adaptations for arboreal climbing (Stern and Susman 1983; Alemseged et al. 2006).

There is also some dispute about the environment occupied by this early hominin (Leakey and Harris 1987; Andrews 1989), particularly the extent and density of woodland cover. Most authors in the first book on Laetoli (Leakey and Harris 1987) concluded that the environment in the past was similar to that of today, but there was little understanding about present day habitats in the region, and what was implied by this conclusion was that the area was dominated by open savanna habitats. In contrast to this, the analysis of community structure of the mammalian fauna (Andrews 1989; Reed 1997; Musiba 1999; Su and Harrison 2007; Kingston and Harrison 2007) suggested a more wooded environment with little grassland. Even a superficial survey of the Laetoli region, however, shows that there are many vegetation associations in the Laetoli area alone (Dale and Greenway 1961; Beentje 1994; Raynal in Leakey and Harris 1987). These associations differ both in their species composition and more greatly in terms of their physical structure and the mammal faunas occupying them, and these variations coincide spatially with variations in topography, soil, altitude and geology. As a result, by making certain assumptions about climate, it is possible to model the past vegetation where these four variables can be reconstructed for the Pliocene. This paper is mainly concerned with setting up a vegetational habitat model for the present day for comparison with paleoecological evidence, for example from sedimentology, isotopes, pollen, phytoliths, wood and vertebrate distributions. The effects of climate change on this reconstruction can then be assessed.

Methods

We have made a collection of plants from 32 collecting localities in the Endulen-Laetoli area (3°08'–3°21' S, 35°06'–35°22' E area; Table 8.1). All species of angiosperms were collected, with woody plants more comprehensively collected as the non-flowering herbs and grasses were difficult to distinguish in the field. All specimens have been identified and stored at the Arusha Herbarium, Tanzania. Duplicate collections have also been housed at the Kew Herbarium, London, under the care of Kaj Volleson, and at the University of the Witwatersrand, Johannesburg. The complete plant list is given in the Appendix, with voucher numbers provided for the collection in the Arusha Herbarium.

The vegetation of the Endulen-Laetoli area has been classified into 7 units and 16 subunits, and these have been mapped at 1:120,000. Mapping was based on direct measurement for the most part, both linear measurements on the ground and by GPS. Data for the eastern (montane) part of the study area were obtained from the topographical map, which shows in broad outline the extent of the montane vegetation. This map was drawn in the 1960s and although land use and vegetation may have shifted since then, the observations we made show that the change was outside the limits of accuracy of our map. Limited areas to the north and south of the study area were not mapped in detail, but were based on long distance sightings combined with spot checks on the ground.

Sampling of the woodland and bushland associations was carried out to document the structures of the woody vegetation in terms of canopy height, canopy density, and relative proportions of tree species within the associations. All plant associations were sampled by transect or sample plot, except for montane forest and grassland, the former because it is so degraded by human action and the latter because most of it was dominated by woody weed species of herbs. Transects were based on a measured line with a standard length of 200 m and measured direction. All trees and shrubs were counted where the vertical projection of the canopy to the ground crossed the line. In riverine bushland and riverine woodland, transects were measured laterally across the river channels as it was in this direction that the greatest variability occurred. Distances across small dry river courses were short, and so the transects measured no more than 50–100 m,

Table 8.1 Collecting localities in the Endulen-Laetoli area

Site	Latitude	Longitude	Location	Altitude (m)	Vegetation
1	3°12'40"S	35° 13' 14"E	Upper Garusi river	1,800	Low *Acacia* riverine woodland
2	3°12'40"S	35° 13' 01"E	Upper Garusi river	1,797	Low *Acacia* riverine woodland
3	3°12'30"S	35° 13' 01"E	Tributary of Garusi	1,802	Riverine bushland
4	3°13'44"S	35° 10' 12"E	Tributary of lower Garusi river	1,735	*Acacia* riverine woodland
5	3°13'44"S	35° 10' 12"E	Lower Garusi river	1,730	Tall *Acacia* riverine woodland
6	3°15'18"S	35° 11' 22"E	Lower Laetoli river	1,680	Mixed tall *Acacia* woodland and bushland
7	3°15'09"S	35° 11' 06"E	Tributary of Olaitoli river	1,685	Dense bushland
8	3°13'06"S	35° 08' 41"E	Emboremony river	1,750	Low *Acacia* riverine woodland
9	3°13'06"S	35° 11' 50"E	Gadjingero river	1,780	Mixed tall *Acacia* woodland and bushland
10	3°13'04"S	35° 16' 06"E	Oldogom river	1,820	Tall gallery forest
11	3°13'12"S	35° 15' 51"E	Oldogom river	1,820	Tall gallery forest
12	3°14'12"S	35° 14' 44"E	Oldogom river	1,675	Tall *Acacia* riverine woodland
13	3°18'38"S	35° 12' 28"E	Noiti	1,560	Low mixed woodland
14	3°10'35"S	35° 13' 14"E	Laetolil bed exposure (Loc.1)	1,800	Mixed woodland
15	3°12'39"S	35° 13' 14"E	Mbuga soil	1,840	Low *Acacia drepanolobium* woodland
16	3°11'12"S	35° 29' 44"E	Secondary grassland, Lemagurut	2,625	Herb-rich grassland
17	3°11'07"S	35° 22' 41"E	Secondary grassland, Lemagurut	2,671	Herb-rich grassland
18	3°10'56"S	35° 22' 33"E	Secondary grassland, Lemagurut	2,715	Herb-rich grassland
19	3°10'56"S	35° 22' 33"E	Secondary grassland, Lemagurut	2,751	Herb-rich grassland
20	3°10'45"S	35° 22' 33"E	Lemagurut forest/grassland	2,830	Montane forest patches disturbed by fire
21	3°10'36"S	35° 22' 34"E	Lemagurut forest/grassland	2,939	Montane forest patches disturbed by fire
22	3°10'33"S	35° 22' 14"E	Lemagurut forest	2,890	Montane evergreen forest
23	3°18'03"S	35° 11' 11"E	Esere	1,590	Tall *Acacia* riverine woodland
24	3°18'49"S	35° 05' 30"E	Esere-Kakesio road	1,610	Edaphic grassland
25	3°21'30"S	35° 06' 06"E	Kakesio river	1,601	Tall *Acacia* riverine woodland
26	3°21'30"S	35° 06' 06"E	Kakesio	1,600	*Commiphora* woodland
27	3°21'30"S	35° 06' 06"E	Kakesio	1,605	*Commiphora* woodland
28	3°13'00"S	35° 14' 06"E	Oldonyo Mati Hills	1,900	Tall mixed woodland
29	3°12'30"S	35° 13' 20"E	Oldonyo Mati Hills	1,825	Tall mixed woodland
30	3°9'31"S	35° 16' 17"E	Oromukeke Gorge	1,827	Tall mixed woodland
31	3°12'40"S	35° 16' 01"E	Endulen/Garusi corner	1,880	Tall escarpment woodland
32	3°13'40"S	35° 18' 21"E	Endulen to Ngorongoro road	2,011	Fringing montane forest

but in deeper valleys with varied associations of riverine woodland, transects of 200 m were measured. All other woodland transects were also 200 m. Sample plots were based on 50 m radius from a marked point, so that the area of the plots was 7,855 m^2. Again this included all plants whose canopy or part of canopy was within the 50 m limits of the sample plot, even if the stem or trunk of the plant was outside the plot.

Soil types and geological substrate were recorded for all transects. Altitude and topographical change were recorded for each transect. The transects and sample plots were positioned in areas of least human disturbance, and no direct measure of human activity was recorded.

Tree heights were measured using a 4 m measure held against the tree trunk and heights estimated from a distance as multiples of the measure. Canopy density was calculated as the percentage of the transect or sample plot beneath tree canopy. All species within transects were identified and provide relative abundance figures of species and groups of species. The transects provide the basis for identifying several catenas within the study area.

The Serengeti Ecosystem and Laetoli

Laetoli is situated on the southern edge of the Serengeti ecosystem, which is a large area of woodland and grassland best defined by the limits of the wildebeest migration (Watson 1967; Metzger et al. 2005). The limits of the study area are shown in Fig. 8.1, and it will be referred to as the Laetoli area since this account was prepared as an adjunct to the paleontological excavations of the Pliocene deposits at Laetoli. It falls within the NCA (Ngorongoro Conservation Area) where the land is protected for conservation, but shared by the Maasai people and their large herds of cattle and goats (Metzger et al. 2005; Boone et al. 2006; see also Lamprey and Reid 2004). The Laetoli area is of generally flat topography sloping down from about 2,000 m in the east to about 1,500 m in the west. Three broad vegetation zones are currently identified: Zone I, grasslands making up approximately 25% of the Serengeti ecosystem; Zone II, *Acacia* woodlands, making up approximately three quarters of the ecosystem (Jager 1982); and Zone III, woodlands with relic forests in the northeast, now mostly replaced by *Combretum-Terminalia* woodland as a result of burning (Schmidt 1975). Descriptions of the woodlands have tended to concentrate on the central and northeastern woodlands, and the ephemeral nature of the vegetation has been documented by Dublin (1991), who showed by means of photographs and written accounts that the Mara grasslands in the north of the ecosystem was covered by *Acacia* woodland and *Croton* thicket in the mid part of the twentieth century. Pre-nineteenth century vegetation is unknown, but in the early nineteenth century, with intensive settlement of the area by pastoralists, the Mara was open grassland, but the Rinderpest epidemic in the late nineteenth century decimated both the human population and their herds, so that the area reverted to woodland by the 1950s (Dublin 1991). Since that time, the spread of people and the immigration into the area of elephants has eliminated the woodland to leave the vegetation as it is today. The vegetation in the southeast corner of the Serengeti ecosystem has never been documented in detail (but see Herlocker and Dirschl 1972), and these southern woodlands will be described here and related to the climate, topography and soil.

The Laetoli area is southeast of the short grass plains, and it is an area with rather greater topographic relief than is present in the central Serengeti. It is an area of low hills and ridges with a generally westerly drainage, trending to the south into Lake Eyasi. The eastern limits of the area are marked by the eastern highlands composed of the extinct volcanoes Lemagurut, Olmoti and Satiman, with Ngorongoro further removed to the east. These rise to over 3,000 m and have patches of degraded montane forest on their upper slopes rapidly diminishing in extent because of human action. They are areas of moderately high rainfall (1,200–1,500 mm per annum or more; TANRIC map), although the presence of abundant epiphytes suggests that the forests owe their existence as much to mist forming over the mountains as to rainfall. The major drainage of the western flanks of the highlands in the Laetoli area are the Garusi and Olaitole Rivers flowing to the west, the Oldogom River flowing to the south and west, the Gadjingero River flowing to the northwest, and the Naisuri River flowing northwards into the Serengeti plains (Fig. 8.1).

Climate

East Africa has generally low rainfall for an area spanning the equator. It has two monsoon systems tracking the movements north and south of the Intertropical Convergence Zone (ITCZ), the southwards movement of the ITCZ bringing the northeastern monsoon with limited moisture content, and the northwards movement bringing the southwestern monsoon off the Indian Ocean, producing the long rainy season in March to June. The area covered by the Serengeti ecosystem is noted for its sporadic and unreliable rainfall (Norton-Griffiths et al. 1975). There is a marked trend from low rainfall in the southeast, around 400 mm mean annual precipitation, increasing to 900 mm in the northeast and 1,200 mm in the northwest. Even more significantly, the pattern of rainfall varies, with the southeast having a single rainy season peaking in December-January while the rest of the Serengeti has the more usual equatorial double rains in November-December (short rains) and April (long rains). Both aspects of rainfall are influenced by the Ngorongoro highlands to the east, which produce a marked rainshadow in the eastern and southeastern parts of the Serengeti. There are no climate records for the Laetoli area, but extrapolating from the very broad scale TANRIC map it appears that most of it falls within the 700–900 mm annual rainfall zone, a level of rainfall more than adequate to support deciduous woodland. The higher land to the east of the Laetoli area appears to have higher rainfall based on the change in vegetation and TANRIC map, 900–1,200 mm although there are no accessible records to confirm this. These values seem rather high and may reflect maximum rainfall only.

Topography

The study area has generally low relief with drainage to the west between low ridges. The drainage channels of the Garusi, Gadjingero and Olaitole Rivers are dry for most of the year. The headwaters of the Garusi are at 1,900 m to the north of Endulen, and the river drops around 200 m over a distance of 16 km, flowing southwest before turning to the south towards Lake Eyasi. Most of the upper part of this valley cuts through Pliocene sediments of the Laetolil Beds, strongly alkaline tuffs and tuffaceous sediments. The sides of the Garusi valley slope up about 40 m above the river channel along most of its length, higher to the south where the ridges are formed of Ogol lavas and lower to the north where the Laetolil tuffs form much of the slopes. There are limited exposures of Ogol lavas to the north as well, for example at Silal Artum to the north of the Gadjingero River. The valley of the Gadjingero has very low relief, from 1,800 m at its origin to 1,720 m over a distance of 8 km to the northwest. The Olaitole River arises from the Oldonyo Mati hills, which rise to heights of 1,886–1,943 m, and the river drops to 1,660 m over a distance of 5 km. The Oldonyo Mati hills are the only area of high relief in the study area apart from the volcanic highlands, and they are formed from Ogol lavas that also form the northwest and southeast ridges bounding the valley of the Olaitole River. Endulen is situated at the eastern end of the Oldonyo Mati hills, and the Oldogom River flows through the town and to the south of the hills. It is a perennial river, fed by springs emerging in the vicinity of Endulen. The Naisuri River rises on the western slopes of Lemagurut, running towards the northwest initially before bending to the north, at which point it forms a gorge 45–60 m in depth cutting vertically through Lemagurut lavas, and although the valley is dry for much of the year, water is present underground at no great depth. Two other unnamed dry river valleys to the east and west of Naisuri are not quite as deep, but are still significant features.

The major features of the study area can be summarized as follows: highlands to the east rising above 3,000 m, west-trending valleys on the lower slopes of the highlands merging into shallow valleys with westwards aspect, low ridges forming the sides of the valleys, the Oldonyo Mati hills to the southeast, and three deep gorges to the north. The distribution of soils, and hence woodland species associations, is related directly to these topographic and geologic features.

Soil

There are four main soil types in the Laetoli area, vertisols (known as black cotton soils or locally as mbugas), shallow to deep dark brown volcanic soils with a high silt content on the Ogol lavas, shallow light brown soils on Lemagurut volcanics, and shallow silty soils on basement rock. These different soils support distinct vegetation associations, but are themselves the products of the underlying rocks. Soils formed in dry river channels are a variation on these as the channels cut through different soil types (Table 8.2). It was found by de Wit (1978), for example, that the distribution of woodland and grassland in the Serengeti ecosystem is largely determined by soil characteristics. Woodland soils were found to have higher infiltration rates and depth of penetration of water, and these were the main factors determining the amount of available moisture. Mbugas were described as heavy clay soils, gray to black with strongly developed prismatic structure resulting in deep cracks. The prismatic structure and strong expansion and stagnating wetness, with low levels of soil oxygen, exclude many plant species from growing on these soils (de Wit 1978). Water availability is low, for although rainfall penetrates deep into the soil through cracks, it is not available in the body of the soil between the cracks (de Wit 1978: 40), and a high proportion of precipitation is lost from run-off into the subsoil (as much as 50%, de Wit 1978: 43). Mbugas form on flat terrain at the foot of gentle slopes in areas with pronounced dry seasons (Retallack 1990).

The mbugas appear to be derived mainly from the Plio-Pleistocene tuffs and are developed on all flat or gently sloping surfaces. No mbuga formation has been observed on the Ogol lavas, but they appear to be developed on Lemagurut lavas in areas of low relief, but not on slopes. An mbuga soil profile described by Anderson and Talbot (1976) 11 km northwest of Endulen was derived from the tuffs, and it was a vertisol of lithomorphic origin. This soil contained free calcium carbonate and soluble salts in the top 1 m of the soil profile, and the upper part of the soil was described as silty clay loam with block structure and pH varying from 6.0 to 6.6 in the top 50 cm of the profile. Below this the pH rises to nearly 8. The vegetation at the soil profile was grasses (i.e., *Pennisetum mezianum, P. stramineum, Cynodon dactylon, Andropogon greenwayi, Chloris pycnothrix, Sporobolus pyramidalis* and *Themeda triandra*). Plants associated with overgrazing are said to be abundant, particularly *Leucas* spp, *Solanum* sp., *Heliotropium steudneri, Crotolaria spinosa*, and *Indigofera* sp., and basal cover averages only about 20% (Anderson and Talbot 1976). The same species are common also in the valleys of the Olaitole, Garusi and Gadjingero Rivers, where the mbuga soils are developed to great depths. Here the vegetation is low woodland with *Acacia drepanolobium* and associated species (see below). This is recognized elsewhere in northern Tanzania as a "natural edaphic savanna on a vertisol" (Schmidt 1975).

Soils with greater silt content occur where the watercourses cut through the mbugas, allowing better drainage locally. Riverine vegetation associations are linked very closely to the drainage channels, which are dry for the greater part of the year, but subsurface water is present as run-off from the highlands and it is a combination of better drainage and greater moisture

Table 8.2 Distribution of soil types in the Endulen-Laetoli area with their associated vegetation and catenas

Soil type	Areas	Edaphic constraints for vegetation	Vegetation association	Catena
Mbuga soils	15	Seasonal water-logging because of impeded drainage, high alkalinity	Low *Acacia drepanolobium* woodland	Bottom of drainage catena
Soils on Ogol lavas	28–29	Good drainage but soils extremely thin on steep slopes	Tall deciduous broad leaved woodland	Top of drainage catena
Lemagurut volcanic soils	13, 16–22, 31–32	Good drainage but soils thin on steep slopes	Tall deciduous broad leaved woodland mixed with *Acacia* woodland	Altitude catena
Basement soils	23–27	Good drainage on generally flat terrain, low fertility because of nutrient-poor source rocks	Low *Acacia*/broad leaved woodland	Altitude catena
Riverine soils	1–12, 30	Increase in drainage where channels cut through mbugas, decreasing water availability at headwaters	*Acacia* woodland and bushland increasing in height down-channel and changing to gallery forest where water is perennial	Riverine catena

content that produces the riverine vegetation. The importance of drainage is seen in the presence of riverine bushland and woodland even in the upper reaches of the river channels, where the water catchments are small, but where the channels are deep enough to cut through the thick mantle of mbuga and allow locally improved drainage. The importance of moisture is demonstrated by the increasing height and species richness of the woodlands lower down the rivers, there being consistent sequences of vegetation on all water courses from riverine bush and thicket along the headwaters of the channels changing first to low woodland and then to tall woodland lower down. This is also where the mbugas are thinner.

The soils developed on the Ogol lavas are "loamy and sandy (and) woody vegetation prevails, because of the good drainage" (de Wit 1978: 37). These soils at Laetoli are always developed on relatively steep erosional slopes and hence are thin and well drained. The lavas are frequently exposed at the surface with pockets of soil in between lava boulders, and these support dense woody vegetation. No differences in soil development have been noted in relation to the different woodland species associations, e.g., between *Acacia tortilis* woodland and *Combretum-Albizia* woodland. It has been observed, however, that the tallest and most diverse woodlands grow in steep-sided valleys that are protected from the fires that are set on the more exposed slopes, and throughout the area there is much physical evidence of burning and clearing of the woodlands.

The soils developed on the Lemagurut volcanics and the basement series have not been examined in detail. They appear thin, with the underlying bedrock never far from the surface, and they are undoubtedly well drained. The basement series is exposed mainly to the west of the Laetoli area, and the vegetation distribution is related more to altitude than to soil (Herlocker and Dirschl 1972): see below.

The Laetoli Woodlands

Seven vegetation units have been recognized in the Laetoli area, ranging from semi-deciduous forest to deciduous woodland, with areas of riverine woodland and bushland, gallery forest, montane forest and edaphic grassland (Fig. 8.3). Some of the units have several different species associations, and these are distinguished here as sub-units. The major vegetation units are classified structurally following Grunblatt et al. (1989) for the most part.

Vegetation Unit 1 – Montane Forest, Tall, Dense, Evergreen

There are remnant patches of montane forest on the upper slopes of Lemagurut, Olmoti and Satiman, all extinct volcanoes forming the part of the highland area that forms the eastern boundary to the southern Serengeti and the Laetoli area in particular. The forest has a main canopy of 22–24 m consisting of *Nuxia congesta, Croton* spp., *Dombeya* spp., *Juniperus procera* and *Hagenia abyssinica*; there is a sparse lower canopy 2–4 m, of *Maytenus heterophylla, Rhamnus prinoides, Sparrmannia ricinocarpa* and a sparse ground flora of *Urtica*, ferns, Rubiaceae species., and forest grasses such as *Snowdenia petitiana.* In places the forest cover has been removed, either from cutting or burning, and these areas are distinctive with large weed species, 2–3 m, of *Leonotis*, Rubiaceae, bracken, *Kalanchoe, Hypericum* but few grasses. The forest has a sharp boundary with lower altitude grasslands at around 2,750 m (Fig. 8.2a), but forest remnants extend to lower altitudes down steep sided valleys. Both of these features suggest that it is fire that has destroyed forest at lower altitudes. In undisturbed conditions there should be a transitional zone bordering the forest, which is vulnerable to fire, and steep sided valleys offer some protection from fire.

The lowest altitude of montane forest that we observed was at 2,750 m on the east and southeast sides of Lemagurut. We have not surveyed any other side of the mountain, but since the prevailing wind, and therefore the highest rainfall, is from the southeast, it is possible that it reaches a lower altitude further round to the south. Herlocker and Dirschl (1972) put the lower limit of montane forest on Lemagurut at 2,450 m, with the montane-edge association of *Acacia lahai* (our unit 5b - see below) ranging down to 2,100 m. Our observations suggest this might still be the case, and it may be significant that the southern part of the mountain is less accessible to people, but the exact altitude is not known. The bedrock is volcanic and the soil is dark brown silty loam, shallow on the ridges but very much deeper and with greater clay content in the valleys.

The parts of the forest that we surveyed have been heavily influenced by human activity. The single *Juniperus* tree that we saw had recently been felled and was still in the process of being cut up for transport. As a result the forest canopy was very open, and we did not measure canopy heights or density. Estimates for the tallest and thickest parts of the forest are 22–24 m in height and ~70% canopy cover.

Vegetation Unit 2 – Deciduous Woodland

Four types of deciduous woodland have been recognized in the Olaitole-Garusi valleys. They form a gradient from high species richness and tall canopies (2a) to low species richness and lower, more open, canopies (2d), and the characteristic tree species are listed below. Full lists are given in the appendix.

2a	Woodland	Tall, closed	*Combretum molle, Dombeya* spp., *Albizia amara, Vangueria* spp., *Celtis africana, Erythrina abyssinica, Ficus lutea, Kigelia africana, Schrebera alata, Pycnostachys deflexifolia*
2b	Woodland	Tall, dense	*Acacia tortilis, A.xanthophloea, A.hockii, A.gerrardii,* and all the species in 2a but scattered and less abundant
2c	Woodland	Tall, open	*A.tortilis, Albizia amara, Balanites aegyptiaca, Ficus cyathistipula, Ziziphus mucronata, Combretum molle, A.hockii, Dombeya* spp.,
2d	Woodland	Low, open	*A. hockii, A.tortilis, A.nilotica, A.drepanolobium, Commiphora* sp., *Combretum molle*

Definitions: Closed > 80% canopy cover
Dense 50–80% canopy cover
Open < 50% canopy cover
Tall >10 m average tree height
Low 6–10 m average tree height

Association 2a is the common one on the Oldonyo Mati hills, and the numbers of species, density of canopy and average height of the trees vary according to the degree of disturbance. The most highly diverse woodlands have closed canopy with tree height between 10 and 15 m, and some areas remain where dense woodlands are still present. The tree crowns form a single but uneven tree canopy, and where the tree canopy is closed the shrub growth is minimal. In places, *Acacia* trees are rare in extensive parts of the woodland, and where they do occur (association 2b) they are less frequent than broadleaved trees (Fig. 8.2b). Five sample plots over the southern and western slopes of the Oldonyo Mati hills showed numbers of *Combretum, Albizia* and *Dombeya* spp. making up 66% of the tree species and *Acacia* species 34% ($N=222$). There is a lower canopy of trees and shrubs (shrubs or bushes are defined as multi-stemmed woody plants as opposed to single-stemmed trees) within the woodlands dominated by *Rhus natalensis, Acacia seyal, Grewia similis, Carissa edulis,* and *Maytenus*, sometimes forming thickets of varying density. Ground vegetation has *Hypoestes verticillaris, Tinnia aethiopica, Indigofera spicatum, Solanum nigrum,* and *Lippia ukambense*. Lianes and creepers are abundant, e.g., *Pterolobium stellatum.* Many of the trees in the dense woodlands have the growth form of closed woodland species, with crown development concentrated in the upper canopy and trunks with few low branches. The more open areas may have dense bushland thickets developed in places, often dominated by low growth of *Ziziphus mucronata.*

A 200 m transect up the west side of one of the Oldonyo Mati hills is shown in Fig. 8.3, an altitude change of 51 m over the 200 m (1,849–1,900 m, ~15° slope). Soils become thinner and more extensive exposures of rock (Ogol lava) are exposed up slope, but tree density and species richness increase up the slope. The lower slopes are dominated by *Combretum molle* and the upper slopes by *Schrebera alata, Mystroxylon aethiopicum* and *Ficus* species. *Albizia amara* is also common, but not present on this transect. The hills are too insignificant to have differential rainfall on the upper slopes, and there is no difference on different sides of the hill other than that caused by the extensive human destruction visible everywhere. The tree species most obviously targeted by human activity (i.e., felling, ring barking, and setting fires around the base of the trees) is *Albizia amara*, which has foliage much liked by domestic animals.

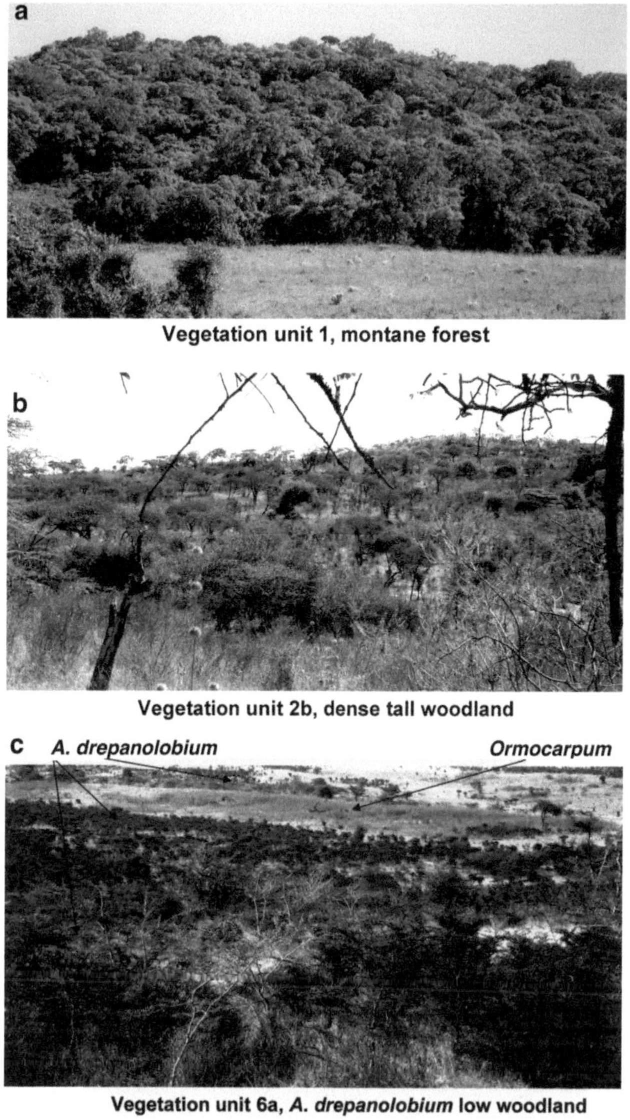

Fig. 8.2 Three views of vegetation: (**a**) montane forest on the eastern highlands, the *upper* slopes of Lemagurut, east-facing slope; (**b**) dense woodland on the Oldonyo Mati hills dominated by *Combretum molle* and *Albizia gummifera*; (**c**) low dense *Acacia drepanolobium* woodland, replaced laterally by *Ormocarpum trichocarpum* due to a change in soil, Garusi valley

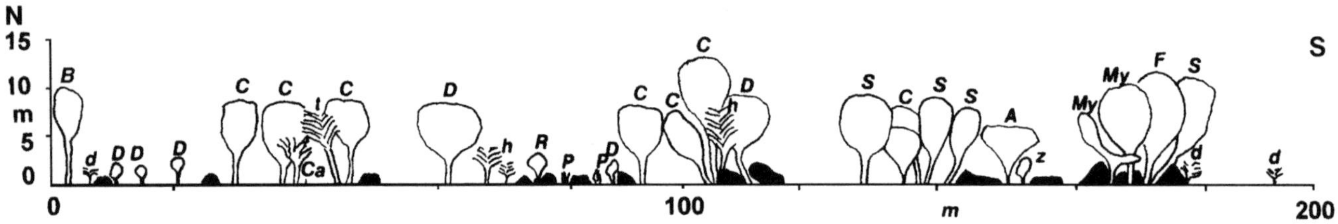

Oldonyo Mati Hill woodlands

A	*Albizia amara*	P	*Pycnostachys deflexifolia*
B	*Balanites aegyptiaca*	R	*Rhus* species
C	*Combretum molle*	S	*Schrebera alata*
Ca	*Commiphora africana*	z	*Ziziphus mucronata*
D	*Dombeya* sp.	d	*A. drepanolobium*
F	*Ficus* sp.	h	*A. hockii*
My	*Mystroxylon aethiopicum*	t	*A. tortilis*

Fig. 8.3 200 m north to south transect from 1,849 to 1,900 m up the west slope of Oldonyo Mati hills: 3°12' 59" S, 35°13' 56" E. Tree identifications are shown by letters above each plant and identified in the key

There are extensive areas of open woodland, especially in the valley of the Olaitole River, and these are dominated by *Acacia tortilis*, with *A. kirkii, A. seyal* and *Balanites aegyptiaca*. Locally this association has greater species richness in areas protected from fire, with the addition of *Cordia ovalis, Ziziphus mucronata, Grewia* spp., *Rhus* spp., *Maerua triphylla* and *Euphorbia candelabrum*. A 200 m transect through *A. tortilis* woodland in the Olaitole valley had 63.5% canopy cover composed almost entirely of this species, but also with 15% *A. kirkii* and scattered *A. drepanolobium*. Generally these woodlands are more open than this. They are widespread along ridge slopes where fires are most frequent.

Further to the west, along the Misigio escarpment above Lake Eyasi, basement rocks are exposed. From the lowest point of the escarpment, 1,100 m at Noiti, the land rises slowly to the west, i.e., away from the eastern highlands, to an altitude of 1,500 m near Kakesio. The whole area is classified as low *Acacia/Combretum* woodland, association 2d, but the associations vary with topography and altitude. At 1,100 m, *Acacia-Commiphora* low woodland also has *A. senegal* and *A. tortilis* with *Adansonia digitata* and *Terminalia brownii* (Herlocker and Dirschl 1972); at 1,300 m, *Combretum* is more abundant, with *Azanza garckeana* and *Sclerocarya birrea*; and at 1,500 m *Erythrina abyssinica, Dombeya* species and *Gardenia latea* produce a flora more like association 2a.

This variation is encapsulated at Noiti based on a combination of geological substrate and topography. On the lahar, with rapid run-off, vegetation is sparse and is dominated by *Commiphora africana* and *Euphorbia cuneata*, with occasional larger *A. tortilis* and *A. drepanolobium*. Ogol lava ridges 4–6 m above the lahar have dense woodland, with *A. tortilis, Balanites aegyptiaca, Ziziphus mucronata, Ximenia americana, Acacia mellifera,* and *Grewia* species. Finally, dry river channels have all of the above species with the addition of *Albizia* species (2), *Maerua parvifolia, Croton dichogamus, Maytenus heterophylla,* and *Cordia ovalis*. All this variation occurs within an area of a few hundred square meters, and spot checks on the area around Noiti suggest that the same pattern is repeated over a much wider area.

The first three types of woodland (2a to 2c) occur exclusively on the Ogol lavas, either on hills or on ridges, and on the basement series. There is often an abrupt transition from low *Acacia* woodland or grassland growing on mbuga soils to tall woodland at the change in geology and soil, a change that is clearly related to edaphic factors rather than to any local climate change, and the factor that is implicated is the improved drainage on the volcanic soils. This comes about both because the volcanic rocks are exposed on moderate to steep slopes and because the soil is thin and sometimes almost absent altogether. The lower valley flats, by contrast, are areas that are seasonally waterlogged due to impeded drainage, deep vertisols with high clay content unable to support tall woodland.

Vegetation Unit 3 – Dense Riverine Bushland

This is a local unit found along the headwaters of the seasonal watercourses, such as the Olaitole and Garusi. It is dominated by shrubs 1–4 m in height, which are sometimes scattered but sometimes forming dense thickets. Taller trees may be present, 5–8 m tall, but they are not common because they have been heavily cleared by local people.

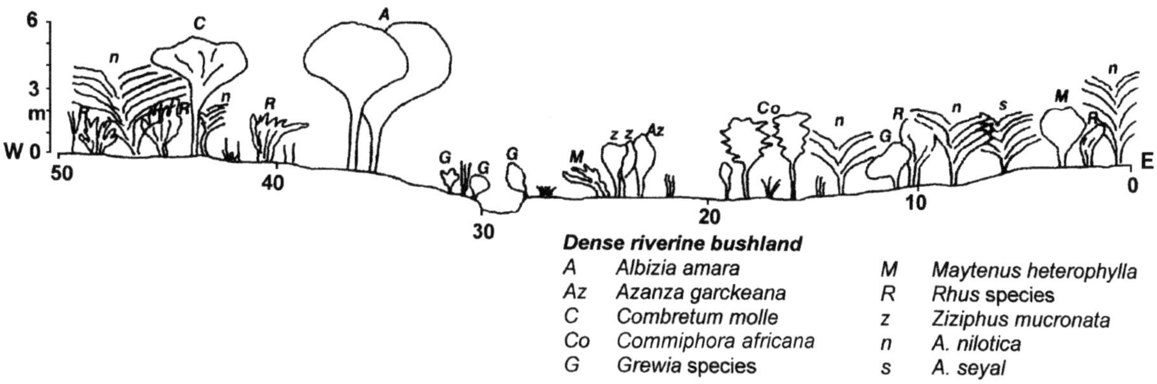

Fig. 8.4 50 m transect through dense riverine bush measured from east to west, altitude 1,800 m, across an *upper* tributary of the Garusi River (3°12' 18" S, 35° 13' 01" E). Water flows irregularly along this watercourse, and for most of the year it is dry. Tree identifications are shown by letters above each plant and identified in the key

Common shrub species are: *Rhus* spp., *Grewia* spp., *Maytenus heterophylla, Ximenia americana,* and many of the tree species have shrubby form, such as *Acacia nilotica, A. kirkii, Balanites aegyptiaca* and *A. drepanolobium.* Some of these are only shrubs because of frequent cutting for boma hedges and browsing by goats. Tree species present are *Albizia amara, Combretum molle* and an occasional *Acacia nilotica.* For the most part, the riverine bushlands are dominated by *Acacia* species, but there are also areas of broadleaved shrubs with no *Acacia* at all. The environmental factors underlying this difference are not clear.

A 50 m transect across the dry river channel of a tributary of the upper Garusi River is shown in Fig. 8.4 (3°12'18" S, 35°13'01" E). This has 83% tree and bush canopy cover, with trees reaching an average height of 5.1 m ($N = 12$), and shrubs having an average height of 2.5 m ($N = 17$). Away from the watercourse the vegetation changes to *Acacia drepanolobium* low woodland (unit 6 below).

Vegetation Unit 4 – Riverine Woodland and Gallery Forest

Most of the watercourses have narrow bands of woodland along their lengths, getting taller, denser and with greater species richness down valley from their headwaters. The one perennial watercourse, the Oldogom River, has gallery forest (association 4d) within 2 km of the springs that are its source, and this changes progressively down stream to tall closed woodland (association 4a) and to tall open woodland (4b). Species associations in the four riverine woodland types are as follows:

4a	Riverine woodland	Tall, closed	*Acacia xanthophloea, A. kirkii, Albizia* spp., *Euclea divinorum, Celtis africana, Combretum molle, Vangueria madagascariensis Zanthoxylum chalybeum*
4b	Riverine woodland	Tall, open	*Acacia kirkii, A. tortilis, Albizia amara, Ziziphus mucronata, Grewia* spp., *Ximenia americana, Euclea divinorum, Combretum molle*
4c	Riverine woodland	Low, open	*Acacia kirkii, A. nilotica, Cordia monoica, Albizia amara, Ximenia americana*
4d	Riverine woodland	Semi-evergreen	*Croton macrostachyus, Lepidotrichilia sp., Ficus* spp., *Celtis africana, Vernonia sp., Vangueria madagascariensis, Rauvolfia caffra, Urtica massaica*

The same criteria for canopy heights and density apply to this vegetation unit as for deciduous woodlands except that canopy cover refers to continuity of cover in one dimension only, along the water courses and not away from them

Definitions: Closed > 80% canopy cover
Dense 50–80% canopy cover
Open < 50% canopy cover
Tall >10 m average tree height
Low 6–10 m average tree height

Fig. 8.5 Three views of *Acacia* woodland: (**a**) *Acacia xanthophloea* lining the river channel in Naisuri Gorge, with an assemblage of *Euclea divinorum, Albizia gummifera* and *Vangueria* species; (**b**) gallery forest along the perennial river Oldogom, with *Acacia xanthophloea* and *Ficus* species, *Croton macrostachyus, Celtis africana*; (**c**) escarpment woodland on the lower slopes of the eastern highlands leading up to Lemagurut, with *Acacia lahai* and *Lepidotrichilia* and *Ficus* species

Tall, closed, riverine woodland (association 4a) is characterized by the tall *Acacia xanthophloea* yellow fever trees that form narrow strips along both sides of the watercourses. The trees range up to 25 m in height in the most striking form of woodland in Naisuri Gorge (Fig. 8.5a). This gorge is 45–55 m deep, with steep, sometimes precipitous slopes, and surface water may be present in places throughout the dry season. The *Acacia* forms a dominant upper story along the river course, with scattered trees of *Albizia amara, Celtis africana* and *Zanthoxylum chalybeum. Euclea divinorum* forms a sometimes-dense under-story together with the other species listed above. In addition to the riverine woodland lining the river course, the slopes and top of the gorge walls are also heavily wooded, as is shown in Fig. 8.6. This is a 100 m north-south transect along the bottom of the gorge on the right, where it crosses the winding river course twice at 20 and 65 m. On the left is the east-west sloping side of the gorge with another 100 m transect up the slope and covering a short distance beyond the edge of the gorge. Offset above is a 70 m transect from south to north along the top of the western edge of the gorge. Species composition changes with topography and aspect of the gorge, with *A. xanthophloea* stopping abruptly, *Euclea divinorum, Albizia amara, Ziziphus mucronata, Ficus* species and *Dombeya* species being the most widespread, and on cliff edges where there is least risk of fire, there are scattered *Euphorbia candelabrum.* Most of these species continue along the top of the gorge, but away from the edge they are replaced by *Balanites aegyptiaca, Acacia seyal* and *A. drepanolobium.* This is a complex and rich vegetation association, much frequented by browsing mammals and two species of monkey, *Papio anubis* and *Chlorocebus aethiops.*

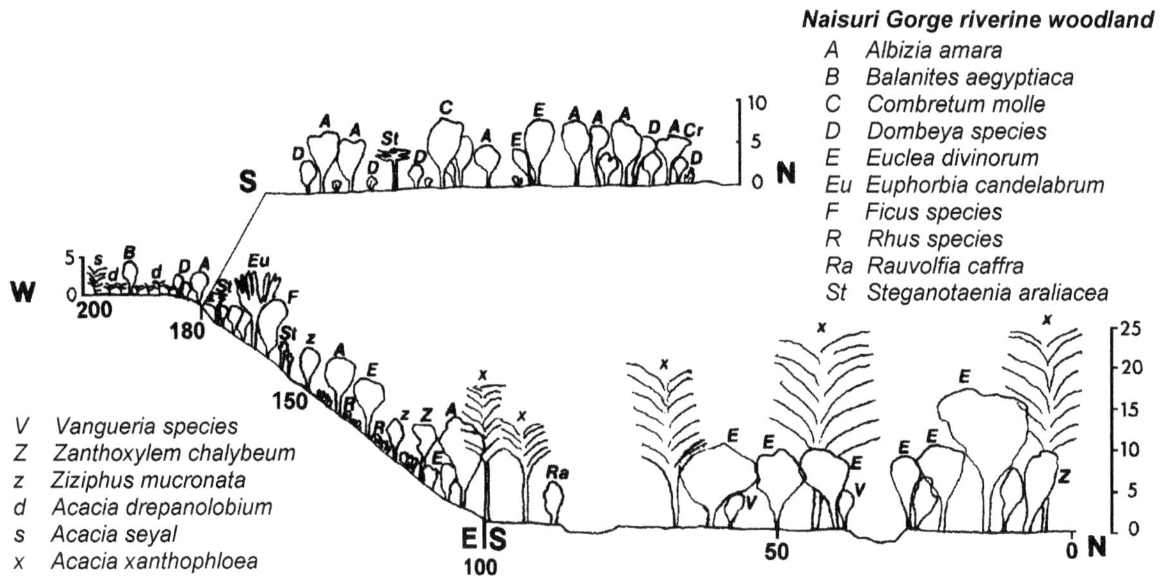

Fig. 8.6 Naisuri Gorge transects (3° 9' 34" S, 35° 14' 30" E): At *bottom right* is a 100 m south to north transect along the *bottom* of the gorge extending into a second 100 m transect up the east-west sloping side of the gorge and extending a short distance beyond the edge of the gorge. Offset above is a 70 m transect from south to north along the *top* of the western edge of the gorge. Tree identifications are shown by letters above each plant and identified in the key

Where conditions are drier, this woodland association becomes more open and it loses some of the species. The dominant *Acacia* species change to *A. kirkii* and *A.tortilis* (association 4b). Some watercourses, such as the Gadjingero River, have lower and open canopies along their length, fewer tree species and reduced ground cover, and the common *Acacia* species is *A. nilotica* (association 4c).

There is a remarkable gallery forest association along the upper reaches of the Oldogom River, which is fed by a perennial spring derived from the eastern highlands (Fig. 8.5b). This has a mixture of evergreen trees, such as the fig species and *Lepidotrichilia volkensii*, and deciduous trees such as the *Croton macrostachyus* and *Celtis africana*, both forest-fringe species. They range up to 20 m in height, with an average of 15.3 m ($N = 17$), and along the river the canopy is closed, with 94% canopy cover. The lower canopy is composed of *Vangueria* spp., *Rauvolfia caffra, Ziziphus mucronata, Maytenus mossambicensis* and *Grewia similis*, with an average height of 3.8 m ($N = 12$). Away from the river and lining both sides of the gallery forest are belts of *Acacia xanthophloea* with some *A. kirkii* similar in structure to the Naisuri Gorge association (4a). The vegetation structure is shown in Fig. 8.7, with a 200 m transect along the river valley. The transect crosses the river several times as it winds about, sometimes crossing the *Acacia* belt and sometimes the narrow line of forest trees alongside the water. The area is much disturbed, for it is the only permanent source of water in the area, so that it is heavily grazed and trampled by cattle, and extensive parts of the gallery forest have been destroyed. This association persists for some distance down the river, and as the water supply diminishes the forest trees disappear and the association changes first to association 4a (above), dominated mainly by *A. xanthophloea*, and further down to 4b, dominated by *A. kirkii* and *A. tortilis*.

The riverine woodlands are clearly linked with the availability of water, and the species composition and richness are closely tied to this. It is also apparent, however, that improved drainage is a factor, and this is seen in the upper reaches of the seasonal rivers, where their channels are cut deep into mbuga soils. Rarely is there free water in these watercourses, but the erosion of the channels has removed most or all of the mbuga soils that mantle the shallow valleys, and where this has happened trees grow along the exposed surfaces.

Vegetation Unit 5 – Escarpment Woodland, Forest Edge

There are two associations that are distinguished from deciduous woodlands because they appear to be derived from areas with stronger forest affinities. They occur from the headwaters of the Olaitoli, Garusi, Naisuri and Oldogom Rivers and extend eastwards to the montane forest of the highland zone, with the species composition changing with increasing altitude. The characteristic species present in the escarpment woodlands are as follows:

5a	Escarpment woodland Altitude 1,900–1,980 m	Tall, open	*Croton macrostachyus, Combretum molle, Dombeya* spp., *Acacia kirkii, Erythrina abyssinica, Albizia* spp., *Ficus* spp, *Vangueria madagascariensis*
5b	Escarpment woodland Altitude > 2,000 m	Tall, dense	*Acacia lahai, Lepidotrichili volkensii, Ficus* spp., *Croton macrostachyus, Albizia amara, Urtica massaica, Buddleia polystachys*

The lower end of escarpment woodlands starts at about 1,900 m, which is similar in altitude to the slopes of the Oldonyo Mati hills and in fact is connected to them by a long ridge extending eastwards from the most eastern of the hills. We measured a 200 m transect from a shallow valley at 1,880 m (Fig. 8.8), and this had riverine woodland association 4c (see above) dominated by *A. kirkii*. Away from the

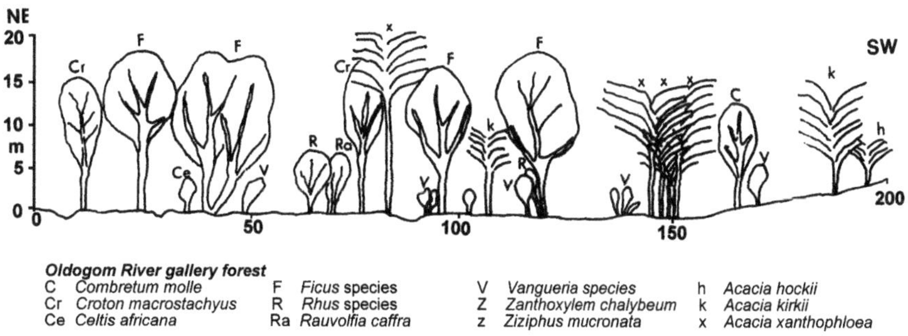

Fig. 8.7 Oldogom River near Endulen with gallery forest (3° 12' 59" S, 35° 16' 07" E) showing a 200 m transect from northeast to southwest along the length of the river. The transect crosses the river four times at 5, 28, 87, and 150 m as it winds across the line of the transect. Tree identifications are shown by letters above each plant and identified in the key

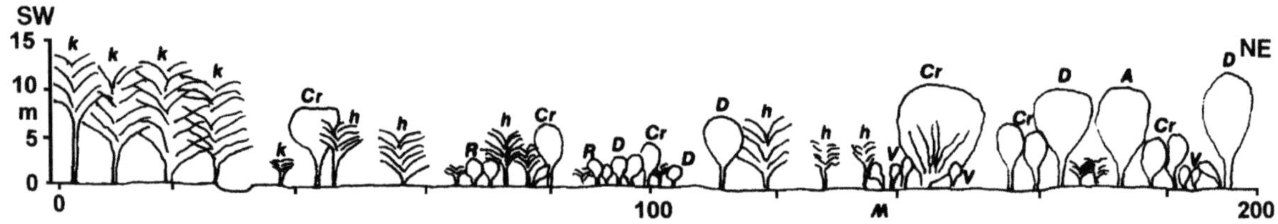

Escarpment woodlands

A	*Albizia amara*	*R*	*Rhus* species
Cr	*Croton macrostachyus*	*V*	*Vangueria* species
D	*Dombeya species*	*h*	*A. hockii*
M	*Maytenus heterophylla*	*k*	*A. kirkii*

Fig. 8.8 Escarpment woodland (3° 12' 37" S, 35° 16' 44" E), altitude 1,880–1,900 m. The 200 m transect runs from the valley bottom (dry river channel is at 30 m) up a moderate slope to the northeast. Tree identifications are shown by letters above each plant and identified in the key

river and increasing in altitude the woodlands were dominated by *Croton macrostachyus* with abundant *Combretum molle, Dombeya* species and *Albizia amara*. The main *Acacia* species was *A. hockii*, but this formed an insignificant part of the tree association. As altitude increases, there is continuous transition of species, probably due to rainfall increasing (Trapnell and Griffiths 1960). At 1,960 m (the height of the tops of the Oldonyo Mati hills) *Croton* has become dominant, with *Albizia amara* and *Albizia petersiana, Celtis africana, Acacia xanthophloea and Acacia senegal*. At 2,011 m, there are dense stands of *Acacia lahai* (Fig. 8.5c), a large forest-edge *Acacia*, together with *Lepidotrichilia volkensii, Ficus* spp., *Senna* spp., *Buddleia* sp., *Ekebergia capensis*, and *Acacia clavigera* subsp. *usambarensis* (association 5b). There is increasing species richness with increasing altitude, and a marked change from dry forest elements to wetter forest elements, and this is no doubt due to increasing rainfall with altitude (Trapnell and Griffiths 1960).

Vegetation Unit 6 – Acacia Drepanolobium Woodland

Acacia drepanolobium woodland covers wide areas of the Garusi and Olaitole River valleys. Canopy height is generally low, but some areas were consistently taller than others. Most of the area is regularly burned by the local Maasai, who also cut it to make hedges, the long thorns being a good deterrent to local lions. This has resulted in some of the trees becoming multi-stemmed, and so strictly should be classified as shrubs rather than trees, but in less disturbed areas the tree form is clearly apparent.

6a	Mbuga woodland	Tall, dense	*Acacia drepanolobium, A. seyal, A. tortilis, Balanites aegyptiaca, A. nilotica, Cordia monoica, Ximenia americana*
6b	Mbuga woodland	Low, dense	*Acacia drepanolobium, A. seyal, A. tortilis, Balanites aegyptiaca, A. nilotica, Cordia monoica*

The *Acacia drepanolobium* woodlands in the Garusi valley are low and dense, and individual trees are rarely more than 2–3 m in height. Average heights for two 200 m transects are 1.2 m ($N = 34$) and 1.1 m ($N = 33$), and they varied in density (Fig. 8.2c). The two transects had canopy densities of 66.4% and 60.9%, well within the criteria for dense woodlands, but most of the area was more open than this as a result of burning and clearing. An area of taller *A. drepanolobium* woodland in the Olaitole valley had an average height of 2.9 m ($N=57$), with individual trees up to 7 m and canopy density of 52.5%. Species associated with the *A. drepanolobium* are *A. seyal* and *Balanites aegyptiaca*, with an occasional *A. tortilis*. There are areas of thicket, where the dominant plants are *Rhus* spp., *Grewia* spp., *Ximenia americana, Balanites aegyptiaca, Carissa edulis* and *Asparagus africanus*.

Acacia drepanolobium is found throughout the Laetoli area, being a minor constituent of most forms of woodland and bushland. It is only found as a dominant species, however, on seasonally waterlogged soils with impeded drainage, i.e., on mbugas. This is not just a feature peculiar to Laetoli, for in western Kenya, for instance, large areas are covered with this soil, and there also the dominant tree is *A. drepanolobium*. At Laetoli it appears that this type of soil forms above the calcareous sediments of the Laetolil fossil beds, for these outcrop all along the Olaitoli and Garusi valleys, and it is likely that the sediments are continuous below the thick mantle of soil. In at least one place the soil was discovered to be thin, and the sediments underlying the soil was discovered by trenching not to be Laetolil Beds but the more recent Ngaloba Beds with very different and less calcareous chemical composition. It is striking that at this place the soil was dark brown rather than black, had a higher silt content, and had no *A. drepanolobium* growing on it. Instead, there was an almost pure stand of *Ormocarpum trichocarpum*, a shrubby legume lacking thorns that was most popular with the local giraffes (Fig. 8.3c). We measured a short transect of 100 m across the contact between the two species (Fig. 8.9), and this shows the abrupt contact between them. The *Acacia* woodland in this case had 72% canopy cover and mean height of 1.8 m ($N = 40$), while the *Ormocarpum* shrubs were more open, 51% canopy cover (N

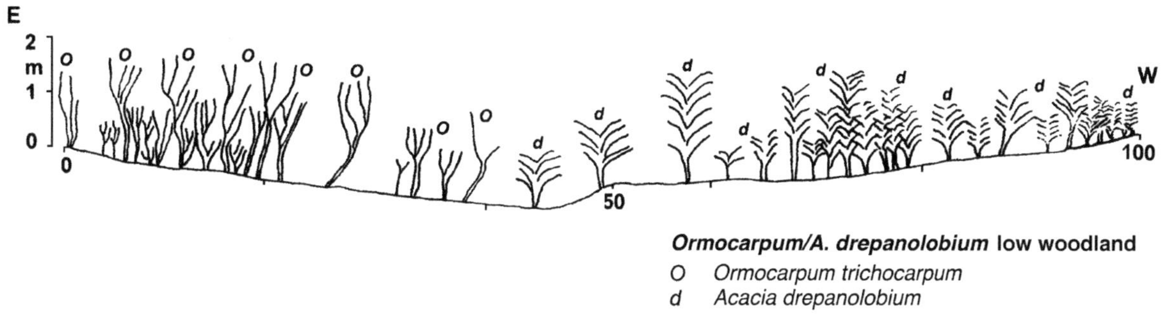

Fig. 8.9 East-west transect of the *Acacia drepanolobium* low woodland replaced by *Ormocarpum trichocarpum* shrubland (3° 12' 17" S, 35° 13' 28" E). Tree identifications are shown by letters above each plant and identified in the key

= 40) and mean height of 0.98 m. The ground cover under the two species was strikingly different also, with a variety of herb species under the *Acacia* (species of *Solanum, Asparagus, Acalypha, Hypoestes, Aristida*), with grass and weed species growing under the *Ormocarpum* (*Themeda triandra, Vernonia, Lippia*). The exact interface between the Ngaloba Beds and the beds underlying the *A. drepanolobium* could not be fixed because of the great depth of soil in the contact area.

Vegetation Unit 7 – Grassland

Parts of the Laetoli area have no woody vegetation, and the first impression is that this consists of grassland, but closer inspection shows that much of the area has little grass and the vegetation is an assemblage of herbaceous weed species. Weeds are species that invade disturbed ground, and they can be extremely abundant and variable depending on the nature of the disturbance (see next section). Where undisturbed grassland does occur in the far west of the study area, it is a medium grassland, intermediate between the short grass of the eastern Serengeti Plains and the tall grasses further west (Anderson and Talbot 1976; Sinclair and Norton-Griffiths 1979). Species associations are as follows:

7a	Grassland	Disturbed	*Bidens schimperi, Lippia javanica,* Chenopodiaceae, *Indigofera, Setaria* spp., *Themeda triandra*
7b	Grassland	Edaphic	*Setaria* spp., *Chloris gayana, Aristida junciformis, Dactylotenium* sp., *Eragrostis* spp.

Disturbance

Most parts of the Laetoli area have been affected by human activity, and the most extreme result of that activity is to produce the disturbed areas described above. Different types of disturbance appear to produce different associations of weed species, and the major weed species that take over areas after human disturbance are listed below. By the term weed, we mean plants that flourish on ground cleared or disturbed by human action and displace the natural vegetation. They are not necessarily invasive or exotic, but are part of the local flora that take advantage of the disturbed growing conditions to take over locally disturbed land. For example, abandoned bomas (Maasai homesteads) are extremely rich in organic remains, and they typically have tall stands of pure *Leonotis nepetifolia* following the limits of the boma very exactly. This species is part of the local flora, but on the disturbed land they regenerate freely and exclude all other plants following abandonment of the bomas. What is seen on the ground are circular patches of pure *L. nepetifolia* sometimes with the decaying remains of the protective fence of *A. drepanolobium* branches forming the perimeter. Abandoned fields also have *L. nepetifolia*, but they also have a greater variety of species, particularly *Hibiscus* species and aromatic *Lippia javanica*. Again there are often the remains of the fence built to keep animals out of the cultivated area. Woodland clearing also leaves a variety of weed species dominated by *Lippia javanica* and *Clerodendron hildebrandtii*. By far the most widespread disturbance, however, is from over grazing, which produces the vast tracts of the *Bidens* weed association described in the previous section. The study by Metzger et al. (2005) has shown that the dry season grazing by livestock increases the proportion of herbs and decreases the amount of dead biomass. The following weed associations are the ones found most commonly after the different human activities:

0	Disturbed - overgrazing	Woody herbs	*Bidens schimperi, Lippia javanica, Solanum incanum*
0	Disturbed - burning	Woody herbs	*Lippia javanica, Clerodendron hildebrandtii*
0	Disturbed - bomas	Woody herbs	*Leonotis nepetifolia, Hibiscus* spp., *Bidens schimperi*

0	Disturbed - agriculture	Woody herbs	*Leonotis nepetifolia, Hibiscus* spp., *Lippia javanica, Clerodendron hildebrandtii, Hypoestes* sp.
0	Disturbed - clearing	Woody herbs	*Hypoestes* sp., *Lippia javanica, Achyranthes aspera, Clerodendron hildebrandti, Leonotis nepetifolia*

Distribution of the Laetoli Woodlands

Figure 8.10 shows the distribution of vegetation types in the Endulen-Laetoli region. The main map shows the detailed distribution of the vegetation associations listed above, and below this are two small-scale maps of the same area showing the source of the mapping evidence (right), and a summary of the main vegetation types. The source of the evidence is mostly based on direct measurement, but the eastern (montane) elements were taken from the topographical map that shows in broad outline the extent of the montane vegetation. The parts of the map marked as 'inferred' are based on long distance sightings combined with spot checks on the ground.

The summary map of the main vegetation types shows a central region of woodlands flanked on three sides by disturbed areas. This is not to imply that the woodlands are not also disturbed, but the disturbance is less and the original vegetation associations are still apparent. The term 'disturbed' on the map signifies that there is very little left of the original vegetation, which has been almost entirely displaced by weed species. To the east are the montane forest (unit 1) and forest fringing *Acacia lahai* woodland (unit 5b), both surrounded by extensively disturbed land. Farther to the west and continuous with the montane elements in at least one place, is escarpment woodland (unit 5a), open to dense woodland heavily depleted by human disturbance.

The distribution of associations is shown in detail in Fig. 8.10. The southern and central areas are composed of woodlands (unit 2), and these are associated with outcrops of the Ogol lava, including the elongated belts of *Acacia tortilis* woodland (unit 2c) extending to the west, which are shallow escarpments exposing the lava. The tall open woodlands in the south of the area are forming on exposures of basement rock beyond the range of the Pliocene and Pleistocene volcanics. The woodlands are transected by riverine woodlands (unit 4), and all are opened up to a certain extent by human action, but still retain enough of the original vegetation to make identification reasonably certain. The north is again heavily disturbed. It has three deep river valleys with open woodland (unit 2d) and dense riverine woodland (units 4a and 4b) that is in the process of being heavily exploited by the local population. The disturbed grasslands across the northern part of the area consist almost entirely of low growing weed species and are heavily exploited for grazing by the local Maasai. Finally, at the western edge of the map, there are areas of grassland that appear to be southern extensions of the Serengeti short grass plains (unit 7b), for the grass species are similar, and the soil has a well developed calcrete (see below) formed 40–60 cm beneath the surface (Anderson and Talbot 1976).

Discussion

If we consider the composition of the dominant species from each of the vegetation types listed here, there are some species that are exclusive to particular vegetation types and many that occur in more than one type (Table 8.3). The average number of species per unit area of the seven vegetation units is 37.7 species, and the change along environmental gradients is relatively high. Some of the weed species are widespread, but they are not considered important here, their abundance being a secondary artifact of human modification of the habitat. No dominant species is ubiquitous. The most distinctive vegetation types are the montane forest (unit 1) and the grassland (unit 7). Riverine woodland and gallery forest (unit 4) have the greatest species richness and spread of taxa, many of which are in common with the deciduous woodland vegetation (unit 2). There are no taxa in common between the escarpment woodland/forest edge species (unit 5) and the mbuga woodland (unit 6). The most widespread species occurring in at least four vegetation types are *Combretum molle, Albizia amara* and *Acacia nilotica,* and six species occur in three vegetation types (*Erythrina abyssinca, Acacia tortilis, Balanites aegyptiaca, Acacia drepanolobium, Ximenia americana,* and *Acacia kirkii.)* Other than the taxa exclusive to vegetation units 1 and 7, there are some other species that occur in one vegetation type only, for example *Ficus lutea, Commiphora* sp., *Juniperus procera, Acacia lahai* and *Acacia seyal* (Table 8.3). If found in the fossil record and recognized (many pollens are not species determinant), any of the restricted taxa would be most useful for interpreting the fossil vegetation. Taxa that occur in more than one vegetation unit would be less useful, for example *Combretum molle,* which occurs in four vegetation units.

The association between soils, topography and vegetation are summarized in Table 8.4. Areas of generally flat topography with impeded drainage, which results in seasonal waterlogging, are uniformly covered with low dense *Acacia drepanolobium* woodland. Slight elevation of land to the east, rising eastwards to the beginning of the eastern highlands, has shallow volcanic soils with tall dense woodland of *Combretum-Albizia* on both the Ogol lavas and the Lemagurut volcanics. Additional forest species come in at higher altitudes and increasing rainfall, such as the *Croton-Celtis-Lepidotrichilia* association (4; Table 8.3).

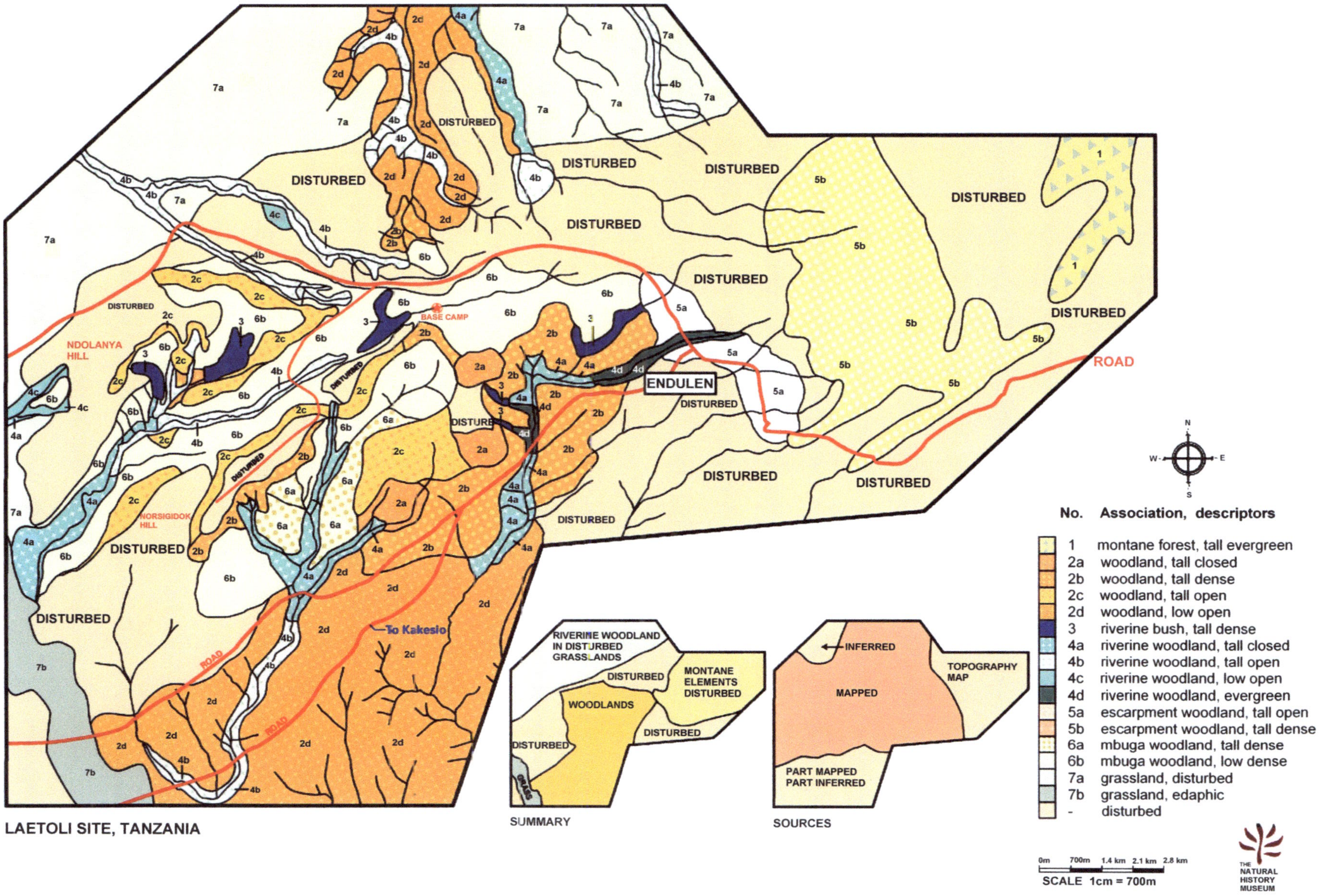

Fig. 8.10 Vegetation map of the Laetoli area showing the distribution of the 16 species associations documented here. Inset *lower right* shows the evidence on which the map is based, being in part measured on the ground and in part inferred from topographic maps. Next to this is a summary map showing the major structural vegetation types

Table 8.3 Occurrence of species in the vegetation units 1–7. Distributions by subunit are shown by the letters A-D representing the subunits in each unit (see Fig. 8.10)

		Montane Forest	Deciduous Woodland	Deciduous Bushland	Riverine Woodland	*A. drep.* Woodland	Escarpment Woodland	Grassland
Family	Taxon	Unit 1	2	3	4	5	6	7
Loganiaceae	*Nuxia congesta*	A						
Euphorbiaceae	*Croton* spp.	A						
Sterculiaceae	*Dombeya* spp.	A	A C			A		
Cupressaceae	*Juniperus procera*	A						
Rosaceae	*Hagenia abyssinica*	A						
Celastraceae	*Maytenus heterophylla*	A						
Rhamnaceae	*Rhamnus prinoides*	A						
Tiliaceae	*Sparmannia ricinocarpa*	A						
Poaceae	*Snowdenia petitiana*	A						
Combretaceae	*Combretum molle*		A B C	A	A B	A		
Mimosaceae	*Albizia amara*		A B C	A	B C	B		
Rubiaceae	*Vangueria* spp.		A B					
Ulmaceae	*Celtis africana*		A B		D			
Papilionaceae	*Erythrina abyssinica*		A B		A	A		
Moraceae	*Ficus lutea*		A B					
Bignoniaceae	*Kigelia africana*		A B					
Oleaceae	*Schrebera alata*		A B					
Lamiaceae	*Pycnostachys deflexifolia*		A B					
Mimosaceae	*Acacia tortilis*		B C D		B		A B	
Mimosaceae	*Acacia xanthophloea*		B		A			
Mimosaceae	*Acacia hockii*		B C D					
Mimosaceae	*Acacia gerrardii*		B					
Balanitaceae	*Balanites aegyptiaca*		C	A			A B	
Moraceae	*Ficus cyathistipula*		C					
Rhamnaceae	*Ziziphus mucronata*		C		B			
Mimosaceae	*Acacia nilotica*		D	A	C		A B	
Mimosaceae	*Acacia drepanolobium*		D	A			A B	
Burseraceae	*Commiphora* spp.		D					
Anacardiaceae	*Rhus* spp.			A				
Tiliaceae	*Grewia* spp.			A	B			
Olacaceae	*Ximenia americana*			A	B C		A	
Mimosaceae	*Acacia kirkii*			A	A B C	A		
Ebenaceae	*Euclea divinorum*				A B			
Mimosaceae	*Albizia* spp.				A	A		
Rubiaceae	*Vangueria madagascariensis*				A	A		
Rutaceae	*Zanthoxylum chalybeum*				A			
Boraginaceae	*Cordia monoica*				C		A B	
Euphorbiaceae	*Croton macrostachus*				D	A B		
Meliaceae	*Lepidotrichilia*				D	B		
Moraceae	*Ficus* spp.				D	A B		
Astereaceae	*Vernonia*				D			
Apocynaceae	*Rauvolfia caffra*				D			
Urticaceae	*Urtica massaica*				D	B		
Mimosaceae	*Acacia lahai*					B		
Loganiaceae	*Buddleia polystachys*							
Mimosaceae	*Acacia seyal*						A B	
Astereaceae	*Bidens schimperi*							A
Verbenaceae	*Lippia javanica*							A
Chenopodiaceae	*Unidentified*							A
Papilionaceae	*Indogofera sp.*							A
Poaceae	*Setaria* spp.							A B
Poaceae	*Themeda triandra*							A
Poaceae	*Chloris gayana*							B
Poaceae	*Aristida junciformis*							B
Poaceae	*Dactyloctenium sp.*							B
Poaceae	*Eragrostis* spp.							B
# Taxa per subunit		9	10, 9, 8, 5	9	8, 8, 5, 7	8, 6	7, 7	6, 5
# Taxa per unit		9	20	9	22	12	7	10

Table 8.4 Relationship of vegetation to soil and topography and the expected change in vegetation under increased rainfall and decreased rainfall

Soils and topography	Present vegetation	Postulated vegetation if increased rainfall	Postulate vegetation if decreased rainfall
Flat or gently sloping terrain, with seasonally waterlogged soils due to impeded drainage; have deep clayey soils, mbugas	*Acacia drepanolobium* low woodland, units 6a & 6b	Woodland becoming more open converting to grassland with dambo formation	Taller and denser woodland
Dry river channels through mbugas provide locally improved drainage where they cut through to bedrock	Riverine woodland and/or bushland dominated by *Acacia* species, especially *A. kirkii*, units 3, 4b & 4c	Greater species richness, change to *A. xanthophloea*, and entry of broad leaved species.	Reduced species richness, change to *A. hockii* or *A. gerardii*
Dark brown non-calcareous soils on Ngaloba beds	*Ormocarpum trichocarpum*, unit 6c	Invasion of other species such as *Combretum molle* to form open woodland	No change
Shallow brown calcimorphic soils on steep slopes on Ogol lavas	Dense deciduous woodland with *Combretum-Albizia-Dombeya* spp., units 2a–c	Greater species richness with species of *Celtis, Ficus* and *Croton*	Reduced species richness, change to *Acacia* woodland, e.g. *A. tortilis*
Shallow brown soils on basement rocks, valleys and hills	Dense deciduous woodland with *Acacia* spp., *Combretum*, *Azanza garkeana* and *Sclerocarya birrea*, unit 2b	*Greater species richness, with Erythrina abyssinica, Dombeya* spp., *Gardenia latea, Kigelia africana,*	*Acacia-Commiphora* low woodland with *A. senegal, A. tortilis, Adansonia digitata* and *Terminalia brownii*
Shallow brown soils on Lemagurut volcanics at 1,800 m on steep slopes	Dense deciduous woodland with *Combretum, Croton macrostachyus, Dombeya-Albizia-Acacia* spp., unit 5a	Taller and denser woodland with acacias becoming less common	More open woodland with *Combretum* becoming common with *Acacia* spp.
Shallow brown soils on Lemagurut volcanics at 1,950 m on steep slopes	Dense deciduous woodland with *Croton* dominant, *Celtis africana*, and *Albizia-Acacia* spp., unit 5a	Taller and denser woodland with *Acacia lahai, Lepidotrichilia, Ficus*	Dense deciduous woodland with *Combretum, Croton macrostachuys, Dombeya- Albizia-Acacia* spp.
Shallow brown soils on Lemagurut volcanics at 2,100 m on steep slopes	Dense tall *Acacia lahai* woodland, some montane forest species, e.g. *Lepidotrichilia* and *Ficus* spp., unit 5b	Montane forest elements such *as Juniperus procera, Croton megalocarpus, Nuxia congesta*	Dense deciduous woodland with *Croton* dominant, *Celtis africana*, and *Albizia-Acacia* spp.
Deep valleys may retain subsurface water even when channels run dry	Riverine woodland with *Acacia xanthophloea*, and *Albizia-Euclea-Celtis* spp., unit 4a	Greater species richness with addition of *Croton, Ficus-Rauvolfia-Lepidotrichilia*	Reduced species richness, with fewer *Celtis* and *Euclea divinorum*
Perennial springs	Gallery forest with evergreen species, *Ficus-Celtis-Lepidotrichilia-Croton*, unit 4d	No change	No change
High altitude above 2,450 m	Montane forest with *Juniperus procera, Olea africana, Nuxia congesta, Croton* spp. *Celtis africana*, unit 1	Similar species but with the forest reaching lower altitudes	Similar species but with the limits of the forest at higher altitudes

Seasonal stream channels flowing from east to west have *Acacia* riverine woodland, with increase in broadleaved trees such as *Celtis-Euclea-Diospyros* with increasing moisture, and becoming gallery forest where perennial springs emerge with *Ficus-Croton-Celtis-Lepidotrichilia* (Tables 8.2 and 8.3). These different plant associations merge into each other *within* the 250 km^2 study area based on variations in soil and topography, and three vegetation catenas can be observed where changes in dominant vegetation are related to these physical changes.

Woodland Drainage Catena

In areas of low relief (e.g., the Garusi River valley), there is a change in woodlands depending largely on soil type and degree of drainage, with elements of moisture availability. Change in altitude is not at issue here, for the topographic change is of the order of a few tens of meters. On flat or gentle slopes, *A. drepanolobium* woodland (vegetation unit 6, Table 8.3) is present on seasonally waterlogged soils. Down slope towards seasonal channels, where drainage is locally improved and moisture availability may be greater, taller riverine woodland (vegetation unit 4) and riverine bushland (vegetation unit 3) are present. Upslope, where soil type changes and topographic relief increases, and therefore better drainage exists, areas of tall deciduous woodland are present, ranging from *Acacia tortilis* woodland on shallow slopes (unit 2c) to *Combretum-Albizia* woodland on steep slopes (units 2a & b) (see Table 8.3). Generally speaking, the greater the degree of topographic relief, the shallower the soils and the better the drainage, resulting in a taller woodland and a more diverse floral association. This is termed the woodland drainage catena (Fig. 8.11).

Riverine Catena

A second catena follows the length of the seasonal river channels and is related to moisture availability. The headwaters of the channels have low riverine bushland with *Acacia-Grewia-Rhus*, unit 3. Down stream in succession are woodlands dominated by *Acacia kirkii* (unit 4c), then *Acacia xanthophloea*, then this same species with an association of *Celtis-Albizia-Euclea* (unit 4a), and finally, if sufficient moisture is available, semi-evergreen gallery forest (unit 4d). The riverine woodlands increase in height, species richness and density down stream with greater water availability (Fig. 8.11).

Altitude Catena

A third catena tracks change in altitude, and it is assumed that the higher the altitude the greater the rainfall. This is a general phenomenon in eastern Africa, well documented for most of the area (Trapnell and Griffiths 1960), although there are no records in the Laetoli area to confirm it. At 1,800 m, in the vicinity of most of the Laetoli fossil sites, the vegetation on the Ogol lava is the *Combretum-Albizia-Dombeya* woodland association (2a or 2b). Moving eastwards on to the Lemagurut lavas, these species are present together with *Croton macrostachyus* at the same altitude. At 1,950 m, further to the east, *Croton* has become dominant. At 2,100 m these lower altitude species have been replaced by dense stands of *Acacia lahai-Lepidotrichilia-Ficus*, all forest edge taxa, forming dense to closed canopy woodlands (Fig. 8.11). And finally, above 2,450 m, montane forest takes over, with *Juniperus procera, Nuxia congesta* and *Croton megalocarpus*.

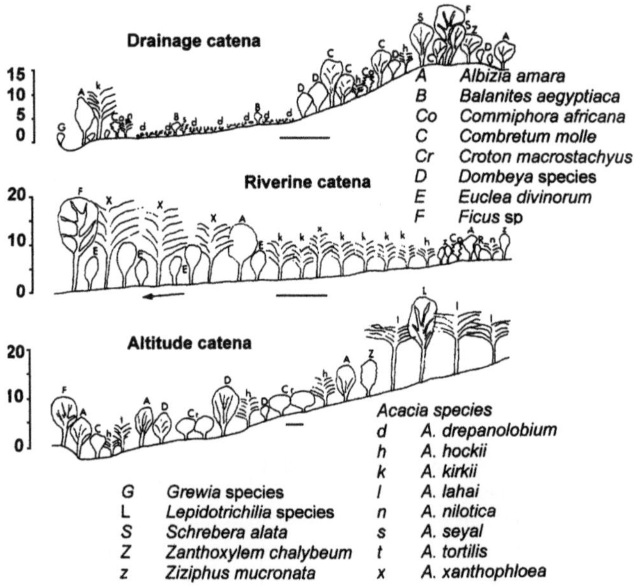

Fig. 8.11 The three vegetation catenas (composite of mapped vegetation data) that document much of the habitat variability at Laetoli at the present time. *Top*: the woodland drainage Catena; *Middle*: the riverine catena; and *Bottom*: the altitude catena. Canopy heights are shown on the vertical axis, and the scale bars beneath each profile signify 1 km horizontal distance. Tree identifications are shown by letters above each plant and identified in the key

Potential Effects of Climate Change

For the pattern of present-day vegetation to be of any use in interpreting past environments, several factors have to be considered. These include past changes in geomorphology, topography and soils, and when these are taken into account it is possible to predict past distributions of vegetation specifically at Laetoli (Andrews and Bamford 2007). These predictions were based initially on a climatic regime similar to today, but it was further possible to predict the effects of changes in climate on the reconstructed environment.

A postulated increase in rainfall can be on a scale of a few tens of years, such as the El Niño, but such an increase is not sustained long enough to affect the vegetation composition or distribution (Woodward 1987). Milankovitch cycles resulting in changes in precipitation may produce shifts in the catenas described here, but a shorter time scale of fluctuations may also be apparent. Such climatic cyclicity may be evident in the sediments. What we do see in the sediments is the evidence of volcanism that is more erratic than regular. If this is linked with climatic cyclicity, a stronger pattern may evolve. More important is the time required for recovery of the vegetation after volcanism and ashfalls. A post-eruptive wet period would reduce the devastating effect of the volcanism.

Long-term changes in rainfall patterns that have occurred from Miocene to present would have had a major effect on the vegetation. Annual rainfall accounts for 79% of the observed variation in plant species richness in southern Africa (O'Brien 1993), whereas temperature has a non-significant relationship with plants in tropical Africa and accounts for only 28% of variation in mammals (Andrews and O'Brien 2000). This arises because energy is in surplus in tropical Africa (O'Brien 1998; O'Brien et al. 2000; Field et al. 2005; Hawkins et al. 2006), and increased solar energy, or temperature, without a corresponding increase in rainfall, would have negative impact on vegetation. This is because temperature-driven increases in photosynthesis of plants could not be sustained if water availability is relatively diminished. On the other hand, lowering of temperature in the Pliocene and Pleistocene could have resulted in the depression of montane vegetation limits, having the effects of bringing montane forest and the fringing escarpment woodlands closer to the Laetoli area where the fossiliferous deposits were accumulating. In addition, the relationship of water availability with temperature may have led to an increased water deficit when temperature increased, despite possible increase in global water availability, because of the effects of increased transpiration of plants, whereas lower temperatures may have increased water availability.

Because plant species richness is positively correlated with rainfall and only indirectly with temperature, increased water availability would have led to more diverse woodland associations at Laetoli, and decreased water to less diverse associations. There is a further complication, however, when climate change impinges on edaphic vegetation associations, for they may have a more complex response to climate change, and each association or catena has to be considered individually. For example, soils affected by waterlogging, e.g., mbugas, may lose their woody vegetation and be transformed into dambo grasslands with increased rainfall (O'Brien and Peters 1999), while free-draining soils on volcanic soils would support taller and denser woodland associations. In the account that follows, we will take these issues into account to predict the nature of the changes in vegetation associations in relation to variations in rainfall.

Taking as a starting point the predicted vegetation distributions based on present day climate, these distributions can now be examined in relation to *increase in rainfall* in the Pliocene modeled on changes that would be apparent today (column 3, Table 8.4). The following predictions can be made:

- *increase* in canopy cover, height of trees and species richness of woodlands on well drained volcanic soils;
- *decrease* in tree cover, canopy height and species richness on poorly drained soils forming on the surface of the Laetolil Beds;
- *increase* in *Acacia* relative to other species, and *increase* in species richness in the riverine woodlands and spread of moisture loving species towards the headwaters.

In other words, the woodland drainage catena described above would become more differentiated, with both ends of the catena having greater species richness and higher woodland and the middle becoming more impoverished. There would also be an *increase* in water availability in the river channels resulting in an increase in species richness, especially in the deep valleys. These might well have perennial water flow, so that there would be increase in species richness in the riverine woodlands, with species of *Celtis-Euclea-Diospyros* and the spread towards the headwaters of moisture loving species. In other words, the riverine catena would shift upstream. There would probably be no change in the vegetation around the perennial springs, where moisture availability is already high. Finally, the montane forest would reach lower altitudes with increased rainfall, possibly with a small increase in species richness, so that the altitude catena would move lower in altitude. In sum, the woodlands would be richer, taller and denser, except in areas of poor drainage where the opposite would occur, perhaps even leading to edaphic grasslands, and gallery forest would be present in the deeper gorges.

Assuming *decreased water availability*, the following effects could be expected (column 4, Table 8.4):

- areas with seasonal waterlogging would support taller and denser woodland, with an *increase* in species richness, because the waterlogging would have less impact;
- well-drained sites would *change* species, losing the ones requiring moister conditions but gaining those able to withstand drier conditions, and the vegetation would become more open and have lower tree canopies.
- riverine woodland species associations would shift downstream as a result both of less moisture and impeded drainage.

As a result of these changes, the woodland drainage catena would become less differentiated. The upper sections of the riverine catena would become drier and more silted up because of reduced water flow, so that the riverine catena would move downstream. While the deep valleys would probably become more open, they would not be likely to lose species because underground water would still be available. Similarly, the sites of perennial springs would change little provided the springs, which are fed by the eastern highlands, did not dry up. Finally, the montane forest would retreat to higher altitudes with less water availability, so that the altitude catena would move up slope.

There is one additional factor that in one way or another affects much of the Serengeti ecosystem. It has been shown by Anderson and Talbot (1976) and de Wit (1978) that the soils covering much of the eastern Serengeti area are immature volcanic soils derived from recent ash falls from the still active volcano Oldonyo Lengai. These soils are highly alkaline, and combined with the strongly seasonal nature of present day rainfall in the region, this has resulted in the formation of thick calcretes 50–70 cm below the surface. The calcretes impede drainage where they form an unbroken layer over much of the eastern Serengeti Plains, and the resulting vegetation is seen in the edaphic short grass plains composed largely of grass species such as *Sporobolus spicatus* and sedges. Grassland is clearly not the climatic climax of this region, for where the calcrete is broken or exposed on the surface the grass/sedge mix disappears to be replaced by low woodland of *Acacia* species and *Commiphora*. This is shown in Fig. 8.12, on which the short grass plains on ash soil contacts the lower slopes of Naibor Soit, an inselberg of basement rock projecting above the mantle of volcanic ash. The vegetation on the basement soils is *Acacia-Commiphora* low woodland, and this extends down onto the exposed calcretes where the ash-derived soil has been eroded away. There is a sharp junction where the ash begins and the woodland is immediately replaced by short grass. This pattern is repeated where ever such a juxtaposition occurs, for example along the lines of faults (First and Second Faults, Hay 1976, 1987), where the loose soil on the elevated slopes of the fault has been eroded and the underlying rocks exposed. In every case, the exposed shallow soils support low *Acacia* woodland cutting across the short grass plains. The granite inselbergs dotted over the Serengeti landscape also have trees, some quite large, growing where they are clear of the ash, as well as having better surface water run-off.

Fig. 8.12 Contrast between the short grass vegetation (i.e., short grass plains) growing on the volcanic immature soils derived from the Oldonyo Lengai ash falls, with no trees present, with the dense low woodland and little grass growing immediately adjacent on the shallow, sloping basement soils of Naibor Soit that are less alkaline and better drained than the plains soils

The soils forming on the ash deposits are immature alkaline soils with thick calcretes just below the surface (Anderson and Talbot 1976). Further to the west, and away from the influence of Oldonyo Lengai, the soils become deeper, less alkaline as the carbonates are leached out. The calcretes start to fragment, and the grasses start to grow taller, classified as medium grassland (Anderson and Talbot 1976). Further still to the west, these changes continue, the soils mature, and first tall grass and then woodland becomes the dominant vegetation. This gradual change is partly due to the greater rainfall further west, which leaches the salts out of the soil more quickly, and partly due to the decreasing influence of the active volcano Oldonyo Lengai. As eruptions continue, the local habitats are kept in a state of flux, quiescent periods resulting in the sequence moving eastwards as the soils develop and mature, and periods of volcanic activity resulting in movement westwards as the volcanic ash falls affect the eastern soils. This, in effect, is a temporal catena since there is a sequence of change over time at any one location due to the volcanic activity (Anderson and Talbot 1976). There are insufficient data to determine the effects of this, for the nature of the evolutionary or ecological change in the plants depends on the response time of the environment to the perturbations arising from volcanism (Woodward 1987).

Oldonyo Lengai is a carbonatite volcano, and the ash falls are rich in sodium and calcium carbonates, hence the alkaline nature of the soils that form on the ash. Several of the other volcanoes formed along the line of the rift valley were also carbonatitic, including in particular Satiman, thought to be the source of much of the sediments at Laetoli. If this is the case, it is more than likely that a temporal catena could have been set up in the Endulen-Laetoli region during the Pliocene. As in the present day this would have brought about continuous change from grassland to woodland regardless of the climatic conditions then existing. In fact, it can be predicted that with rainfall higher than is present today, the temporal catena would change more rapidly as higher rainfall flushed out the salts deposited by the carbonatitic ashes and the soils formed on the tuffs matured more quickly. In contrast, with less rainfall, the temporal catena would change more slowly as soils matured more slowly, so that the earlier stages of bare ground and grassland would persist for longer.

Conclusions

An accurate survey of the modern vegetation in an area is essential before any analogies can be made with past vegetation. This is particularly important where vegetation reconstructions suggest that the Pliocene vegetation preserved in the Laetolil Beds was taxonomically similar to that of today, the conclusion based on pollen data and the evidence of fossil wood. Similarities with modern vegetation structure have also been found based on isotope data and ecomorphological traits of the Laetoli fossil mammals (Kingston and Harrison 2007; Bishop et al. 2011; Kingston 2011; Kaiser 2011; Hernesniemi et al. 2011).

The present study shows that today's vegetation, even though disturbed, is much more complex than the simple types used to date. Studies of mammalian ecomorphology are not able to reconstruct the vegetation to this level of refinement (but see Kovarovic and Andrews 2011). From our observations of the present flora, however, it is apparent that there has been a great deal of disturbance by the past and present people and their livestock. Weeds have been introduced, but in many cases we do not know what has been lost. The botanical fossil record may help in this respect.

We have shown that the vegetation composition in the Endulen-Laetoli area is variable spatially depending on soils and drainage. Fourteen vegetation associations have been mapped (Fig. 8.10) and are grouped into three vegetation catenas as follows:

- woodland-drainage catena dominated by woodlands ranging from low open woodland to closed tall woodland, increasing in density, canopy height and species richness with changes in soil and drainage;
- riverine catena dominated by bushland and woodland, ranging from dense bushland at the headwaters of the dry river channels to tall closed woodland or gallery forest depending on water availability lower down the channels;
- altitude catena, all woodlands but the species composition varying with increased rainfall at higher altitudes.

Two open grassland associations are limited to heavily disturbed areas dominated by weed species and a small area of edaphic grassland at the western margin of the study area.

There is also variability over time, although the scale of fluctuations is difficult to determine. Projections of increased or decreased rainfall and how this may affect the different environments form the basis for predictions about vegetation and habitat change in the past.

The spatial variability of the vegetation is important for the dynamics of the ecosystem because animals and plants have different levels of niche requirements. A variable vegetation structure provides a wide range of habitat availability for early hominins and an ecologically suitable corridors for them to travel through. The combination of evidence for bipedalism with adaptations for climbing in *A. afarensis* suggests that the hominins were able to take advantage of as great a range of vegetation types in the past as occur in the region today.

Acknowledgements We are grateful to Terry Harrison for his invitation to take part in his Laetoli expeditions and to William Mziray for loan of herbarium equipment and facilities for identifying the plant species. We have also profited from discussions with Rosa Albert, Raymonde Bonnefille, Charles Peters, Rob Blumenschine and Terry Harrison. Five reviewers have helped to improve the manuscript, and we thank them. This work was supported by NSF grants BCS-9903434 and BCS-0309513 (to Terry Harrison), a PAST (Palaeoanthropological Scientific Trust, South Africa) grant (to MB) and the Wenner-Gren Foundation for Anthropological Research (to PA).

Appendix 8.1 List of plants collected in 2004 with collection number (EN)[a], site and coordinates

Date	EN	Genus and species	Family	Site	GPS
25-Jun	532	*Aspilia mossambicensis* (Oliv.) Wild	Asteraceae	1	3°12'40"S 35° 13' 14"E
25-Jun	533	*Leonotis nepetifolia* R.Br	Labiatae	1	3°12'40"S 35° 13' 14"E
25-Jun	534	*Monechma debile* (Forsk.) Nees	Acanthaceae	1	3°12'40"S 35° 13' 14"E
25-Jun	535	*Achyranthes aspera* L.	Amaranthaceae	1	3°12'40"S 35° 13' 14"E
25-Jun	536	*Themeda triandra* Forssk.	Poaceae	1	3°12'40"S 35° 13' 14"E
25-Jun	537	*Asparagus africanus* Lam.	Liliaceae	1	3°12'40"S 35° 13' 14"E
25-Jun	538	*Abutilon longicuspe* A. Rich.	Malvaceae	1	3°12'40"S 35° 13' 14"E
25-Jun	539	*Cyathula orthacantha* Schinz.	Amaranthaceae	1	3°12'40"S 35° 13' 14"E
25-Jun	540	*Heracleum abyssinicum* (Boiss) Norman	Apiaceae	1	3°12'40"S 35° 13' 14"E
25-Jun	541	*Cyphostemma adenocaule* (A. Rich.) Wild & Drum	Vitaceae	1	3°12'40"S 35° 13' 14"E
25-Jun	542	*Bidens schimperi* Sch. Bip.	Asteraceae	1	3°12'40"S 35° 13' 14"E
25-Jun	543	*Hirpicium diffusum* (O. Hoffm.) Roess	Asteraceae	1	3°12'40"S 35° 13' 14"E

(continued)

Appendix 8.1 (continued)

Date	EN	Genus and species	Family	Site	GPS
25-Jun	544	*Gutenbergia polycephala* Oliv. & Hiern.	Compositae	1	3°12'40"S 35° 13' 14"E
25-Jun	545	*Setaria* sp.	Poaceae	1	3°12'40"S 35° 13' 14"E
25-Jun	546	*Chloris gayana* Kunth.	Poaceae	1	3°12'40"S 35° 13' 14"E
25-Jun	547	*Setaria pumila* (Poir) Roem & Schult	Poaceae	1	3°12'40"S 35° 13' 14"E
25-Jun	548	*Setaria verticillata* (L.) Beauv.	Poaceae	1	3°12'40"S 35° 13' 14"E
25-Jun	549	*Setaria incrassata* (Hochst.) Hack.	Poaceae	1	3°12'40"S 35° 13' 14"E
25-Jun	550	*Pycnostachys speciosa* Burke	Lamiaceae	1	3°12'40"S 35° 13' 14"E
25-Jun	551	*Setaria verticillata* (L.) P. Beauv.	Poaceae	1	3°12'40"S 35° 13' 14"E
25-Jun	552	*Panicum genuflexum* Stapf.	Poaceae	1	3°12'40"S 35° 13' 14"E
25-Jun	553	*Cyperus rotundus* (L.) subsp. *merkerii* (C. B. Clarke)	Cyperaceae	1	3°12'40"S 35° 13' 14"E
25-Jun	554	*Ocimum gratissimum* subsp. *gratissmium* L.	Lamiaceae	1	3°12'40"S 35° 13' 14"E
25-Jun	555	*Eriochloa meyerana* (Nees.) Pilg	Poaceae	1	3°12'40"S 35° 13' 14"E
25-Jun	556	*Agrostis* sp.	Poaceae	1	3°12'40"S 35° 13' 14"E
25-Jun	557	*Ziziphus mucronata* Willd.	Rhamnaceae	2	3°12'40"S 35° 13' 01"E
25-Jun	558	*Grewia kakothamnus* K.Schum	Tiliaceae	2	3°12'40"S 35° 13' 01"E
25-Jun	559	*Ximenia americana* L.	Olacaceae	2	3°12'40"S 35° 13' 01"E
25-Jun	560	*Cyathula orthacantha*	Amaranthaceae	2	3°12'40"S 35° 13' 01"E
25-Jun	561	*Rhus natalensis* Krauss.	Anacardiaceae	2	3°12'40"S 35° 13' 01"E
25-Jun	562	*Gomphocarpus semilunatus* A. Rich.	Asclepiadaceae	2	3°12'40"S 35° 13' 01"E
25-Jun	563	*Tagetes minuta* L.	Asteraceae	2	3°12'40"S 35° 13' 01"E
25-Jun	564	*Rhynchosia minima* (L.) DC. var. *minima*	Papilionaceae	2	3°12'40"S 35° 13' 01"E
25-Jun	565	*Celosia schweinfurthiana* Schinz.	Amaranthaceae	2	3°12'40"S 35° 13' 01"E
25-Jun	566	*Dregea* sp.	Asclepiadaceae	2	3°12'40"S 35° 13' 01"E
25-Jun	567	*Ipomoea* sp.	Convolvulaceae	2	3°12'40"S 35° 13' 01"E
25-Jun	568	*Vernonia galamensis* (Cass.) Less	Asteraceae	2	3°12'40"S 35° 13' 01"E
26-Jun	569	*Acacia nilotica* (L.) Del.	Mimosaceae	3	3°12'30"S 35° 13' 01"E
26-Jun	570	*Acacia senegal* L. Willd	Mimosaceae	3	3°12'30"S 35° 13' 01"E
26-Jun	571	*Helichrysum panduratum* O. Hoffm.	Asteraceae	3	3°12'30"S 35° 13' 01"E
26-Jun	572	*Conyza pyrrhopappa* Sch. Bip.	Asteraceae	3	3°12'30"S 35° 13' 01"E

(continued)

Appendix 8.1 (continued)

Date	EN	Genus and species	Family	Site	GPS
26-Jun	573	*Secamone parvifolia* (oliv.) Bullock	Asclepiadaceae	3	3°12'30"S 35° 13' 01"E
26-Jun	574	*Acacia kirkii* Oliv.	Mimosaceae	3	3°12'30"S 35° 13' 01"E
26-Jun	575	*Grewia kakothamnos* K. Schum	Tiliaceae	3	3°12'30"S 35° 13' 01"E
26-Jun	576	*Commiphora africana* L.	Burseraceae	3	3°12'30"S 35° 13' 01"E
26-Jun	577	*Balanites aegyptiaca* (L.) Del.	Balanitaceae	3	3°12'30"S 35° 13' 01"E
26-Jun	578	*Rhus vulgaris* Meikle	Anacardiaceae	3	3°12'30"S 35° 13' 01"E
26-Jun	579	*Combretum molle* G. Don	Combretaceae	3	3°12'30"S 35° 13' 01"E
26-Jun	580	*Azanza garckeana* (F.H.) Exell & Hill	Malvaceae	3	3°12'30"S 35° 13' 01"E
26-Jun	581	*Heteromorpha trifoliata* (Wendl.) Eckl. & Zeyh.	Apiaceae	3	3°12'30"S 35° 13' 01"E
26-Jun	582	*Flueggea virosa* (Willd.) Voigt.	Euphorbiaceae	3	3°12'30"S 35° 13' 01"E
26-Jun	583	*Pterocephalus frutescens* Hochst	Dipsacaceae	7	3°15'09"S 35° 11' 06"E
26-Jun	584	*Crossandra* sp.	Acanthaceae	7	3°15'09"S 35° 11' 06"E
26-Jun	585	*Lippia javanica* (Burm. F) Spreng.	Verbenaceae	7	3°15'09"S 35° 11' 06"E
26-Jun	586	*Hibiscus ludwigii* Eckl. & Zeyn.	Malvaceae	3	3°13'44"S 35° 10' 12"E
26-Jun	587	*Indigofera* sp.	Papilionaceae	2	3°12'40"S 35° 13' 01"E
26-Jun	588	*Capparis sepiaria* L.var. *fischeri*	Capparaceae	4	3°13'44"S 35° 10' 12"E
26-Jun	589	*Zanthoxyum chalybeum* (Engl.)	Rutaceae	4	3°13'44"S 35° 10' 12"E
26-Jun	590	*Acacia drepanolobium* Herms ex. Sjostedt.	Mimosaceae	4	3°13'44"S 35° 10' 12"E
26-Jun	591	*Aristida junciformis* Trin. & Rupr	Poaceae	4	3°13'44"S 35° 10' 12"E
26-Jun	592	*Digitaria* sp.	Poaceae	4	3°13'44"S 35° 10' 12"E
26-Jun	593	*Eragrostis* sp.	Poaceae	4	3°13'44"S 35° 10' 12"E
26-Jun	594	*Acacia tortilis* (Forsk) Hyne	Mimosaceae	4	3°13'44"S 35° 10' 12"E
26-Jun	595	*Setaria* sp.	Poaceae	4	3°13'44"S 35° 10' 12"E
26-Jun	596	*Solanum incanum* L.	Solanaceae	4	3°13'44"S 35° 10' 12"E
26-Jun	597	Woody herb	Acanthaceae	5	3°13'44"S 35° 10' 12"E
26-Jun	598	*Maytenus heterophylla* (Eckl. & Zeyh.) N. Robson	Celastraceae	5	3°13'44"S 35° 10' 12"E
26-Jun	599	*Crotolaria laburnifolia* L.	Papilionaceae	5	3°13'44"S 35° 10' 12"E
26-Jun	600	*Turbina stenosiphon* Hall. f.	Convolvulaceae	5	3°13'44"S 35° 10' 12"E
26-Jun	601	*Sphaeranthus steetzii* Oliv. & Hiern.	Asteraceae	5	3°13'44"S 35° 10' 12"E

(continued)

Appendix 8.1 (continued)

Date	EN	Genus and species	Family	Site	GPS
26-Jun	602	*Panicum* sp.	Poaceae	5	3°13'44"S 35° 10' 12"E
26-Jun	603	*Maerua angolensis* DC.	Capparaceae	5	3°13'44"S 35° 10' 07"E
26-Jun	604	*Cordia monoica* Roxb.	Boraginaceae	28	3°13'44"S 35° 10' 12"E
26-Jun	605	*Acacia nilotica* (L.) Del	Mimosaceae	28	3°13'44"S 35° 10' 12"E
26-Jun	606	*Acacia nilotica* (L.) Del	Mimosaceae	1	3°13'44"S 35° 10' 12"E
26-Jun	607	*Kalanchoe lanceolata* (Forssk.) Pers	Crassulaceae	15	3°12'39"S 35° 13' 14"E
26-Jun	608	*Euphorbia* sp.	Euphorbiaceae	14	3°10'35"S 35° 13' 14"E
27-Jun	609	*Maerua triphylla* A. Rich var. *johannis* (Valk. & Gilg.) de Wolf.	Capparaceae	6	3°15'18"S 35° 11' 22"E
27-Jun	610	*Dactylotenium aegyptium* L. Willd	Poaceae	6	3°15'18"S 35° 11' 22"E
27-Jun	611	*Eragrostis* sp.	Poaceae	6	3°15'18"S 35° 11' 22"E
27-Jun	612	*Eragrostis/Panicum*	Poaceae	6	3°15'18"S 35° 11' 22"E
27-Jun	613	*Justicia betonica* L.	Acanthaceae	6	3°15'18"S 35° 11' 22"E
27-Jun	614	*Capparis tomentosa* Lam.	Capparaceae	6	3°15'18"S 35° 11' 22"E
27-Jun	615	Succulent	Asteraceae	6	3°15'18"S 35° 11' 22"E
27-Jun	616	*Chenopodium opulifolium* Schrad.	Chenopodiaceae	6	3°15'18"S 35° 11' 22"E
27-Jun	617	*Heliotropium steudneri* Vatke subsp. *steudneri*	Boraginaceae	6	3°15'18"S 35° 11' 22"E
27-Jun	618	*Helinus integrifolius* (Lam.) Kuntze	Menispermaceae	6	3°15'18"S 35° 11' 22"E
27-Jun	619	Scandent succulent	Aizoaceae	6	3°15'18"S 35° 11' 22"E
27-Jun	620	*Maerua triphylla* A. Rich.	Capparaceae	6	3°15'18"S 35° 11' 22"E
27-Jun	621	*Pavonia* sp.	Malvaceae	7	3°15'09"S 35° 11' 06"E
27-Jun	622	Creeper	Fabaceae	7	3°15'09"S 35° 11' 06"E
27-Jun	623	*Convolvulus* sp.	Convolvulaceae	7	3°15'09"S 35° 11' 06"E
27-Jun	624	*Grewia* sp.	Tiliaceae	7	3°15'09"S 35° 11' 06"E
27-Jun	625	*Opilia amentacea* Roxb.	Opiliaceae	7	3°15'09"S 35° 11' 06"E
27-Jun	626	4–5 m tree	?	7	3°15'09"S 35° 11' 06"E
27-Jun	627	*Acacia mellifera* (Val.) Brenam subsp. *mellifera*	Mimosaceae	7	3°15'09"S 35° 11' 06"E
27-Jun	628	*Albizia amara* (Roxb) var *sericocephala* (Benth.) Brenan	Mimosaceae	7	3°15'09"S 35° 11' 06"E
28-Jun	629	*Acacia senegal* (L.) Willd	Mimosaceae	8	3°13'06"S 35° 08' 41"E
28-Jun	630	*Tragus berteronianus* Schult.	Poaceae	8	3°15'09"S 35° 11' 06"E

(continued)

Appendix 8.1 (continued)

Date	EN	Genus and species	Family	Site	GPS
28-Jun	631	*Chenopodium* sp.	Chenopodiaceae	8	3°15'09"S 35° 11' 06"E
28-Jun	632	*Duosperma quadrangulare* (Klotzsch) Brummit.	Acanthaceae	8	3°15'09"S 35° 11' 06"E
28-Jun	633	*Hibiscus surattensis* L.	Malvaceae	8	3°15'09"S 35° 11' 06"E
28-Jun	634	*Solanum setaceum* Dummer	Solanaceae	8	3°15'09"S 35° 11' 06"E
28-Jun	635	*Hirpicium* sp.	Asteraceae	8	3°15'09"S 35° 11' 06"E
28-Jun	636	*Ipomoea jaegeri* Pilger	Scrophulariaceae	8	3°15'09"S 35° 11' 06"E
28-Jun	637	*Notonia abyssinica* A. Rich.	Asteraceae	8	3°15'09"S 35° 11' 06"E
28-Jun	638	*Pavetta gardeniifolia* A. Rich var. *gardeniifolia*	Rubiaceae	9	3°13'06"S 35° 11' 50"E
28-Jun	639	*Sporobolus spicatus* (Vahl.) Kunth	Poaceae	9	3°13'06"S 35° 11' 50"E
28-Jun	640	*Kalanchoe lanceolata* (Forssk.) Pers.	Crassulaceae	9	3°13'06"S 35° 11' 50"E
28-Jun	641	*Hibiscus trionum* Liun.	Malvaceae	9	3°13'06"S 35° 11' 50"E
28-Jun	642	*Acacia hockii* De Wild.	Fabaceae	9	3°13'06"S 35° 11' 50"E
29-Jun	643	*Allophylus africanus* P. Beauv.	Anacardiaceae	10	3°13'04"S 35° 16' 06"E
29-Jun	644	*Vangueria madagascariensis* Gmel.	Rubiaceae	10	3°13'04"S 35° 16' 06"E
29-Jun	645	*Ocimum gratissimum* (L.) subsp. *gratissimum*	Lamiaceae	10	3°13'04"S 35° 16' 06"E
29-Jun	646	*Lepidotrichilia* sp.	Meliaceae	10	3°13'04"S 35° 16' 06"E
29-Jun	647	*Momordica foetida* Schumach.	Cucurbitaceae	10	3°13'04"S 35° 16' 06"E
29-Jun	648	*Adenostemma perrotteti* DC	Asteraceae	10	3°13'04"S 35° 16' 06"E
29-Jun	649	*Celtis africana* Burm.f.	Ulmaceae	10	3°13'04"S 35° 16' 06"E
29-Jun	650	*Vernonia exsertiflora* Baker	Asteraceae	10	3°13'04"S 35° 16' 06"E
29-Jun	651	*Croton macrostachyus* Del.	Euphorbiaceae	10	3°13'04"S 35° 16' 06"E
29-Jun	652	*Pavonia urens* Cav. var *tomentosa* (Ulbr) Brenan.	Malvaceae	10	3°13'04"S 35° 16' 06"E
29-Jun	653	*Urtica massaica* Mildbr.	Urticaceae	10	3°13'04"S 35° 16' 06"E
29-Jun	654	*Solanum nigrum* L.	Solanaceae	10	3°13'04"S 35° 16' 06"E
29-Jun	655	*Grewia similis* K. Schum.	Tiliaceae	10	3°13'04"S 35° 16' 06"E
29-Jun	656	*Clerodendron hildebrandtii* Vatke	Verbenaceae	10	3°13'04"S 35° 16' 06"E
29-Jun	657	*Pluchea ovalis* DC.	Asteraceae	10	3°13'04"S 35° 16' 06"E
29-Jun	658	*Rauvolfia caffra* var. *caffra* Sond.	Apocynaceae	10	3°13'04"S 35° 16' 06"E
29-Jun	659	*Ficus thonningii* Blume	Moraceae	10	3°13'04"S 35° 16' 06"E

(continued)

Appendix 8.1 (continued)

Date	EN	Genus and species	Family	Site	GPS
29-Jun	660	*Ficus vallis-chondae* Delile	Moraceae	10	3°13'04"S 35° 16' 06"E
29-Jun	661	*Tragia brevipes* Pax	Euphorbiaceae	10	3°13'04"S 35° 16' 06"E
29-Jun	662	*Cyperus involucratus* Rottb.	Cyperaceae	10	3°13'04"S 35° 16' 06"E
29-Jun	663	*Setaria homonyma* (Steud.) Chiov.	Poaceae	10	3°13'04"S 35° 16' 06"E
29-Jun	664	*Hypoestes* sp.	Acanthaceae	10	3°13'04"S 35° 16' 06"E
29-Jun	665	*Maytenus mossambicensis* (Klotzch.) Blakelock var. *ambonensis* (Loes.) Robson	Celastraceae	10	3°13'04"S 35° 16' 06"E
29-Jun	666	*Caesalpinia decapetala* (Roth.) Alston	Caesalpiniaceae	10	3°13'04"S 35° 16' 06"E
30-Jun	667	*Acalypha fruticosa* Forssk. var *villosa* Pax ex Hutch.	Euphorbiaceae	11	3°13'12"S 35° 15' 51"E
30-Jun	668	*Capparis tomentosa* Lam.	Capparaceae	11	3°13'12"S 35° 15' 51"E
30-Jun	669	*Buddleia polystachys* Fresen.	Loganiaceae	11	3°13'12"S 35° 15' 51"E
30-Jun	670	*Eucalyptus sideroxylon* Benth. (Exotic)	Myrtaceae	11	3°13'12"S 35° 15' 51"E
30-Jun	671	*Acacia kirkii* Oliv.	Mimosacaeae	11	3°13'12"S 35° 15' 51"E
30-Jun	672	*Dombeya acutangula* Cav.	Sterculiaceae	12	3°14'12"S 35° 14' 44"E
30-Jun	673	*Syzygium guineense* (Willd.) DC	Myrtaceae	12	3°14'12"S 35° 14' 44"E
30-Jun	674	*Sesbania sesban* (L.) Merrill	Papillionaceae	12	3°14'12"S 35° 14' 44"E
30-Jun	675	*Acacia xanthophloea* Benth.	Fabaceae	12	3°14'12"S 35° 14' 44"E
30-Jun	676	*Pennisetum thunbergii* Kunth	Poaceae	12	3°14'12"S 35° 14' 44"E
30-Jun	677	*Kigelia africana* (Lam.) Benth	Bignoniaceae	12	3°14'12"S 35° 14' 44"E
30-Jun	678	*Vitex doniana* Sweet.	Verbenaceae	12	3°14'12"S 35° 14' 44"E
30-Jun	679	*Combretum molle* G. Don	Combretaceae	12	3°14'12"S 35° 14' 44"E
30-Jun	680	*Senna singueana* (Del.) Lock	Caesalpiniaceae	12	3°14'12"S 35° 14' 44"E
30-Jun	681	*Dichrostachys cinerea* (L.) Wight & Arn.	Mimosaceae	12	3°14'12"S 35° 14' 44"E
30-Jun	682	*Dichrostachys cinerea* (L.) Wight & Arn.	Mimosaceae	12	3°14'12"S 35° 14' 44"E
30-Jun	683	*Commiphora africana* (A. Rich.) Engl.	Burseraceae	12	3°14'12"S 35° 14' 44"E
1-Jul	684	*Commiphora africana* (A. Rich.) Engl.	Burseraceae	13	3°18'38"S 35° 12' 28"E
1-Jul	685	*Euphorbia matabelensis* Pax.	Euphorbiaceae	13	3°18'38"S 35° 12' 28"E
1-Jul	686	*Kleinia kleinioides* (Sch. Bip) M.R.F. Taylor	Asteraceae	13	3°18'38"S 35° 12' 28"E
1-Jul	687	*Sporobolus* sp.	Poaceae	13	3°18'38"S 35° 12' 28"E
1-Jul	688	*Euphorbia* sp.	Euphorbiaceae	13	3°18'38"S 35° 12' 28"E

(continued)

Appendix 8.1 (continued)

Date	EN	Genus and species	Family	Site	GPS
1-Jul	689	*Ansella africana* Lindl.	Orchidaceae	13	3°18'38"S 35° 12' 28"E
1-Jul	690	*Barleria eranthemoides* C.B.Cl	Acanthaceae	13	3°18'38"S 35° 12' 28"E
1-Jul	691	*Eneapogon cenchroides*	Poaceae	13	3°18'38"S 35° 12' 28"E
1-Jul	692	*Phragmanthera* sp.	Loranthaceae	13	3°18'38"S 35° 12' 28"E
1-Jul	693	*Secamone parvifolia* Oliv. Bullock	Asclepiadaceae	13	3°18'38"S 35° 12' 28"E
1-Jul	694	*Maerua parvifolia* Pax.	Capparaceae	13	3°18'38"S 35° 12' 28"E
1-Jul	695	*Aloe macrosiphon* Bak.	Liliaceae	13	3°18'38"S 35° 12' 28"E
1-Jul	696	*Sporobolus pyramidalis* Beauv.	Poaceae	13	3°18'38"S 35° 12' 28"E
1-Jul	697	*Sansevieria kirkii* Baker Vel. Sp. att	Agavaceae	13	3°18'38"S 35° 12' 28"E
1-Jul	698	*Vigna parkeri* Bak.	Papillionaceae	13	3°18'38"S 35° 12' 28"E
1-Jul	699	*Plicosepalus sagittifolius* (Engl.) Danser	Loranthaceae	13	3°18'38"S 35° 12' 28"E
1-Jul	700	*Croton dichogamus* Pax.	Euphorbiaceae	13	3°18'38"S 35° 12' 28"E
1-Jul	701	*Albizia* sp.	Mimosaceae	13	3°18'38"S 35° 12' 28"E
1-Jul	702	*Euphorbia candelabrum* L.	Euphorbiaceae	13	3°18'38"S 35° 12' 28"E
1-Jul	703	*Euphorbia candelabrum* L.	Euphorbiaceae	13	3°18'38"S 35° 12' 28"E
1-Jul	704	*Euphorbia* sp.	Euphorbiaceae	13	3°18'38"S 35° 12' 28"E
2-Jul	705	*Senecio petitianus* A. Rich.	Asteraceae	14	3°10'35"S 35° 13' 14"E
2-Jul	706	*Kalanchoe lanceolata* (Forssk.) Pers	Crassulaceae	14	3°10'35"S 35° 13' 14"E
2-Jul	707	*Lilium* sp.	Liliaceae	14	3°10'35"S 35° 13' 14"E
2-Jul	708	*Osyris lanceolata* Hochst. & Steudal	Santalaceae	14	3°10'35"S 35° 13' 14"E
2-Jul	709	*Lannea schweinfurthii* J. Kokwaro	Anacardiaceae	14	3°10'35"S 35° 13' 14"E
2-Jul	710	*Oliverella hildabrandtii* (Engl.) Van Tiegh	Loranthaceae	14	3°10'35"S 35° 13' 14"E
2-Jul	711	*Carissa edulis* Brenan.	Apocynaceae	14	3°10'35"S 35° 13' 14"E
2-Jul	712	*Azanza gacrkeana* (F.) (Hoffm.) Exell & Hill.	Malvaceae	14	3°10'35"S 35° 13' 14"E
2-Jul	713	*Asplenium* sp.	Pteridaceae	14	3°10'35"S 35° 13' 14"E
2-Jul	714	*Euclea racemosa* Murray	Ebenaceae	14	3°10'35"S 35° 13' 14"E
2-Jul	715	*Rhus natalensis* Berrn. ex Krauss	Anacardiaceae	14	3°10'35"S 35° 13' 14"E
2-Jul	716	*Jasminum meyeri-johannis* Engl.	Oleaceae	14	3°10'35"S 35° 13' 14"E
2-Jul	717	*Rhoicissus tridentata* (L. f.) Wild & Drum.	Vitaceae	14	3°10'35"S 35° 13' 14"E

(continued)

Appendix 8.1 (continued)

Date	EN	Genus and species	Family	Site	GPS
2-Jul	718	*Pterolobium stellatum* (Forssk.) Brenan	Caesalpiniaceae	14	3°10'35"S 35° 13' 14"E
2-Jul	719	*Steganotaenia araliacea* Hochst.	Apiaceae	14	3°10'35"S 35° 13' 14"E
2-Jul	720	*Pappea capensis* Eckl. & Zey.	Sapindaceae	14	3°10'35"S 35° 13' 14"E
2-Jul	721	*Dombeya rotundifolia* L.	Sterculiaceae	14	3°10'35"S 35° 13' 14"E
2-Jul	722	*Vernonia adoensis* Walp.	Asteraceae	14	3°10'35"S 35° 13' 14"E
2-Jul	723	*Lannea schweinfurthii* J. Kokwaro	Anacardiaceae	14	3°10'35"S 35° 13' 14"E
2-Jul	724	*Tinnea aethiopica* Hook.f.	Lamiaceae	14	3°10'35"S 35° 13' 14"E
2-Jul	725	*Commelina benghalensis* L.	Commelinaceae	14	3°10'35"S 35° 13' 14"E
2-Jul	726	*Acacia seyal* L.	Mimosaceae	14	3°10'35"S 35° 13' 14"E
3-Jul	727	*Carsium buchwaldii* O. Hoffm	Asteraceae	15	3°12'39"S 35° 13' 14"E
3-Jul	728	*Leucas stormsii* Gürke	Lamiaceae	15	3°12'39"S 35° 13' 14"E
3-Jul	729	*Andropogon* sp.	Poaceae	15	3°12'39"S 35° 13' 14"E
3-Jul	730	*Sorghum verscolor* Anders.	Poaceae	15	3°12'39"S 35° 13' 14"E
3-Jul	731	*Setaria* sp.	Poaceae	15	3°12'39"S 35° 13' 14"E
3-Jul	732	*Hyparrhenia hirta* (L.) Stapf.	Poaceae	15	3°12'39"S 35° 13' 14"E
3-Jul	733	*Themeda triandra* Forssk.	Poaceae	15	3°12'39"S 35° 13' 14"E
3-Jul	734	*Chloris gayana* Kunth.	Poaceae	15	3°12'39"S 35° 13' 14"E
4-Jul	735	*Cyperus rigidifolius* Steud.	Cyperaceae	21	3°11'12"S 35° 22' 44"E
4-Jul	736	Herb	Campanulaceae	21	3°11'12"S 35° 22' 44"E
4-Jul	737	*Geranium arabicum* Forssk subsp. *arabicum*	Geraniaceae	21	3°11'12"S 35° 22' 44"E
4-Jul	738	*Ageratum conyzoides* L.	Asteraceae	21	3°11'12"S 35° 22' 44"E
4-Jul	739	*Gladiolus dalenii* Van. Geel	Iridaceae	21	3°11'12"S 35° 22' 44"E
4-Jul	740	*Lysimachia ruhmeriana* Vatke	Lamiaceae	21	3°11'12"S 35° 22' 44"E
4-Jul	741	*Nuxia congesta* Fres.	Loganiaceae	21	3°11'12"S 35° 22' 44"E
4-Jul	742	Grass	Poaceae	21	3°11'12"S 35° 22' 44"E
4-Jul	743	*Vernonia pteropoda* Oliv. & Hiern.	Asteraceae	21	3°11'12"S 35° 22' 44"E
4-Jul	744	*Clutia abyssinica* Jaub & Spach var. *abyssinica.*	Euphorbiaceae	21	3°11'12"S 35° 22' 44"E
4-Jul	745	*Crotalaria comanestiana* Volk & Schwein.f.	Papillionaceae	21	3°11'12"S 35° 22' 44"E
4-Jul	746	*Hibiscus urens* Cav.	Malvaceae	21	3°11'12"S 35° 22' 44"E

(continued)

Appendix 8.1 (continued)

Date	EN	Genus and species	Family	Site	GPS
4-Jul	747	*Euphorbia schimperiana* Scheele	Euphorbiaceae	21	3°11'12"S 35° 22' 44"E
4-Jul	748	*Bothriocline tomentosa* S. Moore	Asteraceae	21	3°11'12"S 35° 22' 44"E
4-Jul	749	*Otholobium foliosum* (Oliv.) Stirton	Papilionaceae	20	3°11'12"S 35° 22' 44"E
4-Jul	750	*Cynoglossum coeruleum* Hochst.	Boraginaceae	20	3°11'12"S 35° 22' 44"E
4-Jul	751	*Helichrysum setosum* Harv.	Asteraceae	20	3°11'12"S 35° 22' 44"E
4-Jul	752	*Conyza stricta* Willd.	Asteraceae	20	3°11'12"S 35° 22' 44"E
4-Jul	753	*Bromus unioloides* Kunth.	Poaceae	20	3°11'12"S 35° 22' 44"E
4-Jul	754	*Carduus nyassanus (S. Moore) R.E*	Asteraceae	20	3°11'12"S 35° 22' 44"E
4-Jul	755	*Gomphocarpus fruticosus* (L.) R.Br	Asclepiadaceae	20	3°11'12"S 35° 22' 44"E
4-Jul	756	*Trifolium semipilosum* Fresen.var. *semipilosum*	Papillionaceae	20	3°11'12"S 35° 22' 44"E
4-Jul	757	*Cycnium filicalyx* (E.A. Bruce) O.J.Hansen	Scrophulariaceae	20	3°11'12"S 35° 22' 44"E
4-Jul	758	*Themeda triandra* Forssk	Poaceae	20	3°11'12"S 35° 22' 44"E
4-Jul	759	*Hyparrhenia rufa* (Nees) Stapf	Poaceae	20	3°11'12"S 35° 22' 44"E
4-Jul	760	*Nepeta azurea* Benth	Lamiaceae	20	3°11'12"S 35° 22' 44"E
4-Jul	761	*Leonotis nepetifolia* R.Br	Lamiaceae	20	3°11'12"S 35° 22' 44"E
4-Jul	762	*Aristea alata* Bak	Iridaceae	20	3°11'12"S 35° 22' 44"E
4-Jul	763	*Eleusine jaegeri* Pilg.	Poaceae	20	3°11'12"S 35° 22' 44"E
4-Jul	764	*Bulbostylis* sp.	Cyperaceae	20	3°11'12"S 35° 22' 44"E
4-Jul	765	*Impatiens* sp.	Balsaminaceae	20	3°11'12"S 35° 22' 44"E
4-Jul	766	*Alepidea peduncularis* A. Rich	Asteraceae	20	3°11'12"S 35° 22' 44"E
4-Jul	767	*Eriosema montanum* Baker f.	Papilionaceae	16	3°11'12"S 35° 29' 44"E
4-Jul	768	*Rhynchosia minima* (L.) DC	Papilionaceae	16	3°11'12"S 35° 29' 44"E
4-Jul	769	*Hirpicium angustifolium* (O.Hoffm) Roess	Asteraceae	16	3°11'12"S 35° 29' 44"E
4-Jul	770	*Senecio subsessils* Oliv. & Hiern	Asteraceae	16	3°11'12"S 35° 29' 44"E
4-Jul	771	*Artemisia afra* Willd.	Asteraceae	16	3°11'12"S 35° 29' 44"E
4-Jul	772	*Anisoppus africanus* (Hook. f.) Oliv. & Hiern	Asteraceae	16	3°11'12"S 35° 29' 44"E
4-Jul	773	*Hypericum revolutum* Vahl.	Guttiferae	16	3°11'12"S 35° 29' 44"E
4-Jul	774	*Bothriocline tomentosa* S. Moore	Asteraceae	16	3°11'12"S 35° 29' 44"E
4-Jul	775	*Helichrysum odoratissimum* (L.) Less	Asteraceae	16	3°11'12"S 35° 29' 44"E

(continued)

Appendix 8.1 (continued)

Date	EN	Genus and species	Family	Site	GPS
4-Jul	776	*Veronica glandulosa* Benth.	Scrophulariaceae	16	3°11'12"S 35° 29' 44"E
4-Jul	777	*Satureja punctata* L.	Lamiaceae	16	3°11'12"S 35° 29' 44"E
4-Jul	778	*Guizotia scabra* (Vis.) Chiov.	Asteraceae	16	3°11'12"S 35° 29' 44"E
4-Jul	779	*Eriosema pauciflorum* Klotsch var. *pauciflorum*	Papilionaceae	16	3°11'12"S 35° 29' 44"E
4-Jul	780	*Cryptotaenia calycina* C.C.Townsend.	Apiaceae	16	3°11'12"S 35° 29' 44"E
4-Jul	781	*Chamaecrista hildebrandtii* Vateke	Caesalpiniaceae	16	3°11'12"S 35° 29' 44"E
4-Jul	782	*Agrostis product* Pilg.	Poaceae	16	3°11'12"S 35° 29' 44"E
4-Jul	783	*Festuca abyssinica* A. Rich	Poaceae	16	3°11'12"S 35° 29' 44"E
4-Jul	784	*Cycnium tubulosum* (L.f) Engl.	Scrophulariaceae	16	3°11'12"S 35° 29' 44"E
4-Jul	785	*Indigofera arrecta* A. Rich	Papilionaceae	16	3°11'12"S 35° 29' 44"E
4-Jul	786	*Teramnus* sp.	Papilionaceae	16	3°11'12"S 35° 29' 44"E
4-Jul	787	*Pennisetum sphacelatum* (Nees) Th.	Poaceae	17	3°11'07"S 35° 22' 41"E
4-Jul	788	*Themeda triandara* Forssk	Poaceae	17	3°11'07"S 35° 22' 41"E
4-Jul	789	*Hyperrhenia* sp.	Poaceae	17	3°11'07"S 35° 22' 41"E
4-Jul	790	*Satureja uhligii* Gurke	Lamiaceae	17	3°11'07"S 35° 22' 41"E
4-Jul	791	*Setaria* sp.	Poaceae	17	3°11'07"S 35° 22' 41"E
4-Jul	792	*Maytenus heterophylla* (Ecky. Zeyh)	Celastraceae	22	3°10'33"S 35° 22' 14"E
4-Jul	793	*Stephania abyssinica* (Dill. & A. Rich)	Menispermaceae	22	3°10'33"S 35° 22' 14"E
4-Jul	794	*Rhamnus prinioides* L"Hérit.	Rhamnaceae	22	3°10'33"S 35° 22' 14"E
4-Jul	795	*Impatiens pseudoviola* Gilg.	Balsaminaceae	22	3°10'33"S 35° 22' 14"E
4-Jul	796	*Achyranthes aspera* L.	Amaranthaceae	22	3°10'33"S 35° 22' 14"E
4-Jul	797	*Dombeya kirkii* Mast.	Sterculiaceae	22	3°10'33"S 35° 22' 14"E
4-Jul	798	*Sparrmannia ricinocarpa* (Eckl. & Zeyh.) Kuntze	Tiliaceae	22	3°10'33"S 35° 22' 14"E
4-Jul	799	*Leucas volkensii* var. *volkensii* Gurke	Lamiaceae	22	3°10'33"S 35° 22' 14"E
4-Jul	800	*Juniperus procera* Endl.	Cupressaceae	22	3°10'33"S 35° 22' 14"E
4-Jul	801	*Polystichum fusco-palencum* Alston	Aspleniaceae	22	3°10'33"S 35° 22' 14"E
4-Jul	802	*Galium brenanii* Ehrend & Verdc.	Rubiacea	22	3°10'33"S 35° 22' 14"E
4-Jul	803	*Snowdenia petitiana* (A.Rich) C.E.Hubn	Poaceae	22	3°10'33"S 35° 22' 14"E
4-Jul	804	*Solanum terminale* Forssk	Solanace	22	3°10'33"S 35° 22' 14"E
4-Jul	805	Orchid	Orchidaceae	17	3°11'07"S 35° 22' 41"E

(continued)

Appendix 8.1 (continued)

Date	EN	Genus and species	Family	Site	GPS
4-Jul	806	*Hagenia abyssinica* (Bruce) J. F. Gmel	Rosaceae	18/19	3°10'56"S 35° 22' 33"E
4-Jul	807	*Hypericum revolutum* Vahl.	Hypericaceae	18/19	3°10'56"S 35° 22' 33"E
4-Jul	808	*Kalanchoe aubrevillei* Cuf.	Crassulaceae	18/19	3°10'56"S 35° 22' 33"E
4-Jul	809	*Clutia abyssinica* Jaub & Spach.	Euphorbiaceae	18/19	3°10'56"S 35° 22' 33"E
4-Jul	810	*Rubus pinnatus* Willd.	Rosaceae	18/19	3°10'56"S 35° 22' 33"E
4-Jul	811	*Helichrysum splendidum* (Taub.) C.	Asteraceae	18/19	3°10'56"S 35° 22' 33"E
4-Jul	812	*Pycnostachys meyeri* Gurke	Labiatae	18/19	3°10'56"S 35° 22' 33"E
4-Jul	813	*Ranunculus multifidus*	Ranunculaceae	18/19	3°10'56"S 35° 22' 33"E
4-Jul	814	Herb	Papilionaceae	18/19	3°10'56"S 35° 22' 33"E
4-Jul	815	*Achyranthes aspera* L.	Amaranthaceae	18/19	3°10'56"S 35° 22' 33"E
4-Jul	816	*Nuxia congesta* Fres.	Aquifoliaceae	18/19	3°10'56"S 35° 22' 33"E
4-Jul	817	Grass	Poaceae	18/19	3°10'56"S 35° 22' 33"E
4-Jul	818	Shrub, 2 m, needle-like leaves	Ericaceae	18/19	3°10'56"S 35° 22' 33"E
4-Jul	819	*Psoralea foliosa* Oliv.	Papilionaceae	18/19	3°10'56"S 35° 22' 33"E
4-Jul	820	Fern	Pteridophytes	18/19	3°10'56"S 35° 22' 33"E
4-Jul	821	*Dovyalis* sp.	Flacourtiaceae	18/19	3°10'56"S 35° 22' 33"E
4-Jul	822	Woody climber	Vitaceae	18/19	3°10'56"S 35° 22' 33"E
6-Jul	823	*Chloris gayana* Kunth.	Poaceae	23	3°18'03"S 35° 11' 11"E
6-Jul	824	*Cyperus rotundus* L.	Cyperaceae	23	3°18'03"S 35° 11' 11"E
6-Jul	825	*Typha latifolia* L.	Typhaceae	23	3°18'03"S 35° 11' 11"E
6-Jul	826	*Scirpus aureiglumis* Hooper	Cyperaceae	23	3°18'03"S 35° 11' 11"E
6-Jul	827	*Acacia tortilis* (Forsk.) Hyne	Mimosaceae	24	3°18'49"S 35° 05' 30"E
6-Jul	828	*Tamarix nilotica* (Ehrenb.) Bunge	Tamaricaceae	25	3°21'30"S 35° 06' 06"E
6-Jul	829	*Acacia xanthophloea* Benth.	Mimosaceae	25	3°21'30"S 35° 06' 06"E
6-Jul	830	*Cucumis dipsaceus* Spach.	Cucurbitaceae	25	3°21'30"S 35° 06' 06"E
6-Jul	831	*Commiphora* sp.	Burseraceae	26	3°21'30"S 35° 06' 06"E
6-Jul	832	*Acalypha ornate* Hochst. ex. A. Rich.	Euphorbiaceae	27	3°21'30"S 35° 06' 06"E
6-Jul	833	*Cynodon dactylon* L.	Poaceae	27	3°21'30"S 35° 06' 06"E
6-Jul	834	*Crotolaria* sp.	Papilionaceae	27	3°21'30"S 35° 06' 06"E
6-Jul	835	*Hygrophyla auriculata* (Schum.)Heine	Acanthaceae	27	3°21'30"S 35° 06' 06"E

(continued)

Appendix 8.1 (continued)

Date	EN	Genus and species	Family	Site	GPS
6-Jul	836	*Ziziphus mucronata* Willd.	Rhamnaceae	27	3°21'30"S 35° 06' 06"E
6-Jul	837	*Panicum maximum* L.	Poaceae	27	3°21'30"S 35° 06' 06"E
7-Jul	838	*Schrebera alata* (Hochst.) Welw.	Oleaceae	28	3°13'00"S 35° 14' 06"E
7-Jul	839	*Pycnostachys deflexifolia* Baker	Labiatae	28	3°13'00"S 35° 14' 06"E
7-Jul	840	*Steganotaenia araliacea* Hochst.	Apiaceae	28	3°13'00"S 35° 14' 06"E
7-Jul	841	*Ficus lutea* Vahl.	Moraceae	28	3°13'00"S 35° 14' 06"E
7-Jul	842	*Erythrina abyssinica* DC.	Papilionaceae	28	3°13'00"S 35° 14' 06"E
8-Jul	843	*Ormocarpum trichocarpum* (Taub.) Engl.	Papilionaceae	29	3°12'30"S 35° 13' 20"E
7-Jul	844	*Heliotropium supinum* L.	Lamiaceae	29	3°12'30"S 35° 13' 20"E
7-Jul	845	*Albizia amara* (Roxb.) Boiv.	Mimosaceae	28	3°13'00"S 35° 14' 06"E
7-Jul	846	*Commiphora* sp.	Burseraceae	28	3°13'00"S 35° 14' 06"E
8-Jul	847	*Schrebera alata* (Hochst.) Welw.	Oleaceae	30	3°9'31"S 35° 16' 17"E
8-Jul	848	*Dombeya kirkii* Mast.	Sterculiaceae	30	3°9'31"S 35° 16' 17"E
8-Jul	849	*Albizia petersiana* subsp. *petersiana*	Mimosaceae	30	3°9'31"S 35° 16' 17"E
8-Jul	850	*Momordica foetida* Schumach.	Cucurbitaceae	30	3°9'31"S 35° 16' 17"E
8-Jul	851	*Celtis africana* Burm.f.	Ulmaceae	30	3°9'31"S 35° 16' 17"E
8-Jul	852	*Dombeya* sp.	Sterculiaceae	30	3°9'31"S 35° 16' 17"E
8-Jul	853	*Commiphora* sp.	Burseraceae	30	3°9'31"S 35° 16' 17"E
9-Jul	854	*Entada abyssinica* (Steud. ex A.Rich.)	Mimosaceae	31	3°12'40"S 35° 16' 01"E
9-Jul	855	*Vernonia exsertiflora* Bak.	Asteraceae	31	3°12'40"S 35° 16' 01"E
9-Jul	856	*Acacia senegal* L.	Fabaceae	31	3°12'40"S 35° 16' 01"E
9-Jul	857	*Clerodendrum myricoides* (Hochst.) Vatke	Verbenaceae	31	3°12'40"S 35° 16' 01"E
9-Jul	858	*Laggera alata* (D. Don) Oliv.	Asteraceae	31	3°12'40"S 35° 16' 01"E
9-Jul	859	*Pavetta dolichantha* Brem.	Rubiaceae	31	3°12'40"S 35° 16' 01"E
10-Jul	860	*Acacia lahai* Steud.	Mimosaceae	32	3°13'40"S 35° 18' 21"E
10-Jul	861	*Acacia robusta* Burch	Mimosaceae	32	3°13'40"S 35° 18' 21"E
10-Jul	862	*Senna didymobotrya* (Fresen.) Irwin & Barneby	Papilionaceae	32	3°13'40"S 35° 18' 21"E
10-Jul	863	*Ekebergia capensis* Sparrm.	Meliaceae	32	3°13'40"S 35° 18' 21"E
10-Jul	864	*Ficus cyathistipula* Warb.	Moraceae	32	3°13'40"S 35° 18' 21"E
10-Jul	865	*Mystroxylon aethiopicum* (Thunb.) Loes	Celastraceae	28	3°13'00"S 35° 14' 06"E

[a] Voucher number for the Arusha Herbarium, Tanzania

References

Alemseged, Z., Spoor, F., Kimbel, W. H., Bobe, R., Geraads, D., Reed, D., & Wynn, G. (2006). A juvenile early hominin skeleton from Dikika, Ethiopia. *Nature, 443*, 296–301.

Anderson, G. D., & Herlocker, D. (1973). Soil factors affecting the distribution of the vegetation types and their utilization by wild animals in Ngorongoro crater, Tanzania. *Journal of Ecology, 61*, 627–651.

Anderson, G. D., & Talbot, L. M. (1976). Soil factors affecting the distribution of grassland types and their utilization by wild animals on the Serengeti plains, Tanganyika. *Journal of Ecology, 53*, 33–55.

Andrews, P. (1989). Palaeoecology of Laetoli. *Journal of Human Evolution, 18*, 173–181.

Andrews, P., & Bamford, M. (2007). Past and present vegetation ecology of Laetoli, Tanzania. *Journal of Human Evolution, 58*, 78–98.

Andrews, P., & O'Brien, E. M. (2000). Climate, vegetation and predictable gradients in mammal species richness in southern Africa. *Journal of Zoology, 251*, 205–231.

Andrews, P., Groves, C., & Horne, J. (1975). The ecology of the Tana River flood plain. *Journal of the East African Natural History Society, 151*, 1–31.

Beentje, H. (1994). *Kenya trees, shrubs and lianas*. Nairobi: Kenya National Museum.

Bishop, L. C., Plummer, T. W., Hertel, F., & Kovarovic, K. (2011). Paleoenvironments of Laetoli, Tanzania as determined by antelope habitat preferences. In T. Harrison (Ed.), *Paleontology and geology of Laetoli: Human evolution in context* (Geology, geochronology, paleoecology and paleoenvironment, vol. 1, pp. 355–366). Dordrecht: Springer.

Boone, R. B., Thirgood, S. J., & Hopcroft, J. G. (2006). Serengeti wildebeest migratory patterns modeled from rainfall and new vegetation growth. *Ecology, 87*, 1987–1994.

Burtt, B. D. (1942). Burtt memorial supplement: some east african vegetation communities. *Journal of Ecology, 30*, 65–146.

Dale, I. R., & Greenway, P. J. (1961). *Kenya trees and shrubs*. Nairobi: Buchanan's Kenya Estates.

De Wit, H. A. (1978). Soils and grassland types of the Serengeti Plain (Tanzania): Their distribution and interrelations. Ph.D. dissertation, Agricultural University of Wageningen, Wageningen.

Dublin, H. T. (1991). Dynamics of the Serengeti-Mara woodlands. *Forest Conservation and History, 35*, 169–178.

Field, R., O'Brien, E. M., & Whittaker, R. J. (2005). Global models for predicting the climatic potential for (tree-shrub) plant richness: Development and evaluation. *Ecology, 86*, 2263–2277.

Grunblatt, J., Ottichilo, W. K., & Sinange, R. K. (1989). A hierarchical approach to vegetation classification in Kenya. *African Journal of Ecology, 27*, 45–51.

Hawkins, B. A., Montoya, D., Rodriguez, M. A., Olalla-Tarraga, M. A., & Zavala, M. A. (2006). Global models for predicting woody plant richness from climate: Comment. *Ecology, 87*, 255–259.

Hay, R. L. (1976). *Geology of the Olduvai Gorge*. Berkeley: University of California.

Hay, R. L. (1987). Geology of the Laetolil Beds. In M. D. Leakey, & J. M. Harris (Eds.), *Laetoli: A Pliocene site in northern Tanzania* (pp. 23–61). Oxford: Clarendon.

Herlocker, D., & Dirschl, H. J. (1972). Vegetation of the Ngorongoro Conservation Area, Tanzania. *Canadian Wildlife Services Report Series, 19*, 1–39.

Hernesniemi, E., Giaourtsakis, I., Evans, A. R. & Fortelius, M. (2011). Rhinocerotidae. In T. Harrison (Ed.), *Paleontology and geology of Laetoli: Human evolution in context* (Fossil hominins and the associated fauna, vol. 2, pp. 275–293). Dordrecht: Springer.

Jager, T. (1982). Soils of the Serengeti Woodlands, Tanzania. Ph.D. dissertation, Agricultural University of Wageningen, Wageningen.

Johanson, D., White, T., & Coppens, Y. (1978). A new species of the genus *Australopithecus* (Primates: Hominidae) from the Pliocene of eastern Africa. *Kirtlandia, 28*, 1–14.

Kaiser, T. M. (2011). Feeding ecology and niche partitioning of the Laetoli ungulate faunas. In T. Harrison (Ed.), *Paleontology and geology of Laetoli: Human evolution in context* (geology, geochronology, paleoecology and paleoenvironment, vol. 1, pp. 329–354). Dordrecht: Springer.

Kingston, J. D. (2011). Stable isotopic analyses of Laetoli fossil herbivores. In T. Harrison (Ed.), *Paleontology and geology of Laetoli: Human evolution in context* (Geology, geochronology, paleoecology and paleoenvironment, vol. 1, pp. 293–328). Dordrecht: Springer.

Kingston, J. D., & Harrison, T. (2007). Isotopic dietary reconstructions of pliocene herbivores at Laetoli: Implications for early hominin paleoecology. *Palaeogeography, Palaeoclimatology, Palaeoecology, 243*, 272–306.

Kovarovic, K., & Andrews, P. (2011). Environmental change within the Laetoli fossiliferous sequence: vegetation catenas and bovid ecomorphology. In T. Harrison (Ed.), *Paleontology and geology of Laetoli: Human evolution in context* (Geology, geochronology, paleoecology and paleoenvironment, vol. 1, pp. 367–380). Dordrecht: Springer.

Lamprey, R. H., & Reid, R. S. (2004). Expansion of human settlement in Kenya's Maasai Mara: What future for pastoralism and wildlife? *Journal of Biogeography, 31*, 997–1032.

Leakey, M. D., & Harris, J. M. (Eds.). (1987). *Laetoli: A Pliocene site in northern Tanzania*. Oxford: Clarendon.

Leakey, M. D., & Hay, R. (1979). Pliocene footprints in the Laetolil Beds at Laetoli, northern Tanzania. *Nature, 278*, 317–323.

Lind, E. M., & Morrison, M. E. S. (1974). *East African vegetation*. London: Longman.

Metzger, K. L., Coughenour, M. B., Reich, R. M., & Boone, R. B. (2005). Effects of seasonal grazing on plant species diversity and vegetation structure in a semi-arid ecosystem. *Journal of Arid Environments, 61*, 147–160.

Michelmore, A. P. G. (1939). Observations on tropical African grasslands. *Journal of Ecology, 27*, 282–312.

Musiba, C. M. (1999). Laetoli Pliocene paleoecology: A reanalysis via morphological and behavioral approaches. Ph.D. dissertation, University of Chicago, Chicago.

Norton-Griffiths, M., Herlocker, D., & Pennycuick, L. (1975). The patterns of rainfall in the Serengeti ecosystem, Tanzania. *East African Wildlife Journal, 13*, 347–375.

O'Brien, E. M. (1993). Climatic gradients in woody plant species richness: Towards an explanation based on an analysis of southern Africa's woody flora. *Journal of Biogeography, 20*, 181–198.

O'Brien, E. M., & Peters, C. R. (1999). Climatic perspectives for Neogene environmental reconstructions. In J. Agusti, L. Rook, & P. J. Andrews (Eds.), *Hominoid evolution and climatic change in Europe* (The evolution of Neogene terrestrial ecosystems in Europe, vol. 1, pp. 53–78). Cambridge: Cambridge University.

O'Brien, E. M., Whittaker, R. J., & Fields, R. (1998). Climate and woody plant diversity in southern Africa: Relationships at species, genus and family levels. *Ecography, 21*, 495–509.

O'Brien, E. M., Whittaker, R. J., & Fields, R. (2000). Climatic gradients in woody plant (tree and shrub) diversity: Water-energy dynamics, residual variation, and topography. *Oikos, 89*, 588–600.

Pratt, D. J., Greenway, P. J., & Gwynne, M. D. (1966). A classification of east african rangeland. *Journal of Applied Ecology, 3*, 369–382.

Reed, K. E. (1997). Early hominid evolution and ecological change through the African Plio-Pleistocene. *Journal of Human Evolution, 32*, 289–322.

Retallack, G. J. (1990). *Soils of the past: An introduction to Paleopedology*. Boston: Unwin Hyman.

Schmidt, W. (1975). The vegetation of the northeastern Serengeti National Park, Tanzania. *Phytocoenologia, 3*, 30–82.

Sinclair, A. R. E., & Norton-Griffiths, M. (1979). *Serengeti, dynamics of an ecosystem*. Chicago: University of Chicago.

Stern, J. T., & Susman, R. L. (1983). The locomotor anatomy of *Australopithecus afarensis*. *American Journal of Physical Anthropology, 60*, 279–317.

Su, D. F., & Harrison, T. (2007). The paleoecology of the Upper Laetolil Beds at Laetoli. In R. Bobe, Z. Alemseged, & A.K. Behrensmeyer (Eds.), Hominin environments in the East African Pliocene: An assessment of the faunal evidence (pp. 277–311). Dordrecht: Springer.

Trapnell, C. G., & Griffiths, J. F. (1960). The rainfall–altitude relation and its ecological significance in Kenya. *East African Agricultural Journal, 25*, 207–213.

Watson, R. M. (1967). The population ecology of the wildebeest in the Serengeti. Ph.D. dissertation, University of Cambridge, Cambridge.

White, F. (1983). *The Vegetation of Africa*. Paris: AETFAT/UNESCO.

Woodward, F. I. (1987). *Climate and plant distribution*. Cambridge: Cambridge University.

Chapter 9
Phytoliths and Pollen, the Microscopic Plant Remains in Pliocene Volcanic Sediments Around Laetoli, Tanzania

Lloyd Rossouw and Louis Scott

Abstract We analyzed sediment samples collected from several localities at different stratigraphic levels at Laetoli (i.e., Lower Laetolil Beds [LLB], Upper Laetolil Beds [ULB] and the overlying Upper Ndolanya Beds [UNB]) to establish a record of vegetation succession spanning intermittent periods between 4.3 and 2.66 Ma during the Pliocene. No reliable pollen spectra were found, but phytoliths, especially those of grasses (Poaceae), were investigated. A considerable time interval of deposition for the sequence, combined with a relatively low sample yield, allowed us to present only a low-resolution sequence of environmental changes, but one with marked grass cover variation. Grass was a ubiquitous, but never a dominant vegetation component in the LLB, ULB and the UNB sequences, with a general succession from mainly C_3 grass types in the LLB and older ULB levels to more C_4 grass types in the younger ULB and UNB. The record lends support to fossil herbivore analyses and $\delta^{13}C$ isotope studies, which suggest more heterogeneous habitats and a combination of C_3/C_4 grassland conditions in the ULB and UNB sequences (Andrews 1989; Kingston and Harrison 2007; Kovarovic and Andrews 2007). Productive samples suggest wet, C_3 conditions in the LLB and potentially dry, C_3 conditions in the lower part of the ULB. A shift from drier to more mesic C_4 grass conditions is recorded in the upper part of the ULB. Arid C_4 grassland environments occurred during UNB deposition.

Keywords Grass silica • Palynology • Pollen preservation • Past environment • Vegetation change

L. Rossouw (✉)
Department of Archaeology, National Museum, Bloemfontein, 9300, South Africa
e-mail: lloyd@nasmus.co.za

L. Scott
Department of Plant Sciences, University of the Free State, P.O. Box 339, Bloemfontein, South Africa
e-mail: scottl.sci@ufs.ac.za

Introduction

Reconstructions of past environments associated with fossil faunas can benefit from the availability of macro- and microscopic plant remains. In the case of microscopic remains, however, pollen-bearing sediments are scarce in most dry to semi-dry continental regions because preservation of fossil pollen grains and spores are normally restricted to rare anaerobic lake and swamp deposits. Abundant fossil faunal remains usually do not occur together with pollen in the same layers and often occur in exposures of sediment types like volcanic ash, soils and cave deposits (Klein 1999), which are not ideal for pollen preservation. Consequently, it is not easy to relate paleontological remains to past vegetation types and climate. Raymonde Bonnefille, who pioneered palynological research of East African hominin deposits, had mixed success at different sites, but found enough pollen for palynological analysis in only ca. 10% of samples from several sites including Laetoli, Hadar, Gadeb, Omo, East Turkana and Olduvai (Bonnefille 1977; Bonnefille and Vincens 1985; Bonnefille and Riollet 1987; Bonnefille et al. 2004). Only 11 out of 40 samples of sediments and soils from the volcanic substrates at Laetoli yielded pollen, of which the productive ones were secondary clays and termitaria (Bonnefille and Riollet 1987). Bonnefille and Riolet's results from Laetoli were, nevertheless, important in helping to establish the generally adopted view that the Pliocene vegetation in the Laetoli area was essentially similar to the current savanna and grassland (Bonnefille and Riollet 1987). C_4 grasses were well-established in East Africa during the deposition period of the Laetoli sediments, but may still not have been developed as fully as at present (Ségalen et al. 2007). We attempt to repeat Bonnefille's findings, by studying several samples from different sediment types including ashes, soils, as well as biogenic materials, such as termite mounds and bee broods. We also searched for phytolith microfossils in these sediments in order to complement the pollen study.

Our studies pertain to the pollen and phytolith contents of the fossiliferous Pliocene sediments at Laetoli. The area is situated on the Eyasi Plateau, which ranges in elevation from

T. Harrison (ed.), *Paleontology and Geology of Laetoli: Human Evolution in Context. Volume 1: Geology, Geochronology, Paleoecology and Paleoenvironment*, Vertebrate Paleobiology and Paleoanthropology, DOI 10.1007/978-90-481-9956-3_9,

800 m above sea level in the southwest to 2,500 m above sea level in the northeast (Hay 1987). The age of the sediments from the Laetolil Beds span the period between 4.4 and 3.6 Ma with the ULB sediments deposited over a 200 kyrs interval (Drake and Curtis 1987; Hay 1987; Kingston and Harrison 2007; Deino 2011). The Upper Ndolanya Beds were deposited a million years later, dating to 2.66 Ma (Kingston and Harrison 2007; Kovarovic and Andrews 2007; Deino 2011).

Materials and Methods

We collected separate sets of samples for pollen and phytolith analyses during different field seasons. Both sets of samples generally cover the Lower and Upper Laetolil Beds and the Upper Ndolanya Beds, but do not represent a high-resolution record of the sequence. We do not have close dating control of the samples and expect that considerable temporal gaps may exist between them. Sediment and modern soil samples were collected from the Eyasi Plateau at Laetoli (Localities 1, 3, 4, 5, 6, 7E, 8, 9, 9S,10E, 10W, 12 15, 16, 18, 21, 22S, 22E), Kakesio (Localities 1, 2, 3, 4, 6, 8), Endulen, Noiti (1, 3), Naisuri, Gadjingero River, Naibadad, Oldonyo Mati Hills and Lemagurut. Localities are mapped in Harrison and Kweka (2011) and the geology of the area is described in Ditchfield and Harrison (2011).

Pollen

The pollen samples collected comprised ash layers, soils, termite mounds, bee broods and various modern samples (Table 9.1). There are 42 sediment samples, numerous fossil bee broods and 17 modern samples of various types. In the pollen study we gave special attention to two thin, soft brown clay layers from Loc. 9S and termitaria from Loc. 10W, because Bonnefille and Riollet (1987) recorded pollen in similar sediments from the area. The two soft brown clays from Loc. 9S that we collected showed signs of invertebrate bioturbation. Standard cleaning of all exposure surfaces to at least 10 cm depth was necessary to obtain sediment that was undisturbed or disturbed as little as possible. We cleaned surfaces of the more consolidated materials, like termitaria samples, and consolidated clay slabs in the laboratory but this was not possible for soft crumbly material, like the brown soils.

The general method for pollen extraction included crushing of samples and digestion in acids, HF and HCl, followed by mineral separation with ZnCl2 (specific gravity = 2). We mounted residues on microscope slides in glycerine jelly.

A small degree of modern dust contamination can potentially play a major role in biasing pollen-free or pollen-poor samples (Scott and Bonnefille 1986; Scott 1995; Carrión and Scott 1999). We attempted to identify any contaminants by their fresh appearance and therefore excluded the generally used method of acetolysis (Erdtman 1960), which makes them indistinguishable.

Phytoliths

Opal phytoliths produced by the grass family are effective aids in the reconstruction of paleo-grassland ecology and in acquiring paleoenvironmental and archaeobotanical data in North America (Twiss et al. 1969; Piperno 1988; Mulholland and Rapp 1992; Fredlund and Tieszen 1994; Stromberg

Table 9.1 List of samples used for pollen analysis

Unit	Localities	Material	Samples
Ndolanya Beds	1, 7E, 18, 21, 22S	Soil horizons	13
	18	Root fill	1
	18	Bee broods	3 sets
	15	Biogenic pellet	1
Upper Laetolil Beds	9S, 3, 22E	Clays	11
	8	Bioturbated ash	1
	10W	Termite mounds	4
	1, 3, 6, 12	Bee broods	5 sets
Lower Laetolil Beds	Kakesio, 1, 2–4, 6	Bee broods	3 sets
	Kakesio 2	Calcareous ash	1
	Kakesio 2	Termite mounds	3
	Kakesio 2, Noiti	Clays	4
	Noiti	Tuffs	3
Modern samples	10W, Endulen, Naisuri, Naibadad, Camp	Modern dust scrapings	7
	4, 22, 22S Gadjingero River, Kakesio,	Hyena dung	6
	10W	Other dung	1
	10W, 22S	Owl pellets	2 sets
	7E	Acacia gum	1

2002, 2004), northern and central Africa (Alexandre et al. 1997; Barboni et al. 1999, 2007; Bremond et al. 2005) and in East and South Africa (McLean and Scott 1999; Wooller et al. 2000; Scott 2002; Grab et al. 2005; Scott and Rossouw 2005; Henderson et al. 2006). We restricted the study to the diagnostic phytoliths that originate from the short cells of the grass epidermis (Fig. 9.1). The aim was to establish what vegetation and climatic conditions short-cell grass silica bodies indicate for the Laetoli sequence according to the presence of C_3 and C_4 grasses (Ellis 1977; Vogel et al. 1978; Ellis et al. 1980; Gibbs Russell 1988). Previous studies have shown that the distribution of grass taxa into climatic regions and different environmental conditions correlates with photosynthetic type (Hattersley 1983; Gibbs Russell 1988). Several environmental parameters affect the ratio of C_3 and C_4 plant distribution, favoring C_4 plants under conditions of low atmospheric CO_2 levels, elevated growing temperatures, and increased aridity (Ehleringer et al. 1991, 1997; Sage 2004).

Grass short-cell phytoliths are classified according to the definitions of Twiss et al. (1969), Brown (1984), Mulholland (1989) Fredlund and Tieszen (1994), as well as the descriptions provided by the International Code for Phytolith Nomenclature (Madella et al. 2005).

Fifty-seven sediment samples made up of ash layers, soils, termite mounds and bee brood pellets were collected for phytolith analysis (Table 9.2). The sediment samples were in consolidated blocks to minimize potential contamination. Modern soil samples were collected from the surrounding area including the secondary grasslands on Lemagurut at altitudes between 2,600 and 2,800 m above sea level (Andrews and Bamford 2008), the lower-lying Oldonyo Mati hills and open grassland patches dispersed among the fossil localities at altitudes between 1,700 and 1,800 m, where black, argillaceous soils known as "mbuga" clays, occur. The procedure for recovering phytoliths from sediment samples is based on published techniques (Piperno 1988; Lentfer and Boyd 1998, 1999; Albert and Weiner 2001; Horrocks 2005).

Approximately 20 g of each sample were used for each phytolith extraction. Unlike in the case of pollen samples, we did not use highly corrosive HF that is damaging to the phytoliths. Essential steps included deflocculation, removal of clays by means of sedimentation, and the elimination of carbonates using HCl in low concentration (10%). Phytolith extraction involved mineral separation with a heavy liquid solution of sodium polytungstate (specific gravity = 2.3). Fractions were mounted on microscope slides in glycerin jelly. The phytolith counts are based on systematic scanning following standardized transects for each slide and identification at X500 magnification. Phytolith counts were normalized as percentages of the short-cell sum for each sample, based on the total count of all siliceous microfossils per slide (Fig. 9.4). Phytoliths that could not be classified were also counted. Principal Components Analysis (PCA) of phytolith counts was preformed to aid interpretations.

Application of Modern Phytolith Distribution

A survey of the occurrence of phytoliths in the leaf epidermis of South African grasses shows that deviations from the basic categories do occur (Table 9.3, Fig. 9.2; Rossouw, unpublished data). The mutual relationships between phytoliths and grass taxa are complicated by the production of either a particular phytolith morphotype by many taxa, or a variety of morphotypes by one taxon (Rovner 1971). However, the overall relationship between phytolith distribution and grass ecology appears to be influenced most by photosynthetic variation, growing temperature, and moisture availability. The lobate form, including the bilobate, polylobate and the cross are produced by several subfamilies, but predominantly by the Panicoideae subfamily, which include mostly C_4 grasses adapted to warm growing temperatures and high available soil moisture (Tieszen et al. 1979; Gibbs Russell 1988; Scott 2002). Saddle-shaped types are almost exclusively restricted to C_4 grasses concentrated in the chloridoid subfamily (Twiss et al. 1969; Brown 1984; Mulholland 1989; Fredlund and Tieszen 1994), but also occur in other arid-adapted subfamilies, such as the Aristidoideae and Arundinoideae. Reniform, trapeziform and rondel types are typically common within C_3 grasses reaching their maximum diversity in temperate or moist habitats with cool growing seasons, e.g., winter-rainfall areas (Gibbs Russell 1988, Table 9.3). We consider some of these factors in the interpretation of the fossil phytolith results, and also the role of past atmospheric concentrations of CO_2.

Principal Components Analysis (PCA) for selected short-cell grass phytoliths from South Africa reflects the general relationships between modern short-cell grass phytolith assemblages and grassland conditions. It ordinates data based on the covariance matrix of phytolith percentage between the biogeogaphical distribution of 250 grass species (Gibbs Russell 1985) and seven selected short-cell morphotypes (Fig. 9.3; Rossouw, unpublished data). Although the dataset is from a completely different region, viz. southern Africa, it is included with the intent to demonstrate the extent to which grass short-cell morphotypes can potentially reflect environmental variability and grassland composition. Alignment of the South African short-cell morphotypes in the PCA provides a proxy for climate variability as reflected by the occurrence of C_3 and C_4 grasses.

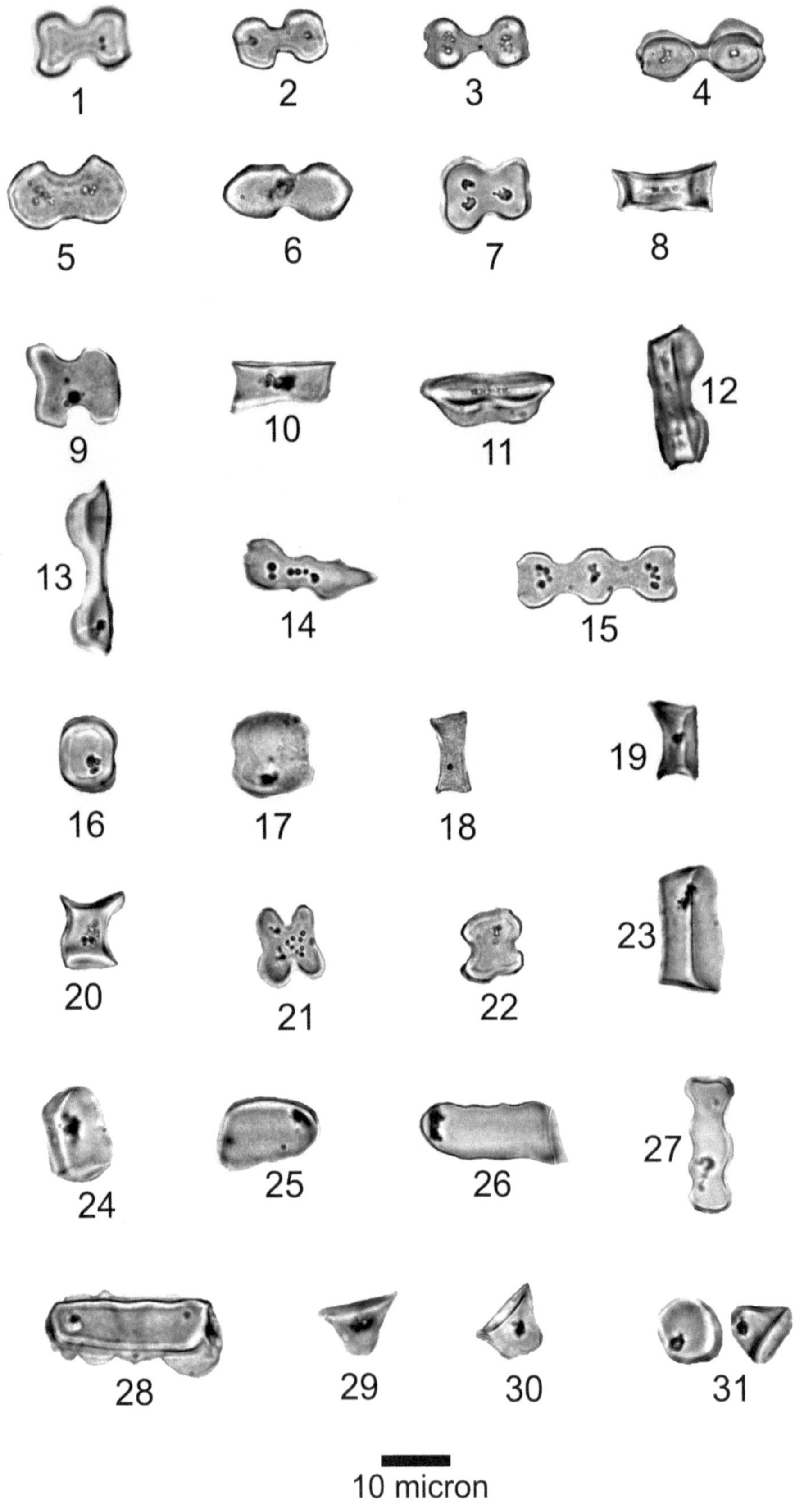

Fig. 9.1 Examples of diagnostic short-cell grass phytoliths selected from modern grass reference collection. 1–13, Bilobate type (8, 10–12 and 13 = rotated); 14, Irregular bilobate type; 15, Polylobate type; 16–20, Saddle type (18–20 = rotated); 21, 22, Cross type; 23–28, Trapeziform type; 29–31, Rondel type

Table 9.2 Occurrence of grass short-cell phytoliths in seven South African grass subfamilies (percentage of species)

Subfamily	Number of species	% C_3 species	Bilobate	Irregular Bilobate	Polylobate	Cross	Saddle	Rondel	Reniform	Trapeziform	Bilobate+ Cross+ Saddle	Polylobate+ Rondel+ Reniform+ Trapeziform
Aristidoideae	9	0.0	88.9	11.1	0.0	0.0	11.1	11.1	0.0	0.0	81.83	18.17
Arundinoideae	27	3.7	37.0	3.7	3.7	7.4	18.5	37.0	14.8	11.1	47.22	52.78
Chloridoideae	60	1.7	20.0	1.7	0.0	3.3	81.7	0.0	0.0	1.7	96.86	3.14
Danthionoideae	68	100.0	54.4	20.6	17.6	7.4	1.5	23.5	22.1	23.5	37.10	62.90
Ehrhartoideae	17	100.0	64.7	70.6	5.9	11.8	5.9	47.1	47.1	41.2	28.00	72.00
Panicoideae	58	6.9	100.0	15.5	25.9	15.5	1.7	1.7	0.0	10.3	68.70	31.30
Pooideae	30	100.0	23.3	23.3	20.0	3.3	0.0	23.3	10.0	90.0	13.77	86.23

Table 9.3 List of samples used for phytolith analysis. Productive samples are marked with an asterisk

Unit	Sample number	Stratigraphic level	Locality	Material
Upper Ndolanya Beds	180*		18	Ash/Soil
	182*		18	Ash/Soil
	183		18	Ash/Soil
	232		18	Bee Brood
Upper Laetolil Beds	126*	35 m below Tuff 6 (below Tuff 1)	1	Calcareous Ash
	127	15 m below Tuff 6 (between Tuffs 3 and 4)	1	Calcareous Ash
	128*	Below Tuff 8	1	Calcareous Ash
	129*	5 m below Tuff 6	1	Calcareous Ash
	130*	10 m below Tuff 6	1	Calcareous Ash
	131	Tuff 8	1	Calcareous Ash
	132	Tuff 6	1	Calcareous Ash
	134*	Yellow Marker Tuff	1	Calcareous Ash
	142*	Between Tuffs 7 & 8	3	Calcareous Ash
	145*	Between Tuffs 7 & 8	3	Calcareous Ash
	147*	Between Tuffs 7 & 8	3	Calcareous Ash
	148*	Below Tuff 7	3	Calcareous Ash
	149	Between Tuffs 7 & 8	3	Calcareous Ash
	152	Tuff 7	10E	Calcareous Ash
	153	Tuff 7	10E	Calcareous Ash
	155*	Below Tuff 7	10E	Calcareous Ash
	157*	Yellow Marker Tuff	10E	Calcareous Ash
	158	Tuff 6	10E	Calcareous Ash
	168	Tuff 2	10W	Calcareous Ash
	170	Tuff 3	10W	Calcareous Ash
	171*	Between Tuffs 1 & 2	10W	Calcareous Ash
	172*	Above Tuff 6	1	Calcareous Ash
	174	Below Tuff 7	1	Calcareous Ash
	208	Between Tuffs 7 & 8	1	Bee Brood
	209	Between Tuffs 7 & 8	3	Bee Brood
	210	Between Tuffs 5 & 7	9	Bee Brood
	211	Between Tuffs 3 & 5	5	Bee Brood
	212	Between Tuffs 7 & 8	16	Bee Brood
	214	Below Tuff 5	5	Calcareous Ash
	215	Above xenolith layer (between Tuffs 3 & 4)	5	Calcareous Ash
Lower Laetolil Beds	177*		Noiti 3	Below laminated silt
	179*		Noiti 3	Fossil plant debris layer
	213		Noiti 1	Fossil woodbearing Lahar
	185		Kakesio 8	Calcareous Ash
	186		Kakesio 8	Calcareous Ash
	187		Kakesio 8	Calcareous Ash
	206		Kakesio 8	Termitary
Modern Soil	277		Open Grassland	Mbuga Clay
	278*		Oldonyo Mati Hills	Topsoil
	279*		Lemagurut	Topsoil
	280*		Lemagurut	Topsoil

Results

Pollen

As expected, the slide preparations of the modern pollen samples from the current surroundings were rich in pollen, but this was not the case in the fossil material. The most prominent types in the modern samples were Poaceae, Asteraceae, Chenopodiaceae, Acanthaceae and lower proportion of arboreal types, varying in proportions in the different samples. Bonnefille and Riollet (1987) provide detailed data on modern assemblages for the study area. In contrast, our fossil materials either contained no or very low numbers of pollen. Unlike in rich fossil pollen spectra, where hard-to-avoid background contamination is negligible, it is problematical in our pollen-poor samples.

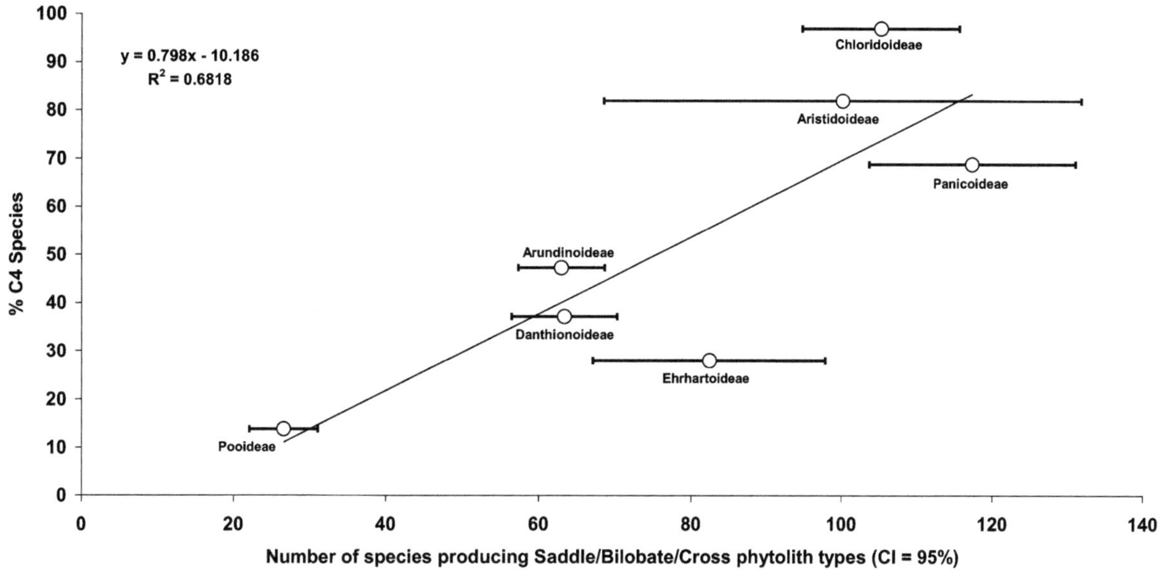

Fig. 9.2 Bivariate plot to demonstrate the relationship between Saddle, Bilobate and Cross types and C_4 species grouped within southern African grass subfamilies

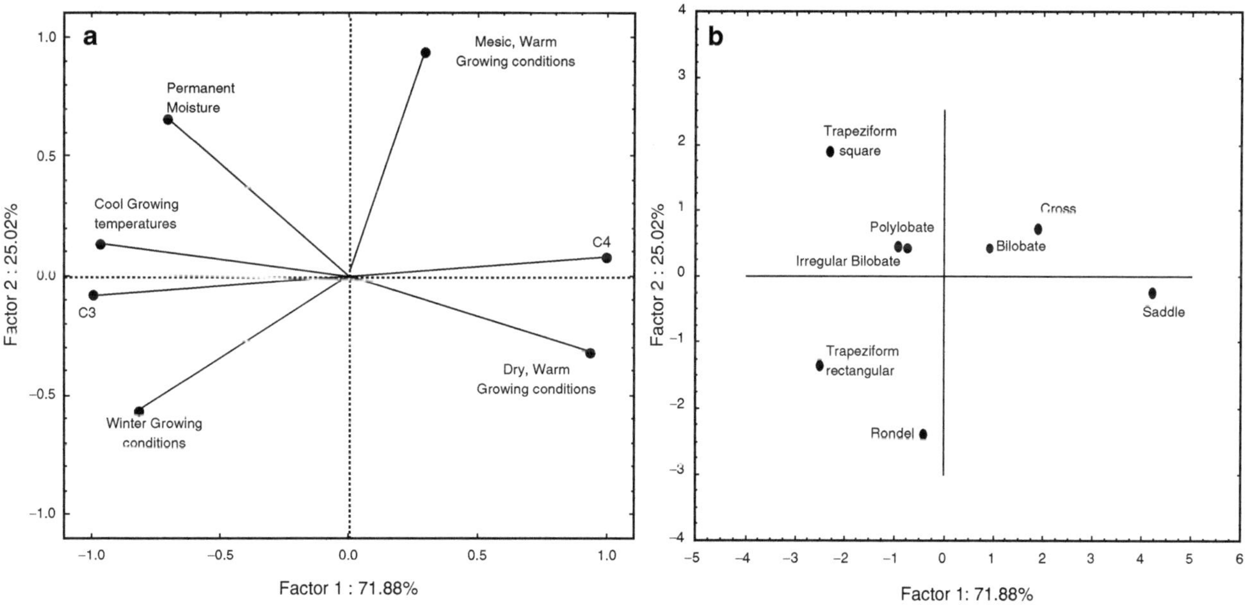

Fig. 9.3 PCA of the covariance matrix of the percentage data between biogeographical distribution of 250 grass species and seven short-cell morphotypes from South Africa. Plot of the case factor coordinates (**a**) and the variable factor coordinates (**b**). The first two axes of the PCA account for almost 97% of the variability in the data (Fig. 9.2). Axis 1 accounts for 72% of the variability and represents a temperature gradient. Axis 2 accounts for 25% of the variability and represents a moisture gradient. In the first component, the high positive loading of the saddle morphotype represents predominantly C_4 grassland conditions, while the negative loadings of the trapeziform morphotypes, in particular, signify mostly C_3 grassland conditions. Photosynthetic type and ambient growing temperature auto-correlate accordingly in the first component. Overall moist conditions in the second component, is demonstrated by the positive loading of square-shaped trapeziform type, the polylobate and the irregular (unsymmetrical) bilobate type. Negative loadings of the rectangular-shaped trapeziform and rondel-type for C_3 grasses represent more xeric, winter growing conditions while the saddle type for C_4 grasses reflect the arid summer rainfall extreme of this component. The first axis generally reflects discrimination of short-cell phytoliths on subfamilial level. The bilobate, cross and saddle morphotypes are abundant in predominantly C_4 subfamilies (Panicoideae, Aristidoideae and Chloridoideae) while the trapeziform, polylobate, irregular bilobate and rondel morphotypes predominate in C_3 subfamilies (Pooideae, Ehrhartoideae and Danthionoideae) (Table 9.2). Generally, the association between the distribution of short-cell morphotypes in grasses and climatic conditions reflects meaningful differences in the geographical distribution of grasses ranging from C_3 cool-growing dry to moist grassy conditions, to C_4 warm-growing dry to mesic grassy conditions

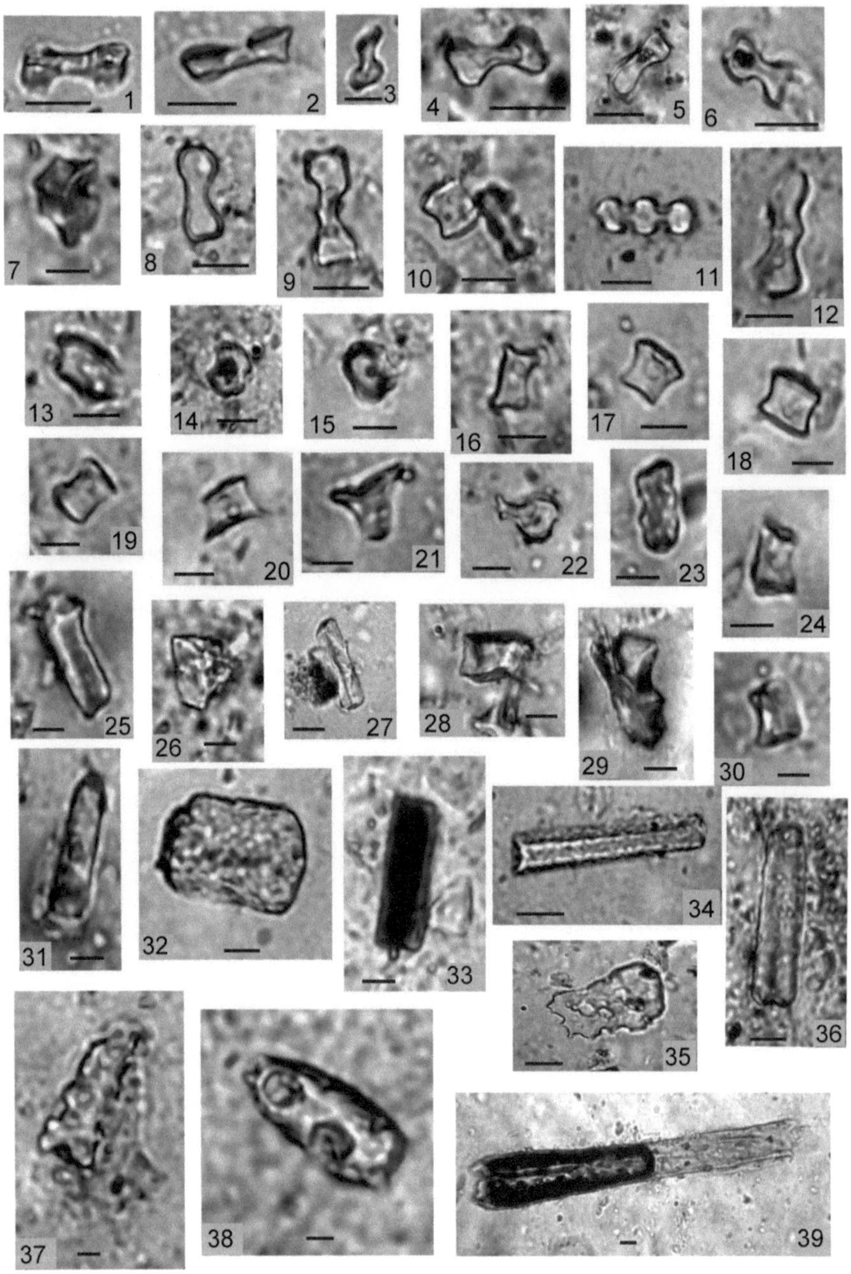

Fig. 9.4 Phytoliths retrieved from Laetoli sediments. Poaceae phytoliths (1–33, ×1,000): 1–8, Bilobate type; 9, Irregular Bilobate; 10, Saddle and Polylobate type; 11–12, Polylobate type; 13–20, Saddle type; 21, Rondel type; 22, Bilobate with one lobe missing; 23–31, Trapeziform type; 32–33, Parallelepipedal Bulliform type. Unclassified phytoliths (34–39, ×500). Scale = 10 micron

Apparent contaminant pollen grains either appeared to be light colored, contained fresh cell contents, or belonged to exotic plants like *Pinus*. Although we cannot exclude the possibility that some Pliocene specimens occurred among these, they are probably rare and difficult to distinguish from modern contaminants. The slide preparations of fossil material further contained only minute proportions of other organic material, including soil fungi, algae, etc., as indicated below.

Lower Laetolil Beds (Kakesio and Noiti)

The basal clays contained rare pollen, apparently contaminations. Termite mounds contained rare organics, including fungi, but no pollen. Bee broods yielded very little organics and fresh-looking blue-green algal cells similar to those living below rock surfaces (Danin 1990).

Upper Laetolil Beds

Clay samples yielded organic material sometimes in minute clumps, various types of pollen grains that are apparently contaminated, rare fern and fungal spores, and insect remains. Termite mounds contained rare organic matter, including microscopic charcoal, a possible sign of burning, while bee broods had blue-green algae, fungal, spores and rare pollen contaminations.

Upper Ndolanya Beds

Soils contained minute clumps of organic material, rare fresh-looking, light-colored contaminant pollen (Chenopodiaceae, Amaranthaceae and Poaceae), insect remains, and soil fungi. Bee broods contained hardly any organic matter and very few pollen grains including Chenopodiaceae, Amaranthaceae and Combretaceae, which are apparently contaminants.

Phytoliths

Samples collected from the calcareous ashy soils between the tuffs generated the best results. A selection of morphotypes from Laetoli is shown in Fig. 9.4.

The bee broods and termitaria produced no phytolith material indicative of grasses. Of the three modern reference localities that were sampled, Lemagurut and Oldonyo Mati hills yielded phytoliths, but the mbuga clays produced very little and are therefore not included in the study. Alkalinity could be an attributing factor in the lack of phytolith preservation in the mbuga soils, as pH values of up to 8 have been recorded previously (Andrews and Bamford 2008). Phytolith results from productive samples are shown in Fig. 9.5.

Lower Laetolil Beds (Kakesio and Noiti)

Diagnostic phytoliths are lacking in the samples collected from the LLB at Kakesio. The lack of phytoliths from a termitarium sample from Kakesio may possibly be attributed to the high alkaline hindgut physiology of termites (Brune and Kuhl 1996; Charles Peters, unpublished manuscript) and the dissolution of phytoliths under high pH conditions (Piperno 1988). Fluvial clays from Noiti 3 yielded grass morphotypes, especially trapeziform short-cells, as well as abundant unclassified phytolith material.

Upper Laetolil Beds

Diagnostic grass morphotypes, including trichomes, bulliform and long-cells were retrieved from the ULB layers. The sediments between Tuffs 6 and 8 and from the Yellow Marker Tuff presented the best preservation of phytoliths compared to the tuffs lower in the sequence. There is a decrease in unclassified phytoliths from below Tuff 8 upwards (Fig. 9.5).

Upper Ndolanya Beds

One sample from the UNB is generally rich in saddles and bilobate forms, but rondel and trapeziform types are also present. A marked difference is observed by the relative decrease in trapeziform morphotypes, which represent only 6.8% of the short-cell phytolith count in the sample (Fig. 9.5).

Interpretation of the Phytoliths

There is a noticeable shift in the ratio of short-cell phytoliths from mainly trapeziform morphotypes in the lower part of

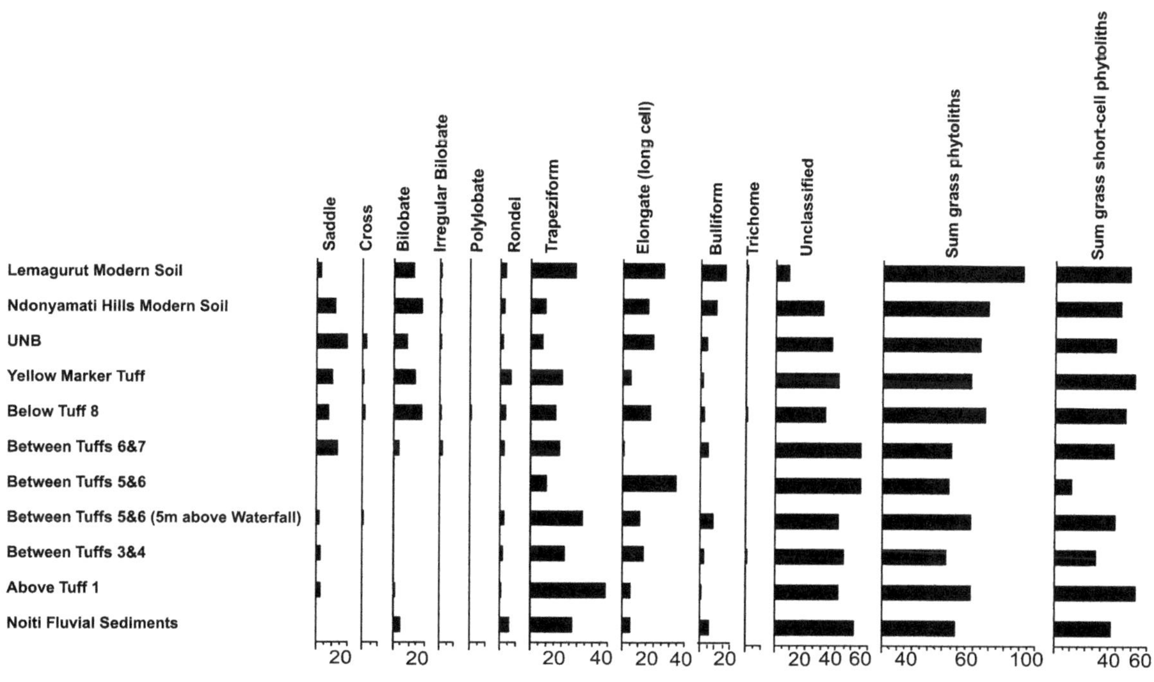

Fig. 9.5 Phytolith percentage diagram of modern and fossil samples from Laetoli. Percentages are calculated on the sum of all the phytolith types

the sequence to a marked increase in saddle and bilobate morphotypes higher up in the sequence (Fig. 9.4). Principal Components Analysis weights of the grass short-cell phytolith types from Laetoli were applied as a C_3/C_4 index along the Factor 1 axis (PC1) and a moisture index along the Factor 2 axis (PC2) as demonstrated in Figs. 9.3 and 9.6. High index values for the fossil data signify an abundance of C_3 grasses (Fig. 9.6a) and generally wetter environmental conditions (Fig. 9.6b), while low index values suggest a majority of C_4 grasses (Fig. 9.6a) and warmer, more arid growing conditions (Fig. 9.6b) (Vogel et al. 1978; Ehleringer et al. 1997). PC1 and PC2 imply wet, C_3 grass conditions in the productive samples from Noiti and potentially dry, C_3 conditions in the lower part of the ULB (between Tuffs 1 and 5). A shift

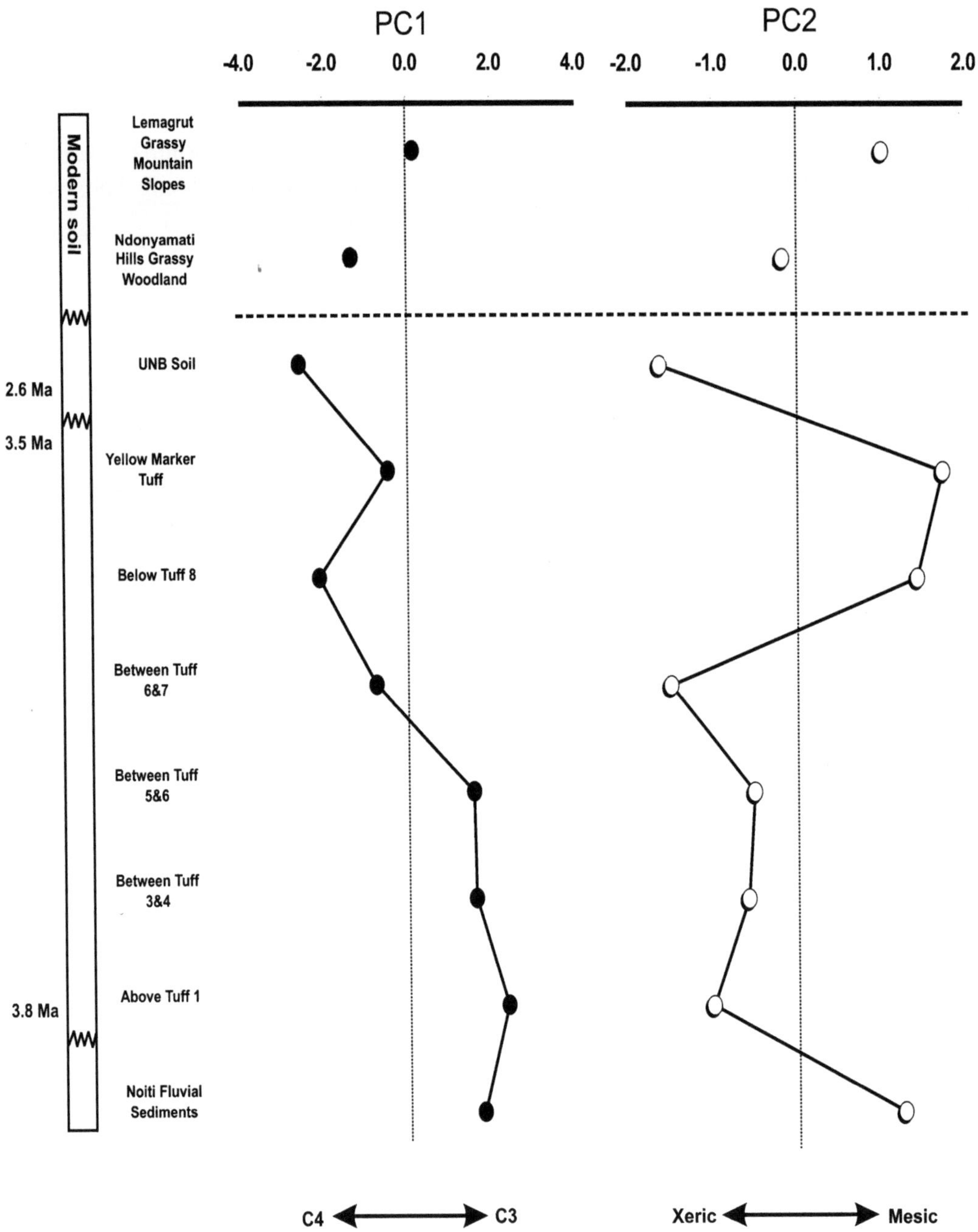

Fig. 9.6 Line graph of the first and second principal component coordinates from a PCA performed on the phytolith sum of the fossil samples to show relative changes of environmental parameters. The sample "5 m above Waterfall between Tuffs 5 and 6" is included in the category "Between Tuffs 5 and 6". The first principal component accounts for 44% and the second principal component accounts for 23% of the variability in the loadings. Contrast in the distribution of the weights was applied as a C_3/C_4 index along the Factor 1 axis and a moisture index along the Factor 2 axis as demonstrated in Fig. 9.3

from drier to more mesic C_4 grass conditions is recorded in the upper part of ULB (between Tuff 7 and the Yellow Marker Tuff). Arid conditions and C_4 grasses occurred in the UNB.

Modern soil samples from Lemagurut collected above 2,600 m represent a relatively high proportion of trapeziform morphotypes compared to the number of saddles and bilobates representative of C_3 grass availability at the cost of C_4 grasses at this high elevation. We expect C_3 growth because of more favorable soil moisture conditions and relatively cooler growing temperatures above 2,600 m. The selective pattern of C_3 and C_4 grass distribution over altitudinal gradients in East Africa clearly shows an increase of C_3 grasses at higher altitudes (Tieszen et al. 1979; Livingstone and Clayton 1980).

Discussion

The general lack of microscopic organic remains of plant origin in the Laetoli sediments suggests that long-term conditions for their preservation were not favorable and that most have been lost through chemical weathering, such as oxidization. This finding is similar to conclusions reached previously about sediments elsewhere in this region (Bonnefille 1977; Bonnefille and Riollet 1987). Although lush plant life must have occurred, as is implied by the rich faunal remains, termitaria and root infillings in the Laetoli region, and solid fossil wood remains at Noiti, virtually no microscopic residues of the Pliocene vegetation in the sediments and paleosols survived, except for opal phytoliths in some layers of the LLB, ULB and the UNB. An interesting question is why macrofaunal and some wood remains survived, but no microscopic pollen? The answer is probably that several factors, including the different susceptibility to oxidization or microbial action of the different materials, play a role. The smallest microscopic particles, like pollen grains and other minute plant remains, are likely to decompose first under unfavorable conditions (Carrión and Scott 1999).

Bonnefille and Riollet (1987) found no pollen in most volcanic layers at Locs. 1, 3, 5, 9, and 11, and some in a few termitaria samples and in clay layers from Locs. 9S and 10. Although we were unable to duplicate Bonnefille's results, we support her finding that fossil pollen concentrations in the Laetoli sediments are generally very poor or absent. It is likely that we did not sample exactly the same levels of termitaria (Bonnefille's samples were provided by W. A. Sands), but we are reasonably confident that we found the thin organic clay layers (brown soils) at Loc. 9S. A possible, but unlikely explanation, for the difference in results is that Bonnefille had access to slightly better exposures that existed two decades ago, which have now disappeared either through exposure, erosion or burial by recent infilling of deep narrow gullies (Bonnefille, personal communication). Although reliable pollen is not available in the exposures that we studied, levels with a better degree of pollen preservation may still be available, but not immediately below currently exposed gully surfaces. Near-surface sediments are experiencing seasonal wetting and drying accompanied by aerobic conditions, microbial action and weathering that can destroy pollen grains. We can also speculate that Bonnefille and Riollet's (1987) few productive pollen assemblages also suffered from a degree of modern contamination, which might have been difficult to recognize. The similarity between modern and fossil pollen assemblages in Bonnefille and Riollet (1987), which have a vast age difference, may imply such contamination.

Regarding preservation in the Laetoli sediments, we believe the pollen and minute organic plants remains originally embedded in the ashes and soils eventually disappeared as results of unfavorable conditions. Their break down could either have been relatively soon after burial, within a few seasons or decades as result of regular wetting, desiccation, aerobic and microbial action, or destruction might have occurred much later below exposed surfaces of gully walls when they were recently exposed. If the latter is true, deep boreholes of sealed-off strata may in future yield productive levels. Given the kind of sediments, however, we believe that pollen is scarce even at deeper levels. Solid coprolites that seal pollen from adverse sedimentary environments have potential (Scott et al. 2003), but we have not looked at any from Laetoli up to now. We therefore conclude that reliable pollen assemblages that potentially reflect the Pliocene vegetation are not available from current exposures, and that it is difficult to replicate the previous results of Bonnefille and Riollet (1987). The discrepancies in the findings of the two studies may lie in different laboratory and field approaches, the details of which are difficult to compare effectively.

Grass phytoliths were more available than pollen, but a considerable time interval of deposition combined with a relatively low phytolith yield, has allowed us at this stage to present no more than a low-resolution sequence of changing grass cover for the LLB, ULB and UNB. The unclassified phytoliths represent a substantial proportion in all of the fossil samples, which suggests more heterogeneous vegetation with grasses as only one of its components (Table 9.4, Fig. 9.5). The unclassified types are possibly represented by either trees or smaller shrubs and herbs and are relatively high in the lower parts of the sequence (Fig. 9.5). The climatic significance of these phytoliths is uncertain.

Phytolith counts of the modern soils from Lemagurut and the Oldonyo Mati Hills generally correspond with local observations that grass density is higher along the slopes of Lemagurut (apparently as a result of earlier deforestation practices, P.J. Andrews, personal communication). In this case, we presume that the processes involved in the turnover

Table 9.4 Detailed counts of phytoliths from Laetoli

Context	Saddle	Cross	Bilobate	Irregular Bilobate	Polylobate	Rondel	Trapeziform	Elongate (long cell)	Bulliform	Trichome	Unclassified phytoliths	Grass phytoliths	Grass short-cell phytoliths	% Total grass phytoliths
Lemagurut Modern Soil	24	1	115	6	4	28	257	235	147	7	76	817	178	91.6
Oldonyo Mati Hills Modern Soil	101	0	152	7	0	19	74	132	88	0	257	573	279	69.0
UNB Soil	140	15	65	5	1	10	50	145	35	0	268	466	236	63.5
Yellow Marker Tuff	62	2	88	2	2	40	130	34	12	0	264	372	196	58.5
Below Tuffs 8	56	7	136	7	7	22	116	138	24	2	247	513	235	67.6
Between Tuffs 6&7	46	0	14	6	0	8	64	4	18	0	196	160	74	44.9
Between Tuffs 5&6	0	0	0	0	0	0	57	186	0	0	307	243	0	44.2
5 m above Waterfall (between Tuffs 5&6)	4	1	1	1	0	6	76	25	18	0	94	132	13	58.4
Between Tuffs 3&4	21	0	3	0	0	14	197	117	23	3	395	375	38	42.2
Above Tuff 1	8	0	4	1	0	2	162	15	5	0	140	197	15	58.5
Noiti Fluvial Sediments	0	0	9	0	0	12	54	11	12	0	105	98	21	48.3

of forest-type silica to predominantly grassland type silica can happen relatively quickly in modern surface soils. Interpreting the unclassified phytolith category is difficult, but it most likely represents woody species as well as non-graminoid herbs and smaller shrubs. The grass phytolith proportion indicates that grasses were a ubiquitous, but never dominant component of the vegetation in the LLB, ULB and the UNB sequences, and this lends support to the current model suggesting a prevalence of woodland to shrubland habitats in the ULB and UNB (Andrews 1989; Kingston and Harrison 2007; Kovarovic and Andrews 2007; Andrews and Bamford 2008). The presence of both C_3 and C_4 grasses agree with dietary classification of large mammalian herbivores based on $\delta^{13}C$ values, that proposes intermediate C_3/C_4 feeding behaviors for fossil herbivores found in the ULB (Kingston and Harrison 2007). The hipparion, *Eurygnathohippus* sp., the predecessor of the white rhinoceros, *Ceratotherium praecox* and two bovid species, *Parmularius pandatus* and a hippotragine, are grazers that exhibit a mixed C_3/C_4 signal in their diet (Kingston and Harrison 2007). Grasses were probably exclusive to their distinctly grazer diets. A mixed C_3/C_4 isotopic signal from the grazers therefore implies that C_3 and C_4 grasses co-occurred intermittently in the ULB. C_3 and C_4 grass distribution may have occurred along an altitudinal gradient analogous to modern grassland conditions in the region (Tieszen et al. 1979; Livingstone and Clayton 1980). Formation of the volcanoes Lemagurut and Ngorongoro occurred after the deposition of the Upper Laetolil Beds, but Satiman is regarded as the probable source of the Upper Laetolil Beds and would have been a considerable landmark at the time (Andrews and Bamford 2008).

Environmental evolution is reflected by a general shift from mainly C_3 grass types in the LLB and older ULB levels to mainly C_4 grass types in the younger ULB and UNB. Potentially drier, C_4 conditions are recorded at Tuff 1, between Tuffs 6 and 7 as well as in the UNB. Marine records show multiple shifts of alternating wetter and drier conditions, driven by orbital variations during the Plio-Pleistocene in subtropical Africa (deMenocal 1995, 2004). This pattern of climate variability was accompanied by a marked increase in aridity between 3.0 and 2.6 million years ago, leading to the inception and intensification of high-latitude glacial cycles (deMenocal 2004).

We must also consider CO_2 pressure in the atmosphere as a factor in the case of the C_3/C_4 status of fossil assemblages. Our results suggest a general increase in C_4 grasses in the upper part of the ULB and in UNB in comparison with the lower part of ULB. However, one must keep in mind the limitations of our wide sample intervals and that the sequence might be more complex. An implied climatic change either from cooler conditions to warmer growing seasons for grasses, or a decrease in CO_2 pressure in the atmosphere are therefore possible scenarios for the change in the sequence, at least in the case of productive phytolith levels. Stable isotope results in the region suggest that C_4 grasses were prominent during cold Last Glacial Maximum times with an altitudinal lowering of vegetation zones (Coetzee 1964), under low atmospheric CO_2 pressures (Ehleringer et al. 1997). In the tropics, the current analog of C_3 and C_4 grass distribution on Mount Kenya along an altitudinal gradient suggests that C_3 grasses become dominant at altitudes above about 2,290 m together with more moisture and cooler temperatures (Tieszen et al. 1979). At higher latitudes, as in South Africa, where Last Glacial Maximum temperatures were generally colder than in the tropics, temperature on the plateau apparently over-ruled the effect of lower CO_2 during "glacial" periods (Scott 2002). However, in order to reconstruct climatic conditions in the study area on the basis of its C_3/C_4-grassland status it will be necessary to disentangle the roles of CO_2, growing-temperature and moisture (Vogel et al. 1978; Tieszen et al. 1979; Ehleringer et al. 1997).

In conclusion, although suitable pollen as evidence of environmental conditions over the studied interval was not preserved, the phytolith data support the presence of heterogenous habitats throughout the sequence. In addition, the upper part of the phytolith sequence indicates a marked change in the grass component toward C_4 dominance and more grassy vegetation. This, together with a decrease in dicotyledonous cover higher up in the sequence, agrees with soil isotopic evidence of a gradual transition towards more C_4 vegetation (Ségalen et al. 2007).

Acknowledgements We thank Terry Harrison for the invitation to take part in the Laetoli project. LS collected most of the pollen samples during the 2000 field season with the help of team members especially John Kingston, while Terry Harrison and LR collected more during following seasons. LR collected the phytolith samples during the 2004 field season with the help of team member John Kingston. Charles Peters kindly provided unpublished data about his research on termitaria and phytolith preservation in East Africa. Petrus Chakane processed the samples chemically in the palynology laboratory at the UFS. The National Research Foundation (NRF) supported laboratory work (Gun 2053236). Any opinions, findings, and conclusions are those of the authors and the NRF does not accept any liability in regard thereto. Travel funds and field support were provided by an NSF grant (BCS-0309513) to Terry Harrison.

References

Albert, R. M., & Weiner, S. (2001). Study of phytoliths in prehistoric ash layers using a quantitative approach. In J. D. Meunier & F. Coline (Eds.), *Phytoliths: Applications in earth sciences and human history* (pp. 251–266). Lisse: Balkema.

Alexandre, A., Meunier, J. D., Lezine, A. M., Vincens, A., & Schwarz, D. (1997). Phytoliths: Indicators of grassland dynamics during the late Holocene in intertropical Africa. *Palaeogeography, Palaeoclimatology, Palaeoecology, 136*, 213–229.

Andrews, P. J. (1989). Palaeoecology of Laetoli. *Journal of Human Evolution, 18*, 173–181.

Andrews, P. J., & Bamford, M. (2008). Past and present vegetation ecology of Laetoli, Tanzania. *Journal of Human Evolution, 54*, 78–98.

Barboni, D., Bonnefille, R., Alexandre, A., & Meunier, J. D. (1999). Phytoliths as palaeoenvironmental indicators, West Side Middle Awash Valley, Ethiopia. *Palaeogeography, Palaeoclimatology, Palaeoecology, 152*, 87–100.

Barboni, D., Bremond, L., & Bonnefille, R. (2007). Comparative study of modern phytolith assemblages from inter-tropical Africa. *Palaeogeography, Palaeoclimatology, Palaeoecology, 246*, 454–470.

Bonnefille, R. (1977). Palynological research at Olduvai Gorge. *National Geographic Society*, 1976 Projects, 227–243

Bonnefille, R., & Riollet, G. (1987). Palynological spectra from the Upper Laetolil Beds. In M. D. Leakey & J. M. Harris (Eds.), *Laetoli: A Pliocene site in northern Tanzania* (pp. 52–61). Oxford: Clarendon.

Bonnefille, R., & Vincens, A. (1985). Apport de la palynologie à l'environnement des hominidés d'Afrique orientale. In M. Beden, A. K. Behrensmeyer, N. T. Boaz, R. Bonnefille, C. K. Brain, H. B. S. Cooke, Y. Coppens, R. Dechamps, V. Eisenmann, A. Gentry, D. Geraads, R. Geze, C. Guerin, J. Harris, J.-C. Koeniguer, R. Letouzey, G. Petter, A. Vincens, & E. Vrba (Eds.), *L'environnement des hominidés au Plio-Pléistocène* (pp. 237–278). Paris: Masson.

Bonnefille, R., Potts, R., Chalié, F., Jolly, D., & Peyron, O. (2004). High-resolution vegetation and climate change associated with Pliocene *Australopithecus afarensis*. *Proceedings of the National Academy of Sciences, 101*, 12125–12129.

Bremond, L., Alexandre, A., Hely, C., & Guiot, J. (2005). A phytolith index as a proxy of tree cover density in tropical areas: Calibration with Leaf Area Index along a forest-savanna transect in southeastern Cameroon. *Global and Planetary Change, 45*, 277–293.

Brown, D. A. (1984). Prospects and limits of a phytolith key for grasses in the central United States. *Journal of Archaeological Science, 11*, 345–368.

Brune, K., & Kuhl, M. (1996). pH profiles of the extremely alkaline hindguts of soil-feeding termites (Isoptera: Termitidae) determined with microelectrodes. *Journal of Insect Physiology, 42*, 1121–1127.

Carrión, J. S., & Scott, L. (1999). The challenge of pollen analysis in palaeoenvironmental studies of hominid beds: The record from Sterkfontein caves. *Journal of Human Evolution, 36*, 401–408.

Coetzee, J. A. (1964). Evidence for a considerable depression of the vegetation belts during the Upper Pleistocene on the East African mountains. *Nature, 204*, 564–566.

Danin, A. (1990). Deterioration of limestone walls in Jerusalem and marble monuments in Rome caused by cyanobacteria and cyanophilous lichens. *International Biodeterioration, 26*, 397–417.

Deino, A. (2011). $^{40}Ar/^{39}Ar$ dating of Laetoli, Tanzania. In: Harrison, T. (Ed.), *Paleontology and geology of Laetoli: Human evolution in context* (Geology, geochronology, paleoecology and paleoenvironment, vol. 1, pp. 77–97). Dordrecht: Springer.

DeMenocal, P. B. (1995). Plio-Pleistocene African climate. *Science, 270*, 53–59.

DeMenocal, P. B. (2004). African climate change and faunal evolution during the Pliocene-Pleistocene. *Earth and Planetary Science Letters, 220*, 3–24.

Ditchfield, P., & Harrison, T. (2011). Sedimentology, litho-stratigraphy and depositional history of the Laetoli area. In T. Harrison (Ed.), *Paleontology and geology of Laetoli: Human evolution in context* (Geology, geochronology, paleoecology, and paleoenvironment, vol. 1, pp. 47–76). Dordrecht: Springer.

Drake, R., & Curtis, G. (1987). K-Ar geochronology of the Laetoli fossil localities. In M. D. Leakey & J. M. Harris (Eds.), *Laetoli: A Pliocene site in northern Tanzania* (pp. 48–51). Oxford: Clarendon.

Ehleringer, J. R., Sage, R. F., Flanagan, L. B., & Pearcy, R. W. (1991). Climate change and the evolution of C_4 photosynthesis. *Trends in Ecology & Evolution, 6*, 95–99.

Ehleringer, J. R., Cerling, T. E., & Helliker, B. R. (1997). C_4 photosynthesis, atmospheric CO_2 and climate. *Oecologia, 112*, 285–299.

Ellis, R. P. (1977). Distribution of the Kranz syndrome in the southern African Eragrostideae and Panicoideae. *Agroplantae, 9*, 73–110.

Ellis, R. P., Vogel, J. C., & Fuls, A. (1980). Photosynthetic pathways and the geographical distribution of grasses in southwest Africa Namibia. *South African Journal of Science, 76*, 307–314.

Erdtman, G. (1960). The acetolysis method. A revised description. *Svensk Botanisk Tidskrift, 54*, 561–564.

Fredlund, G. G., & Tieszen, L. L. (1994). Modern phytolith assemblages from the North American great plains. *Journal of Biogeography, 21*, 321–335.

Gibbs Russell, G. E. (1985). PRECIS: The National Herbarium's computerized information system. *South African Journal of Science, 81*, 62–65.

Gibbs Russell, G. E. (1988). Distribution of subfamilies and tribes of Poaceae in Southern Africa. *Monographs in Systematic Botany, Missouri Botanical Gardens, 25*, 555–566.

Grab, S., Scott, L., Rossouw, L., & Meyer, S. (2005). Holocene palaeoenvironments inferred from a sedimentary sequence in the Tsoaing River Basin, western Lesotho. *Catena, 61*, 49–62.

Harrison, T., & Kweka, A. (2011). Paleontological localities on the Eyasi Plateau, including Laetoli. In T. Harrison (Ed.), *Paleontology and geology of Laetoli: Human evolution in context* (Geology, geochronology, paleoecology and paleoenvironment, vol. 1, pp. 17–45). Dordrecht: Springer.

Hattersley, P. W. (1983). The distribution of C_3 and C_4 grasses in Australia in relation to climate. *Oecologia, 57*, 113–128.

Hay, R. L. (1987). Geology of the Laetoli area. In M. D. Leakey & J. M. Harris (Eds.), *Laetoli: A Pliocene site in northern Tanzania* (pp. 23–47). Oxford: Clarendon.

Henderson, Z. L., Scott, L., Rossouw, L., & Jacobs, Z. (2006). The dating, palaeoenvironments and archaeology of the Sunnyside 1 site, Clarens, South Africa. *Archaeological Papers of the American Anthropological Association, 16*, 139–149.

Horrocks, M. (2005). A combined procedure for recovering phytoliths and starch residues from soils, sedimentary deposits and similar materials. *Journal of Archaeological Science, 32*, 1169–1175.

Kingston, J. D., & Harrison, T. (2007). Isotopic dietary reconstructions of Pliocene herbivores at Laetoli: Implications for early hominin paleoecology. *Palaeogeography, Palaeoclimatology, Palaeoecology, 243*, 272–306.

Klein, R. G. (1999). *The human career*. Chicago: The University of Chicago Press.

Kovarovic, K., & Andrews, P. (2007). Bovid postcranial ecomorphological survey of the Laetoli paleoenvironment. *Journal of Human Evolution, 52*, 663–680.

Lentfer, C. J., & Boyd, W. E. (1998). A comparison of three methods for the extraction of phytoliths from sediments. *Journal of Archaeological Science, 25*, 1152–1183.

Lentfer, C. J., & Boyd, W. E. (1999). An assessment of techniques for the deflocculation and removal of clays from sediments used in phytolith analysis. *Journal of Archaeological Science, 26*, 31–44.

Livingstone, D. A., & Clayton, W. D. (1980). An altitudinal cline in tropical Africa grass floras and its palaeoecological significance. *Quaternary Research, 13*, 392–402.

Madella, M., Alexandre, A., & Ball, T. (2005). International Code for Phytolith Nomenclature. *Annals of Botany, 96*, 253–260.

McLean, B., & Scott, L. (1999). Phytoliths in sediments of the Pretoria Saltpan and their potential as indicators of environmental history at the site. In T. C. Partridge (Ed.), *Tswaing-investigations into the origin, age and palaeoenvironments of the Pretoria Saltpan* (pp. 167–171). Pretoria: Council for Geosciences.

Mulholland, S. C. (1989). Phytolith shape frequencies in North Dakota grasses: A comparison to general patterns. *Journal of Archaeological Science, 16*, 489–511.

Mulholland, S. C., & Rapp, G. J. (1992). A morphological classification of grass silica-bodies. In G. J. Rapp & S. C. Mulholland (Eds.),

Phytolith systematics. Emerging issues. Advances in archeological and museum science 1 (pp. 65–81). New York: Plenum.

Piperno, D. R. (1988). *Phytolith analysis: An archeological and geological perspective*. London: Academic.

Rovner, I. (1971). Potential of opal phytoliths for use in palaeoecological reconstruction. *Quaternary Research, 1*, 343–359.

Sage, R. F. (2004). The evolution of C_4 photosynthesis. *The New Phytologist, 161*, 341–370.

Scott, L. (1995). Pollen evidence for vegetational and climate change during the Neogene and Quaternary in Southern Africa. In E. Vrba, G. Denton, T. C. Partridge, & L. H. Burckle (Eds.), *Paleoclimate and evolution with emphasis on human origins* (pp. 65–76). New Haven: Yale University Press.

Scott, L. (2002). Grassland development under glacial and interglacial conditions in Southern Africa: Review of pollen, phytolith and isotope evidence. *Palaeogeography, Palaeoclimatology, Palaeoecology, 177*, 47–57.

Scott, L., & Bonnefille, R. (1986). A search for pollen from the hominid deposits of Kromdraai, Sterkfontein and Swartkrans: Some problems and preliminary results. *South African Journal of Science, 82*, 380–382.

Scott, L., & Rossouw, L. (2005). Reassessment of botanical evidence for palaeoenvironments at Florisbad, South Africa. *South African Archeological Bulletin, 60*, 96–102.

Scott, L., Fernandez-Jalvo, Y., Carrión, J. S., & Brink, J. S. (2003). Preservation and interpretation of pollen in hyena corprolites: Taphonomical observations from Spain and Southern Africa. *Palaeontologia Africana, 39*, 83–91.

Ségalen, L., Lee-Thorp, J. A., & Cerling, T. (2007). Timing of C_4 grass expansion across sub-Saharan Africa. *Journal of Human Evolution, 53*, 549–559.

Stromberg, C. A. E. (2002). The origin and spread of grassland-dominated ecosystems in the late Tertiary of North America: Preliminary results concerning the evolution of hypsodonty. *Palaeogeography, Palaeoclimatology, Palaeoecology, 177*, 59–76.

Stromberg, C. A. E. (2004). Using phytolith assemblages to reconstruct the origin and spread of grass-dominated habitats in the great plains of North America during the late Eocene to early Miocene. *Palaeogeography, Palaeoclimatology, Palaeoecology, 207*, 239–275.

Tieszen, L., Senyimba, M. M., Imbamba, S. K., & Troughton, J. H. (1979). The distribution of C_3 and C_4 grasses and carbon isotope discrimination along an altitudinal and moisture gradient in Kenya. *Oecologia, 37*, 337–350.

Twiss, P. C., Suess, E., & Smith, R. M. (1969). Morphological classification of grass phytoliths. *Proceedings of the Soil Science Society of America, 33*, 109–115.

Vogel, J. C., Fuls, A., & Ellis, R. P. (1978). The geographical distribution of Krantz grasses in South Africa. *South African Journal of Science, 74*, 209–215.

Wooller, M. J., Street-Perrott, F. A., & Agnew, A. D. Q. (2000). Late Quaternary fires and grassland plaeoecology from charred grass cuticles in Lake sediments. *Palaeogeography, Palaeoclimatology, Palaeoecology, 164*, 207–230.

Chapter 10
Fossil Woods

Marion K. Bamford

Abstract Fossil woods have been recovered from Noiti to the south of the main Laetoli area and are of Lower Laetolil Bed age. Preservation is generally poor because of recrystallization but a variety of taxa have been tentatively identified. Where there is some uncertainty about the identification several possibilities have been given. The fossil woods are from the families Araliaceae, Buddlejiaceae, Caesalpinoideae, Celastraceae, Combretaceae, Dichapetalaceae, Euphorbiaceae, Flacourtiaceae, Meliaceae, Mimosoideae, Myrtaceae, Ochnaceae, Rhamnaceae, Rubiaceae, Rutaceae and Verbenaceae. The flora represents woods from high to low altitudes that grew on the slopes of the volcano and were preserved in a lahar. A variety of soils and vegetation associations are implied from forest, evergreen woodland, dry woodland, bushland, riverine and wooded grassland, in a megathermal environment. Both taxonomic affinities and the vulnerability indices (non-taxonomic) of the woods are used to infer the paleoenvironment.

Keywords Silicified wood • Vegetation associations • Montane • Woodland

Introduction

Locality and Age

The Pliocene site of Laetoli in northern Tanzania comprises fossiliferous sediments and tuffs exposed over a wide area that have been collected and studied since 1935. Vertebrate and invertebrate fossils are relatively common (Harrison and Kweka 2011), but macrobotanical remains are almost unknown. Simon Mataro, the Antiquities Department superintendent at Laetoli, discovered the fossil wood sites and showed them to Terry Harrison. Since 2000 the area has been visited several times. Silicified woods are common at the localities of Noiti 1 and Noiti 2 (Fig. 10.1), which are about 10 km south of Laetoli, in the Nasera valley, northern Tanzania (see Harrison and Kweka 2011).

The woods are preserved in a lahar with no particular orientation or sorting. Lahars are a mix of volcanic ash, debris and water either from rain, snow melt or rivers, that flow down the flank of a volcano and along a river channel or valley until they settle and solidify. The slurry can incorporate and transport any material along its path. These lahars are correlated with the Lower Laetolil Beds (3.8–4.4 Ma; Drake and Curtis 1987; Deino 2011) and the source volcano was possibly Satiman or an undetermined older volcanic center now superimposed by Lemagurut, the extinct volcano that overshadows the Olaitole Valley today (Hay 1987). Mollel et al. (2011) date the phonolite in the lahar at 4.0–4.1 Ma and this provides a maximum age. The lahars in this area do not contain vertebrate fossils, just fossil wood, and they overlie the base of the Lower Laetolil Beds, which do contain rare vertebrate fossils. Vertebrate fossils are common in the Olaitole and Garusi River valleys about 10 km to the north.

At Noiti the stratigraphy is incomplete as the Ogol Lavas, dated to 2.4 Ma, lie uncomformably on the lahars and Lower Laetolil Beds. Pieces of wood weather out of the lahar at the base of the Ogol Lavas and some pieces have been washed down into the gullies. Some pieces remain cemented within the lower part of the lahar and these are randomly aligned (vertical to horizontal). Today, *Acacia, Commiphora* and *Euphorbia* wooded or bushed grasslands are the common vegetation in the Noiti area. Open grasslands occur on interfluves and the upper slopes of the volcanoes are forested (Andrews et al. 2011). A similar gradation most likely occurred in the Pliocene. The lower limit of the upper montane vegetation would have been either lower than that of today when there was more rainfall, or higher when there was less rainfall (Andrews et al. 2011). Nonetheless, the full range of vegetation associations from top to bottom of the volcano would have been in place during the Pliocene.

M.K. Bamford (✉)
Bernard Price Institute for Palaeontological Research,
School of Geosciences, University of the Witwatersrand,
Private Bag 3, WITS 2050, Johannesburg, South Africa
e-mail: marion.bamford@wits.ac.za

T. Harrison (ed.), *Paleontology and Geology of Laetoli: Human Evolution in Context. Volume 1: Geology, Geochronology, Paleoecology and Paleoenvironment*, Vertebrate Paleobiology and Paleoanthropology, DOI 10.1007/978-90-481-9956-3_10,

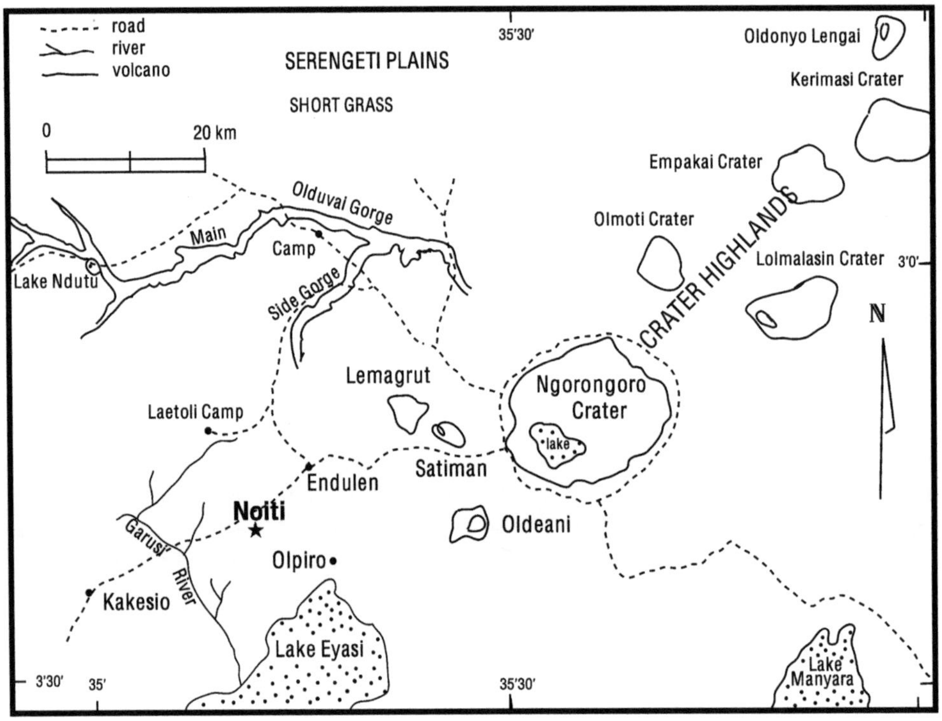

Fig. 10.1 Location of fossil wood site in the Olaitole region

Preservation and Taphonomy

Silica from tephra and rocks is a common source of monosilicic acid that saturates wood buried in a water-logged environment and gradually silicifies the wood (Buurman 1972). The East African rift has an abundance of volcanic material and, under the right conditions of rapid burial and anoxia, wood is preserved. In spite of the abundance of silica in the system, fossil wood is rare in this region implying that some prerequisites are missing, perhaps the temporary water-logged environment of the lahar. Lahar deposits lack the extensive bioturbation seen in paleosols. Termites would certainly have consumed any wood that they encountered during their tunnelling through soils, whereas the relatively high water content of lahars would have caused rapid cementation of the material and prevented bioturbation.

The woods have been silicified, but appear to have either been rapidly and incompletely crystallized or have undergone secondary re-crystallization by calcite. Only in some samples of small diameter woods has the whole piece been well preserved with the central part of the branch as well as the external parts showing vessels, rays, fibers and parenchyma tissue. Most samples are only fairly well preserved around the perimeter. The remaining major portion of the stem contains large crystals that have destroyed the fabric of the once living tissue (Fig. 10.2a). Since the stems have not been distorted and there are ghosts of former cells, vessels in particular, but occasionally rays too, it is assumed that after initial petrification had occurred there has been some secondary chemical process that has modified the primary framework. The final appearance is that of an external rim of well preserved vessels and ground tissue about 2–3 mm thick, with the vessels being the only cells preserved interior to that. Large clear crystal outgrowths of calcite from the vessels have replaced the ground tissue. Often fine details, such as crystals that were deposited in the living wood cells, or the pitting on the vessels walls, are still preserved (Fig. 10.2b). This makes identification of most specimens very difficult. There is no apparent difference in preservation between the two sites.

Paleoenvironmental Approach and Interpretation

Compared with other plant parts, such as flowers, leaves and seeds, wood is relatively conservative as far as evolution is concerned (Wheeler and Baas 1991), with many related species having very similar wood structure. One possible reason for this is that wood is strictly functional and serves for the uptake of water and nutrients from the soil and also for support.

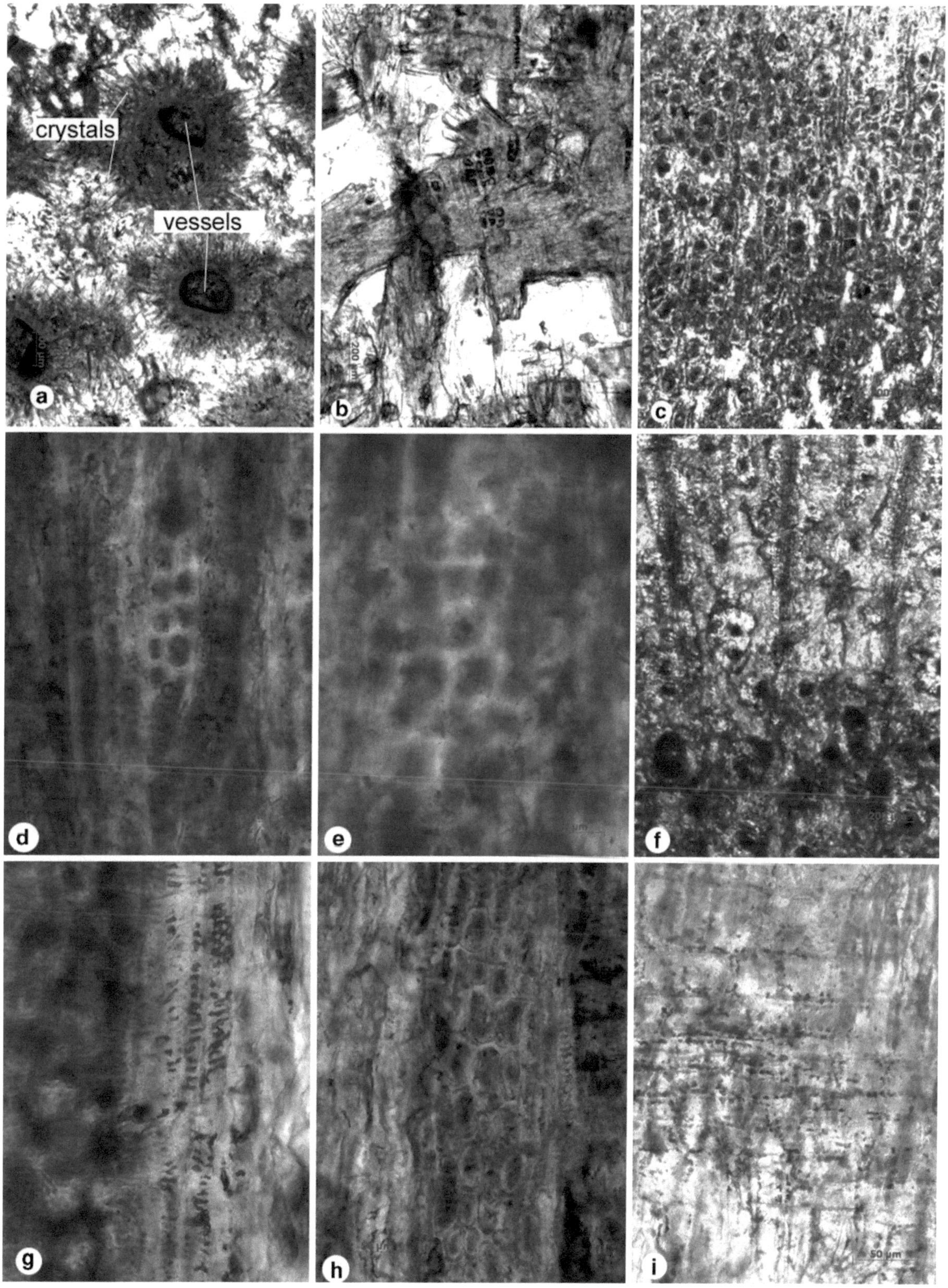

Fig. 10.2 Photomicrographs of thin sections of fossil wood. (**a**) large crystals projecting from the periphery of a vessel and destruction of the ground tissue fabric. (**b**) dark angular crystals in rows but the original encasing ray or axial parenchyma cells have been destroyed. (**c–e**) Photomicrographs of specimen BP/16/1201 from Noiti 1; cf. *Schrebera alata* (Oleaceae). (**c**) transverse section (TS): vessels (larger oval cells) in short radial multiples of 1–4 cells. No parenchyma is visible. Rays are two cells wide (in vertical columns in this view). (**d**) Radial longitudinal section (RLS) with barely visible vessels (horizontal columns in this view). (**e**) square parenchyma ray cells in RLS. (**f–i**) BP/16/1252 from Noiti 1; cf. *Vitex* sp. (Verbenaceae). (**f**) TS, vessels in short radial multiples and a cluster; 1–2 rows of parenchyma cells around the vessels. (**g**) TLS, spiral thickening on walls of tracheids. (**h**) TLS, part of a ray with 3–4 rows of oval cells. (**i**) *RLS*, two rows of square (*top*) and five rows of procumbent (*center*) ray cells

Tall or spreading trees have the advantage of access to more sunlight, but this requires a robust wood type. Lianas have flexible woods with very long "trunks" and use other trees for their support, and so can sacrifice support tissue for more conductive tissue. There are physical restrictions on the height of trees and water conductivity. Conductivity is a function of the vessel diameter, vessel number and distance that the water has to travel from root tips to leaves (Ewers 1985). There is also a risk of embolisms occurring, such as water bubbles or other obstructions, that can stop the conductivity of individual vessels. Such problems can be overcome in a number of ways, for example, by increasing the number of vessels and reducing their diameter, or having vessels and/or tracheids adjacent to one another in radial or longitudinal lines or clusters. With a system of pits between adjacent vessels (and tracheids) such blockages can be by-passed.

In a geographic area that is subject to drought or water shortages the woods tend to have many smaller vessels (Carlquist 1975). In well-watered areas the woods have relatively few large vessels. There are other adaptations to counteract water stress, such as different types of perforation plates and rays. Thus, one can use the functional adaptations of the wood to interpret the environment (Baas et al. 1983; Wheeler and Baas 1991).

It is also possible to identify most plants from their wood anatomy. From the numerous herbarium collections and published Floras one has access to information about the distribution of the various plant taxa. For example, some are restricted to an area (soil, altitude) or to a climate belt (rainfall, temperature, altitude), whereas others are less sensitive and occur in a wide range of conditions. One also knows their growth habits (i.e., large tree, small tree, shrub, liana, herb). Today, there is also a large database of wood anatomy for commercial applications: wood strength, durability, flexibility, weight and hardness determine the best use of the wood for building, furniture, tools, etc. Using the anatomy of fossil woods it is possible to identify some late Tertiary woods to modern taxa, and then one can assume that the modern climatic tolerances and growth habit were the same for the same fossil plants, i.e., the Nearest Living Relative (NLR) method (Mosbrugger and Utescher 1997). Another approach is non-taxonomic ecomorphology, which uses only the physical attributes of the wood. As mentioned above, woods are adapted to cope with environmental stresses and one can assume that fossil woods had the same strategies as modern woods (Wheeler and Baas 1991).

Fossil woods can be used to reconstruct the past vegetation and climate if the modern analogs are known. It is vital, however, to use the whole flora as some taxa or growth forms are non-diagnostic or too general in their habitat requirements to provide useful interpretations. Also, some taxa occur outside of their usual range (i.e., relicts). By studying the whole flora the effects of the problem taxa are reduced.

Materials and Methods

On several visits to the localities of Noiti 1 and 2 fossil wood fragments were collected where they had been weathered out of the sediments. From these the better preserved samples were chosen for sectioning based on observation with a hand lens in the field. Larger pieces encased in the lavas were not retrievable. Wood pieces varied in size from 1–15 cm in diameter. Many showed evidence of knots and branches, but these were not sectioned because branch wood is anomalous and very difficult to identify. Most fossil woods were pale browns and white in color, and the ends were weathered to show raised circles that represent the re-crystallized vessels.

At the Bernard Price Institute for Palaeontological Research, University of the Witwatersrand, petrographic thin sections were made of selected samples for study under a light microscope (Zeiss Axiophot) and photographed with a Zeiss digital camera. Each sample was cut in three planes, transverse, radial longitudinal and tangential longitudinal. Small blocks were cut and the planes were polished with carborundum powder and diamond paste before being glued with the polished surface down, onto glass slides. The rest of the block was cut off using a discoplan, and the resultant thin sections ground and polished to a final thickness of about 40 μm.

Under the light microscope important anatomical features such as vessel size, distribution and frequency, type of wall pitting, ray type and size, parenchyma type, fibers and tracheids, were measured and described. Identification keys such as InsideWood (InsideWood 2006–2007 http://insidewood.lib.ncsu.edu/search/), Wheeler et al. (1986), CSIRO (Ilic 1987) and Lebacq and Dechamps (1964) were used for initial identification, and the results checked with the literature and wood slide collections where possible. Many fossil woods had insufficient features preserved to allow any identification, but some of them were identifiable to genus or species level. Since some of the features critical for distinguishing between species were not preserved, more than one possible identification is given (Table 10.1).

Unless otherwise stated, the details of extant wood anatomy are from the InsideWood database. The habitats for all possible species are included (Table 10.1) and these data come from Beentje (1994), van Wyk and van Wyk (1997) and various Floras (Oliver 1871; Keay 1958; Stanier and Gilbert 1958). Interpretation of the fossil wood flora, therefore, is on different taxonomic levels. Terminology for wood descriptions follows that of the International Association of Wood Anatomists Committee (1989).

Table 10.1 Habit, habitat and distribution of modern woods with possible analogs in the fossil record

Specimen No.	Family	Genus, species	Habit	Habitat	Distribution
1201	Buddlejaceae	*Buddleja parvifolia*	Genus: shrub to small tree	Moist forest, forest margin, bushland; 1,000–2,700 m	Widespread
1201	Caesalpinoideae	*Tetrapterocarpon geayi*			Madagascar
1201	Celastraceae	*Maytenus senegalensis*	Shrub, small tree 1–9 m	Wooded grassland, bushed grassland, riverine bushland in semi-arid areas. 1–2,100 m altitude	Widespread in east and southern Africa
1201	Oleaceae	*Schrebera alata*	Tree 4–24 m	Dry forest edges, evergreen bushland; 1,500–2,400 m	Widespread
1201	Melastomataceae	*Memecylon spp.*	Shrubs, trees to big trees	Evergreen forest, forest, wooded grassland, mostly low altitude but large trees at high altitude	
1252	Melanophyllaceae	*Melanophylla crenata*			Madagascar
1252	Verbenaceae	*Vitex simplicifolia*	Genus: Shrubs or trees; low and high altitude	Wooded grassland, forest edge, evergreen coastal bushland, moist evergreen forest	Nine species in Kenya; widespread
1262	Araliaceae	*Polyscias fulva*	Tree 18–25 m, deciduous	Wet upland forest, riverine forest, 1,400–2,750 m	South west highlands of Kenya, NW of Tanzania
1262	Combretaceae	*Guiera senegalensis*	Small shrub	Widespread in sandy regions and semi-desert areas of the north	Upper Guinea, north central and Lower Guinea to Sudan, Angola
1262	Euphorbiaceae	*Euphorbia tirucalli*	Shrub or tree with succulent branches, 3–6 m	Dry bushland thicket; 0–1,600 m	Widespread but possibly introduced
1262	Rhamnaceae	*Ziziphus mucronata*	Shrub, tree, 2–15 m, thorns	Riverine, bushland, woodland, wooded grassland; 1–1,950 m	Widespread
1262	Sapotaceae	*Manilkara mabokeensis*	Genus: tree	Moist forest, riverine forest, evergreen forest, deciduous bushland, wooded grassland; 1–2,500 m	About five species, widespread
1343	Melastomataceae	*Memecylon* spp.	Shrubs, trees to big trees	Evergreen forest, forest, wooded grassland, mostly low altitude but large trees at high altitude	
1343	Myrtaceae	*Eugenia* spp.	Shrub, tree	Forest, thicket, riverbed; 1–1,500 m	Widespread in Africa, but hybridizes so complicated
1343	Myrtaceae	*Syzygium* spp.	Tree	Riverine, forest, mist forest; 1,050–2,400 m	Widespread
1343	Ochnaceae	*Brackenridgea zanguebarica*	Shrub, tree, deciduous 1–12 m	Coastal bushland, wooded grassland; 1–450 m	East Africa
1343	Ochnaceae	*Ochna calodendron*	Genus: Shrub, small tree or tree	Dry forest, mist forest, woodland, forest, bushland, coastal forest; all altitudes	Widespread
1343	Rubiaceae	*Gardenia* spp.	Shrubs, small trees	Moist forest, wooded grassland, riverine woodland; 1–2,100 m	Widespread; five species
1441	Caesalpinoideae	*Hymenaea verrucosa*	Tree 6–25 m	Dry evergreen forest; 1–400 m, coastal	Northern Tanzania, southeastern Kenya, Mozambique
1441	Meliaceae	*Entandrophragma spp.*	Tree >30 m	Forest	Tropical forest

(continued)

Table 10.1 (continued)

Specimen No.	Family	Genus, species	Habit	Habitat	Distribution
1441	Mimosoideae	*Acacia abyssinica*	Tree 6–20 m	*Acacia* woodland, wooded grassland; forest edge; 1,200–2,300 m	Tanzania, Kenya to southern Africa
1441	Mimosoideae	*Acacia sieberiana*	Tree 3–15 m	Wooded grassland, occasionally riverine; 1–1,250 m; 1,150–1,950 m	Widespread
1441	Rutaceae	*Zanthoxylum sp*	Shrubs, small trees, trees; 1–1,800 m	Semi-evergreen or dry bushland, wooded grassland, moist or dry forest; 1–2,600 m	Widespread
1446	Meliaceae	*Entandrophragma cylindricum*	Large tree	Forest	West Africa
1446	Mimosoideae	*Acacia mellifera*	Shrub, small tree 1.5–9 m	Dry *Acacia* woodland, *Acacia-Commiphora* woodland, *Combretum* woodland; 1–1,800 m	Widespread
1446	Mimosoideae	*Acacia rovumae*	Tree 10–15 m	Riverine, coastal forest; 1–200 m	SE Kenya, NE Tanzania
1446	Mimosoideae	*Acacia senegal*	Tree or shrub, 1–12 m	*Acacia* or *Acacia-Commiphora* bushland; 1–1,900 m	Sudan to South Africa
1446	Mimosoideae	*Acacia seyal*	Tree or shrub, 1–12 m	Dry bushland, grassland, bushland; 200–2,250 m	Widespread, mostly on mbuga
1446	Mimosoideae	*Acacia tortilis*	Tree 2–18 m	Bushland, riverine, dry bushland; 200–1,650 m	Widespread in Africa
1451	Celastraceae	*Cassine sakalava (Elaeodendron)*	Genus: Tree or shrub	Deciduous woodland, riverine, dry evergreen forest, wooded grassland, coastal	Widespread
1451	Clusiaceae	*Catha edulis*	Shrub or tree, 2–12 m, evergreen	Evergreen forest, riverine forest, thicket in *Combretum* wooded grassland; 1,200–2,400 m	Widespread
1451	Flacourtiaceae (Achariaceae)	*Kiggeleria africana*	Tree, 13 m	Evergreen forest among rocks and mountain grasses	Tanzania and southwards
1451	Flacourtiaceae	*Caloncoba ?*			
1453	Dichapetalaceae	*Dichapetalum spp.*	Lianas, shrub, scrambling trees	Moist forest in uplands, lowlands, coastal, evergreen forest	Widespread in woodlands and forests
1456	Caesalpinoideae	*Delonix elata*	Tree 2.5–18 m	*Acacia-Commiphora* bushland, riverine woodland; 100–1,200 m	East Africa
1456	Caesalpinoideae	*Detarium senegalense*	Tree 15 m	Forest, wooded grassland	West Africa
1456	Caesalpinoideae	*Tessmannia africana*	Tree 50 m	Forest	West Africa
1456	Meliaceae	*Entandrophragma spp.*	Tree >30 m	Forest	Tropical forest
1456	Mimosoideae	*Acacia etbaica*	Tree, shrub, 2–10 m	Dry bushland, thicket, wooded grassland; 1–1,800 m	East Africa
1456	Mimosoideae	*Acacia hockii*	Shrub, tree, 1.5–9 m	Wooded grassland, *Acacia* woodland, bushed grassland, deciduous or semi-evergreen forest; 1–1,800 m	Widespread
1456	Mimosoideae	*Acacia polyacantha*	Tree 3.5–18 m	Riverine forest, woodland, impeded drainage; 200–1,800 m	Southern Kenya, Tanzania
1456	Mimosoideae	*Dichrostachys cinerea*	Shrub or tree, 1–12 m	Woodland, bushland, grassland; 1–2,100 m	Widespread

Modern data from Beentje (1994), Oliver (1871), Keay (1958), Coates-Palgrave (2004), van Wyk and van Wyk (1997)

Diversity of Fossil Woods

Of the more than 200 samples collected, 121 have been sectioned (three slides each with the same catalogue number as the macrofossil). Apart from the nine types described and identified here, there are at least another six types with more than one representative specimen and about ten types represented by only one specimen. These 16 types will be described elsewhere. The rest of the samples are too poorly preserved to categorize into informal types. There does not appear to be any difference between woods as far as size, preservation or anatomy are concerned, from Noiti 1 and Noiti 2.

Since several options for the identification of each specimen have been given, the overall diversity appears inflated. Each of the nine specimens described here has at least three other almost identical samples, so it can be assumed that there are more than nine taxa. The total number is not known and cannot be determined because of taphonomic and collecting biases.

Anatomical Descriptions of the Better Preserved Woods

Type A

Catalogue number: BP/16/1201.
Locality: Noiti 1, Laetoli.
Stratigraphy: Lower Laetolil Beds.
Fig. 10.2c–e.

The stem has a diameter of 2 cm and is 4 cm long. There appear to be narrow growth rings on the hand specimen, but there is no evidence of growth rings in the thin section.

Vessels are small, numerous and crowded (average tangential diameter 50 μm and range 30–75 μm, 96 vessels per mm^2) with simple horizontal perforation plates and arranged mostly in short radial multiples of 1–4(-6) vessels. Inter-vessel pits are alternate and less than 5 μm in diameter. Vessel-ray pits were not seen. Vessel element length is unknown. Axial parenchyma is very rare or absent. Rays are biseriate, width 75 μm, average height 341 μm (range 275–450 μm) with uniseriate tails; cells are a similar size in tangential section and square to upright in radial section. Fibers poorly preserved.

Comparison and Identification

Woods with numerous, small vessels in short radial multiples, medium sized alternate inter-vessel pitting and with narrow heterocellular rays occur in members of at least 17 tropical families (Anacardiaceae, Apocynaceae, Celastraceae, Elaeocarpaceae, Erythroxylaceae, Lamiaceae, Lauraceae, LeguminosaeCaesalpinoideae, Lythraceae, Melastomataceae, Myrtaceae, Oleaceae, Rosaceae, Rubiaceae, Salicaceae, Sapotaceae, Scrophulariaceae, Solanaceae).

In the African species of *Maytenus* (Celastraceae) the vessels tend to be somewhat larger than those of the fossil. The species that has the same size vessels has rays more than 1 mm high (*Maytenus senegalensis*). There is no published record of other African woods of *Maytenus. Tetrapterocarpon geayi* of the Caesalpinoideae has matching features but the rays and other elements are storeyed. This tree is from Madagascar. *Memecylon cinnamomoides* (Melastomataceae) is also very similar, but has confluent parenchyma. Vessels in other species of *Memecylon* are mostly exclusively solitary.

Several species of *Buddleja* (Scrophulariaceae) are similar to the fossil wood. *Buddleja parvifolia* is the most similar, but it has large vessels in the mature wood. The fossil wood has a small axis diameter, so it could be juvenile and have smaller vessels.

Several species in the Oleaceae are comparable (Baas et al. 1988; InsideWood database 2006–2007). *Ligustrum* spp. have a greater number of vessels per mm^2; *Schrebera arborea* has larger vessels; *S. trichoclada* has wider and higher rays but *S. alata* is a good match for all the preserved features of the fossil wood. Unfortunately, some of the fine features, such as the fibers, vertical parenchyma strands or crystal inclusions, are not preserved so the identification cannot be easily confirmed.

The extant *Schrebera alata* (common name: wing-leaved wooden pear) is a tree 8–15 m in height, occurring in evergreen forest, dry forest or open woodland from southern Africa to Kenya and Tanzania (Baas et al. 1988; Beentje 1994; Coates-Palgrave 2004). The wood is hard, heavy, durable and makes good fuel (Coates-Palgrave 2004).

Type B

Catalogue number: BP/16/1252.
Locality: Noiti 1, Laetoli.
Stratigraphy: Lower Laetolil beds.
Fig. 10.2f–i.

The fossil wood is yellow-beige in color and is a piece from a trunk of approximately 14 cm diameter. The fragment is 4 cm long and 1.5 × 1 cm wide, not squashed but incomplete. No growth rings were observed. Vessels are solitary and in short radial multiples of 2–3 cells and rare clusters, round to oval in outline with simple, horizontal perforation plates. Mean vessel tangential diameter is 85 μm (range 75–100 μm) and vessel element length is 338 μm (range 105–650 μm). There are 18 vessels per mm^2. Inter-vessel pits are alternate, crowded, and about 5 μm in diameter. Spiral thickening is occasionally preserved. Axial parenchyma is scanty paratracheal to vasicentric with one to two rows of cells around the vessels. The preferential preservation of vessels and cells immediately surrounding them produces a halo, which gives the appearance of aliform parenchyma. No bands of parenchyma are evident. Rays are 4–6 seriate, heterocellular with procumbent, upright and square cells throughout and very high (average height 2,125 µm). Sheath cells are present, but poorly preserved and irregular. Fibers have widely spaced pits on the tangential walls.

Comparison and Identification

Modern woods with the combination of medium diameter vessels that are solitary and in short radial multiples, medium diameter alternate inter-vessel pitting, scanty paratracheal parenchyma and very heterocellular, high rays occur in the Araliaceae, some Euphorbiaceae, Flacourtiaceae, Melanophyllaceae, Myrsinacaeae, Ochnaceae and Verbenaceae. The species within these families that are most similar to the fossil wood are discussed.

Melanophylla crenata has much larger inter-vessel pits than the fossil wood. *Oncostemum pendulum* (Myrsinaceae) is closely comparable but the inter-vessel pitting tends to be larger than that of the fossil. It has well developed sheath cells. Other members of the Myrsinaceae, for example, *Rapanea melanophloeos*, have very similar rays, but the vessels and parenchyma arrangement do not match. *Vitex simplicifolia* (Verbenaceae) is also very similar, but has larger vessels than the fossil.

These genera occur in Africa and surrounding islands today. *Melanophylla* and *Oncostemum* are trees from Madagascar and *Vitex simplicifolia* is a tree or shrub from the Ivory Coast to Central Africa. The latter species is the most similar to the fossil and so the identification rests at *Vitex* sp.

Type C

Catalogue number: BP/16/1262.
Locality: Noiti 1, Laetoli.
Stratigraphy: Lower Laetolil Beds.
Fig. 10.3a–c.

The silicified stem is 1 cm wide and 2.5 cm long with no growth rings visible. Vessels are solitary and in short radial multiples, round to oval in outline with simple horizontal perforation plates and there are 17 vessels per mm^2. Mean vessel tangential diameter is 80 µm, range 70–85 µm, but there are probably two vessel size classes, and the mean vessel element length is 185 µm. Inter-vessel pits are alternate and 10 µm in width. Axial parenchyma is scanty paratracheal to vasicentric. In radial longitudinal view alternating bands of parenchyma and fiber cells are visible. Rays are heterocellular with procumbent, square to upright cells throughout the ray. Average ray width is 55 µm (2–3 cells) and average height is 800 µm.

Comparison and Identification

The combination of mostly solitary medium diameter vessels, scanty paratracheal to vasicentric parenchyma and very heterocellular rays occurs in many woods and in particular some woods of the Araliaceae and Combretaceae, Euphorbiaceae Group B (of Metcalfe and Chalk 1950), Euphorbioideae (of Mennega 2005), some Myrtaceae, Rubiaceae and Sapotaceae.

Polyscias fulva (Araliaceae) has vasicentric parenchyma and rays with sheath cells. Ray cells are also perforated, but the distribution of vessels is not similar to that of the fossil. *Guiera senegalensis* (Combretaceae) is similar and has vessels of two distinct sizes. Important identifying features of the Combretaceae are the vestured pits and included phloem, neither of which is preserved. There are breakdown areas in the fossil transverse section, which may suggest the presence of included phloem originally, but these could also be artifacts of secondary crystallization. *Euphorbia tirucalli* (Euphorbiaceae) shows many similarities, but has significantly fewer vessels. Since the fossil stem shows no growth rings for a tree with a minimum diameter of approximately 14 cm, it cannot be considered as juvenile. *Ziziphus mauritiana* (Rhamnaceae) also shows some similarities, but differences lie in the ray widths as some uniseriate rays are present. It is not possible to tell if the fossil has uniseriate rays. *Manilkara mabokeensis*, *Pouteria heudelotiana* and *P. robusta* of the Sapotaceae have somewhat longer vessel radial multiples than found in the fossil but otherwise are all fairly close. The fossil wood is most similar to that of the Euphorbiaceae, possibly *Euphorbia*, but of an undesignated species because the fine distinguishing details are not preserved. The genus *Euphorbia* exhibits a wide variety in habit, from succulents to herbs to trees.

The more likely identification is a species of *Euphorbia* or *Ziziphus*. The former is a large genus with species occurring in many different habitats, and the latter occurs in dry woodlands.

Type D

Catalogue number: BP/16/1343.
Locality: Noiti 1, Laetoli.
Stratigraphy: Lower Laetolil Beds.
Fig. 10.3d–g.

The silicified stem is coarse-grained, 1.2 cm in diameter and 2.8 mm long. No growth rings are evident. Vessels are exclusively solitary, round to oval in outline, 12 per mm^2 and have

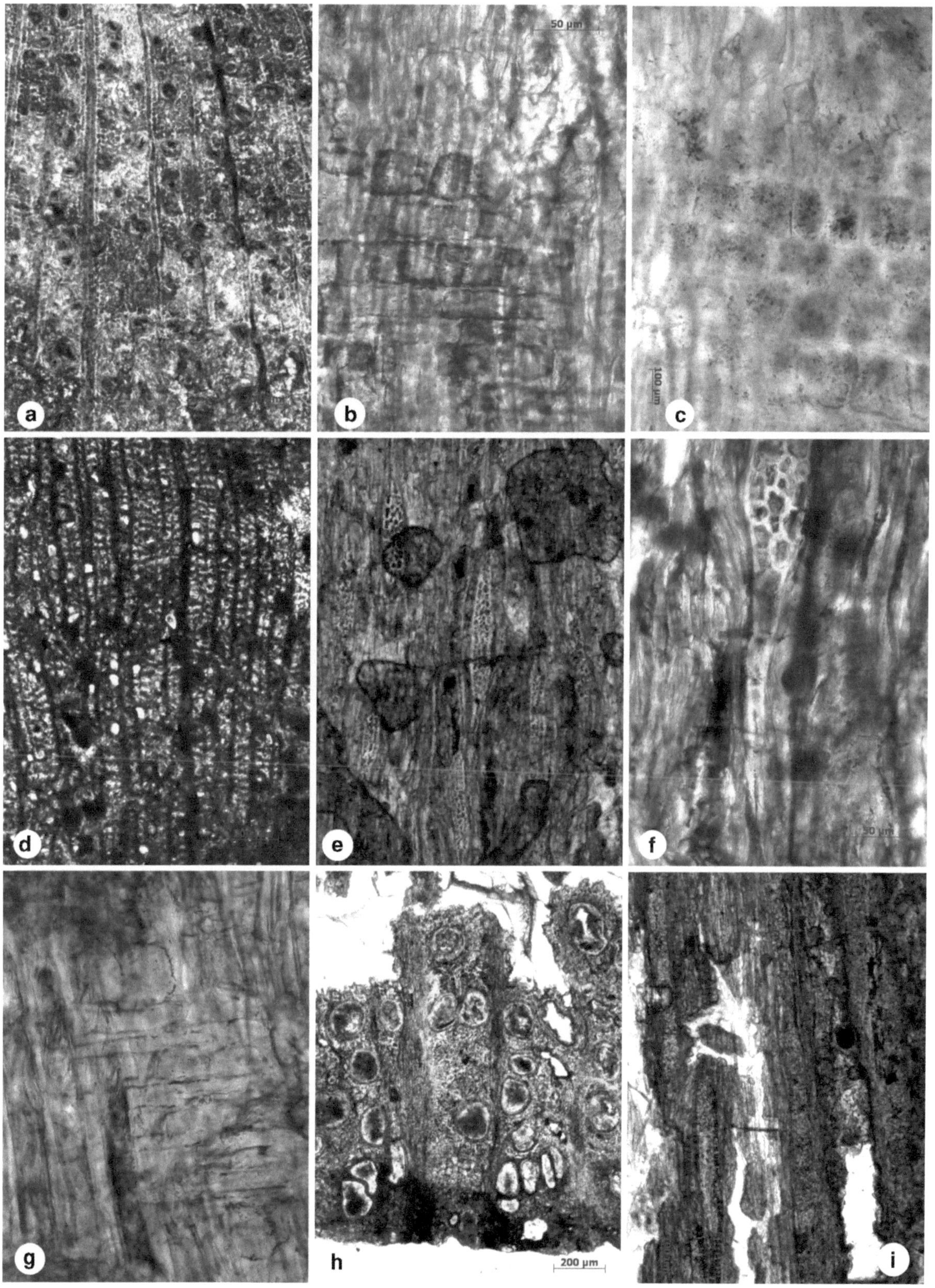

Fig. 10.3 (**a–c**) BP/16/1262 from Noiti 1, A. TS, vessels and scanty paratracheal parenchyma. (**b**) RLS, square and procumbent ray cells. (**c**) RLS, showing square ray cells only. (**d–g**) BP/16/1343 from Noiti 1. (**d**) TS, vessels are light coloured and diffuse parenchyma strands appear as fine horizontal lines between the darker vertical rays. (**e**) TLS, lens-like rays with clear cells, 3–4 cells wide. (**f**) TLS detail of four elongated marginal ray cells. (**g**) RLS, with procumbent body cells of the ray. (**h–i**) BP/16/1441 from Noiti 1. (**h**) TS, vessels large and solitary or in short radial multiples, parenchyma vasicentric to aliform. (**i**) TLS, lens-shaped homocellular rays

an average tangential diameter of 40 µm, range 30–50 µm. Vessel element length and type of perforation plate are unknown. Axial parenchyma is apotracheal, diffuse-in-aggregate with narrow lines of parenchyma between the rays. Rays are 3–4 cells wide (50 µm), up to 1,000 µm high and five rays per linear mm. The rays are markedly heterocellular with long tails of three or more square to upright cells with body cells that are mostly procumbent but also square. Tracheids are common and 20 µm in diameter, but poorly preserved.

Comparison and Identification

Woods with the combination of features of a few small solitary vessels, diffuse-in-aggregate parenchyma and high heterocellular rays occur in the Melastomataceae, Myrtaceae, Ochnaceae and Rubiaceae.

Memecylon spp. (Melastomataceae) are very similar, but the vessels are larger and more numerous than in the small fossilized stem. *Decaspermum* spp. (Myrtaceae) are also similar, but the vessels are bigger. This genus does not occur in Africa, but *Eugenia* and *Syzygium* do. Species of the latter genus tend to have narrower rays, much larger vessels and rarely have diffuse-in-aggregate parenchyma. Some members of *Eugenia* have large or small vessels and diffuse-in-aggregate parenchyma; *E. buxifolia* and *E. mespiloides* are similar, except for the more numerous vessels and possibly narrower rays. Important features to confirm identification to Myrtaceae are the presence of small vestured pits and crystals in parenchyma cells, neither of which is preserved in the fossil. *Brackenridgea* of the Ochnaceae is similar, except for the larger and more numerous vessels and more abundant parenchyma. Other genera in this family, such as *Diporidium, Lophira* and *Ochna,* are also similar, but tend to have very numerous vessels. *Ochna calodendron* is the most similar with only slightly bigger vessels than the fossil. Features to confirm the identification, such as crystals in ray cells, are not preserved. Some *Gardenia* species (Rubiaceae) are similar, but the vessels are much more numerous.

Possible identifications: *Memecylon* sp., *Eugenia* sp., *Gardenia* sp. and *Ochna* sp.

The African species of these taxa occur in forest to wooded grassland and at a variety of altitudes.

Type E

Catalogue number: BP/16/1441.
Locality: Noiti 1.
Stratigraphy: Lower Laetolil Beds.
Figs. 10.3h and 10.4a, b.

Growth rings are almost parallel on the fragment of wood (1.5 cm wide and 8 cm long) so it represents a large diameter trunk. Growth rings are not obvious in the thin section. Vessels are irregular in size and shape and they are arranged in short radial multiples of 1–3 cells and clusters are occasionally present especially in bands, which may represent growth ring boundaries. There are 16 vessels per mm^2 with an average tangential diameter of 110 µm, range 100–150 µm and mean length of 400 µm. No pitting is visible, but the perforation plates are simple and horizontal. Parenchyma is vasicentric to aliform with 3–4 rows of cells around each vessel. Because the tissue between vessels has been recrystallized it is not possible to determine if there is any aliform-lozenge or confluent parenchyma. Rays are wide, 5–6 cells (100 µm) and mostly high (300–) 600–800 µm, with 3 rays per linear mm. Rays are homocellular, comprised of only procumbent body cells. There may be one marginal upright cell in some rays, but the preservation is not good enough to be certain. Uniseriate rays also occur. Crystals occur in the external cells of the rays and form dark staining chains up to 27 cells long.

Comparison and Identification

There are woods with vessels arranged in short radial multiples, vasicentric to aliform parenchyma, rays of medium width and medium-to-high in height with crystal chains in the Leguminosae, Meliaceae and Rutaceae. *Hymenaea verrucosa* (Leguminosae, Caesalpinoideae) is very similar to the fossil, but possibly has more parenchyma in the form of confluent parenchyma. The fossil is small and the vessels are crowded together, so there is little space for parenchyma. *Albizia* species (Leguminosae, Mimosoideae) are also similar, but often the rays are narrower. Some *Acacia* species are very similar, for example, *A. abyssinica,* and *A. sieberiana. Entandrophragma* species (Meliaceae) are also similar with the same type of rays. Some species have storeyed elements (*E. angolense, E. candollei, E. congoense, E. cylindricum, E. utile*). *Zanthoxylum* species (Rutaceae) do not have storeyed rays and the amounts of parenchyma vary from scanty paratracheal to confluent. The ray types also vary in the African species with *Z. decaryi, Z. madagascariense, Z. tsihanimposa* and *Z. thouvenotii* having slightly heterocellular rays with the body cells procumbent and 1–2–4 rows of marginal upright cells. *Z. heitzii* and *Z. macrophyllum* have homocellular rays with all cells procumbent. These differences are very subtle and not discernable in the fossil woods. The legume woods have vestured pits and those of the Meliaceae and Rutaceae do not. The Rutaceae have small inter-vessel pits. Storeyed structures are mostly absent from the Rutaceae, present in the Leguminosae and common in the Meliaceae (Metcalfe and Chalk, 1950).

Possible identifications: *Acacia* cf. *abyssinica, A. sieberiana; Zanthoxylum* spp. The two *Acacia* species occur in dry vegetation types and those of *Zanthoxylum* occur in moist or dry forest and variations of more open woodland.

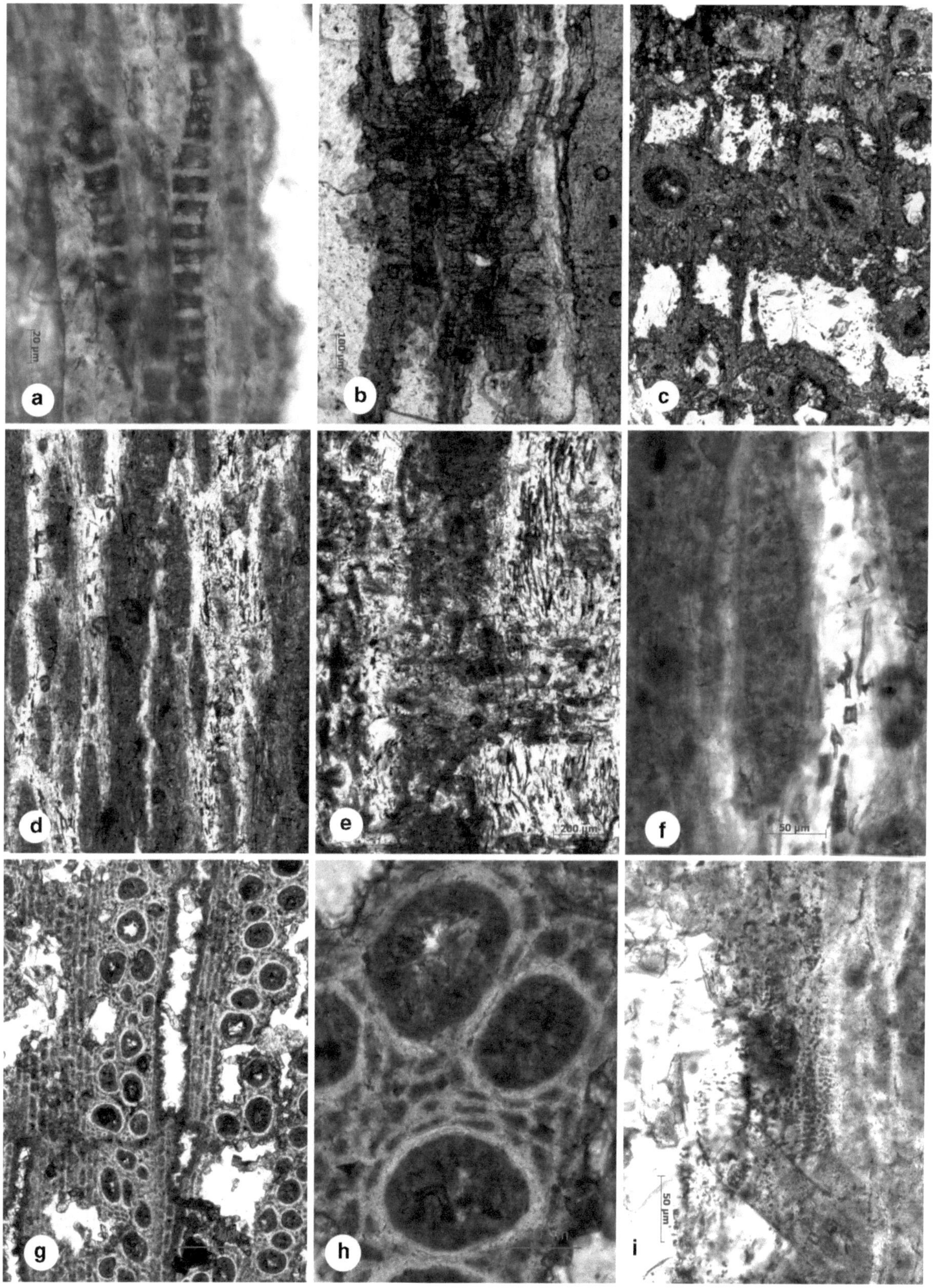

Fig. 10.4 (**a–b**) BP/16/1441 from Noiti 1. (**a**) TLS, dark rectangular crystals in the outer cells of rays. (**b**) RLS, narrow vertical cells are the procumbent ray cells. (**c–f**) BP/16/1446 from Noiti 2. (**c**) TS, large, mostly solitary vessels; light areas are devoid of cells, wide rays. (**d**) TLS, lens-shaped homocellular rays clearer in the upper part of the picture where the surrounding fibers have been lost. (**e**) TLS, detail of one ray with round cells in this view, and (**f**) elongated in the RLS, i.e. procumbent ray cells. (**g–i**) BP/16/1451 from Noiti 2. (**g**) TS, mostly solitary, dark vessels with wide rays and very little ground tissue. (**h**) TLS, inter-vessel pits are very small, dark and alternately arranged (rare because vessels are mostly solitary). (**i**) RLS, black dots are the intervessel pits in alternate arrangement and each one more than 7 µm in diameter

Type F

Catalogue number: BP/16/1446.
Locality: Noiti 2, Laetoli.
Stratigraphy: Lower Laetolil Beds
Fig. 10.4c–f.

Growth rings in the hand specimen indicate that the trunk was at least 24 cm in diameter but the fragment is 5×2 cm wide and 7 cm long, not compressed but incomplete. Vessels are in short radial multiples and solitary with a frequency of eight vessels per mm^2. The mean vessel tangential diameter is 150 μm, range 125–165 μm; vessel element length, perforation plates and pitting are unknown. Axial parenchyma is aliform to confluent and banded; vessels are within the parenchyma bands. These parenchyma bands are more than three cells wide and alternate with fiber bands that are of a similar width. This pattern is also evident in the radial longitudinal section. Rays are wide and homocellular made up of only procumbent cells, 2–4–5 cells wide (100 μm) and 25–40 cells high (375–500 μm). There are 5 rays per linear mm. Crystal chains are preserved as dark squares packed closely together, up to 21 crystals long.

Comparison and Identification

Woods with large vessels in short radial multiples, abundant paratracheal and banded apotracheal parenchyma, homocellular rays and crystal chains associated with the rays occur in the Leguminosae, Mimosoideae and the Meliaceae. *Acacia* and *Parkia* of the Mimosoideae are comparable with the fossil. In the acacias the ray width, height and amount of parenchyma varies, but the fossil is most similar to *Acacia mellifera, A. rovumae, A. seyal, A. senegal, A. sieberiana and A. tortilis.* The differences between these species are very subtle and the presence of vestured pits is necessary to confirm the identification as legumes. The *A. mellifera* samples measured by Robbertse et al. (1980) have an average ray width of 4.6 cells, height 523 μm, 5.8 rays per linear mm and 16 crystals per mm^2 which is almost the same as for the fossil. *Acacia tortilis* is very similar with average ray width 6.2 cells, height 557 μm, 4.5 rays per linear mm and 13 crystals per mm^2 (Robbertse et al. 1980). Höhn (1999) was not able to differentiate between *A. seyal, A. senegal,* and *A. sieberiana*, but they have wide and high rays and the same type of crystal chains. *Parkia biglobosa* is also very similar, but the illustration in Höhn (1999, Fig. 33) shows very low rays. *Entandrophragma cylindricum* (Meliaceae) is the same as the fossil, except for the lack of storeyed rays in the fossil.

Possible identification: *Acacia mellifera, A. rovumae, A. seyal, A. senegal, A. sieberiana and A. tortilis.* These are all common trees in East Africa today in the drier areas.

Type G

Catalogue number: BP/16/1451.
Locality: Noiti 2, Laetoli.
Stratigraphy: Lower Laetolil Beds.
Figs. 10.4g–i and 10.5a.

The fossil wood is very fine grained and is a 1.5 cm diameter stem, 2.5 cm long. No growth rings are visible. Vessels are solitary, but numerous and crowded with 32 vessels per mm^2, round in outline with an average tangential diameter of 104 μm (range 70–125 μm), length unknown. Intervessel pitting is alternate and large. Axial parenchyma is absent or very rare. Rays are 5–8 cells wide (50–250 μm) and very high (2,125 μm), heterocellular with procumbent and square body cells, and square marginal cells up to four cells high. A few crystals are associated with the rays, but it is not possible to see in which cells they actually occur. Spiral thickening is visible on some cells, probably the tracheids.

Comparison and Identification

The combination of small to medium, solitary and numerous vessels, scarce or absent parenchyma, and very high heterocellular rays occurs in woods of the Flacourtiaceae (Achariaceae), Celastraceae and Clusiaceae. *Camptostylus mannii* (Flacourtiaceae) has these features, but the vessel density is slightly greater than the fossil. *Kiggelaria africana*, of the same family, is also very close but the illustration in Kromhout (1975, Fig. 115) shows vessels in short radial multiples and rays that are narrower and much shorter than those of the fossil. This is confirmed by Miller (1975). Another African genus, *Caloncoba*, has more comparable vessel diameter and ray features (Miller 1975). Other African members of the Flacourtiaceae generally have smaller vessels and lower rays (Miller 1975).

Several genera from the Celastraceae are comparable with this fossil wood. The vessels of *Cassine sakalava* are a little on the small size and the rays are a bit shorter. There are two sizes of rays in the extant wood, but the fossil wood is too poorly preserved to determine the presence of uniseriates. *Catha edulis* is similar to *Cassine sakalava. Maytenus senegalensis* is another taxon to be considered, but the rays are only 1–3 cells wide. *Marila* (Clusiaceae) is similar, but has scalariform perforation plates. Other members of the Clusiaceae have larger vessels and more parenchyma (Metcalfe and Chalk 1950).

Possible identifications: *Kiggelaria africana, Caloncoba spp. Maytenus spp.*

Kiggeleria africana is evergreen and occurs in evergreen forests; *Caloncoba* spp. also occur in moister vegetation types, but *Maytenus* species occur in a variety of dry to moist woodlands and riverine bushland.

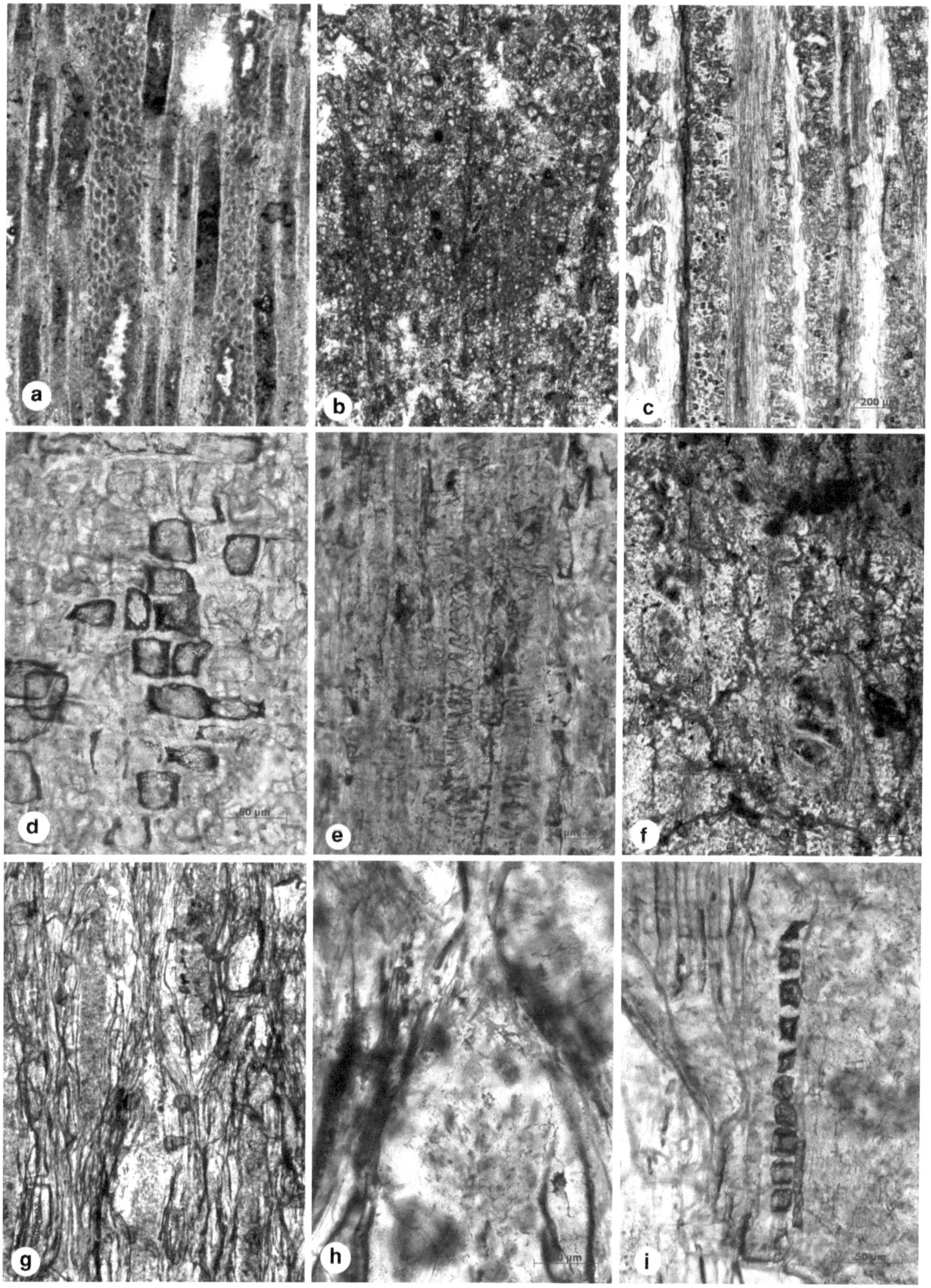

Fig. 10.5 (**a**) BP/16/1451 from Noiti 2, TLS, very long rays, 4–5 cells wide and oval in shape; vessel has a partial infill of dark possibly resinous material. (**b–d**) BP/16/1453 from Noiti 2. (**b**) TS, small vessels, solitary or in short radial multiples with wide rays in between. (**c**) TLS, ray 5–6 cells wide and heterocellular. (**d**) RLS, ray cells are square, procumbent and upright, often containing crystals. (**e**) RLS, spiral thickening on tracheid walls. (**f–i**) BP/16/1456 from Noiti 2. (**f**) TS, three vessels in a radial line, wide rays and vasicentric parenchyma. (**g**) TLS, lens-shaped rays comprising small round cells with sinuous thin-walled fibers surrounding them. (**h**) TLS, detail of ray with small round cells and (**i**) a few dark angular crystals

Type H

Catalogue number: BP/16/1453.
Locality: Noiti 2, Laetoli.
Stratigraphy: Lower Laetolil Beds.
Fig. 10.5b–d.

The stem is 1.5 cm wide and 3 cm long. No growth rings are visible. Vessels are small, numerous and crowded, solitary or in radial multiples of 1–4(-6) cells with a density of more than 50 per mm^2. Average tangential diameter is 40 μm and range 30–50 μm. Length and pitting are unknown. Inter-vessel pits are alternate and less than 5 μm in diameter. Axial parenchyma is absent. Rays are very high, more than 1,000 μm, and 2–4(5) cells wide (25 μm), heterocellular with mixed procumbent and square cells throughout. There are four rays per mm. They are stacked vertically one above the other with only 2–3 curved fibers between them. Crystals are present in some square ray cells. Tracheids with spiral thickening and pitting are present and thin walled fibers form the ground tissue.

Comparison and Identification

Woods with high heterocellular rays, no parenchyma and small, numerous vessels in low radial multiples occur in the Dichapetalaceae, Rhamnaceae, Salicaceae, and some other families. *Berchemia discolor* (Rhamnaceae) differs slightly from the fossil: the ray body cells are all procumbent and there are several rows of upright marginal cells. *Ziziphus mauritiana* (Rhamnaceae) has fewer vessels but more parenchyma than the fossil, and *Z. mucronata* is not at all comparable. *Dovyalis zizyphoides* (Salicaceae) has scalariform perforation plates and the same ray type as *B. discolor. Dichapetalum* species are the most similar with numerous small vessels, little or no parenchyma, tracheids abundant and the same very heterocellular rays as the fossil. Fine details, such as the presence of sheath cells in the rays, are not preserved.

Most of the African members of the Dichapetalaceae are lianas (44 out of 53 species; Hauman 1958) and occur in riverine, gallery and other forest types. Lianas commonly have very high rays and large vessels (Ewers 1985), but not all species of *Dichapetalum* do.

Type I

Catalogue number: BP/16/1456.
Locality: Noiti 2, Laetoli.
Stratigraphy: Lower Laetolil Beds.
Fig. 10.5f–i.

The piece of fossil wood is 2 × 1 cm wide with a longitudinal section missing, and 3.5 cm long and is part of a much larger trunk, beige in color. No growth rings are visible. Vessels are solitary or in short radial multiples of 2–3 cells, nine vessels per mm^2 and with a mean tangential diameter of 150 μm (range 75–180 μm); length and pitting unknown. Axial parenchyma is vasicentric. Rays are 4–6 cells wide (115 μm) and on average 600 μm high. All cells are procumbent and some of the edge cells have dark crystals forming chains of seven or more cells. Often only these crystal chains are preserved. Fibers are thick-walled and curve around the rays in tangential longitudinal view.

Comparison and Identification

Various woods of the three subfamilies of the Leguminosae and the Meliaceae have the combination of large vessels in low radial multiples, vasicentric parenchyma and homocellular rays with only procumbent cells, and with crystal chains. *Delonix regia* and *Detarium senegalense* (Caesalpinoideae), for example, are similar, but they have more parenchyma than the fossil. *Tessmannia africana* (Caesalpinoideae) has less parenchyma. *Dichrostachys cinerea* (Mimosoideae) also has more parenchyma. In general, the rays of the Mimosoideae are narrow, except for those of *Acacia*, which range from 1–10 cells wide (Evans et al. 2006). The species with large vessels (around 150 μm) and rays 4–6 cells wide are *Acacia etbaica, A. hockii, A. polyacantha, A. sieberiana*, and *Acacia tortilis*. The fossil wood is not sufficiently well preserved to distinguish between them, nor can the presence of vestured pits be confirmed, a feature of the Leguminosae, but not of the Meliaceae.

This wood is most likely a species of *Acacia* of dry regions and was a large tree. These species are common in East Africa today in various forms of woodland and occur through a wide range of altitudes.

Vulnerability Index

As mentioned previously, wood functions in the uptake of water and support. Wood anatomical features can be used in a non-taxonomical way to infer climate, particularly water availability and temperature. Wolfe and Upchurch (1987) used Carlquist's Vulnerability Index (1975), which is a measure of the susceptibility or vulnerability of wood to water shortages or drought conditions, and calculated the values for woods from a wide variety of climates. Since the water uptake capacity of a vessel is a function of the fourth power of its radius, then large vessels have considerably more capacity than small vessels. A few large vessels would be more efficient at water uptake than many small ones, but conversely they are more vulnerable. If one large vessel is blocked from embolisms or damage, the effect is much worse than when one small vessel is blocked. The potential uptake of water is also dependent on the number of vessels. The vulnerability index (V) is the mean tangential diameter of vessels (in μm) divided by the mean number of vessels per square mm.

Vessel diameter and density is also related to the size of the plant. Lianas, for example, have higher V values than trees or shrubs, because on average they have wider vessels (Wolfe and Upchurch 1987). Trees generally have bigger vessels than shrubs, whereas shrubs and small trees tend to have more vessels, so large trees have higher V values than small trees and shrubs. The average value for mesic woody plants is 2–3, and megathermal rainforest tress have values greater than this (Wolfe and Upchurch 1987). Trees in tropical arid environments, however, have high V values probably because they have deep tap roots and utilise ground water, whereas those with shallow roots systems would have low V values (Baas et al. 1983). Other wood anatomical features can be correlated with climatic variables, but these are much more complicated (Wiemann et al. 1998, 2001) and require quantitative data relating to some features that have are not been preserved in the Noiti fossil woods.

Table 10.2 shows the vulnerability indices for the nine fossil woods described here, and the interpretations from the modern data generated by Wolfe and Upchurch (1987). The V values for the Noiti woods fall mostly between 3 and 7, with two below 1 and two over 16. The climate inferences show a range from warm mesothermal (13–20°C) to mostly megathermal (20–30°C) and from warm to dry conditions. The environmental interpretations from the vulnerability values are mostly compatible with inferences from the taxonomic affinities. Sample BP/16/1343 does not match. A possible reason could be mis-identification of the wood, or that one other anatomical feature, such as the type of parenchyma or ray type, compensates for the apparent vulnerability of the wood (Wiemann et al. 1998).

For woods from the same region there is rather a lot of variability in the climate signal. The temperature range could represent higher (mesothermal) and lower (megathermal) altitudes on the volcano, but this may be complicated by the amount of insolation. The apparent difference in moisture availability could also be a factor related to altitude, but other causal factors could include soils, aspect and topography (fluves, interfluves).

Discussion

The rather poorly preserved woods cannot be confidently identified to specific taxa because the fine discriminating details are not visible. The most likely candidates for each specimen are given so that all possible identifications can be considered (Table 10.1). The most obvious result is that a wide variety of vegetation types are implied, from moist evergreen forest (*Catha edulis, Entandrophragma* spp., *Gardenia* spp., *Kiggelia africana, Polyscias fulva*) to dry forest (*Ochna calodendron, Schrebera alata*), dry bushland (*Acacia* spp., *Ziziphus mucronata*), wooded grassland (*Acacia* spp., *Maytenus senegalensis, Vitex simplicifolia*) and dry bushland thicket (*Euphorbia tirucalli*). The trees also show a variety of soil types: sandy soils (*Guiera senegalensis*) and heavy seasonally water-logged soils (*Acacia seyal*). There are also a variety of altitudes indicated, from coastal (*Acacia rovumae, Dichapetalum* spp., *Cassine sakalava*) to uplands (*Buddleja parvifolia, Memecylon* spp., *Polyscias fulva*).

Such a diversity of vegetation associations implies either a mixed source or a long time period. The latter is unlikely as the woods are entombed in the same lahar. A mixed source is therefore more likely. Origins would not have been from a wide geographic area, but rather from a great range in altitude, such as that found on a mountain side. The volcanic debris and slurry flowing down the mountain side or volcano would have collected plant material from the source (top) to the bottom where it came to rest and solidified. The debris would include volcanic material, lithic material, older lahars, as well as living vegetation and plant litter. Under suitable conditions of anoxia and free mineral ions (silica, carbonates, iron) in solution, petrifaction of the wood is possible.

The range in sizes of wood fragments, small twigs to parts of trunks over 24 cm in diameter, indicates that large quantities of material were transported in the slurry. Thus, a good yet random representation of the vegetation would be expected. Since the sampling of the wood was not very systematic, based on quality of preservation, it is not possible to provide data on numbers of trunks, twigs, trees, shrubs, etc. that were preserved. There may also be a taphonomic bias towards woods that are more durable, more porous, lighter, heavier or degraded, so abundance and species diversity are unmeasurable. What can be deduced is that vegetation from high mountains (possibly up to 2,700 m) and the whole range down to lowlands was present. This is further supported by the ranges in the vulnerability indices of the woods (Table 10.2).

Some distinctive taxa are missing from the fossil record. Trees that are common in the area today are *Balanites aegyptiaca, Combretum* spp., *Croton* spp. *Euclea* and *Ficus* spp. and the conifer *Juniperus procera*. Their absence in the fossil record, however, does not mean they were absent in the Pliocene. These woods may not have been sufficiently well preserved for identification or may not have been collected. Further collecting and sectioning will only add to our knowledge of the Pliocene vegetation in the Olaitole and Garusi areas.

Table 10.2 Calculated vulnerability indices for the fossil woods described here. Vulnerability index = mean vessel tangential diameter divided by the mean number of vessels per square mm (Carlquist 1975). Implied habit and environment from data from a variety of modern forests and woodlands (Wolfe and Upchurch 1987, Table 2) and from taxonomic affinity. Mesothermal = mean annual temperature from 13–20°C; megathermal = mean annual temperature from 20–30°C. Shrub = multi-stemmed woody plant

Catalogue No.	Size	Vessel diameter (μm)	Vessel number per mm²	Vulnerability = diameter/number	Implied habit and environment from vulnerability index	Implied environment from taxonomic affinity (Table 10.1)
1201	Twig	50	96	0.5	Tree/shrub, mesothermal dry/arid, or mesothermal riparian	Dry forest edge, evergreen bushland
1252	Twig	85	18	4.7	Medium tree, megathermal mesic	Wooded grassland, forest edge, moist evergreen forest
1262	Twig	80	17	4.7	Medium tree, megathermal mesic	Wet upland forest, dry bushland thicket, riverine bushland
1343	Twig	40	12	3.3	Shrub/small tree, megathermal arid	Forest, thicket, riverine, mist forest, woodland
1441	Trunk	110	16	6.9	Medium/small tree, megathermal dry	Wooded grassland, forest edge, riverine
1446	Trunk	150	8	18.7	Large tree, megathermal dry/mesic	Dry woodland, bushland, riverine
1451	Twig	104	32	3.2	Shrub/small tree, megathermal arid	Deciduous woodland, dry evergreen forest
1453	Twig	40	60	0.6	Shrub/small tree, megathermal dry, or tree, mesothermal riparian	Moist forest in uplands, lowlands, coastal, evergreen forest
1456	Trunk	150	9	16.7	Large tree, megathermal dry	Forest, wooded grassland, dry woodland

The diversity of vegetation types indicated by the wood paleoflora is well supported by faunal evidence. Andrews (1989) suggested that Laetoli was a well wooded environment, and since then other authors have agreed (Reed 1997; Kingston and Harrison 2007). The pollen flora (Bonnefille and Riollet 1987) shows that there was montane vegetation, grassland and woodlands with many taxa from today also present in the region during the Pliocene. The wood flora indicates that locally there may have been some more tropical elements and that the vegetation was species rich and structurally complex. Examples of the tropical elements are the West African taxa *Detarium senegalense, Entandrophragma* spp., *Guiera senegalensis* and *Tessmannia africana*, and the possible occurrence of the Madagascan taxa *Melanophyllum crenata, Tetrapterocarpon geayi*, and the liana *Dichapetalum* spp.

Acknowledgements I am grateful to Louis Scott and Terry Harrison for the invitation to take part in the Laetoli expeditions, and to all the team members who helped collect fossil woods. Mr. D. Mbense is thanked for making the thin sections. The Tanzanian Commission for Science and Technology (COSTECH), the Tanzanian Department of Antiquities and the Ngorongoro Conservation Area Authority are also thanked for providing permits to carry out the research. This work was supported by NSF grants BCS-9903434 and BCS-0309513 (to Terry Harrison) and a PAST (Palaeontological Scientific Trust, South Africa) grant (to MKB).

References

Andrews, P. (1989). Palaeoecology of Laetoli. *Journal of Human Evolution, 18*, 173–181.

Andrews, P. J., Bamford, M. K., Njau, E. F., & Leliyo, G. (2011). Past and present vegetation ecology of Laetoli, Tanzania. In T. Harrison (Ed.), *Paleontology and geology of Laetoli: Human evolution in context* (Geology, geochronology, paleoecology and paleoenvironment, vol. 1, pp. 167–200). Dordrecht: Springer.

Baas, P., Werker, E., & Fahn, A. (1983). Some ecological trends in vessel characters. *IAWA Bulletin, n.s. 4*, 141–159.

Baas, P., Esser, P. M., & van der Westen, M. E. T. (1988). Wood anatomy of the Oleaceae. *IAWA Bulletin, n.s. 9*, 103–182.

Beentje, H. J. (1994). *Kenya trees, shrubs and lianas*. Nairobi: National Museums of Kenya.

Bonnefille, R., & Riollet, G. (1987). Palynological spectra from the Upper Laetolil Beds. In M. D. Leakey & J. M. Harris (Eds.), *Laetoli: A Pliocene site in northern Tanzania* (pp. 52–61). Oxford: Clarendon.

Buurman, P. (1972). Mineralization of fossil wood. *Scripta Geologica, 12*, pp. 1–41.

Carlquist, S. (1975). *Ecological strategies of xylem evolution*. Berkeley: University of California Press.

Coates-Palgrave, K. (2004). *Trees of southern Africa* (2nd ed.). Cape Town: Struik Publishers. Revised by M. Coates-Palgrave.

Deino, A. L. (2011). $^{40}Ar/^{39}Ar$ dating of Laetoli, Tanzania. In T. Harrison (Ed.), *Paleontology and geology of Laetoli: Human evolution in context* (Geology, geochronology, paleoecology and paleoenvironment, vol. 1, pp. 77–97). Dordrecht: Springer.

Drake, R., & Curtis, G. H. (1987). K-Ar chronology of the Laetoli fossil localities. In M. D. Leakey & J. M. Harris (Eds.), *Laetoli: A Pliocene site in northern Tanzania* (pp. 48–52). Oxford: Clarendon.

Evans, J. A., Gasson, P. E., & Lewis, G. P. (2006). Wood anatomy of the Mimosoideae (Leguminosae). *IAWA Journal, Suppl. 5*, 117.

Ewers, F. W. (1985). Xylem structure and water conduction in conifer trees, dicot trees, and lianas. *IAWA Bulletin, n.s. 6*, 309–317.

Harrison, T., & Kweka, A. (2011). Paleontological localities on the Eyasi plateau, including Laetoli. In T. Harrison (Ed.), *Paleontology and geology of Laetoli: Human evolution in context* (Geology, geochronology, paleoecology and paleoenvironment, vol. 1, pp. 17–45). Dordrecht: Springer.

Hauman, L. (1958). Dichapetalaceae. In *Flore du Congo Belge et du Ruanda-Urundi, Spermatophytes*, (Vol. VII, pp. 287–348). Bruxelles: INEAC.

Hay, R. L. (1987). Geology of the Laetolil Beds. In M. D. Leakey & J. M. Harris, (Eds.), *Laetoli: A Pliocene site in northern Tanzania* (pp. 23–47). Oxford: Clarendon.

Höhn, A. (1999). Wood anatomy of selected West African species of Caesalpinoideae and Mimosoideae (Leguminosae): A comparative study. *IAWA Journal, 22*, 115–146.

Ilic, J. (1987). The CSIRO family key for hardwood identification. CSIRO Division Chemical and Wood Technology Technical Paper No. 8.

InsideWood, (2006–2007). http://insidewood.lib.ncsu.edu/search/

International Association of Wood Anatomists Committee (1989). IAWA list of microscopic features for hardwood identification. *IAWA Bulletin, n.s. 10*, 219–332.

Keay, R. W. J. (1958). Caesalpiniaceae; Combretaceae. In *Flora of west tropical Africa* (2nd ed., Vol. 1, Part 2, pp. 439–484; Vol. 1, Part 2, pp. 262–281). London: Crown Agents for Overseas Governments and Administrations.

Kingston, J. D., & Harrison, T. (2007). Isotopic dietary reconstruction of Pliocene herbivores at Laetoli: Implications for early hominin palaeoecology. *Palaeogeography, Palaeoclimatology, Palaeoecology, 243*, 272–306.

Kromhout, C. P. (1975). 'n Sleutel vir die mikroskopiese uitkenning van die vernaamste inheemse houtsoorte van Suid-Afrika. *South African Department of Forestry, Pretoria, Bulletin, 50*, 124.

Lebacq, L., & Dechamps, R. (1964). Essai d'identification anatomique des bois d'Afrique centrale. *Annales du Musée Royal de l'Afrique Centrale, Tervuren, Belgique. Serie 8. Sciences Economiques, No 3*, 1–101.

Mennega, A. M. W. (2005). Wood anatomy of the subfamily Euphorbiaceae – a comparison with subfamilies Crotonoideae and Acalyphoideae and the implications for the circumscription of the Euphorbiaceae. *IAWA Bulletin, 26*, 1–68.

Metcalfe, C. R., & Chalk, L. (1950). *Anatomy of the dicotyledons* (vol. 2). Oxford: Clarendon.

Miller, R. B. (1975). Systematic anatomy of the xylem and comments on the relationships of the Flacourtiaceae. *Journal of the Arnold Arboretum, 56*, 20–102.

Mollel, G. F., Swisher, C. C., Feigenson, M. D., & Carr, M. J. (2011). Petrology, geochemistry and age of Satiman, Lemagurut and Oldeani: Sources of the volcanic deposits of the Laetoli area. In T. Harrison (Ed.), *Paleontology and geology of Laetoli: Human evolution in context* (Geology, geochronology, paleoecology, and paleoenvironment, vol. 1, pp. 99–119). Dordrecht: Springer.

Mosbrugger, V., & Utescher, T. (1997). The coexistence approach – a method for quantitive reconstructions of Tertiary terrestrial palaeoclimate data using plant fossils. *Palaeogeography, Palaeoclimatology, Palaeoecology, 134*, 61–86.

Oliver, D. (1871). Caesalpinieae; Combretaceae. In *Flora of tropical Africa. I* (Vol. I, pp. 258–321; Vol. II, pp. 413–436). London: Reeve.

Reed, K. E. (1997). Early hominid evolution and ecological change through the African Plio-Pleistocene. *Journal of Human Evolution, 32*, 289–322.

Robbertse, P. J., Venter, G., & Janse van Rensburg, H. (1980). The wood anatomy of the South African acacias. *IAWA Bulletin, 1*, pp. 93–103.

Stanier, P., & Gilbert, G. (1958). Meliaceae. In *Flore du Congo Belge et du Ruanda-Urundi, Spermatophytes* (Vol. VII, pp. 147–213). Bruxelles: Publications de l'Institut National pour l'Étude Agronomique du Congo Belge (INÉAC).

van Wyk, B., & van Wyk, P. (1997). *Field guide to trees of southern Africa*. Cape Town: Struik.

Wheeler, E. A., & Baas, P. (1991). A survey of the fossil record for dicotyledonous wood and its significance for evolutionary and ecological wood anatomy. *IAWA Bulletin, n.s. 12*, 275–332.

Wheeler, E. A., Pearson, C. A., LaPasha, T., Zack, T., & Hatley, W. (1986). Computer-aided wood identification. Reference manual. *North Carolina Agricultural Research Service Bulletin, 474*, 160.

Wiemann, M. C., Wheeler, E. A., Manchester, S. R., & Portier, K. M. (1998). Dicotyledonous wood anatomical characters as predictors of climate. *Palaeogeography, Palaeoclimatology, Palaeoecology, 139*, 83–100.

Wiemann, M. C., Dilcher, D. L., & Manchester, S. R. (2001). Estimation of mean annual temperature from leaf and wood physiognomy. *Forest Science, 47*, 141–149.

Wolfe, J. A., & Upchurch, G. R. (1987). North American non-marine climates and vegetation during the Late Cretaceous. *Palaeogeography, Palaeoclimatology, Palaeoecology, 61*, 33–77.

Chapter 11
Fossil Leaves, Fruits and Seeds

Marion K. Bamford

Abstract Macroscopic fossil plants from the Upper Laetolil Beds, Laetoli, Tanzania, are described and where possible identified. Leaf impressions were recovered from the top of Tuff 8 at Locality 16 and show only primary and secondary venation. While many cannot be identified, some appear to belong to the Euphorbiaceae. Casts of fruits and seeds occur in almost all localities but exhibit only external features. The most common seed is probably that of *Boscia coriacea* (Capparaceae). Others have been tentatively identified as Papilionaceae, *Ximenia* species (Olacaceae), Lauraceae and *Celtis africana* (Celtidaceae). Monocot leaves, thorns of Rosaceae or Papilionaceae and a large sedge rhizome with basal culms are also described. The macrofossils represent a diverse flora with forest elements (Lauraceae, Celtidaceae) and dry open woodland elements (Capparaceae, Leguminosae) during 3.8–3.5 Ma. The sedge culm implies a wetland or springs at some time in the Upper Laetolil period.

Keywords Dicots • Monocots • Thorns • Capparaceae • Euphorbiaceae • Leguminosae

Introduction

Locality and Geology

Today there are flat-topped *Acacia tortilis* trees, tall and narrow whistling thorn trees (*Acacia drepanolobium*), thorny *Commiphora* bushes and open grasslands between the slightly undulating exposures of the grey tuffs at Laetoli in northern Tanzania. The modern vegetation comprises many units distributed according to the soils, drainage and relief (Andrews et al. 2011). The past vegetation was probably no less complex. Fossil plants have been preserved in the pyroclastic layers (Laetolil Beds) that blanketed the region on a number of occasions during the past 4.3 million years.

The Lower Laetolil Beds are about 4.3 to 3.8 Ma (Drake and Curtis 1987; Deino 2011) and consist mainly of aeolian tuffs interbedded with airfall and water-worked tuffs. Vertebrate fossils from this unit are fragmentary, scattered and rare. The overlying Upper Laetolil Beds (3.8–3.5 Ma; Drake and Curtis 1987; Deino 2011) are much richer with a diverse fauna, including hominins, and flora. A well documented sequence of marker tuffs, Tuffs 1–8, from bottom to top, is exposed in the Olaitole and Garusi River valleys that have been sampled from over 30 localities (Hay 1987; Kingston and Harrison 2007; Harrison and Kweka 2011; see Fig. 11.1). The Laeolil Beds are overlain by the Ndolanya Beds, which are also rich in fossils.

Fossil Flora

Macroscopic plant remains are not common at East African early hominin sites, but Laetoli has produced leaves, seeds, fruits, thorns and a rhizome that are described here. In addition, wood (Bamford 2011) and phytoliths (Rossouw and Scott 2011) are described elsewhere in this volume. Impressions of dicotyledonous and monocotyledonous leaves have been recovered from Loc. 16. Fourteen specimens are described here and only two have been tentatively identified. The leaves are from the top of Tuff 8. By contrast, casts of seeds and fruits have been collected from many of the localities at Laetoli, but are most abundant and diverse from Locality 3 (70 cm above Tuff 7) which has yielded 30% of the seeds. The most abundant morphotype is smooth, spherical with a very small funiculus. Of the 260 seeds collected 132 are of this type (50%) and Loc. 6 has this type exclusively (Table 11.1, Fig. 11.1). The seeds occur between Tuffs 3 and 8, and thus represent a longer time range than do the leaves. For both the leaves and seeds the matrix is coarse-grained, and not many details are apparent, making identification difficult. For this reason most specimens are

M.K. Bamford (✉)
Bernard Price Institute for Palaeontological Research, School of Geosciences, University of the Witwatersrand, Private Bag 3, WITS 2050, Johannesburg, South Africa
e-mail: marion.bamford@wits.ac.za

T. Harrison (ed.), *Paleontology and Geology of Laetoli: Human Evolution in Context. Volume 1: Geology, Geochronology, Paleoecology and Paleoenvironment*, Vertebrate Paleobiology and Paleoanthropology, DOI 10.1007/978-90-481-9956-3_11,

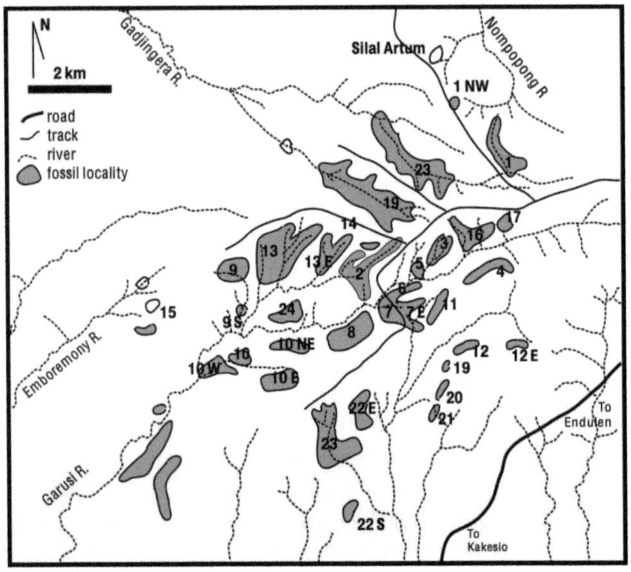

Fig. 11.1 Locality map redrawn from Hay (1987) showing collecting areas

morphotyped rather than identified to taxa. The sample size of the leaves is too small to use for a climate physiognomic analysis, but some environmental features are discussed below. In general, large leaves are typical of wetter environments and small leaves are typical of cold and/or drier climates. Entire margined leaves are more common in evergreen floras and toothed leaves are more common in deciduous floras. The Laetoli leaves are small and include both entire and toothed margined specimens.

Materials and Methods

Small blocks containing the leaf impressions were chipped out of the sediments exposed at Loc. 16 from the top of Tuff 8. Fruits and seeds occurred in many localities and were collected systematically with faunal remains during the surveys carried out by Terry Harrison and his team. These had weathered out of the tuffs. Leaves were photographed with a digital camera and the outlines of the leaves and venation was drawn using a Wacom graphire drawing tablet and Adobe Illustrator software. The reproductive structures were also photographed digitally and a Zeiss Stemi compound microscope was used for the small seeds.

Identification of the seeds and fruits was done from comparative material where available (BPI Palaeobotany Herbarium and Archaeology Department collections, University of the Witwatersrand), but many types of seeds are not identifiable as they lack diagnostic characters. Only some of the seeds are described here.

Preservation

The fruits and seeds are silicified, calcified or both and represent three dimensional casts of the exterior, or of the locule/endosperm cavity of these structures. Broken seed pods are also preserved, but in the absence of data on provenance, they are not described here. Some thorns and spines were preserved. There is no internal structure and little external morphology, which has three possible causes. The fruits and seeds were already worn when buried and preserved; they may have been abraded during transport (allochthonous) or possibly shed and weathered before burial close to the parent plants; or the fossils were weathered by more recent exposure. The last option is more likely because the associated faunal remains are well preserved.

Leaf impressions occur in a coarse-grained tuff and appear to have been incorporated within a wet ashfall. The leaves are not flattened, as might be expected for leaves already lying on the ground prior to burial. Perhaps the ashfall was turbulent and trapped leaves that had just been stripped from the trees. Very thin layers of white material, probably originally calcium carbonate and later modified to a mixture of siliceous and calcitic material, were accumulated on both upper and lower leaf surfaces within the burial matrix. This white mineral has the impression of the leaf but is thin and fragile and absent where pieces of grit occur. Leaves had decomposed completely prior to fossilization, as there is no remaining organic material.

Morphotypes/Identification

Where possible the leaves and seeds have been compared with and identified to modern taxa. However, most of them have been put into morphotypes. For the leaf descriptions, the terminology of Hickey (1973, 1974) and the Leaf Architecture Working Group (1999) have been used. Leaves are distinguished on the following features: margin, laminar shape and venation. There is no similar formal system for the description and databasing of seeds.

Description and Identification of Macrofossils

Dicot Leaves

Morphotype: Leaf with pectinal veins

(Figs. 11.2a, 11.3a)
Suggested Affinity: Euphorbiaceae.
Specimen: EP 126/03.
Locality and stratigraphy: Loc. 16; top of Tuff 8, Upper Laetolil Beds.

Table 11.1 List of plant macrofossils and localities from the Upper Laetolil Beds, Laetoli

EP Number	Year	Locality	Between Tuffs	No.	Description
531	2004	1	7, 8	3	Balls
1911	2003	1	7, 8	2	Round ball; bean
741	2000	2	5, 7	3	Oval = *Ximenia caffra;* 2 balls
2709	2000	2	5, 7	3	Small seeds
2710	2000	2	5, 7	2	Large flat seeds
2711	2000	2	5, 7	1	Large flat seed
233	2001	3	6, 8	1	Short litchi-like seed*
236	2001	3	6, 8 below 7	27	Tiny seeds
353	2003	3	7, 8	3	2 beans = legume; 1 strange curved
355	2003	3	7, 8	1	Rose thorn 4 × 5 mm
663	2004	3	7, 8	6	3 thorns and spines; 2 thorns of Acacia?; 1 rose thorn*
736	2005	3	7, 8	5	Thorns; 3 rose; 2 other*
739	2005	3	7, 8	6	Seeds, boss, beans, stalked, point
1668	2000	3	7, 8	1	Ball
2782	2000	3	below 7	15	Small seeds
2781	2000	3	below 7	12	3 with points; 9 tiny broken seeds
8	2005	4	6, 8	1	Ball
11	2003	4	7, 8	1	Ball
47	2000	4	6, 7	1	Small ball = *Celtis africana?*
207	2001	5	3, 5	3	Small seeds
440	2003	5	3, 5	5	1 small ball with tip; 4 small balls, diameter 4 mm
807	2005	5	3,5	2	Small round seeds 4.5 mm diameter
1366	2004	5	3, 5	1	Small round, black inclusions
2024	2000	5	3, 5	4	Small seeds
158	2001	6	5, 6	30	Balls* = *Boscia coriacea*
691	2005	6	5, 7	13	Balls
1394	2000	6	5, 6	1	Ball
1467	2004	6	7, 7	1	Ball
2070	2003	6	5, 7	12	Balls
2300	2000	6	5, 6	1	Ball
198	2001	7	5, 7	5	Balls
631	2005	7	5, 7	4	Balls
664	2004	7	7, 8	7	3 beans; 2 bosses; 1 layered trilete; 1 bihorned faceted ball
790	2004	7	5, 7	2	Balls
1989	2003	7	5, 7	7	6 balls - 1 shattered; 1 yellow, smooth smaller, flatter ball
2163	2003	7	5, 7	4	Balls
2246	2000	7	5, 7	6	Balls
153	2005	8	5, 7	1	Rough ball, 7 mm diameter, with rim of chemical deposits
343	2000	8	below 6	1	Balls
380	2004	8	5, 7	2	Balls
811	2003	9	5, 7	1	Squashed ball 24 × 26 mm, round = *Cryptocarya latifolia?*
1107	2004	11	7, 8+	3	Balls
1538	2003	12	5, 7	1	Ball
1165	2004	13	6, 8	3	Balls
2347	2003	13	6, 7	1	Elliptic, pointed end = *Ximenia americana*
80	2001	16	7, 8	6	Seeds, various
614	2001	16	above 8	2	Balls
269	2005	17	8, YMT	1	Rough big ball 11 mm diameter
826	2001	18	Ndolanya	3	Small seeds
878	2004	20	6, 8	1	Shell?
914	2005	10E	5, 7	1	Ball
1423	2001	12E	5, 7	1	Ball
1530	2003	12E	5, 7	1	Ball
2506	2003	1NW	below 7	1	Shattered ball
1374	2005	22E	5, 7	1	Petiole and pulvinus?
1375	2005	22E	5, 7	1	Ridged stem with side branches, length 16 mm, diameter 4 mm

(continued)

Table 11.1 (continued)

EP Number	Year	Locality	Between Tuffs	No.	Description
1401	2005	22E	5, 7	3	Narrow stems, 2 ridged like dry sedge culms
1402	2005	22E	5, 7	1	3 incomplete leaf fragments - very small
720	2003	2	5, 7	2	Balls
1889	2000	2	5, 7	6	Balls
1007	2005	2	5, 7	1	Ball
694	2003	2	5, 7	9	Balls
1046	2005	2	5, 7	3	Balls
233	2001	3	6, 8	1	Ball
			Total	260	

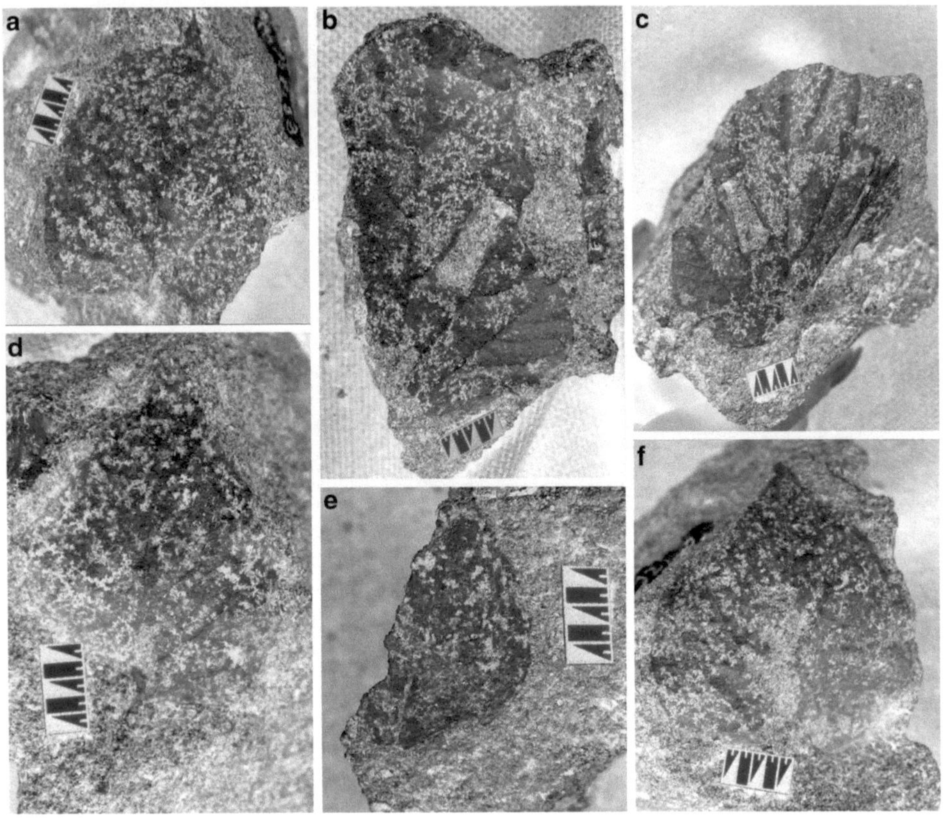

Fig. 11.2 Photographs of fossil dicot leaves from Loc. 16. Scale bars = 10 mm. (**a**) EP 126/03, (**b**) EP 127/03, note pectinal veins on lower right, (**c**) EP 128/03, (**d**) EP 130/03, (**e**) EP 132/03, (**f**) EP 1132/03

Description

The leaf is ovate and symmetrical with an entire margin and an acute to obtuse apex, concave base and no petiole preserved. Lamina length is 44 mm and maximum width is 32 mm. The primary venation is actinodromous suprabasal, with five veins that diverge from a common point just above the point of attachment of the petiole. Pectinal veins (*sensu* Spicer 1986) or agrophic veins (*sensu* Leaf Architecture Working Group 1999) are visible on the left side and curve from the lower two secondary veins. Secondary veins are more crowded in the lower half of the lamina than in the upper half, but the angles of divergence of the secondary veins from the primary vein are all acute. Tertiary venation is extremely faint and partially obscured by the dark minerals.

Identification

Pectinal veins are common in, but not exclusive to, some genera of the Euphorbiaceae, for example *Croton, Macaranga* and *Neoboutonia* or *Alangium* of the Alangiaceae. Leaves of the Moraceae also have small pectinal veins, but the secondary venation is brochidodromous. This leaf is very tentatively identified as Euphorbiaceae.

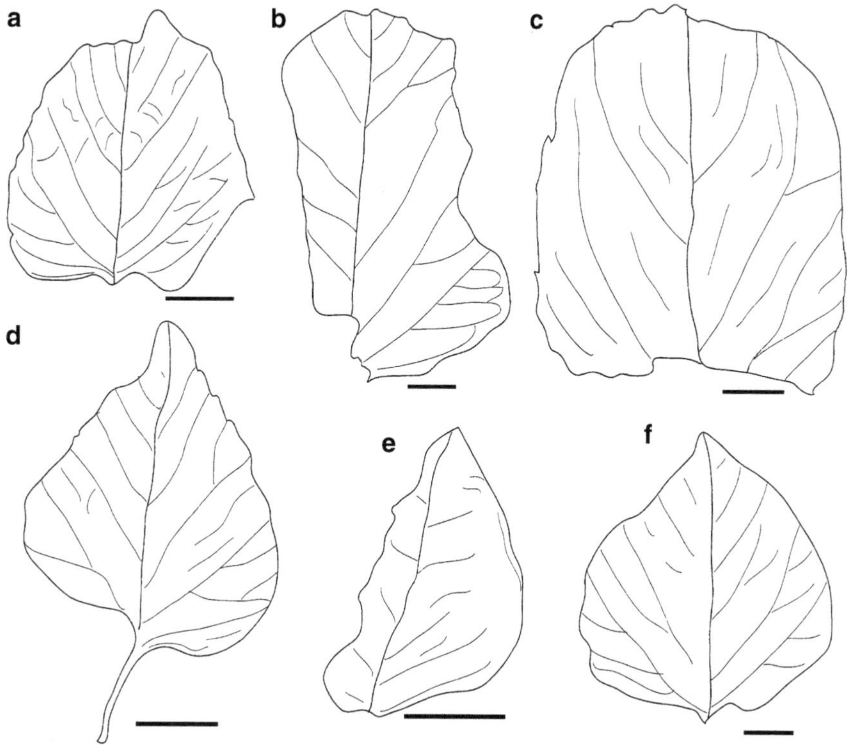

Fig. 11.3 Drawings of leaves in Figs. 11.2a–f respectively. Scale bar = 10 mm. (**a**) EP 126/03, (**b**) EP 127/03, note pectinal veins on lower right, (**c**) EP 128/03, (**d**) EP 130/03, (**e**) EP 132/03, (**f**) EP 1132/03

Morphotype: large leaf with pectinal veins

(Figs. 11.2b, 11.3b).
Specimen: EP 127/03.
Locality and Stratigraphy: Loc. 16; top of Tuff 8, Upper Laetolil Beds.

Description

The leaf is ovate and entire margined, no apex is preserved, but the base is rounded to cordate; length more than 55 mm and width more than 44 mm. The primary venation is simple pinnate but judging by the clustering of the veins near the base (unfortunately incomplete) they could be actinodromous suprabasal with possibly seven veins. Pectinal veins are present and brochidodromous. The secondary vein spacing is acute, irregular and more crowded at the base. Tertiary venation is very poor. This leaf may also be a member of the Euphorbiaceae.

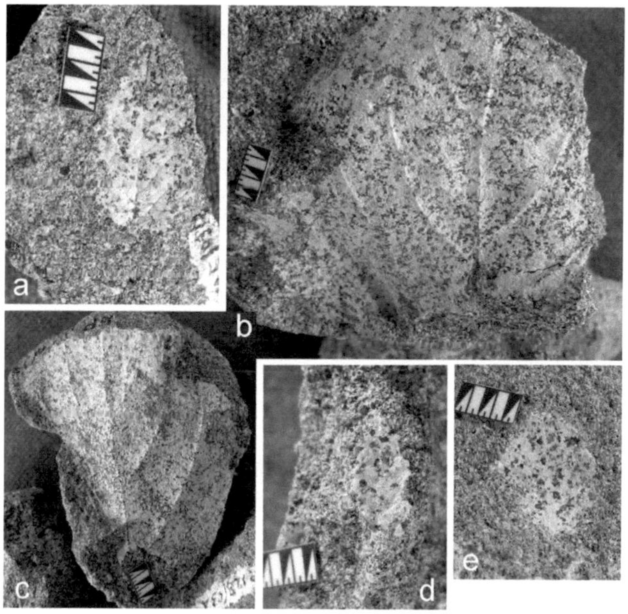

Fig. 11.4 Photographs of leaf specimens from Loc. 16 with crenate margins. Scale bars = 10 mm. (**a**) EP 125/03, (**b**) EP 1131/03, (**c**) EP 1128/03, (**d**) EP 1129/03, (**e**) EP 1133/03

Morphotype: leaf with crowded basal veins

(Figs. 11.2c, 11.4c).
Specimens: EP 128/03 and EP 129/03.
Locality and Stratigraphy: Loc. 16; top of Tuff 8, Upper Laetolil Beds.

Description

Leaf ovate, entire margined, apex unknown, but some form of acute. So much of the base is missing that it could be acute or cordate to round, lamina length >34 mm, width 28 mm. Primary venation is simple craspedodromous or pinnate, or possible actinodromous and secondary venation is pinnate. There are probably seven veins arising from near the base. Secondary veins are more crowded in the lower half of the lamina than in the top. Tertiary veins are very poorly preserved, but a few are just visible and arise from the secondaries at an acute angle and travel towards the apex. Specimen EP129/03 is very similar with pinnate venation, but there is insufficient venation for better comparisons.

Identification

This leaf could be a member of the Moraceae, but is too poorly preserved for any certainty.

Morphotype: pinnate, irregular veined leaf

(Figs. 11.2d, e, 11.4d, e).
Suggested Affinity: Euphorbiaceae?
Specimens: EP 130/03 and EP 132/03.
Locality and Stratigraphy: Loc. 16; top of Tuff 8, Upper Laetolil Beds.

Description

EP130/03: Ovate leaf with rounded to cordate base, long acute apex, entire margined and slender petiole. Laminar length is 44 mm, maximum width 35 mm, petiole is 13 mm long and 1.5 mm wide. Primary venation is pinnate and secondary venation is craspedodromous with the veins forking before reaching the margin, but it is not clear whether or not there is in intramarginal vein or if the veins terminate in the teeth. Tertiary venation is not preserved.

EP132/03: This leaf is the same shape as EP130/03 but no petiole is preserved; length of lamina 52 mm; width 41 mm.

Identification

This leaf shows some similarities with *Macaranga* (see Beentje 1994: 17, 212).

Morphotype: small brochidodromous leaf

(Figs. 11.5 and 11.10).
Specimen: EP 1132/03
Locality and stratigraphy: Loc 16; top of Tuff 8, Upper Laetolil Beds.

Description

The leaf is elliptic with an entire margin, 28 mm long and 24 mm wide. The tip is acute and base slightly rounded. Venation is probably brochidodromous but incomplete because the veins curve back towards the upper adjacent vein, but the preservation is too poor to see if the veins meet the upper adjacent secondary vein. There are inter-secondary veins that are smaller and parallel to the secondary veins. The leaf is too poorly preserved to attempt a comparison.

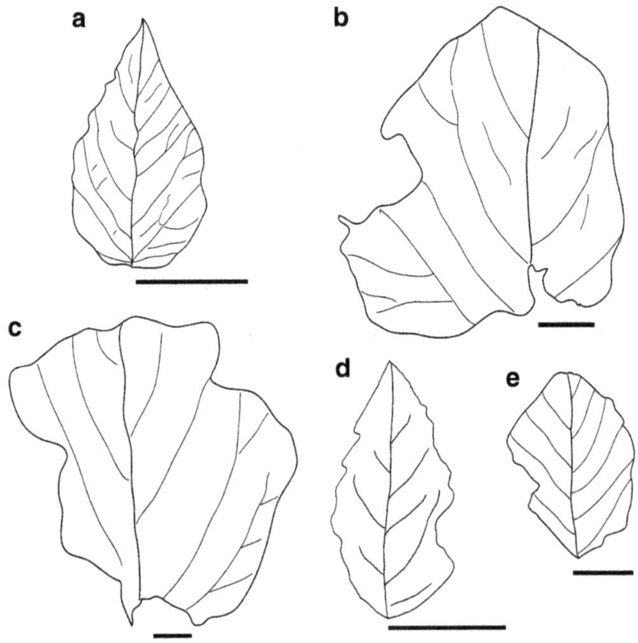

Fig. 11.5 Drawings of leaves in Fig. 11.4a–e respectively. Scale bar = 10 mm. (**a**) EP 125/03, (**b**) EP 1131/03, (**c**) EP 1128/03, (**d**) EP 1129/03, (**e**) EP 1133/03

Morphotype: small, toothed leaf with pectinal veins (Figs. 11.4a, 11.5a).

Specimen: EP 125/03.
Locality and Stratigraphy: Loc 16; top of Tuff 8, Upper Laetolil Beds.

Description

Leaf ovate with a crenate margin; length 28 mm, width 16 mm. The apex is acute and the base convex to truncate. Primary venation is pinnate with craspedodromous secondary venation. Tertiary venation is poor but is generally at right angles to the secondary veins. The crenate margin is poorly preserved. This is a common leaf morphotype among the modern flora, but without more details it cannot be identified.

Morphotype: large serrate leaf (Figs. 11.4b, 11.5b).

Specimen: EP 1131/03.
Locality and Stratigraphy: Loc. 16; top of Tuff 8, Upper Laetolil Beds.

Description

The leaf is incomplete, missing tip, base and one side; length >70 mm, width 80 mm. A small section of the margin is preserved on the upper left and appears to be crenate. Primary venation is probably pinnate and secondary venation craspedodromous. There are two pectinal veins on the lower left. Compare with EP 127/03, which has a serrate margin, and EP1129/03. No identification is possible.

Morphotype: medium crenate

(Figs. 11.4c and 11.5c).
Specimen: EP 1128/03.
Locality and Stratigraphy: Loc. 16; top of Tuff 8, Upper Laetolil Beds.

Description

This small leaf has a crenate margin, elliptic outline and acute to obtuse apex; base unknown. Length is greater than 34 mm; width is 28 mm. Primary venation is pinnate. Secondary venation is simple and goes to the margin without forking. No tertiary veins are visible. No identification is possible.

Morphotype: possibly large serrate leaf

(see EP 1131/03) (Figs. 11.4c, 11.5c).
Specimen: EP1129/03.
Locality and Stratigraphy: 16; top of Tuff 8, Upper Laetolil Beds.

Description

This leaf is poorly preserved and the shape is unknown; length >65 mm, width >66 mm. The venation is either actinodromous or actinodromous suprabasal and there are faint pectinal veins. The venation resembles that of EP 127/03 and EP 1131/03.

Morphotype: small crenate leaf (Figs. 11.4e, 11.5e).

Specimen: EP1133/03.
Locality and Stratigraphy: Loc. 16; top of Tuff 8, Upper Laetolil Beds.

Description

Small elliptic leaf with crenate margin, about 29 mm long and 10 mm wide. Base and tip are acute. Primary venation is pinnate and regular. No other venation is seen.

Monocot Leaves

Morphotype: sedge/grass leaf (Fig. 11.6a, b).

Specimens: EP 1134/03 and EP 1135/03 (2 specimens).
Locality and Stratigraphy: Loc. 16; top of Tuff 8, Upper Laetolil Beds.

Description

The larger specimen (EP 1134/03) is 45 mm long and 10 mm wide, but incomplete lengthwise as it is broken off at both ends. One end is 10 mm wide and the other 8,mm wide, so the leaf appears to taper. Nine parallel, raised, veins (including a central one that is the widest of all) give the slightly convex leaf a minutely corrugated appearance. Between these longitudinal compound veins are fine cross veins. Each of the nine longitudinal veins is composed of extremely fine longitudinal veins or vascular bundles. The smaller specimen is a fragment, only 20 × 4 mm, with parallel veins as described above.

The slightly larger central vein and accompanying convex form are common in both grass and sedge leaves, but there are insufficient features for identification.

Seeds and Fruit

Morphotype: balls (Fig. 11.6c, d).

Specimens: 132 seeds (see Table 11.1 for list).
Localities: see Table 11.1.

Description

The balls are perfectly spherical to sub-spherical, smooth but commonly weathered on one part that has been in contact with the soil (because this always the dirty side).They range in size from 7–11 mm. A very small funiculus is present and opposite this is rarely a tiny point. Most specimens also have one side very slightly flattened. There is no internal structure and fractured seeds are like miniature geodes.

Identification

The balls are interpreted as casts of individual seeds with one slightly flattened side indicating the adpression of an adjacent seed, therefore suggesting derivation from a two-seeded berry or drupe. Such a fruit type occurs in many families, including Anacardiaceae, Apocynaceae, Boraginaceae, Capparaceae, Euphorbiaceae, Lauraceae, Oleaceae, Rhamnaceae, Rubiaceae, Ulmaceae and Vitaceae. Searching among the species that occur in Kenya today, the seeds of Anacardiaceae are too small; Apocynaceae, Boraginaceae and Vitaceae seeds are not round; Euphorbiaceae seeds are usually 3-lobed and grooved, but also compressed, lenticular or angular; Lauraceae and Oleaceae seeds occur singly; Rhamnaceae seeds are elongate (*Berchemia*), or in threes (*Rhamnus*), or single (*Ziziphus*);

Fig. 11.6 (**a**) Monocot leaf with parallel venation and convex curve of lamina. EP 1134/03 from Locality 16. Scale bar = 10 mm. (**b**) fragment of monocot leaf from Locality 16, EP 1135/03. Scale bar = 10 mm. (**c**) the most common seeds from Laetoli, spherical to almost spherical smooth surface. EP 158/01. (**d**) fossil and modern seeds of *Boscia coriacea* (Capparaceae). Fossil seeds are the three on the right. (**e**) two seeds in a berry, loose seeds and leaves of *Boscia coriacea* (Palaeobotany Herbarium modern seeds reference collection No. 47 collected from Ileret, northern Kenya.). (**f**) fossil beans. EP 353/03. (**g**) fossil "melon" pip. EP 2782a/00. (**h**) Morphotype "raspberry", note regular bosses covering the whole seed. EP 2782b/00

Sapindaceae seeds are too big and Rubiaceae seeds tend to have a small residual calyx or scar. That leaves Capparaceae where the fruits are capsules or berries. *Boscia* fruits are one-several seeded; *Cadaba* and *Maerua* fruits are cylindrical, round or ellipsoid and torulose; *Capparis, Cladostemon, Crateva* fruits are round and *Ritchea* fruits are ellipsoidal. From the reference collection of modern seeds at the Bernard Price Institute, University of the Witwatersrand, the fossil seeds most closely resemble those of *Boscia coriacea* Pax. The capsule is completely round, yellow-brown and hairy when young. It breaks open to reveal 1–2 round, smooth seeds with a very faint funiculus and opposite point. (Fig. 11.6d, e).

Boscia coriacea is an evergreen tree or shrub, 1–6 m high and occurs in dry bushland, often along luggas or near wells. It is widespread in East Africa today and occurs from 1 to 1,200 m altitude (Beentje 1994). The fruits are edible when cooked (Beentje 1994).

Morphotype: beans (Fig. 11.6f).

Specimens: EP 353/03 (2 specimens); EP 664/04 (3 specimens).

Locality and Stratigraphy: Locs 3 and 7; between Tuffs 7 and 8, Upper Laetolil Beds.

Description

The specimens are bean-shaped, generally oval with an indentation on one side. Within the indentation are the hilum and micropyle. All except one specimen are smooth.

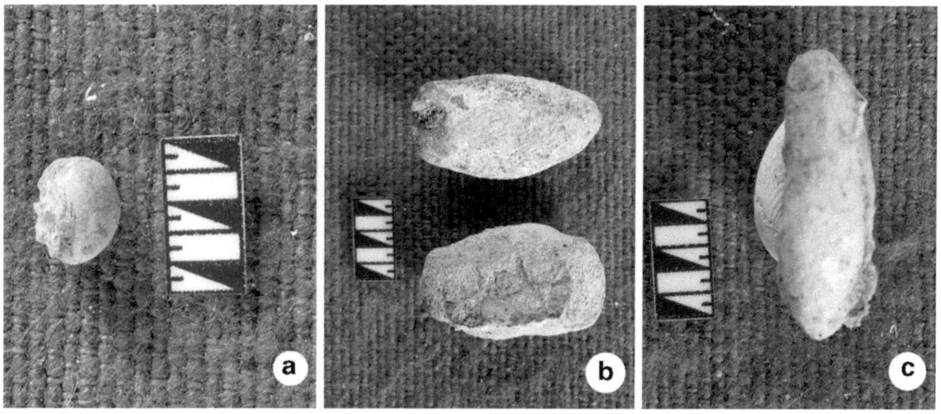

Fig. 11.7 (**a**) Morphotype "tomato"; persistent calyx and stalk on a slightly depressed fruit. EP 2782c/00. (**b, c**) shield-shaped seeds (**b**) with slight bulges below broader part seen in side view, (**c**) EP 2710a/00, EP 2710b/00. Scale bar = 10 mm

The beans from Locality 3 are 14 mm long, 5 mm wide and 4 mm thick. Those from Loc. 7 are 16 mm long, 8 mm wide and 4.5 mm thick.

Identification

The bean-like shape suggests that these could be seeds of the Leguminosae, and the Papilionaceae in particular. This is a very large plant group, and the seeds within the family are very difficult to distinguish. This family is wide ranging in growth form and habitat.

Morphotype: "melon" pip (Fig. 11.6g).

Specimens: EP 27820a/00.
Locality and Stratigraphy: Loc. 3; fine grey clay horizon, 60–70 cm above Tuff 7, Upper Laetolil Beds.

Description

Small, smooth shield-shaped specimens with a small apical structure interpreted as a funiculus, suggesting that these are seeds. The length is 11.5 mm, width 7 mm and thickness 3 mm.

Identification

This seed resembles those of the Cucurbitaceae, but there are very few definitive characteristics.

Morphotype: "raspberry-like" fruits (Fig. 11.6h).

Specimens: EP2782b/00 and EP739a/05.
Locality and Stratigraphy: Loc. 3; fine grey clay horizon 60–70 cm above Tuff 7, Upper Laetolil Beds.

Description

The specimens are 3 mm in diameter, and ovoid with an apical stalk less than 1 mm long and in diameter. It is covered in regular, bosses, round at the base, curved on the surface, each less than 1 mm in diameter. This distinctive specimen resembles the fruit of the Rosaceae, such as *Rubus.*

Morphotype: "tomato"-like fruit (Fig. 11.7a).

Specimens: EP 2782c/00 and EP 2782d/00.
Locality and Stratigraphy: Loc. 3; fine grey clay horizon 60–70 cm above Tuff 7, Upper Laetolil Beds.

Description

The two fossils are almost spherical, but have the top part with the stalk slightly depressed. EP 2782c/00 is 7 mm in diameter and height which includes the short stalk and a persistent structure of four pointed entities that are interpreted as a calyx composed of sepals. EP 2782d/00 is 6 mm in diameter and less depressed on the top as the height is 7 mm, which includes the short stalk and more rounded persistent calyx. Surfaces are smooth. There are no markings on the base.

Identification

Numerous fruits have persistent calyces, so it is not possible to identify these fossils.

Morphotype: large shield seed/fruit (Fig. 11.8a).

Specimens: EP 2710a/00 and EP 2710b/00.
Locality and Stratigraphy: Loc. 2; between Tuffs 5 and 7, Upper Laetolil Beds.

Description

Two flattish fossils with an ovate outline; 13 mm in width at the truncated end, tapering to a rounded point at the other end; 25 mm long. In side view, the top surface is gently convex and the lower side is concave with a slightly enhanced protrusion near the truncated end. The surfaces are smooth

Fig. 11.8 (**a**) bivalved stalked seed with longitudinal groove. EP 739/05b. (**b, c**) fruit with possibly three segments as seen from base of seed (**b**, side view; c=basal view), EP 2347/03. (**d–f**) modern seeds of *Ximenia americana,* Olacaceae, (seed collection of the Archaeology Department, University of the Witwatersrand), side (**d**), top view and point of attachment (**e**), basal view (**f**); note linear contraction lines from the apices. Compare with fossil seed in Figs. 11.8b, c. (**g**) ovoid seed with pointed top and base. EP 741/00a. Note tiny groove or lip on the right at top/bottom end. (**h**) modern seeds of *Ximenia caffra* (Olacaceae; Archaeology Department, University of the Witwatersrand) with both ends pointed and slight puckering or shallow grooves at one end. Compare with fossil in Fig. 11.8g. Scale bar = 10 mm unless otherwise indicated

and unmarked. It is possible that four seeds fitted together in a carpel or that these are fruits.

Morphotype: bivalved, stalked seed/fruit

(Fig. 11.8b).
Specimen: EP739b/05.
Locality and Stratigraphy: Loc. 3; fine grey clay horizon 60–70 cm above Tuff 7, Upper Laetolil Beds.

Description

The seed is ovoid, narrower at the stalk end, rounded at the distal end, 8 mm long and 7 mm in diameter. A longitudinal groove, deeper in the distal part, fading to a mere line near the stalk, divides the seed into two parts. The stalk is poorly preserved, but is broad at the point of attachment and the basal shape is oval. The stalk is broken obliquely, but has a diameter of less than 1 mm. Surface texture is somewhat worn, but there are longitudinal striations parallel to the groove.

Morphotype: ovoid fruit (Fig. 11.8c–f).

Suggested affinity: *Ximenia* (Olacaceae).
Specimen: EP 2347/03.
Locality and Stratigraphy: Loc. 13; between Tuffs 6 and 7, Upper Laetolil Beds.

Description

The fossil is ovoid in shape, 22 mm long and 15 mm in diameter; one end is more pointed than the other. The narrow basal end has two concentric circular scars from the point of attachment. The rounded apical end has a faint trilete or star scar. The surface is smooth and comprises a calcium carbonate layer over a cemented sandy matrix.

Identification

This fruit or seed is similar in size and shape to the fruits of *Ximenia americana* (Fig. 11.8d–f), but some of the surface features are poorly preserved, so it could belong to another member of the Olacaceae.

Morphotype: oval seed with lip (Fig. 11.8g, h).

Specimen: EP741a/00.
Locality and Stratigraphy: Loc. 2; between Tuffs 5 and 7, Upper Laetolil Beds.

Description

The seed is elliptic with both ends pointed; in cross-section it is oval; length 17 mm, width 8 mm, and thickness 10 mm. One end has a short, shallow groove on one surface and close to the edge, so that it resembles a mouth. The surface is slightly rough.

Identification

This seed is very similar in size and shape to *Ximenia caffra* (Fig. 11.8h), but could just as well belong to a member of the Capparaceae.

Morphotype: round fruit (Fig. 11.9a, b).

Specimen: EP 811/03
Locality and Stratigraphy: Loc. 9; between Tuffs 5 and 7, Upper Laetolil Beds.

Description

Round to faintly quadrangular fruit with smooth exterior, 24 mm diameter, and 26 mm high, which includes a short, truncated stalk. The stalk has two remnant sepals of the persistent calyx (Fig. 11.9a). The rounded base (Fig. 11.9b) has three remaining minute ridges of four, implying that it has four-seeds close together, in which case each seed would have a triangular cross-section. (However, many four-carpellate fruits are unilocular, bearing one seed, e.g., Juglandaceae, but this family does not occur in East Africa today).

Identification

This large spherical fruit superficially resembles *Cryptocarya latifolia* (Lauraceae; Fig. 11.9c–e), which has four triangular seeds within a round fruit, but the fossil has insufficient surface details to confirm the identification. *Cryptocarya latifolia* is a large evergreen tree occurring in forests of Natal, South Africa (van Wyk and van Wyk 1997). The East African species is *C. liebertiana* (Beentje 1994) and is also a large forest tree occurring in the Taita Hills of Kenya, at 1,600–1,700 m, with fruits about 20 mm in diameter.

Morphotype: broken bilocular fruit (seed?) (Fig. 11.9f).

Specimen: EP 353c/03.
Locality and Stratigraphy: Loc. 3; between Tuffs 7 and 8, Upper Laetolil Beds.

Description

The fossil is oval, 19 mm long and 13 mm in diameter with an oval cross-section. It is divided in two longitudinally and along the narrow side by a deep groove, making two flat oval seeds. The top of one seed is broken, revealing a hollow interior with calcite crystals projecting from the "seed coat" into the cavity.

Morphotype: short "litchi" (Fig. 11.9g).

Specimen: EP 233/01.
Locality and Stratigraphy: Loc. 3; fine grey clay horizon 60–70 cm above Tuff 7, Upper Laetolil Beds.

Description

The seed is short ovoid, 11 mm long and 8 mm in diameter, with a wide neck, 5 mm in diameter, for attachment or a persistent calyx. Fine concentric layers of substrate are visible within the seed. The exterior is very smooth and featureless.

Morphotype: reticulate-surfaced seed (Fig. 11.9h, i).

Suggested affinity: *Celtis* (Celtidaceae/Ulmaceae).
Specimen: EP 047/00.
Locality and Stratigraphy: Loc. 4; between Tuffs 6 and 7, Upper laetolil Beds.

Description

The seed is small and round, 7 mm in diameter, with fine reticulate markings that have worn off unevenly. This is reminiscent of *Celtis africana* endocarps (Ulmaceae/Celtidaceae; Fig. 11.9i) underneath the seed coat, but as the fossil is rather worn it not possible to identify it with confidence.

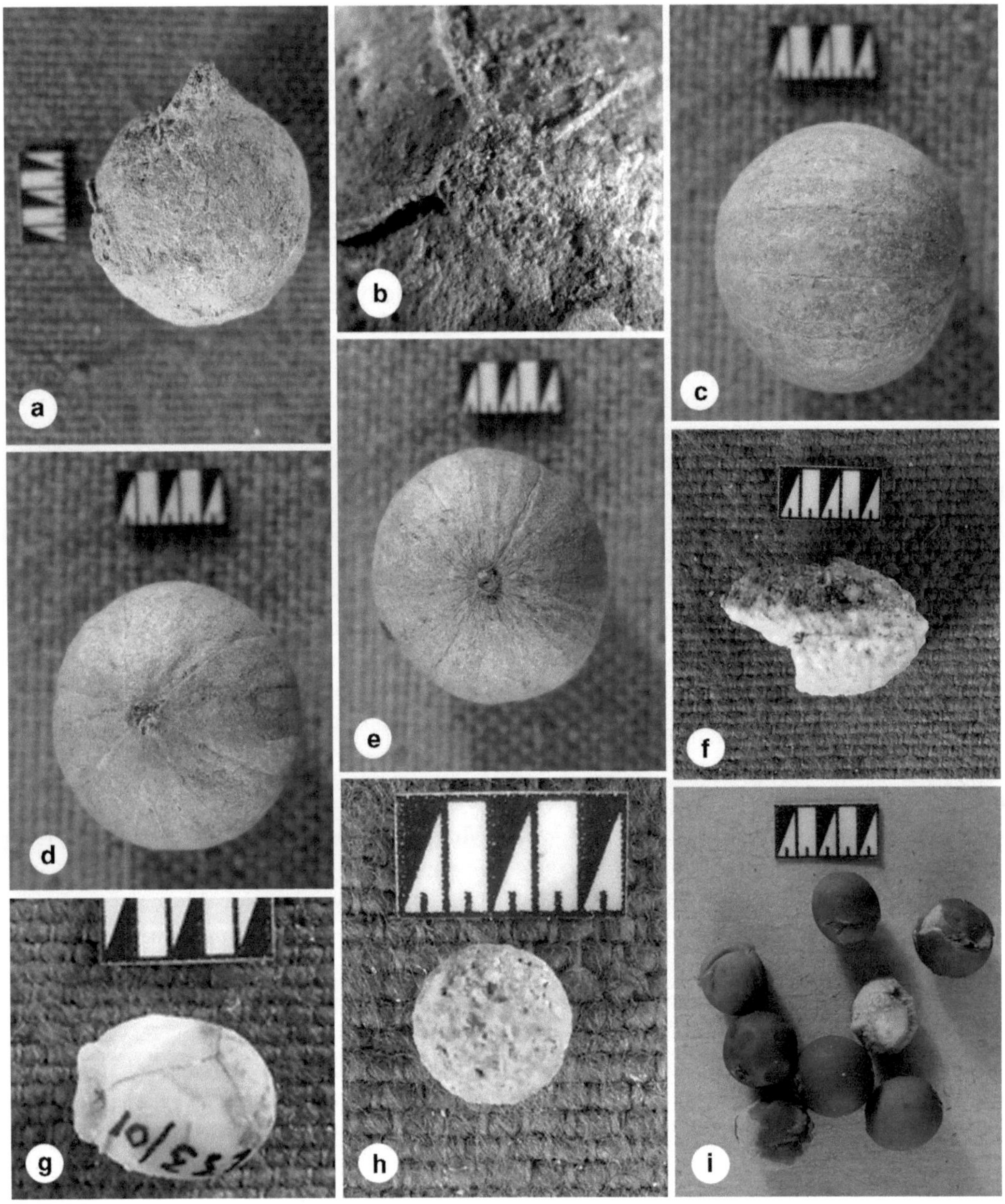

Fig. 11.9 (**a**) Seed EP 2347/03. (**b**) base of seed in (a). (**c–e**) modern seed of *Cryptocarya latifolia* (Lauraceae) from the collection of modern seeds, Archaeology Department, University of the Witwatersrand, (**c**) side view, (**d**) base view, (**e**) top view with very small stalk. (**f**) broken bilocular seed, EP 353c/03; no internal structure, just crystals protruding from the outer shell into the void. (**g**) short ovoid seed with wide point of attachment, EP 233/01. (**h**) round seed with faint reticulate pattern on the surface, EP 047/00. (**i**) modern seeds of *Celtis africana* (Celtidoideae) from the modern seed collection in the Palaeobotany Herbarium, University of the Witwatersrand, No. 156. Examples of whole drupe, outer skin removed exposing spongey flesh and bare seed showing reticulate pattern of the pericarp. Compare with fossil seed in Fig. 11.9h. Scale bar = 10 mm

Thorns and Spines

Morphotype: curved thorns (Fig. 11.10a–c).

Specimens: EP 663a/04, EP 663b/04; EP736b/05; EP736d/05; EP736e/05; EP355/03.

Locality and Stratigraphy: Loc. 3; fine grey clay horizon 60–70 cm above Tuff 7, Upper Laetolil Beds.

Description

Dissociated thorns (curved) and spines (straight) have been collected from Loc. 3. There are six small "rose" thorns. The heights of the rose thorns are 4–5 mm; bases oval, 4–6 mm × 2–3.5 mm (Fig. 11.10a, b). The sharp (?) tips are missing, but the thorns curve inwards from the oval bases to a small oval point.

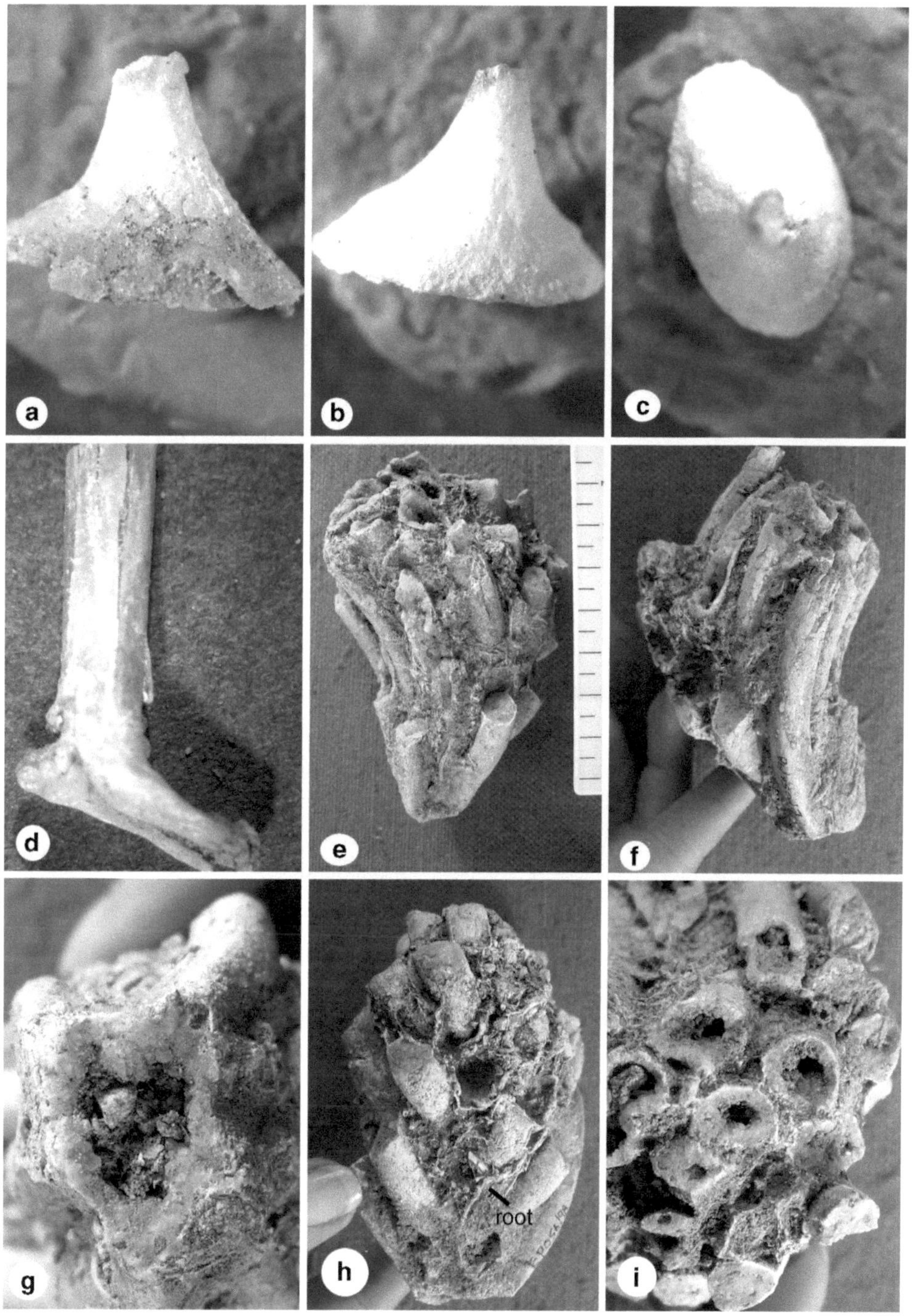

Fig. 11.10 Fossil thorns and monocot rhizomes. (**a–c**) short curved "rose" thorns, EP 736/05a and b, views of both side and top. (**d**) long straight spine EP 663/04, with torn-off 'heel' at the bottom of the figure. (**e–i**) monocot rhizome with attached culms. EP 056/98. (**e**) side view of rhizome (narrow end = base) with culms attached, but broken off about 6 cm from rhizome. Note the lack of leaves but numerous fine roots amongst the culms. (**f**) view of other side. (**g**) cross-section of hollow, woody(?) rhizome (narrow basal end as in Figs. 11.10e, f) 22 × 15 mm and wall thickness 2–5 mm. (**h**) detail of round culms, some are hollow and some have a matrix infill; fine roots indicated. (**i**) detail of hollow culms seen from the top

Morphotype: straight spines (Fig. 11.10d).

Specimens: EP 663c/04; EP 736a/05a; EP 736c/00.

Locality and Stratigraphy: Loc. 3; fine grey clay horizon 60–70 cm above Tuff 7, Upper Laetolil Beds.

Description

The spines are more triangular to oval in cross sectional outline and do not curve inwards (convex). The base of one has been torn from the branch so has a heel; it is oval, 6 × 4 mm and 5 mm high, but the tip is missing. The second spine has a three-lobed base, 4 mm in diameter, and the spine narrows rapidly to a broken point only 6 mm long.

Identification

Although the general impression of East African vegetation is that of thorny trees, relatively few taxa have thorns, but the

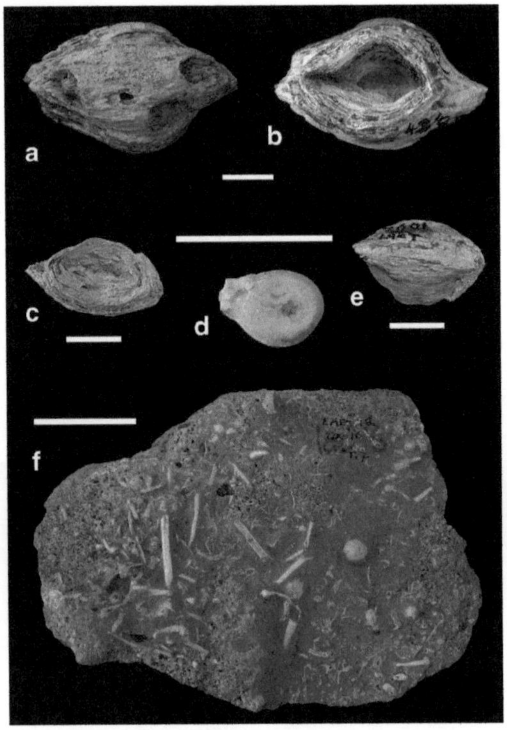

Fig. 11.11 Fossil seeds collected by Mary Leakey. (**a–b**) LAET 75-434 (Loc. 9 S), external and internal views of split seed with pointed ends and thick testa. Scale = 1 cm; (**c** and **e**) LAET 75-2,601 (Loc. 10W), external and internal view of very similar seed. Scale = 1 cm; (**d**) LAET 75-3,506 (Loc. 3), small seed with thin coat. Scale = 1 cm; (**f**) LAET 18-T.7 (Loc. 16, Tuff 7), block of seeds and twigs embedded in tuff. Scale = 5 cm

common taxa are thorny. A few genera in the Apocynaceae, Combretaceae, Cucurbitaceae, Euphorbiaceae, Passifloraceae, Rhamnaceae, Rubiaceae and Rutaceae have spines or thorns. All the acacias have either short or long straight spines and/or hooked thorns. *Acacia nigrescens* (Mimosoideae) also has persistent thorns arising from knobs on the trunk of the tree, and *Zanthoxylum davyi* (Rutaceae) has thorn-tipped knobs on the bark. Several species of *Erythrina* (Papilionaceae) have prickles on the branchlets or along the petioles and midribs. *Erythrina abyssinica* has prickles and also thick thorns on the trunk. In East Africa today there are about nine species of *Rubus* (Rosaceae) that are prickly climbing or scrambling shrubs. The curved thorns are between 2 and 7 mm long. The fossil thorns may belong to any of these taxa. Straight spines are common in the Mimosoideae.

Monocot Rhizome

Specimen: EP 056/98 (Fig. 11.10e–i).
Locality and Stratigraphy: Oleisusu, Upper Laetolil Beds.

Description

A rhizome with attached, very closely spaced basal culms has been preserved as a silicified cast, 8 cm long and 4 cm wide. There are 33 culms attached alternately and closely spaced (Fig. 11.10e, f) to the hollow rhizome (Fig. 11.10g); all culms are curved in one direction. Lots of fine roots are still adhering to the base of the rhizome and twine in between the culms (Fig. 11.10h). The culms are round, 10 mm in diameter. Some are hollow, but others have an irregular infill. All culms have a thin outer layer of dense white material contrasting with the less opaque and more crystalline inner layer. Only about 60 mm remains of the length of the culms and there do not appear to be any leaves associated with them.

Based on the growth form this fossil plant is monocotyledonous with a woody rhizome, so it would belong to the Juncaceae (rushes), Poaceae (grasses) or Cyperaceae (sedges). It is very much larger than most species of these families, except the Bambusoideae in the Poaceae. The East African rushes all have small diameter culms (Haines and Lye 1983). Grasses seldom have such large and closely spaced culms on a rhizome and such densely spaced. Furthermore, grasses usually have leaves, whereas some sedges do not have leaves. Large sedges, such as *Schoenoplectus corymbosus*, are leafless and have a basal culm diameter of 6–8 mm and can attain a height of 100–350 cm. This species grows in lake waters at medium to high altitude (Haines and Lye 1983: 56). Although it is not possible to identify the macrofossil on only the features of the rhizome and culm bases, it is likely that this growth form is common to perennially wet environments, for example lakes, quiet rivers or springs.

Sedges have a very poor macrofossil record, yet pollen is fairly common and nutlets rare. This is probably the earliest record of a large sedge macrofossil and is useful for habit reconstructions even though the taxon is unknown.

Assorted Seeds from Early Collections

Several seeds have been collected prior to the systematic collection described here. Fig. 11.11a, b shows the top and bottom surfaces of a hard seed split in half. The interior is hollow and the outer part made of many layers, ends pointed. The holes visible in the exterior (Fig. 11.11a) may be due to biotic damage, such as insect borings to access the endosperm. The seed in Fig. 11.11c and Fig. 11.11e is very similar, but a bit smaller and less sharp pointed ends. A small seed, split in half longitudinally is shown in Fig. 11.11d with rounded base and thin wall, but mostly infilled. A collection of seeds, twigs and plant fragments, still embedded in the Tuff is shown in Fig. 11.11f. Several occurrences like this were observed, showing the rich assortment of plant matter preserved in ashfalls.

Discussion

East African Plant Record

Angiosperm leaf impressions from the Cenozoic of East Africa are not common but some have been well described and used for paleoclimatic interpretations. The oldest Cenozoic record is from the middle Eocene of Tanzania where leaf impressions have been recovered from kimberlite crater lake facies at Mahenge, Singida Province (Herendeen and Jacobs 2000; Harrison et al. 2001; Jacobs and Herendeen 2004). The collection of about 370 leaves comprises 19–21 morphotypes and includes caesalpinoid and mimosoid legumes, *Aphanocalyx*, cf. *Bauhinia*, cf. *Parinari*, and long, narrow, toothed and untoothed morphotypes. The leaves indicate a rainfall similar to today's (approximately 660 mm per annum) and woodland vegetation (not lowland tropical rainforest).

Oligocene sites in Chilga Woreda, northwestern Ethiopia, have yielded leaf impressions, seeds and silicified woods representing palms and broad and narrow-leaved dicots occurring in different paleoenvironmental facies (Jacobs et al. 2005). The reconstructed environment from the sedimentology and paleobotany is of a river meandering through a wide flat valley with lush vegetation. Not all the woods and leaves have been identified (B. Jacobs, personal communication).

At Bukwa on the slopes of Mt. Elgon, eastern Uganda, Early Miocene (22 Ma) *in situ* plants, grass culms and silicified woods have been recovered that represent Sudano-Zambezian Miombo-like woodlands (Chaney 1933; Hamilton 1968; Pickford 2002). The woods are from the Dipterocarpaceae (Bancroft 1935) and also represent woodland vegetation.

Leaves, seeds, twigs and thorns are abundant from the Early Miocene Rusinga Island deposits in the present-day Lake Victoria and indicate a mixed woodland flora (Chesters, 1957; Collinson 1983; Collinson et al. 2009) with at least 24 families represented, including most of the families identified from Laetoli. From Ethiopia seeds and woods have been recovered from the Early to Middle Miocene aged sites at Fejej (Tiffney et al. 1994; Wheeler et al. 2007). None of the families represented by the woods are represented by the seeds and fruits. Based on the woods, dry deciduous forest and woodlands are indicated (Wheeler et al. 2007).

From Fort Ternan, a Middle Miocene (14 Ma) site in southwestern Kenya, fossil grasses have been recovered. Using isotopic data Cerling et al. (1991) postulate a closed canopy dry woodland or forest, but Retallack and co-workers propose a much more open woodland setting based on paleosols and fossil grasses (Retallack et al. 1990; Dugas and Retallack 1993). Of the three grass genera they described two are comparable with genera from other continents, and one from southern and eastern Africa which occurs in mesophytic open habitats (*Stereochlaena*; Gibbs Russell et al. 1990: 315).

The Ngorora Formation in the Baringo Basin (13–10 Ma) has about 30 leaf species including monocots, dicots and grasses (Hill et al. 1985; Jacobs and Kabuye 1987) that are interpreted as representing forest and possibly riverine vegetation. Leaves of forest trees and herbs were preserved, but no litter was present (Jacobs and Winkler 1992).

The Tugen Hills in western Kenya are Middle to Late Miocene and exhibit a number of sites with leaves, twigs and seeds preserved (Jacobs 2002). Identification of the leaves is incomplete, but reconstructed climates from the dicot leaves indicate high rainfall for Kabarsero (12.6 Ma), medium rainfall for Waril (10–9 Ma) and high rainfall for Kapturo (6.8 Ma). The corresponding vegetation fluctuates from a lowland submontane forest type of central and West Africa, to open vegetation with a pronounced dry season, to woodland and dry forest (Jacobs 2002).

The Laetoli fossil flora falls into the next period of 3.8 to 3.5 Ma and the small sample of leaves and large sample of seeds are described here.

Laetoli Leaves

The leaves are not well preserved and the comparison with the Euphorbiaceae and Moraceae are very tentative, the latter more so. Both families are present in the modern flora and have a long history in the East African region, with Euphorbiaceae leaves recorded from the Middle Miocene Ngorora Formation (Jacobs and Kabuye 1987). With no good leaf identification climate or vegetation reconstructions are not possible.

Using the size classes established by Raunkiaer (1934) and modified by Webb (1959) the leaves are mostly microphylls (11 microphylls, 2 notophylls, 1 mesophyll), in other words, small with only two medium-sized and one large leaf. Small leaves are indicative of dry and/or cold environments, whereas large leaves are more common in mesophytic to humid and high rainfall environments (Dorf 1964; Wolfe 1978; Jacobs and Herendeen 2004). The leaf sample size is too small to use for a statistical analysis, and small leaves may well be preferentially preserved, but the overwhelming small size does point to a drier and/or colder local climate for the short time period that the leaves represent (during deposition of Tuff 8). The fossil leaves are predominantly entire margined or toothed margined and fairly small indicating a fairly dry to mesic environment. This is characteristic of deciduous floras, implying a seasonal climate.

Reproductive Structures

The fossil fruits and seeds occur throughout the area and over most of the Upper Laetolil period. Only some of them have been described and identified because a more comprehensive modern reference collection is needed for comparison. Locality 3 has the greatest number of specimens and the greatest variety (Table 11.1). Locality 6 has the next highest number of seeds, but they are all the same type, *Boscia coriacea*. These seeds occur in almost every locality and throughout the sequence (Tuffs 5–8). The plants may have been abundant, but also the seeds are hard and produced in large numbers, which may bias the apparent relative abundance of this small tree. There are quite a few small lentil-like seeds, but these have not been identified. Beans, most probably from legumes and Papilionaceae in particular, occur in Localities 1, 3, and 7. Other seeds that have been compared with *Ximenia americana* (Loc. 13, between Tuffs 6 and 7), *Ximenia caffra* (Loc. 2, between Tuffs 5 and 7), *Cryptocarya latifolia* (Loc. 9, between Tuffs 5 and 7) and *Celtis* cf. *africana* (Loc. 4, between Tuffs 6 and 7). They may not belong to these taxa, but to closely related species. The members of the Capparaceae (*Boscia coriacea*) are common in drier woodlands and along rivers. The Lauraceae (*Cryptocarya* spp.) and Celtidaceae/Ulmaceae (*Celtis* spp.) are more common in moister environments. The latter are from Tuffs 5–7, whereas the other seeds are from Tuffs 7–8. Based on these affinities the vegetation was moister and more forested in the middle part of the sequence and became somewhat drier during Tuff 7–8 times. The leaves support this evidence.

Thorns and Spines

The dissociated thorns and spines were probably detached from the plants before preservation and have been preserved with the seeds and fruits. If the "rose" thorns belong to *Rubus* or another scrambler it implies that there was dense, multilayered vegetation with plants scrambling over other plants (shrubs, trees?) as is common in many forms of woodlands and forests. If they belong to the other families mentioned then the implication is of armed trees and shrubs.

Monocot Rhizome

The rhizome and basal culms, most likely of a large sedge, are from the Upper Laetolil Beds in the Oleisusu area. Large sedges grow in fresh permanent or seasonal water, for example *Cyperus papyrus, Cyperus dives* and *Schoenoplectus corymbosus* at Ngoitokitok in the Ngorongoro crater, Jipe Lake and other African lakes. The culm height of the fossil sedge would have been over 1 m, probably much more. Its presence implies a reliable source of fresh water, yet *Schoenoplectus corymbosus* grows today in the Ol'Balbal depression at Olduvai Gorge, which is seasonally wet and somewhat saline-alkaline. Nonetheless, there must have been a wetland or some springs present at Oleisusu sometime during the time of deposition of the Upper Laetolil Beds.

Paleoecology

The Upper Laetolil flora is documented here for the period 3.8–3.5 Ma and shows that there was a variety of mesic and dry woodland plants and some possible forest elements. The phytolith record (Rossouw and Scott 2011) demonstrates that grasslands were a ubiquitous, but never dominant part of the vegetation. The fossil woods (Bamford 2011) also show that there were forest and mesic and dry woodlands present. The vegetation models presented by Andrews et al. (2011) show how this mosaic vegetation of higher altitude forest, mesic woodland, dry woodland, wooded grassland and grassland can occur, and how these vegetation bands can shift with changing climate.

The most common and widespread fossil seeds are those of *Boscia coriacea*, which today grows in semi-arid regions of East Africa, but near water courses (Beentje 1994; personal observation). Assuming that its climatic tolerance in the past was the same as that of today, then the past climate would have been hot and seasonally dry. This means that either the species has changed its distribution/climate tolerance, or that there was a very dry period prior to volcanic activity, ashfall and preservation of the local vegetation, such as has been demonstrated at Olduvai Gorge (Bamford et al. 2008; Stollhofen et al. 2008).

Conclusion

If the macrofossils were preserved close to where the plants grew, and there is no evidence of transport of these or the faunal remains, then a complex vegetation distribution can be inferred. The *Boscia coriacea* seeds are common throughout the area and timeframe and may imply some degree of seasonality with fairly dry periods. The legume seeds are too poorly defined, but the Olacaceae seeds also imply a climate on the dry side. The Lauraceae and Celtidoideae seeds indicate more mesic conditions. This mosaic is possible because the soil, drainage and topography would have varied significantly within the Olaitole and Garusi River valleys during the Pliocene.

Acknowledgements Macroplant fossils have been collected over many field seasons by field teams and their assistance is appreciated. Research Permits were granted by the Tanzanian Commission for Science and Technology (COSTECH), Tanzanian Department of Antiquities and the Ngorongoro Conservation Area Authority. The financial assistance of the NSF grants to Terry Harrison (BCS-9903434 and BCS-0308513) and from PAST (the South African Palaeontological Scientific Trust) to the author are gratefully acknowledged.

References

Andrews, P., Bamford, M. K., Njau, E.-F., & Leliyo, G. (2011). The vegetation of the Endulen-Laetoli area in northern Tanzania. In T. Harrison (Ed.), *Paleontology and geology of Laetoli: Human evolution in context* (Geology, geochronology, paleoecology and paleoenvironment, vol. 1, pp. 167–200). Dordrecht: Springer.

Bamford, M. K. (2011). Fossil woods. In T. Harrison (Ed.), *Paleontology and geology of Laetoli: Human evolution in context* (Geology, geochronology, paleoecology and paleoenvironment, vol. 1, pp. 217–233). Dordrecht: Springer.

Bamford, M. K., Stanistreet, I. R., Stollhofen, H., & Albert, R. M. (2008). Late Pliocene grassland from Olduvai Gorge, Tanzania. *Palaeogeography, Palaeoclimatology, Palaeocology, 257*, 280–293.

Bancroft, H. (1935). Some fossil dicotyledonous woods from Mount Elgon, East Africa. *American Journal of Botany, 22*, 164–183. 279–290.

Beentje, H. (1994). *Kenya trees, shrubs and lianas*. Nairobi: National Museums Kenya.

Cerling, T. E., Quade, J., Ambrose, S. H., & Sikes, N. E. (1991). Fossil soils, grasses, and carbon isotopes from Fort Ternan, Kenya: Grassland or woodland? *Journal of Human Evolution, 21*, 295–306.

Chaney, R. W. (1933). A Tertiary flora from Uganda. *Journal of Geology, 41*, 702–709.

Chesters, K. I. M. (1957). The Miocene flora of Rusinga Island, Lake Victoria, Kenya. *Palaeontographica Abt. B, 101*, 30–71.

Collinson, M. E. (1983). Revision of East African Miocene floras. A preliminary report. *International Association of Angiosperm Palaeobotany Newsletter, 8*, 4–10.

Collinson, M. E., Andrews, P., & Bamford, M. K. (2009). Taphonomy of the early Miocene flora, Hiwegi Formation, Rusinga Island, Kenya. *Journal of Human Evolution, 57*, 149–162.

Deino, A. (2011). ^{40}Ar/^{39}Ar dating of Laetoli, Tanzania. In T. Harrison (Ed.), *Paleontology and geology of Laetoli: Human evolution in context* (Geology, geochronology, paleoecology, and paleoenvironment, vol. 1, pp. 77–97). Dordrecht: Springer.

Dorf, E. (1964). The use of fossil plants in palaeoclimate interpretations. In A. E. M. Nairn (Ed.), *Problems in palaeoclimatology. Proceedings of the NATO palaeoclimates conference, University of Newcastle-upon Tyne* (pp. 13–31). London: Interscience.

Drake, R., & Curtis, G. H. (1987). K–Ar geochronology of the Laetolil fossil localities. In M. D. Leakey & J. M. Harris (Eds.), *Laetoli: A Pliocene site in northern Tanzania* (pp. 48–52). Oxford: Clarendon.

Dugas, D. P., & Retallack, G. J. (1993). Middle Miocene fossil grasses from Fort Ternan, Kenya. *Journal of Paleontology, 67*, 113–128.

Gibbs Russell, G. E., Watson, L., Koekemoer, M., Smook, L., Barker, N. P., Anderson, H. M., & Dallwitz, M. J. (1990). Grasses of southern Africa. *Memoirs of the Botanical Survey of South Africa, 58*, 1–437.

Haines, R. W., & Lye, K. A. (1983). *The sedges and rushes of East Africa*. Nairobi: East African Natural History Society.

Hamilton, A. C. (1968). Some plant fossils from Bukwa. *Uganda Journal, 32*, 157–164.

Harrison, T., & Kweka, A. (2011). Paleontological localities on the Eyasi Plateau, including Laetoli. In T. Harrison (Ed.), *Paleontology and geology of Laetoli: Human evolution in context* (Geology, geochronology, paleoecology and paleoenvironment, vol. 1, pp. 17–45). Dordrecht: Springer.

Harrison, T., Msuya, C. P., Murray, A. M., Fine Jacobs, B., Baez, A. M., Mundil, R., & Ludwig, K. R. (2001). Paleontological investigations at the Eocene locality of Mahenge in north-central Tanzania, East Africa. In G. Gunnell (Ed.), *Eocene biodiversity: Unusual occurrences and rarely sampled habitats. Topics in geobiology* (pp. 39–74). New York: Plenum.

Hay, R. L. (1987). Geology of the Laetolil Beds. In M. D. Leakey & J. M. Harris (Eds.), *Laetoli: A Pliocene site in northern Tanzania* (pp. 23–47). Oxford: Clarendon.

Herendeen, P. S., & Jacobs, B. F. (2000). Fossil legumes from the middle Eocene (46.0 Ma) Mahenge flora of Singida, Tanzania. *American Journal of Botany, 87*, 1358–1366.

Hickey, L. J. (1973). Classification of the architecture of dicotyledonous leaves. *American Journal of Botany, 60*, 17–33.

Hickey, L. J. (1974). A revised classification of the architecture of dicotyledonous leaves. In C. R. Metcalfe & L. Chalk (Eds.), *Anatomy of the Dicotyledons* (2nd ed., Vol. 1, pp. 25–39). Oxford: Clarendon.

Hill, A., Drake, R., Tauxe, L., Monaghan, M., Barry, J. C., Behrensmeyer, A. K., Curtis, G., Jacobs, B. F., Jacobs, L., Johnson, N., & Pilbeam, D. (1985). Neogene palaeontology and geochronology of the Baringo Basin, Kenya. *Journal of Human Evolution, 14*, 759–773.

Jacobs, B. F. (2002). Estimation of low-latitude paleoclimates using fossil angiosperm leaves: Examples from the Miocene Tugen Hills, Kenya. *Paleobiology, 28*, 399–421.

Jacobs, B. F., & Herendeen, P. S. (2004). Eocene dry climate and woodland vegetation in tropical Africa reconstructed from fossil leaves from northern Tanzania. *Palaeogeography, Palaeoclimatology, Palaeoecology, 213*, 115–123.

Jacobs, B. F., & Kabuye, C. (1987). A Middle Miocene (12.2 my old) forest in East Africa Rift Valley, Kenya. *Journal of Human Evolution, 16*, 147–155.

Jacobs, B. F., & Winkler, D. A. (1992). Taphonomy of a middle Miocene autochthonous forest assemblage, Ngorora Formation, central Kenya. *Palaeogeography, Palaeoclimatology, Palaeoecology, 99*, 31–40.

Jacobs, B. F., Tabor, N., Feseha, M., Pan, A., Kappelman, J., Rasmussen, T., Sanders, W., Weimann, M., Crabaugh, J., & Massini, J. L. G. (2005). Oligocene terrestrial strata of northwestern Ethiopia: A preliminary report on paleoenvironments and paleontology. *Palaeontologia Electronica, 8*(25A), 1–19.

Kingston, J. D., & Harrison, T. (2007). Isotopic dietary reconstructions of Pliocene herbivores at Laetoli: Implications for early hominin ecology. *Palaeogeography, Palaeoclimatology, Palaeoecology, 243*, 272–306.

Leaf Architecture Working Group (1999). *Manual of Leaf Architecture*. Washington: Smithsonian Institution.

Pickford, M. (2002). Early Miocene grassland ecosystem at Bukwa, Mount Elgon, Uganda. *Comptes Rendus Palevol, 1*, 213–219.

Raunkiaer, C. (1934). *The life forms of plants and statistical plant geography*. Oxford: Clarendon.

Retallack, G. J., Dugas, D. P., & Bestland, E. A. (1990). Fossil soils and grasses of a Middle Miocene East African grassland. *Science, 247*, 1325–1328.

Rossouw, L., & Scott, L. (2011). Phytoliths and pollen, the microscopic plant remains in Pliocene volcanic sediments around Laetoli, Tanzania. In T. Harrison (Ed.), *Paleontology and geology of Laetoli: Human evolution in context* (Geology, geochronology, paleoecology and paleoenvironment, vol. 1, pp. 201–215). Dordrecht: Springer.

Spicer, R. A. (1986). Comparative leaf architectural analysis of Cretaceous radiating angiosperms. In R. A. Spicer & B. A. Thomas (Eds.), *Systematic and taxonomic approaches in palaeobotany*. Systematics Association special vol. 31 (pp. 221–232). Oxford: Oxford University Press.

Stollhofen, H., Stanistreet, I. G., McHenry, L., Hay, R. L., & Swisher, C. C. (2008). Fingerprinting facies of the Tuff IF marker with implications for hominin paleoecology, Olduvai Gorge, Tanzania. *Palaeogeography, Palaeoclimatology, Palaeoecology, 259*, 382–409.

Tiffney, B. H., Fleagle, J. G., & Bown, T. M. (1994). Early to Middle Miocene angiosperm fruits and seeds from Fejej, Ethiopia. *Tertiary Research, 15*, 25–42.

Van Wyk, A., & van Wyk, B. (1997). *Field guide to trees of southern Africa*. Cape Town: Struik.

Webb, L. J. (1959). A physiognomic classification of Australian rain forests. *Journal of Ecology, 47*, 551–570.

Wheeler, E. A., Weimann, M. C., & Fleagle, J. G. (2007). Woods from the Miocene Bakate Formation, Ethiopia. Anatomical characteristics, estimates of original specific gravity and ecological inferences. *Review of Palaeobotany and Palynology, 146*, 193–207.

Wolfe, J. A. (1978). A paleobotanical interpretation of Tertiary climates in the northern hemisphere. *American Scientist, 66*, 694–703.

Chapter 12
Serengeti Micromammal Communities and the Paleoecology of Laetoli, Tanzania

Denné N. Reed

Abstract The fossil deposits at Laetoli provide critical paleoanthropological insights that can be better understood in light of the modern Serengeti ecosystem. Earlier paleoenvironmental reconstructions of the site depicted it as a semi-arid grassland, similar to the modern Serengeti plains. Subsequent analyses, however, have converged around an interpretation emphasizing a greater wooded component in a more humid climate, suggesting that the adaptations of *A. afarensis* remain linked to a more closed and wooded biome. The range of proposed paleoenvironments encompasses the range seen in the modern Serengeti-Masai Mara ecosystem today and thus the Serengeti serves as a natural analogy for comparisons with Laetoli. Our understanding of past environments is informed by the study of modern mammalian communities where it is possible to observe species distributions, adaptations and community composition. As paleoenvironmental indicators, micromammals provide a signal of local paleoenvironments at a smaller spatial grain than large mammals and thus may prove critical to resolving some of the ambiguities of paleoenvironmental interpretations at Laetoli. This chapter presents findings on the modern rodent biodiversity in Serengeti based on material recovered from owl roosting sites distributed throughout the ecosystem. The Serengeti experiences strong ecological gradients, which influence vegetation patterns and the distribution of rodent genera. Examining the interplay between habitats and rodent community composition provides the necessary baseline perspective for interpreting Laetoli paleoenvironments.

Keywords Rodents • Paleoenvironment • Grassland • Woodland • Owl pellets • *Bubo* • *Tyto* • *Australopithecus afarensis*

D.N. Reed (✉)
Department of Anthropology, University of Texas at Austin,
1 University Station C3200, Austin, TX 78712, USA
e-mail: reedd@mail.utexas.edu

Introduction

The fossil site of Laetoli resides within the Ngorongoro Conservation Area (NCA), about 30 km south of Olduvai Gorge, and just outside the southern border of the Serengeti National Park (SNP) in northern Tanzania (Fig. 12.1). Though famous for preserving the lectotype (L.H. 4) of *Australopithecus afarensis* (Leakey et al. 1976; Johanson et al. 1978; Leakey and Harris 1987), as well as the oldest bipedal hominin trackways (Leakey and Hay 1979), Pliocene hominins at Laetoli are relatively rare compared to other mammals and to the abundance of *A. afarensis* at other hominin-bearing sites such as Hadar (Su and Harrison 2008; Reed 2008). At Hadar, *A. afarensis* persists across fluctuating paleoenvironmental regimes in what is generally a wooded to forested lake margin environment (Bonnefille et al. 2004; Campisano and Feibel 2007; Reed 2008) at abundances that range from ca 5–30% of the mammalian fauna depending on the stratigraphic unit examined (Reed 2008), whereas at Laetoli hominins are less than 1% of the fauna (Su and Harrison 2008). The rarity of *A. afarensis* at Laetoli may be explained in part by its taphonomic and paleoecological history. While most other Pliocene hominin localities occur in fluvial and lacustrine settings (White 1988; Behrensmeyer and Hook 1992), Laetoli represents an almost exclusively terrestrial setting (Hay 1987; Leakey 1987; Peters et al. 2008) with an unusual faunal composition and taphonomy. Paleoenvironmental reconstructions of Laetoli range from semi-arid grasslands to more densely wooded to forested habitats (Hay 1987; Su and Harrison 2007; Peters et al. 2008), and it has been suggested that Laetoli perhaps represents a more marginal environment for *A. afarensis* compared with Hadar (Su and Harrison 2008). The paleoenvironment at Laetoli is thus important for understanding the niche breath of *A. afarensis* and for interpreting the degree to which *A. afarensis* retained an arboreal component to its adaptation as is suggested by anatomical studies of this species (Richmond and Jungers 2008).

Paleoenvironmental reconstructions, to be accurate, make use of several independent lines of evidence, including soil

T. Harrison (ed.), *Paleontology and Geology of Laetoli: Human Evolution in Context. Volume 1: Geology, Geochronology, Paleoecology and Paleoenvironment*, Vertebrate Paleobiology and Paleoanthropology, DOI 10.1007/978-90-481-9956-3_12,

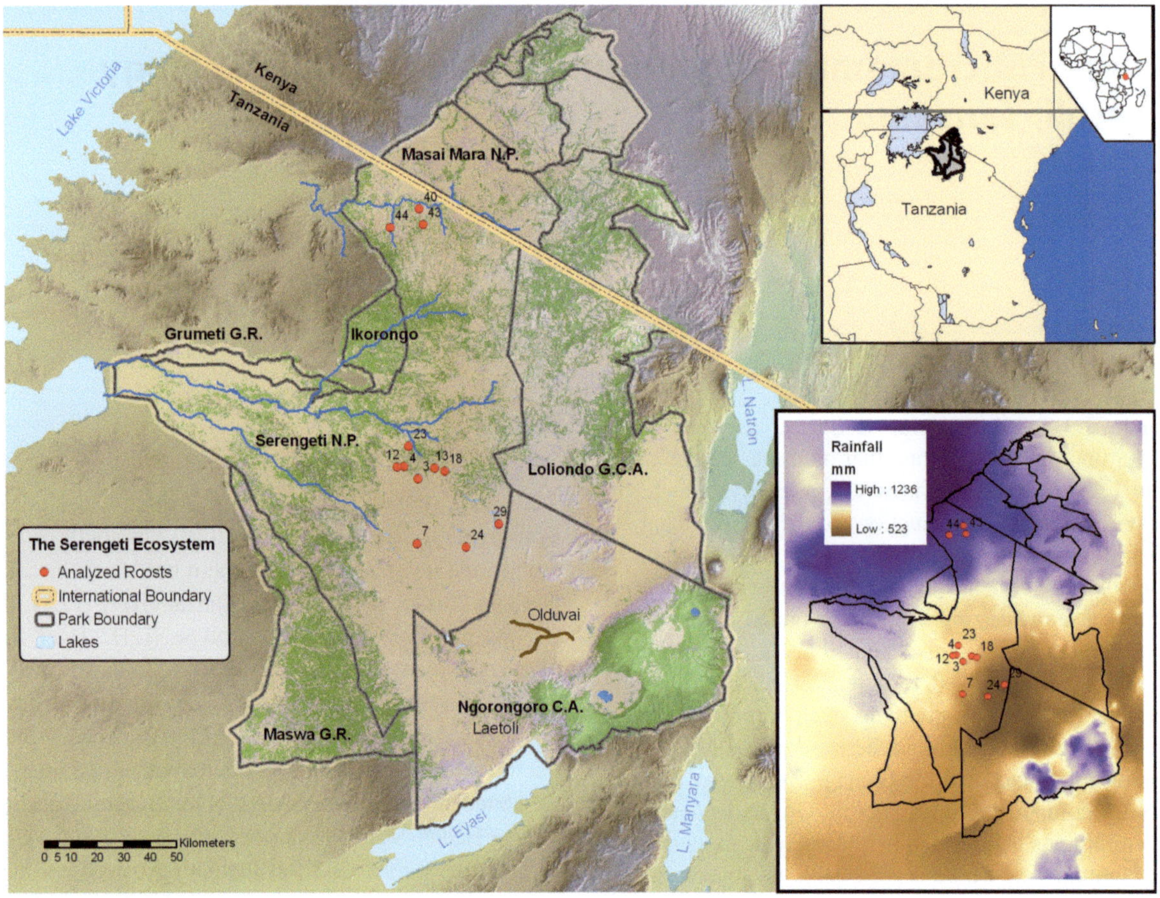

Fig. 12.1 Map of the Serengeti ecosystem showing the location of the 12 sampling sites (roosts) over a vegetation map showing the distribution of grasslands, shrublands, and woodlands. Upper insert shows the location of the Serengeti in East Africa. Lower insert shows a precipitation map and illustrates the orientation of the ecological gradients in Serengeti

isotope and geochemistry, lithology, palynology, large mammal biodiversity and functional morphology, as well as micromammal biodiversity. Concurrence between these methods provides strong evidence for a particular paleoenvironmental interpretation, while discord may point to weaknesses in a method, or may indicate paleoenvironmental heterogeneity at different spatio-temporal scales (Andrews 2006; Kingston 2007). Micromammal analysis provides a paleoecological signal that is sensitive to variation in habitats at spatial scales of 100–1,000 m (Reed 2003, 2007), and is perhaps one of the better tools for resolving habitat ambiguity at finer spatial grains. This chapter supports the analysis of fossil micromammal faunas from Laetoli by providing baseline data on modern vegetation patterns and habitats in the Serengeti ecosystem along with micromammal spatial diversity across those habitats. Data from trapping studies are summarized along with new data from a study of the rodent biodiversity occurring in owl pellets collected from several habitats across the ecosystem.

Ecosystems are dynamic entities and it is critical to recognize that for any fossil locality there is the possibility that no exact modern analog exists (Andrews 2006; Kingston 2007). However, examining the basic structure of modern ecosystems is the foundation for making informed inferences about past mammal communities, and is the best starting point for interpreting the paleoenvironment of Laetoli and how it has changed through time (Soligo and Andrews 2005). Contemporary ecological data abet the analysis of fossil assemblages by: (1) providing data on the distribution and preferred habitats of different taxa throughout the ecosystem; (2) providing baseline data on the competitive dynamics and possible communities structures observed in different habitats; (3) providing baseline data on the variation within and between different habitats within the ecosystem.

The methodological approach to modern faunas taken in this chapter is unique from that of neontologists studying modern rodents in that it relies on collections of bone detritus derived from owl pellets at roosting sites distributed throughout the Serengeti ecosystem. Pellet assemblages such as these are an excellent resource for biodiversity survey (Avery 1992; Avery et al. 2005) because the owl species in Serengeti generating these assemblages, primarily the Barn

Owl, *Tyto alba* and the Spotted Eagle Owl, *Bubo africanus*, take a sufficiently broad diversity of prey species and are generally opportunistic in their selection of prey as long as the prey fall within the constraints of body size and activity pattern suitable to the predator (Reed 2005). The larger Greater Eagle Owl, *Bubo lacteus*, is also present in the Serengeti, but no roosting sites of this species were located in the current study. Comparing pellet data to modern trapping studies reveals strong overlap between the methods in their ability to recover the local biodiversity. Pellet assemblages are useful because owls sample a broad range of the micromammal community and deposit the osteological remnants of prey in dense accumulation at their roosting spots (Glue 1971; Andrews 1990; Avery 2002; Reed 2005). One advantage afforded to the analysis of pellet assemblages is the ability to collect large samples quickly. Trapping success of micromammals in Serengeti average approximately 10% success (D. Reed, unpublished data; A. Sinclair, personal communication) which means that in order to attain a sample of 1,000 specimens an effort of 10,000 trap intervals (i.e., 100 traps set for 100 nights) is needed, whereas samples of 1,000 specimens are readily collected from owl roosting sites at a single visit.

The current study has the following goals. First is to document the modern environments and habitats that occur in the Serengeti ecosystem and the processes that govern the distribution of those habitats across the ecosystem. The second goal is to examine rodent biodiversity across the Serengeti ecosystem for the purpose of analysis of the Laetoli fauna. The third goal is to explore the relationship between habitat distributions and biodiversity to uncover patterns and factors influencing the spatial distribution of rodents in the ecosystem in a manner that may inform us about how rodent communities are assembled and what may be gleaned about the Pliocene environments at Laetoli from the community composition of the Laetoli rodents.

The Serengeti Ecosystem

As a UNESCO World Heritage Site, the Serengeti ecosystem is one of the best-studied and least disrupted tropical savanna ecosystems. Beginning with the first aerial surveys (Grzimek and Grzimek 1960; Talbot and Stewart 1974) interest in Serengeti biodiversity has led to detailed study of most large mammals species. By comparison, Serengeti small mammal ecology and biodiversity have received far less attention. The most recent trapping data for Serengeti rodents comes form the compilation put together by Hendrichs (1969). A few targeted studies on specific rodent species were carried out by Senzota (1978, 1983, 1990, 1992) and Laurie (1971) presented a brief study of barn owl pellets. Since then a wealth of new research has been conducted on the systematics and ecology of East African rodents (Lecompte et al. 2008) as well as further biodiversity studies based on owl pellets (Reed 2003, 2005, 2007).

The Serengeti ecosystem straddles the Tanzania-Kenya border in East Africa between 34 and 36° East longitude and 1–2° S latitude (Fig. 12.1). The Serengeti National Park in Tanzania encompasses an area of 14,800 km^2, but the larger ecosystem–defined as the area covered by the annual migration of nearly two million ungulates–extends to roughly 24,000 km^2 and includes conservation areas that surround the core Serengeti National Park. One of the largest such protected areas is the Ngorongoro Conservation Area, which includes the site of Laetoli.

The Serengeti has a broad range of physical habitats, influenced by a series of interrelated ecological gradients (Sinclair 1995; Reed et al. 2009). A study of rainfall patterns and topography (Norton-Griffiths et al. 1975; Reed et al. 2009) demonstrates that the imposing volcanic highlands of Ngorongoro, Sadiman and Lemagrut to the south of the ecosystem and the chain of low hills called the Gol Mountains along the eastern border, intercept moisture laden winds blowing in from the Indian Ocean during the rainy season (March-June) and cast a rain shadow across the southern portion of the Serengeti. In addition, the northern part of the ecosystem experiences dry season rainfall resulting from weather effects over Lake Victoria and this creates a strong NW-SE rainfall gradient in the ecosystem (Fig. 12.1 inset).

Soils follow a similar gradient as the rainfall. Results from two soil studies (de Wit 1978; Jager 1982) indicate that areas in the driest portion of the rainfall gradient and closest to the volcanic highlands, have the coarsest and most alkaline soils derived from the eruptions of nearby volcanoes. The same prevailing winds that affect rainfall also distributed and size-sorted ash during the eruptions such that heavier, coarser particles landed closest to the volcanic highlands and lighter particles were carried further out toward Seronera. This creates a soil texture and soil alkalinity gradient along roughly the same axis as the precipitation gradient (Anderson and Talbot 1965; Sinclair 1995).

The interaction of these two gradients creates another gradient in soil depth. Precipitation leaches soluble minerals from the soil surface and evaporation causes them to be deposited at a regular depth, leading to the development of extensive calcretes below the surface. The depth of the calcrete depends primarily on the amount of rainfall and the average temperature (and resulting evaporation rate). Since temperatures vary over larger areas than rainfall, it is rainfall that is the primary variable affecting calcrete development at the ecosystem scale. Where the calcrete is continuous, calcrete depth has a great effect on soil depth and the distribution of vegetation communities (de Witt and Jeronimus 1977; de Wit 1978; Sinclair 1995).

Comprehensive mapping of the Serengeti indicates a range of land cover types (Reed et al. 2009), most of which fall in the general category of "savanna", that is a habitat with a continuous grass understory and possibly trees (Cole 1986). Tree densities in Serengeti can range from 0 (grasslands) to greater than 80% (grassed woodlands) while still maintaining a continuous grass understory. There is an active debate among landscape ecologists regarding the mechanisms that regulate tree-grass ratios in savannas and this debate applies to Serengeti and Laetoli as well. Several factors are acknowledged to influence tree-grass ratios, and they can be summarized into three main categories: (1) climatic factors such as rainfall, altitude, temperature, wind (Scholes and Archer 1997; House et al. 2003; Sankaran 2004; Sankaran et al. 2005); (2) disturbance factors such as fire, grazing, browsing, cultivation (Anderson and Talbot 1965; Dublin and Douglas-Hamilton 1987; Dublin et al. 1990; Belsky 1992; Serneels and Lambin 2001; Dempewolf et al. 2007); and (3) edaphic factors including soil chemistry, soil structure, topographic position (Milne 1935; Anderson 1963; Anderson and Talbot 1965; Belsky 1985, 1990; Peters et al. 2008). The most recent studies (Sankaran et al. 2005; Reed et al. 2009) suggest that all of these factors may be active in a hierarchical, scale-dependant fashion. Climate and especially rainfall, influences tree-grass ratios and patterning of vegetation at the broadest scale by setting upper limits to the density of tree cover. Edaphic conditions operate at a more restricted spatial scale, where differences in topography influence soil texture and moisture availability, while disturbance factors may alter vegetation over moderate to small spatial scales (Anderson et al. 2007).

Table 12.1 Details of individual roost locations. Where owls were witnessed at the roost, the species is listed directly. Otherwise best estimates are based on other signs (such as feathers, pellet shape and roost type) and indicated by a question mark. Geographical coordinates are given in decimal degrees in the Arc 1960 datum for Tanzania. Habitat descriptions follow those of Grunblatt et al. (1989)

Roost	Habitat	Owl	Lat	Lon
40	Woodland/forest	*Bubo africanus?*	−1.58351	34.90704
43	Woodland/forest	*Bubo africanus*	−1.63607	34.91956
44	Woodland/forest	*Tyto alba?*	−1.64596	34.80920
3	Wooded grassland	*Tyto alba*	−2.47109	34.89905
13	Wooded grassland	*Tyto alba?*	−2.43625	34.95496
18	Wooded grassland	*Tyto alba?*	−2.44666	34.98977
4	Grassed woodland	*Tyto alba*	−2.43132	34.85326
12	Grassed woodland	*Bubo africanus*	−2.43268	34.82940
23	Grassed woodland	*Bubo africanus?*	−2.36593	34.86813
7	Grassland	*Tyto alba*	−2.68508	34.89518
24	Grassland	*Tyto alba*	−2.69849	35.06356
29	Grassland	*Tyto alba*	−2.62093	35.17648

Methods

Osteological bulk samples collected from 12 roosting sites in Serengeti National Park were analyzed. Three roosts each from four different habitats were analyzed: the short grass plains in the south, wooded-grasslands and grassed-woodlands near the center of the park, the moist, treed-grasslands adjacent to gallery forests of the Mara River system in the north of the park. These roosting locations sample a broad spectrum of "savanna" landscape types, similar to the varieties purported for the Upper Laetolil and Upper Ndolanya Beds. Table 12.1 provides summary information on the roosts. Roost locations were recorded on a consumer grade Garmin XL12 GPS receiver and their distribution is shown in Fig. 12.1.

Taxonomic identification of the rodents appearing in the osteological samples was aided by a reference database and computer application written in the Python programming language (Lutz 2006; Flaig 2008). A database of diagnostic discrete characters was compiled from published keys (Coetzee 1972; Davis 1965; Foster and Duff-Mackay 1966; Delany 1975; Rogers and Stanley 2003), and a character state matrix coded for all genera of East African rodents. A Python program was then written to interactively accept character states present in an unknown specimen and returns a list of matching reference taxa. Voucher reference specimens were drawn from collections in the American Museum of Natural History (AMNH); Field Museum, Chicago (FMNH); Museum National d'Histoire Naturelle (MNHN), National Museum of Natural History, Smithsonian Institution (NMNH) and the Zoologisches Forschungsinstitut und Museum Alexander Koenig, Bonn (ZFMK). The taxonomic classification used here follows that of Musser and Carleton (2005).

Results and analysis are conducted at the generic level because of uncertainty regarding taxonomic assignments at the species level using discrete diagnostic traits on fragmentary cranio-dental remains. The genus level analysis also assures greater comparability with older trapping studies. Voucher specimens from the older studies are not available so it is not possible to verify the taxonomic assignments in older published reports. Taxonomic abundances are reported as the minimum number of individuals (MNI) for each taxon (Grayson 1991; Lyman and Power 2003). Community composition between roosts was analyzed using Non-Metric Multi-Dimensional Scaling (NMDS) (Clarke 1993; McCune and Grace 2002) and other statistical functions in the vegan package of the R statistical programming language (Oksanen et al. 2005; Crawley 2007).

Results

The owl-accumulated faunal assemblages are dominated by small mammals though passerine birds, insects, and reptiles were also observed, supporting other observations that Barn owls and Spotted Eagle owls eat a broad diversity of prey species (Vernon 1972; Andrews 1990; Taylor 1994; Reed 2003, 2005). Previous analyses of prey differences between Barn owls and Spotted Eagle owls indicate that the two owl species have a large overlap in the prey species that they eat and may be considered isotaphonomic with regard to prey selectivity (Reed 2005). In total, the Serengeti owl-accumulated assemblages contain 15 rodent genera (Table 12.2). Sample sizes vary across the roosting sites and range from MNI of 17 to 221. Likewise taxonomic richness varies across sites as well, but the correlation between sample size and species richness is low ($r=0.06168$). The greatest richness (13 taxa) occurs at roosts 3 and 18, both treed grassland sites. The lowest richness values (e.g., richness of 7 at roost 29) occur in the grassland roosts despite these roosts having the larges sample sizes.

A few early studies have explored small mammal biodiversity in Serengeti using live trapping, kill trapping and owl pellet sampling techniques. Tanganyika Game Warden G. H. Swynnerton (1958) compiled one of the first comprehensive lists of mammals for the Serengeti National Park and surrounding areas. This listing features 25 rodent genera (Table 12.3). A later trapping study by Misonne and Verschuren (1966) added three additional genera to this list: *Cricetomys*, *Saccostomus*, and *Zelotomys*. Hendrichs (1969) provides a comprehensive compilation of previous studies but adds no new rodent taxa. The one previous pellet study is a short paper contributed by Laurie (1971).

The correspondence across published studies for the region is good with the early Swynnerton compilation being the most complete, but with the later studies adding new taxa. A total of 28 rodent genera are recorded in the literature for the Serengeti ecosystem. Swynnerton (1958) records 89% of the genera, Misonne and Verschuren (1966) captured 71%, Laurie (1971) 36% and the current pellet study records 54%.

The sampling locations (i.e., owl roosts) are distributed across the ecosystem in four different habitats: grasslands (roosts 7, 24, 29), wooded-grasslands (roosts, 3, 13, 18), grassed-woodlands (roosts, 4, 12, 23) and mesic grasslands beside riverine forest (roosts 40, 43, 44). The distribution of rodent genera is not uniform across these different habitats as is seen in the Non-metric Multi-Dimensional Scaling (NMDS) ordination of the data (Fig. 12.2). The roosts are ordinated based on their taxonomic composition and symbols for the roosts are grouped according to their true habitats (grassland, treed grassland, grassed woodland, mesic grassland/forest) as determined from surveys conducted on foot along with a remote sensing vegetation analysis of the vegetation in the ecosystem as a whole (Reed et al. 2009). The correspondence between the true land cover and position in the ordination plot suggests that taxonomic composition

Table 12.2 Rodent genera occurring in owl-pellet assemblages from Serengeti presented as the minimum number of individuals (*MNI*)

Sampling Unit (Roost Number)												
Taxon	3	4	7	12	13	18	23	24	29	40	43	44
Cricetomyinae												
Staccostomus	6	7	5	1	2	7	11	0	0	0	0	0
Dendromurinae												
Dendromus	24	33	21	2	26	9	4	17	1	2	3	4
Steatomys	32	44	61	4	64	14	2	93	5	2	14	7
Deomyinae												
Acomys	1	0	0	1	2	1	0	0	0	0	0	1
Gerbillinae												
Gerbillus	3	1	44	0	15	3	0	29	30	0	0	0
Gerbilliscus	1	7	10	0	4	3	3	66	122	2	2	3
Murinae												
Aethomys	0	0	0	0	0	0	0	1	0	3	5	5
Arvicanthis	5	21	7	0	3	4	14	1	0	5	10	17
Dasymys	0	0	0	0	0	0	0	0	0	1	0	4
Lemniscomys	4	3	7	1	2	5	6	4	3	0	2	1
Mastomys	9	35	3	1	4	2	21	4	2	3	11	10
Mus	41	27	41	2	24	30	10	6	0	2	10	1
Praomys	1	0	0	0	0	2	0	0	0	0	0	1
Thallomys	7	2	0	5	4	3	26	0	1	0	0	0
Zelotomys	3	6	2	0	5	1	1	0	0	1	1	1
Total MNI	137	186	201	17	155	84	98	221	164	21	58	55
Total Richness	13	11	10	8	12	13	10	9	7	9	9	12

Table 12.3 Checklist of rodent genera occurring in the Serengeti ecosystem

Taxon	Current study	Swynnerton (1958)	Misonne and Verschuren (1966)	Laurie (1971)
Cricetomyinae				
Cricetomys (>400 g)	–	–	*	–
Saccostomus	*	–	*	*
Dendromurinae				
Dendromus	*	*	–	–
Steatomys	*	*	–	*
Deomyinae				
Acomys	*	*	*	–
Gerbillinae				
Gerbillus	*	*	*	–
Gerbilliscus	*	*	*	*
Glirinae				
Graphiurus sp.	–	*	*	–
Hystricidae				
Hystrix (>400 g)	–	*	*	–
Murinae				
Aethomys	*	*	*	–
Arvicanthis	*	*	*	*
Dasymys	*	*	–	–
Grammomys	–	*	*	–
Lemniscomys	*	*	*	–
Lophuromys	–	*	*	–
Mastomys	*	*	*	*
Mus	*	*	*	*
Pelomys	–	*	*	*
Praomys	*	*	*	–
Rhabdomys	–	*	–	–
Thallomys	*	*	*	*
Zelotomys	*	–	*	*
Otomyinae				
Otomys	–	*	–	–
Pedetidae				
Pedetes (>400 g)	–	*	*	–
Tachyoryctinae				
Tachyoryctes	–	*	*	*
Thryonomyidae				
Thryonomys (>400 g)	–	*	–	–
Xerinae				
Heliosciurus	–	*	–	–
Paraxerus	–	*	–	–
Generic Richness	15	25	20	10

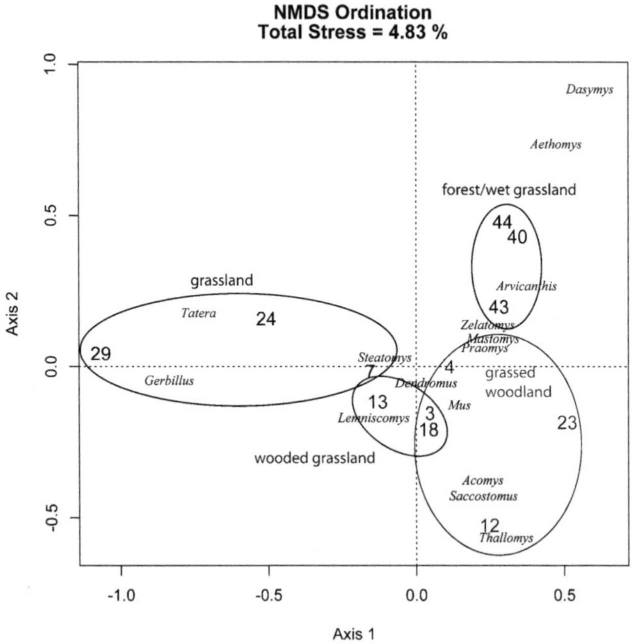

Fig. 12.2 NMDS ordination of roosts based on taxonomic composition. Habitat types: grassland (roosts 7, 24, 29), treed-grassland (roosts 3, 13, 18), grassed-woodland (roosts 4, 12, 23) and mesic grasslands/forest (roosts 40, 43, 44). Low stress indicates a better of fit to the underlying distance matrix

tracks the different habitat types. Each roosting location varies in taxonomic composition and the co-occurrence of taxa in certain habitats is indicated by the taxonomic loadings on the ordination axes, as illustrated by the positioning of the taxa in the ordination space. Many taxa are relatively catholic in their habitat associations and thus fall near the origin, but a few taxa have stronger habitat associations and are positioned at the extremes of the ordination space. *Gerbillus* and *Gerbilliscus,* for example, lie far to the left on Axis 1, along with the open grassland roosts (24 and 29). *Dasymys* and *Aethomys* are associated with the northern mesic grassland/forest roosts (40, 44). *Thallomys* appears in the lower right quadrant, associated with grassed woodland roosts (12, 23). The placement and associations of these taxa correspond well with what is known of the autecology of these taxa, as discussed below.

Discussion

When assessing the biodiversity of Serengeti rodents one encounters several challenges. Traditional methods of live trapping require great effort for an area as large as the Serengeti ecosystem. Large samples can be acquired quickly by examining predator assemblages, and these have the advantage also of being time-averaged assemblages that sample biodiversity over annual to decadal scales. The drawback to owl-accumulated assemblages is that the limitations and habits of the predator will influence representation in the assemblage. This can be seen by comparing results across studies in the Serengeti.

Added to the challenge of sampling is the difficulty of accurately and reliably identifying African rodents. The alpha taxonomy of African rodents has undergone dramatic shifts in recent years as new genetic and cytogenetic analyses reveal morphologically similar but distinct species in some clades, and morphologically diverse populations of a single species in others. Rodent biodiversity is presented here at the generic level, in part because this is more directly relevant to the fossil record at Laetoli, but also in recognition of the uncertainty surrounding the identification of specimens based on fragmentary and incomplete cranio-dental material (Barr and Reed 2009).

There is a 56% overlap between the genera reported by Swynnerton (1958) and the 17 genera present in the owl-accumulated assemblages (Laurie 1971, and the current study). Swynnerton (1958) lists results under five habitat grouping: swamps, plains, bushland, mountain mist forest and heath. The broad range of habitats partly explains the increased taxonomic richness in Swynnerton (1958), since neither swamps nor montane mist forest and heath are sampled by any of the roosts in the pellet studies. An odd result in Swynnerton (1958) is that no rodents are noted for the grassland plains. The current study and others (Senzota 1978, 1983) clearly show rodents inhabiting the grasslands. Subsequently, Misonne and Verschuren (1966) reported the results from trapping conducted at 27 locations in and around Serengeti National Park. These authors express an interest in competition between large and small herbivores (Misonne and Verschuren 1966: 517), an idea that may extend from earlier writings by Verschuren (e.g., 1964 as cited in Senzota 1978). Shortly after, Hendrichs (1969) conducted small mammal trapping as part of an analysis of herbivore biomass and carrying capacity of grassland ecosystems. The faunal list provided in the Hendrichs (1969) report includes summaries of species previously reported in the study area, though curiously it omits the records of *Otomys* and *Thryonomys* reported by Swynnerton (1958). Laurie (1971) produced a small study of owl pellet assemblages in Serengeti. He described three assemblages, one of these sites, Oloserian, is replicated in this study (roost 4).

Considering all the fauna purported to occur in Serengeti, 12 genera were not recorded in the owl assemblages examined as part of the current study: *Hystrix, Pedetes, Thryonomys, Cricetomys, Paraxerus, Graphiurus, Grammomys, Pelomys, Rhabdomys, Otomys, Lophuromys, Tachyoryctes*. Most of these omissions can be attributed to body size or activity patterns. *Hystrix, Pedetes, Thryonomys* and *Cricetomys* are all greater than 400 g and exceed the maximum prey size for barn owls and likely the Spotted Eagle Owls as well. *Paraxerus* lies at the upper limit of the barn owl prey size range and is diurnal. *Graphiurus* was observed at the Serengeti Research Institute (personal observation), in areas near owl roosts, but this is an agile, scansorial species and it may be adept at avoiding predation by owls. *Rhabdomys* is closely related to *Lemniscomys* and *Arvicanthis* (Ducroz et al. 2001; Musser and Carleton 2005), grazing species with which they are presumed to compete. Thus, *Rhabdomys* tends to be limited to highlands and sub-alpine zones that were not sampled by any of the roosting locations (Kingdon 1974). Similar conditions may hold for *Otomys*, which too is associated with more mesic or montane environments.

The absence of the remaining genera is unexplained. *Grammomys* is well within the prey size range for an owl, and though it often prefers moister habitats it has been reported from dry scrub areas, such as the Athi Plains of Kenya (Kingdon 1974). *Lophuromys* is expected to inhabit mesic grasslands free of closed canopy forest (Kingdon 1974). Its absence from roosts 40, 43, 44 in the north is somewhat surprising. The last two genera, *Tachyoryctes* and *Pelomys* were absent from the current study but were recorded by Laurie (1971). *Tachyorcytes* is fossorial and expected to be rare in owl-accumulated assemblages. Laurie (1971) reports only a single specimen out of a combined sample of 1,267 individuals. *Pelomys* inhabits rank grasslands and shrub vegetation associated with river margins (Vesey-Fitzgerald 1964; Kingdon 1974). Its habitat is sampled by the owls in the current study and is expected to occur. It is rare (three specimens in Laurie's sample) and may be absent by chance.

Not surprisingly the current study overlaps greatly with that of Laurie (1971) who also examined owl pellet assemblages. However, six rodent genera appear in the current study that were not recorded in the pellets collected by Laurie (1971): *Acomys, Aethomys, Dasymys, Lemniscomys, Praomys, Dendromus*. *Acomys* and *Lemniscomys*, were absent from pellets in Laurie's study but were caught in live traps near the roosts. As mentioned earlier *Lemniscomys* is semi-diurnal, and even if abundant in the biocoenosis, appears rarely in the owl-accumulated assemblages. *Acomys* is also rare in owl assemblages, perhaps because it has stiff, spine-like dorsal hairs that discourage owls (Ilany and Eilam 2008). Thus, the absence of these species from Laurie's study is not surprising, but it is rewarding to discover them in the current study.

The aforementioned studies are general surveys, providing little detailed information on autecology, community structure, population dynamics or microhabitat use. The first, and really only, in-depth studies on small mammal species in the Serengeti were conducted by Senzota (1978, 1983, 1990, 1992). These studies focused on two abundant rodent taxa *Arvicanthis* and *Gerbilliscus* (referred to then as *Tatera*). Beyond this, the distribution, biodiversity, microhabitat use and community structure of Serengeti small mammals has not been investigated.

The spatial distribution of rodents across the ecosystem suggests that rodents are not randomly distributed and instead track preferred habitats. The grassland roosts (7, 24, 29) feature relatively high abundance of *Gerbillus* and *Gerbilliscus* both members of the subfamily Gerbillinae - and a concomitantly low abundance of murines. Among the gerbils taxonomic names have shifted and this deserves a moment of clarification. *Tatera* was used formerly to refer to both Asian and African species. However, multiple morphological (Pavlinov 2001) and molecular (Chevret and Dobigny 2005) traits separate the African and Asian forms of these gerbils and I follow here the classification of Pavlinov (2001) and Musser and Carleton (2005) in reserving *Tatera* for the Asian forms and raising the subgenus *Gerbilliscus* to genus rank for the African forms. Similarly, specimens here referred to *Gerbillus* may represent members of that genus (e.g., *G. pusillus* or *G. pulvianatus*) or the closely related genus *Dipodillus*, which until recently was treated as a subgenus of *Gerbillus*.

Serengeti specimens of *Gerbillus sensu lato* are smaller than *Gerbilliscus*. There are approximately 60 species according to Musser and Carleton (2005) of which at least three occur in Kenya and Tanzania, *D. harwoodi*, *G. pulvinatus*, and *G. pusillus*. The bulk of the diversity in *Gerbillus* is found in the more arid biomes of North Africa, Arabia, the Levant and into India. *D. harwoodi* and *G. pusillus* are previously reported for Serengeti (Hendrichs 1969). These species prefer loose sandy substrate for burrowing (Hubbard 1972), especially near seasonally flooded areas where deep soil crack form between rains and aid in the construction of their burrows (Kingdon 1974). Burrows are reported to be less complex than in other gerbils. A single specimen of this genus was found (but not trapped) at the edge of the Serengeti short grass plains by Misonne and Verschuren (1966).

Gerbilliscus is a widespread genus with at least seven species occurring in East Africa: *G. boehmi, G. inclusa, G. leucogaster, G. nigricauda, G. philipsi, G. robusta, G. valida*. All species of *Gerbilliscus* prefer loose, sandy ground in open, well-drained areas for constructing elaborate burrows (Avery 1982; Kingdon 1974; Vesey-Fitzgerald 1966). Burrowing is part of their adaptation to generally xeric conditions where these species are most successful (Kingdon 1974; Dauphin et al. 1994). Some species, such as *G. leucogaster* prefer woodlands and bushland areas provided the substrate is suitable for burrowing (Linzey and Kesner 1997; Vesey-Fitzgerald 1966).

Based on biogeographical distributions, Jaeger (1976) suggested the ratio of gerbils to murines as an indicator of openness and aridity. This metric is supported in the current study for local (ecosystem wide) as well as biogeographical (continental) spatial scales. The grassland roosts all feature gerbil to murine ratios near or greater than 1, with the driest roost (29) having the greatest value and the mesic grassland roost (7) having the lowest.

Wooded grasslands and grassed woodlands featuring *Acacia* species consistently have *Thallomys* present in the assemblage, and in one instance (roost 23) *Thallomys* is numerically dominant. *Thallomys* becomes rare or absent in open grassland habitats, although a single specimen is recorded at roost 29. *Thallomys* also becomes rare in evergreen forests to the north suggesting the habitat preference is specifically in Acacia woodlands. The observed distribution of *Thallomys* matches what is described in the literature for this taxon. *Thallomys* is an arboreal species adapted to life in open woodland environments. It possesses adaptations for climbing including longer, broader feet with raised pads on the soles, recurved claws, and partially opposable outer digits. Climbing behavior of captive *Thallomys* reveal it is well adapted to arboreal life and can move up and down trunks head first, while using its tail prehensively (Earl and Nel 1976). Kingdon (1974) and others (Skinner 1990; Hubbard 1972; Vesey-Fitzgerald 1964) describe *Thallomys* as an Acacia woodland specialist, preferring tall *Acacia* such *A. xanthophloea* and *A. tortilis*. However, this may be an oversimplification as there are also reports of *Thallomys paedulcus* being trapped in Miombo woodland, characterized by the genera *Brachystegia* and *Julbernardia,* growing on sandy soils, as well as Mopane woodland, characterized by *Colophopernum mopane,* growing on clays and often with a dense grass understory (Linzey and Kesner 1997). A similar pattern is present for *Saccostomus*, which is present in most of the mixed savanna habitats, but absent from the open grassland roosts and from the northern mesic forest roosts. However, this taxon is never numerically dominant. *Acomys* falls close to *Saccostomus* and in the grassed-woodland part of the plot, but its abundance is too low (1–2 MNI) to have any confidence in its position.

Dasymys is a semi-aquatic specialist that favors mesic savanna environments (Duplantier and Ba 2001; Mullin et al. 2005). It appears at roosts 40 and 44, which are roosts with perennial water that can sustain mesic environments. *Aethomys* has a similar distribution to *Dasymys* across Africa (Mullin et al. 2005) and in its distribution in the Serengeti, occurring primarily in the northern mesic roosts (40, 43, 44). *Aethomys* favors thickets and dense under canopy vegetation so its presence at these locations is not surprising, though its absence in other treed areas with thicket vegetation such as roosts 4, 12, 23 is somewhat surprising.

Though never dominant, *Dendromus* was recovered from every roost in the current study at abundances ranging from 1% to 18%. This catholic distribution suggests it is not a very useful habitat indicator. The molars bear a superficial similarity to *Steatomys*, and confusion with that genus is the only likely explanation for its absence from Laurie's study. In comparing the results with trapping studies it is clear that the owl assemblages are biased toward small, nocturnal rodents, but despite this they still obtain predominantly diurnal species (e.g., *Arvicanthis* and *Lemniscomys*), and manage to capture most of the known diversity at the generic level.

Conclusions

The combined analysis reveals a total of 28 rodent genera inhabiting the Serengeti region. Owl-accumulated assemblages recorded 17 genera or 60% of the total diversity. The most apparent biases are the absence of larger taxa, such as *Pedetes*, *Hystrix* and *Thryonomys*, as well as some diurnal species. These omissions are consistent with the known habits of Barn Owls and Spotted Eagle Owls. Differences appear between roosting sites spread along the ecological gradient in the Serengeti ecosystem. As yet the cause of the differences is not established but the patterning between roosts suggests that vegetation differences have a strong influence on community composition across the ecosystem. Comparing species abundances with their ecological proclivities as summarized in this chapter we find that semi-arid and dry savanna species such as *Gerbillus*, *Gerbilliscus* are associated with roosts 7, 24 and 29, which lie in the plains and experience the lowest mean annual precipitation. Conversely arboreal species such as *Thallomys*, are found in treed-grassland and grassed-woodland locations such as roosts, 3, 12, 23. *Dasymys* is a specialist that favors wet environments. It appears in wet habitats in the north of the park (roosts 40, 44) as does *Aethomys*, which favors thickets and dense vegetation. Most of the Pliocene fossil taxa at Laetoli belong to the same genera, or closely related genera, as those appearing in the modern Serengeti ecosystem (Denys 1987, 2011). Likewise, the structure of the past ecosystem is likely influenced by similar climatic, tectonic and volcanic factors as the modern ecosystem, such that the modern Serengeti ecosystem is the best analog to Laetoli and may be very informative about past environments and ecosystem dynamics. A closer look at the taphonomy and paleoenvironments of the Laetoli rodents is addressed in Chap. 13 of this volume.

References

Anderson, G. D. (1963). Some weakly developed soils of the eastern Serengeti plains, Tanganyika. *African Soils, 8*, 339–347.

Anderson, G. D., & Talbot, L. M. (1965). Soil factors affecting the distribution of the grassland types and their utilization by wild animals on the Serengeti plains, Tanganyika. *Journal of Ecology, 53*, 33–56.

Anderson, T. M., Dempewolf, J., Metzger, K. M., Reed, D., & Serneels, S. (2007). Generation and maintenance of heterogeneity in the Serengeti ecosystem. In A. R. E. Sinclair, C. Packer, S. A. R. Mduma, & J. M. Fryxell (Eds.), *Serengeti III: Human impacts on ecosystem dynamics*. Chicago: Chicago University Press.

Andrews, P. (1990). *Owls, Caves and Fossils*. Chicago: University of Chicago Press.

Andrews, P. (2006). Taphonomic effects of faunal impoverishment and faunal mixing. *Palaeogeography, Palaeoclimatology, Palaeoecology, 241*, 572–589.

Avery, D. M. (1982). Micromammals as palaeoenvironmental indicators and an interpretation of the late Quaternary in the southern Cape Province, South Africa. *Annals of the South African Museum, 85*, 183–374.

Avery, D. M. (1992). Ecological data on micromammals collected by barn owls *Tyto alba* in the West Coast National Park, South Africa. *Israel Journal of Zoology, 38*, 385–397.

Avery, D. M. (2002). Taphonomy of micromammals from cave deposits at Kabwe (Broken Hill) and Twin rivers in central Zambia. *Journal of Archaeological Science, 29*, 537–544.

Avery, D. M., Avery, G., & Palmer, N. G. (2005). Micromammalian distribution and abundance in the Western cape province, South Africa as evidenced by barn owls *Tyto alba* (Scopoli). *Journal of Natural History, 39*, 2047–2071.

Barr, A., & Reed, D. (2009). Coping with taxonomic ambiguity and inter-observer variation in paleontological and paleoanthropological analyses. *American Journal of Physical Anthropology Supplement, 48*, 83.

Behrensmeyer, A. K., & Hook, R. W. (1992). Paleoenvironmental contexts and taphonomic modes. In A. K. Behrensmeyer, J. D. Damuth, W. A. DiMichele, R. Potts, H.-D. Sues, & S. L. Wing (Eds.), *Terrestrial ecosystems through time; evolutionary paleoecology of terrestrial plants and animals* (pp. 15–136). Chicago: University of Chicago Press.

Belsky, J. (1985). Long-term vegetation monitoring in the Serengeti National Park, Tanzania. *African Journal of Ecology, 22*, 449–460.

Belsky, J. (1990). Tree/grass ratios in East African savannas: A comparison of existing models. *Journal of Biogeography, 17*, 483–489.

Belsky, A. J. (1992). Effects of grazing, competition, disturbance, and fire on species composition and diversity of grassland communities. *Journal of Vegetation Science, 3*, 187–200.

Bonnefille, R., Potts, R., Chalie, F., Jolly, D., & Peyron, O. (2004). High-resolution vegetation and climate change associated with Pliocene *Australopithecus afarensis*. *Proceedings of the National Academies of Science USA, 101*, 12125–12129.

Campisano, C., & Feibel, C. (2007). Connecting local environmental sequences to global climate patterns: Evidence from the hominin-bearing Hadar Formation, Ethiopia. *Journal of Human Evolution, 54*, 743–768.

Chevret, P., & Dobigny, G. (2005). Systematics and evolution of the subfamily Gerbillinae (Mammalia, Rodentia, Muridae). *Molecular Phylogenetics and Evolution, 35*, 674–688.

Clarke, K. R. (1993). Non-parametric multivariate analyses of changes in community structure. *Austral Ecology, 18*, 117–143.

Coetzee, C. G. (1972). The identification of southern African small mammal remains in owl pellets. *Cimbebasia, 2*, 54–62.

Cole, M. (1986). *The savannas, biogeography and geobotany*. New York: Academic.

Crawley, M. (2007). *The R Book*. New York: Wiley.

Dauphin, Y., Kowalski, C., & Denys, C. (1994). Assemblage data and bone and teeth modifications as an aid to paleoenvironmental interpretations of the open-air Pleistocene site of Tighenif (Algeria). *Quaternary Research, 42*, 340–349.

Davis, D. H. S. (1965). Classification problems with the African Muridae. *Zoologica Africana, 1*, 121–145.

Delany, M. J. (1975). *Rodents of Uganda*. London: Trustees of the British Museum.

Dempewolf, J., Trigg, S., DeFries, R., & Eby, S. (2007). Burned-area mapping of the Serengeti-Mara region using modis reflectance data. *IEEE Geoscience and Remote Sensing Letters, 4*, 312–316.

Denys, C. (1987). Fossil rodents (other than Pedetidae) from Laetoli. In M. D. Leakey & J. M. Harris (Eds.), *Laetoli: A Pliocene Site in northern Tanzania* (pp. 118–170). Oxford: Clarendon.

Denys, C. (2011). Rodents. In: T. Harrison (Ed.), *Paleontology and geology of Laetoli: Human evolution in context* (Fossil hominins and the associated fauna, vol. 2, pp. 15–53). Springer, Dordrecht.

de Witt, H. A., & Jeronimus, O. D. (1977). *Soil map of the Serengeti plain*. Wageningen: Wageningen Agricultural University.

de Wit, H. A. (1978). Soils and grassland types of the Serengeti plain (Tanzania). Their distribution and interrelations. Ph.D. dissertation, University of Wageningen, Wageningen.

Dublin, H. T., & Douglas-Hamilton, I. (1987). Status and trends of elephants in the Serengeti-Mara ecosystem. *African Journal of Ecology, 25*, 19–23.

Dublin, H. T., Sinclair, A. R. E., & McGlade, J. (1990). Elephants and fire as causes of multiple stable states in the Serengeti-Mara woodlands. *Journal of Animal Ecology, 59*, 1147–1164.

Ducroz, J. F., Volobouev, V., & Granjon, L. (2001). An assessment of the systematics of arvicanthine rodents using mitochondrial DNA sequences: Evolutionary and biogeographical implications. *Journal of Mammalian Evolution, 8*, 173–206.

Duplantier, J., & Ba, K. (2001). Swimming ability in six West-African rodent species under laboratory conditions. In C. Denys, L. Granjon, & A. Poulet (Eds.), *Small African Mammals* (pp. 331–342). Paris: Collection Colloques et Seminaires. IRD Editions.

Earl, Z., & Nel, J. A. J. (1976). Climbing behaviour in three African rodent species. *Zoologica Africana, 11*, 183–192.

Flaig, R.-M. (2008). *Bioinformatics programming in Python: A practical course for beginners*. Weinheim: Wiley.

Foster, J. B., & Duff-Mackay, A. (1966). Keys to the genera of Insectivora, Chiroptera and Rodentia of East Africa. *Journal of the East African Natural History Society, 15*, 189–204.

Glue, D. (1971). Avian predator pellet analysis and the mammalogist. *Mammalogy, 1*, 53–62.

Grayson, D. K. (1991). *Quantitative zooarchaeology*. New York: Academic.

Grunblatt, J., Ottichilo, W. K., & Sinange, R. K. (1989). A hierarchical approach to vegetation classification in Kenya. *African Journal of Ecology, 27*, 45–51.

Grzimek, M., & Grzimek, B. (1960). Census of plains animals in the Serengeti National Park, Tanganyika. *Journal of Wildlife Management, 24*, 27–37.

Hay, R. L. (1987). Geology of the Laetoli area. In M. D. Leakey & J. M. Harris (Eds.), *Laetoli: A Pliocene site in northern Tanzania* (pp. 23–47). Oxford: Clarendon.

Hendrichs, H. (1969). Schatzungen der huftierbiomasse in der dornbuschsavanne nordlich und westlich der serengetisteppe in ostafrika nach einem neuen verfahren un bemerkungen zur biomasse der anderen pflanzenfressenden tierarten. *Saugetierkundliche Mitteilungen, 18*, 237–255.

House, J. I., Archer, S., Breshears, D. D., & Scholes, R. J. (2003). Conundrums in mixed woody-herbaceous plant systems. *Journal of Biogeography, 30*, 1763–1777.

Hubbard, C. A. (1972). Observations on the life histories and behaviour of some small rodents from Tanzania. *Zoologica Africana, 7*, 419–449.

Ilany, A., & Eilam, D. (2008). Wait before running for your life: Defensive tactics of spiny mice (*Acomys cahirinus*) in evading barn owl (*Tyto alba*) attack. *Behavioral Ecology and Sociobiology, 62*, 923–933.

Jaeger, J. J. (1976). Les rongeurs (Mammalia, Rodentia) du Pléistocène Inférieur d'Olduvai Bed I (Tanzanie). In R. J. G. Savage & S. C. Coryndon (Eds.), *Fossil Vertebrates of Africa* (Vol. 4, pp. 57–120). London: Academic.

Jager, T. (1982). Soils of the Serengeti woodlands, Tanzania. Ph.D. dissertation, Agricultural University, Wageningen.

Johanson, D. C., White, T. D., & Coppens, Y. (1978). A new species of the genus *Australopithecus* (Primates: Hominidae) from the Pliocene of eastern Africa. *Kirtlandia, 28*, 1–14.

Kingdon, J. (1974). *East African Mammals* (Vol. 2). Chicago: University of Chicago Press.

Kingston, J. (2007). Shifting adaptive landscapes: Progress and challenges in reconstructing early hominid environments. *Yearbook of Physical Anthropology, 50*, 20–58.

Laurie, W. A. (1971). The food of the barn owl in the Serengeti National Park, Tanzania. *Journal of the East African Natural History Society, 28*, 1–4.

Leakey, M. D. (1987). Introduction. In M. D. Leakey & J. M. Harris (Eds.), *Laetoli: A Pliocene site in northern Tanzania* (pp. 1–22). Oxford: Clarendon.

Leakey, M. D., & Harris, J. M. (Eds.). (1987). *Laetoli: A Pliocene site in northern Tanzania*. Oxford: Clarendon.

Leakey, M., & Hay, R. (1979). Pliocene footprints in the Laetolil Beds at Laetoli, northern Tanzania. *Nature, 278*, 317–323.

Leakey, M. D., Hay, R. L., Curtis, G. H., Drake, R. E., Jackes, M. K., & White, T. D. (1976). Fossil hominids from the Laetolil Beds. *Nature, 262*, 460–466.

Lecompte, E., Aplin, K., Denys, C., Catzeflis, F., Chades, M., & Chevret, P. (2008). Phylogeny and biogeography of African Murinae based on mitochondrial and nuclear gene sequences, with a new tribal classification of the subfamily. *BMC Evolutionary Biology, 8*, 199–220.

Linzey, A. V., & Kesner, M. H. (1997). Small mammals of a woodland-savannah ecosystem in Zimbabwe. I. density and habitat occupancy patterns. *Journal of the Zoological Society, London, 243*, 137–152.

Lutz, M. (2006). *Programming Python* (3rd ed.). Sebastopol: O'Reilly.

Lyman, R. L., & Power, E. (2003). Quantification and sampling of faunal remains in owl pellets. *Journal of Taphonomy, 1*, 3–14.

McCune, B., & Grace, J. B. (2002). *Analysis of ecological communities*. Gleneden Beach: MjM.

Milne, G. (1935). Some suggested units of classification and mapping, particularly for East African soils. *Soil Research, 4*, 183–198.

Misonne, X., & Verschuren, J. (1966). Les rongeurs et lagomorphes de la region du parc national du Serengeti (Tanzanie). *Mammalia, 30*, 517–537.

Mullin, S., Pillay, N., & Taylor, P. (2005). The distribution of the water rat *Dasymys* (Muridae) in Africa: A review. *South African Journal of Science, 101*, 117–124.

Musser, G., & Carleton, M. D. (2005). Superfamily Muroidea. In D. Wilson & D. Reeder (Eds.), *Mammal Species of the World: A taxonomic and geographic reference* (3rd ed., Vol. 2, pp. 894–1531). Baltimore: Johns Hopkins University Press.

Norton-Griffiths, M., Herlocker, D., & Pennycuick, L. (1975). The patterns of rainfall in the Serengeti ecosystem, Tanzania. *East African Wildlife Journal, 13*, 347–374.

Oksanen, J., Kindt, R., & O'Hara, B. (2005). *Vegan: Community ecology package*. CRAN, Technical Report.

Pavlinov, I. (2001). Current concepts of gerbillid phylogeny and classification. In C. Denys, L. Granjon, & A. Poulet (Eds.), *African small mammals* (pp. 141–149). Paris: Collection Colloques et seminaires. IRD Éditions.

Peters, C., Blumenschine, R., Hay, R., Livingstone, D., Marean, C., Harrison, T., Armour-Chelu, M., Andrews, D., Bonnefille, R., & Werdelin, L. (2008). Paleoecology of the Serengeti-Mara ecosystem. In A. R. E. Sinclair, C. Packer, S. A. R. Mduma, & J. M. Fryxell (Eds.), *Serengeti III: Human impacts on ecosystem dynamics* (pp. 47–94). Chicago: University of Chicago Press.

Reed, D. N. (2003). Micromammal paleoecology: Past and present relationships between East African small mammals and their habitats. Ph.D. dissertation, Stony Brook University, Stony Brook.

Reed, D. N. (2005). Taphonomic implications of roosting behavior and trophic habits in two species of African owl. *Journal of Archaeological Science, 32*, 1669–1676.

Reed, D. N. (2007). Serengeti micromammals and their implications for Olduvai paleoenvironments. In R. Bobe, Z. Alemseged, & A. K. Behrenesmeyer (Eds.), *Hominin environments in the East African*

Pliocene: An assessment of the faunal evidence (pp. 217–255). Dordrecht: Springer.

Reed, K. E. (2008). Paleoecological patterns at the Hadar hominin site, Afar Regional State, Ethiopia. *Journal of Human Evolution, 54*, 743–768.

Reed, D. N., Anderson, T. M., Dempewolf, J., Metzger, K. M., & Serneels, S. (2009). The spatial distribution of vegetation types in the Serengeti ecosystem: The influence of rainfall and topographic relief on vegetation patch characteristics. *Journal of Biogeography, 36*, 770–782.

Richmond, B., & Jungers, W. (2008). *Orrorin tugenensis* femoral morphology and the evolution of hominin bipedalism. *Science, 319*, 1662–1664.

Rogers, M., & Stanley, W. (2003). Tanzania Mammal Key. World Wide Web http://www.fmnh.org/tanzania/default.html.

Sankaran, M. (2004). Tree-grass coexistence in savannas revisited–insights from an examination of assumptions and mechanisms invoked in existing models. *Ecology Letters, 7*, 480–490.

Sankaran, M., Hanan, N. P., Scholes, R. J., Ratnam, J., Augustine, D. J., Cade, B. S., Gignoux, J., Higgins, S. I., Roux, X. L., Ludwig, F., Ardo, J., Banyikwa, F., Bronn, A., Bucini, G., Caylor, K. K., Coughenour, M. B., Diouf, A., Ekaya, W., Feral, C. I., Februaiy, E. C., Frost, D. G. H., Hiernaux, P., Hrabar, H., Metzger, K. L., Prins, H. H. T., Ringrose, S., Seal, W., Worden, J. T. J., & Zambalis, N. (2005). Determinants of woody plant cover in African savannas. *Nature, 438*, 846–849.

Scholes, R. J., & Archer, S. R. (1997). Tree-grass interactions in savannas. *Annual Review of Ecology and Systematics, 28*, 517–544.

Senzota, R. M. B. (1978). Some aspects of the ecology of two dominant rodents in the Serengeti ecosystem. M.Sc. thesis, University of Dar es Salaam, Dar es Salaam.

Senzota, R. B. M. (1983). A case of rodent-ungulate resource partitioning. *Journal of Mammalogy, 64*, 326–329.

Senzota, R. B. M. (1990). Activity patterns and social behaviour of the grass rats [*Arvicanthis niloticus* (Desmarest)] in the Serengeti National Park, Tanzania. *Tropical Ecology, 31*, 35–40.

Senzota, R. B. M. (1992). Rodent ectoparasites in the Serengeti National Park, Tanzania. *Tropical Ecology, 33*, 29–33.

Serneels, S., & Lambin, E. (2001). Impact of land-use changes on the wildebeest migration in the northern part of the Serengeti-Mara ecosystem. *Journal of Biogeography, 28*, 391–407.

Sinclair, A. R. E. (1995). Serengeti past and present. In A. R. E. Sinclair & P. Arcese (Eds.), *Serengeti II: Dynamics, management, and conservation of an ecosystem* (pp. 3–30). Chicago: University of Chicago Press.

Skinner, J. D. (1990). *The mammals of the southern African subregion*. Pretoria: University of Pretoria.

Soligo, C., & Andrews, P. (2005). Taphonomic bias, taxonomic bias and historical non-equivalence of faunal structure in early hominin localities. *Journal of Human Evolution, 49*, 206–229.

Su, D., & Harrison, T. (2007). The paleoecology of the upper laetolil beds at Laetoli: A reconsideration of the large mammal evidence. In R. Bobe, Z. Alemseged, & A. Behrenesmeyer (Eds.), *Hominin environments in the East African Pliocene* (pp. 279–313). Dordrecht: Springer.

Su, D., & Harrison, T. (2008). Ecological implications of the relative rarity of fossil hominins at Laetoli. *Journal of Human Evolution, 55*, 672–681.

Swynnerton, G. (1958). Fauna of the Serengeti National Park. *Mammalia, 22*, 435–450.

Talbot, L. M., & Stewart, D. R. M. (1974). First wildlife census of the entire Serengeti Mara region, East Africa. *Journal of Wildlife Management, 28*, 815–827.

Taylor, I. (1994). *Barn Owls*. Cambridge: Cambridge University Press.

Vernon, C. J. (1972). An analysis of owl pellets collected in southern Africa. *Ostrich, 43*, 109–124.

Vesey-Fitzgerald, D. F. (1964). Mammals of the Rukwa Valley. *Tanganyika Notes and Records, 62*, 61–72.

Vesey-Fitzgerald, D. (1966). The habits and habitats of small rodents in the Congo River catchment region of Zambia and Tanzania. *Zoologica Africana, 2*, 111–122.

White, T. (1988). The comparative biology of "robust" *Australopithecus*: Clues from context. In F. Grine (Ed.), *Evolutionary history of the "Robust" australopithecines. Foundations of human behavior* (pp. 449–484). New York: Gruyter.

Chapter 13
The Taphonomy and Paleoenvironmental Implications of the Laetoli Micromammals

Denné N. Reed and Christiane Denys

Abstract Recent fieldwork conducted between 1998 and 2005 significantly increased the sample of fossil rodent specimens from Laetoli, Tanzania, the type locality of *Australopithecus afarensis*, and this allowed the identification of several new micromammal species. This chapter discusses the taphonomy and paleoenvironmental implications of the Laetoli rodents. The taphonomic analysis of the new material looks at element representation, breakage patterns and bone surface modification and finds evidence of predator activity and weathering. In terms of paleoenvironment, the Upper Laetolil Beds assemblage has a very low abundance of murine rodents, a predominantly arboreal taxon (*Thallomys*) alongside an arboreal sciurid (*Paraxerus*), and several fossorial and burrowing taxa, which we interpret to indicate the presence of acacia trees growing on loose, well-drained sediments in a semi-arid environment. The Upper Ndolanya Beds sample remains small, and the species preserved are the same as those found in the Upper Laetolil Beds, with the exception of *Thryonomys*. This provides tentative evidence for a more mesic local environment in Upper Ndolanya times relative to the Upper Laetolil Beds.

Keywords Pliocene • Rodents • Paleoenvironments • Weathering • Breakage patterns • Bone surface modification

Introduction

Hominins at Laetoli are relatively rare compared to other mammals in the Upper Laetolil Beds, and also compared to the abundance of *Australopithecus afarensis* at other hominin-bearing sites, such as Hadar (Su and Harrison 2008). The rarity of *A. afarensis* at Laetoli may be explained in part by the taphonomic and paleoenvironmental history of Laetoli. While most other Pliocene hominin localities occur in fluviatile and lacustrine settings (Behrensmeyer and Hook 1992; White 1998), Laetoli represents a more terrestrial setting (Hay 1987; Leakey 1987; Peters et al. 2008) with an unusual taphonomic history. It has been suggested that Laetoli during the late Pliocene represented a marginal environment for *A. afarensis*, which may have preferred more mesic and closed habitats (Su and Harrison 2008). Obtaining more refined paleoenvironmental reconstructions will aid in understanding the ecology of *A. afarensis*, and micromammals are one method of obtaining a paleoenvironmental signal at a finer spatial grain.

Fossil micromammal assemblages experience taphonomic influences at multiple stages of their preservation and recovery. Most micromammal accumulations are created through the activity of predators such as owls, hawks, civets, foxes etc., which influence what prey appear in the assemblage along with the degree of breakage and quality of preservation (Andrews 1990; Reed 2005, 2007). Micromammals require added effort for recovery and smaller taxa are often overlooked during surface prospecting or coarse dry screening (Stahl 1996). Typically wet screening and careful picking of the screened material under a microscope is necessary to recover the full spectrum of taxa and their relative abundances. Taphonomic influences carry through to the analysis too, where the diagnosis and alpha taxonomy of many groups is challenging because of high species diversity and low levels of morphological distinction between species (Baker and Shaffer 1999; Soligo and Andrews 2005).

All these influences seem to play a role in the Laetoli micromammal collections. Previous taphonomic studies on the Laetoli rodents (Denys 1985a, b, 1986) demonstrated at least two modes of accumulation; there was evidence for predator accumulation producing the bulk of the assemblage, and also *in situ* burial of some animals in their burrows during volcanic eruptions. These findings are substantiated in the analysis of the more recently collected material as described below.

D.N. Reed (✉)
Department of Anthropology, University of Texas at Austin,
1 University Station C3200, Austin, TX 78712, USA
e-mail: reedd@mail.utexas.edu

C. Denys
Department of Systematics and Evolution – CP51, UMR7205 CNRS: Origine structure & évolution de la Biodiversité, MNHN,
55 rue Buffon, 75005, Paris, France
e-mail: denys@mnhn.fr

T. Harrison (ed.), *Paleontology and Geology of Laetoli: Human Evolution in Context. Volume 1: Geology, Geochronology, Paleoecology and Paleoenvironment*, Vertebrate Paleobiology and Paleoanthropology, DOI 10.1007/978-90-481-9956-3_13,

The revised paleontological treatment of the Laetoli rodents provided by Denys (2011) shows a low species diversity compared to other modern and fossil sites, and it has a unique taxonomic composition, unlike those seen in the modern assemblages described by Reed (2011). The aim of this chapter is to describe the taphonomy of the Laetoli micromammals and their paleoenvironmental implications, incorporating data from the newly collected samples.

Material and Methods

New microvertebrate fossil collections were made at Laetoli during a recent campaign led by Terry Harrison from 1998–2005. This fieldwork resulted in the discovery of 295 identifiable rodent remains. These specimens come from localities in the Lower Laetolil Beds (LLB), the Upper Laetolil Beds (ULB) and the Upper Ndolanya Beds (UNB) (Denys 2011). These new collections are compared with published data on micromammal collections from Laetoli made by teams directed by Mary Leakey from 1974–1979 (Davies 1987; Denys 1983, 1985a, 1987).

The Leakey collections at Laetoli were mostly surface finds, however wet screening was conducted by J.-J. Jaeger at Locs. 5 and 6 (Denys 1987). The Harrison campaign made use of surface collection and some dry screening, but did not conduct wet screening (T. Harrison, personal communication). None of the published reports indicate the mesh sizes of the screens used. However, it is common practice that wet screening is conducted at a finer mesh size than dry screening and given the size of the specimens recovered (e.g., isolated molars of small rodents such as *Dendromus*, *Steatomys* and *Mastomys*) it is estimated that Jaeger's wet screening at Locs. 5 and 6 likely employed sieves with 1 mm mesh. The differences in collection methodology complicate direct comparisons of the collections produced by the two campaigns.

The new sample for which taphonomic and surface modification data are available includes 262 specimens, 29 specimens from the Upper Ndolanya Beds, 229 from the Upper Laetolil Beds, and 4 specimens from the Lower Laetolil Beds. Sample sizes for all localities in the recent collections are small. Only eight localities (ULB Locs. 2, 5, 6, 8, 9, 10, 22 and UNB Loc. 15) have micromammal samples sizes greater than 10 specimens, with the largest being Loc. 10 with 63 specimens. With the sample from the Lower Laetolil Beds being very small (four specimens), analysis focuses on material from the ULB and the UNB.

Following the procedures set out by Andrews (1990) and Fernandez-Jalvo and Andrews (1992), each element was examined under a stereo-zoom microscope for signs of incisor digestion, molar digestion, enamel pitting, dentine cracking, general cracking, abrasion, weathering, breakage, and root marking in order to better understand the accumulating agents and taphonomic history of the rodent accumulations at Laetoli. These data also allow us to propose hypotheses about depositional context and history.

Each taphonomic variable was coded on an ordinal scale. The ordinal scores varied depending on the variable. The analysis of digestion used the categories of Fernandez-Jalvo and Andrews (1992) to define grades of digestion intensity (i.e., light, moderate, heavy, extreme) on the incisors only, because for molars there is no modern reference study that documents digestion traces left on East African rodent species. For weathering, we used the categories defined by Behrensmeyer (1978). Breakage patterns were coded following the system of Denys (1983, 1986). Linear trend analysis (Agresti 2007) was used to test for correspondence between the taphonomic variables and stratigraphic level. Linear trend analysis is a test of independence and correspondence, similar to correlation, but designed for ordinal (i.e., ranked) data.

Etching of incisor and molar teeth are an indication of activity by predators, and the degree and frequency of this modification helps identify different classes of predators (Andrews 1990; Fernandez-Jalvo et al. 1998). Enamel pitting, dentine cracking and general cracking are associated either with digestion or with other diagenetic processes such as the corrosion resulting from the chemistry of the interring sediments, or compaction pressure during deep burial. Abrasion, weathering, trampling and root marking are informative about the history of the fossil as it rested on or near the ground surface. Breakage may be informative of predatory activity, trampling or a combination of both. Pictures were taken to illustrate taphonomic features using a Jeol-X25 SEM and a Jeol-JSM 8401. Specimens were cataloged by T. Harrison and C. Denys and are housed in the National Museum of Tanzania, Dar es Salaam, Tanzania.

Taphonomy

A comparison between the earlier (Leakey) and more recent (Harrison) collections provides a test of the impact that different collecting methods have on taxonomic composition and resulting biases.

Figure 13.1 shows the relative abundance of specimens collected by the two campaigns. In these graphs, taxa are arranged according to mean body size as estimated from dental dimensions and comparison to extant species. At both localities the Leakey campaign, where wet screening was conducted, recovered an absolutely larger sample and a greater number and proportion of smaller-bodied taxa, such as *Saccostomus*, *Thallomys*, *Mastomys* and gerbils. Many of

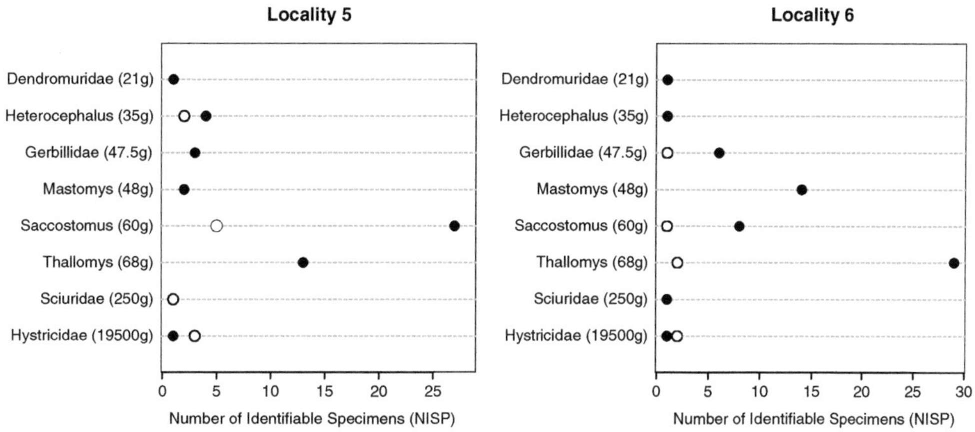

Fig. 13.1 Taxonomic abundances for the main rodent subfamilies present at Locs. 5 and 6. Taxa are arranged by size with body mass estimates given in parentheses. The solid circles indicate the earlier Leakey collections while the open circles indicate the later Harrison collections. The Leakey collections were wet screened, while the Harrison collections were dry screened or surface collected only

the smallest taxa were not recovered at all in the absence of wet screening. Conversely, the larger, more visible taxa, such as the porcupines (Hystricidae) and squirrels (Sciuridae), were absolutely (except at Loc. 5) and relatively more abundant in the non wet-screened collections.

Bone representation could not be systematically evaluated because the assemblage is biased towards identifiable cranio-dental elements. We notice however the presence of well preserved bones (Fig. 13.2a) and skulls (Fig. 13.2b), as well as elements in anatomical connection (Fig. 13.2c) and nearly intact bones (Fig. 13.2d). This exceptional preservation of small mammal elements was noted previously in the Leakey collections (Denys 1986: 99–100, Plates 1 and 2).

Surface Modification

A summary of the taphonomic differences in bone surface modification between the Upper Laetolil Beds (ULB) versus the Upper Ndolanya Beds (UNB) is provided in Table 13.1, while Table 13.2 presents the data as the percentage of specimens showing a given form of modification for each locality. These data do not show a significant difference between the UNB and ULB in incisor digestion, molar digestion or dentine cracking. Significant differences between these stratigraphic units were observed for enamel pitting, cracking, weathering abrasion and breakage. In all of these, save breakage, the pattern is the same; the UNB sample shows a greater number of specimens with higher scores in each of these variables. The reverse is observed for breakage, where the ULB sample shows a greater frequency of higher levels of breakage than the UNB sample.

Rodent incisors are typically the first teeth to show signs of modification as a result of digestion (Andrews 1990). The modification has the effect of etching the enamel surface of the teeth, especially along sharp, salient edges and the effect is often visible at the tip of the tooth. Fig. 13.2f illustrates enamel etching on an upper incisor with traces of digestion, superimposed with other modifications. Digestion traces are not very abundant in the Laetoli sample, but Denys (1986: Plate 3) illustrated a femur head showing a typical moderate grade of digestion. With murid rodent molars, digestion is harder to identify due to their structure, and according to Andrews (1990) it varies considerably between predators. In the Laetoli assemblage molar digestion seems masked by various other alterations, but in some molars (Fig. 13.2i–k, n, o) we see light enamel etching accompanied by dentine cracking and sunken dentine, perhaps reflecting murid molars that passed through the digestive tract of a felid or canid predator (Fig. 13.3).

Most of the samples from the Laetoli localities are very small (NISP <10), but all localities with NISP greater than 2 show some indication of incisor digestion. The NISP counts from Locs. 8 and 22 in the Upper Laetolil Beds are larger, and both localities have percentage incisor digestion in the range of 25–27%, which is consistent with a Level 2 predator, such as an Eagle Owl (Andrews 1990). However, we cannot exclude the presence of different predators in the same level.

The frequency of molar digestion ranges broadly across localities, but concentrating on the localities with greater than 10 specimens, the percentages of molars digested range between 0% and 21%, which are values similar or slightly less than observed for incisor digestion. For a given locality the percentage of molars digested is always less than or equal to the percentage of incisors digested.

Many specimens exhibit fine cracks in the dentine of the tooth crown (Fig. 13.2i, k, o), which may be associated with digestion, weathering or some combination of the two. Values for percentage of specimens showing this modification ranged broadly (0–100%) across all samples and 0–21%

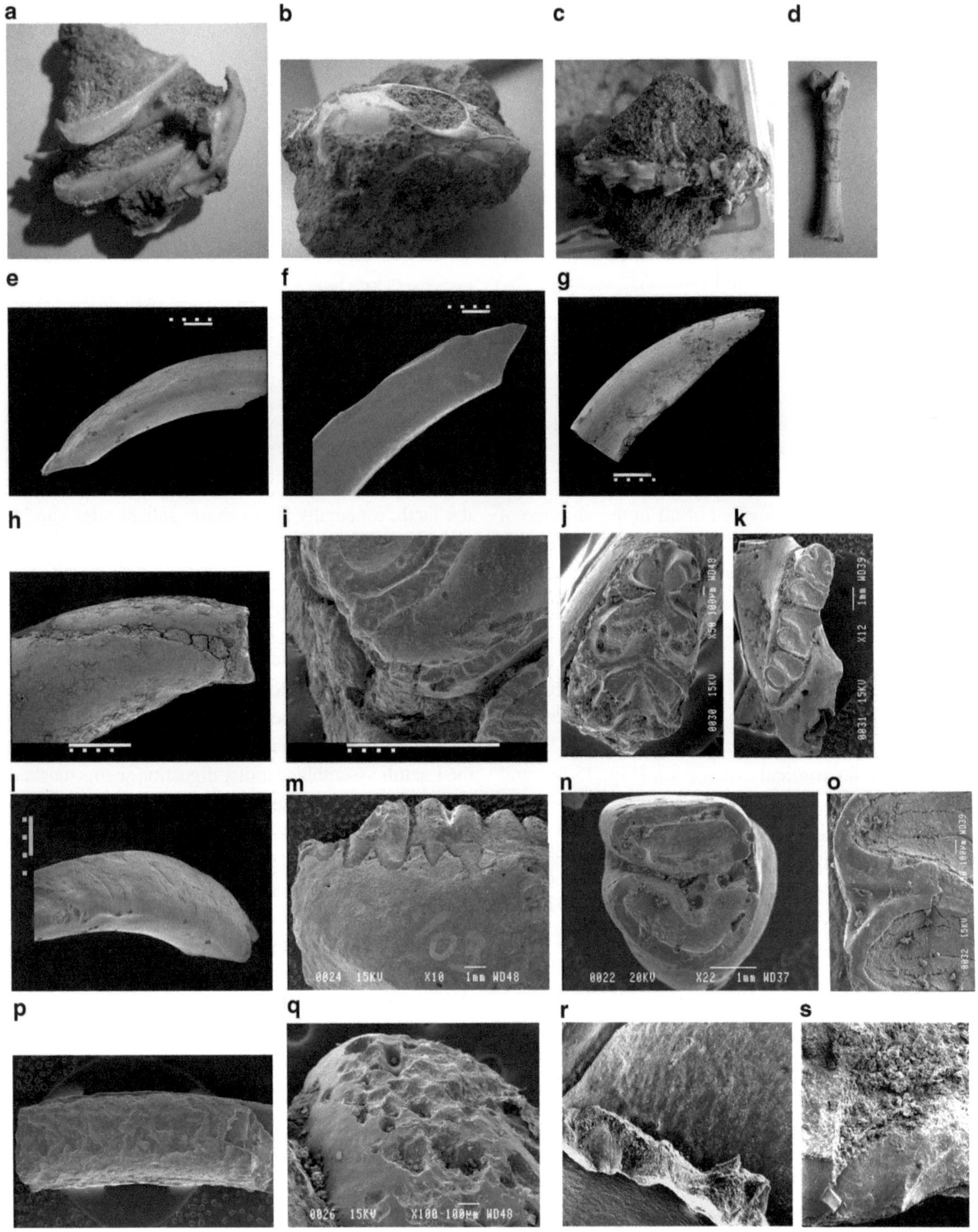

Fig. 13.2 (**a**) Bone breccia (EP 3520/00, Loc. 18, UNB) displaying rodent tibiae and a *Gerbilliscus* mandible embedded in sediment. (**b**) Complete rodent skull embedded in sediment (EP 191/05, Loc. 16, ULB). (**c**) An articulated vertebral column of a small mammal embedded in sediment (EP1902/03, Loc.1, ULB). (**d**) A nearly complete femur of *Pedetes laetoliensis*, the distal articulation is missing while the proximal end does not show any signs of alteration. Rootmarks are visible on the shaft, which also displays very light weathering. (**e**) The tip of a rodent incisor (EP 1171/00, Loc.8, ULB) displaying different alterations, the enamel is nearly removed from the tooth surface except in the middle where it is smooth and polished. Other zones of enamel are more rugose and the whole incisor is affected by longitudinal cracking, which may result from weathering. At the tip of the incisor, the depressed zone, which has sinuous margins, may result form the action of a root mark. Some trampling and pitting are also observed on the dentine. (**f**) The tip of a rodent incisor (EP 1171/00, Loc.8, ULB) showing enamel removal on different parts of the surface, which may result from moderate digestion, but is masked by light longitudinal cracks affecting both dentine and enamel (due to weathering). (**g**) The tip of a rodent incisor (EP 3072/00, Loc.5, ULB) with transverse breakage. Enamel is removed from the tip, except for a small plate on the surface and the dentine is affected by pits, which enter deeply into its structure, possibly resulting from corrosion. Thin longitudinal cracks occur in the dentine. (**h**) The anterior part of a rodent mandible

in the larger samples (NISP >10). Locs. 2, 9 and 10 had the highest rates and values of dentine cracking.

Rodent molar specimens of various species from Laetoli often exhibit small (ca. 0.1 mm) diameter blackened pits on the enamel and dentine surface of teeth, as illustrated in Fig. 13.2m and q. This unusual alteration may be associated with corrosion caused by the very peculiar carbonatite deposits of Laetoli or made by an unknown agent (Fernandez-Jalvo, personal communication). Small black grains of sediments are preserved inside the cavities (Fig. 13.2q). The pitting seems more frequent upon the tooth enamel than in the dentine. This alteration was noted in the Leakey collection and was figured by Denys (1986, Plate 4:1).

Rates of enamel pitting were found to differ significantly between the UNB and ULB levels (Table 13.1). The UNB sample showed a higher number of moderately to heavily pitted specimens than did the ULB sample (Table 13.3), but no significant differences were found between localities within the UNB or the ULB. Every locality had at least some specimens that showed enamel pitting and in most localities the majority of specimens exhibited this surface modification. The percentage of specimens affected in each locality ranged from 33–100% and 67–100% in the larger samples (NISP >10).

Bone cracking is also observed in the assemblage. This form of surface modification includes signs of fracture anywhere on the surface of the specimen, and it is distinct from dentine cracking as described above. The frequency of cracking differs significantly between the UNB and ULB samples (Table 13.1), being greater in the UNB sample, where an unusual number of specimens show moderate and heavy levels of cracking (Table 13.4).

As with the pitting, the highest frequency and the greatest intensity of cracking is seen in the UNB Loc. 15 and Silal Artum samples (Table 13.5). Cracking was the single taphonomic variable that showed significant non-independence between localities. Locs. 7 and 11 have far fewer uncracked bone than expected and the latter has a much greater number of heavily cracked bone. Similarly, Locs. 16 and 21 have higher than expected levels of moderate and heavily cracked bone. Locs. 9 and 15 have substantially more uncracked bone compared to other localities. Cracking may result from the action of various agents occurring at different moments of fossilization. Ingestion by predators causes breakage, as does weathering (Fig. 13.2h), trampling (Fig. 13.2l), compaction (Fig. 13.2g), and root marks (Fig. 13.2e). The differences between localities suggest different taphonomic histories prior to burial, but the exact nature of the differences is not possible to determine at this time given the small samples sizes.

In general, the levels of bone weathering observed in the rodent specimens was light. The weathering classes used here follow the bone weathering classification of Behrensmeyer (1978) and some specimens were classified into intermediary stages if it was ambiguous as to exactly which stage a specimen belonged. An example of the effects of weathering on micromammal bones and teeth was illustrated by Denys (1986: Plate 6: 4, 7).

Significant differences in the degree of weathering were observed between the ULB and UNB samples (Table 13.1), with the UNB sample having substantially more Stage 1 and Stage 2 weathering (Table 13.6). Weathering is evident at every locality with a NISP greater than 2. Frequency of weathering ranges from 33% of specimens weathered to 100% weathered for all localities where weathering is observed. For the larger samples the incidence of weathering ranges between 54% and 92%. The degree of weathering is generally low, with the average weathering stage exceeding

Fig. 13.2 (continued) with a broken incisor *in situ* (EP 3072/00, Loc.5, ULB) showing mosaic and longitudinal bone and dentine cracking. Coating occurs on the enamel and the bone. The entire skeletal element has a rugose appearance. (**i**) Detail of a *Saccostomus major* (EP 1738/04, Loc. 2, ULB) lower M2 showing the enamel of the cusps smooth, with an undulating surface and with some fragments removed. The dentine is smooth and slightly sunken. The roots are not smooth and display an irregular surface. (**j**) *Thallomys laetolilensis* lower M1 (EP 654/03, Loc. 2, ULB) showing strong alteration of the occlusal surface. (**k**) Right mandible of *Saccostomus major* with M/1–3 (EP 1065/03, Loc. 10 W, ULB) showing dentine cracking. Enamel is removed in some parts, especially on the buccal side. (**l**) Rodent upper incisor (EP 1171/00, Loc.8, ULB) showing the enamel removed from most of the surface and the dentine showing a smooth, rounded appearance, perhaps as a result of digestion. At places where the enamel remains, it has a rougher texture, and striations of different sizes and orientations are visible that look like trampling. Small, irregular holes occur on the dentine and the bone structure is visible at the bottom of the holes. (**m**) Lateral view of a *Xerus* sp. mandible (EP 034/03, Silal Artum, UNB) with M/1–3 displaying surface pitting of both enamel and bone. (**n**) *Pedetes* (EP 075/99, Loc. 10 E, ULB) isolated molar showing enamel pitting with a wavy aspect. In some places the enamel is no longer visible. The dentine is smooth, with no cracking and some coating occurring locally. (**o**) Detail of the *Saccostomus* M/2 (detail of Fig. 13.2k) showing enamel alteration and dentine cracking, which may be due to digestion by a mammalian predator or weathering. (**p**) Detail of a broken incisor showing the alteration of the surface, which may result from root etching of the enamel. (**q**) Detail of Fig. 13.2m showing the enamel pitting on the M/2 buccal side. (**r**) Detail of bone breakage displaying both a sharp, beveled breakage and conchoidal breakage (on the left side) (Loc.18–893, Leakey collections, UNB). (**s**) Detail of conchoidal breakage (LAET 75, Loc.5, ULB, Leakey collections). Where the bone was flaked one can see that the bone structure is smooth and abraded, in contrast to the surrounding zones that are rough. In the conchoidal breakage the osteoblasts remain visible, but at higher magnification

Table 13.1 Results of the linear trend analysis comparing the frequency and intensity of each taphonomic variable in the Upper Laetolil Beds (ULB) versus the Upper Ndolanya Beds (UNB)

Variable	Linear trend correlation (r)	M^2	*P*-value	Significance
Incisor digestion	−0.108 LB ≈ UNB	0.907	0.341	NS
Molar digestion	0.002 LB ≈ UNB	0	0.974	NS
Dentine cracking	−0.112 LB ≈ UNB	2.667	0.103	NS
Enamel pitting	0.152 LB < UNB	5.931	0.015	*
Cracking	0.224 LB < UNB	12.873	< 0.001	***
Weathering	0.147 LB < UNB	5.521	0.019	*
Abrasion	0.162 LB < UNB	6.738	0.009	**
Breakage	−0.230 LB > UNB	13.618	< 0.001	***

Positive values in the linear trend correlation; r indicate a trend toward greater values in the UNB, while negative values indicate a trend toward higher values in the ULB; Significance codes: *NS*, not significant; 0.05*; 0.01**; 0.001***

1 in only one small sample (ULB Loc. 21), and the highest observed weathering stage observed for an individual specimen was 2.

There were significant differences observed in the frequency of abrasion markings between the UNB and ULB samples (Tables 13.1 and 13.7). Relative to sample size, the UNB sample had much greater frequency of moderate abrasion (the highest class). No significant difference was observed in the frequency of abrasion between localities within either the ULB or the UNB.

In addition, the Laetoli rodent specimens show evidence of compaction breakage (Fig. 13.2h). Root etching was also present (Fig. 13.2p), and it may alter the surfaces of skeletal elements, as well as enamel of the incisors (Denys 1986, Plate 9:1, 2). Some conchoidal breakage was observed (Fig. 13.2p, q) in both the ULB and UNB samples, and this modification was also noted in the old collection (Denys 1986; Plate 4:3; Plate 9:3). Coating by sediment is seen in nearly all remains, but this has been removed by hand to observe the molars for the taxonomic analyses, consequently we did not attempted to count its frequency. However, this post-burial alteration was illustrated by Denys (1986, Plate 7:4).

Discussion

The difference in collection methodology between the earlier (Leakey) and later (Harrison) collections illustrates the effects that different collecting techniques have on species richness and diversity, yet the details of collection methodology are seldom reported in the paleobiological literature. In the case of Laetoli, for example, it is not documented what screen sizes were used during dry screening or wet screening. Similar problems arise for surface collections where different teams employ different criteria when deciding which fossils are collected and which are not (Alemseged et al. 2007; Eck 2007).

Wet screening is time intensive and may require resources (such as water) that are difficult or dangerous to access in the field, yet the results reported here show that choosing not to wet screen has a dramatic impact on the documented biodiversity and relative abundance of the microvertebrate fossils collected as has already been demonstrated in numerous archaeological experiments (e.g., Stahl 1996 and references therein).

The difference in collection methods and the lack of detailed documentation complicate quantitative comparisons between the earlier and later micromammal collections at Laetoli. It is difficult to know precisely what methodological differences affect the two assemblages, and rarefaction or related techniques that would normally be available to account for differences in samples size cannot be employed because of the size bias. As a result it is impossible to compare relative abundance of fauna, or even presence/absence of taxa effectively because small taxa may be present in wet screened samples yet absent in the dry screened samples.

The analysis of taphonomic bone surface modification between the ULB and UNB samples listed in Table 13.1 reveals a pattern of greater degrees of modification in the UNB sample compared to the ULB sample. The UNB sample shows significantly greater incidence and degree of enamel pitting, general cracking, bone weathering and abrasion. These differences would be expected if the bones recovered from the UNB had spent more time on average at the surface and were not buried as quickly as those in the ULB. Sample sizes at each locality are too small to show the detailed patterns, but the overall difference between the ULB and UNB samples suggests that the two time intervals may have experienced different taphonomic and diagenetic circumstances. This is an hypothesis that can be pursued when more data are available.

Table 13.2 Summary of percentage surface modification for all variables. Cell values show the percentage of specimens at each locality exhibiting each type of surface modification. The NISP column indicates the total number of specimens. Values in parentheses show the ratio of specimens where less than the total NISP is suitable for analysis

Locality	% ID	% MD	% DC	% EP	% CR	%WE	% AB	% BR	NISP
LB L01	(2/2) 100	(1/9) 11	(5/8) 62	67	78	67	11	100	9
LB L02	(1/6) 17	(4/19) 21	(10/17) 59	67	67	67	10	100	21
LB L03	NA	(0/3) 0	(1/3) 33	33	100	100	0	100	3
LB L04	(0/1) 0	(0/3) 0	(2/2) 100	67	67	67	0	100	3
LB L05	(1/5) 20	(0/24) 0	(15/21) 71	88	62	73	31	100	26
LB L06	(1/5) 20	(1/16) 6	(11/15) 73	100	44	62	19	100	16
LB L07	(2/5) 40	(0/8) 0	(3/18) 38	67	89	78	11	100	9
LB L08	(3/11) 27	(1/22) 5	(14/21) 67	96	52	92	12	100	25
LB L09	(1/4) 25	(2/17) 12	(12/18) 67	75	35	60	25	100	20
LB L10	(7/22) 32	(9/50) 18	(38/54) 70	90	46	54	8	100	63
LB L11	(1/2) 50	(0/4) 0	(2/4) 50	60	100	60	0	100	5
LB L12	NA	(0/2) 0	(2/2) 100	100	0	0	0	100	2
LB L15	NA	(0/6) 0	(4/5) 80	100	17	33	0	100	6
LB L16	(0/1) 0	(0/4) 0	(2/4) 50	75	75	100	25	100	4
LB L17	(1/1) 100	(1/1) 100	(0/1) 0	50	100	50	0	100	2
LB L21	(0/1) 0	(0/4) 0	(4/4) 100	75	75	100	50	100	4
LB L22	(1/4) 25	(1/11) 9	(7/9) 78	100	64	73	18	100	11
UNB L07E	NA	(0/1) 0	(0/1) 0	100	0	0	0	100	1
UNB L15	(1/6) 17	(1/9) 11	(1/7) 14	69	85	77	15	77	13
UNB L18	(0/1) 0	(0/7) 0	(2/5) 40	100	57	71	43	100	7
UNB Silal Artum	(0/2) 0	(1/5) 20	(2/3) 67	75	88	100	25	75	8

ID incisor digestion, *MD* molar digestion, *DC* dentine cracking, *EP* enamel pitting, *CR* cracking, *WE* weathering, *AB* abrasion, *BR* breakage, *NISP* Total number of identifiable specimens

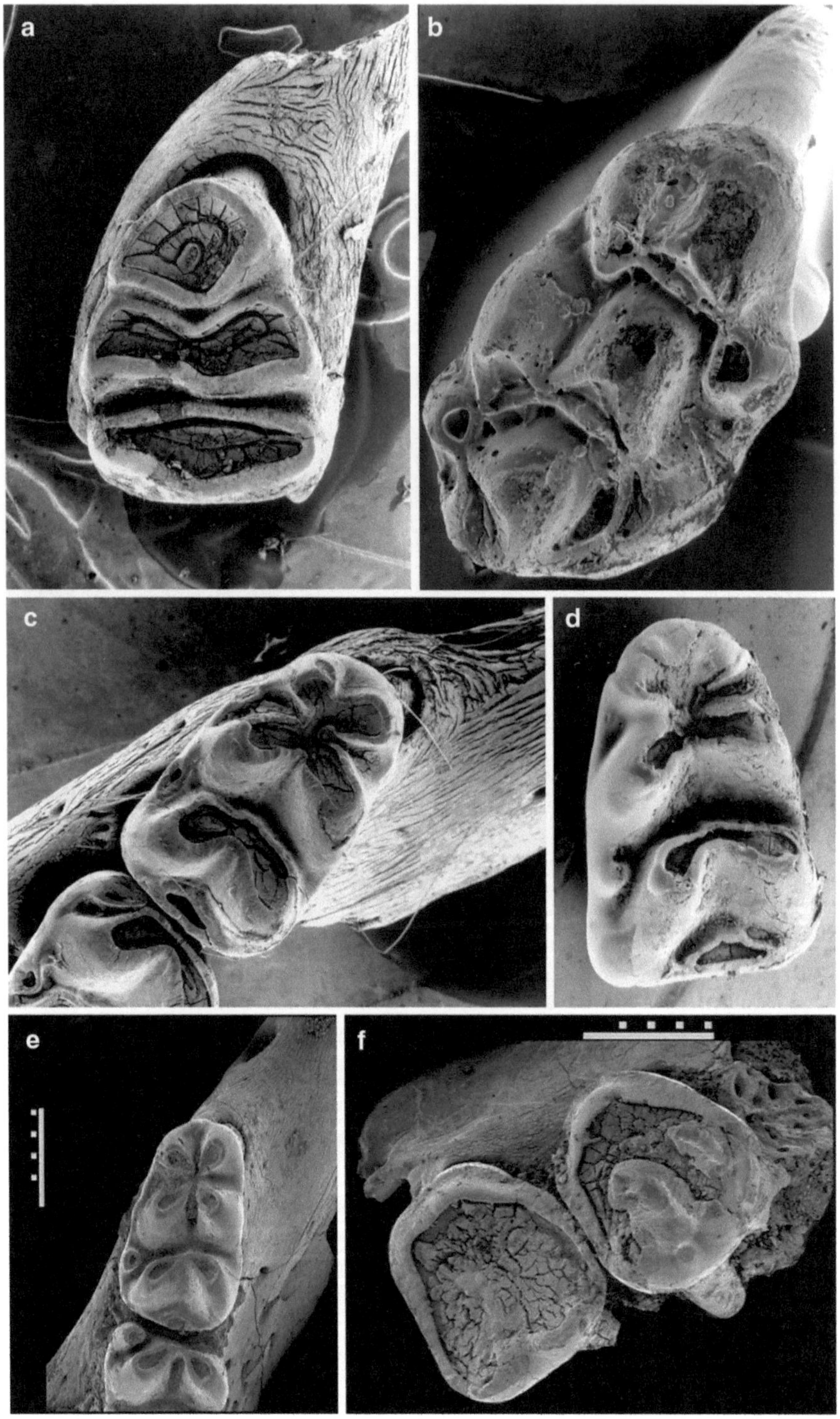

Fig. 13.3 Examples of modern African murid and gerbillid specimens passed through the digestive tracts of mammalian predators (**a–d**) and fossil rodents from Laetoli (**e–f**). (**a**) Left mandible with M/1 of *Gerbilliscus* from modern carnivore feces (Malawi, Karonga) displaying strong dentine and bone splitting and cracking, portions of enamel are removed and the enamel is not cracked. (**b**) Upper M1 of *Grammomys* sp. from modern carnivore feces (Malawi, Karonga) displaying little pits in the enamel, a rugose texture to the enamel, and sunken dentine. (**c**) A left mandible with M/1–2 of *Mastomys natalensis* (Malawi, Karonga) from a modern genet feces accumulation. Notice the mandibular bone splitting, the sunken and cracked dentine, and the little zones of enamel removed. (**d**) An isolated lower M1 of *Mus* sp. (Malawi, Karonga) from indeterminate carnivore feces. Notice the well preserved enamel and the dentine cracking and sinking (**e**) *Thallomys laetolilensis* (EP 2033/03, Loc.6, ULB) lower M1. Note the fine bone splitting at the front of the mandible. Here the dentine is slightly collapsed, but without cracking and the enamel is nearly intact. (**f**) *Paraxerus* sp. (EP 881/005, Loc. 10W, ULB) upper molars showing enamel removal in some parts, dentine mosaic cracking and maxillary bone splitting. The fact that the enamel does not displays cracking is like the situation observed in modern specimens having passed through the digestive tract of a mammalian carnivore rather than weathering

Table 13.3 Summary of rates of enamel pitting between the Upper Ndolanya Beds and Upper Laetolil Beds

Level	None (0)	Light (0.5)	Moderate (1)	Mod/Heavy (1.5)	Heavy (2)	Total
Upper Ndolanya Beds	6	3	14	0	6	29
	(+0.683)	(–3.571)	(–1.812)	(–0.357)	(–2.362)	
Upper Laetolil Beds	36	103	72	1	17	229
	(–0.683)	(+3.571)	(–1.812)	(+0.357)	(–2.362)	
Total	42	106	86	1	23	258

Cell values indicate the NISP (i.e., number of identifiable specimens) falling into each category
Values in parentheses indicate the adjusted residuals for each cell
Positive residuals indicate more specimens appear in a cell than would be expected given the marginal probabilities

Table 13.4 Summary of rates of cracking between the Upper Ndolanya Beds and Upper Laetolil Beds

Level	None (0)	Light (0.5)	Moderate (1)	Heavy (2)	Total
Upper Ndolanya Beds	7	4	15	3	29
	(–2.1)	(–1.8)	(+3.4)	(+1.9)	
Upper Laetolil Beds	102	69	51	7	229
	(+2.1)	(+1.8)	(–3.4)	(-1.9)	
Total	109	73	66	10	258

Cell values indicate the NISP (i.e., number of identifiable specimens) falling into each category
Values in parentheses indicate the adjusted residuals for each cell
Positive residuals indicate more specimens appear in a cell than would be expected given the marginal probabilities

Specimens from the ULB also show a greater degree of breakage, however, and this is contrary to the hypothesis put forth above, but can be accommodated if the ULB sample is suffering greater breakage due to the activity of a more destructive predator. This hypothesis would be corroborated by finding a greater incidence of incisor digestion, molar digestion and dentine cracking in the ULB localities relative the UNB. Again, larger samples are needed to evaluate this hypothesis.

Incisor and molar digestion frequencies are the best indicators of predator activity and the type of predator involved in accumulating fossil micromammal assemblages (Andrews 1990). The sample size at each of the Laetoli localities is too small and biased towards identifiable molars to allow for a detailed determination of accumulating agent, but the relatively low levels of incisor and molar digestion suggest an avian predator. This observation is also consistent with the discovery of bone concentrations recalling fossilized whole owl pellets (see Fig. 13.2a–c; Denys 1986: Plate 1:1–7). At the same time, a few specimens show tooth marks consistent with mammalian predator and signs of molar digestion (e.g., intense dentine cracking; Fig. 13.3), which support the possibility that mammalian carnivores contributed as well. Further actualistic research is needed to establish a comparative baseline for identifying digestion traces left by various predators on African murid rodent molars and to determine whether dentine cracking resulting from intense digestion can be separated from weathering. Similarly, Andrews (1990) pointed out that bone extracted from mammalian carnivore scats (coyote) not exposed to weathering showed extensive splitting and flaking of the surface very similar to that produced by weathering (Fig. 13.3a–d; Andrews 1990, Fig. 3.28).

Concerning post-predation modifications, the surface pitting is an enigmatic type of alteration that has been documented here on enamel, but which is also found on bone and dentine. The fact that it affects all skeletal tissues indicates it results from corrosion probably occurring during burial. The peculiar nature of the carbonatite tuffs in which the Laetoli bones are embedded may be responsible for this alteration, although this was not observed previously at Olduvai Bed I sites (Fernandez-Jalvo et al. 1998).

The low intensity of bone weathering may indicate that bones did not stay long on the surface before burial (less than 2–5 years), or it may indicate a climate with mild temperatures and low levels of seasonality. The presence of root marks indicates an active paleosol. Compaction and conchoidal fractures are quite abundant, and may result from diagenetic alterations.

In the Harrison collections we observe most of the alterations previously observed on the older Leakey collections, and they confirm the rather complex history of the Laetoli small mammals accumulations. A unique alteration, not observed in the new collection, is the little sinuous filaments upon the weathered desquamated bone surface that where observed by Denys (1986, Plates 4:9 and 9:5, 6). These

Table 13.5 Cracking by locality

Level	None (0)	Light (0.5)	Mod (1)	Heavy (2)	Total
LB L01	2	6	1	0	9
	(–1.374)	**(2.437)**	(–0.821)	(–0.544)	
LB L02	7	11	3	0	21
	(–1.084)	**(2.332)**	(–0.923)	(–0.854)	
LB L03	0	3	0	0	3
	(–1.563)	**(2.655)**	(–0.933)	(–0.31)	
LB L04	1	1	1	0	3
	(–0.393)	(0.122)	(0.464)	(–0.31)	
LB L05	10	8	8	0	26
	(–0.663)	(0.075)	(1.106)	(–0.962)	
LB L06	9	1	5	1	16
	(0.977)	**(–2.159)**	(0.895)	(0.769)	
LB L07	1	4	4	0	9
	(–2.059)	(0.955)	(1.631)	(–0.544)	
LB L08	12	8	5	0	25
	(0.369)	(0.216)	(–0.289)	(–0.941)	
LB L09	13	5	2	0	20
	(1.927)	(–0.523)	(–1.381)	(–0.831)	
LB L10	34	13	13	3	63
	(1.768)	(–1.929)	(–0.367)	(0.923)	
LB L11	0	2	2	1	5
	(–2.026)	(0.486)	(0.963)	**(2.225)**	
LB L12	2	0	0	0	2
	(1.585)	(–0.933)	(–0.76)	(–0.252)	
LB L15	5	1	0	0	6
	(1.937)	(–0.728)	(–1.329)	(–0.441)	
LB L16	1	0	3	0	4
	(–0.793)	(–1.325)	**(2.557)**	(–0.358)	
LB L17	0	2	0	0	2
	(–1.273)	**(2.163)**	(–0.76)	(–0.252)	
LB L21	1	1	1	1	4
	(–0.793)	(–0.226)	(0.132)	**(2.572)**	
LB L22	4	3	3	1	11
	(–0.559)	(–0.212)	(0.409)	(1.192)	

Values in parentheses indicate the adjusted residuals for each cell
Positive residuals indicate more specimens appear in a cell than would be expected given the marginal probabilities
Residuals greater than 2 appear in bold

Table 13.6 Summary of weathering rates between the Upper Ndolanya Beds and Upper Laetolil Beds

Level	None (0)	0.5	Stage 1	1.5	Stage 2	Total
Upper Ndolanya Beds	6	3	16	0	4	29
	(−1.405)	(−2.077)	(+2.871)	(−0.620)	(+1.078)	
Upper Laetolil Beds	77	65	66	3	18	229
	(+1.405)	(+2.077)	(−2.871)	(+0.620)	(−1.078)	
Total	83	68	82	3	22	258

Cell values indicate the NISP (i.e., number of identifiable specimens) falling into each category
Values in parentheses indicate the adjusted residuals for each cell
Positive residuals indicate more specimens appear in a cell than would be expected given the marginal probabilities

filaments could be small, fossilized roots or the remains of some other organic matter. The presence of small roots would reinforce the hypothesis of an active soil being present during bone burial, which is consistent with the abundance of the naked mole rat *Heterocephalus* and the presence of urocyclid slugs (Su and Harrison 2007).

Many of the specimens examined here display superimposition of different alterations that imply a rather complex

Table 13.7 Summary of abrasion frequencies between the Upper Ndolanya Beds and Upper Laetolil Beds

Level	None (0)	Light (0.5)	Moderate (1)	Total
Upper Ndolanya Beds	22 (−1.364)	2 (−0.844)	5 (+3.958)	29
Upper Laetolil Beds	196 (+1.364)	28 (+0.844)	5 (−3.958)	229
Total	218	30	10	258

Cell values indicate the NISP (i.e., number of identifiable specimens) falling into each category
Values in parentheses indicate the adjusted residuals for each cell
Positive residuals indicate more specimens appear in a cell than would be expected given the marginal probabilities

scenario of accumulation and fossilization. There are indications of a probable difference in taphonomic histories between the Laetolil Beds and Upper Ndolanya Beds. Further studies on a larger sample and incorporating a broader range of skeletal elements, in addition to the cranio-dental remains analyzed here, will be necessary to further understand the taphonomic history of the Laetoli assemblages.

Paleoenvironment

The taphonomic analysis of the recently collected rodents from Laetoli reveals that the sample suffers from both small sample size and from a bias against smaller-sized taxa. As a result, relative taxonomic abundance estimates, or even rank abundance estimates, using a combined data set for the earlier and later collections are not possible, and this in turn limits the types of paleoenvironmental analyses that are possible.

The recently recovered sample has expanded our knowledge of the biodiversity of the site (Denys 2011). Given the taxa that are present, a few inferences are possible regarding the paleoenvironment of the ULB and UNB. In the ULB, several burrowing taxa are represented, including: *Heterocephalus*, *Gerbilliscus*, *Pedetes*, and *Saccostomus*. From the previous work (Denys 1987), we know at least one species of *Steatomys* is also present. The presence of these burrowing taxa indicates soft sediment, and a relatively high position on the soil catena away from moist and frequently flooded areas, such as river margins. In the modern Serengeti ecosystem, for example, these taxa are prevalent in the grasslands and sparse to open woodlands in the southern part of the ecosystem up to the area just north of Seronera. Further north, the softer soils derived from ashes give way to soils derived from weathered granites of the Precambrian basement rocks (deWit and Jeronimus 1977; Jager 1982). There the terrain becomes more dissected by rivers (Reed et al. 2009) and less suitable for burrowing species. In the north of Serengeti, *Pedetes* is absent (D. Reed, personal observation) and *Gerbilliscus* and *Steatomys* are more rare (Reed 2007, 2011).

Both the old and new fossil rodent collections from Laetoli preserve a significant number of *Thallomys laetolilensis* from the ULB, and in the wet screened samples it is the most abundant murid rodent (Denys 1987). The presence of *Thallomys* is typically a strong indication of wooded vegetation, especially acacia woodland (Reed 2011). This niche is present in all extant forms of *Thallomys*, and is also associated with a stephanodont dental adaptation shared by many other murids that incorporate leaves and bark in their diet. Another climbing genus, *Paraxerus*, is abundant in the Laetoli collections, further supporting the presence of a wooded component to the paleoenvironment.

There is also a good diversity of Gerbillinae (Three species), which are generally indicators of relatively dry and open conditions (see Denys 1999; Reed 2007, 2011). However, gerbils at Laetoli are represented entirely by *Gerbilliscus*, which has a broader ecological tolerance than *Gerbillus* (de Graaff 1981; Linzey and Kesner 1997; Reed 2011).

The genus *Heterocephalus* is not known today from Tanzania, but elsewhere it is associated with loose soils in which it builds elaborate warrens and lives eusocially (Jacobs and Jarvis 1996). In its modern distribution, *Heterocephalus* is associated with arid and open environments (Brett 1991; Jarvis and Bennett 1991; Laden and Wrangham 2005).

Likewise, *Petromus* (dassie rat) are not currently known in East Africa, but appear in the fossil record of Laetoli. Today, they inhabit rocky hills and mountainous areas, living in narrow crevices and large boulders, and occur only in the arid and savannah woodland zones of Southwest Africa (Senegas 2004). Similarly, *Aethomys* is a typical savanna genus often associated with rocky areas.

The most abundant rodent at Laetoli, *Saccostomus*, is not very environmentally informative. This genus is rather catholic in its habits and habitats, occuring in the savannas of East and South Africa. *Saccostomus* is a burrowing species (Hubert 1978; Ellison 1993) that will climb as necessary, but lacks the locomotor adaptations exhibited by more arboreal species, such as *Thallomys* (Earl and Nel 1976). The genus *Xerus,* the African ground squirrel, is also not very informative, but still indicates dry savannas.

The important question for paleoenvironmental reconstruction at Laetoli is how much tree cover was present at the site during ULB times? Many have argued that Laetoli during ULB times was a grassland with perhaps a thin scatter of trees, essentially similar to the modern Serengeti short grass plains (Hay 1987; Bonnefille and Riolet 1987; Gentry 1987). Others (Andrews 1989; Reed 1997; Su and Harrison 2007; Peters et al. 2008) have argued that Laetoli during the Pliocene would have supported a much denser tree canopy, more like the wooded grasslands in the northern Serengeti.

The Laetoli rodent fauna points to a paleoenvironment during ULB times that had low topographic relief and broad expanses of soft sediment that hosted many burrowing taxa. The soft sediments would have been comprised largely of ash from recent volcanic eruptions, such as those that entombed the hominin trackways and the burrowing rodents. These sediments would have been alkaline, as are the modern sediments, a factor that has been shown to limit tree density (Belsky 1990). However if the climatic conditions at the time were less seasonal, then the extensive hardpans that are a physical barrier to tree tap roots may not have existed, removing one major barrier to tree recruitment in the region. The presence of *Thallomys* and the arboreal sciurid *Paraxerus* weigh in favor of tree canopy cover of at least 20% in the area. The presence of *Gerbiliscus* and *Heterocephalus* suggest these trees are growing in a semi-arid environment on loose sediments suitable for borrowing and with sparse to open (0–30%) understory vegetation cover. The presence of *Aethomys* in the ULB and *Petromus* in the LLB points to the presence of some rocky areas, as well, perhaps, evidence for granitic outcroppings, such as kopjes and inselbergs at Laetoli during the Pliocene.

In the Upper Ndolanya Beds, the taphonomic conditions do not seem as good for the preservation of microvertebrates and the samples are thus smaller. Yet we have evidence of a different "environment" with the occurrence of the cane rat, *Thryonomys,* and the abundance of ground squirrel (*Xerus*). The diversity of Gerbillinae falls to one species and the naked mole rat (*Heterocephalus*) is absent which may indicate less suitable soils for this rodent or the absence of tubers and other plants roots they consume. Cane rats do not live in burrows but favor moist grasses for its diet, often in a semi-aquatic context. *Thryonomys* and *Xerus* are both larger rodent species (ca. 6 kg and 320 g respectively) that can be recovered in fossil assemblages by surface prospecting or dry screening. Their absence in the newer collections from the ULB corroborates the hypothesis that these genera are truly absent from the older levels. Given that we did not observe strong differences between the two levels in terms of the accumulating agents, we can attribute the presence of *Xerus* and *Thryonomys* to more mesic conditions in the UNB.

Conclusions

A comparison of the different collection techniques used to recover the fossils confirms the importance of wet screening for recovering the smallest-sized fraction of rodent diversity (e.g. the smallest Murinae and Dendromurinae). These rodents comprise roughy 45% of the species diversity in the Order Rodentia, and because they are generally primary consumers they are the best indicators of paleoenvironment. Efforts must be made in the future to perform systematic wet-screening at fossil sites and to develop new techniques to improve the efficiency of these recovery techniques even under harsh field conditions.

The surface modification study suggests some avian predator as the likely accumulating agent, but with the small sample sizes this can only be a tentative assessment and mammalian carnivores may also be involved as predation agents for some localities. Weathering is low and rootmarking present, which indicates a rather fast burial of the small mammal bones. Other types of alterations that we observed here such as abrasion, trampling, enamel pitting, coating, and cracking may result from soil corrosion during diagenesis. However, a difference of preservation is suggested between ULB and UNB which may explain the low abundance of rodent remains in the UNB.

The soil conditions as well as the environment may have been different in the ULB compared to the UNB, as shown by the taphonomic conditions and the rodent biodiversity. So the relative high proportion of arboreal rodents found in the ULB indicates the presence of acacia trees in non negligeable proportions which are not recovered in the UNB. In both units the abundance of burrowing rodents clearly indicate the presence of soft soils. The presence of the naked mole rat (*Heterocephalus*), the spring hare (*Pedetes*), two species of Gerbillinae are all elements of a semi-arid condition in the ULB; while the occurrence of the sciurid, *Xerus*, the cane rat, *Thryonomys,* the reduction in the diversity of gerbils, and the absence of *Heterocephalus* in the UNB suggests more mesic conditions around 2.7 Ma.

Acknowledgments Thanks to T. Harrison for providing the new rodent material. This work was sponsored by funds from the NSF-RHOI grant under the Taphonomy Working Group. SEM pictures were taken by C. Chancogne Weber and G. Mascarel (MNHN Paris). We thank Yolanda Fernandez-Jalvo and Peter Andrews for taphonomic discussions about this material, as well as two anonymous reviewers whose comments helped significantly in the development of the manuscript.

References

Agresti, A. (2007). *An introduction to categorical data analysis*. Hoboken: Wiley.

Alemseged, Z., Bobe, R., & Geraads, D. (2007). Comparability of fossil data and its significance for the interpretation of hominin environments: A case study in the Lower Omo Valley, Ethiopia. In *Hominin environments in the East African Pliocene: An assessment of the faunal evidence* (pp. 159–181). Dordrecht: Springer.

Andrews, P. (1989). Palaeoecology of Laetoli. *Journal of Human Evolution, 18*, 173–181.

Andrews, P. (1990). *Owls, caves, and fossils*. London: Natural History Museum.

Baker, B. W., & Shaffer, B. S. (1999). Assumptions about species: A case study of tortoise bones from SE Texas. *Journal of Field Archaeology, 26*, 69–74.

Behrensmeyer, A. (1978). Taphonomic and ecologic information from bone weathering. *Paleobiology, 4*, 150–162.

Behrensmeyer, A. K., & Hook, R. W. (1992). Paleoenvironmental contexts and taphonomic modes. In A. K. Behrensmeyer, J. D. Damuth, W. A. DiMichele, R. Potts, H.-D. Sues, & S. L. Wing (Eds.), *Evolutionary paleoecology of terrestrial plants and animals* (pp. 15–136). Chicago: University of Chicago Press.

Belsky, A. (1990). Tree/grass ratios in East African savannas: A comparison of existing models. *Journal of Biogeography, 17*, 483–489.

Bonnefille, R., & Riollet, G. (1987). Palynological spectra from the Upper Laetolil Beds. In M. D. Leakey & J. M. Harris (Eds.), *Laetoli: A Pliocene site in northern Tanzania* (pp. 52–61). Oxford: Clarendon.

Brett, R. A. (1991). The population structure of naked mole-rat colonies. In P. W. Sherman, J. U. M. Jarvis, & R. D. Alexander (Eds.), *The biology of the naked mole-rat* (pp. 97–136). Princeton: Princeton University Press.

Davies, C. (1987). Fossil Pedetidae (Rodentia) from Laetoli. In M. D. Leakey & J. M. Harris (Eds.), *Laetoli: A Pliocene site in northern Tanzania* (pp. 171–189). Oxford: Clarendon.

De Graaff, G. (1981). *The rodents of Southern Africa*. Durban: Butterworth.

Denys, C. (1983). Les rongeurs du Pliocene de Laetoli (Tanzania): Evolution, paléoecologie et paléobiogeographie. Approche qualitative et quantitative. Ph.D. disseration, University Pierre et Marie Curie, Paris.

Denys, C. (1985a). Palaeoenvironmental and palaeobiogeographical significance of the fossil rodent assemblages of Laetoli (Pliocene, Tanzania). *Palaeogeography, Palaeoclimatology, Palaeoecology, 52*, 77–97.

Denys, C. (1985b). Laetoli: A Pliocene southern savanna fauna in the Eastern Rift Valley (Tanzania). Ecological and zoogeographical implications. In K. L. Schuchman (Ed.), *Proceedings of the international symposium on African vertebrates* (pp. 31–35). Bonn: Koenig.

Denys, C. (1986). Le gisement Pliocène de Laetoli (Tanzanie, Afrique de l'Est): Analyse taphonomique des assemblages de microvertébrés. *Palaeontographica Abteilung A-Palaozoologie-Stratigraphie, 194*, 69–98.

Denys, C. (1987). Fossil rodents (other than Pedetidae) from Laetoli. In M. D. Leakey & J. M. Harris (Eds.), *Laetoli: A Pliocene site in northern Tanzania* (pp. 118–170). Oxford: Clarendon.

Denys, C. (1999). Of mice and men. In T. Bromage & F. Schrenk (Eds.), *African biogeography, climate change, and early hominid evolution* (pp. 226–252). Oxford: Oxford University Press.

Denys, C. (2011). Rodents. In T. Harrison (Ed.), *Paleontology and geology of Laetoli: Human evolution in context* (Fossil hominins and the associated fauna, vol. 2, pp. 15–53). Dordrecht: Springer.

deWit, H. A., & Jeronimus, O. D. (1977). *Soil map of the Serengeti plains*. Wageningen: Wageningen Agricultural University.

Earl, Z., & Nel, J. A. J. (1976). Climbing behaviour in three African rodent species. *Zoologica Africana, 11*, 183–192.

Eck, G. G. (2007). The effects of collection strategy and effort on faunal recovery: A case study of the American and French collections from the Shungura Formation, Ethiopia. In R. Bobe, Z. Alemseged, & A. K. Behrensmeyer (Eds.), *Hominin environments in the East African Pliocene: An assessment of the faunal evidence* (pp. 183–215). Dordrecht: Springer.

Ellison, G. T. H. (1993). Group size, burrow structure and hoarding activity of pouched mice (*Saccostomus campestris*: Cricetidae) in Southern Africa. *African Journal of Ecology, 31*, 135–155.

Fernandez-Jalvo, Y., & Andrews, P. (1992). Small mammal taphonomy of Gran Dolina, Atapuerca (Burgos), Spain. *Journal of Archaeological Science, 19*, 407–428.

Fernandez-Jalvo, Y., Denys, C., Andrews, P., Williams, T., Dauphin, Y., & Humphrey, L. (1998). Taphonomy and palaeoecology of Olduvai Bed I (Pleistocene, Tanzania). *Journal of Human Evolution, 34*, 137–172.

Gentry, A. W. (1987). Pliocene Bovidae from Laetoli. In M. D. Leakey & J. M. Harris (Eds.), *Laetoli: A Pliocene site in northern Tanzania* (pp. 378–408). Oxford: Clarendon.

Hay, R. L. (1987). Geology of the Laetoli area. In M. D. Leakey & J. M. Harris (Eds.), *Laetoli: A Pliocene site in northern Tanzania* (pp. 23–47). Oxford: Clarendon.

Hubert, B. (1978). Revision of the genus *Saccostomus* (Rodentia, Cricetomyinae), with new morphological and chromosomal data from specimens from the Lower Omo Valley, Ethiopia. *Bulletin of the Carnegie Museum of Natural History, 6*, 48–52.

Jacobs, D. S., & Jarvis, J. U. M. (1996). No evidence for the work-conflict hypothesis in the eusocial naked mole-rat (*Heterocephalus glaber*). *Behavioral Ecology and Sociobiology, 39*, 401–409.

Jager, T., 1982. Soils of the Serengeti woodlands, Tanzania. Ph.D. dissertation, Agricultural University, Wageningen.

Jarvis, J. U. M., & Bennett, N. C. (1991). Ecology and behavior of the family Bathyergidae. In P. W. Sherman, P. W. Jarvis, & R. D. Alexander (Eds.), *The biology of the naked mole-rat* (pp. 66–96). Princeton: Princeton University Press.

Laden, G., & Wrangham, R. (2005). The rise of the hominids as an adaptive shift in fallback foods: Plant underground storage organs (USOs) and australopith origins. *Journal of Human Evolution, 29*, 482–498.

Leakey, M. D. (1987). Introduction. In M. D. Leakey & J. M. Harris (Eds.), *Laetoli: A Pliocene site in northern Tanzania* (pp. 1–22). Oxford: Clarendon.

Linzey, A. V., & Kesner, M. H. (1997). Small mammals of a woodland-savannah ecosystem in Zimbabwe. I. Density and habitat occupancy patterns. *Journal of Zoology, London, 243*, 137–152.

Peters, C. R., Blumenschine, R. J., Hay, R. L., Livingstone, D. A., Marean, C. W., Harrison, T., Armour-Chelu, M., Andrews, P., Bernor, R. L., Bonnefiller, R., & Werdelin, L. (2008). Paleoecology of the Serengeti-Mara ecosystem. In A. R. E. Sinclair, C. Packer, S. A. R. Mduma, & J. M. Fryxell (Eds.), *Serengeti III: Human impacts on ecosystem dynamics* (pp. 47–94). Chicago: University of Chicago Press.

Reed, K. E. (1997). Early hominid evolution and ecological change through the African Plio-Pleistocene. *Journal of Human Evolution, 32*, 289–322.

Reed, D. N. (2005). Taphonomic implications of roosting behavior and trophic habits in two species of African Owl. *Journal of Archaeological Science, 32*, 1669–1676.

Reed, D. N. (2007). Serengeti micromammals and their implications for Olduvai paleoenvironments. In R. Bobe, Z. Alemseged, & A. K. Behrenesmeyer (Eds.), *Hominin environments in the East African Pliocene: An assessment of the faunal evidence* (pp. 217–255). Dordrecht: Springer.

Reed, D. N. (2011). Serengeti micromammal communities and the paleoecology of Laetoli, Tanzania. In T. Harrison (Ed.), *Paleontology and geology of Laetoli: Human evolution in context* (Geology, geochronology, paleoecology, and paleoenvironment, vol. 1, pp. 253–263). Dordrecht: Springer.

Reed, D. N., Anderson, T. M., Dempewolf, J., Metzger, K., & Serneels, S. (2009). The spatial distribution of vegetation types in the Serengeti ecosystem: The influence of rainfall and topographic relief on vegetation patch characteristics. *Journal of Biogeography, 36*, 770–782.

Sénégas, F. (2004). A new species of *Petromus* (Rodentia, Hystricognatha, Petromuridae) from the Early Pliocene of South Africa and its paleoenvironmental implications. *Journal of Vertebrate Paleontology, 24*, 757–763.

Soligo, C., & Andrews, P. (2005). Taphonomic bias, taxonomic bias and historical non-equivalence of faunal structure in early hominin localities. *Journal of Human Evolution, 49*, 206–229.

Stahl, P. (1996). The recovery and interpretation of microvertebrate bone assemblages from archaeological contexts. *Journal of Archaeological Method and Theory, 3*, 31–75.

Su, D., & Harrison, T. (2007). The paleoecology of the Upper Laetolil Beds at Laetoli: A reconsideration of the large mammal evidence. In R. Bobe, Z. Alemseged, & A. K. Behrenesmeyer (Eds.), *Hominin environments in the East African Pliocene: An assessment of the faunal evidence* (pp. 279–313). Dordrecht: Springer.

Su, D., & Harrison, T. (2008). Ecological implications of the relative rarity of fossil hominins at Laetoli. *Journal of Human Evolution, 55*, 672–681.

White, T. D. (1998). The comparative biology of "Robust" *Australopithecus*: Clues from context. In F. E. Grine (Ed.), *Evolutionary history of the "Robust" australopithecines* (pp. 449–483). New York: Gruyter.

Chapter 14
Coprolites: Taphonomic and Paleoecological Implications

Terry Harrison

Abstract Recent paleontological collections at Laetoli and Kakesio have yielded a number of coprolites of medium- to large-sized carnivores, and a rare collection of ruminant coprolites. The carnivore coprolites appear to belong to a diversity of taxa, including canids, felids and hyaenids. Their occurrence confirms other lines of evidence that carnivores played an important role in the accumulation and composition of the fossil remains at Laetoli. Ruminant coprolites are extremely rare in the African fossil record, and are the result of unusual preservational conditions. The dung can be attributed to medium- to large-sized ruminants, including *Giraffa stillei* and at least two species of bovids. The consistency of most of the ruminant dung and the occasional presence of seeds indicates that deposition occurred primarily during or soon after the rainy season, a finding consistent with the sedimentological evidence. The presence of seeds and of coarse particles of herbaceous and woody plant material in several coprolites supports stable isotope and mesowear studies indicating that the ruminants at Laetoli were predominantly mixed feeders.

Keywords Bovid • Carnivore • Dung pellet • Diet • Ecology • *Giraffa* • Scat

Introduction

The recent paleontological collections at Laetoli and Kakesio have yielded a number of coprolites, and these provide insights into the taphonomy and paleoecology. The material includes a small sample of coprolites of medium- to large-sized carnivores, and a rare collection of ruminant coprolites. All of the coprolites described here were collected by the Harrison expeditions (1998–2005) and are housed in the National Museum of Tanzania in Dar es Salaam.

T. Harrison (✉)
Center for the Study of Human Origins, Department of Anthropology, New York University, 25 Waverly Place, New York, NY 10003, USA
e-mail: Terry.harrison@nyu.edu

Dietrich (1951) briefly described and figured several coprolites from Laetoli (i.e., Garussi-Ursprungsgebeit) collected by Kohl-Larsen in 1938–1939. Mary Leakey's catalog indicates that a number of coprolites were collected ($n=38$) during the 1974–1979 expeditions, mostly from the Upper Laetolil Beds, but, unfortunately, very few specimens have been relocated in the conserved collections. The coprolites are discussed below according to the major taxonomic groups to which they can be assigned – carnivores and ruminants. The carnivore coprolites are described individually, because of their greater diversity in form. The ruminant coprolites are organized into several distinct types, and are described and analyzed collectively.

Although the carnivore coprolite sample is small they appear to belong to a diverse range of taxa, including canids, felids and hyaenids. They provide clues to the significant role that carnivores played in the accumulation and composition of the fossil assemblage at Laetoli, as well as offering some insights into their past behaviors. The collection of ruminant coprolites is much larger than that of carnivores, and belongs to at least three species, a small giraffid and two bovids. The latter specimens were all derived from a single clay horizon in the Upper Laetolil Beds, just above Tuff 7, at Loc. 3 (see Harrison and Kweka 2011). This is the first account of ruminant coprolites from a Plio-Pleistocene locality in East Africa, and as such, it provides a unique perspective on the dietary behavior and taphonomy of the Laetoli ungulates.

Carnivore Coprolites

Fourteen partial or entire coprolites of carnivores have been recovered from Laetoli. These have been primarily recovered from the Upper Laetolil Beds (~3.63–3.85 Ma), but one specimen is from the Lower Laetolil Beds at Kakesio (~3.85–4.36 Ma), and two are from the Upper Ndolanya Beds (~2.66 Ma). A brief description of each of the specimens is presented below, with comments on their possible taxonomic attribution and taphonomic implications. The size of the scat

T. Harrison (ed.), *Paleontology and Geology of Laetoli: Human Evolution in Context. Volume 1: Geology, Geochronology, Paleoecology and Paleoenvironment*, Vertebrate Paleobiology and Paleoanthropology, DOI 10.1007/978-90-481-9956-3_14,

is roughly correlated to the size of the carnivore that produced it, and zoologists commonly use scat size as an important criterion in identifying extant taxa in field studies (e.g., Ray and Sunquist 2001; Breuer 2005). However, there is a good deal of intraspecific variation in scat size according to season, dietary behavior, health, and age of the individual, and carnivore species of different sizes can produce scats that overlap metrically (Farrell et al. 2000). For these reasons, no attempt is made to relate the size of the individual coprolites directly with the fossil taxa represented at Laetoli. Rather, scats of modern carnivores in East Africa are grouped into four sizes classes according to their diameter (i.e., < 15 mm, 15–20 mm, 20–35 mm and>35 mm), which generally correspond to the scats of small, medium, large and very large carnivores (Table 14.1). This allows a more general approximation of the size classes of carnivores that are likely to have deposited the scat at Laetoli during the Pliocene.

Modern carnivores produce scats with varying degrees of bone damage and digestion. Andrews and Evans (1983), Andrews (1990) and Matthews (2002) have shown that scats of small felids (i.e., *Felis wiedi, Felis catus*, and *Felis caracal*) generally contain extremely broken bones with high levels of rounding and polishing produced by digestion. The serval (*Felis serval*) appears to differ from the other small cats in producing scats with lower levels of bone damage and digestion (Matthews 2002). Scats of small canids (i.e., *Otocyon megalotis*, *Canis latrans*, *Vulpes vulpes*, and *Alopex lagopus*) also contain highly digested bones, but the degree of breakage is somewhat less than in small felids. Viverrids, mustelids and herpestids, on the other hand, produce scats in which the bone tends to be less broken and digested than in either small canids or small felids (Andrews 1990). The digestive processes of large carnivores (i.e., hyaenas, large felids, and African wild dogs) produce scats with much

Table 14.1 Size of scats in extant African carnivores in comparison with coprolites from Laetoli

Diameter of feces	Extant African carnivores[a]	Fossil samples[b]	Laetoli carnivores[c] (Corresponding with size of coprolites)[d]
< 15 mm	*Cynictis pencillata* (yellow mongoose)		*Propoecilogale bolti*
	Ichneumia albicauda (white-tailed mongoose)		*Viverra leakeyi*
	Herpestes sanguinea (slender mongoose)		*Genetta* sp.
	Helogale parvula (dwarf mongoose)		*Herpestes palaeoserengetensis*
	Genetta genetta (genet)		*Herpestes ichneumon*
	Galerella nigrata (black mongoose)		*Galerella palaeogracilis*
			Mungos dietrichi
			Mungos sp. nov?
15–20 mm	*Canis mesomelas* (black-backed jackal)	EP 2127/00, EP 4181/00	cf. *Canis* sp. A
	Canis adustus (side striped jackal)		cf. *Canis* sp. B
	Vulpes chama (cape fox)		aff. *Otocyon* sp.
	Otocyon megalotis (bat-eared fox)		*Mellivora* sp.
	Mellivora capensis (honey badger)		*Caracal* sp. or *Leptailurus* sp.
	Felis caracal (caracal)		*Felis* sp.
	Felis serval (serval)		
20–35 mm	*Lycaon pictus* (wild dog)	EP 068/98, EP 1014/00,	?*Nyctereutes barryi*
	Hyaena hyaena (striped hyaena)	EP 1372/00, EP 813/03,	*Crocuta dietrichi*
	Panthera pardus (leopard)	EP 2343/03, EP 599/04,	*Parahyaena howelli*
	Acinonyx jubatus (cheetah)	EP 998/04, EP 1077/04,	*Ikelohyaena* cf. *I. abronia*
		EP 1078/04, EP 1138/04	*Lycyaenops* cf. *L. silberbergi*
			Dinofelis petteri
			Panthera sp. cf. *P. pardus*
			Acinonyx sp.
> 35 mm	*Crocuta crocuta* (spotted hyaena)		?*Pachycrocuta* sp.
	Proteles cristata (aardwolf)		aff. *Proteles* sp.
	Panthera leo (lion)		*Homotherium* sp.
			Panthera sp. aff. *P. leo*

[a]Sources: Horwitz and Goldberg (1989), Andrews (1990), Walker (1996), Stuart and Stuart (1998), Warren et al. (2009), Harrison (unpublished data)
[b]Fossil coprolites – see text for descriptions
[c]From Werdelin and Dehghani (2011)
[d]Size category of fossil taxa based on similarity in dental size to modern carnivores

greater levels of bone breakage and digestion than in small carnivores, and this is reflected in the much lower frequency of occurrence of bone fragments in the former (Andrews 1990; Leakey et al. 1999; Breuer 2005). Given these considerations, it is not possible to assign the coprolites to particular taxa based on the bone content and degree of damage, especially in large carnivore scats where bone fragments are likely to be absent or relatively uncommon.

Description of Specimens

EP 068/98 (Kakesio 4, Lower Laetolil Beds)

The specimen consists of an almost complete coprolite, broken transversely into two segments, and partly eroded (Fig. 14.1a). It is sub-circular in cross-section, with a maximum length of 46.8 mm and a maximum transverse diameter of 24.5 mm. The coprolite has a smooth calcite-rich external surface, with a network of microscopic cracks. The proximal end of the coprolite tapers to a bluntly rounded and roughened protuberance, representing the point of termination by the anal sphincter. The distal end has a roughened and flattened surface, obliquely oriented at 39° to the long axis of the coprolite, presumably produced by impact of the scat with the substrate. Along one margin of the distal end there is a slight outward extrusion at the impact surface. These imply that the bolus was relatively soft and moist when deposited. The exposed internal surface of the coprolite reveals a coarse-grained groundmass of tuffaceous silty sand. No definitive bone fragments are observable, but a few tiny fragments of highly triturated organic hard tissue (bone or shell) are preserved.

This coprolite likely belongs to a large carnivore. In terms of size and form, it is comparable to the scats of wild dog (*Lycaon pictus*), leopard (*Panthera pardus*) and cheetah (*Acinonyx jubatus*) (Table 14.1). Possible candidates among the fossil carnivores from Laetoli are ?*Nyctereutes barryi*, *Dinofelis petteri*, *Panthera* sp. cf. *P. pardus*, and *Acinonyx* sp. (see Werdelin and Dehghani 2011).

EP 1014/00 (Loc. 18, Upper Ndolanya Beds)

This specimen comprises fragments of four associated coprolites (Fig. 14.1c–f). All consist of coarse-grained tuffaceous sediment, mostly without evidence of organic inclusions. They are consistent in size and morphology. (a) A large coprolite with a bluntly rounded apex, a flattened obliquely oriented distal end, a slight longitudinal curvature, and a circular cross-section. There are a few small rounded fragments of highly digested bone, up to 2.5 mm in length. The specimen is 55.2 mm long, with a maximum diameter of 30.0 mm.

Fig. 14.1 Carnivore coprolites. (**a**) EP 068/98 (Kakesio 4, Lower Laetolil Beds); (**b**) EP 599/04 (Loc. 10, Upper Laetolil Beds); (**c–f**) EP 1014/00 (Loc. 18, Upper Ndolanya Beds), four associated coprolite fragments; (**g–h**) EP 1077/04 (Loc. 11, Upper Laetolil Beds), two views of the same coprolite

(b) A smaller segment of coprolite, with bluntly rounded end and a sharp transverse fracture covered with a film of calcite. The coprolite is sub-circular in cross-section. The surface is irregular and granular. The proximal end has a shallow circular depression at its apex. Towards the proximal end is a spiral groove that appears to represent secondary weathering. The length and diameter of the coprolite are 47.9 and 25.8 mm respectively. (c) A coprolite similar in size and form to the previous specimen. It has a rounded apex. The other end has an obliquely oriented transverse fracture covered with a film of calcite. It is 44.5 mm long, with a diameter of 24.3 mm. (d) A short segment of a coprolite, with both ends fractured and coated in calcite. The length is 25.5 mm, with a diameter of 25.2 mm. These specimens belong to a large carnivore about the size of a striped hyaena or a leopard.

EP 1372/00 (Loc. 6, Upper Laetolil Beds, between Tuffs 5 and 6)

A partial coprolite, comprising at least five segments of irregular length, but missing the distal end (Fig. 14.2c). It would appear that the dropping was incomplete at the time of fossilization. The incomplete coprolite is 52.3 long, with a maximum diameter of 22.3 mm. The apex narrows to a rounded point. The coprolite is stained pale orange, with a smooth calcite coat. A number of poorly digested bone fragments of micromammals (or small birds) are exposed on the surface, but none is anatomically or taxonomically identifiable. The largest fragment is a splinter of a limb bone that measures 2.0 by 9.4 mm. Given the size of the coprolite and the amount of undigested bone, this specimen very likely belongs to a large canid or felid, of which ?*Nycteureutes barryi* and *Panthera* sp. cf. *P. pardus* are the most likely candidate taxa (Werdelin and Dehghani 2011).

Fig. 14.2 Carnivore coprolites. (**a**) EP 1138/04 (Loc. 13, Upper Laetolil Beds); (**b**) EP 2127/00 (Loc. 13, Upper Laetolil Beds); (**c**) EP 1372/00 (Loc. 6, Upper Laetolil Beds)

EP 2127/00 (Loc. 13, Upper Laetolil Beds, between Tuffs 6 and 7)

This specimen consists of a small portion of a coprolite, with an apical protuberance produced by the anal sphincter termination (Fig. 14.2b). It is cylindrical in form, with a circular cross-section. The length and diameter of the specimen is 30.9 and 16.9 mm respectively. The external surface is smooth, with no apparent organic inclusions. The groundmass is very fine- grained in a calcite-rich matrix. The coprolite has no distinct segmentation, but it narrows slightly, about ~17 mm below the apex. It is comparable in size, form and content to the droppings of small felids and canids, such as caracal, serval, and jackals. Suitable taxa from Laetoli would be cf. *Canis* spp., *Caracal* sp., and *Felis* sp. (Werdelin and Dehghani 2011).

EP 4181/00 (Loc. 8, Upper Laetolil Beds, between Tuffs 5 and 7)

This specimen consists of a partial coprolite deposited on top of a weathered bone fragment of a large mammal (Fig. 14.3). It is an irregular, flattened mass, somewhat compressed, and tapering towards one side. The flattened contact with the bone and the smoothly rounded contour of the exposed surface of the coprolite suggests a relatively soft stool. It is evident that the bone fragment was heavily weathered at the time of deposition, because the scat penetrates at least 2 mm into wide cracks in the surface. The bone was originally a

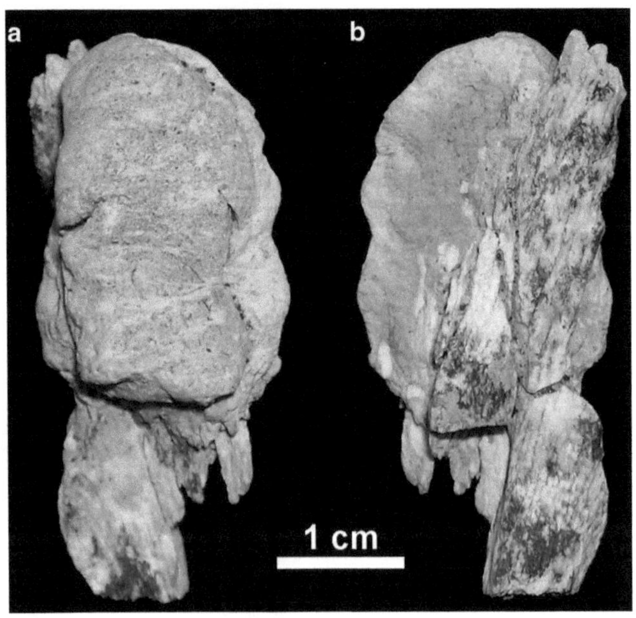

Fig. 14.3 A partial coprolite of a medium-sized carnivore deposited on a weathered bone fragment. EP 4181/00 (Loc 8, Upper Laetolil Beds). (**a**) Top surface; (**b**) Bottom surface

much larger fragment at the time the scat was deposited, but only a small section (43.2 × 16.3 mm) is preserved. The length of the coprolite is 28.2 mm, but originally it may have been as long as ~40 mm. It has a maximum breadth of 17.8 mm, and a thickness of 6.6 mm. The groundmass consists of fine-grained matrix, lacking evidence of organic inclusions. There is a lower band of light colored and coarser-grained matrix, overlain by a thicker mass of pale grey finer-grained matrix.

Deposition of droppings on elevated surfaces, such as on clumps of vegetation and on flat rocks, is a typical territorial marking behavior of extant jackals (Stuart and Stuart 1998) and other medium-sized canids. Large flat bones exposed on the land surface are ideal for this purpose, and the coprolite from Laetoli provides an example of this type of behavior. A similar occurrence has been recorded in the fossil record at the Plio-Pleistocene locality of Kanam East, Kenya, in which a small carnivore deposited a scat on the flattened surface of a weathered cercopithecid bone (Harrison and Harris 1996). The size and form of the coprolite, and its use in territorial marking, indicates that the specimen belongs to a medium-sized canid, comparable in size to the modern jackal.

EP 813/03 (Loc. 9, Upper Laetolil Beds, between Tuffs 5 and 7)

This specimen represents the detached terminal segment of a large coprolite. It is bun-shaped, and circular in cross-section, with a shallow depression on its base for contact with the next segment. It has a maximum diameter of 33.5 mm. It has a roughened irregular surface, with a smooth coat. The groundmass is fine-grained, with a calcite matrix that fills cavities and voids. Several bone fragments and a tooth chip are visible, but not identifiable taxonomically. A thin sliver of bone (12.8 ×4.1 mm) is from a limb bone of an animal about the size of a modern dik-dik (*Madoqua*). Based on its size, form and content, this specimen likely belongs to one of the large hyaenids from Laetoli (Table 14.1).

EP 1246/03 (Loc. 7E, Upper Ndolanya Beds)

This consists of a small fragment of a large coprolite, which is too incomplete to determine its original size or form (Fig. 14.4a). It contains several bone fragments. The two largest fragments (12.2 and 14.6 mm long) are identifiable as pieces of rib, probably of a dik-dik. They are lying side-by-side, and probably represent fragments of the same rib. Other bone fragments are observable, but not identifiable. The bones show no evidence of digestion. Given the size of the included bones and the lack of evidence of digestion, this specimen probably belonged to a medium- to large-sized canid.

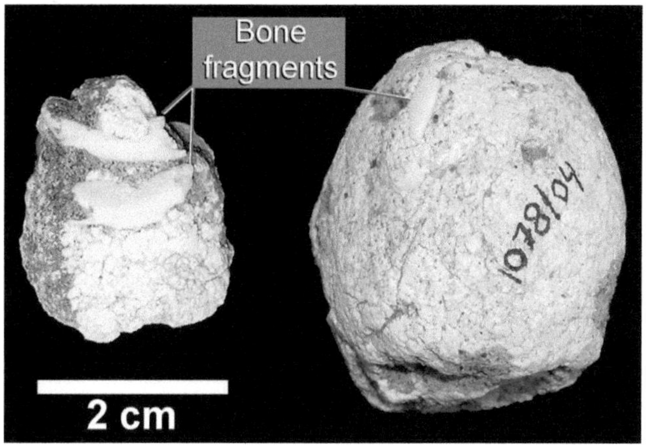

Fig. 14.4 Carnivore coprolites containing bone fragments. (**a**) EP 1246/03 (Loc. 7E, Upper Ndolanya Beds), (**b**) 1078/04 (Loc. 11, Upper Laetolil Beds)

EP 2343/03 (Loc. 13, Upper Laetolil Beds, between Tuffs 6 and 7)

Consists of a partial coprolite that is cylindrical in form, with a circular cross-section. It has a low and broad apical termination, with a shallow and roughened depression at its tip. The distal end is broken. The specimen is 39.5 mm long and 23.2 mm in diameter. The surface is relatively smooth, with a creamy white calcite coat. The groundmass is fine-grained, with a matrix rich in calcite. A few chips of finely triturated and digested bone are visible at the surface. The specimen probably belongs to a hyaenid or canid. Of the carnivores recorded from the Upper Ndolanya Beds, *Crocuta dietrichi*, *Dinofelis petteri*, and *Panthera* sp. cf. *P. pardus* are of a suitable size (Table 14.1).

EP 599/04 (Loc. 10, Upper Laetolil Beds, below Tuff 2)

This consists of a small ovoid coprolite, and a fragment of an associated coprolite (see Fig. 14.1b). The entire coprolite is cylindrical, with rounded ends, and a slight curvature to its long-axis. One side is flattened to slightly concave, and presumably represents the side that settled on the land surface at the time of being deposited. There is a distinct protuberance at one end, which is presumably the result of a large inclusion. The surface was originally coated in a fine film of calcite, but this has been lost in places through weathering. The specimen is 31.8 mm long with a maximum diameter of 20.8 mm. Judging from the fragmentary coprolite, the groundmass was very fine-grained, calcite rich, and without visible organic inclusions. This specimen is consistent in size and form to a medium-sized felid or canid (Table 14.1).

EP 786/04 (Loc. 7, Upper Laetolil Beds, between Tuffs 5 and 7)

A small fragment of a large coprolite. It has a smooth external surface, with a fine-grained cream-colored matrix. There are no visible organic inclusions. It is too incomplete for morphological description or measurement.

EP 998/04 (Loc. 9, Upper Laetolil Beds, between Tuffs 5 and 7)

This specimen consists of an apical segment of a coprolite, with a raised lip surrounding a shallow depression apically. It is cylindrical in form with a sub-circular cross-section. The length of the specimen is 21.2 mm, with a maximum diameter of 21.9 mm. Small shards of bone are exposed on the surface of the coprolite. Given its size and form, this specimen probably belongs to one of the Laetoli hyaenids (Table 14.1).

EP 1077/04 (Loc. 11, Upper Laetolil Beds, between Tuffs 7 and 8)

This specimen represents a fragment of a coprolite segment (see Fig. 14.1g–h). It is cylindrical in form, with a smooth outer coat. The length of the segment is 33.5 mm, and its maximum diameter is 27.5 mm. The coprolite contains a mass of bone fragments of small mammals; the largest piece having a length of 13.2 mm. Several of the bones can be identified anatomically and taxonomically, and these include a distal tibia (without its epiphysis), a vertebral fragment, and a glenoid of the scapula of a juvenile individual of *Serengetilagus*. The size of the coprolite, the identified prey species, and the bone preservation indicate that this specimen was produced by a large-sized canid (see Andrews and Nesbit Evans 1983), such as *?Nyctereutes barryi* (Table 14.1).

EP 1078/04 (Loc. 11, Upper Laetolil Beds, between Tuffs 7 and 8)

Consists of an apical segment of a large, cylindrical coprolite (Fig. 14.4b). One side is slightly flattened, and presumably represents the contact with the land surface after deposition. The proximal end is rounded with a slight apical depression. The other is flattened to slightly concave, where it adjoins the next segment. The surface has a smooth coat, with fine superficial cracking. A number of finely comminuted and corroded bone shards are exposed on the surface; the largest has a length of 8.5 mm. The general size and shape of the coprolite and the nature of the bone fragments suggests that this specimen belongs to the one of the large hyaenids from Laetoli (Table 14.1).

EP 1138/04 (Loc. 13, Upper Laetolil Beds, between Tuffs 6 and 8)

A complete coprolite consisting of four proximo-distally compressed segments (see Fig. 14.2a). Its total length is 45.0 mm, and its maximum diameter is 29.1 mm. Each segment follows a spiral contour in relation to the long-axis of the coprolite. The apical segment is the largest, with a maximum length of 23.5. It tapers and spirals towards a broad apex, with a shallow depression at its center. The remainder of the coprolite is cylindrical, with a circular cross-section. The distal segment has a circular depression in its base, with a rounded rim. The external coat is smooth, stained pale orange and black by the weathering process. The groundmass is fine-grained, with small organic inclusions. These are mostly heavily triturated and digested bone, but several are larger reaching a maximum size of ~5 mm, including one located in the interstitial space between segments 2 and 3. None of the bone fragments can be identified anatomically or taxonomically. Based on its size and form, the coprolite probably belongs to one of the hyaenids from Laetoli (Table 14.1).

Discussion

Relatively few carnivore coprolites have been recovered from Laetoli, especially given the relative abundance and diversity of carnivores throughout the sequence. The specimens are mostly known from the Upper Laetolil Beds, but examples have also been recovered from the Lower Laetolil Beds and Upper Ndolanya Beds. Part of the explanation for the apparent rarity of carnivore coprolites at Laetoli is provided by the observation that they represent a relatively narrow size range (mean = 25.6 ± 4.9 mm; range 16.9–33.5 mm). Droppings of modern African carnivores range in diameter from less than 8 mm to greater than 40 mm (see Table 14.1), and a similar range should be expected at Laetoli given the size range of the carnivores represented. Fossil carnivores at Laetoli range in estimated body mass from animals comparable in size to the dwarf mongoose (*Helogale parvula*) and striped weasel (*Poecilogale albinucha*), with a body mass of just over 200 g, to the size of large felids comparable to the modern-day lion (*Panthera leo*), with an average body mass of almost 200 kg (body mass data from Kingdon 1997). However, the Laetoli coprolites are all from medium- to large-sized taxa, equivalent to animals ranging in body size from a black-backed jackal (~10 kg) to that of a striped hyena (~40 kg). The low incidence and skewed distribution of carnivore coprolites from Laetoli is probably a function of taphonomic influences and collecting bias. Coprolites are likely to be relatively rare in open-air sites with subaerial deposition, such as Laetoli, because animal droppings exposed on the land surface are likely to decompose and weather rapidly, be destroyed or removed by insects, or be eaten by other mammals. Those belonging to smaller carnivores are most likely to be destroyed quickly.

Another important taphonomic consideration is that many of the smaller carnivores and some of the hyaenids deposit their scats in latrines or middens, which would ultimately impact on the distribution of coprolites at fossil sites. Carnivores that deposit their droppings widely increase the chance of representation in the fossil record. In addition, small felids that bury their scats may reduce the chances of preservation as coprolites because of the higher rate of decomposition and destruction prior to fossilization.

Finally, the recovery of coprolites is hampered by the difficulty in recognizing and identifying them in the field, since they do not conform to the typical search image of many fossil collectors, and they often look like weathered nodules of sediment. This collecting bias probably accounts, at least in part, for the low incidence of the recovery of carnivore coprolites at Laetoli, and may also account for the unusual size distribution of coprolites, with those from smaller carnivores being more easily overlooked. This bias is confirmed by the fact that Mary Leakey's team, as noted above, was more successful, recovering more than twice as many carnivore coprolites.

Despite these limitations and biases, the small sample of coprolites does provide some insights into the taphonomy and paleoecology of Laetoli. First, it provides circumstantial evidence that carnivores were important actors on the landscape, in that they had a potentially significant role in influencing the mammalian bone assemblage (Su and Harrison 2008). The coprolites appear to be derived from a diversity of taxa, including felids, canids and hyaenids. In addition to coprolites, we know that a small percentage of the bones and teeth in the Laetoli collection show clear signs of having been digested by the gastric juices of carnivores. For example, 6.7% of the teeth of fossil cercopithecids from Laetoli show evidence of digestion (Harrison 2011). These specimens were presumably regurgitated by carnivores or were left on the land surface after their scats had disintegrated. Similarly, carnivore bite marks occur on 6.1% of large mammal bones (Su 2005; Su and Harrison 2008). The coprolites, when considered in conjunction with digested teeth and bones, and carnivore bite marks, indicate that carnivores had an important impact on the composition of the fossil assemblage as predators and scavengers.

Most of the coprolites (71.4%) do contain organic inclusions in the form of broken and digested bone, but only two specimens contain bone fragments that are taxonomically identifiable. EP 1246/03 and EP 1077/04 contain bones of *Madoqua avifluminis* and *Serentilagus praecapensis* - the two most common mammalian taxa at Laetoli. There is no evidence of hair, seeds or invertebrates, which commonly occur in the scats of small to medium-sized modern African carnivores (Ray 1997; Ray and Sunquist 2001; Kaunda and Skinner 2003; Warren et al. 2009). EP 4181/00 provides evidence that the territorial marking behavior common among extant canids, in which scats are deposited in conspicuous and raised locations, occurred in East Africa during the Pliocene.

Ruminant Coprolites

A collection of coprolites ($n=49$), from a single fossiliferous horizon at Loc. 3, can be attributed to ruminants (see Harrison and Kweka 2011). Harris (1987) made brief reference to coprolites of *Giraffa stillei* in the earlier collections made by Mary Leakey. The Leakey catalogue lists a giraffid coprolite from the Upper Laetolil Beds at Loc. 2 (LAET 74–31), but no detailed stratigraphic information is provided, and the actual specimen could not be relocated in the conserved collections.

Herbivore coprolites are extremely rare in the African fossil record, and their occurrence at Laetoli is a result of the unusual preservational conditions at Loc. 3. The coprolites were recovered from a calcareous clay horizon, located stratigraphically 60–70 cm above Tuff 7. The tuffaceous sediments, derived from the surrounding volcanic ashes, were accumulated in a shallow pond, along with a rich assemblage of organic debris, including bovid dung, insects, twigs and seeds (see Harrison and Kweka 2011). The debris may have been washed into the pond from the surrounding land surface by seasonal run-off, but more likely (given the preservation of the dung) the debris accumulated in a shallow depression that was subsequently flooded and inundated with sediment. It may have been a small waterhole that flooded seasonally. After compaction and consolidation of the deposits, the organic materials decomposed, leaving behind a cavity or void in the sediment. Because the sediments were very fine-grained, the void produced a detailed mold of the external structure of the debris. This was subsequently filled with calcite in solution derived from the calcium carbonate rich carbonatite tuffs of the Laetolil Beds. This produced a natural cast of the organic debris, resulting in remarkably well-preserved fossils. The specimens were recovered as surface finds, having eroded out of the sediments. The fossil insects and plant remains from this locality are described elsewhere in this volume (Bamford 2011; Krell and Schawaller 2011; Kitching and Sadler 2011). The ruminant coprolites are illustrated and briefly described here. The material consists of the following specimens: EP 1669/00, 6 coprolites; EP 2779/00, 29 coprolites; EP 234/01, 3 coprolites; EP 347/03, 4 coprolites; EP 665/04, 3 coprolites; and EP 738/05, 4 coprolites.

Description of Specimens

All of the ruminant coprolites from Loc. 3 consist of pure calcium carbonate that preserve the external morphology in fine detail, but no internal anatomy. Four basic types of coprolites can be recognized, all belonging to large to medium-sized ruminants.

Type A

The largest of the coprolites, comprising about one-third of the collection, are dome-shaped dung pellets with sloping convex sides, which occasionally have a large protuberance at their apex and a shallow depression on their expanded base (Fig. 14.5). The protuberance is produced by contraction of the anal sphincter at the time of expulsion, and the basal depression is caused by contact with the adjacent pellet. The pellets generally have a somewhat flattened base where they have been in contact with the land surface prior to burial; this is referred to here as the undersurface. The diameter of the pellet is greater than its apico-basal height. All of the specimens are deformed by compression, to variable degrees, and this affects the dimensions of the specimens. However, it is possible to estimate that the diameter of the coprolites ranges from ~17–23 mm, with an average diameter of 19.6 mm. Estimation of the height is more difficult because in most cases the pellets have been compressed in the apico-basal plane. Nevertheless, it is possible to estimate a height of ~14–16 mm based on the least distorted specimens.

The surface of the dung was originally smooth, but deformation and compression have typically produced concentric and radial cracking and tearing of the surface. Some cracks are filled with secondary calcite deposition. Several of the coprolites have deep concavities on the undersurface and asymmetrically compressed upper surfaces caused by contact with neighboring dung pellets at the time of compression. The degree of plastic deformation of the pellets indicates that they were relatively soft when buried, probably with high moisture content. Apart from cracking, the outer surface of the pellet is generally featureless. Some of the specimens have areas of the surface coated with an irregular layer of secondary calcite after the initial endocast was formed. This is probably due to slight expansion or accommodation of the void during lithification of the sediments. One specimen (EP 1669/00) has a small fragment of grass blade adhering to the outer surface, and another specimen (EP 2779/00) preserves a fine network of plant rootlets on its undersurface.

Fig. 14.5 Ruminant coprolites of Type A. EP 2779/00 (Loc. 3, Upper Laetolil Beds)

Fig. 14.6 Comparison of coprolite Type B. EP 347/03 (*left*), attributed to *Giraffa stillei*, with dung of extant giraffe, *Giraffa camelopardalis* (*center* and *right*)

These demonstrate that the dung was deposited on a vegetated land surface (and supports the scenario that the dung was autochthonous rather than transported into the pond). Cracks in the surface of the coprolite commonly expose the remains of coarsely triturated plant materials, including fragments of leaves and stems. One specimen (EP 1669/00) preserves a smooth-walled sub-spherical seed with an estimated diameter of less than 2 mm.

These coprolites are generally very similar in shape to the dung of modern *Giraffa camelopardalis*. They fall in the lower end of the size range of the modern species in terms of maximal diameter, but are relatively shorter. They are also similar in diameter to the dung pellets of the eland (*Taurotragus oryx*), the largest pellets produced by extant African bovids, but they differ in shape and proportions. The coprolites can be assigned with confidence to *Giraffa*, and based on their overall size they are most likely attributable to *G. stillei*, the smallest and most common species of giraffid at Laetoli (Robinson 2011). The dung of modern giraffes is known to contain intact and viable seeds of a variety of forbs, shrubs and trees (Milton and Dean 2001).

Type B

An acorn-shaped coprolite that is similar in size and structural characteristics to Type A, but differs primarily in having a height (22.9 mm) that is much greater than its maximum diameter (17.1 mm). Only a single specimen is known (EP 347/03); a well-preserved and relatively undistorted coprolite (Fig. 14.6). It has a high-domed shape with steep, slightly convex sides, a robust protuberance at the apex, and a deep circular pit (with a depth of 2.1 mm) on the base. The external surface is smooth. The estimated diameter of the specimen falls at the lower end of the size range for Type A coprolites. The preservation of this specimen indicates that the original pellet had less moisture content than the dung of Type A coprolites. It is likely that the preservational and structural differences between Type A and Type B coprolites relate to seasonal differences, with the latter being deposited during the dry season, rather than due to a taxonomic difference (Edwards 1991; Stuart and Stuart 1998; Hibert et al. 2008). Sexual dimorphism may be an additional contributing factor, since modern-day giraffe bulls produce relatively longer fecal pellets than do cows (Walker 1996; Harrison, personal observation). This specimen is attributed to *Giraffa stillei*.

Type C

This type of coprolite is the most common, comprising 45% of the collected sample (Fig. 14.7). They are long cylindrical pellets in which the length greatly exceeds the maximum diameter. The average diameter and length of the pellets is 11.0 mm (range, 9.3–12.8) and 17.9 mm (range, 13.1–26.0) respectively, with the greatest variation seen in the length. Occasionally, the pellets have shallow concavities at one or both poles. The mean breadth-length index of the pellets is 63.7 ± 7.6 ($n = 22$). Some pellets are strongly compressed, but generally they are much less deformed than those of Type A coprolites, implying less moisture content in the original pellets (Fig. 14.8). The long, cylindrical shape of the pellets matches those deposited by some extant African bovids during the dry season, rather than the smaller, more squat and misshapen pellets typical of the rainy season (Komers and Brotherton 1997) or by those antelopes that have relatively long large intestines, and produce drier fecal pellets year-round to conserve water (Woodall and Skinner 1993) (Fig. 14.9).

Several pellets have deep concavities on their undersurface produced by compression against other pellets of similar size and form (see Fig. 14.10). The external surface is generally smooth, but in lightly weathered specimens a rough

Fig. 14.7 Ruminant coprolites of Type C. EP 2779/00 (Loc. 3, Upper Laetolil Beds), attributed to a large species of bovid

Fig. 14.8 Comparison of ruminant coprolites Type C (*top row*) and Type A (*bottom row*). EP 1669/00 (Loc. 3, Upper Laetolil Beds)

Fig. 14.9 Latrine of male impala (*Aepyceros melampus*) showing that the individual dung pellets are comparable in form and overall size to ruminant coprolites Type C and Type D (see Fig. 14.7 for comparison). The average length of the pellets is 13.0 mm (range, 11.0–15.1 mm)

Fig. 14.10 Bovid coprolites (EP 1669/00) from Loc. 3 (Upper Laetolil Beds) showing the types of compression and distortion of the dung pellets. (**a**) Two views of the same specimen showing the compression and cracking of the pellet. The large cracks in the specimen expose triturated plant material and small ovoid seeds, (**b**) The undersurface of the pellet shown in Fig. 14.11, illustrating how it has been compressed against another pellet to leave an elongated elliptical depression

fibrous texture is exposed in which the fibers are oriented primarily in the direction of the long-axis of the pellet. These fibers are the remains of finely triturated plant material. Cracks in the coprolite, produced during burial and compression, expose fine particles of vegetation. Occasionally, coarser woody fragments are also observed (Fig. 14.10). Four of the coprolites (EP 1669/00, $n=2$; EP 2779/00, $n=2$) contain a single small seed, all of the same type (Fig. 14.11). It has a flattened discoidal shape with an estimated diameter of 2.8–3.8 mm. It does not match any of the seeds so far recorded from Laetoli (Bamford 2011), so it remains taxonomically unidentified.

Fig. 14.11 Bovid coprolite (EP 1669/00) from Loc. 3 (Upper Laetolil Beds) containing an indeterminate seed

Since fecal pellet weight in African antelopes is correlated with body mass (Coe and Carr 1983; Hashim and Dafalla 1996), it is possible to infer that the coprolites belonged to large- to medium-sized bovids. In terms of their overall size they compare most favorably with the dung pellets of modern bovids ranging in size from lesser kudu (*Tragelaphus strepsiceros*) up to the size of wildebeest (*Connochaetes taurinus*) and gemsbok (*Oryx gazella*), with an average body mass in males ranging from ~80–230 kg (Kingdon 1997). The size and proportions of the sample of coprolites matches well the variation seen in the dung pellets of modern hartebeest (*Alcelaphus buselaphus*) and wildebeest (Stuart and Stuart 1998; Walker 1996; Hibert et al. 2008), although it is possible that the dung pellets of several species of bovids are represented. Of the fossil bovids represented at Laetoli, *Hippotragus* sp., *Oryx deturi*, *Parmularius pandatus* and Alcelaphini larger species (Gentry 2011) are of the appropriate size.

Type D

This is a stout cylindrical coprolite with a rounded apex and a slightly depressed base. It is known from a single specimen (EP 738/05). It differs from Type C in its smaller overall size and in being relatively shorter. The length and diameter are 12.2 and 9.8 mm respectively. As in Type C, the surface exposes a rough fibrous content, with fibers mostly aligned in the direction of the long-axis of the pellet.

This specimen possibly represents a dung pellet of a medium-sized bovid, comparable in size to Grant's gazelle (*Gazella granti*), impala (*Aepyceros melampus*) and lesser kudu (*Tragelaphus strepsiceros*), with a mean body mass in males of ~50–80 kg (Kingdon 1997) (see Fig. 14.8). Several species at Laetoli probably fall into this size range, including *Tragelaphus* sp., *Gazella granti, Aepyceros dietrichi*, and Alcelaphini small sp. (Gentry 2011).

Other Bovid Coprolites

A few coprolites (EP 1669/00, $n=2$; EP 2779/00, $n=5$) represent bovid dung pellets that cannot be readily assigned to any of the types recognized above. They are elliptical, subspherical or irregular in form, and vary considerably in size (with maximum dimension ranging from 12.0 to 22.6 mm).

Discussion

The coprolites from Loc. 3 provide some interesting insights into the taphonomy and paleobiology at Laetoli. The preservation of ruminant dung pellets as coprolites is an unusual occurrence, not previously reported from Plio-Pleistocene sites in Africa, and it serves to highlight the unique preservational circumstances at Laetoli. In most sedimentary settings, ruminant dung would decompose rapidly without trace (unlike carnivore dung, which is usually rich in calcium phosphate and contain a large inorganic component) or be destroyed by termites, but the combination of fine grained and carbonate-rich sediments at Loc. 3 provided ideal conditions for the preservation of ruminant coprolites.

The dung can be attributed to at least three species of medium- to large-sized ruminants. Type A and Type B coprolites most likely represent the dung of *Giraffa stillei*, probably produced during different seasons of the year. The soft and deformable quality of the pellets that produced Type A coprolites suggests that the dung had relatively high moisture content, consistent with their production during or soon after the rainy season. This model also fits with the sedimentary data in which calcareous clays and organic debris were washed into the pond during periods of seasonal run-off. The presence of seeds in several coprolites would also suggest that at least some of the dung was deposited late in the rainy season, since this is when seed and fruit production is synchronize with increased rainfall (Mduma et al. 2007). Type C coprolites, which are the most common at Loc. 3, represent those of medium- to large-sized bovids. The single coprolite assigned to Type D may represent the dung of a different species of bovid, somewhat smaller than those that produced the Type C coprolites.

It is likely that the coprolites belong to some of the commonly represented species in the Laetoli ruminant fauna. For example, *Giraffa stillei* is the most common species of giraffid at Laetoli, comprising about 8% of the ruminant fauna (Su and Harrison 2008). Similarly, it is very likely that the most abundant dung pellets at Loc. 3, those assigned to Type C, belong to the common medium- to large-sized bovids from Laetoli, *Parmularius pandatus*, *Hippotragus* sp. and Alcelaphini indet. (28% of the ruminant fauna combined). It is interesting that there are no examples of coprolites that would be a suitable match for the dung pellets of the most common species at Laetoli, *Madoqua avifluminis* (33% of the ruminant fauna) and *Gazella janenschi* (10%), even though these species are both well-represented by craniodental remains at Loc. 3. This may be the result of a combination of taphonomic factors, including size (i.e., both species are relatively small, with average body weights less than 30 kg), deposition of dung in localized latrines, and the ability to survive without drinking (thereby avoiding ponds and water holes, such as that inferred for Loc. 3).

Finally, the coprolites provide some additional information on the dietary behavior of the bovids at Laetoli. The presence of seeds and coarse particles of herbaceous and woody plant material in at least Type A and Type C coprolites confirms evidence obtained from stable isotope and mesowear data that the Laetoli giraffids and most of the bovids were mixed feeders (Kingston and Harrison 2007; Kaiser 2011) that exploited fruits and leaves of C_3 forbs, shrubs and trees. The occurrence of intact seeds in the dung of modern ruminants is common, especially in browsers, such as *Giraffa* and *Tragelaphus* (Milton and Dean 2001). The higher incidence of seeds in Type C coprolites compared with those of Type A might imply differences in dietary behavior between the two taxa represented.

Conclusions

The recent paleontological collections at Laetoli and Kakesio have yielded a number of coprolites that provide insights into the taphonomy and paleoecology. The material includes coprolites of medium- to large-sized carnivores, and a rare collection of ruminant coprolites. The sample of carnivore coprolites is small, but they appear to belong to a diverse range of taxa, including canids, felids and hyaenids. The low incidence and unrepresentative size distribution of carnivore coprolites from Laetoli is probably a function of taphonomic influences and collecting bias. Carnivore coprolites, in conjunction with evidence of damage to individual teeth and bones, and the skewed composition of the skeletal assemblage (Su and Harrison 2008), confirm that carnivores were important actors on the landscape, and probably played a significant role in the accumulation and composition of the fossil remains at Laetoli. Two coprolites contain bone fragments of *Madoqua avifluminis* and *Serentilagus praecapensis* respectively, the two most common taxa at Laetoli. Another coprolite, deposited on a weathered bone fragment, provides evidence of territorial marking, a common behavior seen today among extant canids.

Ruminant coprolites are extremely rare in the African fossil record, and their occurrence at Laetoli is a result of the unusual preservational conditions. The dung can be attributed to at least three species of medium- to large-sized ruminants. The largest of coprolites can be referred to the small giraffid, *Giraffa stillei* (Robinson 2011). The most abundant ruminant coprolites belong to medium- to large-sized bovids, of which *Parmularius pandatus*, *Hippotragus* sp. and Alcelaphini indet. are of the appropriate size. A single coprolite of a smaller species of bovid likely belongs to *Tragelaphus* sp., *Gazella granti, Aepyceros dietrichi*, or Alcelaphini small sp.

In most cases, the consistency of the ruminant dung indicates high moisture content, which is consistent with deposition during or soon after the rainy season. This inference is supported by the occurrence of seeds in several of the coprolites, as well as the sedimentological evidence that indicates that deposition occurred in a flooded pond. The presence of seeds and coarse particles of herbaceous and woody plant material in several coprolites supports stable isotope and mesowear studies that the giraffids and many of the bovids at Laetoli were mixed feeders.

Acknowledgements The author is grateful to the Tanzania Commission for Science and Technology and the Unit of Antiquities in Dar es Salaam for permission to conduct research in Tanzania. Special thanks to Paul Msemwa (Director) and Amandus Kweka, as well as to all of the staff at the National Museum of Tanzania in Dar es Salaam, for their support and assistance. The Government of Kenya and the National Museums of Kenya are thanked for permission to study the collections in Nairobi. Thanks go to Emma Mbua, Mary Muungu, Meave Leakey (Kenya National Museum) for access to specimens in their care. For their advice, discussion, and help I gratefully acknowledge the following individuals: P. Andrews, P. Ditchfield, A. Kweka, Y. Fernandez-Jalvo, C.P. Msuya, and D. Su. Research at Laetoli was supported by grants from the National Geographic Society, the Leakey Foundation, and NSF (Grants BCS-9903434 and BCS-0309513).

References

Andrews, P. (1990). *Owls, caves and fossils*. Chicago: University of Chicago Press.

Andrews, P., & Nesbit Evans, E. M. (1983). Small mammal bone accumulations produced by mammalian carnivores. *Paleobiology, 9*, 289–307.

Bamford, M. (2011). Fossil leaves, fruits and seeds. In T. Harrison (Ed.), *Paleontology and geology of Laetoli: Human evolution in context* (Geology, geochronology, paleoecology and paleoenvironment, vol. 1, pp. 235–252). Dordrecht: Springer.

Breuer, T. (2005). Diet choice of large carnivores in northern Cameroon. *African Journal of Ecology, 43*, 97–106.

Coe, M. J., & Carr, R. D. (1983). The relationship between large ungulate body weight and faecal pellet weight. *African Journal of Ecology, 21*, 165–174.

Dietrich, W. O. (1951). Koprolithen aus dem Pleistocän Ostafrikas. *Neues Jahrbuch für Geologie und Paläontologie. Monatschefte, 1951*, 310–314.

Edwards, P. B. (1991). Seasonal variation in the dung of African grazing mammals, and its consequences for coprophagous insects. *Functional Ecology, 5*, 617–628.

Farrell, L. E., Roman, J., & Sunquist, M. E. (2000). Dietary separation of symnpatric carnivores identified by molecular analysis of scats. *Molecular Ecology, 9*, 1583–1590.

Gentry, A. W. (2011). Bovidae. In T. Harrison (Ed.), *Paleontology and geology of Laetoli: Human evolution in context* (Fossil hominins and the associated fauna, vol. 2, pp. 363–465). Dordrecht: Springer.

Harris, J. M. (1987). Summary. In M. D. Leakey & J. M. Harris (Eds.), *Laetoli: A Pliocene site in northern Tanzania* (pp. 524–531). Oxford: Clarendon.

Harrison, T. (2011). Cercopithecids (Cercopithecidae, primates). In T. Harrison (Ed.), *Paleontology and geology of Laetoli: Human evolution in context* (Fossil hominins and the associated fauna, vol. 2, pp. 83–139). Dordrecht: Springer.

Harrison, T., & Harris, E. E. (1996). Plio-Pleistocene cercopithecids from Kanam East, western Kenya. *Journal of Human Evolution, 30*, 539–561.

Harrison, T., & Kweka, A. (2011). Paleontological localities. In T. Harrison (Ed.), *Paleontology and geology of Laetoli: Human evolution in context* (Geology, geochronology, paleoecology and paleoenvironment, vol. 1, pp. 17–45). Dordrecht: Springer.

Hashim, I. M., & Dafalla, A. A. (1996). Faecal pellets and biomass of some wild herbivores in Dinder National Park, Sudan. *African Journal of Ecology, 34*, 66–69.

Hibert, F., Fritz, H., Poilecot, P., Abdou, H. N., & Dulieu, D. (2008). Morphological criteria to identify faecal pellets of sympatric ungulates in West African savanna and estimates of associated error. *African Journal of Ecology, 46*, 523–532.

Horwitz, L. K., & Goldberg, P. (1989). A study of Pleistocene and Holocene hyaena coprolites. *Journal of Archaeological Science, 16*, 71–94.

Kaiser, T. M. (2011). Feeding ecology and niche partitioning of the Laetoli ungulate faunas. In T. Harrison (Ed.), *Paleontology and geology of Laetoli: Human evolution in context* (Geology, geochronology, paleoecology and paleoenvironment, vol. 1, pp. 329–354). Dordrecht: Springer.

Kaunda, S. K. K., & Skinner, J. D. (2003). Black-backed jackal diet at Mokolodi Nature Reserve, Botswana. *African Journal of Ecology, 41*, 39–46.

Kingdon, J. (1997). *The Kingdon field guide to African Mammals*. San Diego: Academic.

Kingston, J. D., & Harrison, T. (2007). Isotopic dietary reconstructions of Pliocene herbivores at Laetoli: Implications for hominin paleoecology. *Palaeogeography, Palaeoclimatology, Palaeoecology, 243*, 272–306.

Kitching, I. J., & Sadler, S. (2011). Lepidoptera, Insecta. In T. Harrison (Ed.), *Paleontology and geology of Laetoli: Human evolution in context* (Fossil hominins and the associated fauna, vol. 2, pp. 549–554). Dordrecht: Springer.

Komers, P. E., & Brotherton, P. N. (1997). Dung pellet used to identify the distribution and density of dik-dik. *African Journal of Ecology, 35*, 124–132.

Krell, F.-T., & Schawaller, W. (2011). Beetles (Insecta, Coleoptera). In T. Harrison (Ed.), *Paleontology and geology of Laetoli: Human evolution in context* (Fossil hominins and the associated fauna, vol. 2, pp. 535–548). Dordrecht: Springer.

Leakey, L. N., Milledge, S. A. H., Leakey, S. M., Edung, J., Haynes, P., Kiptoo, D. K., & McGeorge, A. (1999). Diet of striped hyaena in northern Kenya. *African Journal of Ecology, 37*, 314–326.

Matthews, T. (2002). South African micromammals and predators: Some comparative results. *Archaeometry, 44*, 363–370.

Mduma, S. A. R., Sinclair, A. R. E., & Turkington, R. (2007). The role of rainfall and predators in determining synchrony in reproduction of savanna trees in Serengeti National Park, Tanzania. *Journal of Ecology, 95*, 184–196.

Milton, S. J., & Dean, W. R. J. (2001). Seeds dispersed in dung of insectivores and herbivores in semi-arid southern Africa. *Journal of Arid Environments, 47*, 465–483.

Ray, J. C. (1997). Comparative ecology of two African forest mongooses, *Herpestes naso* and *Atilax paludinosus*. *African Journal of Ecology, 35*, 237–253.

Ray, J. C., & Sunquist, M. E. (2001). Trophic relations in a community of African rainforest carnivores. *Oecologia, 127*, 395–408.

Robinson, C. A. (2011). Giraffidae. In T. Harrison (Ed.), *Paleontology and geology of Laetoli: Human evolution in context* (Fossil hominins and the associated fauna, vol. 2, pp. 339–362). Dordrecht: Springer.

Stuart, C., & Stuart, T. (1998). *A field guide to the tracks and signs of Southern and East African Wildlife* (2nd ed.). Cape Town: Southern Book.

Su, D. F. (2005). The paleoecology of Laetoli, Tanzania: Evidence from the mammalian fauna of the Upper Laetolil Beds. Ph.D. dissertation, New York University, New York.

Su, D. F., & Harrison, T. (2008). Ecological implications of the relative rarity of fossil hominins at Laetoli. *Journal of Human Evolution, 55*, 672–681.

Walker, C. (1996). *Signs of the wild* (5th ed.). Cape Town: Struik.

Warren, Y., Cunningham, P., Mbangu, A., & Tutjavi, V. (2009). Preliminary observations of the diet of the black mongoose (*Galerella nigrata,* Thomas, 1928) in the Erongo Mountains, Namibia. *African Journal of Ecology, 47*, 801–803.

Werdelin, L., & Dehghani, R. (2011). Carnivora. In T. Harrison (Ed.), *Paleontology and geology of Laetoli: Human evolution in context* (Fossil hominins and the associated fauna, vol. 2, pp. 189–232). Dordrecht: Springer.

Woodall, P. F., & Skinner, J. D. (1993). Dimensions of the intestine, diet and faecal water loss in some African antelope. *Journal of Zoology, London, 229*, 457–471.

Chapter 15
Stable Isotopic Analyses of Laetoli Fossil Herbivores

John D. Kingston

Abstract In order to further refine early hominin paleoecology at Laetoli, over 500 specimens of fossil enamel and ostrich eggshell fragments collected from the Laetolil Beds and the Upper Ndolanya Beds were analyzed isotopically. The goal was to develop a high-resolution spatio-temporal framework for identifying and characterizing foraging patterns of mammalian herbivore lineages and fossil ostriches that could be used to investigate aspects of plant physiognomy and climate through the Laetoli succession. In general, dietary patterns at Laetoli suggest heterogeneous ecosystems with both C_3 and C_4 dietary plants available that could support grassland, woodland, and forested communities. All large-bodied mammalian herbivores analyzed yielded dietary signatures indicating mixed grazing/browsing strategies or exclusive reliance on C_3 browse, more consistent with wooded than grassland-savanna biomes. Although there were no obvious uniform dietary shifts within specific mammalian herbivore groups in the sequence, the transition from the Upper Laetolil Beds to the Upper Ndolanya Beds documents a significant increase in the representation of grazing bovids. Relative to extant taxa in related lineages, the isotopic ranges of a number of Laetoli fossil herbivores are anomalous, indicating significantly more generalized intermediate C_3/C_4 feeding behaviors, perhaps indicative of dietary niches and habitat types with no close modern analogs. Diets of ostriches as reflected in the isotopic composition of eggshell components indicate predominantly C_3 diets but with discrete isotopic shifts within the sequence linked to taxonomic and possibly environmental change.

Keywords Isotopes • Enamel • Ostrich eggshells • Paleodiet • Paleoecology • Pliocene • Tanzania • Hominin

J.D. Kingston (✉)
Department of Anthropology, Emory University, 1557 Dickey Drive, Atlanta, GA 30322, USA
e-mail: john.kingston@emory.edu

Introduction

Fossil material and stratigraphic interpretations from the site of Laetoli, Tanzania have provided critical insight into the evolution of East African hominin ecosystems during the Pliocene and Pleistocene. In particular, paleoecological reconstructions at the site have been interpreted as indicative of grassland, savanna or open woodland landscapes (Leakey and Harris 1987). These inferences effectively extend the potential habitat range of early hominins to include open grassland/woodland biomes, providing support for adaptive theories linking early hominin evolution with selective pressures encountered in more open, non-forested habitats. However, inferring specific or even generalized aspects of landscapes that dominated the Laetoli area locally during deposition of the Laetolil Beds and the Upper Ndolanya Beds (UNB) has been complicated. These difficulties relate to the recovery of variable environmental proxies from the site that indicate contrasting biomes ranging from more open grassland ecosystems (Leakey 1987; Hay 1987; Denys 1987; Bonnefille and Riollet 1987; Verdcourt 1987) to woodland/bushland habitats (Andrews 1989; Reed 1997; Kingston and Harrison 2007; Andrews and Bamford 2008; Kovarovic and Andrews 2007; Su and Harrison 2007, 2008) to more dense woodland or forests habitats (Walker 1987).

The collective perspective of these studies, as well as that of specific paleoecological studies (Kingston and Harrison 2007; Kovarovic and Andrews 2011; Bamford 2011; Rossouw and Scott 2011), is that of ecological heterogeneity at Laetoli in the past. How this heterogeneity is partitioned temporally or spatially remains unclear. In other words, did the Laetoli area, astride the flanks of the Satiman volcanic edifice, support a variety of contemporaneous habitats differentiated along altitudinal gradients, localized microclimates, and/or variable bedrock/drainage/soils during the Pliocene (i.e., spatial heterogeneity)? Tropical habitats, especially those associated with the topography that characterizes

T. Harrison (ed.), *Paleontology and Geology of Laetoli: Human Evolution in Context. Volume 1: Geology, Geochronology, Paleoecology and Paleoenvironment*, Vertebrate Paleobiology and Paleoanthropology, DOI 10.1007/978-90-481-9956-3_15,

rift systems are complex and may vary greatly within limited geographic areas (Andrews 2006; Andrews et al. 2011). Alternatively, ecosystems may have been more uniform at any given time and instead varied through time in response to orbitally forced climate change and other short-term environmental oscillations occurring during and between deposition of fossiliferous sediments at Laetoli (i.e., temporal heterogeneity). Intermittent volcanic activity that resulted in the deposition of fossil-bearing volcanoclastic sediments at Laetoli would also have disrupted and altered the local ecology periodically, potentially contributing to temporal hetoerogeneity. Processes of accumulation and preservation of fossil sites may have also altered the paleoecological signal, mixing fauna from different habitat types and times or biasing the taxonomic composition through differential selection, artificially mimicking heterogeneity (Behrensmeyer et al. 1992). Although resolution of the fossil record at Laetoli currently remains too coarse to unequivocally establish the relative contribution of the spatial/temporal dimensions to heterogeneity, several chronostratigraphically constrained records support the notion of spatial heterogeneity. For example, fossil wood recovered from a lahar within the Lower Laetolil Beds (LLB), preserving a relatively instantaneous paleovegetational profile, indicates variable and structurally complex contemporaneous ecosystems including forests, evergreen woodlands, dry woodlands, bushlands, riverine and wooded grasslands along the volcanic slopes at Laetoli (Bamford 2011). For less refined intervals, fossiliferous horizons in the Laetoli succession between Tuffs 3 and 5 in the Upper Laetolil Beds (ULB), spanning about 10 kyrs (Deino 2011) have yielded diverse fauna and paleoecologic data, also suggesting mosaic habitats (Kingston and Harrison 2007).

To provide additional insights into the paleoecology of Laetoli during deposition of the Laetolil Beds and the Upper Ndolanya Beds, fossil tooth enamel from multiple taxa were analyzed isotopically. This approach yields direct data on the foraging behavior and dietary guilds of mammalian lineages as well as climatic variables, thereby providing indirect evidence of the floral communities and physiognomy. The stable carbon isotopic composition of herbivore enamel ($\delta^{13}C_{enamel}$) provides a means of: (1) differentiating relative contributions of C_4 and C_3 vegetation to the diet; (2) assessing water stress in dietary plants; (3) examining trophic level effects and prey diet; and (4) potential canopy effects. In general, $\delta^{13}C$ values are utilized to indicate the extent of grazing (C_4 diet), browsing (C_3 diet) or mixed feeding (C_3 and C_4 diet), with additional variation linked to climatic parameters and local conditions. Oxygen isotopic signatures of the tooth enamel ($\delta^{18}O_{enamel}$) primarily reflects aspects of climate, such as evapotranspiration, aridity, and evaporation/precipitation regimes, which are mediated by factors such as water source, altitude and animal behavior and physiology (Kohn et al. 1996). As such, interpretations of the $\delta^{18}O_{enamel}$ signal of fossil assemblage have been limited primarily to comparisons of taxa within assemblages or documenting seasonality. Recently, Levin et al. (2006) have proposed using the oxygen isotopic composition of tooth enamel as a proxy for aridity and this approach is applied here to the Laetoli fossil assemblage.

Given the number of interdependent variables ultimately controlling the isotopic composition of enamel, it is essential to adopt a broad perspective in interpreting these data. This typically requires isotopic analyses of all or most fossil herbivore dietary guilds in the assemblage, comparing the dietary signatures with other relevant fossil and modern isotopic datasets, and developing interpretations in conjunction with other paleodietary indicators such as mesowear studies or analyses of cranio-dental morphology (Kaiser 2011; Hernesniemi et al. 2011; Fortelius and Solounias 2000). This holistic perspective is especially critical in attempting to translate specific dietary signals into general paleoecological indicators. While the notion of utilizing dietary analyses of a limited number of fossil herbivore taxa (and specimens) to reconstruct local plant physiognomy is enticing, emerging fossil studies indicate that foraging strategies of extinct taxa can be unlike modern descendant forms and are highly variable, complicating any direct application of modern analogs (i.e., Zazzo et al. 2000; Cerling et al. 2003; Kingston and Harrison 2007). A number of inferences, related to selectivity of animals, competitive exclusion, migration/immigration, broad habitat tolerance, social behavior, and seasonality, need to be factored into interpretations. Currently, these parameters remain difficult to tease from the fossil record. Alternative paleoecological approaches or lines of evidence, including phytolith/pollen studies, associated fossil wood/leaves/seeds, paleosol analyses, ecomorphology, community structure, and indicator taxa, are necessary to calibrate and establish links between diet and habitat type.

While the carbon isotopic analyses provide a means of documenting the relative contribution of C_3 and C_4 photosynthetic vegetation to herbivore diets, the inference is that they also reflect the extent of grazing, browsing, or mixed feeding in extinct taxa. These correlations are based on the modern distribution of C_3 and C_4 vegetation and the foraging strategies of extant mammalian herbivores in equatorial, lower elevation (currently <2,600 m) African ecosystems. While it is reasonable to assume that these correlations extend to fossil communities, it is worth noting several potential confounding factors. First is the notion that a C_3 isotopic enamel signal reflects predominantly browsing behavior. In modern East African biomes, C_3 grass distribution tends to be restricted to more mesic habitats with cooler growing seasons - within forested or wooded ecosystems or at higher elevations. Grazing guilds in East African mammalian communities currently forage almost exclusively on

C_4 grasses that dominate grassland and more open woodland ecosystems. The expansion of C_4 grasses into African ecosystems over the last 8–5 Ma has been interpreted variably as an adaptive response to increased aridity and seasonality of precipitation, salinity, high temperatures, fires and/or low-CO_2 levels (Cerling et al. 1997; Collatz et al. 1998; Boom et al. 2002). The CO_2 concentrating biochemical and anatomical modifications of C_4 plants represent adaptations that compensate for high rates of photorespiration and carbon deficiency and appear to have evolved independently multiple times to varying conditions, including those listed above, but also potentially in wetlands (Sage 2004). C_3 or C_3/ C_4 intermediate grasses may have comprised significant proportions of grassland or woodland habitats in the past, and the isotopic signature of C_3 grazer would be in effect indistinguishable from that of a C_3 browser. Analysis of phytoliths in the Upper Laetolil Beds indicates that C_3 grasses dominated the grass assemblage below Tuff 7 (Rossouw and Scott 2011). In addition, almost half (2700/7500) of modern C_4 plants are sedges and dicots and it is possible that a C_4 dietary signal may reflect, in part, foraging on C_4 dicots rather than strictly grazing. In a mesowear analysis of fossil ungulates from Laetoli, Kaiser (2011) suggests that a lack of abrasion in taxa yielding C_4 isotopic dietary signatures indicate a diet that included C_4 dicots rather than C_4 grass. Isotopic and corroborative evidence from Laetoli relevant to these issues are included in the discussion sections below.

Fossiliferous sediments at Laetoli span a significant interval of time and a major goal of paleoenvironmental reconstructions is to document environmental change that may mediate or be involved in evolutionary change. The LLB, the ULB and the UNB have yielded fossil assemblages that range in age from 4.36 to 3.85 Ma, 3.85–3.63 Ma, and 2.66 Ma respectively (Deino 2011). A number of faunal changes and environmental change have been associated with transitions between these stratigraphic units, as well as within the ULB (Rossouw and Scott 2011; Hernesniemi et al. 2011; Kovarovic and Andrews 2011; Kaiser 2011; Su 2011). Sedimentation of the ULB spans an interval of about 200 kyrs and the numerous fossil localities within this part of the succession provide an opportunity to assess ecological change at a higher temporal resolution.

In a previous study (Kingston and Harrison 2007), multiple tooth enamel samples of 21 fossil herbivore species from the LLB, ULB and UNB were analyzed isotopically to constrain paleoecological reconstructions of Laetoli during the Pliocene. These data revealed complex and variable foraging patterns with limited representation of obligate grazers and a high diversity of mixed feeders (grazing and browsing) and browsers. While the results overall indicated foraging strategies more representative of closed to open woodlands, the intraspecific isotopic variability complicated any straightforward interpretations as well as attempts to contrast paleodietary patterns through the sequence, particularly comparisons between the ULB and the UNB. To provide further greater representation of the various taxa through the sequence into paleodietary patterns, the original data set of 167 teeth (Kingston and Harrison 2007) was essentially doubled by isotopic analyses of 172 additional mammalian enamel samples, including several primates and carnivores (Tables 15.1 and 15.2).

The new enamel data presented here are also accompanied by preliminary analyses of fossil and modern ostrich eggshells (OES) collected from Laetoli, intended to supplement paleoenvironmental reconstructions based on isotopic analyses of mammalian herbivores. Two components of eggshells were analyzed here, the inorganic or crystalline fraction (predominantly calcite) and the organic fraction (predominantly proteins). The isotopic composition of carbon in eggshell calcite matrix ($\delta^{13}C_{OES-carb}$) and in eggshell organic residue ($\delta^{13}C_{OES-org}$) reflects the isotopic ratios of the birds bulk diet, each offset by a systematic biochemical fractionation (Johnson 1995; Hobson 1995). The oxygen isotopic signal of eggshell carbonate ($\delta^{18}O_{OES}$) depends on that of an organism's metabolic water, which is derived primarily from the isotopic composition of ingested water. For obligate drinkers there is a roughly linear relationship between the $\delta^{18}O$ of drinking water available and $\delta^{18}O_{OES}$, both of which increase with temperature (Folinsbee et al. 1970). For animals that derive most of their water from plant material, like ostriches, the $\delta^{18}O_{OES}$ is a mediated by evapo-transpiration processes which tend to enrich ^{18}O. In general, as aridity increases, so does the $\delta^{18}O$ of plant-leaf water and ultimately the $\delta^{18}O_{OES}$. Finally, the nitrogen isotopic composition of the organic residue ($\delta^{15}N_{OES}$) can yield information about the diet and the ostrich's physiological response to specific environmental factors. Studies have linked $\delta^{15}N_{OES}$ to mean annual precipitation (MAP), providing a proxy to explore relative humidity in the past (Johnson et al. 1997, 1998).

Methods

Mammalian Tooth Enamel

Fossil enamel fragments were collected in the course of field prospecting during the 2000, 2001, and 2004 field seasons with an effort to provide taxonomic identification coupled with specific stratigraphic provenance within the sequence. As most of the fossil material was surface collected rather than excavated directly from the sediments, each tooth or tooth fragment is associated with a stratigraphic range based on the

Table 15.1 Taxa and number of specimens of fossil mammalian herbivore tooth enamel analyzed isotopically at Laetoli from the various geologic units. The table includes numbers from Kingston and Harrison (2007) and this study

Family and tribe for bovids (total number of samples analyzed isotopically)	Taxon	Geologic formation	Number of fossil enamel samples analyzed isotopically	
			Kingston and Harrison (2007)	This study
Bovidae (100)				
Antilopini (17)	*Gazella janenschi*	ULB	9	7
		UNB	1	–
Alcelaphini (43)	*Parmularius pandatus*	LLB	–	1
		ULB	8	17
	Alcelaphini sp. indet	LLB	–	2
		ULB	1	3
	Alcelaphini sp. (medium)	UNB	1	3
	Alcelaphini sp. (large)	ULB	–	4
		UNB	–	3
Hippotragini (18)	*Hippotragus* sp.	ULB	8	10
Neotragini (10)	*Madoqua avifluminis*	ULB	3	7
Tragelaphini (12)	*Tragelaphus sp.*	ULB	4	6
		UNB	2	–
Cercopithecidae (3)	*Parapapio ado*	ULB	–	2
	Cercopithecidae	ULB	–	1
Gomphotheriidae (2)	*Anancus kenyensis*	LLB	1	1
Elephantidae (29)	*Loxodonta exoptata*	LLB	–	1
		ULB	13	8
		UNB	4	3
Deinotheriidae (2)	*Deinotherium bozasi*	ULB	1	1
Suidae (48)	*Notochoerus euilus*	LLB	1	–
		ULB	21	15
	Potamochoerus porcus	ULB	–	5
	Kolpochoerus limnetes	UNB	1	5
Equidae (40)	*Eurygnathohippus* aff. *hasumense*	LLB	1	6
		ULB	9	9
	Eurygnathohippus aff. *cornelianus*	UNB	4	11
Rhinocerotidae (38)	*Ceratotherium efficax*	LLB	1	4
		ULB	8	22
	Ceratotherium sp.	UNB	–	1
	Diceros bicornis	ULB	–	2
Giraffidae (43)	*Giraffa jumae*	ULB	8	9
	Giraffa stillei	ULB	9	8
	Sivatherium maurusium	ULB	2	6
		UNB	1	–

relationship of the fossil find to specific site stratigraphy and exposure rather than a discrete horizon (Table 15.2). In almost all cases, molar enamel (especially third molars) was analyzed, limiting the potentially confounding effects of weaning or juvenile dietary signals present in teeth forming during this developmental stage. Although small amounts of enamel (1–15 mg) were ultimately analyzed for each tooth, longitudinal sections of the teeth perpendicular to growth direction (20–40 mg) were homogenized during preparation to minimize and average out intra-tooth isotopic variability. Isotopic analyses in this study focused on the carbonate in enamel bioapatite.

Enamel was prepared using procedures detailed in Kingston and Harrison (2007). In general, samples were mechanically separated carefully cleaned of adhering sediment, dentine, and weathering rinds with a high speed Z500 Brasseler drilling tool with tungsten-carbide bits, and then ground in an agate mortar. Powdered enamel was reacted for 24 h with ~2% NaOHCl in 2 ml plastic centrifuge tubes to digest any organic material and then rinsed to a pH of 7 by centrifugation with double distilled water. Residue was treated with 0.1 M CH_3COOH for 16 h under a weak vacuum to remove any secondary carbonate, rinsed to neutrality by centrifugation with double distilled water, and freeze dried. In the initial analysis

Table 15.2 Mammalian tooth enamel δ^{13}C and δ^{18}O of fossil and modern samples collected at Laetoli

Lab no.	Taxon	Formation	δ$^{13}C_{enamel}$	δ$^{18}O_{enamel}$	Locality	Stratigraphic level	Element	Field no.
	Bovidae							
	Antilopini							
14	*Gazella janenschi*	ULB	−4.0	0.6	Loc. 13	Between Tuffs 6 & 8	Upper M2	
20	*Gazella janenschi*	ULB	−8.0	−0.1	Loc. 10E	Between Tuffs 5 & 7	Upper M2	
34	*Gazella janenschi*	ULB	−5.5	0.3	Loc. 10E	Between Tuffs 6 & 7	Upper M2	
40	*Gazella janenschi*	ULB	−9.2	2.5	Loc. 10W	Below Tuff 3	Lower molar	
47	*Gazella janenschi*	ULB	−6.4	−1.3	Loc. 10	Below Tuff 3	Upper M3	
120	*Gazella janenschi*	ULB	−4.6	−2.0	Loc. 2	Between Tuffs 5 & 7	Upper molar	
121	*Gazella janenschi*	ULB	−9.8	−0.2	Loc. 2	Between Tuffs 3 & 5	Lower M3	
122	*Gazella janenschi*	ULB	−9.4	0.0	Loc. 8	between Tuffs 5 & 6	Lower M2	
123	*Gazella janenschi*	ULB	−1.9	4.5	Loc. 15	between Tuffs 6 & 8	Lpper molar	
76	*Gazella janenschi*	UNB	−7.8	−2.0	Loc. 18		Upper M2	
LT59B	*Gazella janenschi*	ULB	−9.5	0.2	Loc. 1	Between Tuffs 7 & 8		
LT88B	*Gazella janenschi*	ULB	−10.2	0.6	Loc. 5	Between Tuffs 3 & 5		
LT90B	*Gazella janenschi*	ULB	−10.5	0.9	Loc. 5	Between Tuffs 3 & 5		
LT118B	*Gazella janenschi*	ULB	−6.6	1.3	Loc. 13	Between Tuffs 6 & 8		
LT166B	*Gazella janenschi*	ULB	−8.0	−0.9	Loc. 9S	Below Tuff 2		905/98
LT168B	*Gazella janenschi*	ULB	−5.3	2.4	Loc. 9S	Below Tuff 2		917/98
LT196B	*Gazella janenschi*	ULB	−7.3	0.4	Loc. 10	Below Tuff 2		153/99
	Alcelaphini							
13	*Parmularius pandatus*	ULB	−8.6	−2.0	Loc. 1	Between Tuffs 6 & 8	Upper molar	
19	*Parmularius pandatus*	ULB	−3.1	−0.9	Loc. 10E	Between Tuffs 5 & 7	Upper M1	
33	*Parmularius pandatus*	ULB	−3.8	−0.8	Loc. 10E	Between Tuffs 6 & 7	Lower M3	
39	*Parmularius pandatus*	ULB	−2.2	−0.4	Loc. 10W	Below Tuff 3	Lower M2	
42	*Parmularius pandatus*	ULB	−2.6	0.3	Loc. 10W	Below Tuff 3	Upper M1	
43	*Parmularius pandatus*	ULB	−0.4	−1.6	Loc. 10W	Below Tuff 3	Lower M2	
46	*Parmularius pandatus*	ULB	−1.1	−0.2	Loc. 10	Below Tuff 3	Upper molar	
9	*Parmularius pandatus*	ULB	0.0	3.1	Loc. 2	Between Tuffs 5 & 7	Lower M3	
LT43B	*Parmularius pandatus*	ULB	−3.7	−0.7	Loc. 2	Between Tuffs 5 & 7		
LT45B	*Parmularius pandatus*	ULB	−1.1	−1.7	Loc. 2	Between Tuffs 5 & 7		
LT54B	*Parmularius pandatus*	ULB	−1.4	0.6	Loc. 6	Between Tuffs 5 & 7		
LT61B	*Parmularius pandatus*	ULB	−6.0	0.4	Loc. 7	Between Tuffs 5 & 7		
LT91B	*Parmularius pandatus*	ULB	−4.5	0.8	Loc. 5	Between Tuffs 3 & 5		
LT98B	*Parmularius pandatus*	ULB	−1.6	−3.9	Loc. 8	Between Tuffs 5 & 7		
LT112B	*Parmularius pandatus*	ULB	−0.4	−0.7	Loc. 10	Below Tuff 3		
LT135B	*Parmularius pandatus*	LLB	1.9	−2.2	Kakesio 3			026/98
LT145B	*Parmularius pandatus*	ULB	−2.0	1.6	Loc. 10	Below Tuff 3		416/98
LT167B	*Parmularius pandatus*	ULB	−3.0	1.8	Loc. 9S	Below Tuff 2		916/98
LT169B	*Parmularius pandatus*	ULB	−0.8	−0.9	Loc. 9S	Below Tuff 2		941/98
LT178B	*Parmularius pandatus*	ULB	−1.7	1.5	Loc. 22	Between Tuffs 5 & 7		1200/98
LT182B	*Parmularius pandatus*	ULB	0.4	0.1	Loc. 15	Between Tuffs 7 & 8		1412/98
LT184B	*Parmularius pandatus*	ULB	2.9	2.4	Loc. 9	Between Tuffs 5 & 7		1467/98
LT186B	*Parmularius pandatus*	ULB	−2.9	−0.7	Loc. 9	Between Tuffs 5 & 7		1473/98
LT187B	*Parmularius pandatus*	ULB	−2.6	−1.9	Loc. 10W	Below Tuff 2		1587/98
LT189B	*Parmularius pandatus*	ULB	−9.2	0.1	Loc. 9	Between Tuffs 5 & 7		1645/98
LT193B	*Parmularius pandatus*	ULB	−2.4	−0.2	Loc. 10W	Below Tuff 2		130/99
31	Alcelaphini sp. indet	ULB	−0.1	1.5	Loc. 10E	Between Tuffs 6 & 7	Upper molar	
74	Alcelaphini sp. (medium)	UNB	1.1	−0.4	Loc. 18		Lower M3	
73	Alcelaphini, sp. (large)	UNB	0.6	1.0	Loc. 18		Lower M2	
LT37B	Alcelaphini sp. (medium)	UNB	1.8	1.2	Loc. 7E			
LT39B	Alcelaphini sp. (medium)	UNB	1.0	−0.3	Loc. 7E			
LT76B	Alcelaphini sp. (medium)	UNB	−0.3	2.9	Loc. 14			
LT65B	Alcelaphini sp. (large)	UNB	−0.2	0.8	Loc. 22E			
LT66B	Alcelaphini sp. (large)	UNB	1.1	−0.2	Loc. 22E			
LT110B	Alcelaphini sp. (large)	ULB	−5.6	−0.2	Loc. 21	Between Tuffs 5 & 7		
LT119B	Alcelaphini sp. (large)	ULB	−2.1	0.2	Loc. 4	Between Tuffs 6 & 8		

(continued)

Table 15.2 (continued)

Lab no.	Taxon	Formation	$\delta^{13}C_{enamel}$	$\delta^{18}O_{enamel}$	Locality	Stratigraphic level	Element	Field no.
LT128B	Alcelaphini sp. (large)	UNB	–4.4	–2.4	Silal Artum			
LT139B	Alcelaphini sp. (large)	ULB	–0.6	–1.3	Loc. 10E	Between Tuffs 6 & 7		281/98
LT163B	Alcelaphini sp. (large)	ULB	–2.9	0.5	Loc. 9S	Below Tuff 2		903/98
LT52B	Alcelaphini sp. indet.	ULB	–1.9	–1.5	Loc. 6	Between Tuffs 5 & 7		
LT151B	Alcelaphini sp. indet.	ULB	–2.9	–1.2	Loc. 10W	Below Tuff 2		548/98
LT188B	Alcelaphini sp. indet.	ULB	–1.9	–0.1	Loc. 10W	Below Tuff 2		1595/98
LT190B	Alcelaphini sp. indet.	LLB	–2.4	3.7	Kakesio 6			003/99
LT192B	Alcelaphini sp. indet.	LLB	3.7	–0.9	Esere			090/99
113	*Connochaetes*	Modern	0.6	1.5	Laetoli area			
	Hippotragini							
12	*Hippotragus* sp.	ULB	–9.1	1.7	Loc. 1	Between Tuffs 6 & 8	Upper molar	
21	*Hippotragus* sp.	ULB	–5.0	0.7	Loc. 10E	Between Tuffs 5 & 7	Lower molar	
35	*Hippotragus* sp.	ULB	–4.0	2.3	Loc. 10E	Between Tuffs 6 & 7	Upper molar	
36	*Hippotragus* sp.	ULB	–7.5	2.8	Loc. 10E	Between Tuffs 6 & 7	Upper M1	
45	*Hippotragus* sp.	ULB	–2.3	2.7	Loc. 10W	Below Tuff 3	Upper M3	
141	*Hippotragus* sp.	ULB	–4.9	1.3	Loc. 21	Between Tuffs 5 & 8	Upper molar	
142	*Hippotragus* sp.	ULB	–7.7	1.5	Loc. 21	Between Tuffs 5 & 8	Upper molar	
143	*Hippotragus* sp.	ULB	–6.2	1.9	Loc. 8	Between Tuffs 5 & 8	Upper molar	
LT51B	*Hippotragus* sp.	ULB	–7.0	–0.5	Loc. 6	Between Tuffs 5 & 7		
LT99B	*Hippotragus* sp.	ULB	–6.7	1.0	Loc. 21	Between Tuffs 5 & 7		
LT100B	*Hippotragus* sp.	ULB	–9.8	–0.6	Loc. 21	Between Tuffs 5 & 7		
LT103B	*Hippotragus* sp.	ULB	–5.5	0.9	Loc. 21	Between Tuffs 5 & 7		
LT109B	*Hippotragus* sp.	ULB	–2.3	0.2	Loc. 21	Between Tuffs 5 & 7		
LT115B	*Hippotragus* sp.	ULB	0.8	–2.2	Loc. 13	Between Tuffs 6 & 8		
LT162B	*Hippotragus* sp.	ULB	–7.0	0.2	Loc. 10W	Below Tuff 2		678/98
LT175B	*Hippotragus* sp.	ULB	–9.2	1.7	Loc. 9	Between Tuffs 6 & 7		1165/98
LT180B	*Hippotragus* sp.	ULB	–4.3	–1.1	Loc. 13	Between Tuffs 5 & 7		1344/98
LT181B	*Hippotragus* sp.	ULB	–2.8	–2.1	Loc. 13	Between Tuffs 5 & 7		1345/98
	Neotragini							
38	*Madoqua avifluminis*	ULB	–6.4	–2.7	Loc. 10	Below Tuff 3	Upper molar	
41	*Madoqua avifluminis*	ULB	–10.6	1.6	Loc. 10W	Below Tuff 3	Upper M3	
48	*Madoqua avifluminis*	ULB	–7.3	–1.8	Loc. 10E	Between Tuffs 5 & 7	Mand. M1?	
Md64	*Madoqua kirkii*	Modern	–11.1	–2.2	Laetoli Area			
LT32B	*Madoqua avifluminis*	ULB	–9.6	–1.9	Loc. 10	Below Tuff 3		709/04
LT156B	*Madoqua avifluminis*	ULB	–8.6	0.0	Loc. 10W	Below Tuff 2		553/98
LT179B	*Madoqua avifluminis*	ULB	–6.7	–0.9	Loc. 21	Between Tuffs 5 & 7		
LT183B	*Madoqua avifluminis*	ULB	–10.4	–0.8	Loc. 9	Between Tuffs 5 & 7		1464/98
LT191B	*Madoqua avifluminis*	ULB	–8.6	–0.4	Loc. 10E	Between Tuffs 5 & 7		043/99
LT195B	*Madoqua avifluminis*	ULB	–7.7	–1.4	Loc. 10W	Below Tuff 2		138/99
LT187B	*Madoqua avifluminis*	ULB	1.8	–1.9	Loc. 10	Below Tuff 2		158/99
	Tragelaphini							
32	*Tragelaphus sp.*	ULB	–8.1	–2.5	Loc. 10E	Between Tuffs 6 & 7	Upper M2	
37	*Tragelaphus sp.*	ULB	–7.9	–2.1	Loc. 10	Below Tuff 3	Upper M3	
44	*Tragelaphus sp.*	ULB	–8.2	2.9	Loc. 10W	Below Tuff 3	Upper M1	
177	*Tragelaphus sp.*	ULB	–11.8	–2.2	Loc. 21	Between Tuffs 5 & 8	Upper molar	
LT77B	*Tragelaphus* sp.	ULB	–11.7	–0.1	Loc. 12	Between Tuffs 5 & 8		
LT171B	*Tragelaphus* sp.	ULB	–8.3	–0.3	Loc. 9S	Below Tuff 2		944/98
LT176B	*Tragelaphus* sp.	ULB	–2.8	–0.4	Loc. 9	Between Tuffs 6 & 7		1166/98
LT177B	*Tragelaphus* sp.	ULB	–8.6	1.4	Loc. 22	Between Tuffs 5 & 7		1199/98
LT185B	*Tragelaphus* sp.	ULB	–9.6	–1.1	Loc. 9	Between Tuffs 5 & 7		1470/98
LT194B	*Tragelaphus* sp.	ULB	–4.5	–0.7	Loc. 10W	Below Tuff 2		137/99
72	*Tragelaphus sp.*	UNB	0.1	2.25	Loc. 18		Upper molar frag.	
75	*Tragelaphus sp.*	UNB	–2.1	–1.16	Loc. 18		Upper molar frag.	

(continued)

Table 15.2 (continued)

Lab no.	Taxon	Formation	$\delta^{13}C_{enamel}$	$\delta^{18}O_{enamel}$	Locality	Stratigraphic level	Element	Field no.
	Cercopithecidae							
55	*Parapapio ado*	ULB	−6.6	−1.2	Loc. 1	Between Tuffs 7 & 8		
85	*Parapapio ado*	ULB	−6.5	0.1	Loc. 10E	Between Tuffs 5 & 7		
101	Cercopithecidae	ULB	−6.6	−0.7	Loc. 9	Between Tuffs 5 & 7		
	Gomphotheriidae							
59	*Anancus kenyensis*	LLB	−2.5	−2.4	Kakesio 6		Molar	
LT67P	*Anancus kenyensis*	LLB	−10.5	4.1	Kakesio 8			
	Elephantidae							
LT34P	*Loxodonta* sp.	LLB	−1.7	−1.1	Noiti 3			
1	*Loxodonta exoptata*	ULB	−5.0	−1.9	Loc. 8	Between Tuffs 5 & 7	Molar	
2	*Loxodonta exoptata*	ULB	−2.0	1.4	Loc. 8	Between Tuffs 5 & 7	Molar	
26	*Loxodonta exoptata*	ULB	−0.5	−1.0	Loc. 16	Between Tuffs 7 & just above 8	Molar	
60	*Loxodonta exoptata*	ULB	−5.6	−0.5	Loc. 10E	Between Tuffs 5 & 7	Molar	
61	*Loxodonta exoptata*	ULB	−4.4	−0.9	Loc. 10W	Below Tuff 3	Molar	
62	*Loxodonta exoptata*	ULB	−1.2	0.4	Loc. 10W	Below Tuff 3	Molar	
63	*Loxodonta exoptata*	ULB	−3.8	−0.6	Loc. 10	Below Tuff 3	Molar	
77	*Loxodonta exoptata*	ULB	−4.7	−1.2	Loc. 2	Between Tuffs 5 & 7	Molar	
148	*Loxodonta exoptata*	ULB	−3.5	−2.1	Loc. 3	Between Tuffs 6 & 8	Molar	
150	*Loxodonta exoptata*	ULB	−4.7	−1.7	Loc. 6	Between Tuffs 5 & 6	Molar	
151	*Loxodonta exoptata*	ULB	−3.4	−0.3	Loc. 11	Above Tuff 7	Molar	
152	*Loxodonta exoptata*	ULB	0.9	−1.1	Loc. 10E	Between Tuffs 5 & 7	Molar	
153	*Loxodonta exoptata*	ULB	−2.6	−1.6	Loc. 11	Above Tuff 7	Molar	
LT38P	*Loxodonta exoptata*	ULB	−3.8	−0.5	Loc. 2	Between Tuffs 5 & 7		
LT53P	*Loxodonta exoptata*	ULB	−2.5	0.6	Loc. 20	Between Tuffs 6 & 8		
LT81P	*Loxodonta exoptata*	ULB	−4.2	−0.9	Loc. 21	Between Tuffs 5 & 7		
LT83P	*Loxodonta exoptata*	ULB	−1.4	−0.6	Loc. 10E	Between Tuffs 5 & 7		
LT111P	*Loxodonta exoptata*	ULB	−6.3	−2.1	Loc. 13	Between Tuffs 6 & 8		
LT113P	*Loxodonta exoptata*	ULB	−3.3	−0.8	Loc. 13	Between Tuffs 6 & 8		
LT125P	*Loxodonta exoptata*	ULB	−4.3	−2.5	Loc. 22	Between Tuffs 5 & 7		
LT60	*Loxodonta exoptata*	ULB	−5.9	−1.3	Loc. 10E	Between Tuffs 5 & 7		
71	*Loxodonta exoptata*	UNB	−1.9	−0.3	Loc. 7E		Molar	
146	*Loxodonta exoptata*	UNB	−4.4	0.9	Loc. 18		Molar	
147	*Loxodonta exoptata*	UNB	0.2	0.9	Loc. 7E		Molar	
149	*Loxodonta exoptata*	UNB	−3.1	−0.2	Loc. 18		Molar	
LT33P	*Loxodonta exoptata*	UNB	−1.4	1.2	Loc. 7E			
LT42P	*Loxodonta exoptata*	UNB	−1.2	−2.7	Loc. 14			
LT64P	*Loxodonta exoptata*	UNB	−4.8	−1.5	Loc. 22S			
	Deinotheriidae							
23	*Deinotherium bozasi*	ULB	−11.7	0.0	Loc. 10W	Below Tuff 3	Molar	
LT46P	*Deinotherium bozasi*	ULB	−11.5	0.1	Loc. 6	Between Tuffs 5 & 7		
	Suidae							
10	*Notochoerus euilus*	ULB	−8.2	−2.4	Loc. 2	Between Tuffs 5 & 7	Molar	
18	*Notochoerus euilus*	ULB	−0.1	−1.3	Loc. 3	Between Tuffs 6 & 8	Lower M3	
22	*Notochoerus euilus*	ULB	−1.1	1.4	Loc. 10E	Between Tuffs 5 & 7	Lower M3	
53	*Notochoerus euilus*	ULB	−1.1	0.0	Loc. 10E	Between Tuffs 5 & 7	Molar	
54	*Notochoerus euilus*	ULB	−1.0	−1.3	Loc. 10	Below Tuff 3	Molar	
55	*Notochoerus euilus*	ULB	−2.1	−0.1	Loc. 10	Below Tuff 3	Molar	
154	*Notochoerus euilus*	ULB	−0.5	−1.7	Loc. 9S	Below Tuff 2	Molar	
155	*Notochoerus euilus*	ULB	−3.4	−0.5	Loc. 8	Between Tuffs 5 & 7	Molar	
156	*Notochoerus euilus*	ULB	−4.5	−0.5	Loc. 5N of site	Between Tuffs 6 & 8	Molar	
157	*Notochoerus euilus*	ULB	−0.9	−1.5	Loc. 2	Between Tuffs 5 & 7	Molar	
158	*Notochoerus euilus*	ULB	−3.6	−0.3	Loc. 6	Between Tuffs 5 & 6	Molar	
159	*Notochoerus euilus*	ULB	−0.5	−0.3	Loc. 11	Above Tuff 7	Molar	
162	*Notochoerus euilus*	ULB	−0.3	−0.8	Loc. 10	Below Tuff 3	Molar	
163	*Notochoerus euilus*	ULB	−5.4	−3.1	Loc. 9S	Below Tuff 2	Tusk	
164	*Notochoerus euilus*	ULB	0.0	0.4	Loc. 9	Between Tuffs 6 & 8	Molar	

(continued)

Table 15.2 (continued)

Lab no.	Taxon	Formation	$\delta^{13}C_{enamel}$	$\delta^{18}O_{enamel}$	Locality	Stratigraphic level	Element	Field no.
165	*Notochoerus euilus*	ULB	−2.1	−2.4	Loc. 5	Between Tuffs 3 & 5	Molar	
166	*Notochoerus euilus*	ULB	−2.5	0.3	Loc. 10	Below Tuff 3	Molar	
167	*Notochoerus euilus*	ULB	−1.3	0.0	Loc. 10	Below Tuff 3	Molar	
168	*Notochoerus euilus*	ULB	2.0	−1.6	Loc. 10W	Below Tuff 3	Molar	
LT36S	*Notochoerus euilus*	ULB	−1.6	−2.3	Loc. 2	Between Tuffs 5 & 7		
48	*Notochoerus euilus*	ULB	−3.1	−3.0	Loc. 5	Between Tuffs 3 & 5		
LT49S	*Notochoerus euilus*	ULB	−0.6	−2.4	Loc. 5	Between Tuffs 3 & 5		
LT50S	*Notochoerus euilus*	ULB	−2.0	−2.1	Loc. 5	Between Tuffs 3 & 5		
LT56S	*Notochoerus euilus*	ULB	0.2	−2.8	Loc. 7	Between Tuffs 5 & 7		
LT57S	*Notochoerus euilus*	ULB	−2.3	−1.2	Loc. 7	Between Tuffs 5 & 7		
LT79S	*Notochoerus euilus*	ULB	−3.4	−1.4	Loc. 21	Between Tuffs 5 & 7		
LT92S	*Notochoerus euilus*	ULB	−0.7	−1.4	Loc. 8	Between Tuffs 5 & 7		
93	*Notochoerus euilus*	ULB	−1.5	−2.3	Loc. 8	Between Tuffs 5 & 7		
LT94S	*Notochoerus euilus*	ULB	−3.7	−2.1	Loc. 8	Between Tuffs 5 & 7		
LT95S	*Notochoerus euilus*	ULB	−1.8	−0.8	Loc. 8	Between Tuffs 5 & 7		
LT104S	*Notochoerus euilus*	ULB	−3.1	−3.8	Loc. 9	Between Tuffs 5 & 7		
LT105S	*Notochoerus euilus*	ULB	−4.4	−2.2	Loc. 9	Between Tuffs 5 & 7		
LT117S	*Notochoerus euilus*	ULB	−3.0	−3.1	Loc. 10	Below Tuff 3		
LT121S	*Notochoerus euilus*	ULB	−0.3	1.2	Loc. 10E	Between Tuffs 5 & 7		
LT122S	*Notochoerus euilus*	ULB	0.1	−1.6	Loc. 2	Between Tuffs 5 & 7		
LT123S	*Notochoerus euilus*	ULB	−0.4	−1.2	Loc. 2	Between Tuffs 5 & 7		
160	*Notochoerus euilus*	LLB	−0.4	−1.2	Kakesio 1–6		Molar	
LT35S	*Potamochoerus porcus*	ULB	−4.2	−1.0	Loc. 11	Between Tuffs 7 & 8		
LT152S	*Potamochoerus porcus*	ULB	−1.8	−1.8	Loc. 9	Between Tuffs 6 & 7		1101/98
LT153S	*Potamochoerus porcus*	ULB	−7.7	−2.7	Loc. 22E	Between Tuffs 3 & 8		1343/01
LT154S	*Potamochoerus porcus*	ULB	−0.5	0.8	Loc. 6	Between Tuffs 5 & 6		1357/00
LT155S	*Potamochoerus porcus*	ULB	−2.4	−2.3	Loc. 17	Between Tuffs 7 & YMT		2313/00
145	*Kolpochoerus limnetes*	UNB	−3.7	−1.8	Loc. 22S		Molar	
LT60S	*Kolpochoerus limnetes*	UNB	−0.8	−0.8	Loc. 22S			
LT89S	*Kolpochoerus limnetes*	UNB	−0.3	−2.0	Loc. 9			
LT129S	*Kolpochoerus limnetes*	UNB	0.0	−1.3	Silal Artum			
LT131S	*Kolpochoerus limnetes*	UNB	−1.3	−3.4	Loc. 7E			
LT150S	*Kolpochoerus limnetes*	UNB	−2.0	−0.3	Silal Artum			1520/01
P1	*Phacochoerus africanus*	Modern	−1.1	2.1	Laetoli area		M3	
P2	*Phacochoerus africanus*	Modern	−0.6	2.1	Laetoli area		M3	
WM901 1	*Phacochoerus africanus*	Modern	−0.8	4.9	Laetoli area		M3	
WM901 2	*Phacochoerus africanus*	Modern	−1.0	4.2	Laetoli area		M3	
WM901 3	*Phacochoerus africanus*	Modern	−0.6	3.4	Laetoli area		M3	
WM901 4	*Phacochoerus africanus*	Modern	−1.0	1.8	Laetoli area		M3	
WM901 5	*Phacochoerus africanus*	Modern	−0.8	2.1	Laetoli area		M3	
WM901 6	*Phacochoerus africanus*	Modern	−1.6	2.2	Laetoli area		M3	
WM901 7	*Phacochoerus africanus*	Modern	−3.1	2.7	Laetoli area		M3	
WM901 8	*Phacochoerus africanus*	Modern	−3.4	3.1	Laetoli area		M3	
WM901 9	*Phacochoerus africanus*	Modern	−1.8	3.5	Laetoli area		M3	
WM901 10	*Phacochoerus africanus*	Modern	−0.9	4.2	Laetoli area		M3	
	Equidae							
133	*Eurygnathohippus* aff. *hasumense*	ULB	−0.3	0.9	Loc. 9	Between Tuffs 6 & 8	Molar	
134	*Eurygnathohippus* aff. *hasumense*	ULB	1.2	2.5	Loc. 10W	Below Tuff 3	Upper cheek tooth	
136	*Eurygnathohippus* aff. *hasumense*	ULB	−0.4	0.3	Loc. 5N of site	Between Tuffs 6 & 8	Upper molar	
137	*Eurygnathohippus* aff. *hasumense*	ULB	−4.1	−1.3	Loc. 11	Above Tuff 7	Molar	
139	*Eurygnathohippus* aff. *hasumense*	ULB	0.5	0.3	Loc. 10E	Between Tuffs 5 & 7	Lower cheek tooth	
140	*Eurygnathohippus* aff. *hasumense*	ULB	1.0	0.0	Loc. 7	Above Tuff 7	Molar	

(continued)

Table 15.2 (continued)

Lab no.	Taxon	Formation	$\delta^{13}C_{enamel}$	$\delta^{18}O_{enamel}$	Locality	Stratigraphic level	Element	Field no.
56	*Eurygnathohippus* aff. *hasumense*	ULB	−0.3	−1.5	Loc. 10	Below Tuff 3	Lower molar	
57	*Eurygnathohippus* aff. *hasumense*	ULB	−3.8	−0.5	Loc. 10E	Between Tuffs 5 & 7	Lower molar	
58	*Eurygnathohippus* aff. *hasumense*	ULB	−3.3	−1.8	Loc. 10E	Between Tuffs 5 & 7	Lower molar	
LT44E	*Eurygnathohippus* aff. *hasumense*	ULB	−1.3	0.5	Loc. 3	between 7 & 8		
LT58E	*Eurygnathohippus* aff. *hasumense*	ULB	0.1	−0.5	Loc. 7	between 5 & 7		
LT80E	*Eurygnathohippus* aff. *hasumense*	ULB	−3.0	−1.6	Loc. 21	between 5 & 7		
LT96E	*Eurygnathohippus* aff. *hasumense*	ULB	−2.2	0.4	Loc. 8	Between Tuffs 5 & 7		
LT97E	*Eurygnathohippus* aff. *hasumense*	ULB	−1.2	2.6	Loc. 5	Between Tuffs 3 & 5		
LT102E	*Eurygnathohippus* aff. *hasumense*	ULB	−0.6	−1.3	Loc. 9	Between Tuffs 5 & 7		
LT116E	*Eurygnathohippus* aff. *hasumense*	ULB	−2.7	−1.0	Loc. 10	Below Tuff 3		
LT120E	*Eurygnathohippus* aff. *hasumense*	ULB	1.8	2.4	Loc. 10E	Between Tuffs 5 & 7		
LT124E	*Eurygnathohippus* aff. *hasumense*	ULB	1.4	2.9	Loc. 22	Between Tuffs 5 & 7		
132	*Eurygnathohippus* aff. *cornelianus*	UNB	0.4	−2.0	Loc. 18		Lower molar	
138	*Eurygnathohippus* aff. *cornelianus*	UNB	−0.7	0.4	Loc. 7E		Molar	
69	*Eurygnathohippus* aff. *cornelianus*	UNB	0.3	1.3	Loc. 18		Upper molar	
70	*Eurygnathohippus* aff. *cornelianus*	UNB	0.5	−2.2	Loc. 15		Upper molar	
LT41E	*Eurygnathohippus* aff. *cornelianus*	UNB	−1.2	0.1	Loc. 7E			
LT107E	*Eurygnathohippus* aff. *cornelianus*	UNB	−0.3	−0.5	Loc. 22S			
LT108E	*Eurygnathohippus* aff. *cornelianus*	UNB	−1.3	−2.0	Loc. 15			
LT126E	*Eurygnathohippus* aff. *cornelianus*	UNB	−0.1	0.3	Silal Artum			
LT127E	*Eurygnathohippus* aff. *cornelianus*	UNB	0.6	−2.4	Silal Artum			
LT164E	*Eurygnathohippus* aff. *cornelianus*	UNB	0.3	−0.4	Loc. 18			981/01
LT165E	*Eurygnathohippus* aff. *cornelianus*	UNB	0.5	−1.6	Loc. 18			1033/00
LT170E	*Eurygnathohippus* aff. *cornelianus*	UNB	1.9	−0.2	Loc. 15			4047/00
LT172E	*Eurygnathohippus* aff. *cornelianus*	UNB	1.0	−1.3	Loc. 15			3471/01
LT173E	*Eurygnathohippus* aff. *cornelianus*	UNB	−2.0	−3.5	Loc. 7E			4009/00
LT174E	*Eurygnathohippus* aff. *cornelianus*	UNB	2.7	−0.3	Loc. 1			3044/00
LT70E	*Eurygnathohippus* aff. *hasumense*	LLB	−0.9	−1.0	Kakesio 8			
LT157E	*Eurygnathohippus* aff. *hasumense*	LLB	−3.3	−3.6	Kakesio 4			065/98
LT158E	*Eurygnathohippus* aff. *hasumense*	LLB	0.0	−3.0	Kakesio 1–6			094/98

(continued)

Table 15.2 (continued)

Lab no.	Taxon	Formation	$\delta^{13}C_{enamel}$	$\delta^{18}O_{enamel}$	Locality	Stratigraphic level	Element	Field no.
LT159E	*Eurygnathohippus* aff. *hasumense*	LLB	−1.6	−0.2	Kakesio South			105/98
LT160E	*Eurygnathohippus* aff. *hasumense*	LLB	−1.1	0.2	Kakesio 1–6			033/98
LT161E	*Eurygnathohippus* aff. *hasumense*	LLB	−1.9	−2.5	Kakesio 1–6			085/98
135	*Eurygnathohippus* aff. *hasumense*	LLB	1.9	3.0	Kakesio 1–6		Molar	
EQ14	*Equus burchelli*	Modern	−1.4	−2.0	Laetoli area		Upper molar	
EQ15	*Equus burchelli*	Modern	−1.1	−1.7	Laetoli area		Upper molar	
	Rhinocerotidae							
114	*Ceratotherium efficax*	ULB	−5.9	−1.0	Loc. 9S	Below Tuff 2	Lower cheek tooth	
115	*Ceratotherium efficax*	ULB	−11.0	−1.1	Loc. 21	Between Tuffs 5 & 8	Upper molar	
116	*Ceratotherium efficax*	ULB	−1.3	−1.8	Loc. 6	Between Tuffs 5 & 6	Upper molar	
117	*Ceratotherium efficax*	ULB	−11.0	0.5	Loc. 21	Between Tuffs 5 & 8	Lower molar	
118	*Ceratotherium efficax*	ULB	−11.3	0.0	Loc. 9S	Below Tuff 2	Lower cheek tooth	
119	*Ceratotherium efficax*	ULB	−5.9	−0.4	Loc. 9S	Below Tuff 2	Lower cheek tooth	
LT7R	*Ceratotherium efficax*	ULB	−5.0	4.3	Loc. 9	Between Tuffs 5 & 7	Upper left P2	
LT8R	*Ceratotherium efficax*	ULB	−7.9	−0.9	Loc. 11	Between Tuffs 7 to just above 8	Lower molar frag.	
LT9R	*Ceratotherium efficax*	ULB	−7.3	−1.1	Loc. 8	Between Tuffs 5 & 7	Lower molar	
LT22R	*Ceratotherium efficax*	ULB	−4.2	4.1	Loc. 10	Below Tuff 2	Right M3	865/98
LT23R	*Ceratotherium efficax*	ULB	−5.9	−0.1	Loc. 9S	Below Tuff 2	Molar frag.	1037/98
LT28R	*Ceratotherium efficax*	ULB	−5.3	−1.4	Loc. 3	Between Tuffs 7 & 8	Premolar frag.	454/03
LT31R	*Ceratotherium efficax*	ULB	−3.7	−1.3	Loc. 10	Below Tuff 3		595/04
LT73R	*Ceratotherium efficax*	ULB	−7.8	−1.0	Loc. 11	Between Tuffs 7+just above 8	Lower molar frag.	
LT74R	*Ceratotherium efficax*	ULB	−4.7	−1.6	Loc. 8	Between Tuffs 5 & 7	Lower molar	
170	*Ceratotherium efficax*	ULB	−3.7	−0.8	Loc. 10	Below Tuff 3	Upper molar	
172	*Ceratotherium efficax*	ULB	−5.6	−1.2	Loc. 10	Below Tuff 3	Upper molar	
LT1R	*Ceratotherium efficax*	ULB	−4.4	0.0	Loc. 10W	Below Tuff 3	Lower premolar	
LT2R	*Ceratotherium efficax*	ULB	−8.8	−1.7	Loc. 10E	Between Tuffs 5 & 7	Upper premolar	
LT3R	*Ceratotherium efficax*	ULB	−4.1	−2.4	Loc. 10E	Between Tuffs 5 & 7	Lower molar frag. [2]	
LT4R	*Ceratotherium efficax*	ULB	−5.2	0.3	Loc. 6	Between Tuffs 5 & 7	Lower molar frag. [2]	
LT5R	*Ceratotherium efficax*	ULB	−4.9	−1.1	Loc. 7	Between Tuffs 5 & 7	Lower molar frag. [2]	
LT6R	*Ceratotherium efficax*	ULB	−6.2	−2.7	Loc. 2	Between Tuffs 5 & 7	Lower molar frag.	
LT40R	*Ceratotherium efficax*	ULB	−4.8	−1.5	Loc. 2	Between Tuffs 5 & 7		
LT47R	*Ceratotherium efficax*	ULB	−10.6	−0.1	Loc. 6	Between Tuffs 5 & 7		
LT62R	*Ceratotherium efficax*	ULB	−8.4	−1.7	Loc. 10E	Between Tuffs 5 & 7	Lower molar	
LT63R	*Ceratotherium efficax*	ULB	−7.4	0.2	Loc. 10E	Between Tuffs 5 & 7	Upper molar	
LT71R	*Ceratotherium efficax*	ULB	−4.7	−2.2	Loc. 7	Between Tuffs 5 & 7	Lower molar	
LT72R	*Ceratotherium efficax*	ULB	−8.2	−2.2	Loc. 6	Between Tuffs 5 & 7	Upper molar	
LT78R	*Ceratotherium efficax*	ULB	−6.9	−0.9	Loc. 21	Between Tuffs 5 & 7		
LT68R	*Ceratotherium efficax*	LLB	−6.2	−1.9	Kakesio 8			
LT69R	*Ceratotherium efficax*	LLB	−1.5	−1.6	Kakesio 8			
171	*Ceratotherium efficax*	LLB	−7.0	−1.0	Emboremony 1		Molar	
LT21R	*Ceratotherium efficax*	LLB	−3.5	−0.7	Kakesio 6			103/98
LT30R	*Ceratotherium efficax*	LLB	−10.6	0.2	Kakesio1–6		Lower tooth frag.	1343/03

(continued)

Table 15.2 (continued)

Lab no.	Taxon	Formation	$\delta^{13}C_{enamel}$	$\delta^{18}O_{enamel}$	Locality	Stratigraphic level	Element	Field no.
LT29R	*Ceratotherium sp.*	UNB	−4.0	−1.6	Loc. 11			1298/03
LT25R	*Diceros bicornis*	ULB	−6.0	1.5	Loc. 10E	Below Tuffs 6 & 7	Molar	1554/98
LT27R	*Diceros bicornis*	ULB	−5.2	−1.9	Loc. 3	Between Tuffs 7 & 8	Molar frag.	1611/00
R1010–1	*Ceratotherium simum*	Modern	0.5	4.2	QEP		Upper molar	
R1010–15	*Ceratotherium simum*	Modern	0.4	4.2	QEP		Upper molar	
R1010–30	*Ceratotherium simum*	Modern	0.0	4.2	QEP		Upper molar	
	Giraffidae							
4	*Giraffa jumae*	ULB	−10.9	2.8	Loc. 8	Between Tuffs 5 & 7	Upper molar	
16	*Giraffa jumae*	ULB	−10.6	1.1	Loc. 9	Between Tuffs 6 & 8	Upper molar	
27	*Giraffa jumae*	ULB	−10.0	−0.3	Loc. 1	Between Tuffs 7&YMT	Upper molar	
49	*Giraffa jumae*	ULB	−10.4	−1.1	Loc. 10E	Between Tuffs 5 & 7	Lower molar	
50	*Giraffa jumae*	ULB	−10.2	1.5	Loc. 10E	Between Tuffs 5 & 7	Upper molar	
52	*Giraffa jumae*	ULB	−8.8	8.7	Loc. 10W	Below Tuff 3	Lower M2	
125	*Giraffa jumae*	ULB	−8.3	0.1	Loc. 22E	Between Tuffs 3 & 8	Upper molar	
126	*Giraffa jumae*	ULB	−9.9	0.6	Loc. 7	Between Tuffs 5 & 7	Upper molar	
LT14G	*Giraffa jumae*	ULB	−9.4	0.2	Loc. 8	Between Tuffs 5 & 7		
LT15G	*Giraffa jumae*	ULB	−12.4	−0.6	Loc. 12	Between Tuffs 5 & 8		
LT16G	*Giraffa jumae*	ULB	−10.9	−0.9	Loc.11	Between Tuffs 7 to just above 8		
LT18G	*Giraffa jumae*	ULB	−11.4	−1.2	Loc. 21	Between Tuffs 5 & 7		
LT19G	*Giraffa jumae*	ULB	−12.9	−0.8	Loc. 5	Between Tuffs 3 & 5		
LT87G	*Giraffa jumae*	ULB	−10.6	0.3	Loc. 9S	Below Tuff 2		
LT147G	*Giraffa jumae*	ULB	−9.2	−0.6	Loc. 9S	Below Tuff 2		10436b/98
LT148G	*Giraffa jumae*	ULB	−10.8	−1.2	Loc. 10E	Between Tuffs 6 & 7		061/98
LT149G	*Giraffa jumae*	ULB	−10.7	0.6	Loc. 9	Between Tuffs 6 & 7		207/99
11	*Giraffa stillei*	ULB	−9.6	−0.1	Loc. 2	Between Tuffs 5 & 7	Upper molar	
15	*Giraffa stillei*	ULB	−8.5	2.0	Loc. 2	Between Tuffs 5 & 7	Upper molars (4)	
17	*Giraffa stillei*	ULB	−11.2	0.1	Loc. 9	Between Tuffs 6 & 8	Lower M3	
24	*Giraffa stillei*	ULB	−9.5	1.4	Loc. 21	Between Tuffs 5 & 8	Lower premolar	
25	*Giraffa stillei*	ULB	−9.0	2.7	Loc. 16	Between Tuffs 7 & just above 8	Upper molars (2)	
28	*Giraffa stillei*	ULB	−12.4	2.2	Loc. 8	Below Tuff 6	Upper molar	
30	*Giraffa stillei*	ULB	−9.0	8.9	Loc. 12E	Between Tuffs 5 & 7	Lower molar	
51	*Giraffa stillei*	ULB	−10.2	−0.1	Loc. 10	Below Tuff 3	Lower molar	
127	*Giraffa stillei*	ULB	−10.3	3.5	Loc. 8	Between Tuffs 5 & 7	Upper molar	
128	*Giraffa stillei*	ULB	−11.1	1.1	Loc. 7	Above Tuff 7	Lower molar	
129	*Giraffa stillei*	ULB	−9.6	1.7	Loc. 9S	Below Tuff 2	Upper molar	
130	*Giraffa stillei*	ULB	−10.2	0.1	Loc. 10W	Below Tuff 3	Premolar	
131	*Giraffa stillei*	ULB	−10.5	0.8	Loc. 7	Above Tuff 7	Upper molar	
LT10G	*Giraffa stillei*	ULB	−9.6	1.0	Loc. 16	Between Tuffs 7 to just above 8		
LT13G	*Giraffa stillei*	ULB	−11.3	1.6	Loc. 8	Between Tuffs 5 & 7		
LT134G	*Giraffa stillei*	ULB	−10.0	2.1	Loc. 10E	Between Tuffs 6 & 7		340/98
LT137G	*Giraffa stillei*	ULB	−11.6	2.2	Loc. 10W	Below Tuff 2		604/98
LT138G	*Giraffa stillei*	ULB	−9.2	1.3	Loc. 10W	Below Tuff 2		
LT140G	*Giraffa stillei*	ULB	−10.3	1.9	Loc. 10W	Below Tuff 2		606/98
LT141G	*Giraffa stillei*	ULB	−9.2	0.4	Loc. 10W	Below Tuff 2		674/98
144	*Giraffa stillei*	ULB	−8.6	2.4	Loc. 9S	Below Tuff 2		1004/98
LT131	*Giraffa stillei*	ULB	−10.4	−0.2	Loc. 7	Above Tuff 7		
			−10.2	0.6				
29	*Sivatherium maurusium*	ULB	−12.9	1.0	Loc. 12/12E	Between Tuffs 5 & 7	Upper molar	
174	*Sivatherium maurusium*	ULB	−9.4	1.9	Loc. 21	Between Tuffs 5 & 8	Upper molar	
173	*Sivatherium maurusium*	UNB	−9.6	−0.5	Loc. 15		Upper molar	
LT11G	*Sivatherium maurusium*	ULB	−10.9	−0.4	Loc. 16	Between Tuffs 5 & 7		

(continued)

Table 15.2 (continued)

Lab no.	Taxon	Formation	$\delta^{13}C_{enamel}$	$\delta^{18}O_{enamel}$	Locality	Stratigraphic level	Element	Field no.
LT12G	*Sivatherium maurusium*	ULB	−9.2	0.3	Loc. 7	Between Tuffs 7 to just above 8		
LT17G	*Sivatherium maurusium*	ULB	−10.0	1.2	Loc. 11	Between Tuffs 7 to just above 8		
LT82G	*Sivatherium maurusium*	ULB	−10.7	0.2	Loc. 21	Between Tuffs 5 & 7		
LT132G	*Sivatherium maurusium*	ULB	−10.0	−0.9	Loc. 10E	Between Tuffs 6 & 7		277/98
LT133G	*Sivatherium maurusium*	ULB	−10.3	−0.5	Loc. 9	Between Tuffs 6 & 7		109/98
G1	*Giraffa camelopardalis*	Modern	−12.7	4.0	Laetoli area		M1	
G2	*Giraffa camelopardalis*	Modern	−13.6	6.0	Laetoli area		M2	
G3	*Giraffa camelopardalis*	Modern	−10.6	8.0	Laetoli area		M3	
G4	*Giraffa camelopardalis*	Modern	−9.4	7.9	Laetoli area		M3	
G94 1	*Giraffa camelopardalis*	Modern	−12.4	4.9	Laetoli area		M3	
G94 12	*Giraffa camelopardalis*	Modern	−13.4	4.6	Laetoli area		M3	
GP94 1	*Giraffa camelopardalis*	Modern	−9.8	6.3	Laetoli area		P4	
GP94 20	*Giraffa camelopardalis*	Modern	−8.1	6.8	Laetoli area		P4	
	Hyaenidae + Felidae							
JK86	*Crocuta* sp.	ULB	−7.2	−2.5	Loc. 10E	Between Tuffs 5 & 7		
JK75	*Felis* large sp.	ULB	−8.0	−0.5	Loc 9S	Below Tuff 2		

30–20 mg of the dried samples were reacted with 100‰ phosphoric acid (H_3PO_4) at 90°C in sealed borosilicate reaction y-tubes for 48 h. The liberated carbon dioxide was cleaned, separated cryogenically into 6 mm sealed tubes with silver wire, and sealed. The sealed tubes with the CO_2 and silver were heated to 50°C to remove impurities, such as SO_2 or H_2S, that could contribute interfering masses. Mass spectrometry was performed at Yale Stable Isotope Laboratory using a MAT-Finnegan 251 mass spectrometer interfaced with a tubecracker and at the University of Florida, Department of Geoscience Stable Isotope Lab using a VG Prism 602 also interfaced with a tubecracker. Once it was established that SO_2 or H_2S was not being generated during digestion in the phosphoric acid, samples were analyzed online using a Mountain Mass multi-prep system interfaced with the VG Prism. Results are reported using the conventional notation where:

$$\delta = \left(\left(R_{sample} / R_{standard}\right) - 1\right) \times 1{,}000$$

$$R = {}^{13}C / {}^{12}C, {}^{18}O / {}^{16}O, \text{ or } {}^{15}N / {}^{14}N$$

Carbon and oxygen values are reported relative to PDB (Pee Dee Belemnite) and the nitrogen values relative to air. Precision was ± 0.09‰ for $\delta^{13}C$ ratios and ±0.17‰ for $\delta^{18}O$ ratios for eight replicate pairs of fossil enamel. International and in-house laboratory standards analyzed with the enamel samples at both facilities utilizing the two techniques yielded a standard deviation of ± 0.05‰ ($n = 36$) for C and ± 0.11 for O ($n = 36$). Based on these data, overall analytical precision was better than 0.10‰ for C and 0.22‰ for O.

Diagenesis of the biogenic signal remains an ongoing concern in the analysis and interpretation of the isotopic signature of fossil enamel. Geochemical investigations of the integrity of biogenic isotopic compositions have revealed that alteration and contamination can be pervasive in fossil enamel bioapatite (i.e., Kohn et al. 1999; Sponheimer and Lee-Thorp 2006; Murphy and Bowman 2006; Jacques et al. 2008; Martin et al. 2008). However, no unequivocal criteria have been developed to assess the extent of diagenesis in fossil enamel and how to correct for it (Kohn and Cerling 2002). Perhaps the most convincing evidence for minimal alteration at Laetoli is the preservation of distinct isotopic offset between grazers and browsers documented in modern ecosystems in the fossil assemblages. For example, in the Laetoli fossil data, the giraffid and deinothere enamel consistently yielded distinct browsing isotopic signatures while the equids and alcelaphines have carbon isotopic values consistent with grazing or C_4 dominated diets. These isotopic differences are maintained in enamel fragments recovered from the same stratigraphic horizon meters apart, where alteration would presumably alter all the enamel so isotopic signatures would converge on a diagenetic/matrix value.

Ostrich Eggshell (OES)

Sample Preparation for Isotopic Analyses of Inorganic Calcite Matrix of Ostrich Eggshells

A modified version of the procedure outlined by Johnson (1995) was used here. Eggshell samples are first chemically and mechanically cleaned using a high speed Z500 Brasseler drilling tool with tungsten-carbide bits to remove potential contaminants/diagenetic components. Both the inner and outer discolored eggshell layers were mechanically abraded and the remaining fragment treated in an ultrasound. Samples were extracted from the thickest part of the shell and away from the pore complexes, which tend to be recrystallized and potentially altered. To remove surficial contamination and insure consistent analyses of the core (palisade/spongy layer) of the eggshell, 10–20% of the specimen was dissolved using 2 N HCl. Samples were divided into two fragments for inorganic and organic isotopic analyses. The sample for $\delta^{13}C_{OES-carb}$ and $\delta^{18}O_{OES}$ was powdered in an agate mortar and pestle and loaded onto the mass spectrometer for phosphoric acid digestion.

Sample Preparation for Isotopic Analyses of Organic Residue within Eggshells

Unpowdered samples were prepared using a technique for EA-irms analyses (Fogel, personal communication). 4–7 mg of eggshell fragments or chunks were placed in Ag combustion boats. Using a pipette, 30 μl of 6 N HCl were added to the samples in the boats, letting it sit until the reaction (effervescence) was complete. This procedure was repeated, adding 30 μl each time until there was no more reaction, typically 4–5 times. Samples were dried overnight in a fume hood. Samples were then loaded into the carousel for combustion on an EA-irms.

About 10 cm of silver wool was placed in the bottom of the combustion column to absorb excess HCl.

Results and Discussion

Mammalian Herbivore Enamel

$\delta^{13}C_{enamel}$ Overview

Overall, the analyses presented here support and reinforce previous paleodietary and paleoenvironemental interpretations and conclusions based on isotopic analyses of fossil herbivore enamel (Kingston and Harrison 2007). The new data primarily augments taxa that had been previously analyzed isotopically, in an attempt to refine characterization of the intra- and interspecific isotopic variability as well as situate this variability in a chronostratigraphic context to assess any shifts in dietary patterns through time that might reflect general habitat change (Tables 15.1 and 15.2; Figs. 15.1–15.8). A number of other studies have indicated taxonomic and/or environmental change within the succession, not only between the LLB, ULB, and the UNB but also within the ULB (Kovarovic and Andrews 2011; Kaiser 2011; Hernesniemi et al. 2011; Rossouw and Scott 2011). It is useful to examine any shifts in the isotopic data and potentially integrate interpretations with these parallel paleoecological and taxonomic studies. Despite the extensive nature of the isotopic investigations at Laetoli, involving the analyses of almost 400 enamel samples, detecting shifts within the sequence that are statistically significant or ecologically meaningful remains difficult. A number of factors contribute to these limitations: (1) Almost all of the fossil taxa exhibit a high degree of isotopic variability, especially relative to modern lineages, (2) sampling is uneven through the sequence, (3) a number of the key taxa are not represented at all levels; (4) potential shifts are accompanied by taxonomic changes which complicate direct comparisons; (5) although there was a high degree of stratigraphic control associated with each sample, the temporal uncertainty of the samples range from 10 to 150 kyrs; and (6) the number of analyzed specimens for many of the taxa remain limited, especially when scattered through the succession, typically with only several samples at any given stratigraphic level.

As we attempt to increase the temporal resolution in reconstructing paleodietary patterns, it increasingly becomes important to understand the relationship of specific samples and intra- and inter-population variation in foraging behavior, a parameter that has not been well established even for modern East African herbivore populations. Rather than invoking short term habitat change and resulting foraging behavioral responses, at this point it may be more parsimonious to regard the variability as indicative of generally eclectic foraging strategy by many taxa that were previously interpreted to be more specialized along the grazing-browsing continuum (Kingston and Harrison 2007). The more diverse diets may result from past ecosystems that locally had more variable or distinct microhabitats than modern analogs, leading to more varied diets. The variation in several fossil taxa from the Laetoli succession is significantly greater than that exhibited by modern representatives collected from different ecosystems distributed across equatorial Africa (Kingston and Harrison 2007). Underlying this assessment is the assumption that specimens collected from confined intervals in the Laetoli stratigraphic succession reflect the foraging behavior(s) of individuals in contemporaneous population(s) in the past. However, these stratigraphic intervals typically span tens of thousands of years and potentially involve significant time averaging. Alternatively, the isotopic variability can be interpreted in the framework of general dietary patterns of specific species through time without reducing the perspective to the population level. If the foraging variability is linked to short-term ecological change, assessments are linked more to how the various taxa respond and the range of behavioral flexibility a specific taxon is capable of and/or requires. Ultimately these distinctions are important in framing any sort of foraging

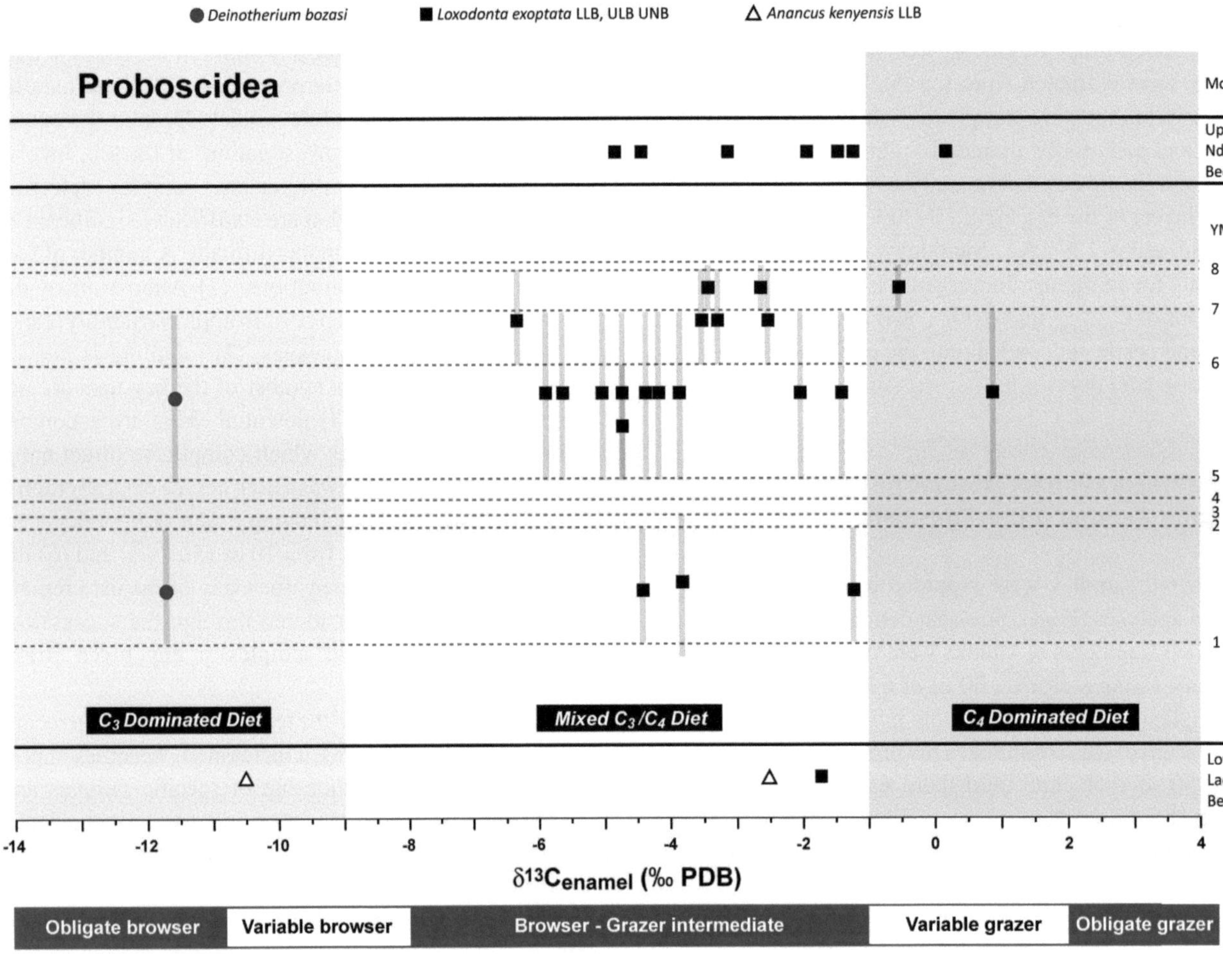

Fig. 15.1 Stable carbon isotopic composition of fossil proboscidean enamel ($\delta^{13}C_{enamel}$) from the Laetolil Beds and Upper Ndolanya Beds, depicting specific stratigraphic provenance of samples. Vertical grey bars represent stratigraphic uncertainty inherent in the collection of the samples, mostly surface finds

conclusions based on isotopic analyses of fossil tooth enamel and eggshell fragments that are scattered through the succession.

Despite these complexities, there are a number of apparent trends or shifts in foraging behavior within the sequence that are included below in discussions of the results for the various herbivore groups analyzed (depicted in Figs. 15.1–15.8 and summarized in Table 15.2). Within the constraints of known stratigraphic ranges, fossil enamel specimens analyzed isotopically form five relatively distinct temporal units. These are: (1) LLB; (2) ULB (below Tuff 3); (3) ULB (between Tuffs 5–7); (4) ULB (at or above Tuff 7); and (5) UNB, with partial overlap between units 3 and 4. Although descriptions of ULB fossils recovered from the interval between Tuffs 3 and 5 suggest that this range is ecologically distinct (i.e., Kovarovic and Andrews 2011; Hernesniemi et al. 2011; Tattersfield 2011), only a few specimens from this interval were analyzed isotopically (due to sampling limitations). While it is unlikely that these stratigraphic units necessarily correspond to differential or variable ecosystems, they provide the best methodological approach at this time to interpret the isotopic data within a chronostratigraphic framework. There are only a few taxa that are represented in all these units and have been systematically analyzed isotopically (i.e. *Loxodonta exoptata, Ceratotherium efficax, Gazella janenschi*), although a series of taxa represented in two or three adjacent units at the genus, family or tribe level were analyzed (i.e., *Notochoerus* sp., Alcelaphini sp., *Sivatherium maurusium*), providing a comparative perspective. Much of the novel data presented here were generated specifically to expand the temporal as well as spatial coverage of dominant herbivore taxa in the fossil assemblages.

The goal of the carbon isotopic analysis is to differentiate foraging patterns of multiple herbivore lineages and ultimately link these behaviors to specific paleoecosystems in the Laetoli area during the Pliocene. This requires partitioning

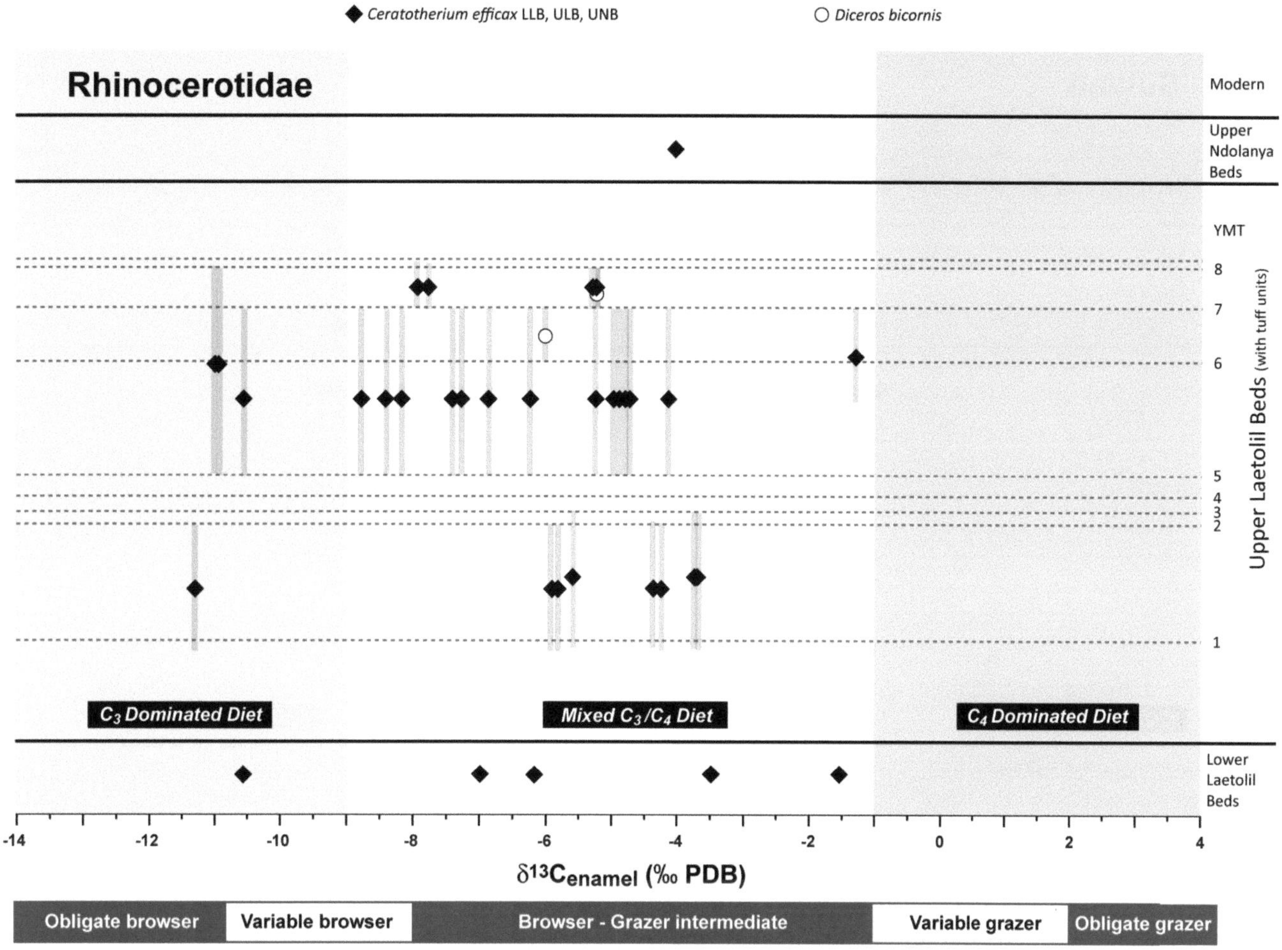

Fig. 15.2 Stable carbon isotopic composition of fossil Rhinocerotidae enamel ($\delta^{13}C_{enamel}$) from the Laetolil Beds and Upper Ndolanya Beds, depicting specific stratigraphic provenance of samples

the $\delta^{13}C$ signals and assigning feeding patterns to discrete ranges of isotopic values. Inherent in this classification is considerable uncertainty, primarily because there is variability in the isotopic composition of C_3, C_4, potential C_3/C_4 intermediates, and CAM (Crassulacean acid metabolism) plants that may have comprised the diet of the extinct fauna. Mixed feeding behavior that may have included variable amounts of these plants further complicates simple interpretations. Given the limited distribution of CAM and C_3/C_4 intermediates in modern habitats and foraging regimes, and no evidence that they comprised any significant proportion of the plant biomass in the past, it is reasonable to interpret the fossil isotopic signatures along a C_3 to C_4 continuum. In following discussions, five foraging classifications are used to describe the Laetoli fauna: (1) obligate grazer, (2) variable grazer, (3) browser-grazer intermediate, (4) variable browser, and (5) obligate browser (Table 15.3). The third category incorporates the greatest range of foraging behavior and can be further modified by indicating if values lie closer to the grazing or browsing end-members. This classification is linked to $\delta^{13}C$ ranges, based on a mixing model that incorporates the isotopic variability associated with plants utilizing the two photosynthetic pathways. These classifications are based on discussions of the diet of modern East African faunas (Tieszen et al. 1979; McNaughton and Georgiadis 1986; Gagnon and Chew 2000) and the isotopic values are adjusted (+1.5‰) for pre-industrial atmospheric $\delta^{13}C$ shifts (Kingston and Harrison 2007).

A final comment critical to interpreting the carbon isotopic data relates to the extent to which C_3 grasses were significant in past habitats at Laetoli. As mentioned earlier, if C_3 grasses were significant at Laetoli in the Pliocene, the browser/grazer distinction based on $\delta^{13}C$ values is seriously compromised. Pollen and phytolith studies (Rossouw and Scott 2011; Bonnefille and Riollet 1987) have documented the presence of grasses in the Laetoli succession. More specifically, Rossouw and Scott (2011) have documented abundant short-cell phytolith morphotypes characteristic of C_3 grasses. Phytolith profiles indicate a general shift from mainly C_3 grass types in the LLB and older ULB levels to

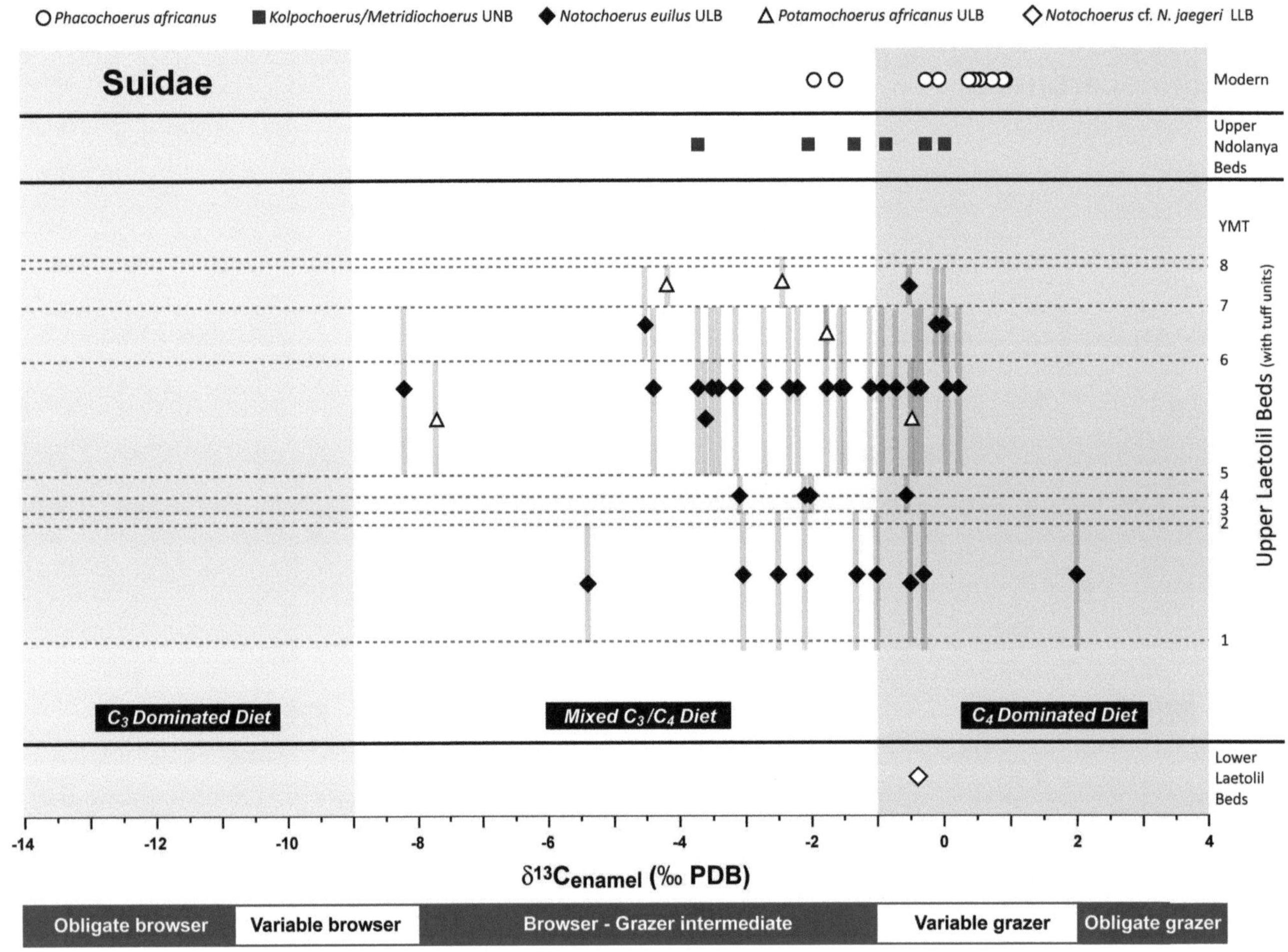

Fig. 15.3 Stable carbon isotopic composition of fossil Suidae enamel ($\delta^{13}C_{enamel}$) from the Laetolil Beds and Upper Ndolanya Beds, depicting specific stratigraphic provenance of samples

mainly C_4 grass types in the younger ULB and UNB, but with C_3 and C_4 grasses potentially co-existing, possibly along altitudinal gradients. Phytolith proportions from LLB, ULB, and UNB indicate that although grasses were ubiquitous in the succession, they were never a dominant component of the vegetation. How can this information be reconciled with isotopic interpretations based on the 'C_3=browse' and 'C_4 = grass' dichotomy? In the LLB and lower levels of the ULB, for example, C_4 grass phytoliths represent a small proportion of the total short-cell grass phytoliths (<1% at some levels). Grass phytoliths in total comprise on average about 54% of the total phytolith component in the fossil samples, with the remaining phytoliths attributed to trees, shrubs and herbs. In addition, phytolith production does not occur in all higher plants, so available browse may be underrepresented in the phytolith records relative to their abundance in the past. Cumulatively, these data suggest that C_4 grasses represented a relatively small component of the plant biomass in lower levels of the succession. However, almost all the large bodied mammalian herbivore lineages from this part of the sequence, with the exception of giraffids, have isotopic signatures consistent with mixed feeding that includes variable degrees of C_4 grazing. $\delta^{13}C$ signatures of some individuals from this interval indicate variable and even obligate C_4 grazing. Relative to C_3 grasses, C_4 grasses may be less nutritious (Caswell et al. 1973; Demment and Van Soest 1985) and herbivores might select differentially for C_3 grasses, further skewing representation of C_4 in the dietary profiles. Overall, it remains difficult to reconcile the ubiquitous C_4 dietary isotopic signatures in the older ULB units with the limited representation of C_4 grass phytoliths. Another perspective in assessing the potential for C_3 grazing is to examine independent paleodietary proxies. Mesowear patterns, for example, have been examined for a number of Laetoli herbivores traditionally interpreted as grazers, such as the rhinocerotid *Ceratotherium efficax*, the alcelaphine *Parmularius pandatus*, and *Hippotragus* sp. (Kaiser 2011; Hernesniemi et al. 2011). These studies indicate mixed feeding strategies that

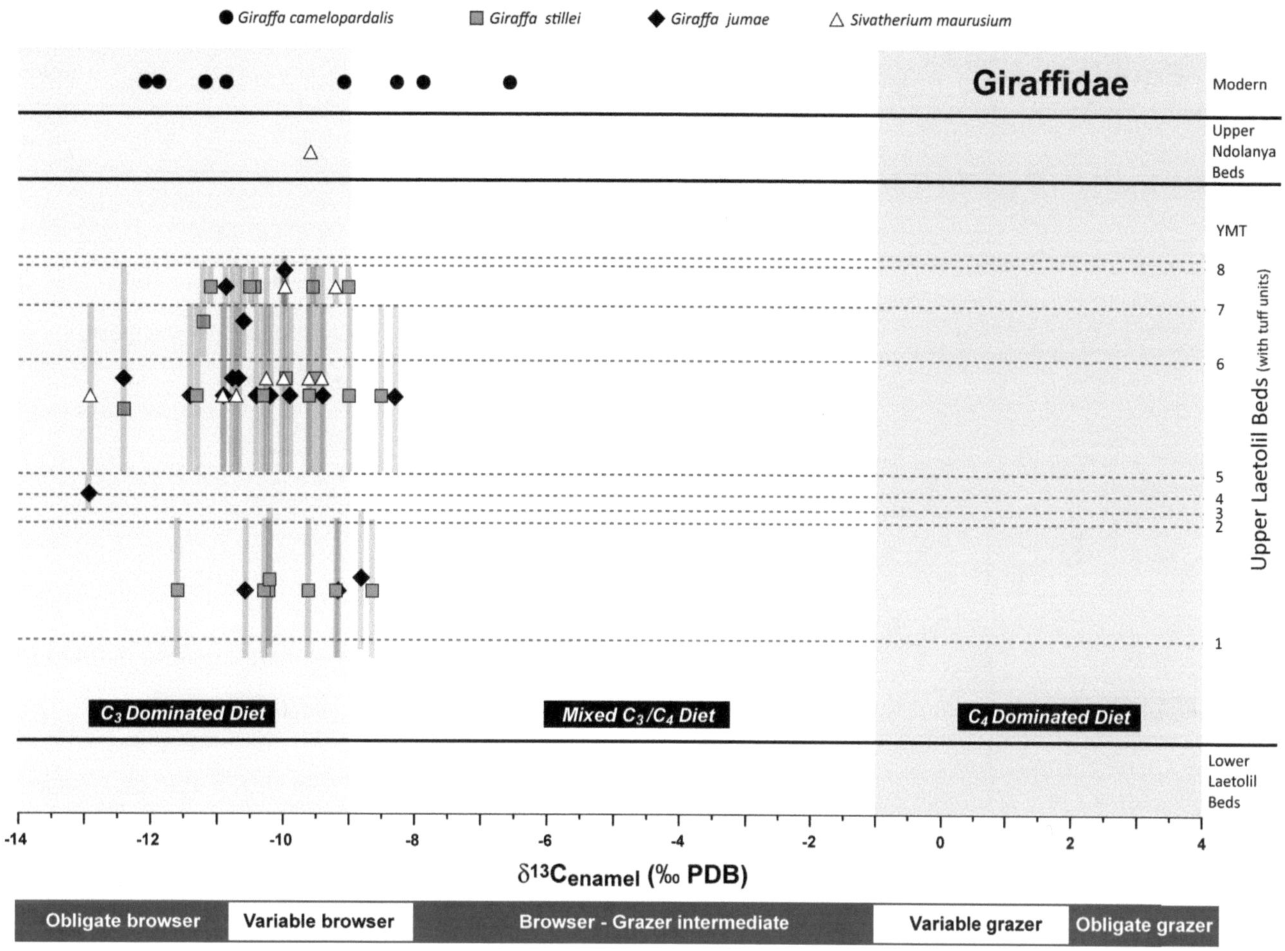

Fig. 15.4 Stable carbon isotopic composition of fossil Giraffidae enamel ($\delta^{13}C_{enamel}$) from the Laetolil Beds and Upper Ndolanya Beds and modern specimens, depicting specific stratigraphic provenance of samples

generally accord well with interpretations based on isotopic data. If a significant component of C_3 component of their diets were C_3 grasses, cusp analyses would typically yield abrasion patterns indicative of significant grazing. Instead, mesowear patterns indicate that the ULB ungulate community (including those from lower ULB horizons) is comprised of 53% mixed feeders and only 6% grazers, an estimate that can accommodate the C_4 grazing isotopic signal in the dietary patterns, but not additional C_3 grazing. In contrast, however, mesowear patterns of the equid *Eurygnathohippus* aff. *hasumense* from lower ULB strata (Kaiser 2011) indicate an abrasive feeding regime that are inconsistent with interpretations of mixed feeding (Kingston and Harrison 2007), instead possibly indicating some C_3 grazing. Overall, the phytolith data are extremely important in documenting the presence of C_3 grasses at Laetoli during the Pliocene and emphasize the limitations of isotopic data for recognizing the potential for C_3 grazing. As discussed later, however, dietary reconstructions based on cranio-dental morphology and mesowear patterns at Laetoli tend to support isotopic interpretations based on the assumption of limited availability of C_3 dietary grass.

$\delta^{18}O_{enamel}$ Overview/Aridity Index

As previously mentioned, the utility of the oxygen isotopic values of fossil enamel ($\delta^{18}O_{enamel}$) for reconstructing terrestrial environments is limited by the complexity of climatic, environmental, physiological and behavioral variables that influence the $\delta^{18}O$ signals. Previous analyses of the $\delta^{18}O_{enamel}$ signatures from Laetoli revealed no apparent patterns through the succession, with inter- and intra-specific $\delta^{18}O_{enamel}$ variability comparable to other African fossil assemblages and modern herbivore populations (Kingston and Harrison 2007). The Laetoli fossil data, together with $\delta^{18}O_{enamel}$ values from other relevant equatorial African fossil sites, reveal a general pattern of ^{18}O depletion relative to modern descendant lineages. While these differences may indicate more humid

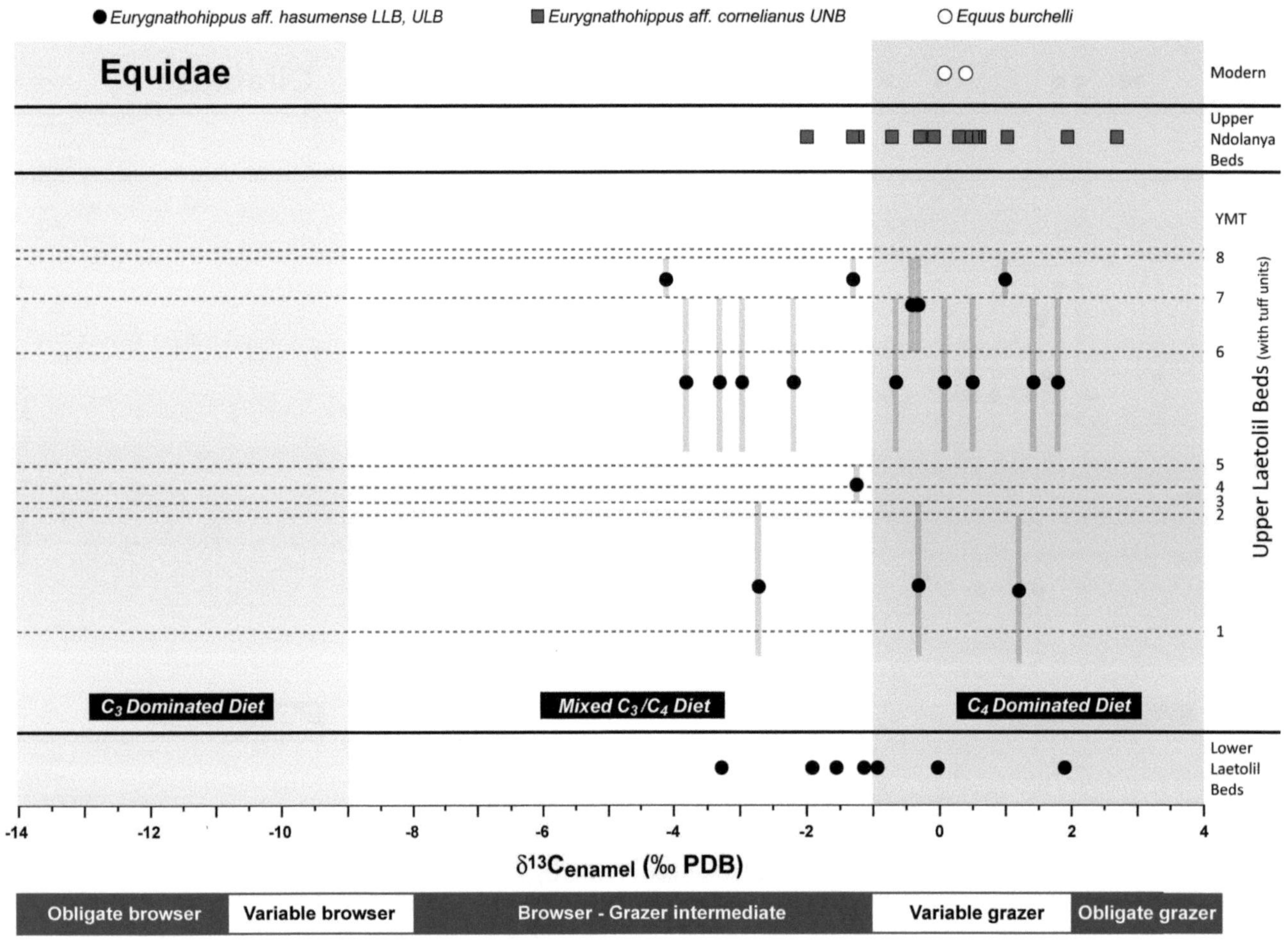

Fig. 15.5 Stable carbon isotopic composition of fossil Equidae enamel ($\delta^{13}C_{enamel}$) from the Laetolil Beds and Upper Ndolanya Beds and modern specimens, depicting specific stratigraphic provenance of samples

conditions associated with fossil localities, simple comparisons are complicated by the great diversity in behaviors, local conditions, and regional and global climatic systems controlling $\delta^{18}O$ in terrestrial ecosystems and ultimately in fossil enamel (Kingston and Harrison 2007: Fig. 6). In recognition of the difficulties in systematically utilizing $\delta^{18}O_{enamel}$ as a proxy for seasonality, rainfall, or humidity at different sites and times in the past, Levin et al. (2006) devised an approach for using the $\delta^{18}O_{enamel}$ of mammals with different foraging behaviors to reconstruct a terrestrial aridity index. The index is based on the notion that the $\delta^{18}O$ of herbivore tooth enamel of some species vary with aridity (evaporation sensitive taxa) due to ingestion of evaporative enrichment in leaf water, while the isotopic values of other species is primarily recording the composition of meteoric water that they drink (evaporation insensitive taxa). The difference between evaporation insensitive and sensitive taxa is minimal in more mesic environments but increases as the aridity does. In theory, this approach eliminates or minimizes the need to control for taxon-specific fractionation factors between the $\delta^{18}O$ of environmental water and body water as well as factors other than aridity that control $\delta^{18}O$ cycling in terrestrial ecosystems. Evaporation sensitive taxa are basically those that can survive by drinking little to no water and get a substantial portion of their water from leaves, such as modern giraffids, oryx, and dik-dik, while evaporation insensitive taxa get their water primarily from drinking. Water deficit is used as a proxy of aridity in this analysis, with water deficit showing a marked increase going from closed canopy forests to arid shrublands (Levin et al. 2006). According to the model, mean annual precipitation (MAP) can be calculated from water deficit, but requires knowledge of the potential evapotranspiration, which in turn is calculated from mean annual temperature and latitude (Thornthwaite 1948; Levin et al. 2006).

In applying this index to the Laetoli fauna (both modern and fossil) (Tables 15.4 and 15.5), it was clear that the calculated water deficit is to a large extent contingent on the taxa assigned the evaporation sensitive or insensitive designation. Noting that isotopic analyses of modern Laetoli fauna are limited, modern giraffes were used as the evaporation

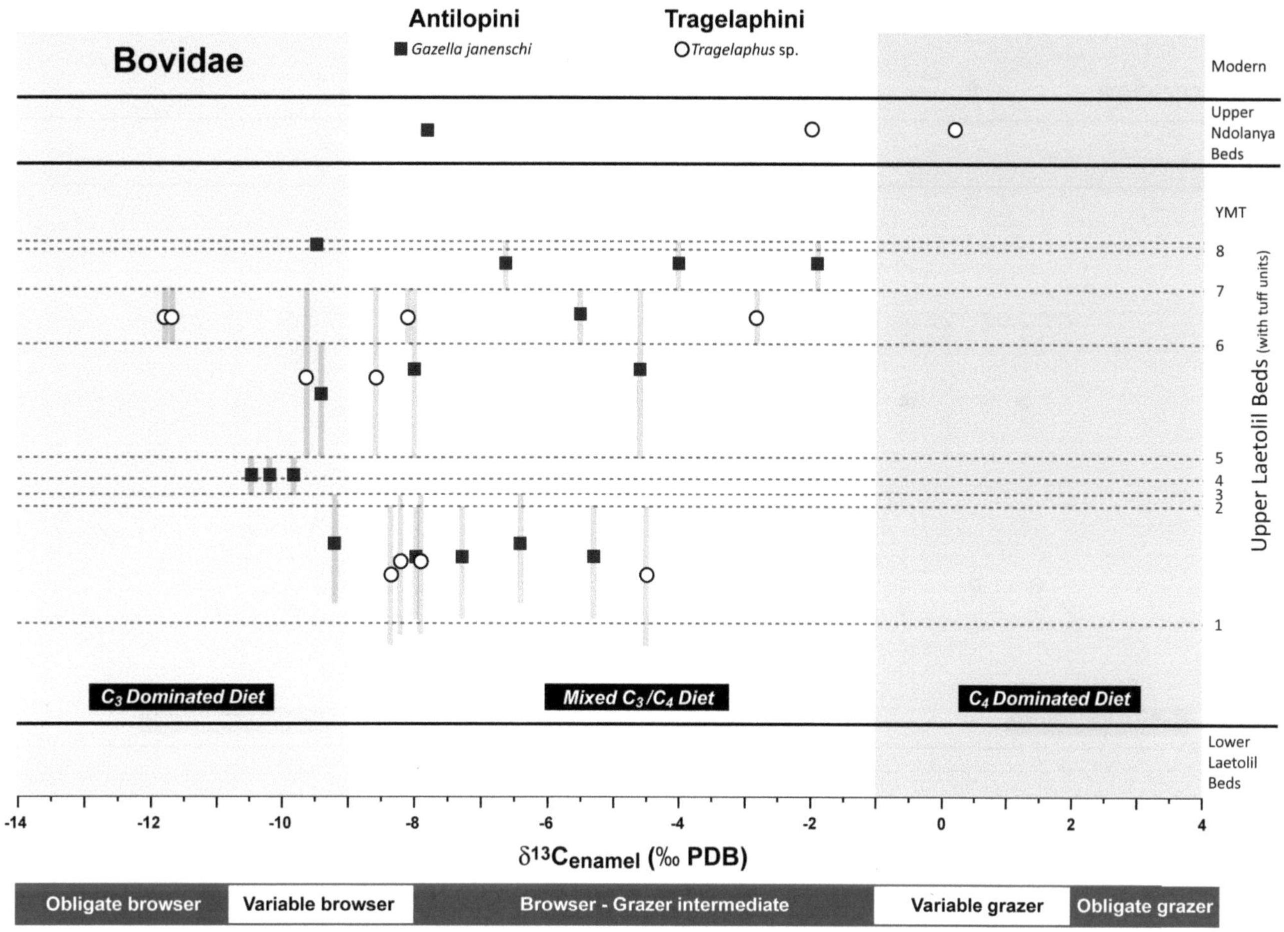

Fig. 15.6 Stable carbon isotopic composition of fossil Bovidae (tribes Antilopini and Tragelaphini) enamel ($\delta^{13}C_{enamel}$) from the Laetolil Beds and Upper Ndolanya Beds, depicting specific stratigraphic provenance of samples

sensitive taxa and the warthog as the evaporation insensitive taxa. Based on modern datasets from elsewhere, this partitioning appears reasonable. The resulting water deficit was 1472 (Table 15.5), a value that lies intermediate between Olorgesailie, Kenya (MAP=417 mm) and Turkana, Kenya (MAP=178 mm) (Levin et al. 2006, Supplementary Table 1). MAP for modern Laetoli has been estimated at 700–900 mm (Andrews et al. 2011), with high local variability related to topography and altitude. The estimated MAP, based on the calculated water deficit, situates modern Laetoli on the low end of the rainfall spectrum, potentially below that typically recorded at Laetoli today but similar to other sites in the Serengeti region such as Olduvai that currently receives ~300 mm per year.

In calculating the aridity index (water deficit) for Laetoli in the past, the focus was on faunal $\delta^{18}O_{enamel}$ recovered from the ULB. Evaporation sensitive taxa were identified based on modern analog comparisons, ^{18}O enrichment relative to other fauna, and $\delta^{13}C_{enamel}$ values indicating significant browsing. Giraffid and hippotragine $\delta^{18}O_{enamel}$ values were most positive and used separately and collectively to calculate enrichment factors relative to evaporation insensitive taxa (*Notochoerus euilus*, *Loxodonta exoptata*, *Ceratotherium efficax*). Depending on specific taxa compared and whether the evaporation sensitive taxa are combined, water deficit calculations range from 744 year to 1,426 mm/year, in general indicating that Laetoli was *more* arid during ULB times than it is currently. The lower water deficit value is most comparable to that calculated for Mpala, Kenya where the MAP is 703 mm and the higher water deficit is similar to Olorgesailie, Kenya which has a MAP of 417 mm (Levin et al. 2006, Supplementary Table 15.1). Assigning an aridity index to the ULB assemblage involves a number of assumptions that ultimately lead to wide range of calculated water deficit values, effectively limiting the resolving potential of this approach to reconstructing humidity in the past at Laetoli. In the application here, criteria used to identify and differentiate evaporation insensitive from sensitive taxa essentially maximized the enrichment factor between the two (Table 15.5) and the resulting water deficit estimate of aridity. If the ULB dik-dik

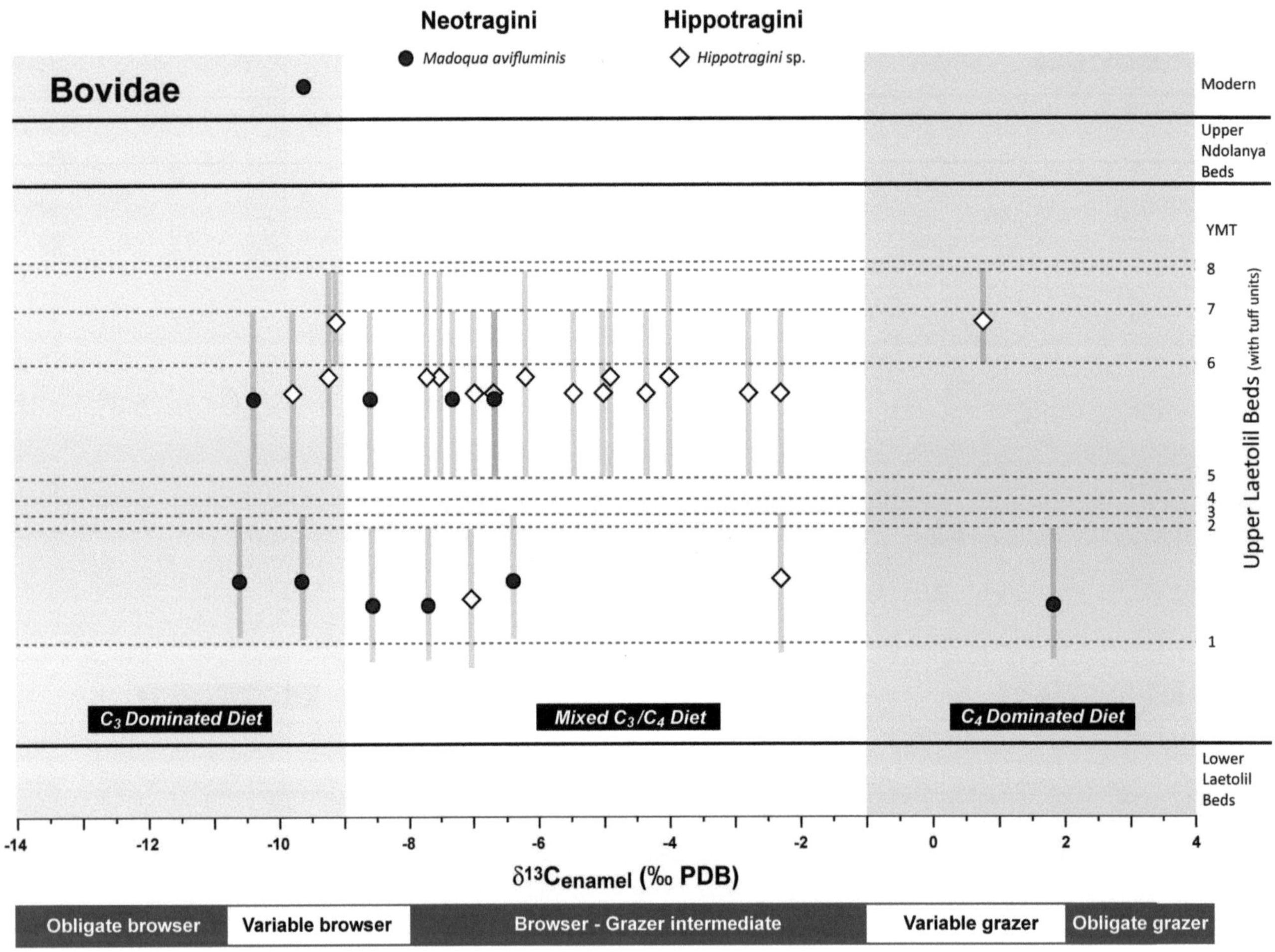

Fig. 15.7 Stable carbon isotopic composition of fossil Bovidae (tribes Neotragini and Hippotragini) enamel ($\delta^{13}C_{enamel}$) from the Laetolil Beds and Upper Ndolanya Beds and a modern sample, depicting specific stratigraphic provenance of samples

(*Madoqua avifluminis*) or the antilopine *Gazella janenschi* are used as evaporation sensitive taxa (a reasonable assignment based on comparisons with modern representatives of these lineages) the enrichment between the $\delta^{18}O_{enamel}$ values from evaporation sensitive and insensitive animals would be significantly reduced as would the water deficit, resulting in more humid interpretations. Specific regressions linking enrichment factors to water deficit are currently unavailable for these taxa. Appropriate application of the aridity index requires that the samples included in the analysis derive from stratigraphically constrained time intervals, primarily to minimize temporal mixing of fauna from oscillating environmental conditions. As applied to Laetoli, these criteria can be difficult to assess since the ~200 kyr potentially represented by the ULB may span fluctuating environmental conditions. Overall, the considerations involved in developing an aridity index provide valuable insights into interpreting the $\delta^{18}O_{enamel}$ of fossil assemblages and, as modern analogs are developed, this may become an increasingly valuable tool.

Isotopic Interpretations of Mammalian Enamel

Previous research at Laetoli included interpretations of isotopic data for each of the fossil herbivore families at Laetoli, including comparisons with relevant fossil and modern data sets (Kingston and Harrison 2007). Discussion here will build on these earlier assessments using new isotopic data and faunal analyses. Where possible, the isotopic data are interpreted relative to the temporal partitioning within the Laetoli succession as discussed earlier.

Proboscidea

The enamel of four of the six proboscideans identified from the Laetolil Beds and the ULB were analyzed isotopically (Fig. 15.1). The bulk of these samples (29/33) were enamel fragments attributed to the elephantid *Loxodonta exoptata* collected from ULB and UNB. In general, the carbon isotopic

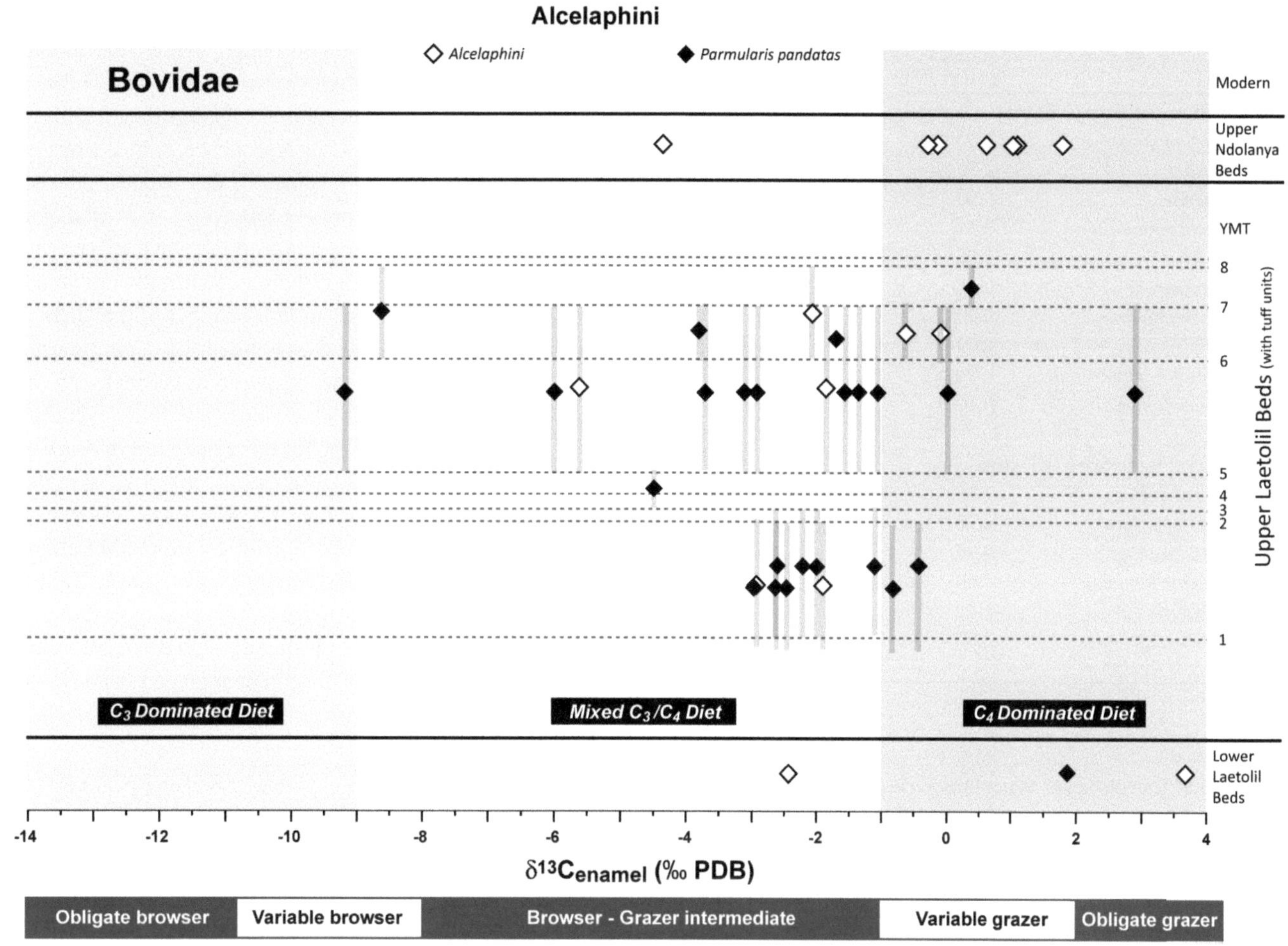

Fig. 15.8 Stable carbon isotopic composition of fossil Bovidae (tribe Alcelaphini) enamel ($\delta^{13}C_{enamel}$) from the Laetolil Beds and Upper Ndolanya Beds, depicting specific stratigraphic provenance of samples

Table 15.3 Dietary classification with $\delta^{13}C$ ranges based on mixing model

Dietary classification	General foraging strategy [a]	Approximate modern $\delta^{13}C$ range	+1.5‰ adjusted pre-industrial fossil $\delta^{13}C$ range [b]
Obligate grazer	Almost exclusively monocots (>95%)	> 0.5‰	>2‰
Variable grazer	Predominance of monocots (75–95%) but with seasonal, geographic, or ecological variability in C_3 dicot consumption	0.5 to −2.5‰	2 to −1‰
Browser-grazer Intermediate	Mixed diets characterized by intermediate proportions of dicots and monocots	−2.5 to −9.5‰	−1 to −8‰
Variable browser	Predominance of dicots (75–95%) with seasonal, geographic or ecological variability in C_4 monocot consumption	−9.5 to −12.3‰	−8 to −10.8‰
Obligate browser	Almost exclusively dicots (>95% C_3 dicots)	<−12.3‰	<−10.8‰

[a]Categories based on discussions of East African herbivore foraging strategies in Tieszen et al. (1979); McNaughton and Georgiadis (1986); and Gagnon and Chew (2000).

[b]All modern enamel data is converted to this range to facilitate comparisons in Figs. 15.1–15.8.

signature of *L. exoptata* yielded highly variable $\delta^{13}C_{enamel}$ signals throughout the sequence. Most carbon isotopic values plotted in the browser-grazer intermediate foraging range, but with a distinct emphasis on C_4 grazing in mixed feeding. Three specimens (two from ULB 1 from UNB) had ^{13}C enriched signatures, consistent with variable grazing characterized by a diet dominated by C_4 grasses (>75%–95%). New data here effectively extended the dietary range of ULB

Table 15.4 Average $\delta^{18}O_{enamel}$ of Laetoli herbivores used to calculate aridity index

	Average $\delta^{18}O$	±	*n*	Formation
Evaporation sensitive (ES) taxa				
Deinotherium bozasi	0.00	0.04	2	ULB
Giraffa stillei	1.50	1.8	22	ULB
Giraffa jumae	0.50	2.4	17	ULB
Sivatherium maurusium	0.34	0.9	8	ULB
all giraffids	1.00	2	47	ULB
Gazella janenschi	0.42	1.6	17	ULB
Madoqua avifluminis	–0.90	1.26	10	ULB
Hippotragus sp.	0.69		18	ULB
ES (Giraffids + *Hippotragus*)	0.90		65	ULB
Giraffa camelopardalis	6.10	1.5	8	Modern Laetoli
Evaporation insensitive (EI) taxa				
Notochoerus euilus	–1.30	1.2	40	ULB
Ceratotherium efficax	–0.70	1.6	30	ULB
Loxodonta exoptata	–0.90	0.9	21	ULB
Kolpochoerus limnetes	–1.60	1.1	6	UNB
Potamochoerus porcus	–1.40	1.4	5	ULB
Phacochoerus africanus	3.00	1	12	Modern Laetoli

L. exoptata toward the C_3 end-member by about 1‰ to an end-member value of –6.3‰, although overall the median $\delta^{13}C_{enamel}$ values shift only slightly (from –3.5‰ to –3.8‰ in the ULB). While there are no statistically significant shifts in isotopic variability within the Laetoli sequence, there is a marked shift in the median $\delta^{13}C_{enamel}$ values of *L. exoptata* between the ULB and the UNB, from –3.8‰ to –1.9‰. If dietary proportions of C_3 and C_4 vegetation reflect that of standing mass in past habitats, these data indicate a shift toward greater C_4 biomass during UNB times relative to that of the ULB.

Analysis of an additional specimen of *Deinotherium bozasi* reinforces the obligate browser assessment of this taxon, with the most ^{13}C depleted values (with the exception of two giraffid specimens). The foraging behavior of *Anancus kenyensis* in the LLB is also based on only two samples. While the earlier $\delta^{13}C_{enamel}$ value of *A. kenyensis* was consistent with that from other fossil localities, indicating almost exclusive C_4 grazing, a second specimen from the LLB indicates obligate browsing. Alternatively, this gomphothere was grazing on C_3 grasses. This dietary signal is 1.5‰ more negative than an *Anancus* sp. fragment analyzed from Nakali

Table 15.5 Calculation of Water Deficit (WD) for modern and fossil Laetoli tooth samples based on $\delta^{18}O_{enamel}$ values (based on Levin et al. 2006). East African localities with comparable WD values and corresponding Mean Annual Precipitation (MAP)

Evaporation sensitive- insensitive Taxon/taxa pairs from Table 15.4 used to calculate enrichment factor (ε_{ES-EI}): *Evaporation sensitive* Taxa/Taxon	*Evaporation insensitive* Taxa/Taxon	ε_{ES-EI}[a]	WD[b] (mm/year)	Modern East African analog localities (Climate station) with similar WD and corresponding MAP[c]
Modern				
Giraffa camelopardalis	*Phacochocperus africanus*	3.09	1,471	Olorgesailie (Magadi) (WD = 1,341 mm/yr; MAP = 417 mm) Turkana (Lodwar) (WD = 1,588 mm/yr; MAP = 178 mm)
Fossils				
All giraffids	*Notochoerus euilus*	2.30	1,347	Olorgesailie (Magadi) (WD = 1,341 mm/yr; MAP = 417 mm)
All giraffids	Rhinocerotids	1.70	821	Mpala (El Karama Ranch) (WD = 751 mm/yr; MAP = 703 mm) Samburu (Isiolo) (WD = 913 mm/yr; MAP = 648 mm)
All giraffids	Elephantids	1.90	870	Mpala (El Karama Ranch) (WD = 751 mm/yr; MAP = 703 mm) Samburu (Isiolo) (WD = 913 mm/yr; MAP = 648 mm)
All *Hippotragus*	*Notochoerus euilus*	1.99	1,298	Amboseli (WD = 1,210 mm/yr; MAP = 400 mm) Olorgesailie (Magadi) (WD = 1,341 mm/yr; MAP = 417 mm)
All *Hippotragus*	Rhinocerotids	1.39	745	Mpala (El Karama Ranch) (WD = 751 mm/yr; MAP = 703 mm)
All *Hippotragus*	Elephantids	1.59	794	Mpala (El Karama Ranch) (WD = 751 mm/yr; MAP = 703 mm)
Giraffa stillei	*Notochoerus euilus*	2.80	1,426	Olorgesailie (Magadi) (WD = 1,341 mm/yr; MAP = 417 mm) Turkana (Lodwar) (WD = 1,588 mm/yr; MAP = 178 mm)
Giraffa stillei	Rhinocerotids	2.20	944	Samburu (Isiolo) (WD = 913 mm/yr; MAP = 648 mm)
Giraffa stillei	Elephantids	2.40	993	Samburu (Isiolo) (WD = 913 mm/yr; MAP = 648 mm) Tsavo (Voi) (WD = 1,059 mm/yr; MAP = 549 mm)
All ES taxa	*Notochoerus euilus*	2.20	1,331	Olorgesailie (Magadi) (WD = 1,341 mm/yr; MAP = 417 mm)
All ES taxa	Rhinocerotids	1.60	796	Mpala (El Karama Ranch) (WD = 751 mm/yr; MAP = 703 mm)

[a] $\varepsilon_{ES-EI} = ([(1{,}000^* \delta^{18}O_{ES})/(1{,}000^* \delta^{18}O_{ES})] - 1) * 1{,}000$

[b] $WD = (\varepsilon_{ES-EI} + 6.21)/0.00632$

[c] Levin et al. (2006)

(10.4–9.6 Ma) (Cerling et al. 1999), at a much earlier time when C_4 grasses were not widespread. Further isotopic sampling and analysis of *Anancus* from the LLB is required to explore this potentially anomalous dietary signal.

Rhinocerotidae

The number of rhinocerotids represented in the isotopic analyses was increased from $n = 9$ to $n = 38$, providing an opportunity to more thoroughly assess the foraging strategy of these taxa in the Laetoli Succession (Fig. 15.2). These data support previous interpretations that *Ceratotherium* in the LLB and ULB was primarily a mixed C_3/C_4 feeder, with the median of ULB rhinos remaining at −5.9‰ with the addition of 29 values. All of the $\delta^{13}C_{enamel}$ values are assigned to a brower-grazer intermediate, variable browser and even obligate browser categories with limited evidence of C_4 dominated diets, in contrast with a number of studies that have linked the evolution of *Ceratotherium* with the spread of (C_4) grasslands (Harris 1983; Carroll 1988). Mesowear studies indicate an intermediate functional state between extant *Diceros* and extant *Ceratotherium,* also suggesting significant amounts of browse (Hernesniemi et al. 2011). These findings support the interpretation that a C_3 dietary signal reflects browsing behavior, rather than C_3 grazing. Mesowear analysis in general, however, does appear to indicate more grazing than is consistent with the isotopic interpretations.

On this basis, it is reasonable to assume that C_3 grass may have comprised a portion of the diet. The fact that equids from the same intervals have yielded almost exclusive C_4 grazing signals suggests that C_4 grasses were readily available and it becomes necessary to envision environmental conditions that could support both C_3 and C_4 grasses. These could be areas of high relief with an altitudinal transition from C_3 to C_4 grasses or more wooded/forested areas with sub-canopy conditions favorable for C_3 grasses in proximity to more open C_4 grass dominated ecosystems. Isotopic analyses of ULB *Ceratotherium* contrast with specimens analyzed from pene-contemporaneous Pliocene sites in Chad, Turkana, and Baringo, Makapansgat and especially modern *Ceratotherium*, all of which tend to indicate significantly more C_4 grazing (Cerling et al. 2003; Kingston 1999; Zazzo et al. 2000; Sponheimer et al. 2001). Two isolated fossil *Ceratotherium* samples from Rift Valley sites (Cerling et al. 2003; Kingston 1999) have yielded $\delta^{13}C_{enamel}$ values similar to those of the ULB. Carbon isotopic signatures of *Ceratotherium* from the LLB also exhibit high variability with a range of over 9‰. Overall, there are no apparent isotopic shifts that would indicate changes of foraging behavior within the LLB and ULB.

Mesowear studies also reflect relative stasis in dietary patterns in the Laetolil Beds, with only minor shifts in grazing/browsing proportions that cannot be distinguished from random noise (Hernesniemi et al. 2011). A major transition to more grazing by UNB times is revealed by significantly more abrasion (Hernesniemi et al. 2011). Unfortunately, only one tooth fragment from the UNB was analyzed. It would be interesting to expand isotopic analyses of UNB rhinos to document any isotopic shifts that might mirror the mesowear pattern transition. Two specimens attributed to *Diceros* sp. were analyzed, yielding $\delta^{13}C_{enamel}$ values indicative of roughly equal dietary proportions of C_3 and C_4 plants (−6.0‰ and −5.2‰). These data contrast with the mean of −10.5‰ for modern *Diceros bicornis*, with ^{13}C enriched vales of −7‰ in arid habitats, and suggest a more mixed feeding niche for *Diceros* in the past. For fragmentary isolated teeth, however, definite identification of *Diceros* and *Ceratotherium* remains difficult (Hernesniemi et al. 2011).

Suidae

Isotopic analyses of suids from Laetolil focused on *Notochoerus euilus* from the ULB, but also included a sample identified as *Notochoerus* cf. *Notochoerus jaegeri* from the LLB, several specimens attributed to *Potamochoerus porcus* from the ULB, and UNB enamel fragments identified as *Metridiochoerus andrewsi* and *Kolpochoerus* sp. Dental and postcranial features of *N. euilus* have been interpreted as adaptations to abrasive grazing diet and foraging in savanna or grasslands biomes (Cooke and Wilkinson 1978; Harris 1983, 1987). $\delta^{13}C_{enamel}$ of 40 individuals ranged from −8.2‰ to 2.0‰ with a median of −1.9‰, indicating eclectic, mixed C_3/C_4 foraging, but with an overall diet dominated by C_4 grasses (Fig. 15.3). These values support the morphologically-based interpretations of grazing, but also are consistent with the notion that these suids were eclectic feeders, incorporating significant amounts of C_3 biomass into their diet. Once again, the question of C_3 grasses or C_3 browse is relevant. Tooth abrasion patterns similar to the forest-dwelling modern giant forest hog *Hylochoerus meinhertzhageni* (Cooke 1985; Kullmer 1999) indicate a browsing component for *N. euilus*. In addition, ecomophological analyses of the postcrania indicate morphological characters associated with woodland habitat preferences (Bishop et al. 1999). Cumulatively these data can be accommodated by inferring paleohabitats in which tropical grasses are more sparsely distributed in the substratum of open woodlands or in the disturbed portions of closed woodland habitats, rather than as grasslands.

Giraffidae

The addition of 18 new isotopic analyses of Laetoli giraffids (bringing the total giraffid sampling to $N=42$) reinforced previous interpretations of this family as obligate to variable browsers with a $\delta^{13}C_{enamel}$ range constrained between −8.3‰

and −12.9‰ (Fig. 15.4). The three giraffids analyzed here - a medium-sized and large giraffine (*Giraffa stillei* and *G. jumae* respectively) and a sivathere (*Sivatherium maurusium*) - were all from the ULB except for a single *S. maurisium* specimen from the UNB. Giraffids comprise an unusually large proportion of the herbivore fauna in the ULB, comprising approximately 15% of the ruminant artiodactyls recovered (Robinson 2011). It is likely that there was ecological partitioning to accommodate a relatively high abundance of three sympatric giraffids, presumably based on different optimum feeding heights, differential selectivity in browse, or foraging in different microhabitats. However, isotopic analyses (C and O) did not differentiate feeding niches in the Laetoli giraffids. This may reflect the complexity of the browsing strategies of these taxa or the limited resolution in differentiating aspects of C_3 vegetation and microhabitat by isotopic analyses. In addition, the data did not indicate shifts in foraging patterns of the three species in the ULB sequence that might correlate with documented changes in size (Robinson 2011).

Equidae

Isotopic analyses of the Laetoli equids indicate variable to obligate grazing. $\delta^{13}C_{enamel}$ values of *Eurygnathohippus* aff. *hasumense* from the LLB and ULB range from −4.1‰ to 1.9‰ with medians of −1.1‰ and 0.5‰ respectively (Fig. 15.5). These relatively ^{13}C depleted values relative to obligate grazers (>2‰) suggests that *Eurygnathohippus* was a C_4 grazer supplementing its diet with either C_3 browse or C_3 grasses. These strategies possibly reflect behaviors or adaptations associated with open C_4 grassland habitats adjacent to more wooded habitats that may have supported C_3 grasses. Mesowear signals of *Eurygnathohippus* sp. show the most abrasion dominated feeding regimes throughout the Laetoli succession (Kaiser 2011), suggesting that the C_3 signal reflects a dietary C_3 grass component. The overall isotopic range for *Eurygnathohippus* from the Upper Laetolil Beds is more positive than in other fossil equid assemblages (except for the KT unit in Chad) and more similar to the adjusted extant equid values. This difference indicates either less C_3 biomass in the diet of equids at other sites or that local C_4 grasses were ^{13}C-enriched because they were dominantly classical-NADP subtypes associated with more mesic environments than the arid adapted Nad-me or PEP-ck C_4 grasses (Chapman 1996). Supporting this later interpretation, $\delta^{18}O_{enamel}$ values for Laetoli *Eurygnathohippus* are similar to or more negative than those from other fossil localites and generally more negative than all sampled modern *Equus burchelli* from East Africa. $\delta^{13}C_{enamel}$ values of *Eurygnathohippus aff. cornelianus* from the UNB range from −1.9‰ to 2.7‰ with a median of 0.3‰. While not statistically significant, these data reflect a shift to increased C_4 grazing consistent with interpretations of UNB environments as more open.

Bovidae

Bovids represent the most abundant mammalian taxa in the Laetoli fossil assemblage (Leakey 1987; Harrison 2011) and provide key insights into paleoecological interpretations within the succession (Kovarovic et al. 2002; Kovarovic and Andrews 2007, 2011; Kaiser 2011). Initial isotopic analyses of representatives of five bovid tribes from the Laetoli assemblage revealed complex, mixed C_3/C_4 feeding patterns for most of the taxa (Kingston and Harrison 2007). These data contrast with the more specialized foraging strategy of modern taxa in these lineages that tend to either graze or browse, but do not typically engage in mixed feeding. The isotopic ranges suggest that ancestral lineages were perhaps more generalized feeders or occupied different niches that do not appear to have modern analogs. Increasing the number of bovid specimens analyzed isotopically by over 60% reinforced these interpretations, but also reveals additional insights into the foraging strategies of extinct Laetoli bovids. Despite the number of bovid teeth analyzed ($n=101$), it remains difficult to define any trends or shifts as statistically significant due to small sample sizes for any given taxon or lineage.

Tragelaphines are not common at Laetoli and may include two differently-sized species in the ULB, although they are described here collectively as *Tragelaphus* sp. (Gentry 2011). Unlike modern tragelaphines (i.e., kudu, eland, bushbuck) that tend to be predominantly browsers, fossil representatives from Laetoli exhibit a significant range of $\delta^{13}C_{enamel}$ values indicating variable inclusion of dietary C_4 grasses along with browse (Fig. 15.6). The median for *Tragelaphus* sp. from the ULB, however, is −8.3‰ indicating a predominance of C_3 plants in the diet, although two specimens indicate that the extinct forms may have also occupied mixed feeding niches with up to 50% of the diet comprised of C_4 grasses. Mesowear scores for ULB tragelaphines yielded the most attrition dominated end of the mesowear continuum for the Laetoli ungulates, indicating exclusive browsing (Kaiser 2011) and prompting the suggestion that the C_4 signal may reflect C_4 dicots rather than grasses. Analysis of tragelaphines from the UNB (*Tragelaphus* sp. cf. *buxtoni*) yielded $\delta^{13}C_{enamel}$ values of −2.1‰ and 0.1‰ indicating a shift to a variable grazing niche. While mesowear scores of UNB tragelaphines indicate intermediate-feeding, both data sets reflect a shift to more grazing at this transition potentially as a result of different ecological settings.

Modern antilopine species (i.e., gazelles, gerenuk, dibatag) tend to inhabit more open habitats and have diets ranging from variable grazers to variable browsers.

An average $\delta^{13}C_{enamel}$ value of –9.2‰ for extant antilopines from South and East Africa (n=78) (Cerling et al. 2003; Sponheimer et al. 2003) reflecting mixed feeding strategies, but with an emphasis on browsing. *Gazella janenschi* is the only taxon from this tribe analyzed isotopically from the Laetolil Beds and the UNB (Fig. 15.6). As with modern gazelles, $\delta^{13}C_{enamel}$ signatures indicate mixed feeding with a range of –10.5‰ to –2‰ but with an emphasis on browsing (median = –7.6‰). Specimens from above Tuff 5 in the ULB appear to include more ^{13}C enriched values suggesting niche expansion to incorporate more C_4 grazing into the foraging strategy. The single sample from the UNB (–7.8‰) is consistent with the median for the ULB, but clearly more sampling is required to characterize the variability that might be inherent in UNB foraging behavior.

Additional analyses of neotragines from the ULB (*Madoqua avifluminis*) shifted the median to a slightly more ^{13}C depleted value (<8‰) than previously reported (Kingston and Harrison 2007), indicating that this extinct species was more of a variable browser than an intermediate feeder (Fig. 15.7). However, values of these extinct neotragines are consistently more ^{13}C enriched, by 2–3‰, than modern dik-diks at Laetoli and various sites in Kenya (Cerling et al. 2003). A single specimen attributed to *M. avifluminis* from the lower part of the ULB yielded one of the most ^{13}C enriched carbon isotopic signals of all the herbivores. The implication that this individual was an obligate grazer is inconsistent with any of the other lines of evidence and contrasts with the other isotopic data from Laetoli and elsewhere. Removing this sample from the analysis shifts the median to almost –9‰. Kaiser (2011) reported no abrasive signal in the mesowea r signal of this species and suggests that either the C_4 component in the isotopic signal is related to feeding on C_4 dicots or was the result of feeding in more humid conditions leading to a less grit-laden environment.

Hippotragus sp. is common in the Laetolil Beds (Gentry 2011) and, based on interpretations of modern representatives of this tribe (i.e., oryxes and sables), potentially represents a major grazing guild in the fossil assemblage. However, isotopic analyses has revealed a wide dietary niche for this extinct taxon that includes primarily mixed feeding and C_3 foraging with no evidence of the obligate grazing that characterizes modern representatives (Fig. 15.7). $\delta^{13}C_{enamel}$ values range from –9.8‰ to 0.8‰, reflecting one of the most eclectic dietary signals analyzed in this study. These data are corroborated by the mesowear signals indicative of a mixed/intermediate feeding strategy (Kaiser 2011), suggesting that the this group may have been formerly more common and widespread, and more ecologically diverse. The availability of browsing niches for hippotragines at Laetoli suggests potentially more wooded habitats during ULB times.

Alcelaphines are represented in the Laetolil Beds by two species – an abundant form with medium-sized teeth (i.e., *Parmularius pandatus*) and a rarer, larger form designated as Alcelaphini larger sp. indet. (Gentry 2011). *Parmularius pandatus* from the ULB yielded the greatest range of $\delta^{13}C_{enamel}$ values (Fig. 15.8). In particular, dietary signals from teeth collected between Tuffs 5 and 7 range from –9‰ to +3‰ indicating eclectic foraging strategies. Although not statistically significant, this range reflects a pronounced expansion in foraging strategy (spanning over 12‰) from earlier ULB times (i.e., below Tuff 3) where $\delta^{13}C_{enamel}$ values are confined to within about 2‰ (–2.6‰ to –0.4‰). An expansion could be related to factors such as increasing ecological heterogeneity, changes in foraging competition, or short-term environmental instability. While this high variability is confined to specimens collected between Tuffs 5 and 7, this stratigraphic interval may represent as much as 130,000 kyrs based on chronostratigraphic extrapolations (Deino 2011), conflating shifting feeding patterns through time. It is also interesting that the one sample of *P. pandatus* collected from the LLB had a carbon isotopic value of 1.9‰, the second most ^{13}C enriched sample of this taxon (n=26), suggesting obligate C_4 grazing at LLB times. Initial isotopic analyses of three specimens identified as Alcelaphini sp. document some of the most convincing evidence for grazing within the bovid groups analyzed (Kingston and Harrison 2007). Fifteen additional samples analyzed here confirm that both the medium and large alcelaphines were essentially dedicated C_4 grazers. In addition, the data reveals a significant shift in the proportions of C_4 graze between the ULB and the UNB alcelaphines with medians of about –1‰ and +2‰ respectively. The more obligate grazing behavior characterizing the UNB assemblage may reflect more open ecosystems with greater proportions of C_4 grass at this time. Alternatively, the UNB fauna, dominated by medium- to large-sized grazing bovids, may represent a mix of resident and migratory communities, documenting seasonal migrations of grazing guilds into the Laetoli communities.

Cercopithecidae

Three cercopithecoid fossil enamel specimens were analyzed isotopically from the upper sequence of the ULB (between Tuffs 5 and 8). Two were identified as *Parapapio ado* and the third as Cercopithecidae gen. and sp. indet. (although it is likely *P. ado* as well). All three samples yielded essentially the same $\delta^{13}C_{enamel}$ values (–6.5 to –6.6‰) and very similar $\delta^{18}O_{enamel}$ signatures (0.0 to –1.2‰). The carbon isotopic signatures indicates an mixed diet that included both C_3 and C_4 dietary components, in the browser-grazer intermediate dietary category. To the extent that modern papionins represent a modern dietary analog for fossil baboons (although see Codron et al. 2005; El-Zaatari et al. 2005), the C_3 category could include the leaves, fruits, nuts, seeds, and corms of trees as well as shrubs, forbs, and tubers and the C_4 component grass leaves, seeds, rhizomes. Insects, C_3 and/or C_4 sedges, and CAM-photsynthesizing succulents are additional dietary components that complicate interpretations based solely on isotopic values. Modern olive baboons (*Papio anubis*), for example, have yielded $\delta^{13}C_{enamel}$ values ranging from −17.1‰ (Ituri Forest)(Cerling et al. 2004) to −4‰ (Laikipia Plateau)(Levin et al. 2008) a range of over 20‰. Although the carbon isotopic composition of fossil and modern baboons reveal that diets of many taxa include varying amounts of a C_4 component, the Laetoli primate $\delta^{13}C_{enamel}$ values are ^{13}C enriched relative to almost all analyzed papionins (Levin et al. 2008; Codron et al. 2005; van der Merwe et al. 2003; Lee-Thorp et al. 1989a, 1994; White et al. 2009), indicating a greater C_4 dietary component. Only fossil enamel of gramnivorous *Theropithicus* from Swartkrans and Sterkfontein in South Africa consistently yield more positive $\delta^{13}C_{enamel}$ values (mean 2.2‰ ± 1.5‰) (Lee-Thorp et al. 1989a; van der Merwe et al. 2003; Codron 2005). The fossil Laetoli data suggest that *P. ado* was a generalist with a diet adapted to exploit C_4 grasses or grass-based foods resources, possibly in more open habitats within a heterogeneous ecosystem. However, studies of modern baboons suggest that they can be highly selective feeders and that their diets do not necessarily reflect the different savanna landscapes in which they forage (Codron et al. 2006).

Carnivora

Factors influencing the $\delta^{13}C_{enamel}$ of carnivore teeth are complex as diets can consist of variable amounts of animal proteins and lipids (as well as some plant material), which in turn have varying carbon isotopic compositions. Despite these differences, the $\delta^{13}C_{enamel}$ of carnivores tends to closely resemble that of their prey (Lee-Thorp et al. 1989b). Carbon isotopic signatures of −7.2‰ and −8.0‰ for *Crocuta* sp. and *Felis* large sp. enamel respectively from the ULB indicates that they were preying on herbivores with a mixed, but predominantly C_3 diet, possibly *G. janenschi*, *Tragelaphus* sp., *Hippotragus* sp., or *M. avifluminis*.

Ostrich Eggshell (OES) Analyses

Struthionid eggshell fragments represent a ubiquitous component of the Laetoli fossil assemblages and >800 specimens have been systematically sampled from the sequence since 1998. Morphology of the eggshells and their taxonomic affinities indicates that two different species of *Struthio* are represented – *Struthio kakesiensis* in the LLB (~4.3–3.8 Ma) and ULB below Tuff 3 (~3.8 Ma) which is replaced higher in the ULB by the modern ostrich, *S. camelus* (Harrison and Msuya 2005). Isotopic analyses of the organic and inorganic components of fossil eggshells provide an archive of diet and prevailing climatic conditions during the past, which can then be compared to morphological attributes the ostrich eggshell, such as thickness and pore density and size which may be linked to environmental conditions (Sauer 1968; Tullett and Board 1977; Mikhailov 1992). Modern ostrich eggshells collected from the Laetoli area were also included in the analysis to provide a comparative dataset that can be linked to known climatic conditions and the foraging behavior of extant ostriches. Isotopic analysis included an assessment of the inorganic fraction of 78 eggshell fragments from the Laetoli succession (Table 15.6). Of these, 22 samples also included analyses of the isotopic composition of the organic phase.

The potential of isotopic analyses of ostrich eggshells for paleodietary and paleoenvironmental reconstruction relies on preservation of the original biogenic and climatic signals during fossilization and burial. One approach to assessing the isotopic integrity of fossil material is to compare the carbon isotopic composition of the organic and inorganic fractions within each sample (Von Schirnding et al. 1982; Johnson et al. 1997). In modern samples, the offset between $\delta^{13}C$ value of the total organic and the inorganic ($\Delta^{13}C$) has been at about 14‰ ± 2‰. For modern ostrich eggshells collected at Laetoli, the average $\Delta^{13}C$ was 13.4‰ ± 1.9‰ ($N = 4$) and for the fossil ostrich eggshells was 12.6‰ ± 0.9‰ ($N = 18$) (Table 15.6). These data imply minimal diagenetic alteration of either of the organic or inorganic fractions for the samples analyzed here.

$\delta^{13}C_{OES}$ and Diet

The $\delta^{13}C_{OES}$ essentially reflects the bulk carbon isotopic signature of the bird's diet, integrated over 3–5 days during shell formation. Modern ostriches are opportunistic herbivores utilizing a wide variety of plant species, although they tend to prefer dicotyledonous C_3 plants to CAM plants and C_4 grasses. Relative to the isotopic composition of the diet, the organic and inorganic fractions in modern ostrich eggshells are 1.5‰ (± 0.8‰) and 16.2‰ (± 0.5‰) enriched in ^{13}C, respectively (Von Schirnding et al. 1982; Johnson et al. 1998). Based on these fractionation factors and the average isotopic signals for C_3, C_4, and CAM plants, specific dietary contributions have been calculated for fossil ratite eggshells (Johnson et al. 1997, 1999; Segalen et al. 2002; Clarke et al. 2006). However, given (1) the environmentally mediated variability inherent in the isotopic composition of C_3, C_4, and CAM plants, (2) deviations in physiological fractionation effects in fossil organisms, and (3) varying climatic and atmospheric parameters influencing isotopic cycling in the past, it remains difficult to

Table 15.6 Isotopic analyses of fossil and modern ostrich eggshells from Laetoli

Lab no.	$\delta^{13}C_{OES-carb}$ (‰)	$\delta^{18}O_{OES}$ (‰)	$\delta^{15}N_{OES}$ (‰)	$\delta^{13}C_{OES-org}$ (‰)	ΔC[a] (‰)	Taxon	Formation	Locality	Stratigraphic interval	Field no.
	Mineral phase		Organic phase							
176	−11.0	2.0	4.9	−22.3	11.3	*Struthio kakesiensis*	LLB	Kakesio 1–6		#60/01
200	−10.5	1.8	6.9	−22.4	11.9	*Struthio kakesiensis*	LLB	Kakesio 8		214/03
201	−9.9	3.3				*Struthio kakesiensis*	LLB	Kakesio 8		214/03
202	−10.1	−0.7	7.0	−22.8	12.6	*Struthio kakesiensis*	LLB	Kakesio 7		222/03
203	−10.1	0.8				*Struthio kakesiensis*	LLB	Kakesio 7		222/03
209	−10.6	−2.2	6.2	−23.0	12.3	*Struthio kakesiensis*	LLB	Kakesio 10		530/03
210	−10.0	−0.2				*Struthio kakesiensis*	LLB	Kakesio 10		530/03
215	−11.4	0.3				*Struthio kakesiensis*	LLB	Kakesio 1–6		1183/01
216	−10.0	3.7				*Struthio kakesiensis*	LLB	Kakesio 1–6		1183/01
243	−11.2	−2.0				*Struthio kakesiensis*	LLB	Kakesio 9		536/03
244	−9.4	5.9				*Struthio kakesiensis*	LLB	Kakesio 9		536/03
253	−9.5	0.9				*Struthio kakesiensis*	LLB	Emboremony 1		2077/03
254	−7.4	−1.0				*Struthio kakesiensis*	LLB	Emboremony 1		2077/03
175	−11.5	2.9	5.6	−23.2	11.7	*Struthio kakesiensis*	ULB	9S	Below tuff 2	#9/01
213	−11.6	−2.1				*Struthio kakesiensis*	ULB	9S	Below tuff 2	1278/01
214	−11.3	2.4	6.2	−23.4	12.0	*Struthio kakesiensis*	ULB	9S	Below tuff 2	1278/01
245	−10.4	1.3				*Struthio kakesiensis*	ULB	10W	Below tuff 2	1108/03
246	−11.1	3.3				*Struthio kakesiensis*	ULB	10W	Below tuff 2	1108/03
64	−10.6	3.1				*Struthio camelus*	ULB	10	Below tuff 3	534/98
219	−12.1	−2.4	4.7	−23.7	11.6	*Struthio kakesiensis*	ULB	10	Below tuff 3	651/01
220	−8.9	3.7				*Struthio kakesiensis*	ULB	10	Below tuff 3	651/01
207	−7.8	3.6				*Struthio camelus*	ULB	5	Between Tuffs 3 & 5	433/03
208	−10.3	4.2				*Struthio camelus*	ULB	5	Between Tuffs 3 & 5	433/03
241	−8.3	3.0				*Struthio camelus*	ULB	19	Between Tuffs 5 & 6	375/00
242	−8.3	3.1				*Struthio camelus*	ULB	19	Between Tuffs 5 & 6	375/00
65	−6.1	4.8				*Struthio camelus*	ULB	2	Between Tuffs 5 & 7	
221	−7.9	2.9				*Struthio camelus*	ULB	10E	Between Tuffs 5 & 7	575/01
222	−10.3	−0.8				*Struthio camelus*	ULB	10E	Between Tuffs 5 & 7	575/01
229	−7.9	3.0				*Struthio camelus*	ULB	13E	Between Tuffs 5 & 7	2096/00
230	−8.0	2.9				*Struthio camelus*	ULB	13E	Between Tuffs 5 & 7	2096/00
235	−8.1	4.0				*Struthio camelus*	ULB	2	Between Tuffs 5 & 7	740/00
236	−8.9	4.9				*Struthio camelus*	ULB	2	Between Tuffs 5 & 7	740/00
239						*Struthio camelus*	ULB	20	Between Tuffs 5 & 7	469/00
240	−8.1	3.8				*Struthio camelus*	ULB	20	Between Tuffs 5 & 7	469/00
251	−7.2	4.8				*Struthio camelus*	ULB	6	Between Tuffs 5 & 7	2041/03
252	−4.9	4.1				*Struthio camelus*	ULB	6	Between Tuffs 5 & 7	2041/04
255	−7.6	2.3				*Struthio camelus*	ULB	7	Between Tuffs 5 & 7	2154/03
256	−9.8	2.3	3.3	−23.1	13.3	*Struthio camelus*	ULB	7	Between Tuffs 5 & 7	2154/04

(continued)

Table 15.6 (continued)

Lab no.	$\delta^{13}C_{OES-carb}$ (‰)	$\delta^{18}O_{OES}$ (‰)	$\delta^{15}N_{OES}$ (‰)	$\delta^{13}C_{OES-org}$ (‰)	ΔC^{a} (‰)	Taxon	Formation	Locality	Stratigraphic interval	Field no.
	Mineral phase		Organic phase							
261	−5.3	5.6				*Struthio camelus*	ULB	12E	Between Tuffs 5 & 7	1572/04
262	−8.5	2.5				*Struthio camelus*	ULB	12E	Between Tuffs 5 & 7	1572/04
263	−8.7	1.0	7.3	−21.7	12.9	*Struthio camelus*	ULB	22	Between Tuffs 3 & 8	1324/01
264	−6.0	4.3				*Struthio camelus*	ULB	22	Between Tuffs 3 & 8	1324/01
265	−10.9	0.2				*Struthio camelus*	ULB	22E	Between Tuffs 3 & 8	1347/01
266	−10.9	0.3				*Struthio camelus*	ULB	22E	Between Tuffs 3 & 8	1347/01
237	−7.1	4.2				*Struthio camelus*	ULB	21	Between Tuffs 5 & 8	532/00
238	−5.8	3.1	4.1	−20.2	14.4	*Struthio camelus*	ULB	21	Between Tuffs 5 & 8	532/00
225	−10.4	−1.8				*Struthio camelus*	ULB	15	Between Tuffs 5 & 7	3369/00
226	−10.0	2.2	6.5	−23.2	13.2	*Struthio camelus*	ULB	15	Between Tuffs 5 & 7	3369/00
257	−6.6	6.3				*Struthio camelus*	ULB	13	Between Tuffs 6 & 7	2344/03
258	−5.0	6.1				*Struthio camelus*	ULB	13	Between Tuffs 6 & 7	2344/03
67	−5.6	9.1				*Struthio camelus*	ULB	1	Between Tuffs 6 & 8	
223	−8.8	3.2				*Struthio camelus*	ULB	9	Between Tuffs 6 & 8	486/01
224	−7.6	3.5				*Struthio camelus*	ULB	9	Between Tuffs 6 & 8	486/01
227	−5.6	8.6				*Struthio camelus*	ULB	1	Between Tuffs 6 & 8	3022/00
228	−5.5	8.6				*Struthio camelus*	ULB	1	Between Tuffs 6 & 8	3022/00
259	−8.2	3.9				*Struthio camelus*	ULB	1NW	Between Tuffs 6 & 8	2495/03
260	−8.2	4.5	8.0	−21.3	13.1	*Struthio camelus*	ULB	1NW	Between Tuffs 6 & 8	2495/03
198	−8.7	2.8				*Struthio camelus*	ULB	21	Between Tuffs 7 & 8	192/03
199	−4.8	4.0				*Struthio camelus*	ULB	21	Between Tuffs 7 & 8	192/03
204	−4.1	4.8				*Struthio camelus*	ULB	16	Between Tuffs 7 & just above 8	304/03
205	−9.8	3.5				*Struthio camelus*	ULB	17	Between tuff 7 & YMT	165/03
206	−7.9	7.1				*Struthio camelus*	ULB	16	Between Tuffs 7 & just above 8	304/03
231	−8.0	2.6				*Struthio camelus*	ULB	3	Between Tuffs 7 & 8	1658/00
232	−5.9	5.5				*Struthio camelus*	ULB	3	Between Tuffs 7 & 8	1658/00
249	−8.2	3.8	4.4	−22.0	13.8	*Struthio camelus*	ULB	8	Between Tuffs 7 & 8	1495/03
250	−7.6	1.9				*Struthio camelus*	ULB	8	Between Tuffs 7 & 8	1495/03
66	−11.7	−0.1	4.9	−22.9	11.2	*Struthio camelus*	UNB	18		

68	−11.2	−2.2				*Struthio camelus*	UNB	15	
211	−11.6	3.5				*Struthio camelus*	UNB	15	1070/01
212	−10.0	1.4	6.0	−23.4	13.4	*Struthio camelus*	UNB	15	1070/01
217	−7.8	3.8				*Struthio camelus*	UNB	18	821/01
218	−11.2	−0.7				*Struthio camelus*	UNB	18	821/01
233	−8.1	2.4				*Struthio camelus*	UNB	22S	1258/00
234	−11.3	1.6	7.1	−23.0	11.7	*Struthio camelus*	UNB	22S	1258/00
247	−6.8	2.2	10.3	−19.7	12.9	*Struthio camelus*	UNB	7E	1262/03
248	−6.1	4.7				*Struthio camelus*	UNB	7E	1262/03
267	−10.4	2.8	4.9	−23.5	13.1	*Struthio camelus*	UNB	Silal Artum	1562/01
268	−9.2	1.5				*Struthio camelus*	UNB	Silal Artum	1562/01
100	−6.3	0.2	7.9	−21.5	15.2	*Struthio camelus*	Modern	Garusi river	
101	−7.1	5.6	8.4	−22.0	14.8	*Struthio camelus*	Modern	Garusi river	
102	−8.3	3.0	7.0	−20.6	12.4	*Struthio camelus*	Modern	Garusi river	
103	−8.7	7.5				*Struthio camelus*	Modern	Garusi river	
104	−10.1	4.4				*Struthio camelus*	Modern		
103	−8.7	8.0	4.4	−22.5	11.3	*Struthio camelus*	Modern		

[a]Offset between the total organic ($\delta^{13}C_{OES-org}$) and inorganic ($\delta^{13}C_{OES-car}$) phases

develop explicit mixing models to quantitatively determine diet. Assuming a maximum $\delta^{13}C$ for C_3 plants of −22‰ and a recent 1.5‰ depletion in atmospheric ^{13}C due to fossil fuels, a conservative estimate for the boundary between an exclusive C_3 diet and a mixed C_3/C4 diet is a $\delta^{13}C_{OES-carb}$ value of about −6‰ or a $\delta^{13}C_{OES-org}$ of about −20‰. As with the enamel carbonate, it is more reasonable to assume a range for this boundary of about −6 to −9‰ for $\delta^{13}C_{OES-carb}$ and −20 to −23‰ for $\delta^{13}C_{OES-org}$.

Based on these criteria, modern and fossil carbon isotopic values of eggshells from Laetoli ostriches indicate foraging predominantly on C_3 plants although, depending on the boundary demarcations chosen, C_4 grasses may comprise a relatively minor dietary component in a number of specimens from the ULB and UNB (Figs. 15.9a, c). If the $\delta^{13}C_{OES}$ values are interpreted as almost exclusively C_3 derived, the variability documented in the succession relates primarily to variations in atmospheric pCO_2 and aridity. The earliest appearance of *S. camelus* in the sequence is marked by a 3–4‰ positive shift relative to eggshell fragments associated with *S. kakesiensis* in the LLB and ULB (below Tuff 3), indicating either increased water stress in C_3 dietary vegetation or/and a greater C_4 dietary contribution associated with this transition, both of which can be linked to more arid conditions. An alternative explanation might be that this isotopic transition results from physiological or behavioral changes associated with the taxonomic shift. However, a later shift to more depleted ^{13}C values for *S. camelus* in the UNB, with values similar to earlier *S. kakesiensis*, suggests that the isotopic changes may be environmentally, rather than taxonomically, mediated. The $\delta^{13}C_{OES}$ values of the UNB indicate more mesic conditions than the upper part of the ULB, contrasting with the enamel isotopic data and other lines of evidence suggesting that UNB ecosystems were relatively more open and arid. Overall, modern ostrich eggshells from the Laetoli area yielded $\delta^{13}C_{OES}$ values consistent with a C_3 dominated diet despite the fact that these ostriches inhabit C_4 grasslands and open woodland ecosystems. These results reinforce the notion that ostriches discriminate appreciably against C_4 grasses and that fossil $\delta^{13}C_{OES}$ data have possibly more limited value in reconstructing past habitats.

$\delta^{18}O_{OES}$ and Aridity/Temperature

Ostriches have been classified as non-obligate drinkers or animals that can access all their water needs through plant-leaf water. In general, the $\delta^{18}O$ of plant-leaf water increases with decreasing relative humidity (Ferhi and Letolle 1977) and increasing temperature (Gonfiantini et al. 1965), and shifting $\delta^{18}O_{OES}$ values may reflect the relative influence of these climatic factors. Within the Laetoli succession, the oxygen isotopic composition of ostrich eggshells mirrors and corroborates interpretations of the carbon isotopic data (Fig. 15.9b). The taxonomic transition from *S. kakesiensis* to *S. camelus* within the ULB is documented by a 1–2‰ ^{18}O enrichment, indicating enhanced evaporative transpiration in dietary plants and possibly higher temperatures and greater evaporative processes. The $\delta^{18}O_{OES}$ from UNB sites reveals a shift back to more negative values and potentially more humid and cooler conditions. The modern Laetoli eggshell fragments have the most ^{18}O-enriched values relative to the fossil samples, indicating that the current conditions are more arid and potentially hotter than in the past.

$\delta^{15}N_{OES}$ and Rainfall

Nitrogen isotope values of ostrich eggshells reflect the isotopic composition of the ostrich's diet. For herbivorous animals, the nitrogen isotope enrichment between diet and consumers is enhanced in arid regions, possibly as the result of certain physiological responses to being drought or protein stressed (Grocke et al. 1997). Research by Murphy and Bowman (2006), however, suggests that the negative relationship is due to links between dietary $\delta^{15}N$ and water availability rather than metabolic factors. A relationship has been described between the $\delta^{15}N$ values of vegetation, modern herbivore collagen, and mean annual precipitation (MAP) (Heaton 1987; Sealy et al. 1987). These links have subsequently been explored for the organic fraction in eggshells (Johnson 1995), and revealed that the $\delta^{15}N$ of the eggshell proteins increases with decreasing rainfall, humidity, and increased physiological stress (Johnson 1995).

Based on studies in South Africa, a significant relationship was quantified between the nitrogen isotopic composition of ostrich eggshells and MAP ($\delta^{15}N_{OES} = 15.2–0.011$[mm/yr]), with an error of about ±25% (Johnson 1995). Direct application of this regression to East African ostrich eggshells is not necessarily appropriate and the relationship needs to be calibrated for the specific environmental factors associated with rift settings. Interestingly, however, application of the South African equation to the modern eggshell Laetoli data yields a MAP of 704 mm, which falls on the low

Fig. 15.9 Box and whisker plots of the stable isotopic composition of fossil and modern ostrich eggshells from the Laetoli succession. Analyses of the calcite (mineral phase) includes determination of the (**a**) carbon isotopic signature ($\delta^{13}C_{OES-carb}$) and (**b**) oxygen isotopic signal ($\delta^{18}O_{OES}$). The inorganic portion of the eggshell was analyzed isotopically for (**c**) carbon ($\delta^{13}C_{OES-org}$) and (**d**) nitrogen ($\delta^{15}N_{OES}$). For the nitrogen analyses, MAP calculated based on the regression developed for South Africa (Johnson et al. 1997)

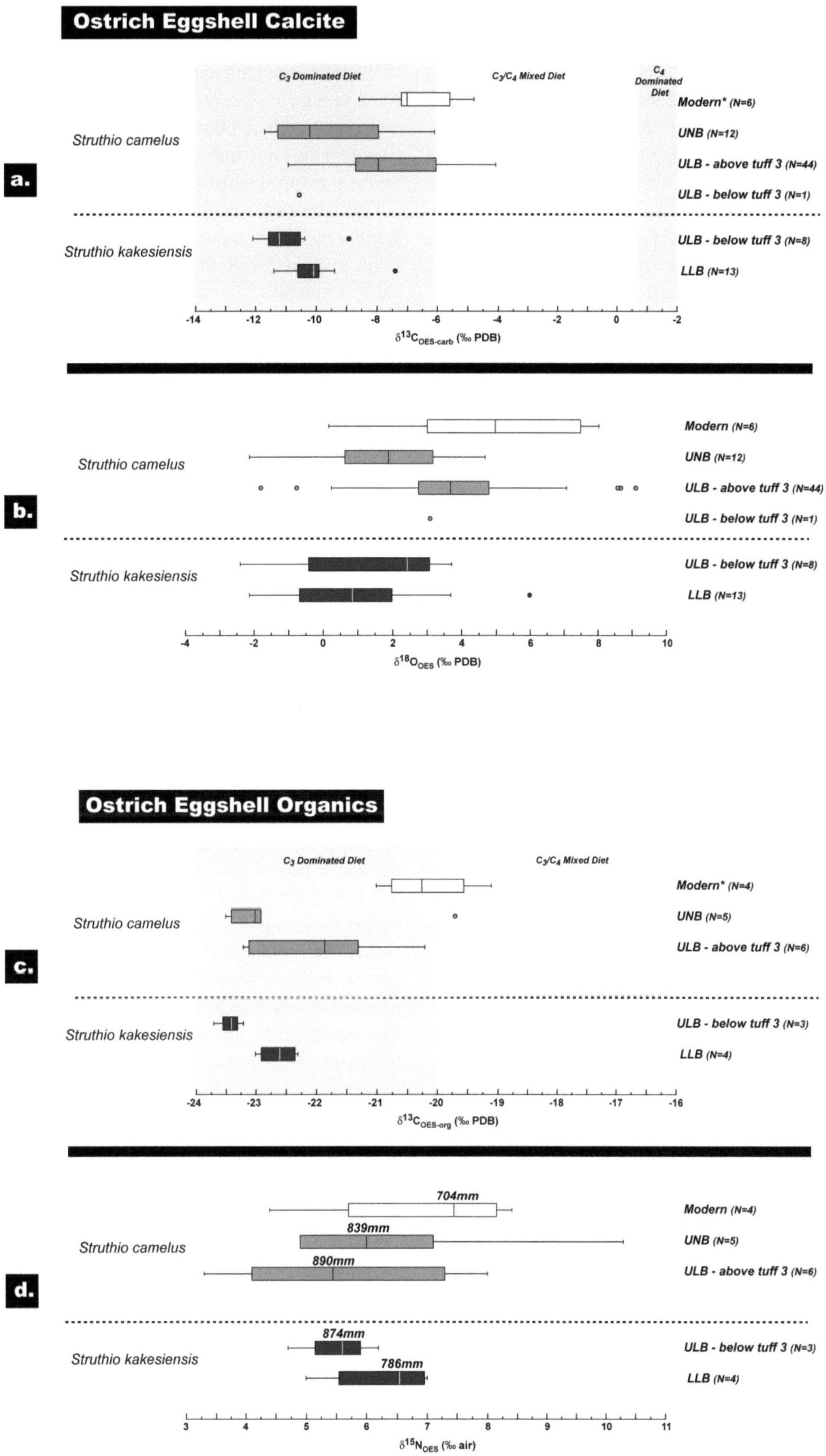

Ostrich Eggshell Calcite
a.
C3 Dominated Diet
C3/C4 Mixed Diet
C4 Dominated Diet
Struthio camelus
Modern* (N=6)
UNB (N=12)
ULB - above tuff 3 (N=44)
ULB - below tuff 3 (N=1)
Struthio kakesiensis
ULB - below tuff 3 (N=8)
LLB (N=13)
-14 -12 -10 -8 -6 -4 -2 0 -2
δ13COES-carb (‰ PDB)
b.
Struthio camelus
Modern (N=6)
UNB (N=12)
ULB - above tuff 3 (N=44)
ULB - below tuff 3 (N=1)
Struthio kakesiensis
ULB - below tuff 3 (N=8)
LLB (N=13)
-4 -2 0 2 4 6 8 10
δ18OOES (‰ PDB)
Ostrich Eggshell Organics
c.
C3 Dominated Diet
C3/C4 Mixed Diet
Struthio camelus
Modern* (N=4)
UNB (N=5)
ULB - above tuff 3 (N=6)
Struthio kakesiensis
ULB - below tuff 3 (N=3)
LLB (N=4)
-24 -23 -22 -21 -20 -19 -18 -17 -16
δ13COES-org (‰ PDB)
d.
704mm
839mm
890mm
874mm
786mm
Struthio camelus
Modern (N=4)
UNB (N=5)
ULB - above tuff 3 (N=6)
Struthio kakesiensis
ULB - below tuff 3 (N=3)
LLB (N=4)
3 4 5 6 7 8 9 10 11
δ15NOES (‰ air)

end of the 700–900 mm range for the local area today. MAP was also calculated for the fossil assemblages, revealing a rather tight range of 786–890 mm for the several levels within the succession (Fig. 15.9d). The nitrogen isotopic data, contrary to interpretations of the carbon and oxygen data, indicate that the ULB (above Tuff 3) had the greatest rainfall and that the UNB only slightly less. However, these calculations were based on median values, which do not necessarily characterize the range of variability in the nitrogen isotopic values. More data, as well as calibration of this approach in East Africa, are needed to assess any significant shifts in the $\delta^{15}N_{OES}$ through the sequence.

Isotopic Interpretations of Ostrich Eggshells

Relative to isotopic analyses of associated fossil mammalian herbivore taxa that that tend to retain overall uniform diets through the sequence, the eggshell isotopic data provide evidence of discrete dietary shifts in ostriches, linked to taxonomic and possibly environmental change. Overall, the data indicate an isotopic shift in the ULB at about Tuff 3 and also between the ULB and the UNB. The earlier isotopic transition is related to the replacement of the more primitive *S. kakesiensis* by the more competitive *S. camelus*, which presumably had a diet that was more specialized for coping with C_4 grasses or C_3 vegetation in more water stressed habitats. The carbon and oxygen isotopic ostrich eggshell signatures suggest the possibility of a shift to higher evaporative and/or warmer conditions above Tuff 3 in the ULB. This transition also involves a substantial decrease in the eggshell thickness (from an average of 3.2 mm for *S. kakesiensis* to 2.1 mm for early *S. camelus*) (Harrison and Msuya 2005), which is likely linked to diet and water conservation in modern ostriches. Sauer (1968) reports empirical data for thicker shells in ostriches with higher quality food or higher rainfall. Within the Laetoli succession, there is a negative, linear correlation between $\delta^{13}C_{OES}$ and eggshell thickness ($R^2 = 0.82$), suggesting that eggshell thickness and carbon isotopes are tracking related dietary or climatic phenomena. While the $\delta^{13}C_{OES}$ values may indicate increasing proportions of dietary C_4 grasses, it is possible that changes in the carbon isotopic values are more directly a function of varying environmental factors. An assessment of links between eggshell thickness and $\delta^{13}C_{OES}$ from fossil localities in Turkana, Namibia and Laetoli yielded an R^2 value of 0.25 (Harrison and Msuya 2005; Segalen et al. 2002), suggesting that this relationship cannot be extrapolated between sites. There is also a marked reduction in pore size between the two Laetoli ostrich taxa at the Tuff 3 transition (Harrison and Msuya 2005) that may represent an important adaptation for reducing vapor conductance in more arid environments (Tullett and Board 1977). However, later changes in $\delta^{13}C_{OES}$ in the sequence are not accompanied by corresponding changes in pore diameter. The ostrich eggshell isotopic data indicates the UNB environments were more mesic than the upper ULB (above Tuff 3) and perhaps more similar to the LLB and lower part of the ULB. This interpretation is at odds with reconstructions of the ecology based on the large mammals, but consistent with those based on the terrestrial gastropods (Tattersfield 2011).

An important consideration in the interpretation of the fossil eggshell isotopic data for paleoecological reconstructions is the implications of the timing and duration of eggshell formation. In natural environments, ostriches are opportunistic breeders and tend to lay eggs when resources are abundant, typically at the end of the rainy season (Sinclair 1978). The eggs, formed in three to five days, thus reflect environmental conditions and foraging behavior of a particular portion of the year. Depending on the degree of seasonality, this may not be representative of the total environment.

Conclusions

Carbon isotopic analyses of fossil enamel and ostrich eggshell fragments from the Laetoli succession reveal substantial intraspecific variability for a number of the 23 herbivore species included in this study. The variability relates primarily to different proportions of C_3 and C_4 vegetation in feeding strategies, with most taxa engaged in various degrees of mixed feeding. These dietary patterns contrast with modern representatives of these lineages that tend to specialize as grazers or browsers (Kingston and Harrison 2007). Fossil herbivore taxa from Laetoli occupy variable ranges along this C_3/C_4 continuum, suggesting multiple niches within the mixed feeding guild. Equids and alcelaphines anchor the C_4 end of this continuum with dietary signals indicating obligate to variable grazing behavior. Within these two grazing groups there is evidence of C_3 dietary intake, although these may represent C_3 grasses. Feeding pattern in the suids, primarily *Notochoerus euilus*, indicate mixed feeding, but with a considerable C_4 grazing component, although not to the extent documented for modern warthogs. In contrast to interpretations based on modern analogs, elephantids, rhinocerotids, and various bovid tribes (i.e., hippotragines, antilopines, and tragelaphines) have $\delta^{13}C_{enamel}$ patterns reflecting mixed C_3/C_4 feeding with extensive ranges of variation. Neotragines range from mixed feeders to variable browsers, with a distinct emphasis on C_3 feeding. The giraffids represent the only obligate to variable browsers in the assemblage, with no apparent isotopic niche differentiation of the three species recovered from Laetoli. Phytolith analysis documents the presence of C_3 grasses, especially in the lower part of the sequence (Rossouw and Scott 2011), but overall concordance of isotopic and mesowear studies (Kaiser 2011; Hernesniemi et al. 2011) suggest that the C_3 dietary signal primarily represents a browsing component.

Although Laetoli chronostratigraphy is relatively well resolved and provenance of the enamel samples tightly constrained within this framework, deciphering the nature of this variability remains complex. It is unclear whether the isotopic variability is inherent in fossil populations or a result of time-averaging of shifting dietary patterns in response to short term ecological change driven by Milankovitch cycling and/or volcanic disturbance. In general, feeding patterns for the various lineages are consistent through the sequence, certainly within the Laetolil Beds, suggesting that extinct forms are eclectic feeders and that the overall foraging behavior of these extinct herbivore communities has no close modern analog. There is evidence of shifting dietary patterns for some of the lineages within the succession, relating primarily to apparent changes in the relative breadth of mixed C_3/ C_4 diets, although these transitions are not statistically significant. Analysis of the fossil ostrich eggshells document dietary changes within the ULB, relating primarily to taxonomic change. The transition between the ULB and the UNB, spanning a hiatus of almost 1 million year, remains difficult to characterize isotopically. This relates primarily to changes in taxonomic representation and assumptions required in comparing dietary patterns of variable species. For herbivore lineages where there was reasonable sampling in both units, such as the suids and elephantids, the range and location of the isotopic variability remains consistent. For the equids, some alcelaphine species, and tragelaphines there is an apparent shift in the variability to an increased grazing component, suggesting that the UNB ecosystems were perhaps more open and arid. Interpretations of the carbon and oxygen isotopic signatures of the ostrich eggshells, however, suggests the opposite trend, with a transition to more mesic conditions in the UNB. Attempts to use the isotopic data to constrain estimates of precipitation/humidity in the sequence yielded estimates roughly similar but generally more arid than current conditions at Laetoli. Application of the aridity index, based on oxygen isotopic signals of enamel, yielded water deficit values comparable to modern East African habitats with MAP ranging from 420 to 750 mm, although these values likely represent minimal rainfall predictions of a potentially broad range of estimates. Nitrogen isotopic analyses of organic residues in ostrich eggshells yielded signals consistent with MAP estimates of 800–900 mm for the fossil levels based on regressions developed for South Africa.

Overall, the dietary patterns revealed by the isotopic analysis provide a comprehensive perspective on the collective foraging behavior of the herbivore community through time and by inference the paleoecology of Laetoli in the Pliocene. The limited representation of obligate grazers and the high diversity of mixed feeders and browsers in the assemblage are most consistent with a heterogeneous woodland ecosystem. Although C_4 grasses represent a dietary component for most of the fossil herbivore taxa, the obligate grazing guild that characterizes modern East African communities is not apparent. These data do not support the notion of extensive grassland-savanna biomes at Laetoli in the past. The interpretive value of the isotopic data is greatly enhanced by corroborating or independent studies that provide insights into the foraging behavior and ecology of the fossil Laetoli assemblages. Not surprisingly, as the resolution of paleoecological reconstructions at Laetoli is increased, the data collectively reveal the complexity of past ecosystems and the need to move beyond the application of simple modern analogs. While there may be broad similarities in faunal and floral composition between the fossil and modern assemblage, specific dietary patterns within these ecosystems are probably quite different. This recognition is especially relevant when attempting to constrain the niche of early hominins within this context and develop adaptive scenarios based on generalized reconstructions.

Acknowledgements I thank Terry Harrison for the opportunity to participate in research conducted by the Eyasi Plateau Paleontological and Geological Project. The following project members contributed to the recovery of the fossil material analyzed here: P. Abwalo, P. Andrews, E. Baker, M. Bamford, R. Chami, S. Cooke, P. Ditchfield, M. Duru, C. Feibel, T.S. Harrison, T. Harrison, S. Hixson, K. Kovarovic, A. Kweka, M. Lilombero, M.L. Mbago, K. McNulty, C. Msuya, S. Odunga, C. Robinson, L. Rossouw, W.J. Sanders, L. Scott, D. Su, M. Tallman, and S. Worthington. I thank the Tanzania Commission for Science and Technology and the Unit of Antiquities in Dar es Salaam for permission to conduct research in Tanzania. Special thanks go to N. Kayombo (Director General), P. Msemwa (Director), Amandus Kweka and all of the curators and staff at the National Museum of Tanzania in Dar es Salaam and Arusha for their support and assistance. Fieldwork at Laetoli and isotopic analyses were supported by grants from National Geographic Society, the Leakey Foundation, and NSF (Grants BCS-9903434 and BCS-0309513). Matt Sponheimer and an anonymous reviewer provided comments and suggestions that improved the manuscript.

References

Andrews, P. J. (1989). Palaeoecology of Laetoli. *Journal of Human Evolution, 18*, 173–181.

Andrews, P. (2006). Taphonomic effects of faunal impoverishment and faunal mixing. *Palaeogeography, Palaeoclimatology, Palaeoecology, 241*, 572–589.

Andrews, P., & Bamford, M. (2008). Past and present vegetation ecology of Laetoli, Tanzania. *Journal of Human Evolution, 54*, 78–98.

Andrews, P., Bamford, M. K., Njau, E. F., & Leliyo, F. (2011). The ecology and biogeography of the Endulen-Laetoli area in northern Tanzania. In T. Harrison (Ed.), *Paleontology and geology of Laetoli: Human evolution in context* (Geology, geochronology, paleoecology and paleoenvironment, vol. 1, pp. 167–200). Dordrecht: Springer.

Bamford, M. K. (2011). Fossil wood. In T. Harrison (Ed.), *Paleontology and geology of Laetoli: Human evolution in context* (Geology, geochronology, paleoecology and paleoenvironment, vol. 1, pp. 217–233). Dordrecht: Springer.

Behrensmeyer, A. K., & Hook, R. W. (1992). Paleoenvironmental contexts and taphonomic modes. In A. K. Beherensmeyer, J. D. Damuth, W. A. DiMichele, R. Potts, J. Sues, & S. L. Wing (Eds.), *Terrestrial ecosystems through time: Evolutionary paleoecology of terrestrial plants and animals* (pp. 15–136). Chicago: University of Chicago Press.

Bishop, L. C., Hill, A., & Kingston, J. D. (1999). Paleoecology of Suidae from the Tugen Hills, Baringo, Kenya. In P. Andrews & P. Banham (Eds.), *Late Cenozoic environments and hominid evolution: A tribute to Bill Bishop* (pp. 99–112). London: Geological Society of London.

Bonnefille, R., & Riollet, G. (1987). Palynological spectra from the Upper Laetolil Beds. In M. D. Leakey & J. M. Harris (Eds.), *Laetoli: A Pliocene site in northern Tanzania* (pp. 52–61). Oxford: Clarendon.

Boom, A., Marchant, R. A., Hooghiemstra, H., & Damste, J. S. S. (2002). CO_2 and temperature-controlled altitudinal shifts of C_4- and C_3-dominated grasslands allows reconstruction of ρCO_2. *Palaeogeography, Palaeoclimatology, Palaeoecology, 177*, 29–45.

Carroll, R. L. (1988). *Vertebrate paleontology and evolution*. San Francisco: Freeman.

Caswell, H., Reed, R., Stephenson, S. N., & Werner, P. A. (1973). Photosynthetic pathways and selective herbivory: A hypothesis. *The American Naturalist, 107*, 465–480.

Cerling, T. E., Harris, J. M., MacFadden, B. J., Leakey, M. G., Quade, J., Eisenmann, V., & Ehleringer, J. R. (1997). Global vegetation change through the Miocene/Pliocene boundary. *Nature, 389*, 153–158.

Cerling, T. E., Harris, J. M., & Leakey, M. G. (1999). Browsing and grazing in elephants: The isotope record of modern and fossil proboscideans. *Oecologia, 120*, 360–374.

Cerling, T. E., Harris, J. M., & Passey, B. H. (2003). Diets of East African Bovidae based on stable isotope analysis. *Journal of Mammalogy, 84*, 45–470.

Cerling, T. E., Hart, J. A., & Hart, T. B. (2004). Stable isotope ecology in the Ituri Forest. *Oecologia, 138*, 5–12.

Chapman, G. P. (1996). *The biology of grasses*. Wallingford: CAB International.

Clarke, S. J., Miller, G. H., Fogel, M. L., Chivas, A. R., & Murray-Wallace, C. V. (2006). The amino acid and stable isotope biogeochemistry of elephant bird (*Aepyornis*) eggshells from southern Madagascar. *Quaternary Science Reviews, 25*, 2343–2356.

Codron, D., Luty, J., Lee-Thorp, J. A., Sponheimer, M., de Ruiter, D., & Codron, J. (2005). Utilization of savanna-based resources by Plio-Pleistocene baboons. *South African Journal of Science, 101*, 245–248.

Codron, S., Lee-Thorp, J. A., Sponheimer, M., de Ruiter, D., & Codron, J. (2006). Inter- and Intrahabitat dietary variability of chacma baboons (*Papio ursinus*) in South African savannas based on fecal $\delta^{13}C$, $\delta^{15}N$, and %N. *American Journal of Physical Anthropology, 129*, 204–214.

Collatz, G. J., Berry, J. A., & Clark, J. S. (1998). Effects on climate and atmospheric CO_2 partial pressure on the global distribution of C_4 grassland: Past, present, and future. *Oecologia, 114*, 441–454.

Cooke, H. B. S. (1985). Plio-Pleistocene Suidae in relation to African hominid deposits. In Y. Coppens (Ed.), *L'Environments des hominides au Plio-Pleistocene* (pp. 101–117). Paris: Masson.

Cooke, H. B. S., & Wilkinson, Q. F. (1978). Suidae and Tayassuidae. In V. J. Maglio & H. B. S. Cooke (Eds.), *Evolution of African mammals* (pp. 435–482). Cambridge: Harvard University Press.

Deino, A. L. (2011). $^{40}Ar/^{39}Ar$ dating of Laetoli, Tanzania. In T. Harrison (Ed.), *Paleontology and geology of Laetoli: Human evolution in context* (Geology, geochronology, paleoecology and paleoenvironment, vol. 1, pp. 77–97). Dordrecht: Springer.

Demment, M. W., & van Soest, P. J. (1985). A nutritional explanation for body-size patterns of ruminant and nonruminant herbivores. *The American Naturalist, 125*, 641–672.

Denys, C. (1987). Rodentia and Lagomorpha. In M. D. Leakey & J. M. Harris (Eds.), *Laetoli: A Pliocene site in northern Tanzania* (pp. 118–170). Oxford: Clarendon.

El-Zaatari, S., Grine, F. E., Teaford, M. F., & Smith, H. R. (2005). Molar microwear and dietary reconstruction of fossil Cercopithecoidea from the Plio-Pleistocene deposits of South Africa. *Journal of Human Evolution, 49*, 180–205.

Ferhi, A., & Letolle, R. (1977). Transpiration and evaporation as the principal factors in oxygen isotope variations of organic matter in land plants. *Physiologie Végétale, 15*, 363–370.

Folinsbee, R. E., Fritz, P., Krouse, H. R., & Robblee, A. R. (1970). Carbon-13 and oxygen-18 in dinosaur, crocodile, and bird eggshells indicate environmental conditions. *Nature, 168*, 1353–1356.

Fortelius, M., & Solounias, N. (2000). Functional characterization of ungulate molars using the Abrasion-Attrition wear gradient: A new method for reconstructing paleodiets. *American Museum of Natural History Novitates, 3302*, 1–36.

Gagnon, M., & Chew, A. E. (2000). Dietary preferences in extant African Bovidae. *Journal of Mammalogy, 81*, 490–511.

Gentry, A. W. (2011). Bovidae. In T. Harrison (Ed.), *Paleontology and geology of Laetoli: Human evolution in context* (Fossil hominins and the associated fauna, vol. 2, pp. 363–465). Dordrecht: Springer.

Gonfiantini, G. R., Gratziu, S., & Tongiorgi, E. (1965). Oxygen isotope composition of water in leaves. In *Isotopes and radiation in soil-plant nutrition studies* (pp. 405–410). Vienna: International Atomic Energy Agency.

Grocke, D. R., Bocherens, H., & Mariotti, A. (1997). Annual rainfall and nitrogen-isotope correlation in macropod collagen: Application as a paleoprecipitation indicator. *Earth and Planetary Science Letters, 153*, 279–285.

Harris, J. M. (1983). Family Rhinocerotidae. In J. M. Harris (Ed.), *Koobi Fora research project – The fossil ungulates: Proboscidea, Perissodactyla, and Suidae* (pp. 130–155). Oxford: Clarendon.

Harris, J. M. (1987). Fossil Suidae from Laetoli. In M. D. Leakey & J. M. Harris (Eds.), *Laetoli: A Pliocene site in northern Tanzania* (pp. 349–358). Oxford: Clarendon.

Harrison, T. (2011). Laetoli revisited: Renewed palaeontological and geological investigations at localities on the Eyasi Plateau in northern Tanzania. In T. Harrison (Ed.), *Paleontology and geology of Laetoli: Human evolution in context* (Geology, geochronology, paleoecology and paleoenvironment, vol. 1, pp. 1–15). Dordrecht: Springer.

Harrison, T., & Msuya, C. P. (2005). Fossil struthionid eggshells from Laetoli, Tanzania: Taxonomic and biostratigraphic significance. *Journal of African Earth Sciences, 41*, 303–315.

Hay, R. L. (1987). Geology of the Laetoli area. In M. D. Leakey & J. M. Harris (Eds.), *Laetoli: A Pliocene site in northern Tanzania* (pp. 23–47). Oxford: Clarendon.

Heaton, T. H. E. (1987). The $^{15}N/^{14}N$ ratio of plants in South Africa and Namibia: Relationship to climate and coastal/saline environment. *Oecologia, 74*, 236–246.

Hernesniemi, E., Giaourtsakis, I., Evans, A., & Fortelius, M. (2011). Rhinocerotidae. In T. Harrison (Ed.), *Paleontology and geology of Laetoli: Human evolution in context* (Fossil hominins and the associated fauna, vol. 2, pp. 275–293). Dordrecht: Springer.

Hobson, K. A. (1995). Reconstruction avian diets using stable-carbon and nitrogen isotope analysis of egg components: Patterns of isotopic fractionation and turnover. *The Condor, 97*, 752–762.

Jacques, L., Ogle, N., Moussa, I., Kalin, R., Vignaud, P., Brunet, M., & Bocherens, H. (2008). Implications of diagenesis for the isotopic analysis of Upper Miocene large mammalian herbivore tooth enamel from Chad. *Palaeogeography, Palaeoclimatology, Palaeoecology, 266*, 200–210.

Johnson, B. J. (1995) The stable isotope biogeochemistry of ostrich eggshell and its application to late *Quaternary paleoenvironmental reconstructions in South Africa*. Ph.D. dissertation. University of Colorado, Boulder.

Johnson, B. J., Miller, G. H., Fogel, M. L., & Beaumont, P. B. (1997). The determination of late Quaternary paleoenvironments at Equus Cave, South Africa, using stable isotopes and amino acid racemization in ostrich eggshell. *Palaeogeography, Palaeoclimatology, Palaeoecology, 136*, 121–137.

Johnson, B. J., Fogel, M. L., & Miller, G. H. (1998). Stable isotopes in modern ostrich eggshell: A calibration for paleoenvironmental applications in semi-arid regions of southern Africa. *Geochimica et Cosmochimica Acta, 62*, 2451–2461.

Johnson, B. J., Miller, G. H., Fogel, M. L., Magee, J. W., Gagan, M. K., & Chivas, A. R. (1999). 65,000 Years of vegetation change in central Australia and the Australian summer monsoon. *Science, 284*, 1150–1152.

Kaiser, T. M. (2011). Feeding ecology and niche partitioning of the Laetoli ungulate faunas. In T. Harrison (Ed.), *Paleontology and geology of Laetoli: Human evolution in context* (Geology, geochronology, paleoecology and paleoenvironment, vol. 1, pp. 329–354). Dordrecht: Springer.

Kingston, J. D. (1999). Environmental determinants in early hominid evolution: Issues and evidence from the Tugen Hills, Kenya. In P. Andrews & P. Banham (Eds.), *Late Cenozoic environments and hominid evolution: A tribute to Bill Bishop* (pp. 69–84). London: Geological Society.

Kingston, J. D., & Harrison, T. (2007). Isotopic dietary reconstructions of Pliocene herbivores at Laetoli: Implications for early hominin paleoecology. *Palaeogeography, Palaeoclimatology, Palaeoecology, 243*, 272–306.

Kohn, M. J., & Cerling, T. E. (2002). Stable isotope compositions of biological apatite. *Reviews in Mineralogy and Geochemistry, 48*, 455–488.

Kohn, M. J., Schoeninger, M. J., & Valley, J. W. (1996). Herbivore tooth oxygen isotope compositions: Effects of diet and physiology. *Geochimica et Cosmochimica Acta, 60*, 3889–3896.

Kohn, M., Schoeninger, M., & Barker, W. (1999). Altered states: Effects of diagenesis on fossil tooth chemistry. *Geochimica et Cosmochimica Acta, 63*, 2737–2747.

Kovarovic, K., & Andrews, P. (2007). A bovid postcranial ecomorphological survey of the Laetoli palaeoenvironment. *Journal of Human Evolution, 52*, 663–680.

Kovarovic, K., & Andrews, P. (2011). Environmental change within the Laetoli fossiliferous sequence: Vegetation catenas and bovid ecomorphology. In T. Harrison (Ed.), *Paleontology and geology of Laetoli: Human evolution in context* (Geology, geochronology, paleoecology and paleoenvironment, vol. 1, pp. 367–380). Dordrecht: Springer.

Kovarovic, K., Andrews, P., & Aiello, L. (2002). The palaeoecology of the Upper Ndolanya Beds at Laetoli, Tanzania. *Journal of Human Evolution, 43*, 395–418.

Kullmer, O. (1999). Evolution of African Plio-Pleistocene suids (Artiodactyla: Suidae) based on tooth pattern analysis. *Kaupia, 9*, 1–34.

Leakey, M. D. (1987). Introduction. In M. D. Leakey & J. M. Harris (Eds.), *Laetoli: A Pliocene site in northern Tanzania* (pp. 1–21). Oxford: Clarendon.

Leakey, M. D., & Harris, J. M. (Eds.). (1987). *Laetoli: A Pliocene site in northern Tanzania*. Oxford: Clarendon.

Lee-Thorp, J. A., Sealy, J. C., & van der Merwe, N. J. (1989a). Stable carbon isotope ratio differences between bone collagen and bone apatite, and their relationship to diet. *Journal of Archaeological Science, 16*, 585–599.

Lee-Thorp, J. A., van der Merwe, N. J., & Brain, C. K. (1989b). Isotopic evidence for dietary differences between two extinct baboon species from Swartkrans. *Journal of Human Evolution, 18*, 183–190.

Lee-Thorp, J. A., van der Merwe, N. J., & Brain, C. K. (1994). Diet of *Australopithecus robustus* at Swartkrans from stable carbon isotopic analysis. *Journal of Human Evolution, 27*, 361–372.

Levin, N. E., Cerling, T. E., Passey, B. H., Harris, J. M., & Ehleringer, J. R. (2006). A stable isotope aridity index for terrestrial environments. *Proceedings of the National Academy of Sciences of the United States of America, 103*, 11201–11205.

Levin, N. E., Simpson, S. W., Quade, J., Cerling, T. E., & Frost, S. R. (2008). Herbivore enamel carbon isotopic composition and the environmental context of *Ardipithecus* at Gona, Ethiopia. *Geological Society of America Special Paper, 446*, 215–234.

Martin, C., Bentaleb, I., Kaandorp, R., Iacumin, P., & Chatri, K. (2008). Intra-tooth study of modern rhinoceros enamel $\delta^{18}O$: Is the difference between phosphate and carbonate $\delta^{18}O$ a sound diagenetic test? *Palaeogeography, Palaeoclimatology, Palaeoecology, 266*, 183–189.

McNaughton, S. J., & Georgiadis, N. J. (1986). Ecology of African grazing and browsing mammals. *Annual Review of Ecology and Systematics, 17*, 39–66.

Mikhailov, K. E. (1992). The microstructure of avian and dinosaurian eggshell: Phylogenetic implications. In: K. E. Campbell, (Ed.), *Papers in avian paleontology honoring Pierce Brodkorb, Science Series* (pp. 361–373). Los Angeles: Natural History Museum of Los Angeles County.

Murphy, B. P., & Bowman, D. M. J. S. (2006). Kangaroo metabolism does not cause the relationship between bone collagen $\delta^{15}N$ and water availability. *Functional Ecology, 20*, 1062–1069.

Reed, K. E. (1997). Early hominid evolution and ecological change through the African Plio-Pleistocene. *Journal of Human Evolution, 32*, 289–322.

Robinson, C. (2011). Giraffidae. In T. Harrison (Ed.), *Paleontology and geology of Laetoli: Human evolution in context* (Fossil hominins and the associated Fauna, vol. 2, pp. 339–362). Dordrecht: Springer.

Rossouw, L., & Scott, L. (2011). Phytoliths and pollen, the microscopic plant remains in Pliocene volcanic sediments around Laetoli, Tanzania. In T. Harrison (Ed.), *Paleontology and geology of Laetoli: Human evolution in context* (Geology, geochronology, paleoecology and paleoenvironment, vol. 1, pp. 201–215). Dordrecht: Springer.

Sage, R. F. (2004). The evolution of C_4 photosynthesis. *The New Phytologist, 161*, 341–370.

Sauer, E. G. F. (1968). Calculations of struthious egg sizes from measurements of shell fragments and their correlation with phylogenetic aspects. *Cimbebasia Series A, 1*, 27–55.

Sealy, J. C., van der Merwe, N. J., Thorp, J. A. L., & Lanham, J. L. (1987). Nitrogen isotopic ecology in southern Africa: Implications for environmental and dietary tracing. *Geochimica et Cosmochimica Acta, 51*, 2707–2717.

Segalen, L., Renard, M., Pickford, M., Senut, B., Cojan, I., Callonnec, L. L., & Rognon, P. (2002). Environmental and climatic evolution of the Namib Desert since the Middle Miocene: The contribution of carbon isotope ratios in ratite eggshells. *Comptes Rendus Geoscience, 334*, 917–924.

Sinclair, A. R. E. (1978). Factors affecting the food supply and breeding season of resident birds and movements of Palaearctic migrants in a tropical African savannah. *Ibis, 120*, 480–497.

Sponheimer, M., & Lee-Thorp, J. A. (2006). Enamel diagenesis at South African australopith sites: Implications for paleoecological reconstruction with trace element. *Geochimica et Cosmochimica Acta, 70*, 1644–1654.

Sponheimer, M., Reed, K., & Lee-Thorp, J. A. (2001). Isotopic plaeoecology of Makapansgat limeworks Perissodactyla. *South African Journal of Science, 97*, 327–329.

Sponheimer, M., Lee-Thorp, J. A., DeRuiter, D. J., Smith, J. M., van der Merwe, N. J., Reed, K., Grant, C. C., Ayliffe, L. K., Robinson, T. F., Heidelberg, C., & Marcus, W. (2003). Diets of southern African Bovidae: Stable isotope evidence. *Journal of Mammalogy, 84*, 471–479.

Su, D. F. (2011). Large mammal evidence for the paleoenvironment of the Upper Laetolil and Upper Ndolanya Beds of Laetoli, Tanzania. In T. Harrison (Ed.), *Paleontology and geology of Laetoli: Human evolution in context* (Geology, geochronology, paleoecology and paleoenvironment, vol. 1, pp. 381–392). Dordrecht: Springer.

Su, D. F., & Harrison, T. (2007). The paleoecology of the Upper Laetolil Beds at Laetoli: A reconsideration of the large mammal evidence. In R. Bobe, Z. Alemseged, & A. K. Behrensmeyer (Eds.), *Hominin environments in the East African Pliocene: An assessment of the faunal evidence* (pp. 279–313). Dordrecht: Springer.

Su, D., & Harrison, T. (2008). Ecological implications of the relative rarity of fossil hominins at Laetoli. *Journal of Human Evolution, 55*, 672–681.

Tattersfield, P. (2011). Gastropoda. In T. Harrison (Ed.), *Paleontology and geology of Laetoli: Human evolution in context* (Fossil hominins and the associated fauna, vol. 2, pp. 567–587). Dordrecht: Springer.

Thornthwaite, C. W. (1948). An approach toward a rational classification of climate. *Geographical Review, 38*, 55–94.

Tieszen, L. L., Hein, D., Qvortrup, S. A., Troughtonn, J. H., & Imbama, S. K. (1979). Use of $\delta^{13}C$ values to determine vegetation selectivity in East African herbivores. *Oecologia, 37*, 351–359.

Tullett, S. G., & Board, R. G. (1977). Determinants of avian eggshell porosity. *Journal of Zoology, 183*, 203–211.

van der Merwe, N. J., Thackeray, J. F., Lee-Thorp, J. A., & Luyt, J. (2003). The carbon isotope ecology and diet of *Australopithecus africanus* at Sterkfontein, South Africa. *Journal of Human Evolution, 44*, 581–597.

Verdcourt, B. (1987). Mollusca from the Laetolil and Upper Ndolanya Beds. In M. D. Leakey & J. M. Harris (Eds.), *Laetoli: A Pliocene site in northern Tanzania* (pp. 438–450). Oxford: Clarendon.

Von Shirnding, Y., van der Merwe, N. J., & Vogel C. J. (1982). Influence of diet and age on carbon isotope ratios in ostrich eggshell. *Archaeometry, 24*, 3–20.

Walker, A. C. (1987). Fossil Galaginae from Laetoli. In M. D. Leakey & J. M. Harris (Eds.), *Laetoli: A Pliocene site in northern Tanzania* (pp. 88–90). Oxford: Clarendon.

White, T. D., Ambrose, S. H., Suwa, G., Su, D. F., DeGusta, D., Bernor, R. L., Boisserie, J. -R. Brunet, M., Delson, E., Frost, S. Garcia, N., Giaourtsakis, I. X., Haile-Selassie, Y., Howell, F. C., Wehmann, T., Likius, A., Pehlevan, C., Saegusa, H., Semprebon, G., Teaford, M., & Vrba, E. (2009). Macrovertebrate paleontology and the Pliocene habitat of *Ardipithecus ramidus*. *Science, 326*, 87–93.

Zazzo, A., Bocherens, H., Brunet, M., Beauvilain, A., Billiou, D., Mackaye, H. T., Vignauch, P., & Mariotti, A. (2000). Herbivore paleodiet and paleoenvironmental changes in Chad during the Pliocene using stable isotope ratios of tooth enamel carbonate. *Paleobiology, 26*, 294–309.

Chapter 16
Feeding Ecology and Niche Partitioning of the Laetoli Ungulate Faunas

Thomas M. Kaiser

Abstract The Pliocene ungulate fauna from the hominid-bearing Laetolil succession (Southern Serengeti, Tanzania) is investigated with regard to dietary adaptation, niche segregation and change over time. The fossiliferous Upper Laetolil Beds (ULB) (3.63–3.85 Ma) are unconformably overlain by the Upper Ndolanya Beds (UNB) (2.66 Ma). Both stratigraphic units contain a rich mammalian fauna, with ungulates predominating. Analysis of dental mesowear is applied to 23 ungulate taxa from both units, including Equidae, Bovidae and Giraffidae, and the results are compared to extant species. The equids at Laetoli represent the only specialized grazers throughout the succession. All Upper Laetolil Alcelaphini and Hippotragini have mesowear signatures that indicate intermediate feeding strategies, different from their modern counterparts that are mostly specialized grazers. This indicates a dietary shift in these lineages, a finding that is also supported by isotope studies. Mesowear data of ungulates from the ULB also suggest that extant ungulates representing closely related lineages in the same genus or even tribe may not serve as actualistic model taxa in faunal reconstructions using taxonomic uniformitarianism. The three species of giraffids and the remaining bovid taxa were either browsers or intermediate feeders, but not grazers. The almost complete absence of grazing guilds, and the heavy reliance on browse by most fossil herbivores, do not support the inference that the Laetoli environment was dominated by grassland. Within the Laetoli succession it appears that fundamental feeding niches converged over time, with grazers increasingly engaged in feeding on less abrasive components and intermediate-feeders closing the dietary gap by exploiting more abrasive feeding niches. Niche partitioning in the Laetoli ungulates appears to reflect environmental change and evolutionary trajectories in the major lineages. This distribution of feeding niches may serve as an overall indicator of niche diversity. Within the succession it appears that the diversity of feeding niches generally decreased. A decrease in feeding niches would suggest that the diverse habitat structure, which was typical of the ULB environment, no longer existed after the faunal and environmental transition that occurred after deposition of the ULB. After a hiatus of 1.0 million years, the UNB environment was more or less free of forest and woodland patches and can be characterized as more or less open grassland.

Keywords Ruminantia • Bovidae • Giraffidae • Equidae • Paleodiet • Mesowear • Tooth wear • Pliocene • Paleoecology

Introduction

Laetoli is located in Tanzania at the southern part of the eastern branch of the East African Rift Valley. Sediments are exposed along the margin of the Eyasi Plateau that forms the divide between the Lake Eyasi and Olduvai Gorge drainage basins. Stratigraphic units investigated in this study are the Upper Laetolil Beds (ULB) and the Upper Ndolanya Beds (UNB) (Fig. 16.1). The Laetoli succession rests unconformably on the Precambrian basement. The Laetolil Beds, which consist of two lithologic units, form the base of the succession. The lower unit consists of aeolian tuffs interbedded with airfall and waterworked tuffs. Radiometric dates indicate an age of 4.4–3.85 Ma (Drake and Curtis 1987; Deino 2011). Fossils have been recovered from this unit, but they were not considered in this study due to their scarcity. The upper unit is comprised of a series of aeolian and airfall tuffs, including eight distinctive marker tuff horizons. Mean K/Ar dates (Drake and Curtis 1987) indicate an age of 3.76 ± 0.03 Ma for the base of the upper unit, and 3.46 ± 0.12 Ma for Tuff 8, close to the top of the fossiliferous sediments. Deino (2011) provides $^{40}Ar/^{39}Ar$ dates of 3.85–3.63 Ma for this unit. The fauna from the ULB is diverse and spans more than 0.2 million years, with much time averaging. Fossils have been recovered from more than 30 localities from throughout the ULB. The ULB are separated from the overlying UNB by a depositional hiatus that lasted 1.0 million years (Fig. 16.1).

T.M. Kaiser (✉)
Biozentrum Grindel und Zoologisches Museum, Universität Hamburg, Martin-Luther-King-Platz 3, D-20146 Hamburg, Germany
e-mail: thomas.kaiser@uni-hamburg.de

T. Harrison (ed.), *Paleontology and Geology of Laetoli: Human Evolution in Context. Volume 1: Geology, Geochronology, Paleoecology and Paleoenvironment*, Vertebrate Paleobiology and Paleoanthropology, DOI 10.1007/978-90-481-9956-3_16,

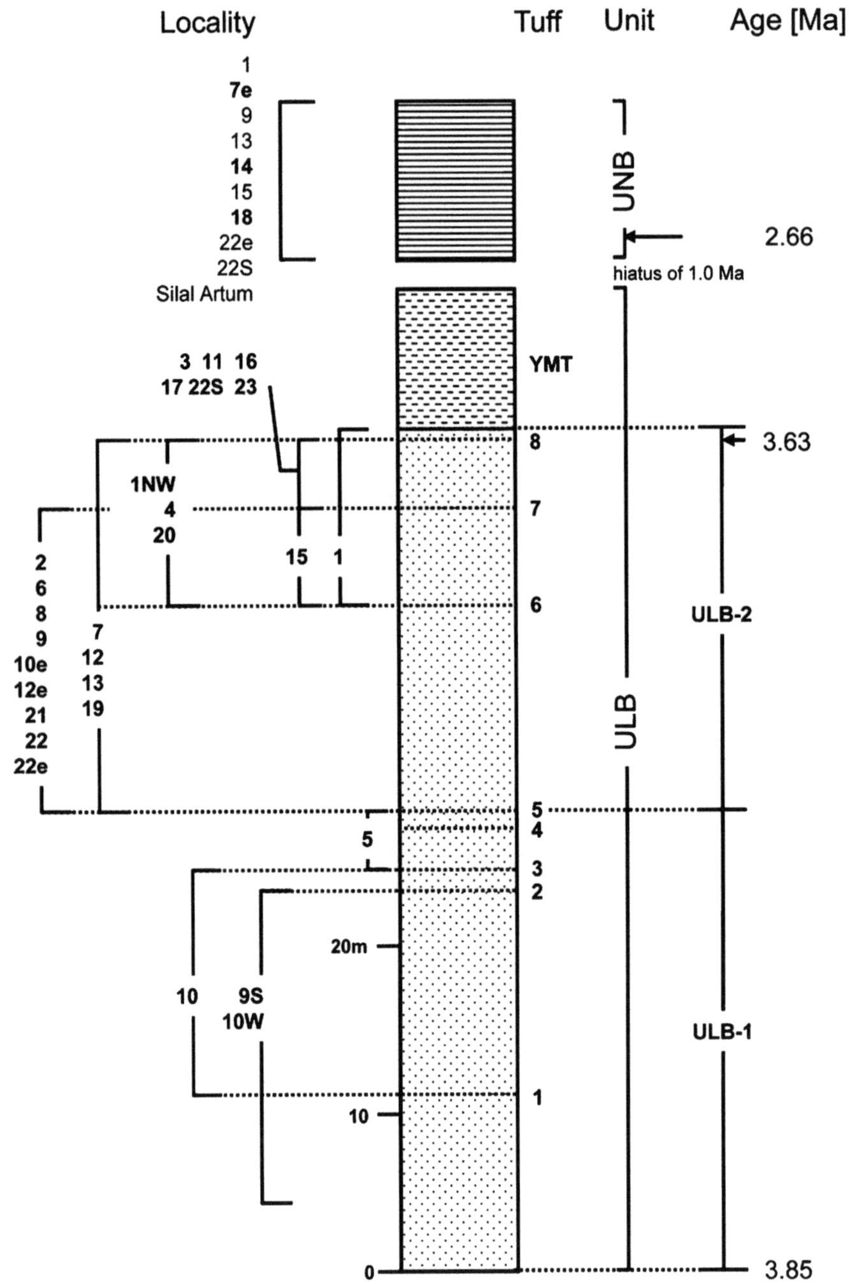

Fig. 16.1 Section of the Laetoli stratigraphic sequence investigated (after Drake and Curtis 1987; Deino 2011). The section is bracketed into three major time slices. *ULB*-1 Upper Laetolil Beds below Tuff 5, *ULB*-2 Upper Laetolil Beds above Tuff 5, *UNB* Upper Ndolanya Beds

The Upper Ndolanya Beds overly the Upper Laetolil Beds and are older than the overlying Ogol Lavas, which are dated to 2.41 ± 0.12 Ma (Drake and Curtis 1987; Deino 2011). The Upper Ndolanya Beds have yielded a rich mammalian fauna that is consistent with a radiometric date of 2.66 Ma (Gentry 1987; Harris 1987a; Ndessokia 1990; Manega 1993; Deino 2011). Fossils at Laetoli derive primarily from extensive volcanogenic sediments deposited on a low-relief land surface distal to major sources of water (Hay 1981, 1987). This results in relatively autochthonous assemblages in which the dominant taphonomic biases are related to carnivore activity and differential preservation (Kaiser, unpublished; Su 2005; Su and Harrison 2008).

Only stratigraphically well-provenienced ungulate specimens were included in the study. Stratigraphic information attached to fossils either refers to locality numbers or to the relative position (below or above) one of the marker tuffs 1–8 and Yellow Marker Tuff (YMT). Laetoli fossil localities expose a section of the ULB column, but they overlap each other in their stratigraphic range. This makes it difficult to define time slices and to assign fossil specimens to one of these slices without time overlap. In fact, the only consistent system of time bracketing based on the specimen information available and covering the entire fossil record from Laetoli would be a three-slice system with Tuffs 3 and 5 as delimitations (Fig. 16.1). The comparatively low number of specimens represented between Tuffs 3 and 5, a part of the section which is only well-exposed and fossiliferous at Locality 5, would favor a two-slice system with Tuff 5 as the boundary between an upper and a lower division of the ULB. Here this two-slice bracketing system is employed, which covers two time zones in the ULB a lower division or ULB-1 below Tuff 5 and an upper division, ULB-2, above Tuff 5. The Upper Ndolanya Beds (UNB) are treated as the third stratigraphic unit in this system.

Reconstructing fossil biome conditions is still largely based on two major approaches: taxonomic uniformitarianism (e.g., Vrba 1975; Shipman and Harris 1988; Wilkinsin 2002) and ecomorphology (e.g., Köhler 1993; Bernor et al. 1999). In a strict sense, only paleobiomes with stable habitats saturated with well-adapted and stable species would be accessible for study by either of these techniques, but at least in local communities this situation is hypothetical (e.g., Morris 1998). Using the dietary interface as a taxon-independent pathway may allow a more precise comparison of biomes independent from species historical circumstances.

In recent years, the mesowear method (Fortelius and Solounias 2000; Kaiser et al. 2000) has been employed to evaluate the average abrasiveness of the food of fossil herbivorous mammals by comparing macroscopic dental wear with extant representatives with known diet and habitat preference. Because mesowear provides an immediate long-term dietary signal compared to the individual's lifespan, it allows comparison of biomes somewhat independent of differences in species historical circumstances and time averaging (Kaiser 2003). The mesowear method measures the relative abrasiveness or relative attritional nature of a diet, and therefore reflects the composition of the diet as reflected by inherent mechanical properties of food items, as well as exogenous grit content. Mesowear indicates, therefore, the specific dietary behavior of individuals or populations, as well as environmental conditions. This approach is employed here, because, as a taxon-independent pathway to paleoecology, it should reflect major habitat properties independent of slowly evolving characters such as the degree of hypsodonty.

Food availability and habitat structure are closely linked parameters in terrestrial ecosystems. Because dental durability in ungulates is linked to the abrasiveness of foods comminuted, dental wear analysis evaluates a major habitat interface, which in turn reflects the overall abrasiveness of foods eaten. In general, many angiosperms, both monocotyledons and dicotyledons (Piperno et al. 2002), impregnate their vegetative and reproductive organs with phytoliths (solid particles of amorphic SiO_2). By increasing abrasiveness, phytoliths constitute an important system of mechanical defense (McNaughton et al. 1985). Since opalines are the only substance hard enough to grind and abrade the tooth enamel of herbivorous mammals (Baker et al. 1959), the evolution of hypsodont dentitions during the late Miocene, when open woodlands and grasslands expanded globally at the expense of forests, has been hypothesized to be linked to the high phytolith content of grass leaves in particular (Cerling et al. 1998). Recently, Sanson et al. (2007) challenged hardness measurement data given by Baker et al. (1959) and report hardness values of phytoliths lower than those of sheep dental enamel. However, this study uses indentation hardness measurement, which probably does not allow immediate conclusions about brittle materials. Their results therefore should be considered with caution. Airborne grit and grit derived from attached soil certainly make up the second major source of abrasives eaten by herbivorous mammals (Kay and Covert 1983). Additionally, some general habitat-driven mechanisms can increase abrasiveness of food available to herbivores.

First, if the growth period is short, dry (dead) vegetation represents a more frequent forage. This forage will cause increased tooth wear, because it accumulates more grit. Ruminants, however, tend to reduce their consumption under these conditions and extract more energy from their food. Second, in drier environments plants have more water stress, growth is slow, and they need better protection against herbivores. Phytoliths are thus expected to be more abundant in both dicotyledons and monocotyledons under these circumstances. Third, if the vegetation cover is thin more airborne grit will accumulate on plants. Fourth, low precipitation washes off less grit. Fifth, scarce resources are more intensively exploited, and forage is thus collected closer to the ground, causing grit from the ground to increasingly contaminate the forage.

All these mechanisms are expected to cause increased tooth wear and shift the faunal mesowear equilibrium towards the abrasion dominated end of the spectrum. Because mesowear equilibria need a certain degree of wear to be established, increased absolute wear rates (as typical for hypsodont species) would increase the dietary resolution of mesowear (Rivals et al. 2007). Fortelius and Solounias (2000) consider this bias to be not strong enough to seriously affect analyses like those undertaken here. Fortelius et al. (2002) have indicated that there is a general interrelationship between precipitation and hypsodonty in ungulate faunas. Since ecomorphological parameter and mesowear are both a result of food interaction at the occlusal surface, it is likely that mesowear reflects the same habitat parameters on an individual scale as hypsodonty does on an evolutionary scale.

A central goal of this study concerns the interrelationship between the environmental water availability and food abrasiveness, and to refine the understanding of the paleoenvironmental setting at Laetoli as indicated by dental mesowear equilibria. It is to be expected that habitats including both open and closed components would support more species feeding on abrasive food components compared to those inhabiting closed habitat settings. If this is correct, one would expect dietary niche segregation between herbivorous ungulates to reflect the proposed vegetation structures. This in turn will allow an examination of ecological variability between the major stratigraphic units, and comparisons of the dietary signatures with modern relatives of the fossil herbivore taxa. It will further allow one to address the question of whether or not there are apparent dietary shifts corresponding to ecological transitions within the Laetoli sequence. In particular, does the transition between the ULB and UNB, which is represented by a sedimentary hiatus of 1.0 million years, reflect differences in the mesowear signatures of ruminants and equids. During this interval there was a faunal turnover in the ungulate fauna, in which most of the younger elements made their first appearance. Another goal of this study is to compare dietary assessments based on mesowear analyses to paleodiet reconstructions based on stable isotopes (Kingston and Harrison 2007), and to identify taxa that are opportunistic or eclectic feeders that shift diet with ecological change, thereby providing the most sensitive proxies for characterizing Pliocene ecosystems.

Materials

Over 448 representative upper molars of 24 herbivore taxa, almost all identified to species, were analyzed in terms of their mesowear signatures. Specimens derive from multiple stratigraphic horizons within the Laetoli succession. Investigated specimens representing the Upper Laetolil Bed biome derive from Laetoli Localities 1, 1NW, 2, 3, 4, 5, 6, 7, 8, 9, 9 S, 10, 10E, 10W, 11, 12, 12E, 13, 15, 16, 17, 19, 21, 22, and 22E as indicated in Fig. 16.1. Specimens representing the UNB biome are from Laetoli Localities 7E, 9, 14, 15, 18, 22E, 22S and Silal Artum. All these materials are curated at the National Museum of Tanzania in Dar es Salaam. The ungulate sample investigated comprised all fossil ungulate teeth collected until 01/2006 by the two major research projects at Laetoli directed by Mary Leakey and Terry Harrison respectively. Earlier collections by Margit and Ludwig Kohl-Larsen, housed in the Museum für Naturkunde (Berlin) were evaluated, but due to the lack of reliable provenience data could not be considered. Also, a number of well-provenienced, yet unstudied, specimens collected by T. Bromage, F. Schrenk and T. M. Kaiser (mentioned in Kaiser et al. 1995) could not be included.

The fossil ungulate fauna from Laetoli is predominantly represented by six major groups, the Rhinocerotidae (Guerin 1987; Hernesniemi et al. 2011), Equidae (Hooijer 1987a, b; Armour-Chelu and Bernor 2011), Proboscidea (Beden 1987; Harris 1987a; Sanders 2011), Bovidae (Gentry 1987; Gentry 2011), Giraffidae (Harris 1987b; Robinson 2011), Camelidae (Harris 1987b). The Bovidae, Giraffidae and Equidae are well-represented by postcranial and dental specimens, and these taxa form the basis of dietary evaluation in this study.

From the ULB assemblage 13 taxa of Bovidae, three Giraffidae and one member of the Equidae were available for mesowear analysis. The Rhinocerotidae are analyzed in a separate study (Hernesniemi et al. 2011). From the UNB assemblage six members of the Bovidae, and one member of the Equidae were analyzed (Table 16.1).

Because dental specimens cannot always be confidently assigned to species in the Bovidae (Alan Gentry, personal communication), one UNB sample (*mecon*) is likely to comprise two distinct bovid species *Megalotragus kattwinkeli* and *Connochaetes isaaci* of Gentry (2011). A similar bias applies to the large Bovidae in the ULB assemblage. According to Alan Gentry (personal communication), the (*alsp*) sample includes the two large bovids *Parmularius pandatus* and Alcelaphini (larger) sp. indet. after Gentry (2011). Specimens clearly assigned to *Parmularius pandatus* are in the (*papan*) sample, which results in this species being subdivided into two individual datasets. The remaining bovid datasets are likely to represent one species each (Table 16.1).

In addition to being analyzed as one unit, the Upper Laetolil Beds are subdivided into two stratigraphic sub-units, ULB-1 = below Tuff 5 and ULB-2 = above Tuff 5 (Fig. 16.1). This results in four ungulate communities investigated (Table 16.2). One represents the total ULB community with 17 ungulate species available for study. The second community represents only the older ULB-1, where eight fossil ungulate species were selected for analysis, because they are consistently represented in both sub-units of the ULB. The third community represents the younger ULB-2, also with eight ungulates analyzed. The youngest community represents the Upper Ndolanya Beds (UNB), with seven species analyzed.

Although the two biomes (Upper Laetolil Beds and Upper Ndolanya Beds) are separated by 1.0 million years, the bovids *Gazella janenschi* and *Madoqua avifluminis* are represented in both units. The Equidae are represented by two distinct species of hipparionines (three toed horses; Armour-Chelu and Bernor 2011), each of which represents one of the major biomes. The giraffid *Giraffa stillei* seems to be the third ungulate species common to both biomes.

With few exceptions all materials investigated derive from surface collections and have undergone a variable degree of post-depositional alteration due to secondary weathering. In fact, taphonomic alteration has rendered the vast majority of dental material unsuitable for any kind of dental wear analysis, because of broken cusp apices and abraded occlusal surfaces. All specimens showing abrasion to a degree that would obliterate obvious microwear scars

Table 16.1 Mesowear scorings of the Laetoli Bovidae, Giraffidae and Equidae from the Upper Laetolil Beds (*ULB*) and the Upper Ndolanya Beds (*UNB*). N = number of dental specimens, %h percentage of high occlusal relief, %s = percentage of sharp cusps, %r percentage of round cusps, %b percentage of blunt cusps (after Fortelius and Solounias 2000)

Unit	Spec-short	Order	Family	Tribe	Genus	Species	N	SCORE	EXTREF
UNB	*mecon*	Artiodactyla	Bovidae	Alcelaphini	*Megalotragus/ Connochaetes*	*kattwinkeli* or *isaaci*/sp.	8	1.88	ct
UNB	*paalt*	Artiodactyla	Bovidae	Alcelaphini	*Parmularius*	*? altidens*	4	1.75	hn
UNB	*anrec*	Artiodactyla	Bovidae	Antilopini	*Antidorcas*	*recki*	9	0.78	Om
UNB	*gajan*	Artiodactyla	Bovidae	Antilopini	*Gazella*	*janenschi*	4	0.25	OH
UNB	*masp*	Artiodactyla	Bovidae	Neotragini	*Madoqua*	*?avifluminis*	3	1.00	ke
UNB	*trbux*	Artiodactyla	Bovidae	Tragelaphini	*Tragelaphus*	sp. cf. *buxtoni*	5	1.20	Me
UNB	*eusp*	Perissodactyla	Equidae		*Eurygnathohippus*	sp.	7	3.14	cs
ULB-all	*alsp*	Artiodactyla	Bovidae	Alcelaphini	*Parmularius/Alcelaphini* larger	*pandatus*/sp. indet.	31	0.80	Me
ULB-all	*papan*	Artiodactyla	Bovidae	Alcelaphini	*Parmularius*	*pandatus*	41	0.83	Me
ULB-all	*aedie*	Artiodactyla	Bovidae	Antilopini	*Aepyceros*	*dietrichi* sp. nov.	25	0.24	OH
ULB-all	*gajan*	Artiodactyla	Bovidae	Antilopini	*Gazella*	*janenschi*	35	0.40	GC
ULB-all	*gakoh*	Artiodactyla	Bovidae	Antilopini	*"Gazella"*	*kohllarseni*	11	0.27	OH
ULB-all	*sikoh*	Artiodactyla	Bovidae	Bovini	*Simatherium*	*kohllarseni*	5	0.60	Me
ULB-all	*brnan*	Artiodactyla	Bovidae	aff. Cephalophini or Bovini	*Brabovus*	*nanincisus*	8	0.25	OH
ULB-all	*bocep*	Artiodactyla	Bovidae	Cephalophini		sp. or spp.	3	1.00	Gg
ULB-all	*hisp*	Artiodactyla	Bovidae	Hippotragini	*Hippotragus*	*sp.*	27	0.54	Cec
ULB-all	*ordet*	Artiodactyla	Bovidae	Hippotragini	*Oryx*	*deturi*	9	0.75	Me
ULB-all	*maavi*	Artiodactyla	Bovidae	Neotragini	*Madoqua*	*avifluminis*	24	0.13	OJ
ULB-all	*rasp*	Artiodactyla	Bovidae	Neotragini	*? Raphicerus*	sp.	2	0.50	To
ULB-all	*trsp*	Artiodactyla	Bovidae	Tragelaphini	*Tragelaphus*	sp.	5	0.00	AA
ULB-all	*gijum*	Artiodactyla	Giraffidae		*Giraffa*	*jumae*	6	1.00	Gg
ULB-all	*gisti*	Artiodactyla	Giraffidae		*Giraffa*	*stillei*	33	0.45	To
ULB-all	*simau*	Artiodactyla	Giraffidae		*Sivatherium*	cf. *maurusium*	6	0.50	To
ULB-all	*eusp*	Perissodactyla	Equidae		*Eurygnathohippus*	sp.	32	2.55	dl
ULB-2	*alsp*	Artiodactyla	Bovidae	Alcelaphini	*Parmularius*/Alcelaphini larger	*pandatus*/sp. Indet.	16	0.94	Me
ULB-2	*papan*	Artiodactyla	Bovidae	Alcelaphini	*Parmularius*	*pandatus*	31	0.90	Me
ULB-2	*aedie*	Artiodactyla	Bovidae	Antilopini	*Aepyceros*	*dietrichi* sp. nov.	13	0.31	OH
ULB-2	*gajan*	Artiodactyla	Bovidae	Antilopini	*Gazella*	*janenschi*	31	0.42	GC
ULB-2	*hisp*	Artiodactyla	Bovidae	Hippotragini	*Hippotragus*	*sp.*	24	0.57	Ca
ULB-2	*maavi*	Artiodactyla	Bovidae	Neotragini	*Madoqua*	*avifluminis*	10	0.00	AA
ULB-2	*gisti*	Artiodactyla	Giraffidae		*Giraffa*	*stillei*	26	0.38	GC
ULB-2	*eusp*	Perissodactyla	Equidae		*Eurygnathohippus*	sp.	23	2.39	ab
ULB-1	*alsp*	Artiodactyla	Bovidae	Alcelaphini	*Parmularius*/Alcelaphini larger	*pandatus*/sp. Indet.	13	0.58	Ca
ULB-1	*papan*	Artiodactyla	Bovidae	Alcelaphini	*Parmularius*	*pandatus*	9	0.56	Ca

(continued)

Table 16.1 (continued)

Unit	Spec-short	Order	Family	Tribe	Genus	Species	N	SCORE	EXTREF
ULB-1	*aedie*	Artiodactyla	Bovidae	Antilopini	*Aepyceros*	*dietrichi* sp. nov.	9	0.11	Ov
ULB-1	*gajan*	Artiodactyla	Bovidae	Antilopini	*Gazella*	*janenschi*	4	0.25	OH
ULB-1	*hisp*	Artiodactyla	Bovidae	Hippotragini	*Hippotragus*	*sp.*	2	0.50	To
ULB-1	*maavi*	Artiodactyla	Bovidae	Neotragini	*Madoqua*	*avifluminis*	10	0.20	DS
ULB-1	*gisti*	Artiodactyla	Giraffidae		*Giraffa*	*stillei*	6	0.67	Me
ULB-1	*eusp*	Perissodactyla	Equidae		*Eurygnathohippus*	sp.	9	3.00	dl

SCORE score value, *EXTREF* Extant reference species as identified in dendrograms Fig. 16.4. Symbols as in Fig. 16.4

Table 16.2 Statistics based on mesowear scores. *ULB*-1 lower section of the Upper Laetolil Beds profile (Fig. 16.1), *ULB*-2 upper section of the Upper Laetolil Beds profile, *ULB*-all Entire Upper Laetolil Beds. SPECSHORT species as in Table 16.1

	SCORE				SCORE-diff	TRAIT			
Unit	ULB-1	ULB-2	ULB-all	UNB	ULB	ULB-1	ULB-2	ULB-all	UNB
Stratum									
SD-all	0.94	0.74	0.57	0.94	0.34				
SD-BR	0.07	0.19	0.32						
SD-IM	0.07	0.21	0.19	0.30					
SD-GR				0.89					
AV-score-all	0.73	0.74	0.64	1.43					
AV-score-BR	0.19	0.28	0.33	0.25	−0.06	3	4	7	1
AV-score-IM	0.58	0.80	0.66	0.99	−0.12	4	3	9	2
AV-score-GR	3.00	2.39	2.55	1.94	0.61	1	1	1	4
SPECSHORT									
mecon				1.88					GR
paalt				1.75					GR
anrec				0.78					IM
eusp				3.14					GR
trbux				1.20					IM
masp				1.00					GR
ordet			0.75					IM	
bocep			1.00					BR	
brnan			0.25					BR	
gakoh			0.27					BR	
gijum			1.00					IM	
rasp			0.50					IM	
sikoh			0.60					IM	
simau			0.50					IM	
trsp			0.00					BR	
aedie	0.11	0.31	0.24		−0.20	BR	BR	BR	
alsp	0.58	0.94	0.80		−0.35	IM	IM	MI	
hisp	0.50	0.57	0.54		−0.07	IM	IM	IM	
eusp	3.00	2.39	2.55		0.61	GR	GR	GR	
gajan	0.25	0.42	0.40	0.25	−0.17	BR	BR	BR	BR
gisti	0.67	0.38	0.45		0.28	IM	BR	IM	
maavi	0.20	0.00	0.13		0.20	BR	BR	BR	
papan	0.56	0.90	0.83		−0.35	IM	IM	IM	

SD standard deviation, *Av* average, *SCORE* mesowear score, *SCORE-diff* mesowear score difference between unites *ULB*-1 and *ULB*-2, *TRAIT* feeding trait of identified extant reference species, *BR* browser, *IM* intermediate feeder, *GR* grazer on dental enamel were not considered in this study. Furthermore, all specimens were excluded where stratigraphic information was uncertain. The largest number of specimens, however, had to be excluded, because no taxonomic assignment could be established based on isolated teeth. The dental sample was further reduced to 343 specimens because teeth in early and late wear had to be excluded (see scoring protocol below). Of the resulting 40 datasets representing a taxon in its stratigraphic context 21 were based on a critically low number of teeth (less than 10), and therefore had to be considered as statistically less meaningful, while the remaining datasets range between 11 and 41 dental specimens and can be considered as fairly stable in mesowear analysis according to feasibility tests undertaken by Fortelius and Solounias (2000).

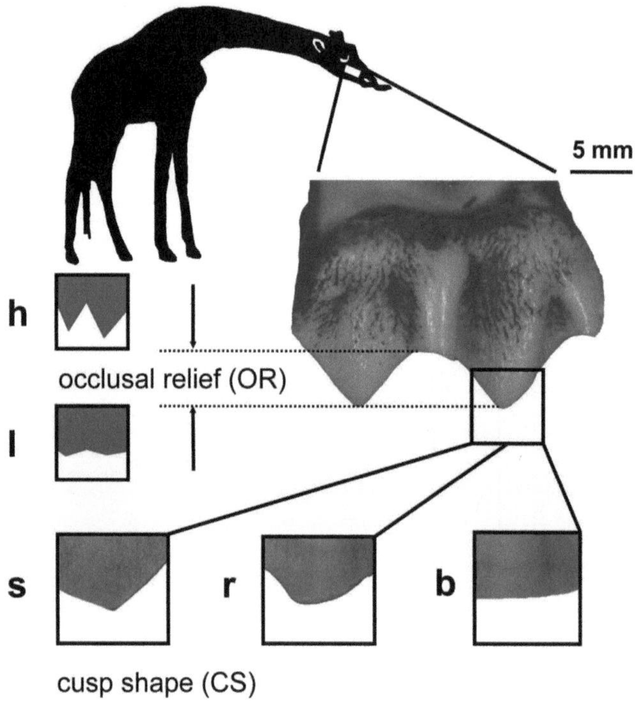

Fig. 16.2 The mesowear variables of a brachydont giraffe (*Giraffa camelopardalis*) cheek tooth (upper right M2). Arrows indicate how the occlusal relief (*OR*) is evaluated. It may be scored "high" (*h*) or "low" (*l*), the cusp shape (*CS*) is classified as "sharp" (*s*), "round" (*r*) and "blunt" (*b*). Scale bar = 5 mm

Methods

Museum specimens were carefully cleaned where needed, and a negative mold of the occlusal aspect was made using PROVIL novo Putty regular set (Heraeus Kulzer, Hanau, Germany) polysiloxane dental molding putty. Subsequently, positive casts of the teeth were produced by filling the molds with epoxy resin Injektionsharz EP (Reckli-Chemiewerkstoff, Herne, Germany).

The mesowear method of dietary analysis is based on facet development of cheek tooth occlusal surfaces. The degree of facet development reflects the relative proportions of tooth-to-tooth contact (attrition) and food to tooth contact (abrasion). Attrition creates facets while abrasion obliterates them. The entire surface of the teeth is affected by tooth wear, but mesowear analysis has focused on the buccal cutting edges of the enamel surfaces where the buccal wall (ectoloph) meets the occlusal surface. The mesowear method treats ungulate tooth mesowear as two variables: occlusal relief and cusp shape. Occlusal relief (OR) is classified as high (h) or low (l), depending on how high the cusps rise above the valley between them. The second mesowear variable, cusp shape, includes three scored attributes: sharp (s), round (r) and blunt (b) according to the degree of facet development (Fig. 16.2).

In addition to established mesowear convention, a combined mesowear score was computed from each population similar to Mihlbachler and Solounias (2006), Rivals and Semprebon (2006) and Semprebon and Rivals (2007). A combination of high relief and sharp cusps was assigned a score of "0", a combination of high relief and round cusp was assigned a score of "1", a combination of low relief and sharp cusp was assigned a score of "2", a combination of low relief and round cusp was assigned a score of "3" and a combination of low relief and blunt cusp was assigned a score of "4". In this convention, a score of 0 represents the most attrition dominated mesowear signature, while a score of 4 would represent the most abrasion dominated signature. Individual scores were averaged, and a mean score was calculated for each fossil species. Thus, scores indicate the overall abrasiveness of the diet a species had to cope with.

Only permanent upper postcanine dentitions were investigated, because up to now there is no consistent comparative mesowear dataset available for the lower dentition. Unworn teeth and teeth in early wear (occlusal surface not yet entirely exposed to wear) were excluded from this study, because when too little wear is involved, no stable mesowear equilibrium can be established in the early stages of tooth wear. Also, teeth in very late wear were excluded, following Fortelius and Solounias (2000). In ruminants, both the second and the third molar were scored, following feasibility tests on three extant ruminant species (Franz-Odendaal and Kaiser 2003), and the procedure applied to giraffes (Clauss et al. 2007a) and other extant free ranging and captive ruminants (Kaiser et al. in press). For the Equidae, the extended scoring convention (Kaiser and Solounias 2003) was applied, which involves inclusion of the upper P4, P3, M1, M2 and M3 into the mesowear model. Only the sharpest of the two cusps of a cheek tooth was scored in order to be consistent with the comparative data presented by Fortelius and Solounias (2000).

Data Analysis and Statistics

Cluster statistics

The mesowear parameter frequencies were calculated for each fossil population of specimens. In cluster analysis (Fig. 16.3) extant comparative species comprise a set of 27 ungulates reported as "typical" dietary categories by Fortelius and Solounias (2000). Dietary traits of extant species are adapted from the conservative (CONS) classification by Fortelius and Solounias (2000) with one exception. The bushbuck (*Tragelaphus scriptus*; TS) was recognized as a browser instead of an intermediate-feeder. This species is classified as a 'browser' by Kingdon (1979), Janis and Ehrhardt (1988), Simbotwe and Sichone (1989) and Gagnon and Chew (2000), and a 'concentrate selector' by Hofmann (1989), while Simpson (1974), MacLeod et al. (1996), and Fortelius and Solounias (2000) consider it a mixed feeder. The browse/graze ratio of 70/10% as indicated by Van Wieren (1996), the Dicot/Monocot ratio of 80/10% given by Gagnon and Chew (2000), the high percentage of fruit (10%) as reported by Gagnon and Chew (2000) and the lack of C_4 grass in the diet as reported by Sponheimer et al. (2003) indicate that the bushbuck should be assigned as a browser or a browse dominated mixed feeder rather than a typical mixed feeder.

Subsequently, cluster statistics were performed using Systat 11.0 (SYSTAT Software Inc., Richmond, CA) software and using default settings. Hierarchical cluster analysis with complete linkage (furthest neighbors) was applied following the standard hierarchical amalgamation method of Hartigan (1975). The algorithm of Gruvaeus and Weiner (1972) was then used to order the cluster tree using the three cusp shape variables (% high,% sharp and% blunt). The resulting trees show the relationship of the datasets by joining them in clusters. The closer the data the smaller is the normalized Euclidean distance (NED) at the branching point. The exact sequence and direction of species arrangement in the dendrogram, however, may not be interpreted as an expression of sequential differences because clusters may flip. Using the cluster trees we identify extant specimens as classifying closest to fossil species. These extant species are thus regarded as dietary analogs to the fossil population and dietary trait identification was taken from these extant species as actualistic analogs (Table 16.2).

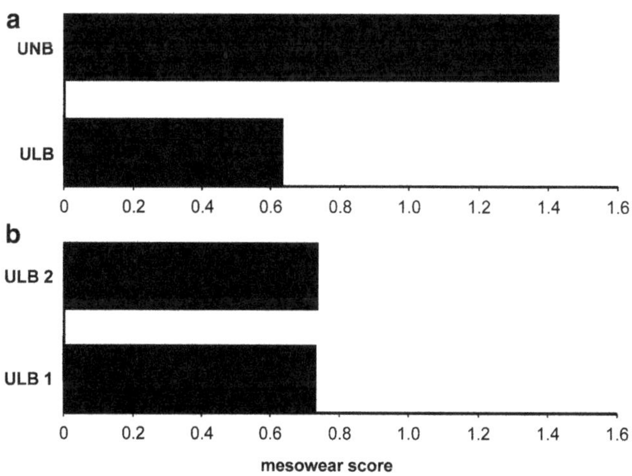

Fig. 16.3 (**a**) Average scores of *ULB* and *UNB* (all species investigated). Note that average scores are highest in the *UNB* (1.43), and lowest in the total of *ULB* (0.64). (**b**) Average scores of *ULB*-1 and *ULB*-2 (8 species investigated). Note that there is no notable difference between *ULB*-1 (0.73) and *ULB*-2 (0.74)

Mesowear scores

Because cluster analysis is an exploratory technique and there is no quantitative information associated, a combined mesowear score was computed for each fossil population and for a set of extant African Bovidae, Giraffidae and Equidae. Individual scores were averaged and a mean score was calculated for each taxon as related to the detailed stratigraphic context. Scores therefore indicate the overall abrasiveness of the diet a species had to cope with in a given context. The higher the score, the greater amount of abrasive components was in the diet of the animal. The scores are plotted together with extant taxa in related lineages (Fig. 16.4). In order to better resolve patterns in the two ULB sub-units (ULB-1 and ULB-2), differences of average scores between feeding trait (SCORE_diff) were computed (Table 16.2; Fig. 16.5), as well as standard deviations of score differences in each of the three feeding traits: browsers, intermediate-feeders and grazers (Table 16.2, SD_diff). Standard deviations are calculated for scores in all species per unit investigated (Table 16.2, SD_all).

Results and Interpretations

Mesowear scorings are listed by species in Table 16.1. Extant reference species of fossil populations as identified by cluster analysis are also given. In all Ruminantia, high occlusal relief prevails over low relief (63–100% high). In the ULB community, the lower range of high relief ratios in the Ruminantia is slightly higher (67%; *bocep*). In the two sub-units of the ULB, the Ruminantia show slightly higher ratios of high relief in the older Unit (ULB-1), where all have 100% high reliefs, while in the ULB-2 two species (*gajan* and *alsp*) have slightly lower (94%) ratios of high reliefs. In the UNB Ruminantia, high occlusal relief ranges from 63% (*mecon*) to 100% (*gajan* and *masp*).

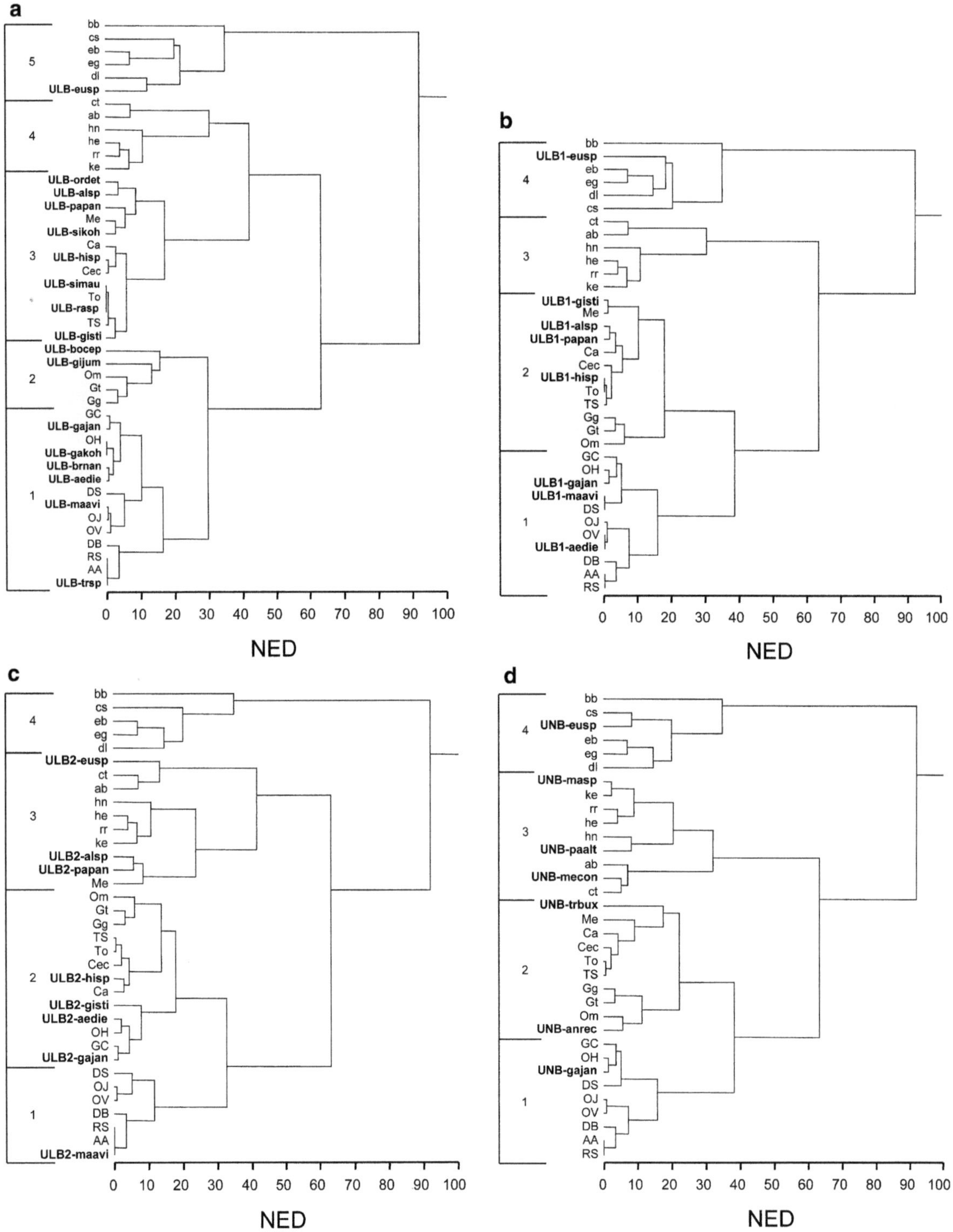

Fig. 16.4 Hierarchical dendrograms based on a set of 27 "typical" extant comparison species and fossil ungulates from Laetoli. The closer the species are in their mesowear signature, the smaller is the normalized Euclidean distance at the branching point. With the exception of *Tragelaphus scriptus* (*TS*) the dietary classification of extant species is according to the "CONS" classification of Fortelius and Solounias (2000). Fossil species from Laetoli are given in Table 16.1 are indicated in bold. *ULB* Upper Laetolil Beds, *UNB* Upper Ndolanya Beds. (**a**) All 17 fossil species investigated from the Upper Laetolil Beds. (**b**) Eight fossil species from the lower unit of the Upper Laetolil Beds (*ULB*-1). (**c**) Eight fossil species from the lower unit of the Upper Laetolil Beds (*ULB*-2). (**d**) Seven fossil species investigated from the Upper Ndolanya Beds. Symbols of extant comparison species: UPPER CASE=BROWSER, lower case=grazer, Mixed case=intermediate-feeder. Browsers*:* *Alces alces* (AA), *Dicerorhinus sumatrensis* (DS), *Diceros bicornis* (DB), *Giraffa camelopardalis* (GC), *Odocoileus hemionus* (OH), *Odocoileus virginianus* (OV), *Okapia johnstoni* (OJ), *Rhinoceros sondaicus* (RS), *Tragelaphus scriptus* (TS). Intermediate-feeders*:* *Aepyceros melampus* (Me), *Capricornis sumatraensis* (Ca), *Cervus elaphus canadensis* (Cec), *Gazella granti* (Gg), *Gazella thomsoni* (Gt), *Taurotragus oryx* (To). Grazers*:* *Alcelaphus buselaphus (ab), Bison bison (bb), Ceratotherium simum (cs), Connochaetes taurinus* (ct), *Damaliscus lunatus* (dl), *Equus burchellii* (eb), *Equus grevyi* (eg), *Hippotragus equinus* (he), *Hippotragus niger* (hn), *Kobus ellipsiprymnus* (ke), *Ovibos moschatus* (Om), *Redunca redunca* (rr)

Sharp cusps prevailed in nine species from the ULB and two species from the UNB. In the two sub-units of the ULB, sharp cusps prevail in three ULB-1 populations and four ULB-2 populations. Blunt cusps have generally low frequencies. In the UNB four of the seven taxa investigated have considerably high blunt ratios (>19%), while only two of the 17 ULB populations have as high blunt ratios. The Equidae consistently have more than 28% blunt cusps and in ULB-1 *eusp* shows the highest blunt ratio recorded (50%). Mesowear scores range between 1.0–3.1 in the UNB and 0–2.6 in the ULB assemblage. The two Equidae consistently have the highest scores in all units (Table 16.1). Average scores of the ungulate community are highest in the UNB (1.43), and lowest in the ULB (0.64) (Table 16.2, Fig. 16.6a). Score averages are only slightly higher in ULB-2 (0.74) compared to ULB-1 (0.73) (Fig. 16.6b).

Cluster analysis of mesowear variables classifies the 27 extant reference species and 17 ULB species into five major clusters (Fig. 16.3a). Cluster 1 contains all extant browsers and six ULB species. Species in this cluster have the most attrition dominated mesowear signatures. The extant intermediate-feeders fall in clusters 2 and 3, which also contain the majority (10 species) of the ULB ungulate community. Clusters four and five (Fig. 16.3a) correspond to the most abrasion dominated mesowear signatures, and among all extant grazers contain the equid (*eusp*) from the ULB, which is classified close to the extant topi (*Damaliscus lunatus*). Fig. 16.3a identifies six ULB species as browsers, 10 as intermediate-feeders and one as a grazer (also see Table 16.2).

When ULB units 1 and 2 are analyzed separately (Figs. 16.3b–c) clustering patterns are slightly different. All extant intermediate-feeders now fall within one major cluster 2, which also contains four ULB-1 species. Three species from the ULB-1 are classified as browsers and one (*eusp*) as a grazer. In the ULB-1, *eusp* is closest to the extant zebra *Equus quagga (burchellii)* (eb) and *Equus grevyi* (eg). In the ULB-2 (Fig. 16.6c) four species classify as browsers, three as intermediate-feeders, and again the equid (*eusp*) is the only grazer.

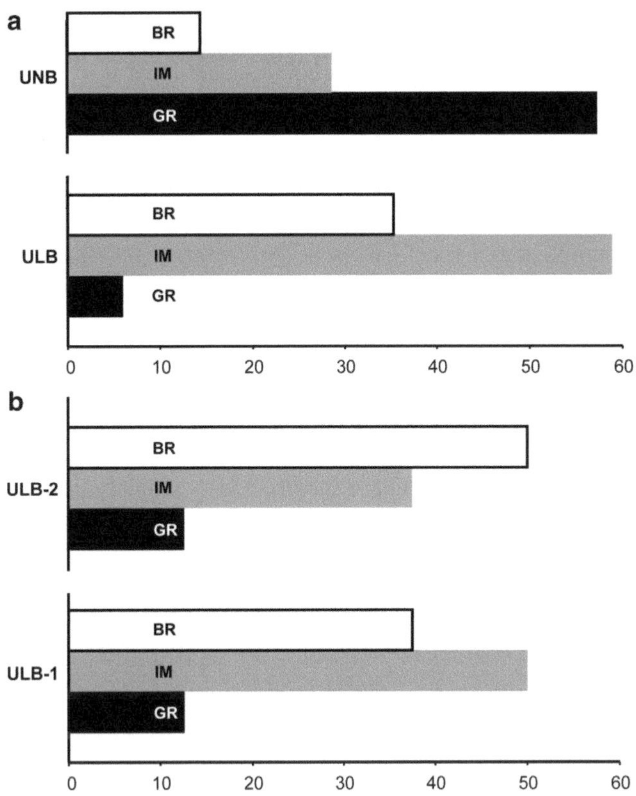

Fig. 16.5 Percentages of fossil ungulate species representing a certain dietary trait. *BR* browsers, *IM* intermediate-feeders, *GR* grazers. (**a**) Upper Laetolil Beds (*ULB*) versus Upper Ndolanya Beds (*UNB*). Note that in the *ULB* intermediate-feeders dominate the community and grazers are at a minimum (one species), while in the *UNB* grazers dominate and browsers are at a minimum. (**b**) Comparison of two sub-units of the Upper Laetolil Beds. *ULB*-1 below Tuff 5, *ULB*-2 above Tuff 5. Note that one flexible species (*Giraffa stillei*) is an intermediate-feeder in *ULB*-1 and a browser in *ULB*-2

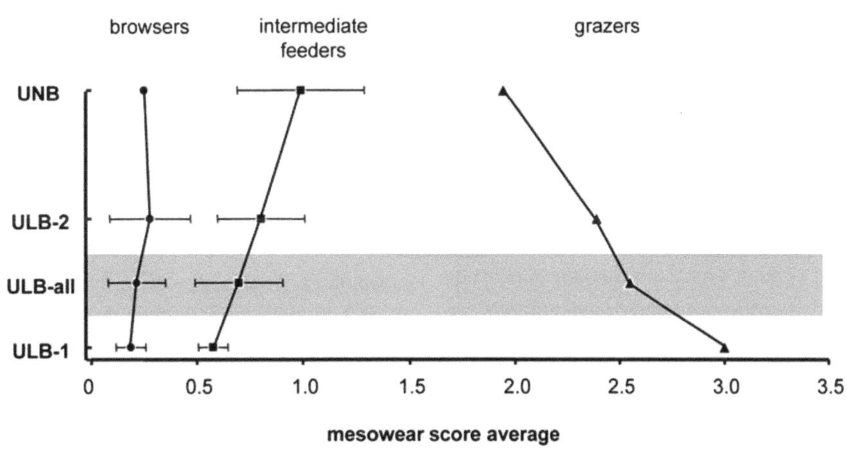

Fig. 16.6 Score average per feeding type plotted against stratigraphic units. *ULB*-all all Upper Laetolil Beds species, *ULB*-1 lower section of *ULB* (below Tuff 5), *ULB*-2 upper section of *ULB* (above Tuff 5), *UNB* Upper Ndolanya Beds. Grey shaded bar indicates the total of all *ULB* species investigated (17 species), while in *ULB*-1 and *ULB*-2 data base on a reduced number of 8 species. Note that browsers show little shift in their signature within the sequence and intermediate-feeders show a consistent shift towards more abrasion control in the mesowear equilibrium. The reverse trend is seen in grazers

Within the ULB succession we find *Giraffa stillei* to be the only species showing a shift in its dietary trait through time. In ULB-1 the species behaves like a browser, while in ULB-2 its feeding signature gets more abrasion dominated, and the signal falls within the extant intermediate-feeders (Table 16.2).

Cluster analysis of UNB species (Fig. 16.3d) results in four major clusters, which represent feeding traits in the same pattern as Fig. 16.3b. Only *Gazella janenschi* (*gajan*) from the UNB classifies in cluster 1, and is identified as the only browser in the UNB. *Antidorcas recki* (*anrek*) and *Tragelaphus* sp. cf. *buxtoni* (*trbux*) are intermediate-feeders classified in cluster 2, while four species (*mecon*, *paalt*, *masp* and *eusp*) classify at the most abrasion dominated end of the mesowear continuum together with extant grazers.

Lineage Based Analysis of Mesowear Scores

Bovidae

Bovids are the most abundant mammalian taxa in the Laetoli assemblage, comprising 35% of the Upper Laetolil specimens and 81% of the Upper Ndolanya material (Leakey 1987, Kingston and Harrison 2007). Here mesowear analyses are focused on taxa from seven of the eight bovid tribes represented in the Laetoli material, those that are relatively well-represented in the Laetoli collections - Alcelaphini, Antilopini, Bovini, Cephalophini, Hippotragini, Neotragini, and Tragelaphini.

Alcelaphini

The extant members of tribe Alcelaphini include the wildebeest, hartebeest, bontebok, topi and hirola. All are variable to obligate grazers (Gagnon and Chew 2000), having diets dominated by C_4 grass (Cerling et al. 2003a; Sponheimer et al. 2003). They inhabit open woodland or edaphic/secondary grasslands (Scott 1979; Estes 1991), and are postcranially adapted for cursoriality (Kappelman et al. 1997).

In the ULB, alcelaphines are represented by *Parmularius pandatus*, a species estimated to be the size of the extant bontebok (*Damaliscus dorcas*) and a slightly larger species, Alcelaphini sp. indet. about the size of the extant hartebeest (*Alcelaphus buselaphus*) (Alan Gentry, personal communication). Due to earlier difficulties of species assignment, the latter sample may be, to a limited extent, mixed with *P. pandatus* (Alan Gentry, personal communication). The two ULB alcelaphines are very similar in their mesowear signatures (scores = 0.83/0.8) and show a mixed/intermediate feeding strategy, close to the modern-day impala (*Aepyceros melampus*) (Fig. 16.3a). Kingston and Harrison (2007) find a $\delta^{13}C$ enamel signal ranging from −8.6‰–0.6‰, with a median of about −2‰. This indicates, that *P. pandatus* also had a significant C_3 browse component in its diet and occupied a grass dominated mixed feeding niche. This is in complete accordance with the mesowear signature. Alcelaphini sp. indet. falls in the same transitional $\delta^{13}C$ range close to 0‰ (Kingston and Harrison 2007). This would suggest a higher ratio of C_4/C_3 grasses in the diet of this grass dominated mixed feeder without increasing the general proportion of abrasive grasses compared to *Parmularius pandatus*. The mixed feeding signatures of the ULB alcelaphines imply that some early alcelaphine lineages may not have been as committed to a grazing niche as their extant counterparts.

Four species of alcelaphines are known from the UNB (Alan Gentry, personal communication). Mesowear data are available for *Megalotragus kattwinkeli*/*isaaci* or *Connochaetes* sp. (*mecon*), which may include the *Parestigorgon gadjingeri* of Dietrich (1950), Gentry (1987) and Kingston and Harrison (2007) (Alan Gentry, personal communication) and *Parmularius* ? *altidens* (*paalt*), which certainly includes the *Parestigorgon gadjingeri* of Gentry (1987) and Kingston and Harrison (2007). The two UNB alcelaphines investigated have mesowear scores of 1.9 (*mecon*) and 1.8 (*paalt*) (Fig. 16.4) and classify at the abrasion-dominated end of the mesowear continuum together with modern-time grazers only. Specimens that Kingston and Harrison (2007) indicate as Alcelaphini sp. (no species given) in their Fig. 16.7 have $\delta^{13}C$ enamel signals ranging between 0.5‰ and 2 ‰ are among the most positive $\delta^{13}C$ enamel values, and these would complement the grazing signature of some UNB alcelaphines by indicating the C_4 nature of the graze.

Modern-day alcelaphines, the hartebeest (*Alcelaphus buselaphus*) and the wildebeest (*Connochaetes taurinus*), have almost the same mesowear scores (1.89/1.8) as the two UNB species investigated (Fig. 16.3). This may signify that by Upper Ndolanya times the modern grazing trait was already established in the lineage.

Antilopini

Modern gazelles have dietary traits ranging from variable grazing in Grant's gazelle (*Gazella granti*; score = 0.65), Thomson's gazelles (*G. thomsoni*; score = 0.66) and impala (*Aepyceros melampus*; score = 0.647) to browsing in the gerenuk (*Litocranius walleri*; score = 0.75). Antilopines from the ULB available for this study include *Gazella janenschi* (*gajan*), "*Gazella*" *kohllarseni* (Gentry 2011), *Aepyceros dietrichi* sp. nov. (*aedie*) (Gentry 2011). All three ULB-species have scores ranging between 0.4 (*gajan*) and 0.24 (*aedie*), which signifies that all ULB antilopines were browsers. No modern-day representative has a mesowear score as low as the three ULB gazelles (Fig. 16.4). *Gazella janenschi* seems to be the only species with a certain amount of abrasive component

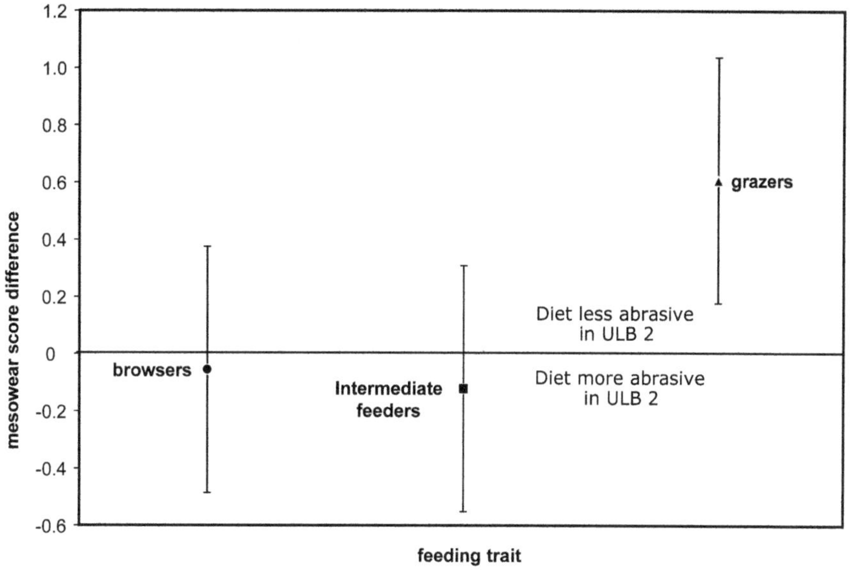

Fig. 16.7 Deviation of average scores per feeding trait computed for *ULB*-units 1 and 2. Note that browsers and intermediate-feeders have a slightly more abrasion dominated average score, while the only grazer in the *ULB* ungulate community (i.e., *Eurygnathohippus*) has a less abrasion-dominated signature in *ULB*-2

in its diet. As in the extant giraffe (*Giraffa camelopardalis*), the dietary reference species of ULB *G. janenschi* (Fig. 16.3a), the forage of *G. janenschi* may have been grit laden (Clauss et al. 2007a; Fortelius and Solounias 2000). Alternatively, this species may have foraged a larger proportion of grass or other abrasive foods compared to the remaining ULB antilopines.

Kingston and Harrison (2007) give $\delta^{13}C$ enamel signals for *G. janenschi* that indicate a variable diet (−2‰ to −10‰), probably focused more on browsing than modern antilopines. This isotope signal again allows one to identify the abrasive component in the diet of the species as C_4 grass, and indicates that the ULB environment was probably less grit-laden compared to the modern-day giraffe's feeding regime.

Antilopines from the UNB investigated are *Gazella janenschi* (*gajan*) and *Antidorcas recki* (*anrec*) (Gentry 2011). *Gazella janenschi* from the UNB classifies as the only browser in this youngest studied unit of the Laetoli sequence, and compared to the ULB population of the same taxon it has a less abrasion dominated feeding regime (score=0.24). Isotope data for this population given by Kingston and Harrison (2007) fall in the C_3/C_4 Mixed diet range, but close to C_3 dominated browsing and shifted more towards the browsing range than the median of ULB *Gazella janenschi.* Both methods independently indicate that *G. janenschi* from the UNB was engaged in more leaf browsing compared to the ULB population. However, since both studies had few specimens available, these differential signals should be regarded as preliminary. It is noteworthy that the signature of *G. janenschi* shifts from more attrition dominated in ULB-1 to slightly more abrasion dominated in ULB-2 and back again to the most attrition dominated signal in the UNB. The ULB assemblage is signified by the appearance of *Antidorcas recki* (anrek), a species with a browse dominated intermediate feeding signature (score=0.78). As a possible competitor in a probably narrow browsing niche in the UNB environment, the presence of *A. recki* could be responsible for the shift towards the even more browse dominated feeding niche of *G. janenschi*. Gentry (1987) notes that the teeth of *G. janenschi* are rather generalized and fairly hypsodont. This would indicate that *G. janenschi* had undergone an adaptation towards more abrasive food in the Pliocene, which would have enabled this species to occupy a more opportunistic feeding niche compared to taxa that retained brachydonty. A typical leaf browsing signal in the UNB, as indicated by the close linkage of the species to the mule deer (Fig. 16.3d) would then reflect the species capability to behave like a "hypsodont browser" and opportunistically exploit a high energy feeding niche. This remarkable flexibility may be the major pre-adaptation allowing the species to survive the environmental transition.

Bovini

The Bovini have only one modern-day representative in Africa, the African buffalo (*Syncerus caffer*). The African buffalo is the largest bovid, and is most consistently considered a variable grazer (Gagnon and Chew 2000). A dicot/monocot ratio of 22.5/77.5% has also led some authors to suggest that the species is an intermediate-feeder (Janis 1988; Fortelius and Solounias 2000). Mesowear scores of *Syncerus caffer* are low (0.94) for a typical grazer, and they could well correspond with a more variable feeding strategy.

Sponheimer et al. (2003) indicate 100% of C_4 graze for this species in East Africa and only 88% in South African environments, illustrating the flexible feeding behavior of the species. The extinct *Simatherium kohllarseni* (*sikoh*) from the ULB is a moderate to large-sized bovine, which is considered ancestral to *Pelorovis*, the extinct long-horned buffalo of the African Pleistocene (Gentry 1987). The mesowear signature of *S. kohllarseni* indicates an intermediate feeding strategy (score = 0.6), while the extant dietary reference species is identified as the impala (*Aepyceros melampus*; Fig. 16.3a). The almost balanced ratio of sharp and round cusps, together with only high occlusal relief is frequently found in typical intermediate-feeders (Fortelius and Solounias 2000). Compared with extant buffalos, *S. kohllarseni* certainly occupied a more browse dominated feeding niche and did not share the same feeding trait as its modern counterpart in the ULB paleoenvironment.

Cephalophini

The modern-day duikers are predominantly small-bodied bovids almost strictly bound to forested environments and are never found in grasslands. With a slightly arched body and the front legs a little shorter than the hind legs, they are well-adapted to penetrate thickets. The diet of extant duikers seems to be rather diverse. Although, they are sometimes simply considered browsers (Hofmann 1989; Janis 1990a), other authors recognize the duikers as highly frugivorous, and are thus not typical browsers (Gautier-Hion et al. 1980; Lumpkin and Kranz 1984; Feer 1989). For 18 extant species, Gagnon and Chew (2000) list fruit as representing more than 70% of the diet, while monocot forage never makes up more than 2.5%, the remainder being dicots. This would correspond to a rather consistent dietary trait for all extant species that would be best characterized as a fruit dominated browsing niche. Mesowear scores of five modern-day cephalophines, *C. dorsalis, C. natalensis, C. nigrifrons, C. niger* and *C. silvicultor* range from 0.9 in *C. niger* to 1.46 in *C. silvicultor*, and reflects the close dietary traits within the genus today. Due to a significant abrasive component in their diet, they fall within the score spectrum of typical intermediate feeders and some attrition dominated grazing species (Fig. 16.4). In terms of mesowear, the cephalophines are therefore considered "mabra"(minute abraded brachydonts) in the convention by Fortelius and Solounias (2000) because their mesowear signal is more abrasion dominated than would be expected for a brachydont browser. Solounias and Fortelius (2000) attribute the rounding of their cusps to the "tip crushing wear" typically associated with frugivory (Janis 1990b).

The fossil record of cephalophines at Laetoli is restricted to one species, assigned to Cephalophini sp. or spp. (Gentry 2011) and a second one (*Brabovus nanincisus* from the ULB), which shows affinities either to the Cephalophini or Bovini (Gentry 2011). Here, for comparative reasons, *B. nanincisus* is listed as a member of the Cephalophini. Mesowear scores of Cephalophini sp. or spp. are 1.0 and place the species in the range of intermediate-feeders, occupied by modern-day gazelles like Grant's gazelle (*Gazella granti*). As a members of the "mabra" group the Cephalophini cannot be accommodated within the classical browser/grazer spectrum, and therefore comparison must be restricted to the extant members of the tribe. The score of Cephalophini sp. or spp. is almost identical with modern-day Cephalophini, which indicates that there is no major difference in the feeding trait of these extinct Pliocene taxa as compared to the majority of extant species. We therefore have to consider Cephalophini sp. or spp. as fruit dominated browser in the ULB habitat. *Brabovus nanincisus,* the second ULB species that may be a member of Cephalophini, has a score of 0.25 that classifies closest to the North American mule deer (*Odocoileus hemionus*; Fig. 16.3a), a browser feeding on soft items. Compared to all extant and fossil cephalophines investigated, *B. nanincisus* has the least abrasion dominated mesowear signature and implies that this species was certainly not engaged in the type of frugivory typical of modern-day cephalophines. It thus is characterized as a leaf browser rather than as a frugivore. Because of the uncertain taxonomic status of the species, these observations can only tentatively be regarded as relevant to the question of niche segregation within the Cephalophini. The dietary trait identified may even hint at a less close relationship of this species with the Cephalophini.

Hippotragini

The extant Hippotragini (i.e., oryxes, roan and sable antelopes) are primarily obligate grazers, associated with habitats ranging from desert grasslands to open woodlands (Estes 1991; Gagnon and Chew 2000). Fossil Hippotragini from the ULB include *Hippotragus* sp. (hisp) (the *Praedamalis deturi* of Dietrich (1950), Gentry (1987), and Kingston and Harrison (2007) and *Oryx deturi* (ordet), a species conceived as rare by Alan Gentry (personal communication). No Hippotragini from the UNB have been investigated here. The ULB hippotragines show a mesowear signal indicative of a mixed/intermediate feeding strategy, with the impala (*Aepyceros melampus*) and the wapiti (*Cervus elaphus canadensis*) as extant reference species (Fig. 16.3a). The Laetoli Hippotragini show significantly lower mesowear scores compared to the extant *Hippotragus equinus* (score = 1.27) and *H. niger* (score = 1.45). Kingston and Harrison (2007) find isotopic signals ranging from about −2‰ to −9‰, reflecting a mixed C_3/C_4 diet, but with an emphasis on browse. These relatively negative values also contrast with all modern Hippotragini isotopic data from East and South Africa that indicate a C_4 dominated diet (Sponheimer and Lee-Thorp 2003).

Mesowear and isotope proxies concordantly indicate that hippotragines were formerly more ecologically diverse compared to their extant representatives and that their dietary diversity declined during the Plio-Pleistocene. Kingston and Harrison (2007) suggest that this decline could be related to hippotragines being forced into increasingly more marginal environments by the radiation of alcelaphines. Also, there may have been a broader availability of browsing and intermediate-feeding niches in the ULB environment that would have accommodated a greater taxonomic diversity. By ULB times, there would have been less competition for access to grazing niches, which obviously were occupied only by the equid (*Eurygnathohippus*) at Laetoli. Expanding into a grazing niche in the late Pliocene, however, may have forced several groups of ruminants (including alcelaphines and hippotragines) to compete with equids in order to compensate for decreasing niche diversity in the attrition (browse) dominated part of the dietary spectrum.

Neotragini

The Neotragini are represented by two species in the ULB, referred to *Madoqua avifluminis* and ?*Raphicerus* sp. (Gentry 1987; Gentry 2011). Modern *Madoqua kirkii* (Kirk's dikdik), which is a common species in the southern Serengeti ecosystem today, is a browser that inhabits dense, dry thickets dominated by evergreen scrub. Gagnon and Chew (2000) give a monocot/dicot ratio of 80/17% for this species, which indicates that some grass is being eaten, especially during the wet season (Kingdon 1982). $\delta^{13}C$ enamel signatures of *M. kirkii* and *M. guentheri* (Guenther's dikdik) suggest exclusive C_3 browsing (Cerling et al. 2003b; Kingston and Harrison 2007). Mesowear scores of *Madoqua avifluminis* (maavi) from the ULB are as low as 0.13, and the modern-day reference species is the okapi (*Okapia johnstoni*) a dedicated browser (Fig. 16.3a). Within the ULB assemblage, *M. avifluminis* has among the most attrition dominated feeding traits, only slightly exceeded by *Tragelaphus* sp.

$\delta^{13}C$ enamel signatures of ULB *Madoqua avifluminis* published by Kingston and Harrison (2007) are more positive than the modern taxa, which leads Kingston and Harrison to suggest that *M. avifluminis* incorporated 15–20% more grass in its diet. Since there is no abrasive signal in the mesowear data of this extinct species, it must be considered more likely that the C_4 component derived from either C_4 dicots or was the result of feeding in a less grit laden environment, perhaps indicative of more humid conditions in the wet season. In the UNB, only *Madoqua avifluminis* has been recorded, which probably represent the same species as in the ULB. *Madoqua avifluminis* survived the faunal turnover that took place after the deposition of the ULB. The UNB population has a mesowear score of 1.0, and equals the extant *Kobus ellipsiprymnus* (common waterbuck), a variable grazer which has a dicot/monocot ratio of 15/84% in its diet (Gagnon and Chew 2000). What makes the UNB population of *M. avifluminis* different from other taxa with similarly low scores is the complete lack of sharp cusps (Table 16.1), a mesowear feature that is never found in extant browsers or even intermediate-feeders. On the other hand, intermediate-feeders, in particular those representing the no (no particular class) group of Fortelius and Solounias (2000), may sometimes have mesowear scores higher than 1.0. Finding *M. avifluminis* to score like an abrasive feeder is surprising. It may illustrate a flexible and opportunistic feeding behavior that may be, at least in part, responsible for the survival of the species into the late Pliocene when browsing niches declined. However, it may also indicate a change in food availability associated with general ecological change and imply a highly flexible opportunistic feeding strategy.

Tragelaphini

Modern tragelaphines are mostly browsers and browse dominated intermediate-feeders with low-crowned teeth and digestive tracts adapted for nutritious vegetation and fruit (Clauss et al. 2007b; Kingdon 1982, 1997). A mixed feeding strategy, including a grazing component, has been reported for some taxa (Gagnon and Chew 2000). Isotopic analyses of keratin and tooth enamel of modern tragelaphines from South and East Africa indicate diets dominated by browse, although there is evidence for mixed C_3/C_4 feeding (e.g., eland, *Taurotragus oryx*) (Lee-Thorp 1989; Cerling et al. 2003a; Sponheimer et al. 2003). $\delta^{13}C$ enamel values of *Tragelaphus cf. kyaloae* from the Pliocene of Gona in Ethiopia (Semaw et al. 2005) (−0.1 to 11.0‰) indicate that the foraging strategy of some taxa can span the entire browse/graze spectrum.

The one known ULB tragelaphine (*Tragelaphus* sp.) has the lowest mesowear score of all Laetoli ungulates (0.0), which equals that of the moose (*Alces alces*) in being at the most attrition dominated end of the mesowear continuum. Carbon isotope signatures of the species range from −7.8 to −11.8‰ (Kingston and Harrison 2007). These values would indicate a C_3 dominated diet with some C_4 forage. Since the mesowear data do not indicate an abrasive component in the diet of *Tragelaphus* sp., this C_4 component is most likely due to C_4 dicots rather than grasses. Kingston and Harrison (2007: 280), however, consider C_4 dicots unlikely to comprise a significant component of the equatorial African biomass during the Pliocene. The low level of abrasiveness in the diet of *Tragelaphus* sp. from the ULB may merit reconsideration of this assumption.

The only known tragelaphine from the UNB is *Tragelaphus* sp. cf. *buxtoni*, which is larger than *Tragelaphus* sp. from the ULB (Gentry 1987). The extant *T. buxtoni* (mountain nyala) from Ethiopia is a browser that prefers diversely structured environments including woodland, bushland, but also grassland (Kingdon 1997). A mesowear score of 1.2 classifies *T.* sp. cf. *buxtoni* from the UNB as an intermediate-feeder, closest in its mesowear signal to the extant impala (*Aepyceros melampus*; Fig. 16.3a), and close to the transition between intermediate-feeders and grazers. Kingston and Harrison (2007) find their UNB *Tragelaphus* sample to be ^{13}C-enriched, indicating variable grazing behavior; a dietary signal consistent with the mesowear signature.

Giraffidae

At Laetoli, giraffid material has been attributed to three species from both the Upper Laetolil and Upper Ndolanya Beds - a medium-sized giraffid (*Giraffa stillei*) and a large sized species (*G. jumae*) and a sivathere (*Sivatherium* cf. *maurusium*). In addition, Robinson (2011) recognizes *Giraffa pygmaea* in the Upper Ndolanya Beds. Compared to other East African Plio-Pleistocene localities, Laetoli is unusual in that giraffids comprise a significant proportion of the mammalian specimens (8.7%) and that *G. stillei* is numerically the most abundant species (comprising 65% of *Giraffa* specimens) (Leakey 1987; Harris 1987a; Robinson 2011). It is likely that there was ecological partitioning to accommodate a relatively high abundance of at least three sympatric giraffids, presumably based on different optimum feeding heights, differential selectivity in browse, or foraging in different microhabitats. Harris (1987a) notes that there are some functional differences between the two giraffines, and suggests that competition for dietary resources was reduced by differences in size. This may have resulted in niche partitioning related to different vegetation levels. Based on a study of the feeding ecology of the Masai giraffe, *G. camelopardalis tippelskirchi*, in the Serengeti National Park (Pellow 1983), Harris (1991) proposed that the larger *G. jumae* may have specialized in foraging in a habitat equivalent to mid-slope open *Acacia–Commiphora* woodland and *G. stillei* in shorter ridge-top or riverine woodlands. Sivatheres have been associated with habitats ranging from forest to open woodland (Churcher 1978; Geraads 1985). Kingston and Harrison (2007) did not obtain carbon isotopic results to allow differentiation of feeding niches in the Laetoli giraffids. They infer that this may reflect the complexity of the browsing strategies of these taxa or the limited resolution in differentiating aspects of C_3 vegetation and microhabitat by isotopic analyses.

The mesowear data add complementary information here. *Giraffa stillei* shows mesowear scores ranging from 0.67 in ULB-1–0.45 in ULB-2. In ULB-2 *G. stillei* has extant *G. camelopardalis* as its closest dietary reference species (Fig. 16.3c), while in the lowermost part of the ULB succession (ULB-1), the signal is more abrasion dominated, closest to the extant impala (*Aepyceros melampus*; Fig. 16.3b). In the ULB-all assemblage, *G. stillei* shows an intermediate feeding signal close to the extant eland (*Taurotragus oryx*; Fig. 16.3a). It is also close to *Sivatherium* cf. *maurusium*, which has the same extant analog. *Giraffa jumae* differs from all other giraffids in showing 17% low occlusal reliefs (Table 16.1). This indicates a typical intermediate-feeding signature, close to that of Grant's gazelle (*Gazella granti*), with the most abrasion dominated signal of the three giraffids from Laetoli. *Giraffa jumae* displays a more abrasion dominated dietary regime than the extant *G. camelopardalis*, which never has blunt cusps in the wild (Clauss et al. 2007a; Kaiser et al. in press). The mesowear signal supports the hypothesis of niche segregation put forward by Harris (1991). The larger *G. jumae* was adapted to more open woodland with comparatively more grit laden browse than the closed riverine woodlands inferred to be the preferred habitat of *G. stillei*. The observation that the mesowear signal of *G. stillei* is closest to the extant *G. camelopardalis* in the upper section of the ULB (Fig. 16.3c) indicates that the Laetoli habitat may have been slightly more arid and grit laden compared to modern *Acacia-Commiphora* woodlands.

Features of sivatheres, including a relatively short neck, have suggested to some a grazing or mixed low-level leaf browsing/grazing strategy (Meladze 1964; Hamilton 1973; Churcher 1978; Geraads 1985). Based on premaxillary shape and tooth microwear, Solounias et al. (1988) suggested that the Miocene *Samotherium boissieri* (a potential ancestor of *Sivatherium*) could have occupied a grazing niche. Application of three dietary reconstruction tools - hypsodonty, microwear and mesowear - indicate a mixed feeding strategy for *Sivatherium hendeyi* from the early Pliocene of South Africa (Franz-Odendaal and Solounias 2004). Kingston and Harrison (2007) could not differentiate *Sivatherium* cf. *maurusium* isotopically from *Giraffa* spp. in the Laetoli sequence. The same applies to the mesowear signatures, which are essentially the same as in *G. stillei*. Harris and Cerling (1998) report that late Miocene and early Pliocene sivatheres from East Africa have exclusive browsing isotopic signatures, but they became grazers during the late Pliocene. Analyses of *S. maurusium* enamel from the Upper Laetolil and Upper Ndolanya Beds by Kingston and Harrison (2007), however, reveal no evidence of a grazing component in the diet. In fact, in their analysis the δ^{13}C signature of one specimen yielded the most negative carbon isotopic value of all herbivores sampled, an observation that leads Kingston and Harrison (2007) to suggest browsing in more wooded or forested portions of the habitat for *S. maurusium* where water stress and/or irradiance were reduced. The mixed feeding dietary signature found for *Sivatherium* cf. *maurusium* in the

ULB is in accordance with these previous interpretations. It confirms that *Sivatherium* cf. *maurusium* was not a grazer, but that it foraged a significant proportion of browse with some abrasive components. Considering the carbon isotope evidence (Kingston and Harrison 2007), this component would most likely be attributed to grit rather than a substantial proportion of grass in the diet. A similar grit component is also considered likely to be responsible for the wear signature of the extant giraffe, which, if compared to other pure browsers like the moose (*Alces alces*) is comparatively abrasion-dominated (Fortelius and Solounias 2000: 26).

Equidae

Two hipparionine equids were recognized at Laetoli by Hooijer (1987a, b). *Hipparion* sp. from the ULB and *Hipparion* cf. *ethiopicum* from the UNB (Leakey 1987). Following Bernor and Armour-Chelu (1999) and Bernor and Harris (2003), these species are provisionally included in *Eurygnathohippus*. Mesowear and stable isotope investigation of *Cormohipparion* sp. from Chorora (Ethiopia; 10.7–10.1 Ma) independently indicate an intermediate feeding regime (Bernor et al. 2004) for this oldest known hipparion (s.l.) from Africa. The only member of the *Eurygnathohippus* clade analyzed to date in terms of paleodiet is *E. hooijeri* (Bernor and Kaiser 2006). This was shown to be a grazing species at Langebaanweg (South Africa) by Franz-Odendaal et al. (2003), where C_4 grasses were not established by ~5 Ma (Franz-Odendaal et al. 2002). Equid enamel younger than about 7 Ma in tropical and sub-tropical Africa, including all modern samples, have previously yielded $\delta^{13}C$ enamel values in the range of −2‰–3‰, indicating C_4 dominated to exclusive C_4 grass diets (Morgan et al. 1994; Bocherens et al. 1996; Cerling et al. 1997; Kingston 1999; Zazzo et al. 2000; Cerling et al. 2003a, b; Kingston and Harrison 2007). Kingston and Harrison (2007) find the overall isotopic range for *Eurygnathohippus* sp. from the ULB to be more positive than other fossil assemblages and more similar to the adjusted extant equid values. $\delta^{13}C$ values of two specimens of *Eurygnathohippus* from the ULB indicate a mixed C_3/C_4 diet (Kingston and Harrison 2007). Kingston and Harrison infer that Laetolil equids were either C_4 grazers supplementing their diet with C_3 browse or C_3 grasses or they were more eclectic feeders.

Mesowear signals of *Eurygnathohippus* sp. show the most abrasion dominated feeding regimes throughout the Laetolil succession. All extant reference species are dedicated grazers. Scores decrease slightly during the ULB section (3.00 in ULB-1 and 2.39 in ULB-2), indicating a slightly less abrasive feeding regime in the upper portion of the section (ULB-2) (Fig. 16.4). *Eurygnathohippus* sp. from the UNB has the most abrasive feeding regime of all taxa from Laetoli (score = 3.14), and corresponds to the extant white rhinoceros (*Ceratotherium simum*). This score indicates a more abrasive feeding regime compared to extant Plains zebra (*Equus quagga*; score = 2.1). The ULB equids are thus closest to the extant *Equus zebra* (score = 2.6) in their mesowear signatures, and they do not show evidence of a mixed feeding behavior in the sense that typical browse was consumed. Supplementing the stable isotope data (Kingston and Harrison 2007), their diets most likely consisted of C_4 and C_3 grasses only, rather than C_3 browse as suggested by Kingston and Harrison (2007). In the ULB assemblage, *Eurygnathohippus* sp. is the only definitive grazing species in the ungulate community.

The Laetoli Ungulate Community Structure

In quantitative terms, the ULB community comprises 41% browsing, 53% intermediate-feeding, and 6% grazing species. In the youngest stratigraphic unit, the UNB, grazers are the most frequent feeding type (57%), while browsers are rare (14%) (Fig. 16.7a). The two sub-units of the ULB are only distinguished by a shift in the dietary trait of *Giraffa stillei* (Fig. 16.7b). This shift results in an increase of the browser ratio at the expense of the intermediate-feeders in the upper unit (ULB-2). Comparison of average scores per feeding trait in the two ULB-units (Table 16.2; Fig. 16.5) indicates that browsers and intermediate-feeders have a more abrasion dominated average score, while the only grazer in the ULB ungulate community (i.e., *Eurygnathohippus* sp.) shifted towards less abrasion in ULB-2.

Ecological and Environmental Trends in the Laetoli Succession

The Upper Laetolil Beds Environment and Its Transitions

The Laetoli mesowear data indicate a range of foraging strategies, characterized in general by intermediate feeding on variable food sources and by browsing (Figs. 16.8 and 16.9). Similar inferences are indicated by the isotope data presented by Kingston and Harrison (2007). Grasses certainly represented a component of herbivore forage at Laetoli during ULB times, but only equids occupied a typical grazing niche. Kingston and Harrison (2007) assume the availability of C_3 grass in the ULB habitat to have been extremely limited, and interpret relatively negative isotope signals for *Eurygnathohippus* sp. to reflect a significant amount of browse in the diet of some individuals. Mesowear data of the species, however, are close to the extant African Equidae,

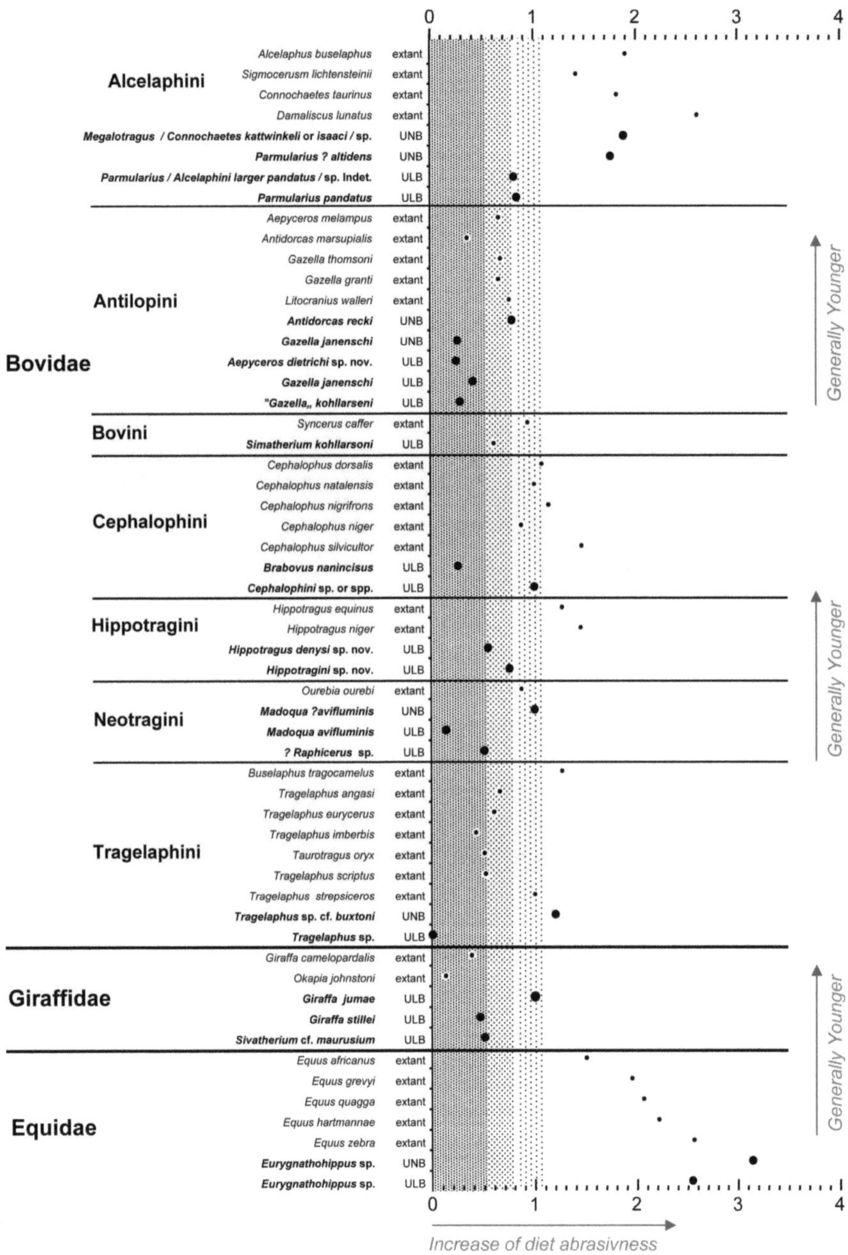

Fig. 16.8 Mesowear scores of extant African ungulates (*small symbols*) and fossil ungulates from Laetoli (*bold symbols*). Scores of extant Bovidae and Giraffidae were calculated based on published data from Solounias and Fortelius (2000). Samples are sorted in stratigraphic order with extant samples above. Scores ranging from 0–0.5 are generally associated with dedicated leaf browsing behavior in extant species (*hatched dark grey*). Scores ranging from 0.5–0.8 represent the attrition dominated intermediate-feeders and some of the abrasion dominated browsers (*hatched medium dark grey*). Scores ranging from 0.8–1.1 (*hatched light grey*) represent the abrasion dominated intermediate-feeders, most of the fruit browsers "minute abraded brachydonts" after Fortelius and Solounias (2000) as well as some attrition dominated grazers. All species eating more abrasive diets (predominantly grass or grit laden vegetation) have scores ranging from 1.1–3.2 (no hatching). The latter group includes all grazers and some of the fruit browsers. Delimitations were calculated based on species classified as "typical" by Fortelius and Solounias (2000), because their mesowear signatures represent conservative classifications of feeding traits. Abbreviations: *ULB* Upper Laetolil Beds, *UNB* Upper Ndolanya Beds

which all have less than 5% typical browse in their diet, but instead graze on C_4 and C_3 grasses. Mesowear data indicate that the ULB habitat included a significant amount of C_3 grasses that would be responsible for a more negative $\delta^{13}C$ signature in those taxa engaged in some degree of grazing.

Representatives of ULB Hippotragini and Alcelaphini have mesowear signatures that indicate intermediate feeding strategies, different from their modern counterparts that are mostly dedicated grazers. This implies a dietary shift within these lineages, which is also reflected in the isotope data

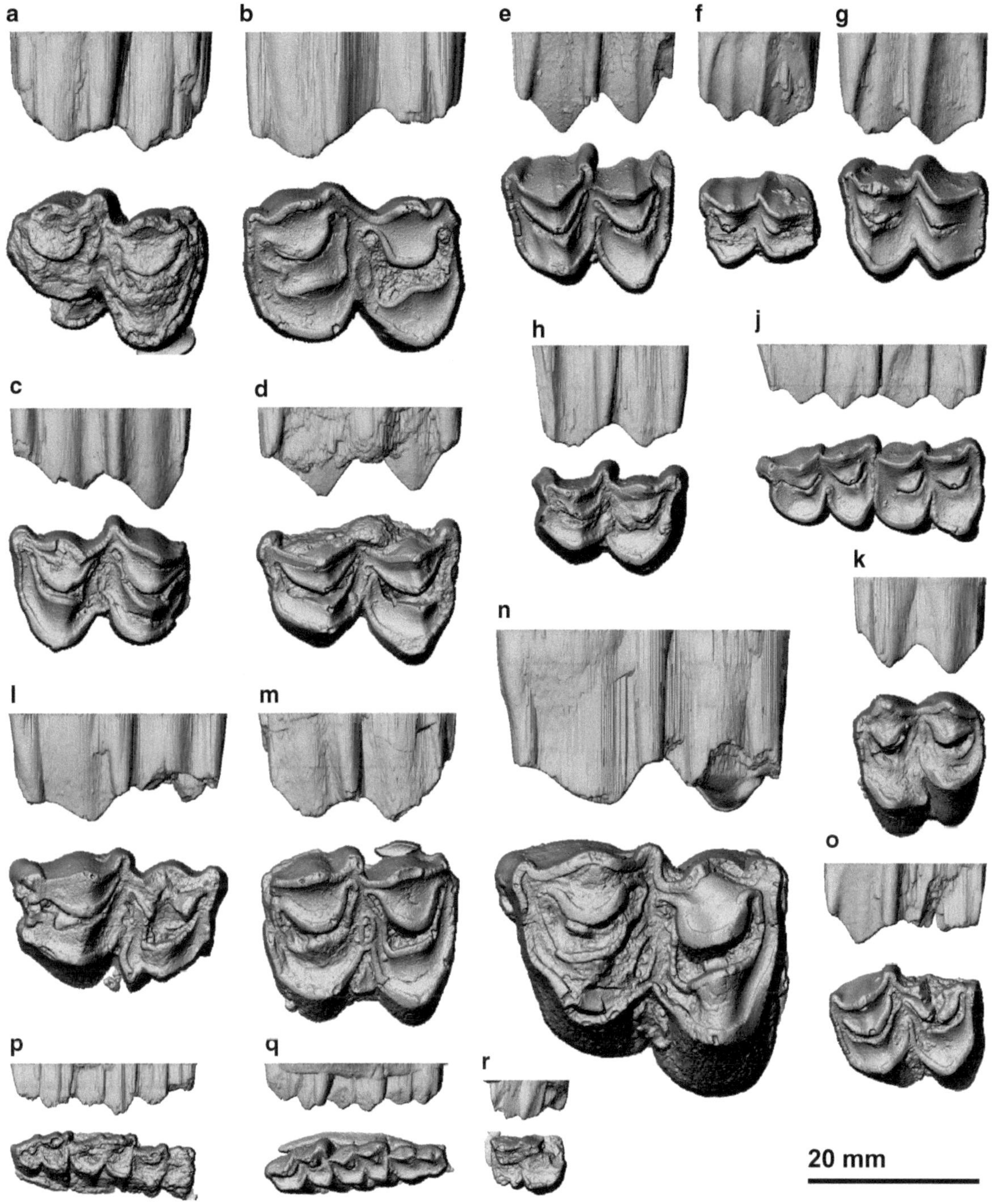

Fig. 16.9 A selection of upper first and second molars of Laetoli Bovidae. Upper view=buccal aspect (ectoloph as evaluated for mesowear), lower view=occlusal aspect. All figures are virtual 3-D models made of plastic casts with a ROLAND Modela MDX 20 scanner at a grit resolution of 0.1 mm. (**a**) LAET 76-7E-23 *Parmularius* ? *altidens* (Alcelaphini), UNB; (**b**) LAET 75-739 *Megalotragus/ Connochaetes kattwinkeli or isaaci/sp*. (Alcelaphini), UNB; (**c**) EP 1286/00 *Parmularius pandatus* (Alcelaphini) ULB; (**d**) EP 1139/03 *Tragelaphus sp*. cf. *buxtoni* (Tragelaphini) UNB; (**e**) EP 1143/03 *Antidorcas recki* (Antilopini) UNB; (**f**) LAET 78-5198 *Gazella janenschi* (Antilopini) UNB; (**g**) LAET 74–117 *Aepyceros dietrichi* sp. nov. (Antilopini) ULB; (**h**) EP 230/03 "*Gazella*" *kohllarseni* (Antilopini) UNB; (**j**) EP 1290/00 *Gazella janenschi* (Antilopini) ULB; (**k**) LAET 75-3052 Cephalophini sp. or spp. ULB; (**l**) EP 088/98 *Oryx deturi* (Hippotragini) ULB; (**m**) EP 1156/98 *Hippotragus* sp. (Hippotragini) ULB; (**n**) EP-393-04 *Simatherium kohllarseni* (Bovini) ULB; (**o**) EP 1894/00 *Brabovus nanincisus* (aff. Cephalophini or Bovini) ULB; (**p**) EP 113/03 *Madoqua ? avifluminis* (Neotragini) UNB; (**q**) EP 927/98 *Madoqua avifluminis* (Neotragini) ULB; (**r**) EP 1542/98 ? *Raphicerus* sp. (Neotragini) ULB

(Kingston and Harrison 2007). The extant Tragelaphini, which currently inhabit a diverse set of ecosystems ranging from forests to arid, seasonal bush, are mostly browsers or browse dominated intermediate-feeders. None of the extant taxa is found at the very attrition dominated end of the mesowear spectrum (Fortelius and Solounias 2000), unlike the ULB tragelaphine investigated (*Tragelaphus* sp.; Fig. 16.4). Kingdon (1982) suggests that ancestral members of the lineage may have preferred more humid woodlands and forests, and the postcranial morphology of fossil tragelaphines implies environments with light to heavy vegetation cover (Kappelman et al. 1997). The attrition dominated feeding trait of *Tragelaphus* sp. from the ULB may reflect an ancestral condition of the tribe not represented in the modern Tragelaphini, or alternatively may indicate that ULB environments have no immediate modern analogs. The ULB Antilopini display a similar dietary pattern. All three ULB-species have lower mesowear scores than any of the extant representatives investigated.

Compared to modern African herbivore mesowear, the ULB dietary spectrum is thus heavily weighted towards the attrition dominated (browsing) end of the mesowear continuum. Most of the ULB ungulates appear to be more generalist compared to related extant taxa, and they must have occupied fundamentally different feeding niches. Alternatively, they may have inhabited environments in ULB times for which there are no modern analogs in the range of their extant conspecifics. Kingston and Harrison (2007) find concordant isotopic signatures indicating more feeding on C_3 biomass compared to extant African herbivore. It is difficult to reconcile these data, independently acquired using complementary methodology, with earlier interpretations that Laetoli represents a grassland-savanna habitat (Leakey and Harris 1987).

The feeding traits identified for Pliocene Hippotragini, Alcelaphini, Tragelaphini and Antilopini have implications for our understanding of actualistic approaches in reconstructing past paleoenvironments. Taxonomic uniformitarianism is a widely used approach to Neogene mammal ecology (Vrba 1975; Shipman and Harris 1988) and faunal analysis of herbivores (including the Laetoli fauna) have often been employed as immediate environmental proxies, whereby the interpretation of adaptations of extinct organisms hinge on analogy with related modern forms. In particular, evolutionary patterns of the Bovidae have been specifically associated with environmental change and faunal turnover during the Neogene of Africa (Vrba 1985, 1995; Bobe and Eck 2001; Alemseged 2003). High percentages of alcelaphines and antilopines have been interpreted to indicate comparatively arid grassland habitats (Vrba 1980, 1985; Greenacre and Vrba 1984). Antilopines play a role in this notion, because many modern taxa have cursorial adaptations, and are particularly drought resistant (Plummer and Bishop 1994). However, the tribe also contains water-bound species as well as a wide spectrum of dietary traits.

Faunal approaches widely applied in the past, involved taxonomic uniformitarianism to a great degree. Habitat preferences of Neogene bovids in Africa were inferred from relative representations of bovid tribes (Vrba 1980, 1985; Shipman and Harris 1988). Faunal analyses of bovids have in fact been an integral approach to reconstructing hominin paleoenvironments, and these have dramatically influenced our picture of hominin evolution in East Africa (WoldeGabriel et al. 1994; Sponheimer et al. 1999; Pickford and Senut 2001; WoldeGabriel et al. 2001; Vignaud et al. 2002; Sponheimer and Lee-Thorp 2003). Studies using this approach, supplemented with data on the flora and geology, have led to a reconstruction of the ecology of the hominin-bearing ULB that was inferred to be an arid to semi-arid seasonal grasslands with scattered bush and tree cover, but generally a non-wooded environments, much like some areas of the modern Serengeti ecosystem (Hay 1981; Harris 1987a; Leakey and Harris 1987). The significance of Laetoli was therefore partly seen in that it provides some of the earliest evidence for the establishment of modern savanna ecosystems in East Africa (Kingston and Harrison 2007). This reconstruction has contributed essentially to the establishment of a framework for understanding and interpreting the evolutionary history and paleobiology of East African faunas, including hominins, during the Plio-Pleistocene (Leakey and Harris 1987).

Andrews (1989) was the first to reinterpret aspects of the structure of the fossil mammalian community to indicate more wooded ecosystems than previously considered. Utilizing a faunal approach and minimizing time-averaging, community reconstructions by Reed (1997) also indicate a closed to medium density woodland. Subsequently, Andrews and Humphrey (1999) supported the idea that the ULB paleoenvironment tends more toward the woodland end of the spectrum than toward grassland. Also, stable isotope data of paleosol carbonates indicate that woodland habitats predominated (Cerling 1992). Recent studies provide further support for this interpretation (Musiba 1999; Hicks 1999; Kingston and Harrison 2001, 2002, 2005, 2007; Musiba et al. 2002; Su and Harrison 2003, 2005; Bamford and Harrison 2004; Harrison and Su 2004; Kovarovic et al. 2005; Harrison 2005; Su 2005). Finally, the mesowear results clearly support an environmental reconstruction of the ULB favoring a more complex mosaic of habitats with greater representation of bush cover and woodland.

Differences in mesowear data between lower ULB-1 and upper ULB-2 did not reveal significant changes or trends in foraging strategies of ungulates in any of the lineages and general feeding traits. This observation is shared by Kingston and Harrison (2007), who also do not find evidence for changes within the stratigraphic subdivisions they investigated.

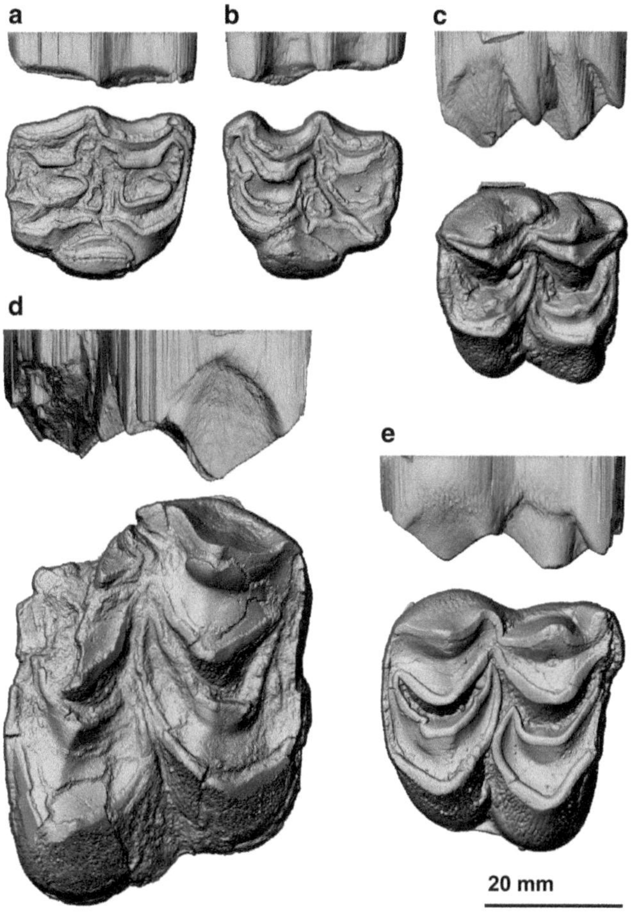

Fig. 16.10 A selection of upper first and second molars of Laetoli Equidae and Giraffidae. Upper view=buccal aspect (ectoloph as evaluated for mesowear), lower view=occlusal aspect. All figures are virtual 3-D models made of plastic casts with a ROLAND Modela MDX 20 scanner at a grit resolution of 0.1 mm. (**a**) EP 765/03 *Eurygnathohippus* sp. (Equidae) *ULB*; (**b**) LAET 75–1685 *Eurygnathohippus* sp. (Equidae) *UNB*; (**c**) EP 4831/01 *Giraffa stillei* (Giraffidae) *ULB*; (**d**) EP 1112/98 *Sivatherium* cf. *maurusium* (Giraffidae) *ULB*; (**e**) EP 1753/03 *Giraffa jumae* (Giraffidae) ULB

However, they do not rule out the possibility of subtle shifts in ecology or community composition through the ULB interval. Such subtle shifts are indicated by: (1) an increase in browsers at the expense of intermediate-feeders between ULB1 and ULB2 (Fig. 16.7b); (2) a shift of average scores of feeding traits (Fig. 16.10); and (3) ULB2 grazers are shifted towards less abrasive diet and browsers and intermediate-feeders display the reverse trend (Fig. 16.5).

The ULB-UNB Transition and the Upper Ndolanya Beds Paleoenvironment

Differential mesowear assessment across the ULB-UNB transition revealed significant changes in the foraging strategies of several ungulate groups (Fig. 16.4), as well as in the proportion of feeding traits (Fig. 16.7a) and the position of the abrasion/attrition equilibrium as represented by feeding trait main scores (Fig. 16.10). These results are consistent, although limited sample sizes may compromise the result in some taxa.

Mesowear average scores per feeding type (Fig. 16.10), indicate that browsers show only little shift in their signature across the ULB-UNB transition. Intermediate-feeders, however, show a consistent shift towards more abrasion in the mesowear equilibrium (0.58 in ULB-1–0.99 in the UNB). The reverse trend is seen in the grazers, which consistently have lower score averages in the younger units (3.0 in ULB-1 to 1.94 in the UNB). The gap between intermediate-feeders and grazers is successively reduced towards the top of the youngest members of the Laetoli section (Fig. 16.10). Within the succession, it appears that fundamental feeding niches continuously converged, with grazers getting more engaged in feeding on less abrasive components and intermediate-feeders exploiting more abrasive feed (Fig. 16.10). Within the succession it appears that the spectrum of feeding niches generally decreased, because the most attrition dominated section of the browsing niche, the pure leaf browsing niche, was almost abandoned and seems to have been taken over by more opportunistic intermediate feeding species, which did not rely on leaf browse only. The obvious decrease of the diversity of browsing species further suggests that the formerly diverse and heterogeneous habitat structure, which for the ULB environment seems to be supported, did no longer exist in UNB times. This in turn would make it likely that the UNB environment more or less lacked substantial forest patches that would have supported a more diverse browser community, including both leaf browsers and fruit browsers. It thus should to be characterized as more or less open grassland.

The faunal transition between the ULB and UNB is interpreted as reflecting a shift towards drier and more open, semi-arid scrub or bushland habitats (Leakey 1987; Harris 1987a; Kovarovic et al. 2002, Kingston and Harrison 2007). Also, stable isotope data of paleosol carbonates (Cerling 1992) seem to indicate that woodland habitats predominated during ULB times, and then gave way to more open grassland-savanna habitats in the UNB.

This resulted in a UNB fauna that was dominated by large bovids (mainly alcelaphines), which comprise more than 80% of the mammalian taxa (Kingston and Harrison 2007). The two alcelaphines analyzed here (*mecon* and *paalt*) show a grazing mesowear signal, which is close to the signal of the extant counterparts of the lineage (Fig. 16.4) and demonstrate modern characteristics in terms of their dietary niches. The grazing traits demonstrated by independent paleodietary reconstruction have also resulted in morphological adaptations of the chewing system in alcelaphines. In *mecon* folding occurs along the joins of R2 and R3, creating additional ridges, further increasing the amount of cutting edges perpendicular to the movement of the jaw. Such foldings are distinctive of most extant alcelaphines and of most extant grazing bovids in general (Archer and Sanson 2002; Clauss et al. 2007b), (Fig. 16.8b). Archer and Sanson (2002) also find evidence for shearing optimisation in the transverse (antero-posterior) orientation of the teeth in grazers that are clearly developed in the UNB alcelaphines, but also in the ULB hippotragines (*Hippotragus;* Fig. 16.8m) and bovines (*Simatherium;* Fig. 16.8n), and this may be of equal importance to the function of a grazer's molars.

The evidence shows that most taxa analyzed across the ULB/UNB transition, at the tribe (Alcelaphini, Tragelaphini, Antilopini) and genus levels (*Eurygnathohippus*), reflect more abrasive feeding regimes and more distinct functional adaptations to cope with abrasive forage as indicated by higher mesowear scores in the younger UNB. This may simply indicate a greater grazing component, but could also be the result of increased aridity leading to either increased graze availability or increased grit load of foliage or both, thereby resulting in generally increased food abrasiveness. The validity of this general interrelationship, between the aridity of the habitat and dental wear signatures, has been documented by Kaiser and Rössner (2007) and Kaiser and Schulz (2006). The general shift observed would also be supported by the slightly more ^{13}C-enriched median carbon isotopic values in the UNB presented by Kingston and Harrison (2007). The coincidence of mesowear and stable isotope data reconciles both methods and demonstrate their analytical and complementary potential for assessments of environmental change.

The ecological shift towards more arid and open conditions after the transition obviously caused intermediate-feeders to shifted more towards the grazing end of the continuum, and display a continuous extension of their feeding range in the younger members. This does not mean that this feeding niche was more diverse after the transition, but coincident with the reverse trend in grazers, it reflects differences in food availability and an incremental environmental shift that allowed increasingly more exploitation of an increasingly broader abrasive intermediate feeding niche. This would mean that the formerly diverse browsing niche would have narrowed and *G. janenschi*, the only browser identified in the of the UNB, could then again be understood as a highly flexible form that this time extended its niche even more towards the attrition dominated end of the mesowear continuum, while the remainder of species abandoned the niche. The shift within grazing niche can then be understood as reflecting a broad set of resources in a general intermediate feeding regime that was exploited by the more browse dominated grazers in the succession.

Conclusions

The almost complete absence of a dedicated grazing guilds and the heavy reliance on browse by most fossil herbivores do not support the inference that Laetoli was dominated by extensive grassland habitats. This makes it difficult to reconcile dietary traits of the community with earlier notions that Laetoli represents a grassland-savanna habitat. Mesowear data of ULB ungulates suggest that extant ungulates representing the same lineages may not serve as actualistic model taxa in faunal reconstructions using taxonomic uniformitarianism on a tribe or even genus level. Kingston and Harrison (2007) emphasize, that "faced with incremental environmental shifts, many taxa are able to maintain foraging niches, and ecological change may only be revealed by combining isotopic analyses of many herbivore lineages with other dietary or ecological proxies such as ecomorphology and micro-/mesowear analysis. Alternatively, dietary shifts may reflect niche jostling within the community independent of any environmental or large-scale ecological changes". The presented mesowear data add important complementary environmental information. δ^{13}C enamel signatures of ULB led Kingston and Harrison (2007) to suggest that *M. avifluminis* and *Tragelaphus* sp. incorporated some C_4 grass into their diet. Since there is no abrasive signal in the mesowear data of either of these extinct species, it now must be considered more likely that the C_4 component was derived from C_4 dicots, which would have been a component of the Laetoli habitat. On the other hand, the ULB equid, *Eurygnathohippus* sp., is the only dedicated grazer at Laetoli, even exceeding extant African equids in the abrasiveness of its diet. Supplementing the isotope data (Kingston and Harrison 2007), their diets most likely consisted of C_4 and C_3 grasses only, but no C_3 browse, as alternatively suggested by Kingston and Harrison (2007). The apparent presence of C_4 dicots on the one hand and C_3 grasses on the other serve to emphasize the great ecological diversity of the Laetoli habitat.

Mesowear and isotope proxies concordantly indicate that the Laetoli ungulates were formerly more ecologically diverse compared to their extant representatives and that their dietary diversity declined during the Plio-Pleistocene. In particular, in the ULB hippotragines, alcelaphines and antilopines, compared to modern African conspecifics, the dietary spectrum is heavily weighted towards attrition dominated (browsing and intermediate-feeding) traits. Across the transition with the Upper Ndolanya Beds modern dietary traits appear as distinct functional dental adaptations, strictly related to a specialized and less flexible feeding behavior.

Acknowledgements I thank the Tanzania Commission for Science and Technology (COSTECH) and the Department of Antiquities in Dar es Salaam for permission to conduct research in Tanzania. Thanks also to the curators and staff at the National Museum of Tanzania (Dar es Salaam) and for their hospitality during the related museum work. Alan Gentry is most sincerely acknowledged for sharing his new Laetoli bovid taxonomy and database. Without Alan's tremendous contribution concerning the clarification of taxonomic assignments of bovid specimens, this manuscript would probably not have been possible. Thanks to Christoph Zahn for laboratory assistance in sampling, labeling, databasing and molding specimens. Regina Mertins did the 3-D scan of dental specimens at the 3-D morphometry lab of the Mammalogy Section of the Zoological Museum Hamburg. I thank two anonymous reviewers for their very constructive comments and suggestions, that improved the manuscript and I sincerely thank Terry Harrison for his invitation to conduct this study and contribute to this volume. Fieldwork at Laetoli was supported by grants to Terry Harrison from the National Geographic Society, the Leakey Foundation, and the National Science Foundation (Grants BCS-0216683 and BCS-0309513). This contribution further benefited from grants to TMK from the German Research Foundation KA 1525/4-1/4-2, KA 1525/6-1.

References

Alemseged, Z. (2003). An integrated approach to taphonomy and faunal change in the Shungura Formation (Ethiopia) and its implication for hominid evolution. *Journal of Human Evolution, 44*, 451–478.

Andrews, P. J. (1989). Palaeoecology of Laetoli. *Journal of Human Evolution, 18*, 173–181.

Andrews, P. J., & Humphrey, L. (1999). African Miocene environments and the transition to early hominines. In T. Bromage & F. Schrenk (Eds.), *African biogeography, climate change, and human evolution* (pp. 282–300). Oxford: Oxford University Press.

Archer, D., & Sanson, G. (2002). Form and function of the selenodont molar in southern African ruminants in relation to their feeding habits. *Journal of Zoology, 257*, 13–26.

Armour-Chelu, M., & Bernor, R. L. (2011). Equidae. In T. Harrison (Ed.), *Paleontology and geology of Laetoli: Human evolution in context* (Fossil hominins and the associated fauna, vol. 2, pp. 295–326). Dordrecht: Springer.

Baker, G., Jones, L. H. P., & Wardrop, I. D. (1959). Cause of wear in sheep's teeth. *Nature, 184*, 1583–1584.

Bamford, M. K., & Harrison, T. (2004). Pliocene vegetation and palaeoenvironment of Laetoli, Tanzania evidence from fossil woods. *PaleoAnthropology Selected Issue*: PAS 2004 Abstracts: A02.

Beden, M. (1987). Fossil Elephantidae from Laetoli. In M. D. Leakey & J. M. Harris (Eds.), *Laetoli: A Pliocene site in northern Tanzania* (pp. 259–294). Oxford: Clarendon.

Bernor, R. L., & Armour-Chelu, M. (1999). Toward an evolutionary history of African hipparionine horses. In T. Bromage & F. Schrenk (Eds.), *African biogeography and human evolution* (pp. 189–215). Oxford: Oxford University Press.

Bernor, R. L., & Harris, J. M. (2003). Systematics and evolutionary biology of the late Miocene and early Pliocene hipparionine equids from Lothagam, Kenya. In M. G. Leakey & J. M. Harris (Eds.), *Lothagam: The dawn of humanity in eastern Africa* (pp. 387–438). New York: Columbia University Press.

Bernor, R. L., & Kaiser, T. M. (2006). Systematics and paleoecology of the earliest Pliocene equid, *Eurygnathohippus hooijeri* n. sp. from Langebaanweg, South Africa. *Mitteilungen aus dem Hamburgischen Zoologischen Museum und Institut, 103*, 149–186.

Bernor, R. L., Kaiser, T. M., Kordos, L., & Scott, R. S. (1999). Stratigraphic context, systematic position and paleoecology of *Hippotherium sumegense* Kretzoi, 1984 from MN 10 (Late Vallesian of the Pannonian basin). *Mitteilung der Bayerischen Staatssammlung für Paläontologie und historische Geologie, 39*, 1–35.

Bernor, R. L., Kaiser, T. M., & Nelson, S.V. (2004). The oldest Ethiopian Hipparion (Equinae, Perissodactyla) from Chorora: Systematics, paleodiet and paleoclimate. *Courier Forschungsinstitut Senckenberg, 246*, 213–226.

Bobe, R., & Eck, G. G. (2001). Responses of African bovids to Pliocene climatic change. *Paleobiology, 27*, 1–47.

Bocherens, H., Koch, P. L., Mariotti, A., Geraads, D., & Jaeger, J. (1996). Isotopic biogeochemistry (^{13}C, ^{18}O) of mammalian enamel from African Pleistocene hominid sites. *Palaios, 11*, 306–318.

Cerling, T. E. (1992). Development of grasslands and savannas in east Africa during the Neogene. *Palaeogeography, Palaeoclimatology, Palaeoecology, 97*, 241–247.

Cerling, T. E., Harris, J. M., MacFadden, B. J., Leakey, M. G., Quade, J., Eisenmann, V., & Ehleringer, J. R. (1997). Global vegetation through the Miocene/Pliocene boundary. *Nature, 389*, 153–158.

Cerling, T. E., Ehleringer, J. R., & Harris, J. M. (1998). Carbon dioxide starvation, the development of C_4 ecosystems, and mammalian evolution. *Philosophical Transactions of the Royal Society of London, 353*, 159–171.

Cerling, T. E., Harris, J. M., Leakey, M. G., & Mudida, N. (2003a). Stable isotope ecology of northern Kenya, with emphasis on the Turkana basin. In M. G. Leakey & J. M. Harris (Eds.), *Lothagam: The dawn of humanity in eastern Africa* (pp. 583–603). New York: Columbia University Press.

Cerling, T. E., Harris, J. M., & Passey, B. H. (2003b). Diets of east African Bovidae based on stable isotope analysis. *Journal of Mammalogy, 84*, 456–470.

Churcher, C. S. (1978). Giraffidae. In V. J. Maglio & H. B. S. Cooke (Eds.), *Evolution of African mammals* (pp. 509–535). Cambridge: Harvard University Press.

Clauss, M., Franz-Odendaal, T., Brasch, J., Castell, J., & Kaiser, T. M. (2007a). Tooth wear in captive giraffes (*Giraffa camelopardalis*): Mesowear analysis classifies free-ranging specimens as browsers but captive ones as grazers. *Journal of Zoo and Wildlife Medicine, 38*, 433–435.

Clauss, M., Kaiser, T. M., & Hummel, J. (2007b). The morphophysiological adaptations of browsing and grazing mammals. In I. J. Gordon & H. H. T. Prins (Eds.), *The ecology of browsing and grazing* (pp. 47–88). Heidelberg: Springer.

Deino, A. (2011). $^{40}Ar/^{39}Ar$ dating of Laetoli, Tanzania. In T. Harrison (Ed.), *Paleontology and geology of Laetoli: Human evolution in context* (Geology, geochronology, paleoecology, and paleoenvironment, vol. 1, pp. 77–97). Dordrecht: Springer.

Dietrich, W. O. (1950). Fossile antilopen und rinder aquatorialafrikas. *Palaeontographica, 99*, 1–62.

Drake, R., & Curtis, G. H. (1987). K–Ar geochronology of the Laetoli fossil localities. In M. D. Leakey & J. M. Harris (Eds.), *Laetoli: A Pliocene site in northern Tanzania* (pp. 48–52). Oxford: Clarendon.

Estes, R. D. (1991). *The behavior guide to African mammals, including hoofed mammals, carnivores, primates*. Berkeley: University of California Press.

Feer, F. (1989). Comparison des régimes alimentaires de *Cephalophus callipygus* et *Cephalophus dorsalis*, Bovidés sympatriques de la forêt sempervirente africaine. *Mammalia, 53*, 563–604.

Fortelius, M., & Solounias, N. (2000). Functional characterization of ungulate molars using the Abrasion-Attrition wear gradient: A new method for reconstructing paleodiets. *American Museum of Natural History Novitates, 3302*, 1–36.

Fortelius, M., Eronen, J., Jernvall, J., Liu, L., Pushkina, D., Rinne, J., Tesakov, A., Vislobokowa, I., Zhang, Z., & Zhou, L. (2002). Fossil mammals resolve regional patterns of Eurasian climate change over 20 million years. *Evolutionary Ecology Research, 4*, 1005–1016.

Franz-Odendaal, T. A., & Kaiser, T. M. (2003). Differential mesowear in the maxillary and mandibular cheek dentition of some ruminants (Artiodactyla). *Annales Zoologici Fennici, 40*, 395–410.

Franz-Odendaal, T. A., & Solounias, N. (2004). Comparative dietary evaluations of an extinct giraffid (*Sivatherium hendeyi*) (Mammalia,

Giraffidae, Sivatheriinae) from Langebaanweg, South Africa (early Pliocene). *Geodiversitas, 26*, 675–685.
Franz-Odendaal, T. A., Lee-Thorp, J. A., & Chinsamy, A. (2002). New evidence for the lack of C_4 grassland and expansions during the early Pliocene at Langebaanweg, South Africa. *Paleobiology, 28*, 378–388.
Franz-Odendaal, T. A., Kaiser, T. M., & Bernor, R. L. (2003). Systematics and dietary evaluation of a fossil equid from South Africa. *South African Journal of Science, 99*, 453–459.
Gagnon, M., & Chew, A. E. (2000). Dietary preferences in extant African Bovidae. *Journal of Mammalogy, 81*, 490–511.
Gautier-Hion, A., Emmons, L. H., & Dubost, G. (1980). A comparison of the diets of three major groups of primary consumers of Gabon (primates, squirrels, and ruminants). *Oecologia, 45*, 182–199.
Gentry, A. W. (1987). Pliocene Bovidae from Laetoli. In M. D. Leakey & J. M. Harris (Eds.), *Laetoli: A Pliocene site in northern Tanzania* (pp. 378–407). Oxford: Clarendon.
Gentry, A. W. (2011). Bovidae. In T. Harrison (Ed.), *Paleontology and geology of Laetoli: Human evolution in context* (Fossil hominins and the associated fauna, Vol. 2, pp. 363–465). Dordrecht: Springer.
Geraads, D. (1985). *Sivatherium maurusium* (Pomel) (Giraffidae, Mammalia) du Pléistocène de la République de Djibouti. *Paläontologische Zeitschrift, 59*, 311–321.
Greenacre, M. J., & Vrba, E. S. (1984). Graphical display and interpretation of antelope census data in African wildlife areas, using correspondence analysis. *Ecology, 65*, 984–997.
Gruvaeus, G., & Weiner, H. (1972). Two additions to hierarchical cluster analysis. *The British Journal of Mathematical and Statistical Psychology, 25*, 200–206.
Guerin, C. (1987). Fossil Rhinocerotidae (Mammalia, Perissodactyla) from Laetoli. In M. D. Leakey & J. M. Harris (Eds.), *Laetoli: A Pliocene site in northern Tanzania* (pp. 320–348). Oxford: Clarendon.
Hamilton, W. R. (1973). The lower Miocene ruminants of Gebel Zelten, Libya. *Bulletin of the British Museum (Natural History) Geology Series, 21*, 73–150.
Harris, J. M. (1987a). Summary. In M. D. Leakey & J. M. Harris (Eds.), *Laetoli: A Pliocene site in northern Tanzania* (pp. 524–531). Oxford: Clarendon.
Harris, J. M. (1987b). Fossil Giraffidae and Camelidae from Laetoli. In M. D. Leakey & J. M. Harris (Eds.), *Laetoli: A Pliocene site in northern Tanzania* (pp. 358–377). Oxford: Clarendon.
Harris, J. M. (1991). Family Giraffidae. In J. M. Harris (Ed.), *Koobi Fora research project* (Vol. 3, pp. 93–138). Oxford: Clarendon.
Harris, J. M., & Cerling, T. E. (1998). Isotopic changes in the diet of giraffids. *Journal of Vertebrate Paleontology, 18*, 49A.
Harrison, T. (2005). Fossil bird eggs from the Pliocene of Laetoli, Tanzania: Their taxonomic and paleoecological relationships. *Journal of African Earth Sciences, 41*, 289–302.
Harrison, T., & Su, D. F. (2004). Laetoli paleoecology reconsidered: Implications for early hominin habitats. *PaleoAnthropology Selected Issue*: PAS 2004 Abstracts: A33.
Hartigan, J. A. (1975). *Clustering algorithms*. New York: Wiley.
Hay, R. L. (1981). Paleoenvironment of the Laetolil beds, northern Tanzania. In G. Rapp Jr. & C. F. Vondra (Eds.), *Hominid sites: Their geological settings* (AAAS Selected Symposia Series, Vol. 63, pp. 7–24). Boulder: Westview.
Hay, R. L. (1987). Geology of the Laetoli area. In M. D. Leakey & J. M. Harris (Eds.), *Laetoli: A Pliocene site in northern Tanzania* (pp. 23–47). Oxford: Clarendon.
Hernesniemi, E., Giaourtsakis, I. X., Evans, A. R., & Fortelius, M. (2011). Rhinocerotidae In T. Harrison (Ed.), *Paleontology and geology of Laetoli: Human evolution in context* (Fossil hominins and the associated fauna, vol. 2, pp. 275–293). Dordrecht: Springer.
Hicks, M. M. (1999). Paleoenvironmental interpretation of the Upper Laetolil Beds, Laetoli archaeological site, Tanzania. Abstracts, Geological Society of America. Boulder: Geological Society of America (GSA), p. 401.
Hofmann, R. R. (1989). Evolutionary steps of ecophysiological adaptation and diversification of ruminants: A comparative view of their digestive system. *Oecologia, 78*, 443–457.
Hooijer, D. A. (1987a). Hipparions of the Laetolil beds, Tanzania. In M. D. Leakey & J. M. Harris (Eds.), *Laetoli: A Pliocene site in northern Tanzania* (pp. 301–312). Oxford: Clarendon.
Hooijer, D. A. (1987b). Hipparion teeth from the Ndolanya Beds. In M. D. Leakey & J. M. Harris (Eds.), *Laetoli: A Pliocene site in northern Tanzania* (pp. 312–315). Oxford: Clarendon.
Janis, C. M. (1988). An estimation of tooth volume and hypsodonty indices in ungulate mammals, and the correlation of these factors with dietary preference. In D. E. Russell, J. P. Santoro, & D. Sigogneau-Russell, (Eds.), *Teeth Revisited.* Proceedings of the 7th International Symposium Dental Morphology (pp. 367–387), Paris 1986. Mémoires Museum Nationale d'Histoire Naturelle, Série C, Paris.
Janis, C. M. (1990a). Correlation of cranial and dental variables with body size in ungulates and macropodoids. In J. Damuth & B. J. MacFadden (Eds.), *Body size in mammalian paleobiology* (pp. 255–300). Cambridge: Cambridge University Press.
Janis, C. M. (1990b). The correlation between diet and dental wear in herbivorous mammals, and its relationship to the determination of diets of extinct species. In J. Boucot (Ed.), *Evolutionary paleobiology of behaviour and coevolution* (pp. 241–260). Amsterdam: Elsevier.
Janis, C. M., & Ehrhardt, D. (1988). Correlation of relative muzzle width and relative incisor width with dietary preference in ungulates. *Zoological Journal of the Linnean Society, 92*, 267–284.
Kaiser, T. M. (2003). The dietary regimes of two contemporaneous populations of *Hippotherium primigenium* (Perissodactyla, Equidae) from the Vallesian (upper Miocene) of Southern Germany. *Palaeogeography, Palaeoclimatology, Palaeoecology, 198*, 381–402.
Kaiser, T. M., & Rössner, G. E. (2007). Dietary resource partitioning in ruminant communities of Miocene wetland and karst palaeoenvironments in southern Germany. *Palaeogeography, Palaeoclimatology, Palaeoecology, 252*, 424–439.
Kaiser, T. M., & Schulz, E. (2006). Tooth wear gradients in zebras as an environmental proxy – a pilot study. *Mitteilungen aus dem Hamburgischen Zoologischen Museum und Institut, 103*, 187–210.
Kaiser, T. M., & Solounias, N. (2003). Extending the tooth mesowear method to extinct and extant equids. *Geodiversitas, 25*, 321–345.
Kaiser, T., Bromage, T. G., & Schrenk, F. (1995). Hominid corridor research project update: New Pliocene fossil localities at Lake Manyara and putative oldest early stone age occurrences at Laetoli (Upper Ndolanya Beds), northern Tanzania. *Journal of Human Evolution, 28*, 117–120.
Kaiser, T. M., Solounias, N., Fortelius, M., Bernor, R. L., & Schrenk, F. (2000). Tooth mesowear analysis on *Hippotherium primigenium* from the Vallesian Dinotheriensande (Germany) – a blind test study. *Carolinea, 58*, 103–114.
Kaiser, T. M., Brasch, J., Castell, J., & Clauss, M. (2009). Tooth wear in captive wild ruminant species differs from that of free-ranging conspecifics. *Mammalian Biology, 74*, 425–437.
Kappelman, J., Plummer, T., Bishop, L., Duncan, A., & Appleton, S. (1997). Bovids as indicators of Plio-Pleistocene paleoenvironments in East Africa. *Journal of Human Evolution, 32*, 229–256.
Kay, R. F., & Covert, H. H. (1983). True grit: A microwear experiment. *American Journal of Physical Anthropology, 61*, 33–38.
Kingdon, J. (1979). *East African mammals: An atlas of evolution in Africa* (Vol. III (B)). London: Academic.
Kingdon, J. (1982). *East African mammals: An atlas of evolution in Africa (Vol. III. Parts C and D (Bovids))*. London: Academic.
Kingdon, J. (1997). *The Kingdon field guide to African mammals*. San Diego: Academic.
Kingston, J. D. (1999). Environmental determinants in early hominid evolution: Issues and evidence from the Tugen hills, Kenya.

In P. Andrews & P. Banham (Eds.), *Late Cenozoic environments and hominid evolution: A tribute to Bill Bishop* (pp. 69–84). London: Geological Society.

Kingston, J. D., & Harrison, T. (2001). High-resolution middle Pliocene landscape reconstructions at Laetoli, Tanzania. *Journal of Human Evolution, 38*, A11.

Kingston, J. D., & Harrison, T. (2002). Isotopically based reconstructions of early to middle Pliocene paleohabitats at Laetoli, Tanzania. *American Journal of Physical Anthropology, 119*(S35), 95–96.

Kingston, J. D., & Harrison, T. (2005). Ostrich eggshells as paleoenvironmental indicators in the Pliocene Laetoli succession, N. Tanzania. *PaleoAnthropology A37* (PAS 2005 Abstracts).

Kingston, J. D., & Harrison, T. (2007). Isotopic dietary reconstructions of Pliocene herbivores at Laetoli: Implications for hominin paleoecology. *Palaeogeography, Palaeoclimatology, Palaeoecology, 243*, 272–306.

Köhler, M. (1993). Skeleton and habitat of recent and fossil ruminants. *Münchner Geowissenschaftliche Abhandlungen A, 25*, 1–88.

Kovarovic, K., Andrews, P., & Aiello, L. (2002). The palaeoecology of the Upper Ndolanya Beds at Laetoli, Tanzania. *Journal of Human Evolution, 43*, 395–418.

Kovarovic, K., Andrews, P., & Aiello, L. (2005). Palaeoenvironmental reconstruction of Laetoli, Tanzania using a bovid community ecomorphological survey. *PaleoAnthropology A50* (PAS 2005 Abstracts).

Leakey, M. D. (1987). Introduction. In M. D. Leakey & J. M. Harris (Eds.), *Laetoli: A Pliocene site in northern Tanzania* (pp. 1–21). Oxford: Clarendon.

Leakey, M. D., & Harris, J. M. (1987). *Laetoli: A Pliocene site in northern Tanzania*. Oxford: Clarendon.

Lee-Thorp, J. A. (1989). Stable carbon isotopes in deep time: The diets of fossil fauna and hominids. Ph.D. dissertation, University of Cape Town, Cape Town.

Lumpkin, S., & Kranz, K. R. (1984). *Cephalophus sylvicultor*. *Mammalian Species, 225*, 1–7.

MacLeod, S. B., Kerley, G. I. H., & Gaylard, A. (1996). Habitat and diet of bushbuck *Tragelaphus scriptus* in the woody cape nature reserve: Observations from faecal analysis. *South African Journal of Wildlife Research, 26*, 19–25.

Manega, P. C. (1993). Geochronology, geochemistry and isotopic study of the Plio-Pleistocene hominid sites and the Ngorongoro volcanic highlands in northern Tanzania. Ph.D. dissertation, University of Colorado, Boulder.

McNaughton, S. J., Tarrants, J. L., McNaughton, M. M., & Davis, R. H. (1985). Silica as a defense against herbivory and a growth promotor in African grasses. *Ecology, 66*, 528–535.

Meladze, G. K. (1964). About the paleobiological study of Sivatheriinae. *Soobshcheniia Akademii nauk Gruzinskoi SSR, 33*, 567–600.

Mihlbachler, M. C., & Solounias, N. (2006). Coevolution of tooth crown height and diet in oreodonts (Merycoidodontidae, Artiodactyla) examined with phylogenetically independent contrasts. *Journal of Mammalian Evolution, 13*, 11–36.

Morgan, M. E., Kingston, J. D., & Marino, B. D. (1994). Carbon isotopic evidence for the emergence of C_4 plants in the Neogene from Pakistan and Kenya. *Nature, 367*, 162–165.

Morris, S. C. (1998). The evolution of diversity in ancient ecosystem. *Philosophical Transactions: Biological Sciences, 353*(1366), 327–345.

Musiba, C.M. (1999). Laetoli Pliocene paleoecology: A reanalysis via morphological and behavioral approaches. Ph.D. dissertation, University of Chicago, Chicago.

Musiba, C. M., Vogt, M., & Branting, S. (2002). Laetoli Pliocene landscape reconsidered: A reanalysis via functional morphology and taphonomic data. *Journal of Human Evolution, 42*, A25.

Ndessokia, P. N. S. (1990). The mammalian fauna and archaeology of the Ndolanya and Olpiro Beds, Laetoli. Tanzania. Ph.D. dissertation, University of California, Berkeley.

Pellow, R. A. (1983). The feeding ecology of a selective browser, the giraffe (*Giraffa camelopardalis tippelskirchi*). *Journal of Zoology, 202*, 57–81.

Pickford, M., & Senut, B. (2001). The geological and faunal context of Late Miocene hominid remains from Lukeino, Kenya. *Comptes Rendus de l'Académie des Sciences Paris. Série 2a 332*, 145–152.

Piperno, D. R., Holst, I., Wessel-Beaver, L., & Andres, T. C. (2002). Evidence for the control of phytolith formation in cucurbita fruits by the hard rind (Hr) genetic locus: Archaeological and ecological implications. *Proceedings of the National Academy of Sciences of the United States of America, 99*, 10923–10928.

Plummer, T. W., & Bishop, L. C. (1994). Hominid paleoecology at Olduvai Gorge, Tanzania, as indicated by antelope remains. *Journal of Human Evolution, 27*, 47–75.

Reed, K. E. (1997). Early hominid evolution and ecological change through the African Plio-Pleistocene. *Journal of Human Evolution, 32*, 289–322.

Rivals, F., & Semprebon, G. (2006). A comparison of the dietary habits of a large sample of the Pleistocene pronghorn *Stockoceros onusrosagris* from the Papago Springs Cave in Arizona to the modern *Antilocapra americana*. *Journal of Vertebrate Paleontology, 26*, 495–500.

Rivals, F., Mihlbachler, M. C., & Solounias, N. (2007). Effect of ontogenetic-age distribution in fossil and modern samples on the interpretation of ungulate paleodiets using the mesowear method. *Journal of Vertebrate Paleontology, 27*, 763–767.

Robinson, C. (2011). Giraffidae. In T. Harrison (Ed.), *Paleontology and geology of Laetoli: Human evolution in context* (Fossil hominins and the associated fauna, vol. 2, pp. 339–362). Dordrecht: Springer.

Sanson, G. D., Kerr, S. A., & Gross, K. A. (2007). Do silica phytoliths really wear mammalian teeth? *Journal of Archaeological Science, 34*, 526–531.

Scott, K. M. (1979). Adaptation and allometry in bovid postcranial proportions. Ph.D. dissertation, Yale University, New Haven.

Semaw, S., Simpson, S. W., Quade, J., Renne, P. R., Butler, R. F., McIntosh, W. C., Levin, N., Dominguez-Rodrigo, M., & Rogers, M. J. (2005). Early Pliocene hominids from Gona, Ethiopia. *Nature, 433*, 301–305.

Semprebon, G., & Rivals, F. (2007). Was grass more prevalent in the pronghorn past? An assessment of the dietary adaptations of Miocene to recent Antilocapridae (Mammalia: Artiodactyla). *Palaeogeography, Palaeoclimatology, Palaeoecology, 253*, 332–347.

Shipman, P., & Harris, J. M. (1988). Habitat preference and paleoecology of *Australopithecus boisei* in eastern Africa. In F. Grine (Ed.), *The evolutionary history of the "robust" australopithecines* (pp. 343–381). New York: Aldine de Gruyter.

Simbotwe, M. P., & Sichone, P. W. (1989). Aspects of behavior of bushbuck in Kafue national park, Zambia. *South African Journal of Wildlife Research, 19*, 38–41.

Simpson, C. D. (1974). Food studies on the Chobe bushbuck, *Tragelaphus scriptus arnatus* Pocock 1900. *Arnoldia, 6*(32), 1–9.

Solounias, N., Teaford, M., & Walker, A. (1988). Interpreting the diet of extinct ruminants: The case of a non-browsing giraffid. *Paleobiology, 14*, 287–300.

Sponheimer, M., & Lee-Thorp, J. A. (2003). Using carbon isotope data of fossil bovid communities for palaeoenvironmental reconstruction. *South African Journal of Science, 99*, 273–275.

Sponheimer, M., Reed, K. E., & Lee-Thorp, J. A. (1999). Combining isotopic and ecomorphological data to refine bovid paleodietary reconstruction: A case study from the Makapansgat Limeworks hominin locality. *Journal of Human Evolution, 36*, 705–718.

Sponheimer, M., Lee-Thorp, J. A., DeRuither, J. D. J., Smith, J. M., Van Der Merwe, N. J., Reed, K., Grant, C. C., Ayliffe, L. K., Robinson, T. F., Heidelberg, C., & Marcus, W. (2003). Diets of southern African Bovidae: Stable isotope evidence. *Journal of Mammalogy, 84*, 471–479.

Su, D. F. (2005). The paleoecology of Laetoli, Tanzania: Evidence from the mammalian fauna of the Upper Laetolil Beds. Ph.D. dissertation, New York University, New York.

Su, D. F., & Harrison, T. (2003). Faunal differences in the sequence at Laetoli: Implications for taphonomy and paleoecology. *American Journal of Physical Anthropology, 120*, 203.

Su, D. F., & Harrison, T. (2005). Large mammal differences between fossil and modern communities: Implications for the reconstruction of hominin paleoenvironments *PaleoAnthropology, PAS 2005* Abstracts: A22.

Su, D., & Harrison, T. (2008). Ecological implications of the relative rarity of fossil hominins at Laetoli. *Journal of Human Evolution, 55*, 672–681.

Van Wieren, S. E. (1996). *Digestive strategies in ruminants and nonruminants*. Den Haag: Landbouwuniversiteit Wageningen.

Vignaud, P., Duringer, P., Mackaye, H. T., Likius, A., Blondel, C., Boisserie, J. R., de Bonis, L., Eisenmann, V., Etienne, M. E., Geraads, D., Guy, F., Lehmann, T., Lihoreau, F., Lopez-Martinez, N., Mourer-Chauvire, C., Otero, O., Raje, J. C., Schuster, M., Viriot, L., Zazzo, A., & Brunet, M. (2002). Geology and palaeontology of the Upper Miocene Toros–Menalla hominid locality, Chad. *Nature, 418*, 152–155.

Vrba, E. S. (1975). Some evidence of chronology and palaeoecology of Sterkfontein, Swartkrans and Kromdraai from the fossil Bovidae. *Nature, 254*, 301–304.

Vrba, E. S. (1980). The significance of bovid remains as indicators of environment and predation patterns. In A. K. Behrensmeyer & A. Hill (Eds.), *Fossils in the making: Vertebrate taphonomy, and paleoecology* (pp. 247–272). Chicago: Chicago University Press.

Vrba, E. S. (1985). Ecological and adaptive changes associated with early hominid evolution. In E. Delson (Ed.), *Ancestors: The hard evidence* (pp. 63–71). New York: Alan R. Liss.

Vrba, E. S. (1995). On the connections between paleoclimate and evolution. In E. S. Vrba, G. H. Denton, T. C. Patridge, & L. H. Burckle (Eds.), *Paleoclimate and evolution, with emphasis on human origins* (pp. 1–45). New Haven: Yale University Press.

Wilkinsin, D. M. (2002). Ecology before ecology: Biogeography and ecology in Lyell's 'Principles'. *Journal of Biogeography, 29*, 1109–1115.

WoldeGabriel, G., White, T. D., Suwa, G., Renne, P., de Heinzelin, J., Hart, W. K., & Heiken, G. (1994). Ecological and temporal placement of early Pliocene hominids at Aramis, Ethiopia. *Nature, 371*, 330–333.

WoldeGabriel, G., Haile-Selassie, Y., Renne, P. R., Hart, W. K., Ambrose, S. H., Asfaw, B., Heiken, G., & White, T. (2001). Geology and palaeontology of the Late Miocene Middle Awash Valley, Afar Rift, Ethiopia. *Nature, 412*, 175–178.

Zazzo, A., Bocherens, H., Brunet, M., Beauvilain, A., Billiou, D., Mackaye, H. T., et al. (2000). Herbivore paleodiet and paleoenvironmental changes in Chad during the Pliocene using stable isotope ratios of tooth enamel carbonate. *Paleobiology, 26*, 294–309.

Chapter 17
Paleoenvironments of Laetoli, Tanzania as Determined by Antelope Habitat Preferences

Laura C. Bishop, Thomas W. Plummer, Fritz Hertel, and Kris Kovarovic

Abstract We examined the fossil remains of antelope (Mammalia: Bovidae) postcrania recovered from the Upper Laetolil Beds and the overlying Upper Ndolanya Beds in northern Tanzania. We used analyses of the ecomorphology of extinct antelopes to determine their habitat preferences. The most common postcranial elements - the phalanges, astragali, and distal radii - were examined. A total sample of 446 specimens was analyzed. Changes in the relative proportion of habitat preferences through time are suggestive of temporal fluctuations in habitat availability. As has been noted elsewhere, the mammals from Laetoli do not show the expected distribution of body sizes, so this paper also examined how such size biases, assumed to be taphonomic in origin, might have affected the reconstructions proposed here. We conclude that, on the basis of habitat preferences of the antelopes recovered from Laetoli, there is evidence for the continuous regional presence of woodland and forest throughout deposition of the Upper Laetolil Beds. Antelopes preferring forest and heavy cover habitats dominate the assemblage. Antelopes that locomoted principally in open and light cover habitats are largely in the minority, with a combined frequency never exceeding around 25% of the total in each stratigraphic unit of the Upper Laetolil Beds. The relative proportions of heavy cover and forest preferring antelopes in particular are likely affected by body size biases caused by taphonomic processes active during the deposition of the Upper Laetolil Beds, which favoured the preservation of the postcrania of smaller antelopes. During the formation of the Upper Ndolanya Beds, the proportion of antelopes preferring more open habitats is greatly increased, although forest-preferring antelopes are still present. The conclusion that forest and woodland were always present throughout the sequence is robust.

Keywords Ecomorphology • Bovidae • Postcranials • Pliocene

Introduction

The importance of Laetoli as a Pliocene paleontological site and as a catalyst for paradigm change within the field of paleoanthropology cannot be overestimated. Located south of the equally famous and important Olduvai Gorge in Tanzania, Laetoli has been generating important data and insight since the 1930s. This paper is concerned with the reconstruction of local environments during the formation of the Upper Laetolil Beds, and to a lesser extent the overlying Upper Ndolanya Beds. Composition and paleobiology of the mammalian fauna have become established proxies for paleoecological reconstructions of fossil sites (e.g., Andrews et al. 1979; Reed 1997, 1998). We used the ecomorphology of fossil antelopes (Mammalia: Bovidae) in order to reconstruct their habitat preferences. The relative proportions of antelopes having particular habitat preferences were then used to examine the availability of preferred habitats in the region of the deposits, and how these have varied through time. Finally, in view of taphonomic biases affecting the size distributions of Laetoli mammals relative to their presence in modern ecosystems (Su and Harrison 2007), we examined how body size distributions may alter the interpretation of these results.

L.C. Bishop (✉)
School of Natural Sciences and Psychology, Research Centre in Evolutionary Anthropology and Palaeoecology, Liverpool John Moores University, Byrom Street, Liverpool, L3 3AF, UK
e-mail: L.C.Bishop@ljmu.ac.uk

T.W. Plummer
Department of Anthropology, Queens College, CUNY & NYCEP, 65–30 Kissena Blvd, Flushing, NY 11367, USA
e-mail: Thomas.Plummer@qc.cuny.edu

F. Hertel
Department of Biology, California State University, Northridge, CA 91330-8303, USA
e-mail: fritz.hertel@csun.edu

K. Kovarovic
Department of Anthropology, Durham University, Dawson Building, South Road, Durham, DH1 3LE, UK
e-mail: kris.kovarovic@durham.ac.uk

T. Harrison (ed.), *Paleontology and Geology of Laetoli: Human Evolution in Context. Volume 1: Geology, Geochronology, Paleoecology and Paleoenvironment*, Vertebrate Paleobiology and Paleoanthropology, DOI 10.1007/978-90-481-9956-3_17,

Ecomorphology?

Ecomorphology, or ecological morphology, is the study of the relationship between morphology and ecology, and examines the interface between an organism's phenotype and its environment (Van der Klaauw 1948). Ecomorphological studies can address the relationships between any aspect of morphology and ecology, but paleoecological studies have focused principally on locomotor (postcranial) and dietary (craniodental) reconstructions.

Although studies of ecomorphology are common in modern neontological sciences, applications of this method to extinct life are now becoming more established in the literature (see Plummer et al. 2008 for a review). Ecomorphological studies of extinct taxa give us the opportunity to reconstruct their paleobiology using analogy to modern taxa, which is neither restrictive nor proscriptive. Ecological similarities to modern, sometime distant, relatives are not assumed. The methods used here can be applied to postcrania that cannot be identified to genus or species and are thus ideal for expanding the information that can be gained from the commonly disarticulated fossil remains found at paleontological and archaeological sites.

Here we focus on the relationship between postcranial locomotor morphology and habitat preference. Building upon pioneering work by Gentry (1970), which examined skeletal specializations in the limb bones of cursorial antelopes, this work has been expanded by Kappelman (1988, 1991), Kappelman et al. (1997) and numerous other authors (Bishop 1994, 1999; Plummer and Bishop 1994; Kappelman et al. 1997; Spencer 1997; Lewis 1997; Bishop et al. 1999; Sponheimer et al. 1999; Werdelin and Lewis 2001; DeGusta and Vrba 2003, 2005; Kovarovic and Andrews 2007; Plummer et al. 2008).

Why Antelopes?

At Laetoli, as at other African paleontological sites, Bovidae is a common, if not the most common, fossil taxon recovered. The Neogene radiation of African antelopes was broad in geographical and ecological scope and the taxonomic diversity of this group is still evident today. Modern antelopes are distributed across all African habitat types and vary in body size, locomotor preference, and diet, as well as in other morphological, behavioral, and ecological characteristics. Modern antelope species are also relatively habitat-specific and, with the exception of some ecotonal species, prefer to live largely in a single habitat type. These are preferences, not restrictions, and antelopes may be found in more than one of these habitat types during their lifetime. Their relative habitat specificity makes them most valuable for paleoenvironmental reconstructions. Habitat specific morphologies have been identified in many taxa that transcend phylogenetic relationships – they are shared by taxa similar ecologically yet relatively distant phylogenetically. These morphologies seem to relate principally to the compromises between joint stability and mobility associated with locomotion across variable substrates, and movement patterns in habitats with different vegetation densities (Kappelman 1988).

Materials and Methods

Extant Antelopes

We examined modern antelope postcrania from several museum collections (American Museum of Natural History, New York, NY; National Museum of Natural History, Washington, DC; The Natural History Museum, London, UK) (Table 17.1). Adult, wild-shot specimens were measured in preference to zoo specimens, when available. Less than 5% of the individuals were zoo specimens, many of which had been captured in the wild. Thirty-six modern African species were studied. The largest antelope species, weighing more than 250 kg (genera *Taurotragus* and *Syncerus*), were excluded from the analysis because for very large antelopes, scaling differences and behavioral specializations related to large body size may override any morphological habitat signal (Scott 1979, 1985; Kappelman 1988; Estes 1991).

Here we present results on the most common well-preserved skeletal elements found at Laetoli – the phalanges, astragali, and distal radii. Due to variations in completeness of museum specimens, the sample sizes for these different bones varied from a minimum of 188 for the third (distal) phalanx to a maximum of 312 for the first (proximal) phalanx (Table 17.1). The minimum sample size for each species was two adult specimens per taxon, with the exceptions of the distal and intermediate phalanges of *Kobus kob* and the proximal phalanx of *Raphicerus campestris*. The maximum sample size for a species was 27 for proximal phalanges of *Tragelaphus scriptus*.

Each antelope species was assigned to a habitat preference category based upon previous ethological research (Scott 1979, 1985; Kappelman 1986, 1988, 1991, Kappelman et al. 1997) (Table 17.1). The environmental categories used here were open (e.g., grassland, arid country, ecotones bordering open country), light cover (e.g., light bush, tall grass), heavy cover (e.g., heavy bush, woodland, densely vegetated swamp), and forest. These general terms artificially partition the continuous vegetational spectrum - from those generally lacking trees and bush to those with a continuous tree canopy - into discrete categories. This categorization puts complicated but

Table 17.1 Modern African antelope taxa used in this analysis. This table shows the average body mass and the habitat category assigned for each of the 36 taxa used for this study. The sample sizes for each taxon for the DFA models for each skeletal element are shown, as are the percentages of their successful classification by the models used here. Although most species were classified correctly by the models, a discussion of those that consistently misclassify can be found in the text

Taxon	Body mass (kg)	Habitat category	Astragalus	AST percent misclassified	Phalanx 1	PHA1 percent misclassified	Phalanx 2	PHA2 percent misclassified	Phalanx 3	PHA3 percent misclassified	Radius	RAD percent misclassified
Alcelaphus buselaphus	155	Open	14	0.00%	23	4.35%	13	0.00%	10	0.00%	10	0.00%
Damaliscus dorcas	68	Open	8	0.00%	13	7.69%	8	0.00%	7	0.00%	8	12.50%
Damaliscus lunatus	136	Open	10	0.00%	16	0.00%	12	0.00%	8	0.00%	11	9.09%
Connochaetes taurinus	214	Open	6	0.00%	3	0.00%	3	0.00%	3	0.00%	6	0.00%
Connochaetes gnou	148	Open	5	0.00%	5	0.00%	3	0.00%	3	0.00%	4	0.00%
Hippotragus niger	228	Open	9	0.00%	16	0.00%	11	0.00%	9	0.00%	4	0.00%
Addax nasomaculatus	96	Open	4	25.00%	4	0.00%	4	0.00%	4	0.00%	4	0.00%
Oryx gazella	169	Open	13	0.00%	12	8.33%	11	0.00%	8	0.00%	5	0.00%
Hippotragus equinus	270	Open	4	0.00%	5	0.00%	3	0.00%	3	0.00%	4	0.00%
Antidorcas marsupialis	38	Open	8	12.50%	12	8.83%	5	20.00%	5	20.00%	7	0.00%
Gazella thomsoni	21	Open	11	45.45%	15	40.00%	7	57.14%	7	57.14%	10	10.00%
Gazella granti	55	Open	10	20.00%	13	7.69%	6	33.33%	4	25.00%	10	30.00%
Sylvicapra grimmia	20	Light cover	17	17.65%	9	0.00%	9	0.00%	8	25.00%	8	50.00%
Aepyceros melampus	53	Light cover	13	7.70%	19	0.00%	11	0.00%	10	10.00%	12	16.67%
Ourebia ourebia	17	Light cover	7	0.00%	11	0.00%	7	14.29%	7	0.00%	7	28.57%
Raphicerus campestris	11	Light cover	7	28.57%	1	0.00%					8	12.50%
Kobus kob	79	Light cover	10	20.00%	4	0.00%	1	0.00%	1	0.00%	10	20.00%
Redunca redunca	45	Light cover	10	10.00%	6	0.00%	4	0.00%	3	33.33%	10	10.00%
Redunca arundinum	58	Light cover	7	0.00%	2	0.00%	2	0.00%	1	100%	7	14.29%
Redunca fulvorufula	30	Light cover	8	0.00%	5	0.00%	5	0.00%	4	0.00%	6	0.00%
Tragelaphus spekei	78	Heavy cover	11	9.09%							11	18.18%
Tragelaphus euryceros	270	Heavy cover	6	0.00%	18	0.00%	13	7.69%	17	11.76%	8	0.00%

(continued)

Table 17.1 (continued)

Taxon	Body mass (kg)	Habitat category	Astragalus	AST percent misclassified	Phalanx 1	PHA1 percent misclassified	Phalanx 2	PHA2 percent misclassified	Phalanx 3	PHA3 percent misclassified	Radius	RAD percent misclassified
Tragelaphus strepsiceros	214	Heavy cover	7	0.00%	11	0.00%	10	0.00%	13	0.00%	7	14.29%
Tragelaphus imberbis	82	Heavy cover	4	0.00%	5	0.00%	7	0.00%	4	0.00%	4	25.00%
Madoqua kirkii	5	Heavy cover	11	0.00%	15	0.00%	11	0.00%	8	12.50%	10	10.00%
Kobus ellipsiprymnus	210	Heavy cover	10	0.00%	4	0.00%	4	50.00%	4	75.00%	10	10.00%
Kobus megaceros	90	Heavy cover	7	42.90%							7	57.14%
Tragelaphus scriptus	43	Forest	17	5.88%	27	0.00%	14	0.00%	7	0.00%	11	9.09%
Cephalophus natalensis	13	Forest	4	0.00%	2	0.00%	2	0.00%	2	0.00%	4	0.00%
Cephalophus leucogaster	18	Forest	4	25.00%	5	0.00%	4	0.00%	4	25.00%	4	0.00%
Cephalophus silvicultor	63	Forest	7	0.00%	8	12.50%	8	50.00%	8	12.50%	7	0.00%
Cephalophus monticola	6	Forest	4	25.00%	11	0.00%	5	0.00%	5	0.00%	5	0.00%
Cephalophus nigrifrons	16	Forest	4	0.00%	4	0.00%	4	0.00%	3	0.00%	4	0.00%
Cephalophus weynsi	20	Forest	5	0.00%	4	0.00%	4	0.00%	4	0.00%	5	0.00%
Cephalophus dorsalis	22	Forest	5	0.00%	4	0.00%	4	0.00%	4	25.00%	5	0.00%
Neotragus moschatus	5	Forest	4	25.00%							4	0.00%
Total/misclassified			291	8.93%	312	4.49%	215	7.44%	188	11.17%	257	10.12%

Table 17.2 Success rates for the models used in this analysis. This table summarizes the successful classification rates for the different DFA models created for antelope skeletal elements most commonly found at Laetoli. The overall success rates of the models, for both straight resubstitution and jackknifed tests of accuracy are presented by habitat type. Success rates are generally high and similar across skeletal elements and habitat types

Anatomical part		Number of variables	Overall	Open	Light cover	Heavy cover	Forest
Radius	Resubstitution	11	89%	93.90%	80.88%	82.46%	97.96%
	Jackknifed	11	84%	89.02%	77.94%	75.44%	95.92%
Astragalus	Resubstitution	11	93%	93.94%	90.91%	92.86%	92.59%
	Jackknifed	11	87%	91.92%	85.71%	83.93%	83.33%
Proximal phalanx	Resubstitution	11	95%	92.20%	98.21%	94.23%	98.46%
	Jackknifed	11	90%	89.36%	87.50%	90.38%	95.38%
Intermediate phalanx	Resubstitution	10	92%	90.91%	94.87%	93.18%	91.11%
	Jackknifed	10	88%	87.50%	84.62%	86.36%	91.11%
Distal phalanx	Resubstitution	7	89%	91.55%	82.35%	86.96%	91.89%
	Jackknifed	7	87%	91.55%	76.47%	84.78%	89.19%

environmentally-relevant data within a simplified framework allowing statistical analysis, and as such is a compromise between descriptive reality and analytical utility [as discussed in Kappelman et al. (1997) and DeGusta and Vrba (2003)]. Within each model we attempted to ensure that sample sizes were as consistent as possible across habitat categories.

Numerous measurements were collected on each bone. The logarithms of these measurements were then subjected to a quadratic discriminant function analysis (DFA) as described fully elsewhere (Plummer et al. 2008). DFA is a classification algorithm that separates individual cases into naturally occurring, predefined groups (James 1985). In this procedure, an equation (or function) is derived using the metrics and indices from the anatomical part or portion in question that best discriminate among habitat preference groups (hence the name discriminant function). Not all of the measurements taken, and the ratios calculated to reflect shape differences, were used in the final analysis. For each skeletal element a subset of these variables having the best predictive power was selected using a manual experimental approach combined with the stepwise procedure in SYSTAT v. 9 (Table 17.2). Within-group covariance matrices showed significant differences, so the matrices were not pooled to calculate a linear discriminant function, but instead were used to compute quadratic discriminant functions (James 1985; Reyment, 1991). The number of variables required for accurate prediction varied among seven for the distal phalanx and 11 for the proximal phalanx, radius, and astragalus. For the definitions of the ratios and measurements used in the DFA models, see Appendix.

Discriminant function analyses of our morphological variables allowed separation of the modern antelopes into their habitat preference groups with very high levels of accuracy. This was determined by model success rates as indicated by multivariate statistics testing the significance of differences among group means, as well as by how well the discriminant function classified specimens of known habitat. Model accuracy was further tested using jackknife analysis (Table 17.2). For the five skeletal elements used in this analysis, the highest success rates were for the proximal phalanx ($n=312$, 95% resubstitution, 90% jackknifed) and the lowest for the radius ($n=257$, 89% resubstitution, 84% jackknifed). Although no DFA is perfect, these models all perform more than three times better than chance for modern antelopes and have high rates of successful reclassification.

We also examined accuracy of classification within individual species samples. With two exceptions, the bones of each species were reclassified correctly in the tests of model accuracy, with the majority of the sample being assigned to their pre-assigned habitat category. The exceptions were the Thomson's Gazelle (*Gazella thomsoni*), for which the second and third phalanges were misclassified from open to light cover in approximately 57% of 7 cases. The third (distal) phalanx of the waterbuck (*Kobus ellipsiprymnus*) also was misclassified in 3 of 4 cases, from heavy cover to open ($n=4$). Model accuracy was thus high both across skeletal elements and within individual taxa, allowing use of this method on the fossil material from Upper Laetolil Beds and environs.

Fossils

Fossils from Laetoli have been recovered by several expeditions; this study examined specimens collected by the Eyasi Plateau Paleontological Expedition (1998–2005), which employed a sampling strategy geared towards collecting fossils for both paleoecological reconstruction and more traditional paleontological studies. For ecomorphological studies, the extent of the articular surfaces must be observable and well preserved, and the preservation of shaft dimensions should not be altered by shattering, deformation, predepositional

Table 17.3 Distribution of skeletal parts by stratigraphic subdivision. The skeletal sample is shown as distributed through the stratigraphic section, which is divided into categories based on upper and lower limits determined by marker tuffs. This division allows us to examine how habitat distributions at Laetoli have changed through time. The category 'Strat indet' refers to specimens whose potential stratigraphic assignment is to a larger or different grouping of levels than the subdivisions used here for the purposes of analysis

Stratigraphic subdivision	Astragalus	Proximal phalanx	Intermediate phalanx	Distal phalanx	Distal radius	Sample size
Below Tuff 3	9	27	20	6	1	63
Between Tuff 3–Tuff 5	7	3	9	4	2	25
Between Tuff 5–Tuff 7	31	11	24	20	2	88
Above Tuff 7	7	14	21	10	0	52
Laetolil Beds – Strat indet	22	25	21	14	3	85
Upper Ndolanya Beds	19	24	45	22	23	133
Total	95	104	140	76	31	446

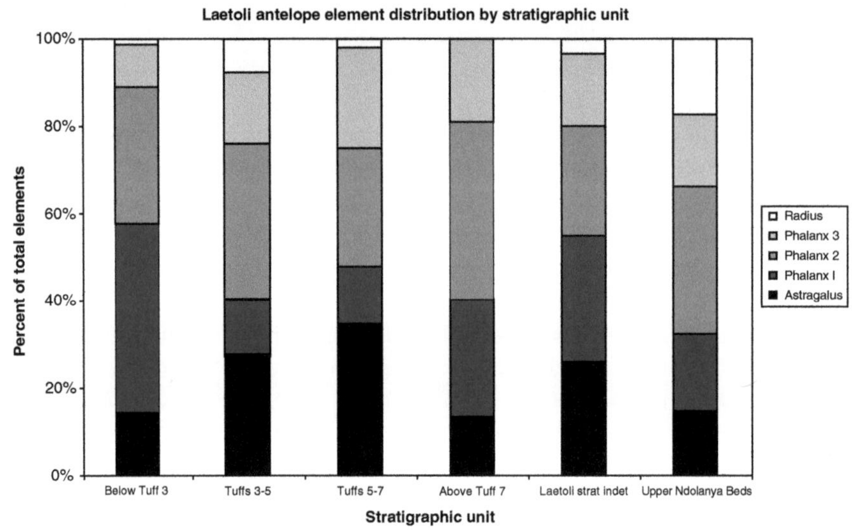

Fig. 17.1 Laetoli antelope element distribution by stratigraphic unit. The data presented in Table 17.3 are provided here in figurative form to allow comparison of each skeletal element by stratigraphic interval. Relative frequencies of elements differ slightly. However, the success rates for the different DFA models are sufficiently similar to allow us to examine results of all bones together when considering habitat reconstructions and environmental change

(e.g., carnivore gnawing) or other distortion of the surface of the cortical bone. Both post-depositional and post-erosional processes can limit the utility of a fossil for this type of analysis. We measured all of the long bones as well as the phalanges, calcanei and astragali at the National Museum of Tanzania in Dar es Salaam. Here, we present data on the most commonly preserved elements, the astragali, phalanges, and distal radii.

Four hundred and sixty four fossils from the Upper Laetolil Beds and the Upper Ndolanya Beds were sufficiently well preserved for use in this study (Table 17.3). These fossils were assigned to habitat category using the DFA model developed for each skeletal element using the modern sample as described above. The complete sample contained fossils that were relatively evenly distributed across stratigraphic levels and all skeletal elements (Table 17.3, Fig. 17.1). For analyses of taphonomy and habitat preference, we divided the sample into six different stratigraphic categories as outlined in Table 17.3. There were four stratigraphic levels within the Upper Laetolil Beds, plus one Upper Laetolil Beds category, termed 'indeterminate' here, which comprises specimens that could not be assigned to one of the more common stratigraphic units delimited for this study. Results for the Upper Ndolanya Beds are shown separately. The success rates for the DFA models used to classify different skeletal elements were so similar that we combined the results for the different bones to increase sample sizes in each stratigraphic level (Plummer and Bishop 1994; Kovarovic and Andrews 2007). Here we have not used a statistical 'threshold' to eliminate specimens with low estimated probabilities for their habitat preferences from the analysis (e.g., DeGusta and Vrba 2003, 2005). That technique can unnecessarily and artificially lower sample sizes, and alter the natural pattern of results (Plummer et al. 2008).

The measurements taken on the fossil bones also allow us to examine the effects of body size on habitat category representation within the Laetoli sample. Numerous postcranial measurements scale well with body size (Scott 1979, 1985). We examined body mass proxies from the different fossil and modern skeletal elements studied, bivariate plots between body size and habitat preference, and differences in body size distribution among different stratigraphic units. We used these data to examine the effects of taphonomic alteration of the body size distribution, and the relative abundance of animals of different body sizes, on environmental reconstructions using ecomorphological methods.

Results

Habitat Preference

The results of the reclassification DFA of fossil antelope postcrania from Laetoli showed two main patterns (Table 17.4; Fig. 17.2). The first was an overall preponderance of specimens that showed morphologies related to preferences for forest and heavy cover habitats. Throughout the Laetolil Beds sequence, from the lowest stratigraphic subdivision to the highest, nearly 75% or more of the sample represents antelopes that prefer the more densely vegetated part of the African habitat spectrum. Only in the Upper Ndolanya Beds do open and light-cover preferring antelopes begin to dominate the assemblage. In the Upper Laetolil Beds, light-cover and open habitat preferring antelopes only represent about one quarter of the analyzed assemblage.

When examined in sequence, the variations in antelope habitat preference abundances can be interpreted as a proxy to show fluctuations in the presence of different habitats. Forest habitat antelopes vary from a low of 19% of the sample in the lowest stratigraphic unit below Tuff 3 to a maximum of 34% of the sample during the interval deposited between Tuffs 5 and 7. However, open habitat preferring antelopes reached their maximum relative abundance of 12.5% of the sample during this same interval. It is the heavy cover specimens that are at their lowest relative frequency of 5.7% during this time. Overall, the pattern is one of slight fluctuations in the relative proportions of antelopes having more closed and more open habitat types but the ranges of variation are not very high.

Table 17.4 Combined results of the ecomorphological analyses for each stratigraphic subdivision, showing sample sizes and distributions of habitat preferences. This table shows the number and percentage of skeletal elements assigned to each habitat from each stratigraphic unit

Stratigraphic unit	Subdivision	Sample size	Open	Light cover	Heavy cover	Forest
Laetolil Beds	Below Tuff 3	63	5	11	35	12
			7.94%	17.46%	55.56%	19.05%
Laetolil Beds	Between Tuff 3–Tuff 5	25	3	3	13	6
			12.00%	12.00%	52.00%	24.00%
Laetolil Beds	Between Tuff 5–Tuff 7	88	11	10	37	30
			12.50%	11.36%	42.05%	34.09%
Laetolil Beds	Above Tuff 7	52	5	3	28	16
			9.62%	5.77%	53.85%	30.77%
Laetolil Beds	Strat indet	85	7	15	39	24
			8.24%	17.65%	45.88%	28.24%
Ndolanya Beds	Upper	133	41	46	21	25
			30.83%	34.59%	15.79%	18.80%
Total		446	72	88	173	113
			16.14%	19.73%	38.79%	25.34%

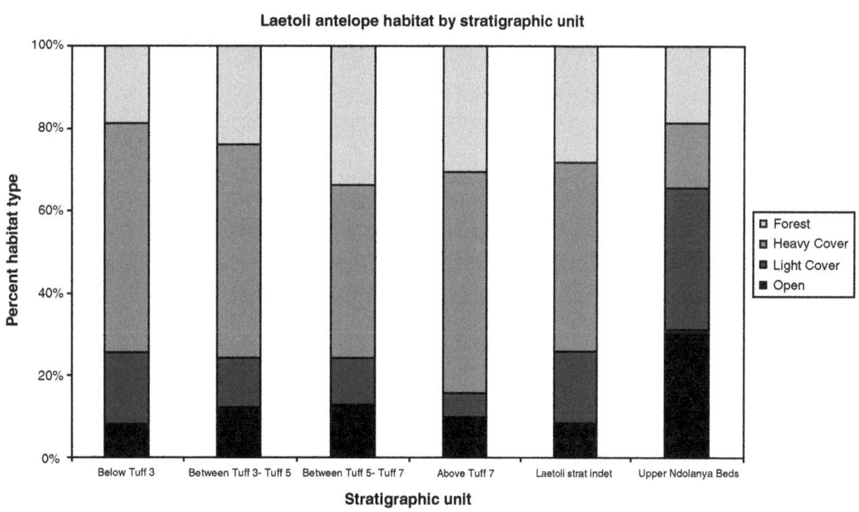

Fig. 17.2 Laetoli antelope habitat predictions by stratigraphic unit. This figure presents the data from Table 17.4 in graphic format. Some fluctuations in proportions of antelopes preferring different habitats are shown. The overall dominance of heavy cover and forest ecomorphs is apparent here

Due to the type of data used in these analyses, demonstrating statistical significance in any meaningful way is challenging. MANOVA and paired t-tests were performed on the data in Table 17.4 to assess whether there were statistically significant differences among layers in the number of specimens per habitat preference category. These show statistically significant differences between some pairs of stratigraphic units. For example, the distributions of habitat types are significantly different between Tuffs 3–5 and the Upper Ndolanya Beds ($p<0.05$). These units are also the most divergent in overall sample size, being the smallest and largest samples, respectively. To test the effect of sample size on the analysis, we standardized the sample size and found no significant differences among any of the stratigraphic units. So, we cannot state that the differences in distributions of antelope habitat preferences among the levels used for this analysis are statically significant, merely that there appear to be differences, however minor, among the proportions of different habitat types present. When the Upper Laetolil Beds are considered together, however, there is a statistically significant difference among the proportions of more open habitat categories between the Upper Laetolil Beds and the Upper Ndolanya Beds. The values for the Upper Ndolanya Beds for open, light cover, and heavy cover habitat preferences are greater than two standard deviations away from those of the Upper Laetolil Beds. Thus, there is strong evidence for a continuous presence of heavily wooded (heavy cover and forest) habitats throughout the Laetolil Beds sequence. This will be discussed in light of taphonomic considerations below.

Table 17.5 Size distribution of skeletal elements by stratigraphic subdivision. The contrast between the numbers of larger (sizes three and four) and smaller (sizes one and two) antelopes is shown. For the samples having stratigraphic provenance, the ratio of smaller to larger animals is up to 20:1. Some stratigraphic units (between Tuffs 3 and 5) have very few larger antelopes due to an overall smaller sample size. The discrepancy evens out somewhat in the Upper Ndolanya Beds sample (less than 3–1). Mammal size classes follow Brain (1981) and Bunn (1986)

Stratigraphic unit	Model	Size 1, 2	Size 3, 4	Total
Below Tuff 3	Astrag	9		9
	Pha1	25	2	27
	Pha2	19	1	20
	Pha3	5	1	6
	Radius		1	1
	Total	58	5	63
Between Tuff 3 – Tuff 5	Astrag	5	2	7
	Pha1	3		3
	Pha2	9		9
	Pha3	4		4
	Radius	2		2
	Total	23	2	25
Between Tuff 5 –Tuff 7	Astrag	27	4	31
	Pha1	10	1	11
	Pha2	21	3	24
	Pha3	19	1	20
	Radius	1	1	2
	Total	78	10	88
Above Tuff 7	Astrag	4	3	7
	Pha1	12	2	14
	Pha2	20	1	21
	Pha3	8	2	10
	Total	44	8	52
Laetolil Beds strat indet	Astrag	15	7	22
	Pha1	24	1	25
	Pha2	18	4	22
	Pha3	12	4	16
	Radius	3		3
	Total	72	16	88
Upper Ndolanya Beds	Astrag	16	3	19
	Pha1	18	6	24
	Pha2	32	13	45
	Pha3	17	5	22
	Radius	15	8	23
	Total	98	35	133

Body Size

Taphonomic bias, particularly carnivore ravaging of carcasses, may have led to a disproportionate destruction of postcranial elements from medium-sized mammals in the Upper Laetolil Beds (Su 2005; Su and Harrison 2007, 2008). Small mammals less than 5 kg in mass exhibit better postcranial preservation, because their remains were more likely to be buried and preserved in ash falls or by reworked tuffs. There are known to be disproportionately high frequencies of small mammals found in these assemblages, whereas medium-sized animals (i.e., wildebeest sized antelopes) were relatively poorly represented (Su 2005). The contrast between the presence of smaller and larger antelopes in the postcranial fossil assemblage is easily seen (Table 17.5, Fig. 17.3). This taphonomic bias potentially impacts the reliability of antelope habitat preferences as a simple proxy for the availability of habitats whilst the deposits were formed. The actual proportional representation of different habitats is difficult to ascertain. However, the fact that ecomorphs indicative of well-wooded settings are present throughout the entire Upper Laetolil Beds sequence indicates that these settings were present throughout their 200 kyr long deposition, as well as during Upper Ndolanya Beds deposition, albeit to a lesser degree.

We examined the relationship between habitat preference and body size by using a series of body size proxies, determined by examining modern antelope postcrania, and examining their relationship to habitat preferences in the modern and

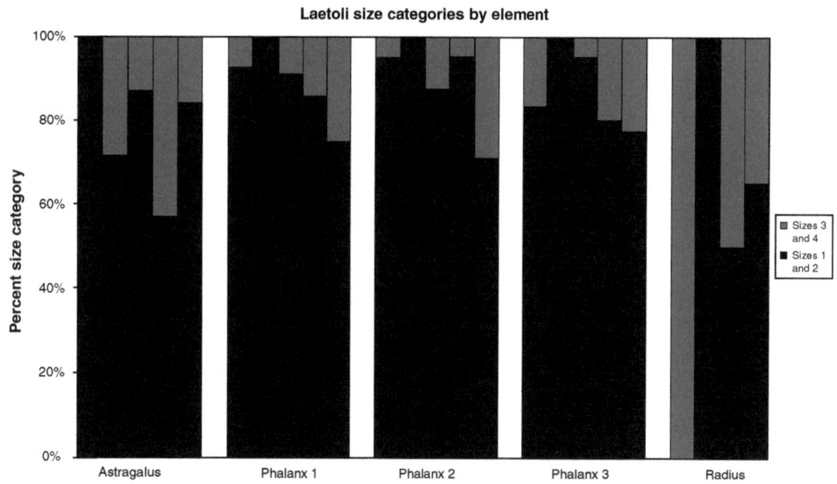

Fig. 17.3 Distribution of antelope body size classes by element, further divided by stratigraphic subdivision. This figure presents the data from Table 17.5 graphically. The dominance of smaller antelopes throughout the sequence is clearly seen. Very few size three and four antelopes are found in the small samples. The bars represent the stratigraphic divisions used throughout this chapter, from *left* to *right* – below Tuff 3, between Tuffs 3–5, between Tuffs 5–7, above Tuff 7, and stratigraphic position indeterminate. There are no radii from above Tuff 7 so that bar is omitted

fossil samples. Femoral length is a good predictor of body size in the Bovidae (Scott 1979, 1985; Kappelman 1988), and other measurements that are isometric with femoral length can also be used as body size proxies. We analyzed a variety of these proxies relative to habitat preference assignments to assess the extent to which our habitat reconstructions might be altered by size biases in the fossil assemblage.

We can then examine these body size abundances using size approximations for antelope species known from Laetoli. The fossil *Madoqua avifluminis* is approximately the same size as the modern dik dik (*M. kirkii*). The most common gazelle in the assemblage, *Gazella janenschi*, is also about the same size as the modern Thomson's gazelle (*G. thomsoni*). There are two fossil Alcelaphini known, one of which (*Parmularius pandatus*) is roughly the same size as the hartebeest (*Alcelaphus buselaphus*). As can be seen from the distribution of body sizes, the Upper Laetolil Beds are dominated by smaller antelopes, which are frequently classified as heavy cover or forest preferring. The model for the third (distal) phalanx classifies many more of the smaller phalanges to the forest rather than to the heavy cover category. This may reflect differences in the morphology of fossil versus modern *Madoqua* and potentially allows us to infer that extinct dik diks or their ancestors preferred more heavily forested environments than their modern representatives.

This is not to say that the habitat preferences determined here are biased; there are real biological and ecological reasons that animals that prefer more closed habitats tend to be small. Conversely, there are exceptions to this in the antelope world where some bovids preferring heavier cover are large bodied, particularly in the Reduncini, which are underrepresented at Laetoli. The preponderance of smaller bodied animals in the Laetoli assemblage is not due purely to a post-depositional winnowing; Su and Harrison (2007) demonstrated a bias in the preservation of postcranial remains of medium-sized bovids relative to their cranio-dental remains. However, the proportions of craniodental remains clearly indicate that *Madoqua* and *Gazella* were among the commonest bovids at Laetoli; small bovids dominated the original bovid community despite the apparent preservational bias (Su and Harrison 2007). This must, to some extent, reflect the reality of a more heavily vegetated setting for this Pliocene locality than was originally proposed (Leakey and Harris 1987; see Andrews 1989).

Discussion

Several conclusions can be inferred from this study. Ecomorphological analyses of antelope postcrania yield clear indications of habitat availability and how it changed through time when viewed within a stratigraphic and temporal analytical framework. Antelopes preferring forest and heavy cover habitats dominate the assemblage from the Upper Laetolil Beds. Although the overall proportion of forest and heavy cover antelopes might be altered by taphonomic biases (Su and Harrison 2007), the continuous presence of wooded and forested habitats during Upper Laetolil Beds deposition is clearly documented. The Upper Ndolanya Beds also show a relatively high proportion of antelopes preferring forest habitats. There are fewer heavy cover-preferring antelopes and an expanded number of light-cover and open habitat antelopes relative to the proportions during Laetolil Beds deposition.

The results of this analysis are in general agreement with other studies of paleoenvironments using a variety of indicators and proxies. First, studies of phytoliths from the Laetolil and Upper Ndolanya Beds confirm that grass cover, whilst common, is not a dominant feature of the Laetoli region throughout the sequence (Rossouw and Scott 2011). This accords well with the antelope ecomorphology, which shows the presence of antelopes preferring more closed habitats throughout the sequence. The continuous presence of a variety of habitat preferences also is supported by work on ungulate diets, which suggests that browsing was the dominant herbivore diet of the Laetoli succession; this is true even for ungulates that are today known to be obligate grazers (e.g., Alcelaphini) (Kaiser 2011). The rhinoceros taxon *Ceratotherium efficax* shows both temporal changes and significant variability in tooth morphology throughout the sequence suggesting that they had mixed grazing diets with variable amounts of browse (Hernesniemi et al. 2011). Results of studies using similar faunal indicators (Kovarovic and Andrews 2007, 2011) also support the presence of relatively heavy vegetation throughout the sequence, with an increase in more open habitats in the Upper Ndolanya Beds. When compared with modern habitat variability in similar geographic areas today, it would seem that regional variation and heterogeneity in habitat type was considerable during the deposition of the Laetoli sequence (see also Su and Harrison 2007; Kingston and Harrison 2007). Su and Harrison (2008) suggest that the high frequency of *Gazella, Madoqua,* and *Raphicerus*, with the presence of alcelaphines indicate a woodland-bushland-grassland mosaic; this is supported here. Our results show that environments at Laetoli always had the presence of forest and heavy cover habitats. This conclusion is inescapable; even the potential for taphonomic bias altering the relative portions of these more densely vegetated habitat types does not impact this view. The presence of these habitats is also in accordance with sedimentological studies suggesting that riverine environments ranging from ephemeral braided streams to larger channels were present during the formation of the Upper Laetolil Beds, potentially supporting riparian woodlands and forest (Ditchfield and Harrison 2011).

The main advantages of paleoenvironmental reconstructions using ecomorphological methods are three-fold. First, these methods allow us to decouple phylogeny and ecology to a certain degree; we can thus examine and reconstruct the adaptations of extinct species rather than assume them. Second, this approach requires minimal (family-level) taxonomic identification, allowing use of evidence that would not be exploited in more traditional paleontological and paleoecological reconstructions, because the postcranial fossils used are not typically associated with the craniodental remains that make genus or species-level identifications possible. Third, ecomorphological studies allow us to take relative abundance into account, as long as taphonomic biases are duly considered. The proportional representation of different ecomorphs may provide information about relative habitat abundance during the formation of a fossil assemblage, and document shifts in habitat representation through time in a sequence of isotaphonomic assemblages.

Acknowledgements The authors thank Terry Harrison for the invitation to work on this material and to contribute to this volume. Research permission was obtained from the Tanzania Commission for Science and Technology. LCB and KK acknowledge generous support from The Leverhulme Trust. TP acknowledges support from the Professional Staff Congress-City University of New York Research Award Program. Thanks also to the staff of the numerous museums where we measured antelopes for this study, particularly the American Museum of Natural History, the National Museum of Natural History, and The Natural History Museum (London). At the National Museum of Tanzania in Dar es Salaam, Amandus Kweka provided endless support and hospitality. Thanks to two anonymous reviewers whose comments helped us improve the manuscript. We are grateful to Julien Louys for assistance in data collection and analysis.

Appendix: Measurements and Ratios Used in This Study

Distal Radius

Measurements

LDARTML	log(mediolateral breadth of distal articular surface)
LDMAXAP	log(maximum anteroposterior diameter, distal end)
LDCENTAP	log(anteroposterior length of central facet, distal end)
LDCENML2	log(mediolateral breadth, most constricted point of the medial facet, distal end)
LDLATAP	log(anteroposterior length of lateral facet, distal end)
LDLATML	log(mediolateral length of lateral facet, distal end)
LDTUBML	log(distance between tuberosities, dorsal side of distal end)

Ratios

LDISI3	ldmaxml/ldcresap = log(maximum mediolateral breadth, distal end)/log(anteroposterior length of distal crest)
LDISI8	ldcentap/ldlatap = log(anteroposterior length of central facet, distal end)/log(anteroposterior length of lateral facet, distal end)
LDISI24	ldmaxap/ldtubml = log(maximum anteroposterior diameter, distal end)/log(distance between tuberosities, dorsal side of distal end)
LDISI25	ldtubml/ldcresap = log(distance between tuberosities, dorsal side of distal end)/log(anteroposterior length of distal crest)

Astragalus

Measurements

TARSMLT	Tarsal articulation M-L
TAMAP	Depth of medial portion of tarsal articulation
MINLEN	Minimum astragalus length
TUBTIBA	Tibial articulation to fibular tuberosity

Ratios

LENRAT11	Flangeml/medlen*100=ML across flange of medial trochlea/medial length
LENRAT21	Timap/medlen*100=Length of tibial articulation, medial side/medial length
PCFRAT24	Pcflap/tubml*100=posterior calcaneal facet, lateral length/mediolateral width at tuberosity
TICRAT2	Ticap/tilap*100=tibial articulation, minimum length/tibial articulation, lateral length
SIRAT3	Lcfdsi/tubtiba*100=distal lateral calcaneal facet, depth/tibial articulation to fibular tuberosity
TARAT5	Tarsmlm/Tarsmlt*100=tarsal articulation ML, medial side/tarsal articulation ML
TARAT9	Tarsmlm/tamap*100=tarsal articulation ML, medial side/depth of medial portion of tarsal articulation

Proximal Phalanges

Measurements

LPARTBR	log of the proximal articular breadth
LMAP	log of midshaft anteroposterior diameter
LDMAXAP	log of maximum anteroposterior dimension of the distal end
LDMINML	log of the minimum mediolateral dimension of the distal end

Ratios

LPROX1	lmaxlen/lminlen=log(maximum length)/log (minimum length)
LPROX2	lpmaxht/lpmaxbr=log(maximum height of proximal end)/log(articular breadth of the proximal end)
LPROX7	ldmaxml/ldminml=log(maximum mediolateral breadth of distal end)/log(minimum mediolateral breadth of distal end)
LPROX12	lminlen/lpmaxht=log(minimum length)/log (maximum height of the proximal end)
LPROX18	lminlen/ldminap=log(minimum length)/log (minimum anteroposterior dimension, distal end)
LPROX20	ldmaxml/lmml=log(maximum mediolateral breadth of distal end)/log(midshaft mediolateral diameter)
LPROX21	lmaxlen/lmml=log(maximum length)/log (midshaft mediolateral diameter)

Intermediate Phalanges

Measurements

LMAXLEN	log(maximum length)
LMML	log(midshaft mediolateral diameter)
LDARHMAX	log(maximum height, distal articular surface)

Ratios

LINT1	lmaxlen/lminlen=log(maximum length)/log (minimum length)
LINT4	lmap/lmml=log(midshaft anteroposterior diameter)/log(medshaft mediolateral diameter)
LINT5	ldmaxml/ldminm=log(maximum breadth, distal end)/log(minimum breadth, distal end)
LINT8	ldarhmax/ldminml=log(maximum height, distal articular surface)/log(minimum breadth, distal end)
LINT11	lmaxlen/lLS=log(maximum length)/log(superior length)
LINT17	lpmaxbr/ldmaxml=log(maximum breadth, proximal end)/log(maximum breadth, distal end)
LINT21	lmaxlen/lmml=log(maximum length)/log (midshaft mediolateral diameter)

Distal Phalanges

Measurements

LMAXLEN	log(maximum length)
LPMAXBR	log(maximum breadth, proximal end)
LLENSUP	log(superior length)
LMAXHT	log(maximum height)
LMPSB	log(breadth of the plantar surface at its midpoint)

Ratios

LDIS8	llensup/lpartht=log(superior length)/log(proximal articular height)
LDIS17	lpartht/lmpsb=log(proximal articular height)/log(breadth of the plantar surface, at its midpoint)

References

Andrews, P. J. (1989). Palaeoecology of Laetoli. *Journal of Human Evolution, 18*, 173–181.

Andrews, P., Lord, J., & Nesbit Evans, E. M. (1979). Patterns of ecological diversity in fossil and modern mammalian faunas. *Biological Journal of the Linnean Society, 11*, 177–205.

Bishop, L. C. (1994). Pigs and the ancestors: Hominids, suids, and environments during the Plio-Pleistocene of East Africa. Ph.D. dissertation, Yale University, New Haven.

Bishop, L. C. (1999). Suid paleoecology and habitat preference at African Pliocene and Pleistocene hominid localities. In T. G. Bromage & F. Schrenk (Eds.), *African biogeography, climate change and early hominid evolution* (pp. 216–225). Oxford: Oxford University Press.

Bishop, L. C., Hill, A., & Kingston, J. (1999). Paleoecology of Suidae from the Tugen Hills, Baringo, Kenya. In P. Andrews & P. Banham (Eds.), *Late Cenozoic environments and hominid evolution: A tribute to Bill Bishop* (pp. 99–111). London: Geological Society.

Brain, C. K. (1981). *The hunters or the hunted*. Chicago: University of Chicago Press.

Bunn, H. T. (1986). Patterns of skeletal representation and hominid subsistence activities at Olduvai Gorge, Tanzania, and Koobi Fora, Kenya. *Journal of Human Evolution, 15*, 673–690.

DeGusta, D., & Vrba, E. S. (2003). A method for inferring paleohabitats from the functional morphology of bovid astragali. *Journal of Archaeological Science, 30*, 1009–1022.

DeGusta, D., & Vrba, E. S. (2005). Methods for inferring paleohabitats from the functional morphology of bovid phalanges. *Journal of Archaeological Science, 32*, 1099–1113.

Ditchfield, P., & Harrison, T. (2011). Sedimentology, lithostratigraphy and depositional history of the Laetoli area. In T. Harrison (Ed.), *Paleontology and geology of Laetoli: Human evolution in context* (Geology, geochronology, paleoecology, and paleoenvironment, vol. 1, pp. 47–76). Dordrecht: Springer.

Estes, R. D. (1991). *The behavior guide to African mammals*. Berkeley: University of California Press.

Gentry, A. W. (1970). The Bovidae (Mammalia) of the Fort Ternan fossil fauna. In L. S. B. Leakey & R. J. G. Savage (Eds.), *Fossil vertebrates of Africa* (vol. 2, pp. 243–323). London: Academic.

Hernesniemi, E., Evans, A. R., Giaourtsakis, I. X., & Fortelius, M. (2011). Rhinocerotidae. In T. Harrison (Ed.), *Paleontology and geology of Laetoli: Human evolution in context* (Fossil hominins and the associated fauna, vol. 2, pp. 275–293). Dordrecht: Springer.

James, M. (1985). *Classification algorithms*. New York: Wiley.

Kaiser, T. M. (2011). Feeding ecology and niche partitioning of the Laetoli ungulate faunas. In T. Harrison (Ed.), *Paleontology and geology of Laetoli: Human evolution in context* (Geology, geochronology, paleoecology and paleoenvironment, vol. 1, pp. 329–354). Dordrecht: Springer.

Kappelman, J. (1986). The paleoecology and chronology of the middle Miocene hominoids from the Chinji Formation of Pakistan. Ph.D. dissertation, Harvard University, Cambridge.

Kappelman, J. (1988). Morphology and locomotor adaptations of the bovid femur in relation to habitat. *Journal of Morphology, 198*, 119–130.

Kappelman, J. (1991). The paleoenvironment of *Kenyapithecus* at Fort Ternan. *Journal of Human Evolution, 20*, 95–129.

Kappelman, J., Plummer, T. W., Bishop, L. C., Duncan, A., & Appleton, S. (1997). Bovids as indicators of Plio-Pleistocene paleoenvironments of East Africa. *Journal of Human Evolution, 32*, 95–129.

Kingston, J., & Harrison, T. (2007). Isotopic dietary reconstructions of Pliocene herbivores at Laetoli: Implications for early hominin paleoecology. *Palaeogeography, Palaeoclimatology, Palaeoecology, 243*, 272–306.

Kovarovic, K., & Andrews, P. (2007). Bovid postcranial ecomorphological survey of the Laetoli paleoenvironment. *Journal of Human Evolution, 52*, 663–680.

Kovarovic, K., & Andrews, P. (2011). Environmental change within the Laetoli fossiliferous sequence: Vegetation catenas and bovid ecomorphology. In T. Harrison (Ed.), *Paleontology and geology of Laetoli: Human evolution in context* (Geology, geochronology, paleoecology and paleoenvironment, vol. 1, pp. 367–380). Dordrecht: Springer.

Leakey, M. D., & Harris, J. M. (Eds.). (1987). *Laetoli: A Pliocene site in northern Tanzania*. Oxford: Clarendon.

Lewis, M. E. (1997). Carnivoran paleoguilds of Africa: Implications for hominid food procurement strategies. *Journal of Human Evolution, 32*, 257–288.

Plummer, T. W., & Bishop, L. C. (1994). Hominid paleoecology at Olduvai Gorge, Tanzania as indicated by antelope remains. *Journal of Human Evolution, 29*, 321–362.

Plummer, T. W., Bishop, L. C., & Hertel, F. (2008). Habitat preference of extant African bovids based on astragalus morphology: Operationalizing ecomorphology for palaeoenvironmental reconstruction. *Journal of Archaeological Science, 35*, 3016–3027.

Reed, K. E. (1997). Early hominid evolution and ecological change through the African Plio-Pleistocene. *Journal of Human Evolution, 32*, 289–322.

Reed, K. E. (1998). Using large mammal communities to examine ecological and taxonomic organization and predict vegetation in extant and extinct assemblages. *Paleobiology, 32*, 384–408.

Reyment, R. A. (1991). *Multidimensional palaeobiology*. New York: Pergamon.

Rossouw, L., & Scott, L. (2011). Phytoliths and pollen, the microscopic plant remains in Pliocene volcanic sediments around Laetoli, Tanzania. In T. Harrison (Ed.), *Paleontology and geology of Laetoli: Human evolution in context* (Geology, geochronology, paleoecology and paleoenvironment, vol. 1, pp. 201–215). Dordrecht: Springer.

Scott, K. M. (1979). Adaptation and allometry in bovid postcranial proportions. Ph.D. dissertation, Yale University, New Haven.

Scott, K. (1985). Allometric trends and locomotor adaptations in the Bovidae. *Bulletin of the American Museum of Natural History, 197*, 197–288.

Spencer, L. (1997). Dietary adaptations of Plio-Pleistocene Bovidae: Implications for hominid habitat use. *Journal of Human Evolution, 32*, 201–228.

Sponheimer, M., Reed, K. E., & Lee-Thorp, J. A. (1999). Combining isotopic and ecomorphological data to refine bovid paleodietary reconstruction: A case study from the Makapansgat Limeworks hominin locality. *Journal of Human Evolution, 36*, 705–718.

Su, D. F. (2005). The paleoecology of Laetoli, Tanzania: Evidence from the mammalian fauna of the Upper Laetolil Beds. Ph.D. dissertation. New York University, New York.

Su, D. F., & Harrison, T. (2007). The paleoecology of the Upper Laetolil Beds at Laetoli: A reconsideration of the large mammal evidence. In R. Bobe, Z. Alemseged, & A. K. Behrensmeyer (Eds.), *Hominin environments in the East African Pliocene: An assessment of the faunal evidence* (pp. 279–313). Dordrecht: Springer.

Su, D. F., & Harrison, T. (2008). Ecological implications of the relative rarity of fossil hominins at Laetoli. *Journal of Human Evolution, 55*, 672–681.

Van der Klaauw, C. J. (1948). Ecological studies and reviews. IV. Ecological Morphology. *Bibliotheca Biotheoretica, 4*, 27–111.

Werdelin, L., & Lewis, M. E. (2001). A revision of the genus *Dinofelis* (Mammalia, Felidae). *Zoological Journal of the Linnean Society, 132*, 147–258.

Chapter 18
Environmental Change within the Laetoli Fossiliferous Sequence: Vegetation Catenas and Bovid Ecomorphology

Kris Kovarovic and Peter Andrews

Abstract Here we investigate changes in Pliocene paleoecological conditions between the Upper Laetolil Beds (3.85–3.63 Ma) and Upper Ndolanya Beds (2.66 Ma) at Laetoli, Tanzania. We assess the ecomorphology (i.e., post-cranial adaptations that indicate habitat preferences) of all measureable specimens of the family Bovidae in both beds in order to reconstruct the likely habitat distribution during the deposition of both major geological units. Predicted habitat categories are defined and discussed in terms of extant vegetation associations found in the modern day Laetoli region. Our analyses indicate that the Upper Ndolanya Beds represent a more open environment than was present during deposition of the Upper Laetolil Beds one million years earlier. Additionally, we identify a trend during the 200 kyr deposition of the Upper Laetolil Beds, during which the proportion of woodland, bushland and grass-dominated habitat types shifts throughout the sequence.

Keywords Habitat • Paleoecology • Bovid • Ecomorphology • Vegetation • Catena

Introduction

Laetoli is an important Pliocene site located 36 km south of Olduvai Gorge, consisting of a series of fossil-bearing outcrops that cover an area of over 1,600 km^2. The sedimentary sequence at Laetoli consists of the Lower and Upper Laetolil Beds and the Lower and Upper Ndolanya Beds (Hay 1987). The Lower Laetolil Beds are mainly aeolian tuffs, but they are poorly fossiliferous and it has been difficult to reconstruct the paleoenvironment of this unit (Hay 1987). The most richly fossiliferous units comprise the Upper Laetolil Beds, which consist of aeolian and airfall tuffs dated to 3.85–3.63 Ma (Hay 1987; Drake and Curtis 1987; Deino 2011). They are subdivided into nine stratigraphic units by eight marker tuffs (Fig. 18.1), which enable local correlations of sedimentary sections that are separated from each other by vegetation. The marker tuffs within these beds are also useful subdivisions for contextualizing the fossil fauna within a temporal framework. The Upper Laetolil Beds are overlain unconformably by the lower unit of the Ndolanya Beds, which contains no fossils. Above this, and also separated by a disconformity, is the upper unit of the Ndolanya Beds which has a fauna distinct from those of the Upper Laetolil Beds and is dated to 2.66 Ma (Ndessokia 1990; Deino 2011).

Both beds have yielded hominin fossils. Numerous remains of *Australopithecus afarensis*, including the lectotype in the Upper Laetolil Beds (Leakey 1987; Leakey and Harris 1987; Harrison 2011a), and two specimens, one of which is identified as *Paranthropus aethiopicus*, in the Upper Ndolanya Beds (Harrison 2002, 2011). Of particular interest in the Upper Laetolil Beds is Tuff 7, which has yielded valuable information about the mammalian paleocommunity in the form of tracks preserved in ashfall deposits (Leakey and Hay 1979).

Australopithecus afarensis is the best known species of early hominin and it possesses a unique suite of characteristics indicating adaptations for both bipedal and arboreal movement in the locomotor repertoire and an unspecialized diet of herbivorous resources (e.g., Teaford and Ungar 2000). Because of its age and particularly well-studied morphology, it has been the focus of many questions pertaining to the evolution of novel hominin behaviors, including bipedalism (e.g., Ward 2002). Evidence from the Laetolil Beds figures heavily into these arguments, notably the Tuff 7 tracks, which feature footprints from three hominin individuals presumed to be made by *A. afarensis*, pointing unequivocally to the establishment of bipedal locomotion in this taxon (Leakey and Hay 1979; White and Suwa 1987). *Paranthropus aethiopicus*, the earliest of the robust australopiths, is a lesser known taxon, but available material, including an edentulous maxilla from the Upper Ndolanya Beds (Harrison 2002, 2011a), indicates

K. Kovarovic (✉)
Department of Anthropology, Durham University, Dawson Building, South Road, Durham, DH1 3LE, UK
e-mail: kris.kovarovic@durham.ac.uk

P. Andrews
Palaeontology Department, Natural History Museum, Cromwell Road, London, SW7 5BD, UK
e-mail: pjandrews@uwclub.net

T. Harrison (ed.), *Paleontology and Geology of Laetoli: Human Evolution in Context. Volume 1: Geology, Geochronology, Paleoecology and Paleoenvironment*, Vertebrate Paleobiology and Paleoanthropology, DOI 10.1007/978-90-481-9956-3_18,

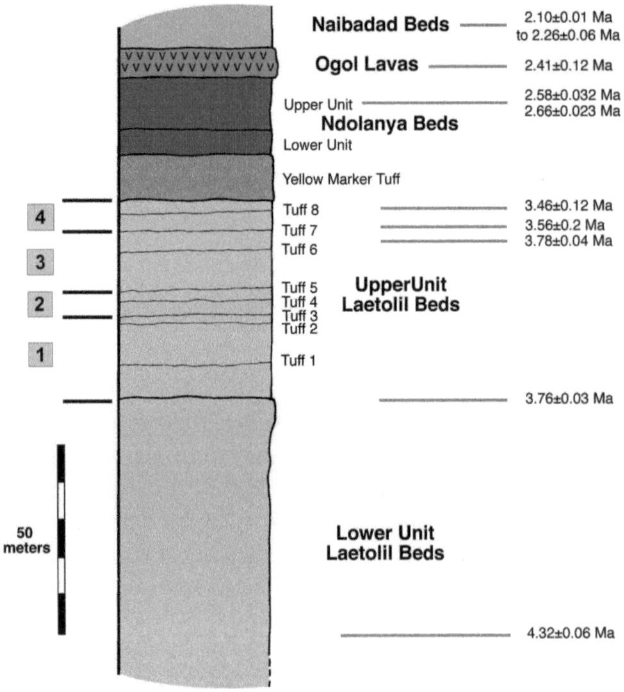

Fig. 18.1 General Laetoli stratigraphy (after Hay 1987; Drake and Curtis 1987; Ndessokia 1990; Manega 1993; Harrison 2005). Numbers *1–4* (on *left* side of column) indicate the units that have been considered in the ecomorphological analyses: Unit *1* = below Tuff 3, Unit *2* = between Tuffs 3 and 5, Unit *3* = between Tuffs 5 and 7, Unit *4* = above Tuff 7

that it was adapted to a specialized diet of tougher, more fibrous plant matter. There are no postcranial remains assigned to the taxon, but a proximal tibia, possibly attributable to this species, is known from the Upper Ndolanya Beds (Harrison 2002, Harrison 2011a). The interpretation of this species is that bipedalism was firmly established in this species, as it was in later members of the robust lineage.

Contrary to initial interpretations that the Upper Laetolil Beds represent an open "savanna" habitat (Leakey and Harris 1987), this time period at Laetoli is now understood to have been characterized by a mosaic of wooded habitats (Andrews 1989; Reed 1997; Andrews and Humphrey 1999; Harrison 2005; Kingston and Harrison 2007; Kovarovic and Andrews 2007; Su and Harrison 2007; Su 2011). The Upper Ndolanya Beds appear to have been quite different, with the area supporting a mammalian community adapted to more arid climates and less wooded habitats (e.g., Kovarovic et al. 2002; Kovarovic 2004; Kingston and Harrison 2007; Kaiser 2011; Rossouw and Scott 2011). Considering our revised views on the adaptive habitats of *A. afarensis* at Laetoli, and the difference in habitats one million years later when the robust *P. aethiopicus* ranged in the region, it is desirable that we investigate habitat change at a finer temporal resolution than a comparison of the Laetolil and Ndolanya Beds allow. We also take the view that environments in the past at Laetoli would have shown comparable variability of habitat to that present in the Laetoli region today (Andrews and Bamford 2008; see Andrews et al. 2011). The aim of this chapter is to investigate the variability and change through time of the habitat at Laetoli, combining fossil bovid postcranial ecomorphology with a catena-based habitat classification to take account of habitat variability.

Methods

Ecomorphological Analysis

Ecomorphological analyses relate characteristics of skeletal anatomy directly to the exploitation of particular habitat types (cf. Kappelman 1988). Bovidae have been particularly useful in such studies; this diverse family inhabits a wide range of global habitat types providing ample comparative material, and their remains are abundant at Pliocene hominin sites. The entire bovid appendicular skeleton was evaluated for each element's ability to predict the habitat type exploited by the individual, based on a sample of 205 extant adult bovids. This was supplemented by five tragulids and 14 cervids from Africa, Asia, Europe and North America, which together comprised the comparative dataset (Kovarovic and Andrews 2007). Linear measurements were taken on 21 elements (all long bones, phalanges, carpals and tarsals) from each bovid. These collections were housed in The Natural History Museum, London; Powell-Cotton Museum, Kent; the American Museum of Natural History, New York, and the National Museum of Natural History, Smithsonian Institution, Washington, DC

A discriminant function analysis (DFA) was conducted on each element, as well as the distal and proximal end of each long bone. This type of analysis predicts group membership (i.e., habitat) on the basis of a set of variables (i.e., anatomical measurements). The numbers of individuals that are correctly assigned to their known habitats are reported as a final percentage of the total. This percentage of correct classifications serves as a useful benchmark for "good" versus "bad" predictors. A total of 24 good predictors were identified (Kovarovic 2004). These analyses all yielded percentages of correct classification over the "baseline of accuracy", the percentage of individuals that would be classified correctly on the basis of chance alone (cf. DeGusta and Vrba 2003).

The 24 predictors from the Upper Laetolil and Upper Ndolanya Beds were analyzed. In contrast to DFAs on the modern data, which investigate whether the measurements could be used to correctly assign the individuals to their *known* habitat categories, DFAs of the fossil data predicted the habitats of the fossil specimens, which are of unknown habitat affiliation. Small sample sizes may obscure patterns in the data, so combining the results of all of the analyses regardless of the element, the total number of predictions in each habitat group can be calculated for each stratum in order to better observe a trend in habitat affiliation.

The Fossil Sample

The fossil collections are held at the National Museums of Tanzania, Dar es Salaam, Tanzania and were studied July-September 2002 and July-August 2005. The majority of the material is part of the collections accumulated during the 1998–2005 seasons of the Eyasi Plateau Paleontological Expeditions. A smaller number of the fossil specimens are part of the Mary Leakey collections that were originally housed at the camp at Olduvai Gorge, but now transferred to the National Museums of Tanzania in Dar es Salaam (Harrison 2011b).

A total of 360 specimens from the Upper Laetolil Beds and 170 from the Upper Ndolanya Beds were initially analyzed. However, although these elements form the basis for the predictive DFA models that fall above the baseline of accuracy, here we report only on a subset of these analyses, those that yielded percentages of correct classification over 50%. This still allows for adequate sample sizes and, more importantly, because the percentage of correct classification is higher, the probabilities associated with the predictions are likewise higher and therefore greater confidence can be placed in them (Kovarovic 2004). In total, there were 242 specimens from the Upper Laetolil Beds and 44 from the Upper Ndolanya Beds. Table 18.1 summarizes the fossil bovid data sample, and lists the sample size for each element measured, as well as providing a breakdown of the elements derived from the stratigraphic units described below. Complete long bones are rare and include only one humerus, two radii and three metatarsals, while carpals and phalanges in particular dominate both assemblages.

The Upper Laetolil Beds are thought to span approximately 200 kyr (Hay 1987; Deino 2011), and much faunal change could potentially have occurred during that time. The present analysis therefore investigates the fauna on an horizon-by-horizon basis. Unfortunately, sample sizes are too low to obtain representative faunas from every individual horizon, so the nine horizons have been grouped into four units as indicated in Fig. 18.1 and Table 18.1 (derived from Su 2005). The four units (and sample sizes) are: unit 1, below Tuff 3 ($n=77$); unit 2, between Tuffs 3 and 5 ($n=18$); unit 3, between Tuffs 5 and 7 ($n=88$); and unit 4, above Tuff 7 ($n=59$). The Upper Ndolanya Beds are approximately one million years later in time than the Laetolil Beds. They cannot be divided into smaller, discrete time intervals. The Upper Ndolanya fauna is therefore considered as one 2.66 Ma sample of 44 fossil elements.

Habitats and Vegetation Associations

The habitat classification used here is based on 16 modern plant associations present in the Laetoli area today, and they are linked with each other through three catenas. The catenas track vegetation changes due to variations in soil, topographic relief, altitude and moisture availability, and together they provide a fuller picture of the variability in habitat present in the area today. The emphasis here is on habitat variability rather than on one habitat type as opposed to another, and we consider that it provides a more accurate reconstruction of past environments when linked with the evidence available on ecomorphology of the bovid fauna.

Rationale for the Habitat Classification

Most ecomorphological analyses use a simple system of three or four habitat types: open, closed and intermediate or open, closed, light cover and heavy cover (Plummer and Bishop 1994; Kappelman et al. 1997; DeGusta and Vrba 2003, 2005; Plummer et al. 2008). These are not very precise, and Kovarovic (2004; Kovarovic and Andrews 2007) originally proposed a division of habitats into seven categories that were designed to apply to a wide geographic range. What we will attempt to do here is relate these categories to the local vegetation associations at Laetoli (however, here we have combined the two montane categories into one for a total of six habitats). There are two reasons for doing this. Firstly, the vegetation distribution of the Laetoli area is well documented (Andrews et al. 2011) and includes habitats ranging from evergreen forest to localized edaphic grassland, but the most common vegetation types are woodlands varying from low *Acacia* woodland to tall mixed species closed woodland, although all areas have been disturbed to some extent by clearing, burning and grazing domestic stock. Predictions have been made about the nature and extent of variation of these habitat types under differing climatic regimes (Andrews and Bamford 2008). In addition, the relationship between vegetation and landforms has been established for the region today, and the geomorphology proposed by Hay (1987) indicates similar landforms and altitude to those existing today, with generally flat topography cut by valleys.

The second reason for relating ecomorphology to the local environment at Laetoli is that available fossil evidence suggests that the distribution of Pliocene habitats at Laetoli was similar to what we see today, with possibly greater emphasis on wooded habitats. The evidence for this is based on a wide range of data: the taxonomic diversity of fossil mammals (contributions in Leakey and Harris 1987; Harrison 2011), the pollen spectra from the Upper Laetolil

Table 18.1 Summary of Laetoli fossil samples

HUMERUS (n=1)		DISTAL PHALANGES (n=47)	
Laetolil Beds	0	Laetolil Beds	43
Below Tuff 3	0	Below Tuff 3	6
Between Tuffs 3 & 5	0	Between Tuffs 3 & 5	3
Between Tuffs 5 & 7	0	Between Tuffs 5 & 7	24
Above Tuff 7	0	Above Tuff 7	10
Ndolanya Beds	1	Ndolanya Beds	4
RADIUS (n=2)		MAGNUM (n=19)	
Laetolil Beds	1	Laetolil Beds	15
Below Tuff 3	1	Below Tuff 3	3
Between Tuffs 3 & 5	0	Between Tuffs 3 & 5	1
Between Tuffs 5 & 7	0	Between Tuffs 5 & 7	7
Above Tuff 7	0	Above Tuff 7	4
Ndolanya Beds	1	Ndolanya Beds	4
METATARSAL (n=3)		UNCIFORM (n=10)	
Laetolil Beds	3	Laetolil Beds	6
Below Tuff 3	1	Below Tuff 3	1
Between Tuffs 3 & 5	0	Between Tuffs 3 & 5	0
Between Tuffs 5 & 7	2	Between Tuffs 5 & 7	4
Above Tuff 7	0	Above Tuff 7	1
Ndolanya Beds	0	Ndolanya Beds	4
PROXIMAL PHALANGES (n=85)		LUNAR (n=15)	
Laetolil Beds	76	Laetolil Beds	12
Below Tuff 3	42	Below Tuff 3	1
Between Tuffs 3 & 5	4	Between Tuffs 3 & 5	1
Between Tuffs 5 & 7	13	Between Tuffs 5 & 7	5
Above Tuff 7	17	Above Tuff 7	5
Ndolanya Beds	9	Ndolanya Beds	3
INTERMEDIATE PHALANGES (n=104)			
Laetolil Beds	86		
Below Tuff 3	22		
Between Tuffs 3 & 5	9		
Between Tuffs 5 & 7	33		
Above Tuff 7	22		
Ndolanya Beds	18		

Beds (Bonnefille and Riollet 1987) and the paleoflora, which is taxonomically similar to the present flora (Andrews and Bamford 2008; Bamford 2011). In addition, evidence from plant phytoliths indicates that although grasses were ubiquitous in the Laetoli succession, as they are today, they were never dominant, and it also indicates a range of C_3 and C_4 plants in the Pliocene environments at Laetoli with a shift towards the latter in the upper levels of the Upper Laetolil Beds and the Upper Ndolanya Beds (Rossouw and Scott 2011). In contrast to the phytoliths, however, isotopic studies of fossil dentition do not indicate any significant change in the availability of grassland, woodland and forest throughout the Laetoli succession, although sample sizes are small (Kingston and Harrison 2007). Finally, extensive evidence of termite activity in the Upper Laetolil Beds (Sands 1987; Darlington 2011) suggests that conditions in the Pliocene were comparable to those of today, with an indication that the climate might have been somewhat warmer and moister.

There are, however, several ways in which Pliocene environments could have differed from present day ones due to changes in climate, altitude, geomorphology and soils. Altitude probably was similar to that of today, but there may have been some uplift associated with the formation of the Eyasi fault, although it is not certain exactly when this tectonic event occurred (Hay 1987). Based on Hay's (1987) reconstruction paleotopographic differences within the study area appear to have been less than today, and the volcanic highlands to the east were probably still at an early stage of formation so that the rain shadow effect would have been less than it is now. The area had generally low relief, with seasonal river gullies extending westwards from the eastern highlands, and there was at least one deep valley, ~250 m in depth, which was cut into the basement rocks and which is deeper than the ones today having tall riverine woodland (Hay 1987). The overall effects of these differences would have produced minor changes if applied to modern vegetation distribution.

One major difference between the Pliocene and Present is seen in the soils. Much of the area today has black clay soils on areas of low relief, and on these areas drainage is impeded, which has major impact of the vegetation (Jager 1982). In the Pliocene, there is a series of weakly developed soils with calcretes, suggesting formation over relatively brief periods of time in a seasonal climate (Hay 1987). The sediments are derived from wind blown tuffs from the eastern volcanoes, and frequent ash falls would have buried the newly forming soils and caused some degree of habitat disturbance (Hay 1987). Periods of soil stability are indicated, however, by the presence of massive termite nests, as termite populations require periods of 5–8 thousands of years to build up and stabilize (Sands 1987). At Loc. 10W there are two successive termite nests, one above the other and separated by an estimated 8 kyr, and the overall period of stability indicated was an estimated 17 kyr. Periods of this length are consistent with the poorly developed soils, but they are more than sufficient for fully developed vegetation to become established according to climatic and edaphic constraints (Lind and Morrison 1974; de Wit 1978). Parts of the Laetoli area today are also subjected to deposition of wind blown tuffs from the active volcano Oldonyo Lengai, and where these ashes accumulate on flat surfaces, the distinctive grass/sedge flora of the short grass plains of the Serengeti develops (Anderson and Talbot 1965).

Pliocene environments may also have differed from the present as the result of climate change, which could have had two major impacts on habitats at Laetoli in the past. Temperature has a non-significant relationship with plants and accounts for only 28% of variation in mammals in southern Africa (Andrews and O'Brien 2000), but higher temperature during the Pliocene and Pleistocene could have resulted in the raising of montane vegetation limits, thus moving montane forest and the fringing escarpment woodlands further away from the area of deposition of the fossiliferous deposits. In contrast to temperature, annual precipitation accounts for 79% of the observed variation in plant species richness in southern Africa (O'Brien 1993), and any change in rainfall regime, whether annual amount or degree of seasonality, would have had a major impact on local environments. Higher rainfall would have depressed montane limits and increased woodland cover except on areas of impeded drainage, where woodland diversity and density would have decreased. Similarly, lower rainfall would have raised montane limits and decreased woodland diversity and density, with opposite effects again on areas of impeded drainage.

In summary, there is sufficient similarity in both flora and fauna between the Pliocene and Present to allow a greater degree of precision in their comparison. We will start from the general habitat descriptions used in Kovarovic and Andrews (2007) and then combine them with the actual habitat variation present in the Laetoli area today.

Present Habitats

The following is a summary of six habitat categories modified from Kovarovic (2004) and Kovarovic and Andrews (2007). These form the basis for the ecomorphological predictions.

1. *Grassland/tree-less - open grassland, tundra, steppe, desert and some semi-desert habitats*. Dominated by grasses and sedges not usually exceeding 1 m in height, although occasionally a height of 2–3 m can be obtained. Scattered woody cover does not usually exceed 2% of the ground cover.
2. *Wooded-bushed grassland* – Grasses dominate the ground vegetation and may grow to a height of 1–3 m. There is always some contribution of herbaceous growth, but epiphytes are uncommon. Open canopy of trees and shrubs, with 3–40% cover (White 1983). Trees have a single bole while shrubs have short, low, multi-stemmed branches (Aubréville 1963). This category includes semi-desert habitats, which present the same proportion of vegetation to open areas, but the ground vegetation includes more annual grasses and herbs.
3. *Light woodland-bushland* – Grasses and herbs dominate the ground vegetation, with 40–60% shrub and tree canopy cover. This category includes open woodland and open bushland.
4. *Heavy woodland-bushland* – Ground vegetation is lighter and with higher proportions of herbs to grasses, and shrub and tree cover is denser, with 61–75% canopy cover. This category includes dense thickets, woodland and bushland.
5. *Forest* – Forest habitats have continuous or nearly continuous canopy of interlocking tree crowns providing 76–100% cover. They generally have more than one storey and the dominant ground cover comprises herbs and shrubs if it is not bare (Menaut 1983; White 1983). Lichens, lianas and epiphytic plants such as mosses and ferns are also common (Pratt et al. 1966; Greenway 1973). The uppermost canopy height ranges from 6 or 7.5 m in dwarf forests to 40–50 m in true forests (Pratt et al. 1966; White 1983; Grunblatt et al. 1989).
6. *Montane habitats* – Habitats range from open, lightly covered by trees or other woody vegetation, including alpine habitats above the tree-line, to denser woodland and forest habitats in the mountains. Vegetation species differ from lowland forms by their adaptations to low temperatures and night-time frosts.

These six habitat types can now be matched against the actual vegetation variation observed at present in the Laetoli area. Sixteen vegetation associations have been mapped in the Laetoli area (Andrews et al. 2011), and they are grouped into three vegetation catenas.

1. *Woodland drainage catena.* In areas of low relief at Laetoli, there is a change in woodlands depending largely on soil type and degree of drainage, with elements of moisture availability but little change in altitude. Low woodland is present on areas of low relief with seasonally waterlogged soils. Down slope towards seasonal channels, where drainage is locally improved and moisture availability may be greater, taller riverine woodland and riverine bushland are present; and upslope, where soil type changes and topographic relief increases, and therefore better drainage exists, areas of tall deciduous woodland are present. In terms of the habitat classification above, the catena ranges from light woodland-bushland in the middle with heavy woodland-bushland both downslope in the dry river channels and upslope on the areas of greater topographic relief. A variation of this catena, which is so much disturbed at Laetoli that it can only be estimated, is the shift from wooded- bushed grassland on recent volcanic ash soils to low woodland on basement soils.
2. *Riverine woodland catena.* A second catena follows the length of the seasonal river channels at Laetoli and is related to moisture availability. The headwaters of the channels have low riverine bushland changing in succession down stream to taller riverine woodland with increasing numbers of species, until finally, if sufficient moisture is available, semi-evergreen gallery forest is developed. This catena ranges from light woodland-bushland at the headwaters of the river channels to heavy woodland-bushland down stream. In some cases, gallery forest may be present in deep valleys.
3. *Altitude catena.* A third catena tracks change in altitude eastwards from Laetoli towards the volcanic highlands, and it is assumed that the higher the altitude the greater the rainfall, a feature well documented for eastern Africa (Trapnell and Griffiths 1960). At 1,800 m, in the vicinity of most of the Laetoli fossiliferous sites, the deciduous woodlands at the top end of the woodland catena increase in species richness as altitude increases, and it changes dramatically to forest edge *Acacia* at around 2,100 m forming dense to closed canopy woodlands. Finally, above 2,450 m montane forest takes over. The altitude catena ranges from heavy woodland-bushland at lower altitudes to denser and taller woodlands with increasing altitude and eventually to montane forest.

Note that true grasslands do not form part of the variation, except in places where recent wind-blown volcanic ashes have accumulated. Where the ashes end and the underlying rock is exposed, there is an abrupt transition to low woodland and wooded bushed grassland (Anderson and Talbot 1965; Jager 1982; Andrews 1989).

Results of Ecomorphology Related to the Modern Laetoli Catenas

Table 18.2 presents the results of the ecomorphological habitat predictions from the four units in the Upper Laetolil Beds (units are shown in Fig. 18.1 and Table 18.1), the combined predictions from these units, and predictions from the Upper Ndolanya Beds. Considering the bovid fauna from the Upper Laetolil Beds as a composite sample, it shows a range of ecomorphologies, but the greatest proportion, 39.7%, is found in the heavy woodland-bushland category, followed by lower proportions in light woodland-bushland (17.4%) and wooded-bushed grassland (21.1%). A forest component is also represented (13.6%). This pattern is distinct from that of the Upper Ndolanya Beds, where the greatest number of predictions belongs to wooded-bushed grassland and represents 43.2% of the sample, followed by light woodland-bushland (29.5%) and heavy woodland-bushland (18.2%) with a minor forest component (4.5%).

Figure 18.2 illustrates the habitat predictions in the Laetoli sequence as a percentage of the total sample from each analyzed unit. The last two bars represent the combined Upper Laetolil Beds sample and the Upper Ndolanya Beds, where the distinction between the two beds is clear. It is particularly noticeable in the greater percentage of wooded-bushed grassland (black bar) and decreased heavy woodland-bushland (white bar) in the Upper Ndolanya Beds.

Looking at habitat predictions *within* the Laetolil Beds, heavy woodland-bushland is represented by the highest proportion in each unit (Fig. 18.2) and this varies by 13%; the highest is 46.8% below Tuff 3 and the lowest is 33.9% above Tuff 7 (Table 18.2). Similar variation is found in the other two best-represented categories, wooded-bushed grassland, ranging from 14.3% below Tuff 3 to 27.1% above Tuff 7 and light woodland-bushland ranging from 13.6% between Tuffs 5 and 7 to 22.2% between Tuff 3 and 5. There is also a much smaller but more stable forest component, comprising anywhere from 11.1% to 15.3% of the predictions.

Comparing these ecomorphological categories with the present-day vegetation catenas indicates that the Upper Laetolil environments formed part of a woodland catena that was similar to that present today, but probably with less differentiation. Local topography was flatter (Hay 1987) and drainage was probably better at this time, so that the catena would have ranged from tall open woodland to tall closed woodland. Acacias would have formed a minor element in these woodlands (Bonnefille and Riollet 1987), for evidence from the floras indicate they would have been composed of Sudano-Zambezian broad-leaved tree species growing to ~20 m (Bonnefille and Riollet 1987, Andrews and Bamford 2008, Bamford 2011). The river channels, probably dry for

Table 18.2 Number and percentage of Laetoli specimens predicted to belong to each habitat by unit

	Grassland/ tree-less	Wooded-bushed grassland	Light woodland-bushland	Heavy woodland-bushland	Forest	Montane	Total # of specimens
	n, %	n, %	n, %	n, %	n, %	n, %	n
LAETOLIL BEDS							
Below Tuff 3	0, 0%	11, 14.3%	15, 19.5%	36, 46.8%	11, 14.3%	4, 5.2%	77
Between Tuffs 3 & 5	0, 0%	3, 16.7%	4, 22.2%	8, 44.4%	2, 11.1%	1, 5.6%	18
Between Tuffs 5 & 7	0, 0%	21, 23.9%	12, 13.6%	32, 36.4%	11, 12.5%	12, 13.6%	88
Above Tuff 7	2, 3.4%	16, 27.1%	11, 18.6%	20, 33.9%	9, 15.3%	1, 1.7%	59
All tuffs combined	2, 0.8%	51, 21.1%	42, 17.4%	96, 39.7%	33, 13.6%	18, 7.4%	242
NDOLANYA BEDS							
	1, 2.3%	19, 43.2%	13, 29.5%	8, 18.2%	2, 4.5%	1, 2.3%	44

most of the year as they are today, would have had bushland in the headwaters extending down stream to *Acacia* riverine woodland and further down to tall closed riverine woodland, which would have included both tall *Acacia* species and broad-leaved trees, which by analogy with today could have included species of *Combretum* and *Euclea*. In at least one instance, the deep Olpiro valley identified by Hay (1987) would most likely have had perennial water, as do the deep valleys today, and this would have supported gallery forest in the deepest parts of the valley. Finally, the altitude catena would have been much like today, passing from woodland in the lower altitudes, similar to those of the woodland catena, to montane forest at higher altitudes, although the location of the transition zones is not known because relative altitudes are not known for this period. The proportions of these vegetation categories that are suggested by the proportions of ecomorphologies indicate that the woodland catena was most extensive in the Laetoli area during the time of deposition of the Upper Laetolil Beds. The areas of forest indicated by the ecomorphology could either have been riverine gallery forest or the montane forests in the east of the Laetoli area.

Ecomorphological Proportions within the Laetoli Succession

The ecomorphological analysis of the subdivisions of the Laetoli bovid faunas show that all faunas have summed ecomorphologies denoting woodland environments (i.e., light heavy woodland-bushland together) ranging from approximately 50–67% of the fauna in the four units of the Upper Laetolil Beds compared to 47.7% for the Upper Ndolanya fauna (Table 18.2). As noted above, this is a clear difference between the Upper Laetolil and the Upper Ndolanya Beds where the largest single habitat type (43.2%) is the wooded-bushed grassland category (Fig. 18.2), suggesting an increase in habitats at the more open end of the vegetation cover spectrum. A 2-tailed Z-test for two proportions was conducted for three habitat categories (sample sizes are too small in the grassland/tree-less, forest and montane categories) in order to obtain some idea as to the importance of these observed shifts in the habitat proportions between the beds (Table 18.3a). The increase in wooded-bushed grassland ($p = .002$) and the decrease in heavy woodland-bushland ($p = .006$) are both significant (where significance is set at $p < .05$) and the increase in light woodland-bushland is approaching significance ($p = .059$). These results suggest that the habitat shift towards less woodland was both statistically and ecologically significant.

The Upper Ndolanya Beds have previously been interpreted to indicate open woodland and bushland with more extensive grass cover (Kovarovic et al. 2002; Andrews 2006; Kovarovic and Andrews 2007). This shift of habitat in the Ndolanya Beds could well be due to change in climate either regionally with changing global climate, or locally due to the expansion of the eastern volcanic highlands, reducing the effects of the south eastern monsoon. Grass is, of course, the dominant ground cover of most woodland or wooded grassland associations, but generally speaking the denser the woodland canopy the less grass and more herbs.

Within the four Upper Laetolil Beds units, the ecomorphology of the bovid faunas suggest a shift in the woodland catena after Tuff 5, with a proportional change to more open vegetation. All four units have he highest proportion of vegetation represented by the heavy-woodland-bushland category, and in the lowest two units, light woodland -bushland also has a higher proportion than wooded-bushed-grassland. Above Tuff 5, however, the wooded-bushed grassland elements are greater than the light woodland-bushland, and above Tuff 7 there is a marked increase in wooded-bushed-grassland and reduction in the heavy and light woodlands (Fig. 18.2). Investigating the statistical significance of the trends between each chronological unit in the Laetolil Beds is hampered by low sample sizes in the habitat categories between Tuffs 3 and 5, as these violate the assumptions of a number of relevant

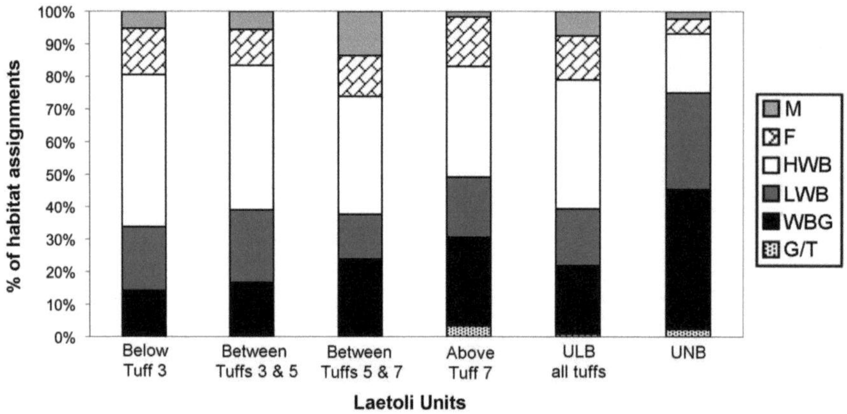

Fig. 18.2 Habitat predictions of the Laetoli bovid sample indicate a range of ecomorphologies across the open to closed gradient of vegetation cover in both the Upper Laetolil and Upper Ndolanya Beds. The proportions of habitat predictions indicate that the Upper Ndolanya Beds (*UNB*) had a higher percentage of more open adapted bovids, while the Laetolil Beds were biased towards heavy woodland-bushland. However, the proportion of wooded-bushed grassland increases up through the Laetolil sequence. *G/T* = grassland/tree-less, *WBG* = wooded-bushed grassland, *LWB* = light woodland-bushland, *HWB* = heavy woodland-bushland, *F* = forest, *M* = montane

Table 18.3 (a) Results of the 2-tailed Z-tests for two proportions comparing the differences in the percentage of individuals in three habitat categories between the Upper Laetolil and Upper Ndolanya Beds. The increase in wooded-bushed grassland and the decrease in heavy woodland-bushland are found to be significant (significance = $p < .05$ and significant p-values are in bold). The increase in light woodland-bushland is approaching statistical significance. (b) Results of the 2-tailed Z-tests for two proportions comparing differences in the percentage of individuals in two habitat categories at the beginning and end of the Laetolil Beds sequence. Neither are significant (significance = $p < .05$), but the increase in wooded-bushed grassland approaches it

(a)	Laetolil Beds	Ndolanya Beds
Wooded-bushed grassland		
n, %	51, 21.1%	19, 43.2%
Z	−3.137	
p	.002	
Light woodland-bushland		
n, %	42, 17.4%	13, 29.5%
Z	1.887	
p	.059	
Heavy woodland-bushland		
n, %	96, 39.7%	8, 18.2%
Z	2.726	
p	.006	
(b)	**Laetolil Beds below Tuff 3**	**Laetolil Beds above Tuff 7**
Wooded-bushed grassland		
n, %	11, 14.3%	16, 27.1%
Z	1.859	
p	.063	
Heavy woodland-bushland		
n, %	36, 46.8%	20, 33.9%
Z	1.51	
p	.131	

tests. However, a Z-test for two proportions can be applied to the ecomorphologies represented in the two habitat categories that exhibit the greatest amount of change from the beginning to the end of the Laetolil Beds sequence: wooded-bushed grassland (14.3% below Tuff 3 and 27.1% above Tuff 7) and heavy woodland-bushland (46.8% below Tuff 3 and 33.9% above Tuff 7). At 95% confidence threshold, the differences in these proportions are not quite significant, with $p = .063$ for the shift in wooded-bushed grassland and $p = .131$ for the shift in heavy woodland-bushland (Table 18.3b).

All of the Upper Laetolil Bed faunas have a low and stable forest signal varying by only 4.2% throughout the sequence. This suggests that the riverine or montane elements of the catenas, both of which include elements of forest, changed little through the Upper Laetolil sequence. This in turn could indicate little climate change during the time of deposition of the Upper Laetolil Beds, for variations in the occurrence of forest in both the riverine catena (more moisture, more forest) and the altitude catena (lower temperatures and/or more moisture, more forest) would be the product of variations in climate. It would take a major climatic shift to extend the dominant woodland catena into forest, and since the forest proportions do not appear to have changed through the Laetolil Bed sequence it would appear to rule out any *major* climatic change at this time. However, the shift in the woodland catena in the later part of the sequence indicates that a gradual directional change towards more open woodlands and greater grass cover occurred, and this is likely to have been accompanied by increased aridity. This trend continued into the Upper Ndolanya Beds as shown above.

Faunal Differences in the Laetoli Sequence

The results of our analysis highlight two main points about the paleoenvironment in the Laetoli region between 3.85 Ma and 2.66 Ma. Firstly, there is a distinct difference in the faunas of the Upper Laetolil Beds and the Upper Ndolanya Beds one million years later, both in terms of species composition (Su and Harrison 2007, 2008) and ecomorphology. Secondly, there is some evidence for a shift in ecomorphology within the Upper Laetolil Beds.

The Laetolil-Ndolanya Transition

The transition between the Upper Laetolil and Upper Ndolanya Beds has long been recognized as having been characterized by faunal change (Leakey and Harris 1987). Taxonomically the mammalian communities differ and the Laetolil Beds have a higher diversity of taxa. At the family level, bovids and lagomorphs dominate the Laetolil Beds. Bovids alone dominate the Ndolanya fauna, which is generally characterized by a high number of medium and large-sized herbivores, particularly grazers (Kovarovic et al. 2002). How much of this is due to taphonomic bias is difficult to assess, and there is clearly some loss of smaller species in both beds. However, recent total surface collection methods that have been practiced at the site since the publication of the original Laetoli monograph (Leakey and Harris 1987) would mitigate in some respects against the "invisibility" of the smaller taxa that are less likely to survive (see Harrison 2011b).

Mammal community-based analyses of the two beds have indicated a clear difference between the trophic and locomotor adaptations of their representative faunas (Andrews 1989; Reed 1997; Andrews and Humphrey 1999). The Laetolil community structure bears the greatest resemblance to the wooded end of the spectrum of extant Serengeti habitats, whilst the Upper Ndolanya fauna (bovids, in particular) appears adapted for the exploitation of habitats dominated by short scrubby growth and is interpreted to indicate a semi-arid bushland environment with a small woodland component. There is no significant evidence for permanent water at Laetoli at any time period (Hay 1987), although seasonal streams and rivers would have supported heavy wooded growth in the Laetolil Beds, as well as the lesser component of such in the Ndolanya Beds (Ditchfield and Harrison 2011).

These conclusions are qualified by other sources of evidence. A number of woodland adapted taxa, frugivorous species and a diverse browsing community are present in the Laetolil Beds including tragelaphines, cephalophines, the bush squirrel *Paraxerus*, the large elephant shrew *Rhynchocyon*, and chalicotheres, which although quadrupedal are understood to have used their hindlegs for balance as they reached for and fed on a browsing diet of young shoots and leaves (Chavanon 1962; Butler 1987; Denys 1987; Guérin 1987; Harris 1987). The incidence of *Madoqua*, which is abundant at Laetoli and requires dense thicket and undergrowth in which to retreat when alarmed, further indicates wooded conditions (Gentry 1987), as does the diversity of the primate community, which includes a bushbaby, two colobines and two cercopithecines, and which would have required a significant tree canopy in which to seek shelter and safety (Leakey and Delson 1987; Walker 1987; Harrison 2011c, d). On the other hand, the presence of gazelles and alcelaphines indicates some more open habitat and, like the present-day habitats, the Laetoli environments are best indicated by the woodland catenas described above. This allows for the transition from open woodland (light woodland-bushland or even wood-bush-grassland) to closed woodland along edaphic or topographic transects. There is also evidence from the invertebrates that there was thick vegetation or forest in the region (Verdcourt 1987; Tattersfield 2011). Additionally, the pollen signature presents a higher proportion of Afro-montane elements than is known in the modern pollen rain, indicating that wooded and/or forested regions were more important during the Laetolil time period (Bonnefille and Riollet 1987).

The Upper Ndolanya mammal species are related to those today living in bushland or bush-woodland mosaic habitats. There is a dominance of open country antilopines and alcelaphines in the bovid community and the equid present is more hypsodont than the earlier Laetolil form, indicating an increased reliance on grazing (Harris 1987; Hooijer 1987a, b; Kaiser 2011). Isotopic signatures of the paleosol carbonates in fact indicate grass dominated habitats, in contrast to the woodland signatures from the Laetolil Beds (Cerling 1992). Furthermore, allowing for some taphonomic loss of micromammals, the number of species present in the Upper Ndolanya Beds approaches the number often found in tropical semi-arid bushlands that lack a permanent water source (Andrews et al. 1979). However, although not common in the Upper Ndolanya fauna, the presence of primates and the rodent genus *Thallomys*, which lives in *Acacia* woodland (Kingdon 1997), indicate that some limited tree canopy would have been available to the Upper Ndolanya community. The presence of *Thryonomys*, which has a diet of grass in wetland habitats (Kingdon 1997), also suggests that the seasonal rivers may have been enough to support the presence of small populations of this species.

In contrast to this, recent isotopic reconstructions of the dietary regimes of Laetoli herbivores found limited support for significant change (Kingston and Harrison 2007). However, the two Ndolanya alcelaphines included in the data sample were among the highest $\delta^{13}C_{enamel}$ values, and a number of other taxa also presented values that were ^{13}C-enriched, albeit not statistically significant. These are possible indications of an increased grazing component and/or increased aridity, although it has been suggested that it may relate to an influx of migratory grazing species from beyond the Laetoli area (T. Harrison, personal communication). Ecomorphological analysis cannot inform on this issue, for if a resident community of bovids from a woodland habitat were combined with a group of migratory grazing herbivores, it could produce a signal comparable to a bushland. Analysis of the Upper Ndolanya herbivore species indicates that they are divided into 56% grazers and 44% browsers (Kovarovic et al. 2002), but this would not of itself change the ecological signal from woodland to semi-arid bushland, for which a mixture of 75% or more grazing species would be necessary (Andrews 2006). Additionally, a number of Laetolil taxa are $\delta^{18}O$ depleted, suggesting greater humidity during the deposition of these beds in contrast to the Ndolanya Beds (Kingston and Harrison 2007).

Environmental Change within the Upper Laetolil Beds

The ecomorphological analysis presented here suggests a transition in the vegetation resource base exploited by the bovid community. While the forest postcranial ecomorphologies of the Laetolil bovids remain relatively equally represented throughout the sequence, there is a shift in the percentage of ecomorphologies indicating less wooded habitats half-way though the succession. In particular, heavy woodland-bushland steadily decreases and the light woodland-bushland gives way to a greater proportion of wooded-bushed grassland in unit three (Fig. 18.2). This implies a shift in the woodland vegetation from more closed to more open woodland, both forms still being present but the former becoming more widespread at the expense of the latter. On the other hand, it has been shown that mammalian species diversity remains relatively constant throughout the sequence (Su 2005), as does the woody vegetation (Bamford and Harrison 2004). Furthermore, both $\delta^{13}C_{enamel}$ and $\delta^{18}O_{enamel}$ values indicate that there is a consistently high amount of variability, more so than that observed in extant lineages, in the foraging regimes of the Laetolil herbivores regardless of the stratigraphic unit from which they are derived (Kingston and Harrison 2007). Only giraffes and equids adhered to any specific feeding pattern, with the former yielding obligate browsing and the latter variable to obligate grazing signals, but no other significant trends within the succession could be identified. Within some taxa, however, there is evidence of change; for example, in the dietary regime of Laetolil tragelaphines. Those early in the sequence yield both browsing and mixed diet signals, but higher in the sequence they change to more grazing diets (Kingston and Harrison 2007).

In order to illustrate the ecomorphological trends more clearly, we removed the montane predictions from consideration (Fig. 18.3). The species on which the montane predictions are based, mainly members of Caprinae, were included in the discriminant function analyses in order to capture the morphological diversity of the entire bovid family and create predictive models applicable across a global array of ecologies (Kovarovic and Andrews 2007). The absence of both mountains and caprines at Laetoli renders it unlikely that the Laetoli bovids are, in fact, mountain dwellers, so that viewing the five remaining habitat categories in isolation is more appropriate to Plio-Pleistocene East Africa.

The pattern identified in the ecomorphologies is now clearer: above Tuff 5 there is a shift towards the wooded-bushed grassland component of the woodland catena, replacing light woodland-bushland as the second habitat component after heavy woodland. What is also apparent is a trend across all of the units in which the heavy woodland steadily diminishes towards the more open woodland, approaching the more open end of the woodland catena. Approximately 47% of the predictions below Tuff 3 are closed woodland, and this decreases to just below 35% above Tuff 7 (Fig. 18.3). Bovids adapted for grassland/tree-less habitats appear only after Tuff 5 where they add to the general trend towards more open habitats in the Upper Laetolil Beds. The forest component remains stable throughout the sequence of the Upper Laetolil Beds at roughly 15%. There is a slight decrease in the number of forest predictions between Tuffs 3 and 5, but this is the smallest sample size of any unit in the analysis and it is difficult to know how robust that observed pattern is. The forest component indicated by the ecomorphology may be attributed in part to the likely presence of gallery forest in the valleys and possibly also if there were springs emerging in the area as they do today, and in both cases they form parts of the riverine catena described above, merging into woodland as moisture availability decreases away from the valley or spring source.

The identification of these trends in the Upper Laetolil Beds may be confounded by a small sample size in unit two, between Tuffs 3 and 5, which amounts to only 18 specimens across all six of the habitat types, as opposed to the larger samples sizes (between 59 and 88) in the other units. The addition of new fossils derived from sediments between Tuffs 3, 4 and 5 may change the overall representation of the habitat categories half way through the sequence and the shift we observe at this point may prove to be less severe than how it currently appears (Fig. 18.3). Regardless of this, the sample sizes in unit one and four, at the beginning and end of the Laetolil Beds deposition, are greater and the distributions indicate a decrease in heavy woodland-bushland and an increase in wooded-bushed grassland over the 200 kyrs of sediment deposition. The lack of statistical significance between the differences in the proportions of individuals

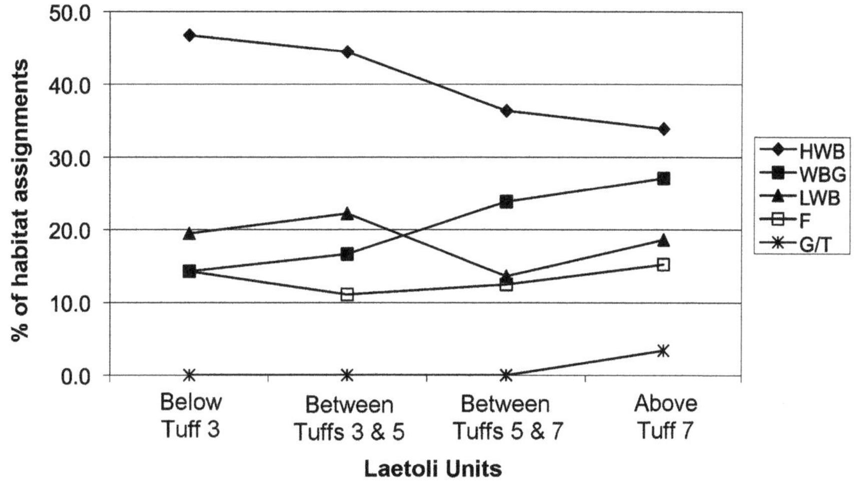

Fig. 18.3 Line graph showing the habitat predictions of the bovid sample in each chronological unit of the Upper Laetolil Beds. The montane category has been removed in order to make trends within the most relevant habitats more clear. After Tuff 5 the wooded-bushed grassland component becomes greater than the light woodland-bushland component. In addition to this, it can be seen that the heavy woodland-bushland component gradually decreases in each unit over time. *G/T* = grassland/tree-less, *WBG* = wooded-bushed grassland, *LWB* = light woodland-bushland, *HWB* = heavy woodland-bushland, *F* = forest

in these habitats in these units (as determined by the Z-tests above) is, in these instances, perhaps not as meaningful as one might normally attribute to it. The addition of just one more individual in the wooded-bushed grassland category above Tuff 7 makes the increase in the representation of this habitat type statistically significant with a p-value of.038, opposed to.063 (Table 18.3b) and the removal of three individuals from heavy woodland-bushland above Tuff 7 would likewise make the decrease in this habitat category significant so that $p = .036$ rather than .131 (Table 18.3b). With the attainment of statistical significance simply a matter of the addition or removal of such a small number of specimens, we argue that the lack of statistical significance in these cases does not equate to a corresponding lack of ecological significance. The transition in the catenas that is suggested by the bovid ecomorphologies is supported by other lines of evidence, and we conclude that it was real and ecologically meaningful, although the nuances of the transition may not be clear in the middle of the sequence. Insofar as ecomorphological indications are concerned, only an increased sample size in unit two would resolve this.

Laetoli Hominins

The majority of the paleoecological evidence from *A. afarensis* sites indicates that this species frequented "edge" areas between the forests and dense woodlands that fringe rivers and lakes and the more open woodland and bushland beyond them (Reed 1997, 2008). Reed (2008) has shown that in the ~1.5 million year succession at Hadar, *Australopithecus afarensis* has been recovered from all of the collection units below the BKT-2, that is before 2.94 Ma. Habitats with medium density tree and/or bush cover dominated the landscape through most of the earlier time period, with the Sidi Hakoma Member having the most closed habitat. The overlying Denen Dora Member had higher frequency of reduncines, possibly indicating the presence of flood plain grasslands, while the Kada Hadar Member had an influx of ungulates that indicate a more arid environment. The Hadar Formation had open environments, wooded grassland and some floodplain environments (Reed 2008). *Australopithecus afarensis* was common in the Sidi Hakoma Member, but it formed part of the fauna in all four members so that it appears to have had a relatively broad habitat preference. *Australopithecus afarensis* has also been recovered from Maka, Ethiopia, and the associated fauna indicates a habitat that is similar to that of the upper Denen Dora Member, namely bushland or woodland with floodplain grassland. In all cases, the nearby presence of water is indicated (Reed 1997, 2008).

The only site that does not entirely agree with this scenario is Laetoli, which has no permanent water source, although seasonal streams would have existed for periods of time during the year (Hay 1987). Water may have been present all year in the deep valley mentioned above (Hay 1987), and there may have been springs fed by rainfall on the eastern volcanic highlands, as there are today, but clearly this aspect of the environment at Laetoli was very different from that throughout the Hadar sequence. The reconstruction presented here of the Laetoli environment indicates that vegetation cover was also an important aspect of the habitat of *A. afarensis* (see also Su and Harrison 2008).

Most of the Laetoli hominins were found in the deposits between Tuffs 5 and 8, although the exact level is not always clear. Leakey (1987) gives the following figures: 2 specimens below Tuff 3, 1 specimen between Tuffs 3 and 5, 16 specimens between Tuffs 5 and 7 and 7 specimens above Tuff 7. Hominins therefore appear to span the environmental shift in the Laetolil sequence, but are most abundant between Tuffs 5 and 8, and this appears to indicate that *A. afarensis* may not have experienced the long term habitat stability that is indicated by species diversity indices and other evidence (Su 2005; Kingston and Harrison 2007). Faunal change occurred through the sequence of the Laetolil Beds, and analyzing these faunas as a composite sample representing a homogenous environment may not be appropriate.

Paranthropus aethiopicus sites are intermediate in vegetation cover between those of earlier australopithecines, which are moderately to densely vegetated, and those of later robust species, which are generally more open (Reed 1997). Often found in deltaic environments with nearby woodland-bushland and edaphic grasslands, the exception is once again Laetoli where *P. aethiopicus* survived in a region without a large permanent water source or edaphic grasslands (Reed 1997). However, the Upper Ndolanya habitat was similar to other *P. aethiopicus* sites in terms of the amount of vegetation cover it provided. The Upper Ndolanya Beds do not represent the same kind of time depth as the Laetolil Beds, so there is no way to evaluate if there were ecological shifts around this time period or how *P. aethiopicus* might have fared. But, for this one brief window at 2.66 Ma, we know that the ecological conditions in the Laetoli region supported at least a small population of this taxon.

Acknowledgements We would like to extend our sincere thanks to Leslie Aiello, Marion Bamford, Laura Bishop, Tom Plummer, Christophe Soligo and Brian Villmoare. Thank you to Terry Harrison and Denise Su for access to Laetoli material and for figures used herein. Thanks also to the 2001 Eyasi Plateau Paleontological and Geological Project field crew, the National Museums of Tanzania, American Museum of Natural History, New York, National Museum of Natural History, Washington, DC, The Natural History Museum, London, and the Powell-Cotton Museum, Kent. This research was supported by an Overseas Research Studentship, the UCL Graduate School (to K. Kovarovic), Wenner-Gren Foundation (to P. Andrews) and the NSF (to T. Harrison, BCS-9903434 and BCS-0309513).

References

Anderson, G. D., & Talbot, L. W. (1965). Soil factors affecting the distribution of the grassland types and their utilization by wild animals on the Serengeti plains, Tanganyika. *Journal of Ecology, 53*, 33–56.

Andrews, P. J. (1989). Palaeoecology of Laetoli. *Journal of Human Evolution, 18*, 173–181.

Andrews, P. J. (2006). Taphonomic effects of faunal impoverishment and faunal mixing. *Palaeogeography, Palaeoclimatology, Palaeoecology, 241*, 572–589.

Andrews, P., & Bamford, M. (2008). Past and present vegetation ecology of Laetoli, Tanzania. *Journal of Human Evolution, 241*, 572–589.

Andrews, P. J., & Humphrey, L. (1999). African Miocene environments and the transition to early hominines. In T. G. Bromage & F. Schrenk (Eds.), *African biogeography, climate change, and human evolution* (pp. 282–300). Oxford: Oxford University Press.

Andrews, P. J., & O'Brien, E. M. (2000). Climate, vegetation, and predictive gradients in mammal species richness in southern Africa. *Journal of Zoology, 251*, 205–231.

Andrews, P. J., Lord, J. M., & Nesbit Evans, E. M. (1979). Patterns of ecological diversity in fossil and modern mammalian faunas. *Biological Journal of the Linnean Society, 11*, 177–205.

Andrews, P., Bamford, M. K., Njau, E. F., & Leliyo, F. (2011). The ecology and biogeography of the Endulen-Laetoli area in northern Tanzania. In T. Harrison (Ed.), *Paleontology and geology of Laetoli: Human evolution in context* (Geology, geochronology, paleoecology and paleoenvironment, vol. 1, pp. 167–200). Dordrecht: Springer.

Aubréville, A. (1963). Classification des formes biologiques des plantes vasculaires en milieu tropical. *Adansonia, 3*, 221–226.

Bamford, M. K. (2011). Fossil wood. In T. Harrison (Ed.), *Paleontology and geology of Laetoli: Human evolution in context* (Geology, geochronology, paleoecology and paleoenvironment, vol. 1, pp. 217–233). Dordrech: Springer.

Bamford, M. K., Harrison, T. (2004). Pliocene vegetation and palaeoenvironment of Laetoli, Tanzania evidence from fossil woods. *PaleoAnthropology Selected Issue*: PAS 2004 Abstracts: A02.

Bonnefille, R., & Riollet, G. (1987). Palynological spectra from the Upper Laetolil Beds. In M. D. Leakey & J. M. Harris (Eds.), *Laetoli: A Pliocene site in northern Tanzania* (pp. 52–61). Oxford: Clarendon.

Butler, P. M. (1987). Fossil insectivores from Laetoli. In M. D. Leakey & J. M. Harris (Eds.), *Laetoli: A Pliocene site in northern Tanzania* (pp. 85–87). Oxford: Clarendon.

Cerling, T. E. (1992). Development of grasslands and savannas in East Africa during the Neogene. *Palaeogeography, Palaeoclimatology, Palaeoecology, 97*, 241–247.

Chavanon, S. (1962). Les Chalicotheriidae du Bassin Aquitain, Observation morphologiques, biométriques et éthologiques. Ph.D. dissertation, University of Bordeaux, Bordeaux.

Darlington, J. P. E. C. (2011). Trace fossils interpreted in relation to the extant termite fauna at Laetoli, Tanzania. In T. Harrison (Ed.), *Paleontology and geology of Laetoli: Human evolution in context* (Fossil hominins and the associated fauna, vol. 2, pp. 555–565). Dordrecht: Springer.

DeGusta, D., & Vrba, E. S. (2003). A method for inferring paleohabitats from the functional morphology of bovid astragali. *Journal of Archaeological Science, 30*, 1009–1022.

DeGusta, D., & Vrba, E. S. (2005). Methods for inferring paleohabitats from the functional morphology of bovid phalanges. *Journal of Archaeological Science, 32*, 1099–1113.

Deino, A. L. (2011). $^{40}Ar/^{39}Ar$ dating of Laetoli, Tanzania. In T. Harrison (Ed.), *Paleontology and geology of Laetoli: Human evolution in context* (Geology, geochronology, paleoecology and paleoenvironment, vol. 1, pp. 77–97). Dordrecht: Springer.

Denys, C. (1987). Fossil rodents (other than Pedetidae) from Laetoli. In M. D. Leakey & J. M. Harris (Eds.), *Laetoli: A Pliocene site in northern Tanzania* (pp. 118–170). Oxford: Clarendon.

De Wit, H.A. (1978). *Soils and grassland types of the Serengeti Plain (Tanzania)*. Ph.D. dissertation, University of Wageningen, Wageningen.

Ditchfield, P., & Harrison, T. (2011). Sedimentology, lithostratigraphy and depositional history of the Laetoli area. In T. Harrison (Ed.), *Paleontology and geology of Laetoli: Human evolution in context* (Geology, geochronology, paleoecology and paleoenvironment, vol. 1, pp. 47–76). Dordrecht: Springer.

Drake, R. E., & Curtis, G. H. (1987). K–Ar geochronology of the Laetoli fossil localities. In M. D. Leakey & J. M. Harris (Eds.), *Laetoli: A Pliocene site in northern Tanzania* (pp. 48–52). Oxford: Clarendon.

Gentry, A. W. (1987). The bovids from Laetoli. In M. D. Leakey & J. M. Harris (Eds.), *Laetoli: A Pliocene site in northern Tanzania* (pp. 378–408). Oxford: Clarendon.

Greenway, P. J. (1973). A classification of the vegetation of East Africa. *Kirkia, 9*, 1–68.

Grunblatt, J., Ottichilo, W. K., & Sinange, R. K. (1989). A hierarchical approach to vegetation classification in Kenya. *African Journal of Ecology, 27*, 45–51.

Guérin, C. (1987). Chalicotheriidae (Mammalia, Perissodactyla) remains from Laetoli. In M. D. Leakey & J. M. Harris (Eds.), *Laetoli: A Pliocene site in northern Tanzania* (pp. 315–320). Oxford: Clarendon.

Harris, J. M. (1987). Fossil Suidae from Laetoli. In M. D. Leakey & J. M. Harris (Eds.), *Laetoli: A Pliocene site in northern Tanzania* (pp. 349–358). Oxford: Clarendon.

Harrison, T. (2002). The first record of fossil hominines from the Ndolanya Beds, Laetoli, Tanzania. *American Journal of Physical Anthropology Supplement, 34*, 83.

Harrison, T. (2005). Fossil bird eggs from the Pliocene of Laetoli, Tanzania: Their taxonomic and paleoecological relationships. *Journal of African Earth Sciences, 41*, 289–302.

Harrison, T. (2011a). Hominins from the Upper Laetolil and Upper Ndolanya Beds, Laetoli. In T. Harrison (Ed.), *Paleontology and geology of Laetoli: Human evolution in context* (Fossil hominins and the associated fauna, vol. 2, pp. 141–188). Dordrecht: Springer.

Harrison, T. (2011b). Laetoli revisited: Renewed paleontological and geological investigations at localities on the Eyasi Plateau in northern Tanzania. In T. Harrison (Ed.), *Paleontology and geology of Laetoli: Human evolution in context* (Geology, geochronology, paleoecology and paleoenvironment, vol. 1, pp. 1–15). Dordrecht: Springer.

Harrison, T. (2011c). Cercopithecids (Cercopithecidae, Primates). In T. Harrison (Ed.), *Paleontology and geology of Laetoli: Human evolution in context* (Fossil hominins and the associated fauna, vol. 2, pp. 83–139). Dordrecht: Springer.

Harrison, T. (2011d). Galagidae (Lorisoidea, Primates). In T. Harrison (Ed.), *Paleontology and geology of Laetoli: Human evolution in context* (Fossil hominins and the associated fauna, vol. 2, pp. 75–81). Dordrecht: Springer.

Hay, R. L. (1987). Geology of the Laetoli area. In M. D. Leakey & J. M. Harris (Eds.), *Laetoli: A Pliocene site in northern Tanzania* (pp. 23–47). Oxford: Clarendon.

Hooijer, D. A. (1987a). Hipparions of the Laetolil Beds, Tanzania. In M. D. Leakey & J. M. Harris (Eds.), *Laetoli: A Pliocene site in northern Tanzania* (pp. 301–312). Oxford: Clarendon.

Hooijer, D. A. (1987b). Hipparion teeth from the Ndolanya Beds. In M. D. Leakey & J. M. Harris (Eds.), *Laetoli: A Pliocene site in northern Tanzania* (pp. 312–315). Oxford: Clarendon.

Jager, T. (1982). *Soils of the Serengeti Woodlands, Tanzania*. Agricultural Research Report 912. Wageningen: Agricultural University.

Kaiser, T. M. (2011). Feeding ecology and niche partitioning of the Laetoli ungulate faunas. In T. Harrison (Ed.), *Paleontology and geology of Laetoli: Human evolution in context* (Geology, geochronology, paleoecology and paleoenvironment, vol. 1, pp. 329–354). Dordrecht: Springer.

Kappelman, J. (1988). Morphology and locomotor adaptations of the bovid femur in relation to habitat. *Journal of Morphology, 198*, 119–130.

Kappelman, J., Plummer, T., Bishop, L., Duncan, A., & Appleton, S. (1997). Bovids as indicators of Plio-Pleistocene paleoenvironments in East Africa. *Journal of Human Evolution, 32*, 229–256.

Kingdon, J. (1997). *The Kingdon field guide to African mammals*. London: Academic.

Kingston, J. D., & Harrison, T. (2007). Isotopic dietary reconstructions of Pliocene herbivores at Laetoli: Implications for early hominin paleoecology. *Palaeogeography, Palaeoclimatology, Palaeoecology, 243*, 272–306.

Kovarovic, K.M. (2004). Bovids as palaeoenvironmental indicators: An ecomorphological analysis of bovid postcranial remains from Laetoli, Tanzania. Ph.D. dissertation, University College London, London.

Kovarovic, K., & Andrews, P. (2007). A bovid postcranial ecomorphological survey of the Laetoli palaeoenvironment. *Journal of Human Evolution, 52*, 663–680.

Kovarovic, K., Andrews, P., & Aiello, L. (2002). An ecological diversity analysis of the Upper Ndolanya Beds, Laetoli, Tanzania. *Journal of Human Evolution, 43*, 395–418.

Leakey, M. D. (1987). The Laetoli hominid remains. In M. D. Leakey & J. M. Harris (Eds.), *Laetoli: A Pliocene site in northern Tanzania* (pp. 108–117). Oxford: Clarendon.

Leakey, M. G., & Delson, E. (1987). Fossil Cercopithecidae from the Laetolil Beds. In M. D. Leakey & J. M. Harris (Eds.), *Laetoli: A Pliocene site in northern Tanzania* (pp. 91–107). Oxford: Clarendon.

Leakey, M. D., & Harris, J. M. (Eds.). (1987). *Laetoli: A Pliocene site in northern Tanzania*. Oxford: Clarendon.

Leakey, M. D., & Hay, R. L. (1979). Pliocene footprints in the Laetolil Beds, northern Tanzania. *Nature, 278*, 317–323.

Lind, E. M., & Morrison, M. E. S. (1974). *East African vegetation*. London: Longman.

Manega, P. C. (1993). Geochronology, geochemistry and isotopic study of the Plio-Pleistocene hominid sites and the Ngorongoro Volcanic Highland in northern Tanzania. Ph.D. dissertation, University of Colorado, Boulder.

Menaut, J. (1983). The vegetation of African savannas. In F. Bouliére (Ed.), *Ecosystems of the world: Tropical savannas* (Vol. 13, pp. 109–149). Amsterdam: Elsevier.

Ndessokia, P. N. S. (1990). The mammalian fauna and archaeology of the Ndolanya and Olpiro Beds, Laetoli, Tanzania. Ph.D. dissertation, University of California, Berkeley.

O'Brien, E. (1993). Climatic gradients in woody plant species richness: Towards an explanation based on an analysis of southern Africa's woody flora. *Journal of Biogeography, 20*, 181–198.

Plummer, T. W., & Bishop, L. C. (1994). Hominid paleoecology at Olduvai Gorge, Tanzania as indicated by antelope remains. *Journal of Human Evolution, 27*, 47–75.

Plummer, T. W., Bishop, L. C., & Hertel, F. (2008). Habitat preference of extant African bovids based on astragalus morphology: Operationalizing ecomorphology for palaeoenvironmental reconstruction. *Journal of Archaeological Science, 35*, 3016–3027.

Pratt, D., Greenway, P. J., & Gwynne, M. D. (1966). A classification of East African rangeland, with an appendix on terminology. *Journal of Applied Ecology, 3*, 369–382.

Reed, K. E. (1997). Early hominid evolution and ecological change through the African Plio-Pleistocene. *Journal of Human Evolution, 32*, 289–322.

Reed, K. E. (2008). Paleoecological patterns at the Hadar hominin site, Afar Regional State, Ethiopia. *Journal of Human Evolution, 54*, 743–768.

Rossouw, L., & Scott, L. (2011). Phytoliths and pollen, the microscopic plant remains in Pliocene volcanic sediments around Laetoli, Tanzania. In T. Harrison (Ed.), *Paleontology and geology of Laetoli: Human evolution in context* (Geology, geochronology, paleoecology and paleoenvironment, vol. 1, pp. 201–215). Dordrecht: Springer.

Sands, W. A. (1987). Ichnocoenoses of probable termite origin from Laetoli. In M. D. Leakey & J. M. Harris (Eds.), *Laetoli: A Pliocene site in northern Tanzania* (pp. 409–433). Oxford: Clarendon.

Su, D.F. (2005). The paleoecology of Laetoli, Tanzania: Evidence from the mammalian fauna of the Upper Laetolil Beds. Ph.D. dissertation, New York University, New York.

Su, D. F. (2011). Large mammal evidence for the paleoenvironment of the Upper Laetolil and Upper Ndolanya Beds of Laetoli, Tanzania. In T. Harrison (Ed.), *Paleontology and geology of Laetoli: Human evolution in context* (Geology, geochronology, paleoecology and paleoenvironment, vol. 1, pp. 381–392). Dordrecht: Springer.

Su, D. F., & Harrison, T. (2007). The paleoecology of the Upper Laetolil Beds at Laetoli: A reconsideration of the large mammal evidence. In R. Bobe, Z. Alemseged, & A. K. Behrensmeyer (Eds.), *Hominin environments in the East African Pliocene: An assessment of the faunal evidence* (pp. 279–313). Dordrecht: Springer.

Su, D., & Harrison, T. (2008). Ecological implications of the relative rarity of fossil hominins at Laetoli. *Journal of Human Evolution, 55*, 672–681.

Tattersfield, P. (2011). Gastropoda. In T. Harrison (Ed.), *Paleontology and geology of Laetoli: Human evolution in context* (Fossil hominins and the associated fauna, vol. 2, pp. 567–587). Dordrecht: Springer.

Teaford, M. F., & Ungar, P. S. (2000). Diet and the evolution of the earliest human ancestors. *Proceedings of the National Academy of Sciences of the United States of America, 97*, 13506–13511.

Trapnell, C. G., & Griffiths, J. F. (1960). The rainfall–altitude relation and its ecological significance in Kenya. *East African Agricultural Journal, 25*, 207–213.

Verdcourt, B. (1987). Mollusca from the Laetolil and Upper Ndolanya Beds. In M. D. Leakey & J. M. Harris (Eds.), *Laetoli: A Pliocene site in northern Tanzania* (pp. 438–450). Oxford: Clarendon.

Walker, A. (1987). Fossil Galaginae from Laetoli. In M. D. Leakey & J. M. Harris (Eds.), *Laetoli: A Pliocene site in northern Tanzania* (pp. 88–91). Oxford: Clarendon.

Ward, C. V. (2002). Interpreting the posture and locomotion of *Australopithecus afarensis*: Where do we stand? *American Journal of Physical Anthropology, 119*, 185–215.

White, F. (1983). *The vegetation of Africa*. Switzerland: UNESCO.

White, T. D., & Suwa, G. (1987). Hominid footprints at Laetoli: Facts and interpretations. *American Journal of Physical Anthropology, 72*, 485–514.

Chapter 19
Large Mammal Evidence for the Paleoenvironment of the Upper Laetolil and Upper Ndolanya Beds of Laetoli, Tanzania

Denise F. Su

Abstract There has been much debate on the environment of Pliocene Laetoli. These disagreements reflect the complexity of the paleoenvironment and the difficulties in reconciling contradictory evidence. In this paper, the community structure of the large mammal fauna at Laetoli is compared to that of modern faunal communities and the relative abundances of bovid tribes are examined. The results of these analyses are interpreted within the context of other lines of evidence, including those based on rodents, gastropods, phytoliths, stable isotopes and mesowear. The balance of evidence suggests that the ecology of the Upper Laetolil Beds was a mosaic of grassland-shrubland-open woodland habitats with extensive woody vegetation in the form of shrubs, thickets, and bush. There was also a significant presence of dense woodland and possibly riverine forest habitats. The results also indicate that the ecological conditions in the Upper Laetolil Beds became progressively drier and less wooded through time. There is no clear consensus as to the paleoenvironment of the Upper Ndolanya Beds. While there is evidence to suggest that it was drier and more open than the Upper Laetolil Beds, there is contrary evidence indicating that it was at least as humid and wooded as the Upper Laetolil Beds.

Keywords Community analysis • Pliocene • Bovidae • Indicator species • Relative abundances.

Introduction

There has been much debate regarding the paleoenvironmental reconstructions of the Upper Laetolil Beds of Laetoli, Tanzania. Past reconstructions have ranged in a continuum from open, dry savanna habitats (Leakey and Harris 1987) to woodland-grassland habitats (Su and Harrison 2007, 2008) to medium-density woodland (Reed 1997) and to dense woodland and forest habitats (Kovarovic and Andrews 2007). Based on previous faunal and paleoecological analyses (papers in Leakey and Harris 1987; Reed 1997; Kovarovic and Andrews 2007; Su and Harrison 2007, 2008), it is clear that Pliocene Laetoli was most likely ecologically heterogenous, but it is unclear as to the proportion and the changes in the geographical and temporal distribution of the different habitats. Taphonomic and geologic analyses suggest that the heterogeneity seen in the Laetoli large mammal fauna was not the result of time-averaged and transported assemblages, but rather a reflection of the mosaic nature of the environment at the time of deposition (Su and Harrison 2007, 2008). The lack of higher-level resolution in the data and the difficulty in resolving and integrating the contradictory inferences from different lines of evidence are major factors in the debate. In this paper, the large mammal evidence will be examined using a variety of methods, including indicator species (specifically species of the family Bovidae), relative abundances of bovid taxa, and community analysis using ecovariable proportions. The results are then integrated with other lines of evidence to provide a better understanding of the paleoenvironment of Pliocene Laetoli.

Materials and Methods

The analyses presented in this paper are derived from fossil specimens recovered from the Upper Laetolil Beds (ULB; 3.85–3.6 Ma) and Upper Ndolanya Beds (UNB; 2.66 Ma) at Laetoli by Terry Harrison and his team from 1998–2005. All specimens are from surface collections. All anatomically identifiable specimens were collected; however, only dental specimens were included in these analyses to minimize biases in sampling and element representation due to taphonomic factors (Su and Harrison 2008).

Community Structure

In this analysis, an attempt is made to compare and contrast the Laetoli faunas to known extant community structures

D.F. Su (✉)
Department of Anthropology, Bryn Mawr College,
101 N. Merion Avenue., Bryn Mawr, PA 19072, USA
e-mail: dfayshensu@brynmawr.edu

T. Harrison (ed.), *Paleontology and Geology of Laetoli: Human Evolution in Context. Volume 1: Geology, Geochronology, Paleoecology and Paleoenvironment*, Vertebrate Paleobiology and Paleoanthropology, DOI 10.1007/978-90-481-9956-3_19,

from different habitat types. The goal of this analysis is to determine to which modern habitat types the Upper Laetolil and Upper Ndolanya Beds are most similar in community structure. The method is based on the observation that differences in community structure reflect differences in habitats, and represents an approach that is commonly used in paleoecological analyses (for examples, see Reed 1997; Kovarovic et al. 2002; Su and Harrison 2008; Su et al. 2009). Andrews et al. (1979) have shown that patterns of community structure based on dietary and locomotor variables (=ecovariables) are similar in similar habitats, regardless of species composition. This is a method for interpreting the paleoecology of fossil communities based on general ecological principles, rather than on using inferences from closely related modern taxa (Andrews et al. 1979; Reed 1997). The method is based on faunal list composition, and is subject, therefore, to the same biases associated with presence and absence data.

All species greater than 500 g were included in the analysis. Definition and categorization of habitat types are based on and modified from those of Lind and Morrison (1974), Pratt and Gwynne (1977), and White (1983). The modern communities are categorized into nine habitat types: forest, woodland, open woodland, riparian woodland, bushland, shrubland, grassland, floodplain grassland, and desert (Table 19.1). The ecovariables used to characterize community structure in this analysis include dietary and locomotor adaptations. There are 10 dietary and five locomotor variables (Table 19.2). For the Laetoli faunas these are assigned from published studies on inferred locomotor and dietary behaviors for fossil genera (Lewis 1995; Spencer 1995; Bishop 1999; Cerling et al. 1999; Sponheimer and Lee-Thorp 1999; Zazzo et al. 2000; papers in Leakey and Harris 2003; Werdelin and Lewis 2001; Harris and Cerling 2002; Kingston and Harrison 2007) and inferences from their extant relatives (see Su and Harrison 2007 for designations). Locomotor and dietary variables for fauna from comparative modern communities are taken from published behavioral and carbon isotopic studies (see Su and Harrison 2007 for references). In order to demonstrate which modern faunal community is most similar to that of Laetoli, Hierarchical Clustering Analysis was run on ecovariable frequencies using SPSS version 17.0 (Clustering method: furthest neighbor, chi-square between frequency). The faunas from the Upper Laetolil and Upper Ndolanya Beds are treated as two separate units in the analysis.

Indicator Species and Their Relative Abundance

Indicator species analysis is based on the assumption that closely related species are behaviorally similar from the past

Table 19.1 Modern African localities and vegetation types

Vegetation	Locality	References
Forest	Congo rainforest	Rahm (1966)
	E. of River Niger	Happold (1987)
	W. of River Niger	Happold (1987)
	E. of River Cross	Happold (1987)
	Kibale	Struhsaker (1997)
Riparian woodland (with swamps and grasslands)	Chobe National Park	Smithers (1971)
	Okavango	Smithers (1971)
	Moremi Game Reserve	Smithers (1971)
	Linyanti swamp	Smithers (1971)
Woodland	Zambia southern woodland	Ansell (1960, 1978)
	Guinea savanna woodland	Smithers (1983)
	Amboseli National Park	Behrensmeyer et al. (1979)
Open woodland	Tarangire National Park	Lamprey (1962)
	Southern savanna woodland	Smithers (1983)
	Kalahari thornveld	Rautenbach (1978a, b)
	Sudan savanna	Smithers (1983)
	Southwest arid	Smithers (1983)
Bushland	Mkomazi Game Reserve	Coe et al. (1999)
	Serengeti bush	Swynnerton (1958)
Shrubland	Sahel savanna	Smithers (1983)
	Karoo-Nama	Vernon (1999)
	Karoo-succulent	Vernon (1999)
Floodplain grassland	Kafue flats	Sheppe and Osborne (1971)
	Makgadikgadi Pan	Smithers (1971)
	Rukwa valley	Vesey-FitzGerald (1964)
Grassland	Central Kalahari	Rautenbach (1978a, b)
	Serengeti plains	Swynnerton (1958)
	Southern savanna grassland	Smithers (1983)
Desert	Namib desert	Rautenbach (1978a, b)

Table 19.2 Locomotor and dietary variable categories (following Reed 1997; Su and Harrison 2007)

Code	Locomotor adaptations	Code	Trophic adaptations
T	Terrestrial	G	Grazer
T-A	Terrestrial-Arboreal	FG	Fresh Grass Grazer
A	Arboreal	B	Browser
AQ	Aquatic	MF	Mixed Feeder
AQ-T	Aquatic-Terrestrial	GuI	Gumnivore-Insect
F	Fossorial	TF	Total Frugivory
		TC	Total Carnivory
		I	Insectivore
		O	Omnivore
		RT	Root and Tuber

to the present, so that the habitat preferences of living members of a taxon are extended to its fossil relatives. Thus, the presence of a species with extant members that are ecologically constrained can be used to infer paleoenvironment (for examples, see Gentry 1970, 1978; Shipman and Harris 1988; Vrba 1980). This is supported by recent ecomorphological and stable carbon isotopic studies that show that certain fossil taxa generally shared similar diets with their extant relatives (Sponheimer et al. 1999; Kingston and Harrison 2007). However, species presence can be based on single or rare specimens that may have been stratigraphically or ecologically intrusive, and as such, would not be accurate indicators of paleohabitat. This bias can be ameliorated by incorporating relative abundance data, so that rare species do not take on disproportionate importance for paleoecological inferences. By combining indicator species and their relative abundance data, a more precise inference of the paleoenvironment can be obtained, as this takes into account the correlation between species abundance and habitat preferences (Su et al. 2009). It has been shown that there is differential abundance of bovid taxa in different habitats (Western 1973; Greenacre and Vrba 1984) and that habitat preferences of living species are reflected in surface bone assemblages (Behrensmeyer et al. 1979; Behrensmeyer and Dechant Boaz 1980; Behrensmeyer 1993). Thus, it should be possible to make the same correlation and inferences for habitat-specific fossil taxa.

Bovids are the most common mammals at Laetoli and are the focus of this analysis. Their abundance and specificity in habitat preferences render them ideal for using relative abundances to infer habitat types. Premolars and molars were counted as identifiable specimens (NISP). Associated teeth, such as those in a mandible or maxilla, were counted only as a single NISP. Abundance data were collected at the tribal level only. Due to the uncertainty in their tribal affiliations (Gentry 2011), *Brabovus nanincisus* and *"Gazella" kohllarseni* are treated separately from other bovid tribes as "Brabovus" and "aff. Antilopini?", respectively. Bovid relative abundances are compared between localities and stratigraphic levels (ULB bovid dental NISP=1888; UNB bovid dental NISP=283). Locality designations follow those of Harrison and Kweka (2011). Marker tuffs are used to define the basic stratigraphic units within the Upper Laetolil Beds. However, because several horizons are exposed within a single locality (see Harrison and Kweka 2011), fossils may derive from strata that span several marker tuffs. This did not allow for the subdivision of fossils according to individual horizons separated by consecutive tuffs. However, it was possible to divide the Upper Laetolil fauna into three stratigraphic units, i.e. below Tuff 3 (BT3), between Tuff 3 and Tuff 5 (T3T5), and Above Tuff 5 (AT5) (see Table 2.2 in Harrison and Kweka (2011), for the localities associated with each unit). Most of the Upper Laetolil fossils derive from above Tuff 5. The Upper Ndolanya Beds were treated as one unit. Spatial and temporal variations in the relative abundances of the Laetoli bovids were examined through Correspondence Analysis, a multivariate analysis that can be used to examine the relationship between bovid tribal frequencies and localities/stratigraphic units. The analysis was conducted on weighted bovid dental NISP using SPSS version 17.0 (symmetrical normalization and standardization by removing row and column means).

Table 19.3 Comparative hominin-bearing Plio-Pleistocene sites used in the bovid relative abundance analysis

Locality	Age (Ma)	References
Lothagam, Kenya		Leakey and Harris (2001)
Apak Member	~5.0–4.2	
Kaiyumung Member	<3.9	
Hadar, Ethiopia		Reed (2008)
Basal member	3.8–3.4	
Sidi Hakoma	3.42–3.26	
Denen Dora	3.26–3.18	
Kada Hadar	3.2	
Omo, Ethiopia		Bobe (1997); Alemesged (2003)
Shungura Member B	3.36–2.95	
Shungura Member C	2.95–2.6	
Middle Awash, Ethiopia		
Aramis	4.4	White et al. (2009)

Relative abundances of bovid tribes from the Upper Laetolil Beds and Upper Ndolanya Beds are compared to those from other Plio-Pleistocene sites using Correspondence Analysis (Table 19.3). The count data for the comparative Plio-Pleistocene sites were derived from the published literature (see Table 19.3 for references). While the data from different sites are not directly comparable due to different collecting methodologies and taphonomic factors, the use of only dental material minimizes the effects of these factors as these elements tend to be more systematically collected.

Results and Discussion

Community Structure

Relative frequencies of the locomotor and dietary variables indicate that terrestrial, carnivorous, and mixed-feeding species dominate the faunal list for the Upper Laetolil Beds (86%, 30%, and 22%, respectively) and that terrestrial, mixed-feeding, and grazing species dominate the faunal list for the Upper Ndolanya Beds (87%, 32%, and 21%, respectively) (Table 19.4). The high proportion of mixed-feeders among the herbivorous species indicates the availability of not only wooded habitats, but also of grasslands. It is unclear, however, what the proportion of these two major types of habitats is in relation to each other. To examine the locomotor and dietary

Table 19.4 Frequencies of species locomotor and dietary variables of the Upper Laetolil Beds (*ULB*) and the Upper Ndolanya Beds (*UNB*). See Table 19.2 for Locomotor and Trophic codes

	ULB (%)	UNB (%)
Locomotor		
T	86	87
T-A	6	8
F	5	3
A	3	3
Aq	0	0
Trophic		
G	8	21
FG	0	0
B	16	18
MF	22	32
TF	10	8
TC	30	18
RT	5	3
I	3	0
O	6	0

variable frequencies further, and to put them in relation to the faunal communities of known habitats, they are compared to those of modern communities using Hierarchical Clustering. In the results of the Hierarchical Clustering analysis, the modern faunal communities are found to be divided into three distinct clusters: (1) those found in wet and/or wooded habitats (i.e., floodplain grassland, open woodland, riparian woodland, and woodland), (2) those found in more arid and less wooded habitats (i.e., desert, shrubland, grassland), and (3) those found in forests (Fig. 19.1). The faunal community of the Serengeti Plains does not fall into any of these three groups, but instead clusters with the faunas of Upper Laetolil Beds and Upper Ndolanya Beds, which group most closely with each other (Fig. 19.1). This suggests that the Laetoli fossil communities may share similar ecological parameters with that of the Serengeti and that the types of habitat in which they were found may have been comparable. It is possible that the similarity between the faunal communities of Pliocene Laetoli and

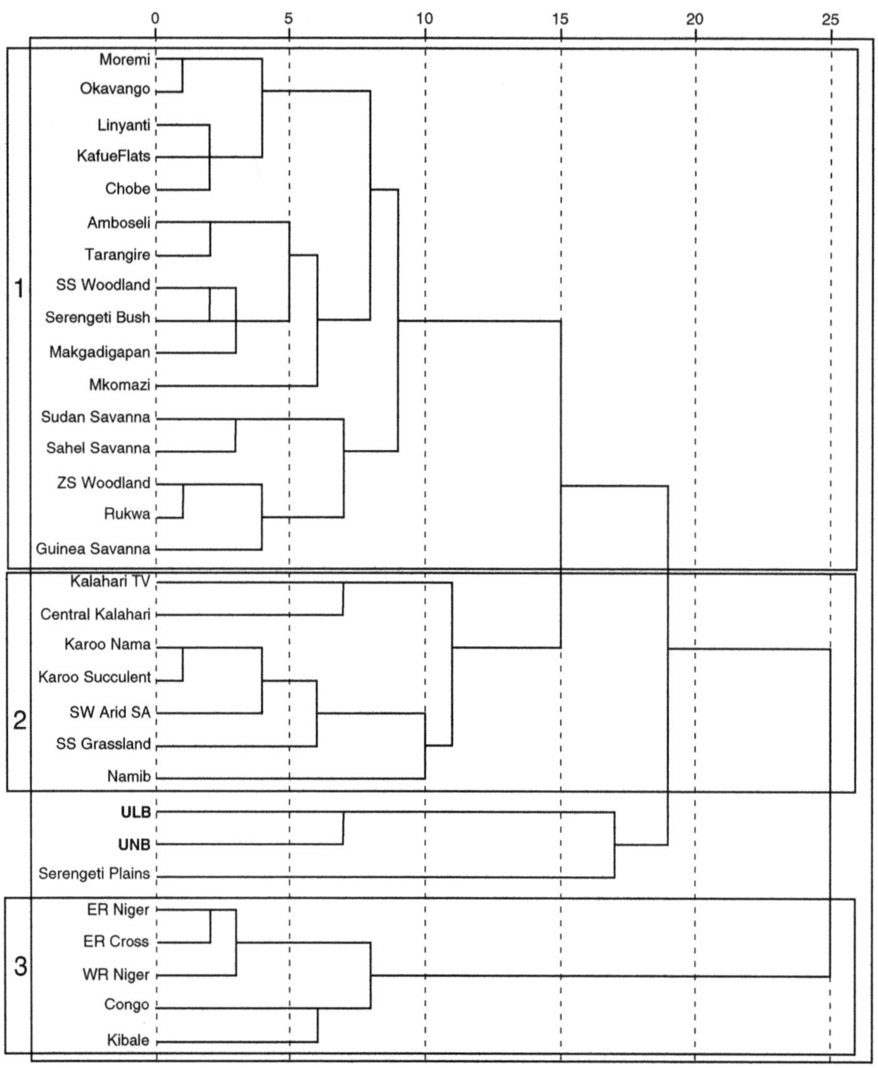

Fig. 19.1 Dendrogram from the hierarchical clustering analysis of the Laetoli and modern faunal communities. Upper Laetolil Beds (*ULB*) and Upper Ndolanya Beds (*UNB*) are in bold. There are three distinct clusters: (*1*) faunal communities of wet and/or wooded habitats (i.e., floodplain grassland, open woodland, riparian woodland, and woodland), (*2*) faunal communities of more arid and less wooded habitats (i.e., desert, shrubland, grassland), and (*3*) faunal communities of forests. Note that ULB and UNB are clustered with Serengeti Plains, and grouped with cluster 2 (i.e., faunal communities of more arid and less wooded habitats)

that of modern Serengeti is due to geographic proximity, rather than strictly ecological similarity. If geographic proximity is the driving factor behind the distinct grouping of the faunal communities of Laetoli and Serengeti Plains, then the faunal community of the Serengeti Bush might be reasonably expected to also be found within the ULB-UNB-Serengeti Plains cluster. However, it appears to be distinct from that of the Serengeti Plains, and is instead found in Cluster 1 (Fig. 19.1). This suggests that the distinctiveness of the faunal communities of ULB-UNB-Serengeti Plains cluster may indeed be reflective of shared ecological parameters. The clustering of forest communities apart from all other modern communities indicate that the faunal communities that inhabit forests are distinctive and may be more easily detectable in the fossil record than those of other habitat types (Su et al. 2009).

Indicator Species and Their Relative Abundances

The Upper Laetolil Beds (ULB) is dominated by alcelaphines (28%) and neotragines (29%) (Table 19.5). Extant alcelaphines are committed grazers found mostly in open habitats, with many species requiring regular access to water (Kingdon 1982, 1997; Smithers 1983; Sponheimer et al. 2003). The high proportion of neotragine is atypical, as they are usually one of the rarest elements of the bovid fauna in East African Plio-Pleistocene sites. Laetoli neotragines are made up overwhelmingly of *Madoqua*. As a group, extant neotragines, particularly *Madoqua*, are mostly arid-adapated and dependent on low-level thickets and succulents (Kingdon 1982, 1997). The classification of *Madoqua* habitat preferences can be problematic. Because *Madoqua* inhabits dense thickets and bushes and they browse almost exclusively, they are often classified as heavy cover animals (for example see Kovarovic and Andrews 2007). However, the dense thickets and bushes that modern dik-diks inhabit are often situated in open habitats, since they prefer an unobstructed view of predators (Kingdon 1982). Thus, while *Madoqua* is a heavy cover animal in relation to its immediate surroundings, it is often situated within a greater ecological context of more open habitats. This distinction is important since the interpretation of ecological preferences of modern relatives can dramatically affect inferences of the paleoenvironment based on the fossil taxa, especially in the case of Laetoli where *Madoqua* is particularly abundant. The prevalence of *Madoqua* may be indicative of relatively open habitats with an abundance of thickets, shrubs, and bush.

Table 19.5 Percentage (%) of bovid tribes in the Upper Laetolil Beds (*ULB*) and Upper Ndolanya Beds (*UNB*). NISP=number in parentheses

	ULB	UNB
	(1888)	(283)
Aepycerotini	7	0
Aff. antilopini?	5	0
Alcelaphini	28	54
Antilopini	11	24
Bovini	0.5	2
Cephalophini	0.7	0
Hippotragini	16	2
Brabovus	3	0
Neotragini	29	12
Tragelaphini	0.4	6
?Reduncini	0.1	0

Hippotragines (16%) also make up a significant proportion of the Upper Laetolil bovid fauna. All species of the modern hippotragine tribe are arid-adapted animals; however, some are better adapted to desert conditions than others (Kingdon 1982, 1997). Modern *Hippotragus*, the genus to which most of the Laetoli hippotragines are identified (Gentry 2011), requires regular water and prefers grassland-woodland ecotones or open woodland habitats, while avoiding closed woodland and forest habitats (Joubert 1976; Kingdon 1982, 1997; Smithers 1983). Antilopines comprise 11% of the bovid fauna in ULB. As a group, modern gazelles prefer open habitats, such as short- to medium-grasslands and open bushlands, and have significant browse in their diet, so that they range from mixed-feeders to browsers in their dietary preferences (Kingdon 1982, 1997; Estes 1991). The dominance of these tribes, particularly alcelaphines and neotragines, signals the abundance of their preferred habitats, which may consist of open grasslands, dense thickets and bushes, open woodlands, and grassland-woodland ecotones. While there is no geological evidence for permanent sources of large bodies of water at Laetoli, there were ephemeral streams and permanent springs present (Su and Harrison 2008) and these would have supplied those animals that required regular access to water.

Bovid tribes associated with wet and/or wooded conditions, i.e., Tragelaphini, Cephalophini, Aepycerotini, Reduncini (Kingdon 1982, 1997; Smithers 1983), are rare in the Upper Laetolil Beds. Tragelaphines and cephalophines together comprise less than 1% of the ULB bovid fauna, and aepycerotines are 7% of the ULB bovid fauna (Table 19.5). It is worth noting here that the presence of cephalophines is unusual. They are rarely found in African fossil localities (Gentry 2011), with only two other recorded instances at Lukeino (Thomas 1980) and Koobi Fora (Harris 1991). Most species of extant cephalophines are found in woodland and forest habitats, except for *Sylvicapra*, the bush duiker (Kingdon 1982; Newing 2001). *Sylvicapra* is not found in forests, but in savannas and open woodland habitats, where there are bush, thickets, and dense underbrush (Kingdon 1982). It has also been hypothesized that forest cephalophines are secondarily adapted to forest habitats (Kingdon 1982; Heckner-Bisping 2001), suggesting that *Sylvicapra* may exhibit the habitat preferences of ancestral cephalophines. It is conceivable that the Laetoli cephalophine (gen. et sp. indet.) may have been ecologically more similar to *Sylvicapra* than to the forest forms. Two possible reduncine teeth were recovered, but their precise taxonomic attribution is uncertain (Gentry 2011). Regardless, they comprise only 0.1% of the bovid fauna (Table 19.5). The low abundance of these taxa suggest that while densely wooded and/or wet habitats were present, they were probably not the dominant habitats on the Upper Laetoli paleolandscape. Of course, taphonomic factors may have impacted the abundance of the bovids that are often associated with densely wooded habitats. Previous taphonomic analysis showed that medium-sized bovids in the range of 25–100 kg are probably under-represented in the Laetoli fauna, due to a combination of carnivore activity and lower probability for immediate burial (Su and Harrison 2008). Specimens of tragelaphines, cephalophines, and aepycerotines from the Upper Laetolil Beds fall within this under-represented weight category (Su and Harrison 2008). It is likely that the original proportions of these bovids were higher than what is preserved and collected. However, antilopines, which also fall within this weight class and are likely similarly under-represented, are more abundant than tragelaphines, cephalophines, and aepycerotines combined (Table 19.5). This suggests that tragelaphines, cephalophines, and aepycerotines may indeed have been relatively rare in the original Upper Laetolil bovid fauna.

The Upper Ndolanya Beds is dominated by alcelaphines (54%), followed by antilopines (24%) (Table 19.5). The overwhelming abundance of alcelaphines in the Upper Ndolanya Beds suggest that it was drier and more open than the Upper Laetolil Beds. However, tragelaphines are much more common in the Upper Ndolanya Beds (6%) compared to the Upper Laetolil Beds (0.4%). This apparent contradiction is reinforced by evidence from ostrich eggshell stable isotopic data and gastropods that suggest that the Upper Ndolanya Beds was cooler, wetter, and more wooded than the Upper Laetolil Beds (Kingston 2011; Tattersfield 2011). Furthermore, studies of enamel carbon isotopic signatures and mesowear of Upper Ndolanya Beds tragelaphines indicate that they were mixed feeders with significant graze in their diet (Kaiser 2011; Kingston 2011), while studies of phytoliths suggest the prevalence of arid C_4 grasses in the Upper Ndolanya Beds (Rossouw and Scott 2011). There does not appear to be any way to reconcile these contradictory lines of evidence at this time.

The association of bovid tribes with localities and stratigraphic units is examined through the use of Correspondence Analysis. Frequencies of bovid tribes in each locality and stratigraphic unit is presented in Tables 19.6 and 19.7, respectively. The results of the locality analysis show that the first dimension explains 34.4% of the inertia and the second dimension explains 20% of the inertia. There is no clear association of any bovid taxon to any particular Upper Laetolil locality. However, when bovid abundances are analyzed based on stratigraphic position, clear associations of bovid tribes to stratigraphic units (as described in Materials and Methods) are seen. The results of the stratigraphic analysis show that the first dimension explains 69.4% of the inertia and Dimension 2 explains 25.4% of the inertia. The first dimension separated UNB from the ULB stratigraphic units, such that UNB is distinct from all ULB stratigraphic units (Fig. 19.2). Although alcelaphines represent more than half of the UNB bovids, tragelaphines and bovines are most closely associated with UNB (Fig. 19.2), probably due to the fact that tragelaphines and bovines are relatively more common in UNB compared to ULB. While tragelaphines are most closely associated with UNB compared to ULB stratigraphic units, they are actually distinct from any stratigraphic unit. This is probably a reflection of their overall rarity in the bovid fauna. Overall, Upper Ndolanya Beds is most closely associated with alcelaphines, antilopines, bovines, and tragelaphines, a mix of taxa that have habitat preferences that range from grassland to woodland. Alcelaphines, antilopines, aff. antilopines, hippotragines, and ?reduncines associate most closely with AT5, while neotragines, aepycerotines, *Brabovus*, and cephalophines associate most closely with BT3 (Fig. 19.2). The bovids associated with AT5 can generally be classified as those that are commonly found in habitats with less woody vegetation, such as grassland, shrubland, and wooded grassland. The bovids associated with BT3 can generally be classified as those that are most commonly found in habitats with more woody vegetation, such as bushland, open woodland, closed woodland, and forest. T3T5 is distinct from the other ULB stratigraphic units, but more closely associates with the "open" habitat bovids (Fig. 19.2). While this analysis is relatively coarse-grained in resolution, it does suggest that there was a shift in ecology from the lower part (below Tuff 3) to the upper part (above Tuff 5) of the Upper Laetolil Beds, as has been suggested by other studies (Kingston 2011; Kovarovic and Andrews 2011; Rossouw and Scott 2011; Tattersfield 2011). The bovid relative abundance suggests that there was a greater proportion of woody vegetation (including trees, shrubs, and bushes) in the lower part of the Upper Laetolil sequence (below Tuff 3) and decreased over time.

In order to place the relative abundances of the ULB and UNB bovids into context, they are compared to those from other Plio-Pleistocene sites using Correspondence Analysis (see Table 19.3 for a list of sites). The first and second dimensions account for 58.8% and 27.3% of the inertia, respectively. Three distinct clusters can be seen in the plot of Dimensions 1 and 2: (1) Laetoli and its associated bovids, (2) Aramis and its associated bovids, and (3) all other Plio-Pleistocene sites and their associated bovids (Fig. 19.3). The Upper Laetolil Beds are considered as one unit in this analysis and, along with the Upper Ndolanya Beds, they are most closely associated with bovid tribes that are usually found in drier habitats with less woody vegetation cover,

Table 19.6 Percentage (%) of bovid tribes in the Upper Laetolil Beds by locality. NISP = number in parentheses

	Loc. 1 (86)	Loc. 1NW (6)	Loc. 2 (188)	Loc. 3 (44)	Loc. 4 (19)	Loc. 5 (75)	Loc. 6 (98)
Aepycerotini	8	33	8	7	11	23	2
aff. antilopini?	3	0	4	7	0	3	2
Alcelaphini	10	0	34	18	21	8	35
Antilopini	0	17	10	14	5	7	12
Bovini	2	0	0.5	0	0	0	0
Celphalophini	0	0	1	5	0	1	0
Hippotragini	14	33	18	14	26	3	22
Brabovus	2	0	1	9	0	9	1
Neotragini	59	17	24	25	32	47	23
Tragelaphini	0	0	0	2	5	0	2
?Reduncini	0	0	0	0	0	0	0
	Loc. 7 (128)	Loc. 8 (121)	Loc. 9 (108)	Loc. 9 S (108)	Loc. 10 (109)	Loc. 10E (196)	Loc. 10 W (174)
Aepycerotini	2	7	6	10	12	4	8
aff. antilopini?	5	7	7	10	8	2	3
Alcelaphini	32	32	30	26	20	18	36
Antilopini	9	9	20	6	15	17	5
Bovini	0	1	0	0	1	0.5	0
Celphalophini	0	1	0	0	1	0	1
Hippotragini	21	18	24	2	8	20	7
Brabovus	2	2	2	0	3	0.5	4
Neotragini	29	21	10	45	32	37	36
Tragelaphini	0	1	1	1	0	0	0
?Reduncini	0	0	0	0	0	0.5	0
	Loc. 11 (56)	Loc. 12 (15)	Loc. 12E (32)	Loc. 13 (52)	Loc. 13SG (32)	Loc. 13E (13)	Loc. 15 (28)
Aepycerotini	4	0	3	6	3	0	0
aff. antilopini?	2	0	3	4	9	0	7
Alcelaphini	29	33	13	71	50	62	36
Antilopini	13	7	9	2	13	15	21
Bovini	0	7	0	0	0	0	0
Celphalophini	0	0	0	0	0	0	0
Hippotragini	25	27	13	6	9	15	21
Brabovus	2	7	16	0	3	0	0
Neotragini	27	20	44	12	13	8	14
Tragelaphini	0	0	0	0	0	0	0
?Reduncini	0	0	0	0	0	0	0
	Loc. 16 (71)	Loc. 17 (11)	Loc. 19 (6)	Loc. 20 (11)	Loc. 21 (47)	Loc. 22 (45)	Loc. 22E (9)
Aepycerotini	8	9	0	9	4	11	0
aff. antilopini?	4	0	0	0	4	2	0
Alcelaphini	24	9	50	18	38	29	11
Antilopini	14	18	0	9	17	11	0
Bovini	4	0	0	0	2	2	11
Celphalophini	1	0	0	0	0	0	11
Hippotragini	27	18	17	36	21	20	56
Brabovus	3	18	0	0	0	7	0
Neotragini	13	27	33	27	13	18	11
Tragelaphini	0	0	0	0	0	0	0
?Reduncini	1	0	0	0	0	0	0

such as alcelaphines, antilopines, and hippotragines. Since Laetoli is the only site in the comparative sample to have recorded occurrences of cephalophines, it is not surprising that this bovid tribe associates most closely with ULB.

The results of this analysis show that the pattern of bovid tribal frequencies at Laetoli (ULB and UNB) differs from those found at other Plio-Pleistocene sites. This signifies important ecological differences. Hadar paleoenvironments

Table 19.7 Percentages (%) of bovid tribes in the Upper Laetolil Beds by stratigraphic units. See text for discussion of stratigraphic units. NISP = number in parentheses

	below T3	T3–T5	above T5
	(391)	(75)	(1422)
Aepycerotini	10	23	6
aff. antilopini?	7	3	4
Alcelaphini	29	8	29
Antilopini	8	7	12
Bovini	0.3	0	0.8
Celphalophini	0.5	1	0.5
Hippotragini	6	3	20
Brabovus	3	9	2
Neotragini	38	47	26
Tragelaphini	0.3	0	0.4
?Reduncini	0	0	0.1

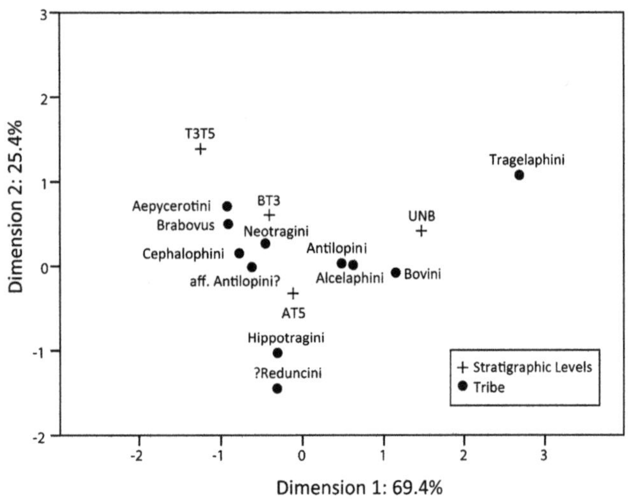

Fig. 19.2 Results of a correspondence analysis of bovid tribes from the Upper Laetolil Beds (*ULB*) and Upper Ndolanya Beds (*UNB*). The Upper Laetolil Beds are divided into three stratigraphic units: below Tuff 3 (*BT3*), between Tuff 3 and Tuff 5 (*T3T5*), and above Tuff 5 (*AT5*). See text for discussion

appear to have varied through time, but were generally wooded, with varying proportions of floodplain grassland (Bonnefille et al. 2004; Reed 2008). It is not until the upper part of the Denen Dora Member that there was a shift to more open woodland and wooded grassland (Reed 2008). Omo, during the deposition of Shungura Members B and C, was likely to have been a wooded and wet environment with closed woodland, riverine forest, and edaphic grassland (Bonnefille and DeChamps 1983; Wesselman 1985; Reed 1997; Bobe and Eck 2001; Bobe et al. 2002), which changed to one that was more open and arid after 2.5 Ma (Bobe et al. 2002; Alemseged 2003). The Lothagam fauna indicates that the paleoecological setting in the Apak Member was predominantly woodland with abundant grassland nearby, which transitioned into a more open habitat with a relative increase in grassland and bushland in the Kaiyumung Member (Leakey and Harris 2003). Aramis is distinct from other Plio-Pleistocene sites in

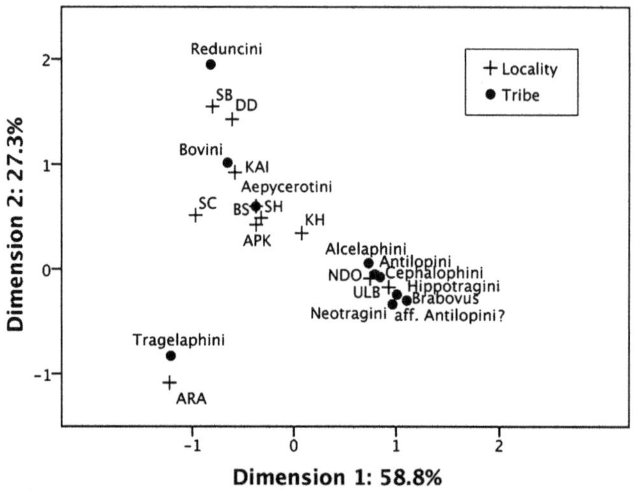

Fig. 19.3 Results of a correspondence analysis of bovid tribes from Laetoli in comparison with other Plio-Pleistocene fossil sites. Abbreviations: *ULB*, Upper Laetolil Beds; *UNB*, Upper Ndolanya Beds; *ARA*, Aramis; *APK*, Apak Member, Lothagam; *KAI*, Kaiyumung Member, Lothagam; *SH*, Sidi Hakoma Member, Hadar; *DD*, Denen Dora Member, Hadar; *SB*, Shungura Member B; *SC*, Shungura Member C. See Table 19.3 for the age and references for each site. See text for discussion

its pattern of bovid relative abundance (Fig. 19.3), mostly due to the overwhelming dominance of tragelaphines (White et al. 2009). Geological, isotopic, and faunal data indicate that Aramis was most likely densely wooded during the Pliocene (White et al. 2009; WoldeGabriel et al. 2009). Given the results of the Correspondence Analysis and the pattern of bovid frequencies at each site, it is likely that Laetoli was less wooded than the other fossil sites included in this study. The difference in vegetation cover may have been due to the differential presence of permanent bodies of water. While all of the comparative sites had either a river or a lake (Bobe et al. 2002; Feibel 2003; Campisano and Feibel 2007; WoldeGabriel et al. 2009), there is no evidence to indicate that either was present at Laetoli during the Pliocene. Instead, water sources were apparently limited to small springs and seasonal watercourses (Harris 1987; Hay 1987; Ditchfield and Harrison 2011).

Paleoenvironmental Implications

The cumulative evidence from the analyses presented here indicates that while densely wooded habitats were present and more prevalent at Laetoli in the Pliocene than today, they were unlikely to have been the dominant vegetation type. Pliocene Laetoli (ULB and UNB) is most similar in ecovariable structure to modern-day shrubland and grassland habitats, dominated by bovid species that are associated with more arid and less wooded habitats. This is corroborated by the rodent fauna, which is dominated by taxa whose extant relatives are found in arid, open habitats, such as *Pedetes*, *Saccostomus*, and *Heterocephalus* in the Upper Laetolil Beds and *Xerus* and

Gerbilliscus in the Upper Ndolanya Beds (Denys 2011). While it is impossible to say with certainty that these fossil rodents share similar ecological preferences as their extant relatives, it has been shown that there is a strong correlation between rodent taxa and vegetation beginning as early as 6 Ma (Denys 1985, 1999). Thus, it is not unreasonable to infer that fossil relatives of modern arid, open habitat rodents may have had similar ecological preferences and that their dominance may indicate the prevalence of such habitats during the Pliocene at Laetoli. This is not to say, however, that Laetoli was a grassland or savanna. Other lines of evidence for the Upper Laetolil Beds indicate that the paleoenvironment was much more complex. Terrestrial gastropod composition in the Upper Laetolil Beds suggests that the paleoenvironment was heavily vegetated, with woodlands and forests (Tattersfield 2011). Analyses of phytoliths indicate that while grasses were ubiquitous and common (grass=54% of total phytoliths), they were never dominant in the ULB sequence (Rossouw and Scott 2011). This inference is supported by enamel carbon isotopic (Kingston and Harrison 2007; Kingston 2011) and mesowear (Kaiser 2011) analyses that reveal that the ULB bovids were much more generalized in their dietary preferences than their extant relatives. Even alcelaphines, which are commonly classified as dedicated C_4 grazers, were consuming significant portions of C_3 vegetation (Kingston and Harrison 2007; Kingston 2011) and this suggests that woody vegetation was abundant during the deposition of the Upper Laetolil Beds at Laetoli. However, evidence from phytoliths indicate that C_3 grasses were present and may have contributed to the C_3/C_4 signal seen in many bovid taxa (Rossouw and Scott 2011), so they might have been more dedicated to grazing than it seems based on carbon isotopic data. Thus, the balance of evidence suggests that while there were significant proportions of dense woodland, and perhaps even riverine forest habitats along ephemeral watercourses, the Upper Laetolil paleohabitat was likely dominated by a mosaic of grassland, shrubland, and open woodland.

By examining bovid abundances in different stratigraphic units (below Tuff 3, between Tuffs 3 and 5, above Tuff 5) separately, it is possible to see a shift in ecological conditions from the lower part (below Tuff 3) to the upper part (above Tuff 5) of the Upper Laetolil Beds. The lower part of ULB appeared to have had a greater proportion of woody vegetation compared to the upper part of ULB. Similar ecological shifts from wetter and more wooded to drier and less wooded conditions are suggested by the stable oxygen isotopic signature of ostrich eggshell fragments, phytoliths, and gastropod composition (Kingston 2011; Rossouw and Scott 2011; Tattersfield 2011). Interestingly, phytolith analysis indicates that mesic C_4 grass was the dominant grass phytolith above Tuff 7; the two possible reduncine specimens (?Reduncini) are also from above Tuff 7, suggesting that there might have been limited areas of wet grasslands during the latest part of the Upper Laetolil Beds. However, this is directly contradicted by the paleoecological inference based on the gastropod fauna. Tattersfield (2011) indicates that there was a slight shift to drier conditions above Tuff 5, which intensified above Tuff 7. It is difficult to reconcile these contradictory lines of evidence, but there is general agreement that there was an ecological shift to drier and more open habitats in the upper part of the Upper Laetolil Beds.

There is no clear inference that can be drawn about the paleoenvironment of Upper Ndolanya Beds based on the bovid abundance data. While alcelaphines are the dominant bovid taxon in the Upper Ndolanya Beds, suggesting an open, arid habitat, tragelaphines are relatively more abundant in the Upper Ndolanya Beds compared to the Upper Laetolil Beds. However, Upper Ndolanya tragelaphines were apparently variable grazers (Kingston 2011), which may indicate the dominance of arid grassland habitats that necessitated animals that emphasized browse in their diet to consume large amounts of graze. Studies of ecomorphology, mesowear, enamel carbon isotopes, and phytoliths all indicate that the Upper Ndolanya Beds was more arid and open than the Upper Laetolil Beds (Kovarovic et al. 2002; Kingston and Harrison 2007; Kingston 2011; Kaiser 2011; Rossouw and Scott 2011). Contradictory evidence is derived from ostrich eggshell oxygen isotope and gastropods, which indicate that the Upper Ndolanya paleoenvironment was cooler, more humid, and more wooded than that of the Upper Laetolil Beds (Kingston 2011; Tattersfield 2011). Furthermore, the Upper Ndolanya woodland/forest gastropod community is found at all UNB localities, suggesting that these habitats were relatively widespread, rather than constrained to microhabitats (Tattersfield 2011). Taphonomic factors may be at play here, because the preservation of large mammals differs in the Upper Ndolanya Beds and Upper Laetolil Beds. Until detailed taphonomic analyses are conducted for the Upper Ndolanya Beds, it is difficult to determine the role it played in influencing the paleoecological inferences drawn from different lines of evidence. Other considerations include the possibility that the large bovid fauna (e.g., alcelaphines) in the Upper Ndolanya Beds may have been migratory forms that were non-residents in the Laetoli area, thus inflating their proportions.

Summary

The results of the analyses highlight the complexity of the Laetoli paleohabitat and the reasons why there has been so much debate. The ecology of the Upper Laetolil Beds was most likely a mosaic environment dominated by grassland, shrubland, and open woodland, with dense woodland and possible riverine forest along ephemeral watercourses. However, the paleoenvironment was not static, and there was a transition between the lower part and the upper part of the sequence in which ecological conditions became generally drier and less wooded. The mammalian community of the Upper Ndolanya Beds is most similar to those of modern-day shrubland and grassland habitats. When combined with the dominance of alcelaphine bovids and evidence from community structure, mesowear, enamel carbon isotopes, and phytoliths, the Upper Ndolanya Beds was likely dominated by semi-arid to arid grasslands. However, contradictory evidence from stable oxygen isotopes and gastropods implies that the ecology of the Upper Ndolanya Beds was cooler, wetter, and more wooded than that of the Upper Laetolil Beds. Unfortunately, it is not possible at this time to reconcile the conflicting evidence. It is evident that much of the ongoing debate that surrounds paleoenvironmental reconstructions of Laetoli is due to the ecological complexity and to the difficulties in reconciling contradictory evidence. This study highlights the importance of using different lines of evidence to reconstruct the paleoenvironment so that a more nuanced and finer-grained interpretation can be made.

Acknowledgments I am grateful to Terry Harrison for inviting me to contribute to this volume and for the numerous discussions on Laetoli and paleoecology (in general and of Laetoli in particular) over the years, which have been invaluable in the development of this paper. I thank all team members who participated in the expeditions to Laetoli that contributed to the recovery of the material discussed here and Terry Harrison for leading the expeditions. I am grateful to the Tanzania Commission for Science and Technology, the Unit of Antiquities in Dar es Salaam, and the National Museum of Tanzania for permission to conduct research in Tanzania. Special thanks to the curators and staff at the National Museum of Tanzania for their support and assistance throughout the years of research at the museum and in the field. I thank A. Gentry, Y. Haile-Selassie, J. Kingston, and W. Sanders for discussions during the formulation of this paper and two anonymous reviewers for their comments. These discussions and comments greatly improved the quality and clarity of the paper.

References

Alemseged, Z. (2003). An integrated approach to taphonomy and faunal change in the Shungura Formation (Ethiopia) and its implication for hominid evolution. *Journal of Human Evolution, 44*, 451–478.

Andrews, P., Lord, J. M., & Nesbit Evans, E. M. (1979). Patterns of ecological diversity in fossil and modern mammalian faunas. *Biological Journal of the Linnean Society, 11*, 177–205.

Ansell, W. F. H. (1960). *Mammals of northern Rhodesia*. Lukasa: Government Printer.

Ansell, W. F. H. (1978). *The mammals of Zambia*. Chilanga, Zambia: National Park and Wildlife Service.

Behrensmeyer, A. K. (1993). The bones of Amboseli. *National Geographic Research and Exploration, 9*, 402–421.

Behrensmeyer, A. K., & Dechant-Boaz, D. E. (1980). The recent bones of Amboseli National Park, Kenya, in relation to East African paleoecology. In A. K. Behrensmeyer & A. P. Hill (Eds.), *Fossils in the making: Vertebrate taphonomy and paleoecology* (pp. 72–92). Chicago: University of Chicago Press.

Behrensmeyer, A. K., Western, D., & Dechant Boaz, D. E. (1979). New perspectives in vertebrate paleoecology from a recent bone assemblage. *Paleobiology, 5*, 12–21.

Bishop, L. (1999). Suid paleoecology and habitat preferences at African Pliocene and Pleistocene hominid localities. In T. G. Bromage & F. Schrenk (Eds.), *African biogeography, climate change, and human evolution* (pp. 216–225). Oxford: Oxford University Press.

Bobe, R. (1997). Hominid environments in the Pliocene: An analysis of fossil mammals from the Omo Valley, Ethiopia. Ph.D. dissertation, University of Washington, Seattle.

Bobe, R., & Eck, G. G. (2001). Responses of African bovids to Pliocene climatic change. *Paleobiology, 27*, 1–47.

Bobe, R., Behrensmeyer, A. K., & Chapman, R. (2002). Faunal change, environmental variability and late Pliocene hominin evolution. *Journal of Human Evolution, 42*, 475–497.

Bonnefille, R., & DeChamps, R. (1983). Data on fossil flora. In: de Heinzelin, J. (Ed.), *The Omo Group: Archives of the International Omo Research Expedition*, Annales, S. 8, Sciences Geologiques, Musée de l'Afrique Centrale, Tervuren, pp. 191–207.

Bonnefille, R., Potts, R., Chalié, F., Jolly, D., & Peyron, O. (2004). High-resolution vegetation and climate change associated with Pliocene *Australopithecus afarensis*. *Proceedings of the National Academy of Sciences of the United States of America, 101*, 12125–12129.

Campisano, C. J., & Feibel, C. S. (2007). Connecting local environmental sequences to global climate patterns: Evidence from the hominin-bearing Hadar Formation, Ethiopia. *Journal of Human Evolution, 53*, 515–527.

Cerling, T. E., Harris, J. M., & Leakey, M. G. (1999). Browsing and grazing in elephants: The isotope record of modern and fossil proboscideans. *Oecologia, 120*, 364–374.

Coe, M., McWilliam, N., Stone, G., & Packer, M. (1999). *Mkomazi: The ecology, biodiversity and conservation of a Tanzanian savanna*. London: Royal Geographical Society (with The Institute of British Geographers).

Denys, C. (1985). Paleoenvironmental and paleobiogeographical significance of the fossil rodent assemblages of Laetoli (Pliocene, Tanzania). *Palaeogeography, Palaeoclimatology, Palaeoecology, 52*, 77–97.

Denys, C. (1999). Of mice and men. Evolution in East and South Africa during Plio-Pleistocene times. In T. Bromage & F. Schrenk (Eds.), *African biogeography, climate change and human evolution* (pp. 226–252). Oxford: Oxford University Press.

Denys, C. (2011). Rodents. In T. Harrison (Ed.), *Paleontology and geology of Laetoli: Human evolution in context* (Fossil hominins and the associated fauna, vol. 2, pp. 15–53). Dordrecht: Springer.

Estes, R. D. (1991). *The behavior guide to African mammals*. Berkeley: University of California Press.

Ditchfield, P., & Harrison, T. (2011). Sedimentology, lithostratigraphy and depositional history of the Laetoli area. In: T. Harrison (Ed.), *Paleontology and geology of Laetoli: Human evolution in context* (Geology, geochronology, paleoecology, and paleoenvironment, vol. 1, pp. 47–76). Dordrecht: Springer.

Feibel, C. S. (2003). Stratigraphic and depositional history of the Lothagam sequence. In M. G. Leakey & J. M. Harris (Eds.), *Lothagam: The dawn of humanity in eastern Africa* (pp. 17–30). New York: Columbia University Press.

Gentry, A. W. (1970). Revised classification for *Makapania broomi* Wells and Cooke (Bovidae, Mammalia) from South Africa. *Palaeontologia Africana, 13*, 63–67.

Gentry, A. W. (1978). Bovidae. In V. Maglio & H. S. B. Cooke (Eds.), *Evolution of African mammals* (pp. 540–572). Cambridge: Harvard University Press.

Gentry, A. W. (2011). Bovidae. In T. Harrison (Ed.), *Paleontology and geology of Laetoli: Human evolution in context* (Fossil hominins and the associated fauna, vol. 2, pp. 363–465). Dordrecht: Springer.

Greenacre, M. J., & Vrba, E. S. (1984). Graphical display and interpretation of antelope census data in African wildlife areas, using correspondence analysis. *Ecology, 65*, 984–997.

Happold, D. C. D. (1987). *The mammals of Nigeria*. Oxford: Clarendon.

Harris, J. M. (1987). Summary. In M. D. Leakey & J. M. Harris (Eds.), *Laetoli: A Pliocene site in northern Tanzania* (pp. 524–531). Oxford: Clarendon.

Harris, J. M. (1991). *Koobi Fora research project* (Vol. 3). Oxford: Clarendon.

Harris, J. M., & Cerling, T. E. (2002). Dietary adaptations of extant and Neogene african suids. *Journal of Zoology, London, 256*, 45–54.

Harrison, T., & Kweka, A. (2011). Paleontological localities on the Eyasi Plateau, including Laetoli. In T. Harrison (Ed.), *Paleontology and geology of Laetoli: Human evolution in context* (Geology, geochronology, paleoecology, and paleoenvironment, vol. 1, pp. 17–45). Dordrecht: Springer.

Hay, R. L. (1987). Geology of the Laetoli area. In M. D. Leakey & J. M. Harris (Eds.), *Laetoli: A Pliocene site in northern Tanzania* (pp. 23–47). Oxford: Clarendon.

Heckner-Bisping, U. (2001). Appendix E – The walking patterns of duikers in respect to their origin in evolution. In V. J. Wilson (Ed.), *Duikers of Africa* (pp. 743–752). Bulawayo: Chipangali Wildlife Trust.

Joubert, S. C. J. (1976). The population ecology of the roan antelope, *Hippotragus equinus equinus* (Desmarest, 1804) in the Kruger National Park. Ph.D. dissertation, University of Pretoria, Pretoria.

Kaiser, T. M. (2011). Feeding ecology and niche partitioning of the Laetoli ungulate faunas. In T. Harrison (Ed.), *Paleontology and geology of Laetoli: Human evolution in context* (Geology, geochronology, paleoecology and paleoenvironment, vol. 1, pp. 329–354). Dordrecht: Springer.

Kingdon, J. (1982). *East African mammals Volume IIIC: Bovidae*. Chicago: University of Chicago Press.

Kingdon, J. (1997). *The Kingdon field guide to African mammals*. San Diego: Academic.

Kingston, J. (2011). Stable isotopic analyses of Laetoli fossil herbivores. In T. Harrison (Ed.), *Paleontology and geology of Laetoli: Human evolution in context* (Geology, geochronology, paleoecology and paleoenvironment, vol. 1, pp. 293–328). Dordrecht: Springer.

Kingston, J., & Harrison, T. (2007). Isotopic dietary reconstructions of Pliocene herbivores at Laetoli: Implications for early hominin paleoecology. *Palaeogeography, Palaeoclimatology, Palaeoecology, 243*, 272–306.

Kovarovic, K., & Andrews, P. (2007). Bovid postcranial ecomorphological survey of the Laetoli paleoenvironment. *Journal of Human Evolution, 52*, 663–680.

Kovarovic, K., & Andrews, P. (2011). Environmental change within the Laetoli fossiliferous sequence: Vegetation catenas and bovid ecomorphology. In T. Harrison (Ed.), *Paleontology and geology of Laetoli: Human evolution in context* (Geology, geochronology, paleoecology and paleoenvironment, vol. 1, pp. 367–380). Dordrecht: Springer.

Kovarovic, K., Andrews, P., & Aiello, L. (2002). An ecological diversity analysis of the Upper Ndolanya Beds, Laetoli, Tanzania. *Journal of Human Evolution, 43*, 395–418.

Lamprey, H. F. (1962). The Tarangire Game Reserve. *Tanganyika Notes and Records, 60*, 10–22.

Leakey, M. D., & Harris, J. M. (Eds.). (1987). *Laetoli: A Pliocene site in northern Tanzania*. Oxford: Clarendon.

Leakey, M. G., & Harris, J. M. (Eds.). (2001). *Lothagam: The dawn of humanity in eastern Africa*. New York: Columbia University Press.

Leakey, M. G., & Harris, J. M. (2003). Lothagam: Its significance and contributions. In M. G. Leakey & J. M. Harris (Eds.), *Lothagam: The dawn of humanity in eastern Africa* (pp. 625–660). New York: Columbia University Press.

Lewis, M. E. (1995). Plio-Pleistocene carnivoran guilds: Implications for hominid paleoecology. Ph.D. dissertation, State University of New York, Stony Brook.

Lind, E. M., & Morrison, M. E. S. (1974). *East African vegetation*. Bristol: Longman.

Newing, H. (2001). Bushmeat hunting and management: Implications of duiker ecology and interspecific competition. *Biodiversity and Conservation, 10*, 99–118.

Pratt, D. J., & Gwynne, M. D. (1977). *Rangeland management and ecology in East Africa*. London: Hodder and Stoughton.

Rahm, U. (1966). Les mammifères de la forêt équatoriale de l'est du Congo. *Musée Royal de l'Afrique Centrale Annales, Serie 8, 149*, 9–121.

Rautenbach, I. L. (1978a). A numerical re-appraisal of southern African biotic zones. *Bulletin of the Carnegie Museum of Natural History, 6*, 175–187.

Rautenbach, I. L. (1978b). Ecological distribution of the mammals of the Transvaal (Vertebrata: Mammalia). *Annals of the Transvaal Museum, 31*, 131–153.

Reed, K. E. (1997). Early hominid evolution and ecological change through the African Plio-Pleistocene. *Journal of Human Evolution, 32*, 289–322.

Reed, K. E. (2008). Paleoecological patterns at the Hadar hominin site, Afar Regional State, Ethiopia. *Journal of Human Evolution, 54*, 743–768.

Rossouw, L., & Scott, L. (2011). Phytoliths and pollen, the microscopic plant remains in Pliocene volcanic sediments around Laetoli, Tanzania. In T. Harrison (Ed.), *Paleontology and geology of Laetoli: Human evolution in context* (Geology, geochronology, paleoecology and paleoenvironment, vol. 1, pp. 201–215). Dordrecht: Springer.

Sheppe, W., & Osborne, T. (1971). Patterns of use of a flood plain by Zambian mammals. *Ecological Monographs, 41*, 179–205.

Shipman, P., & Harris, J. M. (1988). Habitat preference and paleoecology of *Australopithecus boisei* in eastern Africa. In F. E. Grine (Ed.), *Evolutionary history of the "robust" australopithecines* (pp. 343–382). New York: Aldine de Gruyter.

Smithers, R. H. N. (1971). The mammals of Botswana. *Museum Memoirs of the National Museums and Monuments of Rhodesia, 4*, 1–340.

Smithers, R. H. N. (1983). *The mammals of the southern African subregion*. Pretoria: University of Pretoria.

Spencer, L. M. (1995). Antelopes and grasslands: Reconstructing African hominid environments. Ph.D. dissertation, State University of New York, Stony Brook.

Sponheimer, M., & Lee-Thorp, J. A. (1999). Isotopic evidence for the diet of an early hominid, *Australopithecus africanus*. *Science, 283*, 368–370.

Sponheimer, M., Reed, K. E., & Lee-Thorp, J. A. (1999). Combining isotopic and ecomorphological data to refine bovid paleodietary reconstruction: A case study from the Makapansgat Limeworks hominin locality. *Journal of Human Evolution, 36*, 705–718.

Sponheimer, M., Lee-Thorp, J. A., DeRuiter, D. J., Smith, J. M., van der Merwe, N. J., Reed, K., Grant, C. C., Ayliffe, L. K., Robinson, T. F., Heidelberg, C., & Marcus, W. (2003). Diets of southern African Bovidae: Stable isotope evidence. *Journal of Mammalogy, 84*, 471–479.

Struhsaker, T. T. (1997). *Ecology of an African rain forest*. Gainesville: University Press of Florida.

Su, D. F., & Harrison, T. (2007). The paleoecology of the Upper Laetolil Beds at Laetoli: A reconsideration of the large mammal evidence. In R. Bobe, Z. Alemseged, & A. K. Behrensmeyer (Eds.), *Hominin environments in the East African Pliocene: An assessment of the faunal evidence* (pp. 279–313). Dordrecht: Springer.

Su, D. F., & Harrison, T. (2008). Ecological implications of the relative rarity of fossil hominins at Laetoli. *Journal of Human Evolution, 55*, 672–681.

Su, D. F., Ambrose, S. H., DeGusta, D., & Haile-Selassie, Y. (2009). Paleoenvironment. In Y. Haile-Selassie & G. WoldeGabriel (Eds.), *Ardipithecus kadabba: Late Miocene evidence from the Middle Awash* (pp. 521–547). Berkeley: University of California Press.

Swynnerton, G. H. (1958). Fauna of the Serengeti National Park. *Mammalia, 22*, 435–450.

Tattersfield, P. (2011). Gastropoda. In T. Harrison (Ed.), *Paleontology and geology of Laetoli: Human evolution in context* (Fossil hominins and the associated fauna, vol. 2, pp. 567–587). Dordrecht: Springer.

Thomas, H. (1980). Les bovidés du Miocène supérieur des couches de Mpesida et de la Formation de Lukeino (District de Baringo, Kenya). In: R. E. F. Leakey, & B. A. Ogot (Eds.), *Proceedings of the 8th Pan-African Congress of Prehistory*, Nairobi, 1977 (pp. 82–91).

Vernon, C. J. (1999). Biogeography, endemism and diversity of animals in the Karoo. In W. R. J. Dean & S. J. Milton (Eds.), *The Karoo: Ecological patterns and processes* (pp. 57–85). Cambridge: Cambridge University Press.

Vesey-Fitzgerald, D. F. (1964). Mammals of the Rukwa Valley. *Tanganyika Notes and Records, 62*, 61–72.

Vrba, E. S. (1980). The significance of bovid remains as indicators of environment and prediction patterns. In A. K. Behrensmeyer & A. P. Hill (Eds.), *Fossils in the making* (pp. 247–271). Chicago: University of Chicago Press.

Werdelin, L., & Lewis, M. E. (2001). A revision of the genus *Dinofelis* (Mammalia, Felidae). *Zoological Journal of the Linnean Society, 132*, 47–258.

Wesselman, H. B. (1985). Fossil micromammals as indicators of climatic change about 2.4 Myr ago in the Omo Valley, Ethiopia. *South African Journal of Science, 81*, 260–261.

Western, D. (1973). The structure, dynamics and changes of the Amboseli ecosystem. Ph.D. dissertation, University of Nairobi, Nairobi.

White, F. (1983). *The vegetation of Africa: A descriptive memoir to accompany UNESCO/AETFAT/UNSO vegetation maps of Africa*. Paris: UNESCO.

White, T. D., Ambrose, S. H., Suwa, G., Su, D. F., DeGusta, D., Bernor, R. L., Boisserie, J.-R., Brunet, M., Delson, E., Frost, S., Garcia, N., Giaourtsakis, I. X., Haile-Selassie, Y., Howell, F. C., Lehmann, T., Likius, A., Pehlevan, C., Saegusa, H., Semprebon, G., Teaford, M., &Vrba, E. (2009). Macrovertebrate paleontology and the Pliocene habitat of *Ardipithecus ramidus*. *Science, 326*, 87–93.

WoldeGabriel, G., Ambrose, S. H., Barboni, D., Bonnefille, R., Bremond, L., Currie, B., DeGusta, D., Hart, W.K., Murray, A. M., Renne, P. R., Jolly-Saad, M. C., Stewart, K. M., &White, T. D. (2009). The geological, isotopic, botanical, invertebrate, and lower vertebrate surroundings of *Ardipithecus ramidus*. *Science, 326*, 65–70.

Zazzo, A., Bocherens, H., Brunet, M., Beauvilain, A., Biliou, D., Mackaye, H. T., Vignaud, I., & Mariotti, A. (2000). Herbivore paleodiet and paleoenvironmental changes in Chad during the Pliocene using stable isotope ratios of tooth enamel carbonate. *Paleobiology, 26*, 294–309.

Index

T. Harrison (ed.), *Paleontology and Geology of Laetoli: Human Evolution in Context. Volume 1: Geology, Geochronology, Paleoecology and Paleoenvironment*, Vertebrate Paleobiology and Paleoanthropology, DOI 10.1007/978-90-481-9956-3,

O

P

MIX
Papier aus verantwortungsvollen Quellen
Paper from responsible sources
FSC® C105338

If you have any concerns about our products,
you can contact us on
ProductSafety@springernature.com

In case Publisher is established outside the EU,
the EU authorized representative is:
Springer Nature Customer Service Center GmbH
Europaplatz 3, 69115 Heidelberg, Germany

Printed by Libri Plureos GmbH
in Hamburg, Germany